HANDBOOK OF VITAMINS

CLINICAL NUTRITION IN HEALTH AND DISEASE

Additional Volumes in Preparation

HANDBOOK OF VITAMINS

Third Edition
Revised and Expanded

edited by

Robert B. Rucker
*University of California
Davis, California*

John W. Suttie
*University of Wisconsin
Madison, Wisconsin*

Donald B. McCormick
*Emory University
Atlanta, Georgia*

Lawrence J. Machlin
*Nutrition Research and Information, Inc.
Livingston, New Jersey*

MARCEL DEKKER, INC.   NEW YORK · BASEL

The previous edition was published as *Handbook of Vitamins*, edited by Lawrence J. Machlin, 1991, (Marcel Dekker, Inc.).

ISBN: 0-8247-0428-2

This book is printed on acid-free paper.

Headquarters
Marcel Dekker, Inc.
270 Madison Avenue, New York, NY 10016
tel: 212-696-9000; fax: 212-685-4540

Eastern Hemisphere Distribution
Marcel Dekker AG
Hutgasse 4, Postfach 812, CH-4001 Basel, Switzerland
tel: 41-61-261-8482; fax: 41-61-261-8896

World Wide Web
http://www.dekker.com

The publisher offers discounts on this book when ordered in bulk quantities. For more information, write to Special Sales/Professional Marketing at the headquarters address above.

Current printing (last digit):
10 9 8 7 6 5 4 3 2 1

PRINTED IN THE UNITED STATES OF AMERICA

QP
771
.H35
2001

In Memoriam

Lawrence J. Machlin was truly distinguished in his field. A world-renowned researcher on vitamin E, he published widely, chaired conferences, directed the Department of Clinical Nutrition of Hoffmann-La Roche, and taught nutrition at Cornell University Medical College, Washington University, and New York University, among many other accomplishments. He will be missed.

Preface

Knowledge of vitamins in health and disease has increased markedly in recent years due to extraordinary advances in analytical and separation methodologies (1,8,9). Although each of the past six decades has been exciting from the standpoint of improved approaches in the isolation, identification, and synthesis of vitamins, the past 10 years have been particularly important. In addition, improved methodology has never been so available to the typical researcher. With improved methodologies and sources of inexpensive vitamins, it is now possible to study the functions of vitamins in clinical settings. Improved methods have led, in turn, to a better understanding of the roles that vitamins play as catalysts, cellular regulators, and co-substrates. Moreover, there is now a better appreciation of vitaminlike accessory factors, which may be beneficial to overall health or required on a conditional basis.

As a result of the continued interest in vitamin-related research, as might be expected, a large body of information has accumulated in the past 10 years. One goal for the *Handbook of Vitamins: Third Edition, Revised and Expanded*, is to provide basic and fundamental background material to aid the reader in assessing the importance of new findings regarding vitamin function. In addition to the *Handbook of Vitamins*, the reader is directed to other books published by Marcel Dekker, Inc. (5–7, 10–13), to the Methods in Enzymology series (e.g., Ref. 8), edited by two of the editors of the *Handbook of Vitamins*, and to other related textbooks (2–4).

Much of the interest in vitamins stems from an appreciation that there remain sizable populations at risk for vitamin deficiencies. In addition, the need for vitamins in some individuals is known to be influenced by genetic polymorphisms and related genetic factors. Some of the material in the *Handbook of Vitamins* addresses such points and relationships that have significant public health impact, e.g., the relationship of B_{12} and folate metabolism to homocysteine regulation and, in turn, the connections that homocysteine may have to vascular diseases and developmental defects.

Indeed, for each of the vitamins, there now exists a broader understanding of function and requirement. In the chapters dedicated to fat-soluble vitamins, new roles for vitamin A, K, and D are described. An overview of vitamin E's role as an antioxidant is included, as well as a separate chapter on various flavinoids that play roles in oxidant defense. The section on water-soluble vitamins begins with a brief discussion and review of bioorganic mechanisms. Each of the subsequent chapters focusing on water-soluble vitamins was written to provide a general understanding of chemical, physiological, and nutritional relationships. Some sections highlight methods for evaluating deficiencies or nutritional requirements and the interaction of vitamins with environmental factors. Also, the efficacy and safety of vitamin use at high levels is addressed.

The chapters were written with a varied audience in mind: the clinician, the biochemist whose physiological background may be limited, the advanced nutrition student, and the dietician. The *Handbook of Vitamins* is a single source, a summary of vitamin functions, that is complete, yet accessible. In this kind of endeavor, the challenge is to deal with an expanding factual information base. The contributors have often chosen to develop the material in a conceptual, rather than in a highly detailed manner. The *Handbook of Vitamins* serves as an authoritative and comprehensive source.

We are indebted to the staff at Marcel Dekker, Inc., for their patience and assistance. The *Handbook of Vitamins* is part of a serious and significant effort by Dekker to provide seminal works that speak broadly to micronutrient function. In addition to those referenced below, other offerings that complement this book may be found at http://www.dekker.com.

We also wish to acknowledge three colleagues who died before this edition was completed. Graham Garratt, Executive Vice President and Publisher, was very helpful in the early stages of assembling the handbook. James Olson, one of the eminent leaders in the field of vitamin A metabolism, and Lawrence Machlin, who edited the first two editions of *Handbook of Vitamins*, passed away during the past year. Each man was a dear friend, truly distinguished in his field, and will be missed.

Robert B. Rucker
John W. Suttie
Donald B. McCormick
Lawrence J. Machlin

REFERENCES

1. G. F. M. Ball, *Bioavailability and Analysis of Vitamins in Foods*, Chapman & Hall, New York, 1998, pp. 1–569.
2. J. Basu and T. Kumar, *Vitamins in Human Health and Disease*, Wallingford: CAB International, New York, 1996, pp. 1–345.
3. G. F. Combs, *The Vitamins: Fundamental Aspects in Nutrition and Health*, Academic Press, San Diego, CA, pp. 1–618.
4. P. J. Quinn and V. E. Kagan, eds., *Fat-Soluble Vitamins*, Plenum, New York, 1998, pp. 1–533.
5. C. Rice-Evans and L. Packer, eds., *Flavonoids in health and disease*, Marcel Dekker, New York, 1997.
6. L. B. Bailey, ed., *Folate in Health and Disease*, Marcel Dekker, New York, 1994.
7. A. Bendich, C. E. Butterworth, *Micronutrients in Health and in Disease Prevention*, Marcel Dekker, New York, 1991.

8. D. B. McCormick, J. W. Suttie, C. Wagner, eds., *Vitamins and Coenzymes, Parts I, J, K, and L*, Academic Press, San Diego, CA, Methods in Enzymology series, Vol. 279–283, 1997.

9. Sauberlich, H. E., *Laboratory Tests for the Assessment of Nutritional Status, 2nd ed.*, CRC Press, Boca Raton, 1999, pp. 1–486.

10. R. Blomhoff, *Vitamin A in Health and Disease*, Marcel Dekker, New York, 1994.

11. L. Packer, J. Fuchs, eds., *Vitamin C in Health and Disease*, Marcel Dekker, New York, 1997.

12. L. Packer, J. Fuchs, eds., *Vitamin E in Health and Disease*: Biochemistry and Clinical Applications, Marcel Dekker, New York, 1992.

13. S. K. Gaby, A. Bendich, V. N. Singh, L. J. Machlin, eds., *Vitamin Intake and Health: A Scientific Review*, Marcel Dekker, New York, 1990.

Contents

Contributors

William S. Beck, M.D. Department of Medicine and Department of Biochemistry and Cell Biology, Harvard University, Cambridge, Massachusetts

Richard A. Berg, Ph.D. Biomaterials Research, Collagen Corporation, Palo Alto, California

Tom Brody, Ph.D. Department of Nutritional Sciences and Toxicology, University of California, Berkeley, California

Ching K. Chow, Ph.D. Graduate Center for Nutritional Sciences, University of Kentucky, Lexington, Kentucky

Elaine D. Collins, Ph.D. Department of Chemistry, San José State University, San José, California

Jeffrey C. Geesin, Ph.D. Wound Healing Technology Resource Center, Johnson and Johnson, Skillman, New Jersey

J. Bruce German, Ph.D. Department of Food Science and Technology, University of California, Davis, California

Minnie Holmes-McNary, Ph.D. Lineberger Comprehensive Cancer Center, School of Medicine, University of North Carolina at Chapel Hill, Chapel Hill, North Carolina

Carol S. Johnston, Ph.D.   Department of Nutrition, Arizona State University East, Mesa, Arizona

James B. Kirkland, Ph.D.   Department of Human Biology and Nutritional Sciences, University of Guelph, Guelph, Ontario, Canada

James E. Leklem, Ph.D.   Department of Nutrition and Food Management, Oregon State University, Corvallis, Oregon

Donald B. McCormick, Ph.D.   Department of Biochemistry, School of Medicine, Emory University, Atlanta, Georgia

Donald M. Mock, M.D., Ph.D.   Departments of Biochemistry and Molecular Biology, University of Arkansas for Medical Sciences, Little Rock, Arkansas

Anthony W. Norman, Ph.D.   Division of Biomedical Sciences, Department of Biochemistry, University of California, Riverside, California

James Allen Olson, Ph.D.†   Department of Biochemistry, Biophysics, and Molecular Biology, Iowa State University, Ames, Iowa

John Thomas Pinto, Ph.D.   Nutrition Research Laboratory, Department of Medicine, Memorial Sloan-Kettering Cancer Center, New York, New York

Nora S. Plesofsky, Ph.D.   Department of Plant Biology, University of Minnesota, St. Paul, Minnesota

Jean M. Rawling, M.D., Ph.D.   Department of Family Medicine, Faculty of Medicine, University of Calgary, Calgary, Alberta, Canada

Richard S. Rivlin, M.D.   Weill Medical College of Cornell University, and Department of Medicine, Clinical Nutrition Research Unit, Memorial Sloan-Kettering Cancer Center, New York, New York

Robert B. Rucker, Ph.D.   Department of Nutrition, University of California, Davis, California

Barry Shane, Ph.D.   Department of Nutritional Sciences and Toxicology, University of California, Berkeley, California

Francene M. Steinberg, Ph.D., R.D.   Department of Nutrition, University of California, Davis, California

John W. Suttie, Ph.D.   Department of Biochemistry, University of Wisconsin, Madison, Wisconsin

---

† Deceased.

Vichai Tanphaichitr, M.D., Ph.D., F.A.C.P., F.R.A.C.P., F.R.I. Department of Medicine, Ramathibodi Hospital, and Faculty of Medicine, Mahidol University, Bangkok, Thailand

Maret G. Traber, Ph.D. Linus Pauling Institute, Oregon State University, Corvallis, Oregon

Steven H. Zeisel, M.D., Ph.D. Department of Nutrition, School of Public Health and School of Medicine, University of North Carolina at Chapel Hill, Chapel Hill, North Carolina

HANDBOOK OF VITAMINS

1

Vitamin A

JAMES ALLEN OLSON

Iowa State University, Ames, Iowa

I. HISTORY

Probably the first nutritional deficiency disease to be clearly recognized was night blindness. The ancient Egyptians, as indicated in the Papyrus Ebers and later in the London Medical Papyrus, recommended that juice squeezed from cooked liver be topically applied to the eye to cure night blindness. These writings date from 1500 BC, but the observations probably are of much earlier origin. The Greeks, who depended heavily on Egyptian medicine, recommended both the ingestion of cooked liver and its topical application as a cure for night blindness, a tradition that has persisted in many societies to this day (1).

Although interesting references to vitamin A deficiency diseases and their cure can be found throughout history, the modern science of nutrition is only about a century old. The observation that experimental animals lose weight and die on purified diets was noted by many investigators toward the end of the nineteenth century. In the early part of this century, specific factors necessary for growth and survival were beginning to be identified. Frederick Gowland Hopkins in England, for example, during the period 1906–1912 found that a growth-stimulating principle from milk was present in an alcoholic extract of milk rather than in the ash.

During the same period, Stepp in Germany identified one of these "minimal qualitative factors" as a lipid. Soon thereafter, E. V. McCollum and Marguerite Davis in Wisconsin showed that butter or egg yolk, but not lard, contained a lipid-soluble factor necessary for the growth of rats. In 1913 they coined the term "fat-soluble A" and thereby attributed for the first time the growth-stimulating property of these extracts to a single compound. Approaching the problem in a very different way, Osborne and Mendel at Yale concomitantly found that cod liver oil or butter was an essential growth-promoting food for rats. The year 1913, therefore, was the beginning of the modern age of vitamin A exploration.

Many outstanding findings have been made during the past eight decades, of which only the most notable will be mentioned here. Inasmuch as both colorless extracts of liver as well as colored plant lipids showed biological activity, Steenbock in Wisconsin postulated in 1919 the interconversion of two forms of the vitamin. A decade later, Moore in England showed that β-carotene, the plant pigment, was converted to the colorless form of the vitamin in liver tissue. In the early 1920s, experimental vitamin A deficiency was well characterized by Wolbach and Howe in Boston, and the existence of vitamin A deficiency in children was noted by Bloch in Denmark. In 1930, Karrer and his colleagues in Switzerland determined the structure both of β-carotene and of vitamin A and shortly thereafter synthesized some of its derivatives. One of these derivatives was retinaldehyde, and in 1935 George Wald of Harvard, while working in Germany, proved that the retinene found in visual pigments of the eye was identical to Karrer's chemical compound retinaldehyde.

Of great importance for applied nutrition was the elegant synthesis of all-*trans* vitamin A from the inexpensive precursor β-ionone by Otto Isler and his collaborators in Basel in the late 1940s. Within a few years the price of vitamin A, which earlier had been painstakingly isolated from fish liver oils by molecular distillation, fell 10-fold, and the possibility of using vitamin A more generally in foods at a reasonable cost was assured.

With the availability of radioactive isotopes in the 1950s, many studies were initiated on the metabolism of vitamin A, and much of our present knowledge concerning it was defined. The details of the visual cycle involving rhodopsin in the eye were in large part worked out, and attention began to focus on the possible somatic function of vitamin A in growth and cellular differentiation. In subsequent sections, the present state of knowledge relative to vitamin A will be summarized. Much of the early history of vitamin A is discussed in Moore's highly readable monograph (2).

II. CHEMISTRY

A. Isolation

In nature vitamin A is largely found as an ester and, consequently, is highly soluble in organic solvents but not in aqueous solutions. The major provitamin carotenoid, β-carotene, has similar solvent properties. One of the richest sources of vitamin A is liver tissue, in particular the liver oils of marine fish and mammals. The esters can be directly isolated from these oils by molecular distillation at very low pressure, a procedure that has been used extensively for the commercial preparation of vitamin A–rich oils. Alternatively, vitamin A might be directly extracted with chloroform or with some other solvent combination, such as hexane together with ethanol, followed by purification of vitamin A by chromatographic means. To hydrolyze esters, not only of vitamin A and carotenoids but also of triglycerides and other lipids, saponification with KOH is commonly used, followed by extraction with organic solvents. Retinol or its esters can be readily crystallized at low temperature from a variety of organic solvents, including ethyl formate, propylene oxide, and methanol.

B. Structure and Nomenclature

The structure and recommended names of major compounds in the vitamin A group of substances are depicted in Fig. 1. The acknowledged reference compound is called all-*trans* retinol, whose formula and numbering system are given in Fig. 1a (3).

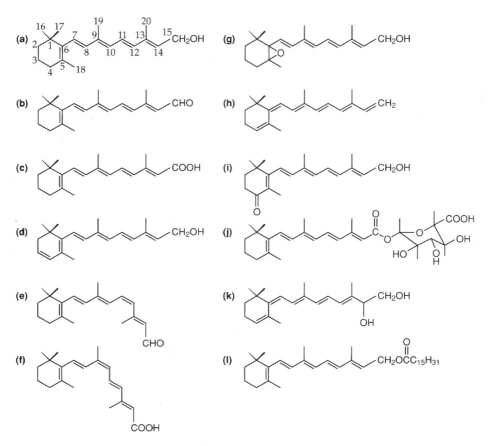

Fig. 1 Formulas for retinol and its derivatives. All are all-trans isomers except as noted. (a) retinol, (b) retinal, (c) retinoic acid, (d) 3,4-didehydroretinol, (e) 11-*cis* retinal, (f) 9-*cis* retinoic acid, (g) 5,6-epoxyretinol, (h) *retro* anhydroretinol, (i) 4-oxoretinol, (j) retinoyl β-glucuronide, (k) 14-hydroxy-4,14-*retro* retinol, (l) retinyl palmitate.

In addition to the all-trans form of vitamin A, all 15 of the other possible isomers have been prepared and characterized (4–6). The most interesting and important isomers (4–6) are 11-*cis* retinal, depicted in Fig. 1e, which is the chromophore of the visual pigments rhodopsin and iodopsin; 9-*cis* retinoic acid (Fig. 1f), which is the ligand for the retinoid X receptor (RXR) transcription factors; and 13-*cis* retinal, which serves light-dependent functions in halobacteria. As the number of cis bonds increase, both the absorption maximum of the isomer and its absorbance tend to decrease. Thus, whereas all-*trans* retinaldehyde has an absorption maximum and molecular extinction coefficient of 368 nm and 48,000 in hexane, respectively, the 9,11,13-cis isomer shows corresponding values of 302 nm and 15,500 (5,6). The tetra-cis isomer, 7,9,11,13-cis, of ethyl retinoate, retinal, and retinol show similar hypsochromic and hypochromic shifts relative to their all-trans analogs (4,5). Although the terms "cis" and "trans" have traditionally been used to denote various isomers of vitamin A, the current chemical notation is "Z" for "cis" and "E" for "trans." Thus, 11-*cis* retinal might equally well be termed 7E,9E,11Z,13E-retinal.

The major commercial forms of vitamin A are all-*trans* retinyl acetate and all-*trans* retinyl palmitate (Fig. 1l). The esters are generally produced because of their increased stability and better solubility in oils and other commercial preparations. The esters can also be incorporated into a gelatin matrix that protects them from oxidation over reasonable periods of time, even when subjected to cooking. These latter forms have been extensively used in the fortification of foods and animal feeds.

Carotenoids not only serve as precursors of vitamin A but also can quench singlet oxygen and act as both antioxidants and prooxidants. Of over 600 carotenoids that have been isolated from nature, only about 50 possess provitamin A activity. An excellent summary of many aspects of carotenoid isolation, structure, and synthesis is given in the fine treatise edited by Isler (7) and in a recent comprehensive work (8). Thus, the term provitamin A is used as a generic indicator for all carotenoids that show the biological activity of vitamin A. Carotenoids commonly found in ingested fruits and vegetables are depicted in Fig. 2. The most active and quantitatively the most important of these provitamins is all-*trans* β-carotene (Fig. 2:5). Generally, carotenoids must contain at least one β-ionone ring that is not hydroxylated to show vitamin A activity.

C. Structures of Important Synthetic Analogs of Vitamin A

A very large number of analogs of vitamin A have been synthesized during the past 15 years. The procedures for synthesizing these retinoids have been comprehensively reviewed (9). The initial impetus for this vast effort was to identify analogs of all-*trans* retinoic acid with a better efficacy/toxicity ratio ("therapeutic index") in the prevention of chemically induced cancers in experimental animals (10). With the discovery of the retinoid receptors, RAR and RXR, the focus has shifted to the synthesis of analogs that bind strongly and specifically to various forms of these receptors, thereby either activating them or inhibiting their activation by retinoic acid (9). Some retinoids with interesting therapeutic or binding properties are given in Fig. 3. In brief, 13-*cis* retinoic acid (Accutane) (Fig. 3a) has been used effectively in the treatment of acne, etretinate (Fig. 3b) and acitretin (Fig. 3c) in the treatment of psoriasis, and 4-hydroxyphenylretinamide (fenretinide) (Fig. 3d) in preventive studies of breast cancer. TTNPB (Fig. 3e) is one of the most toxic retinoids as well as one of the most physiologically active (11). TTNPB interacts well with RARs but poorly with RXRs. Its 3-methyl derivative (Fig. 3f), in contrast, reacts well with RXRs but not with RARs. The naphthylamide derivative Am-580 (Fig. 3g) is specific for RARα, CD-417 (Fig. 3h) for RARβ, CD-666 (Fig. 3i) for RARγ, and SR-11237 (Fig. 3j) for RXR (12,13). The latter compound has a fixed cisoid configuration at the position equivalent to 9-*cis* retinoic acid.

D. Synthesis

Since the complete elucidation of the structure of vitamin A by Karrer in 1930, intense effort has been expended in synthesizing vitamin A from a variety of precursors. Of many possibilities, the two major synthetic procedures used commercially are those of Hoffmann-La Roche and of Badische Anilin- und Sod-Fabrik (BASF). The Roche procedure involves as a key intermediate a C14 aldehyde and further requires the efficient reduction of acetylenic to olefinic bonds near the end of the synthesis. The BASF procedure, on the other hand, depends heavily on the Wittig reaction, by which a phosphonium ylid reacts with an aldehyde or ketone to give an olefin and phosphine oxide. The major steps in

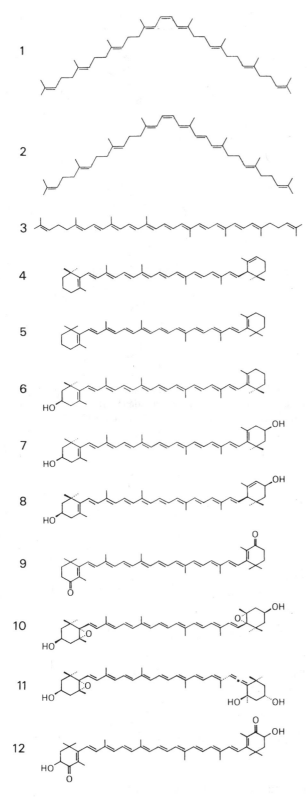

Fig. 2 Polyenes and carotenoids in foods that may also be found in animal tissues. 1, phytoene; 2, phytofluene; 3, lycopene; 4, α-carotene; 5, β-carotene; 6, β-cryptoxanthin; 7, zeaxanthin; 8, lutein; 9, canthaxanthin; 10, violaxanthin; 11, neoxanthin; 12, astaxanthin. (From Ref. 139.)

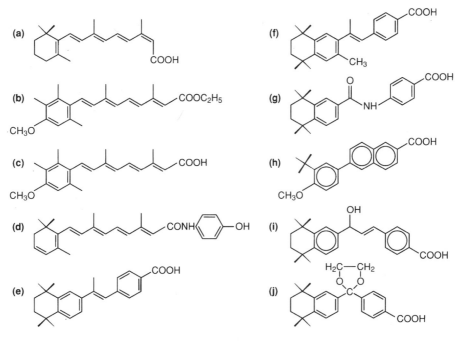

Fig. 3 Some synthetic retinoids with specific therapeutic or protein-binding properties. (a) 13-*cis* retinoic acid, (b) etretinate, (c) acitretin, (d) 4-hydroxyphenylretinamide (HPR, Fenretinide), (e) tetrahydrotetramethylnaphthalenylpropenylbenzoic acid (TTNPB), (f) 3-methyl TTNPB, (g) AM-580, (h) CD-417, (i) CD-666, (j) SR-11237.

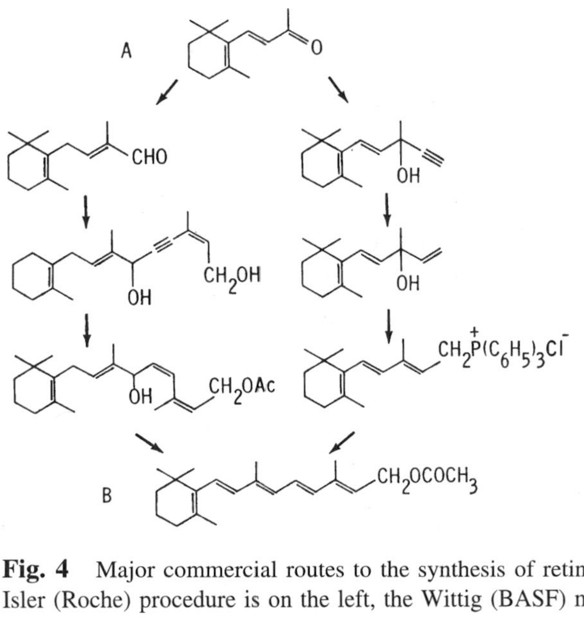

Fig. 4 Major commercial routes to the synthesis of retinyl acetate (B) from β-ionone (A). The Isler (Roche) procedure is on the left, the Wittig (BASF) method on the right.

Table 1 Physical Properties of All-*trans* Retinol and Its Esters

Property	Retinol	Retinyl acetate	Retinyl palmitate
Formula	$C_{20}H_{30}O$	$C_{22}H_{32}O_2$	$C_{36}H_{60}O_2$
Formula weight	286.46	328.50	524.88
Melting point (°C)	63–64	57–59	28–29
UV absorption[a]			
λ_{max}	325	326	326
$E_{1cm}^{1\%}$	1820	1530	960
ϵ	52,140	50,260	50,390
Fluorescence			
Excitation λ_{max}	325	325	325
Emission λ_{max}	470	470	470

[a]In isopropanol. Values are similar in ethanol but differ in chloroform and other solvents. Absorbance values for retinyl esters in hexane are about 3% higher than for retinol in hexane.

these two procedures are presented in Fig. 4. Major synthetic routes to vitamin A and the retinoids have been exhaustively reviewed (7,9,14).

E. Chemical and Physical Properties

In concentrated solution, retinol and its esters are light yellow to reddish hued oils that solidify when cooled and have a mild pleasant odor. As already mentioned, they are insoluble in water or glycerol but are readily miscible with most organic solvents. The formula, formula weights, melting points, and characteristics of the absorption and fluorescent spectra are summarized in Table 1. Indeed, many other physicochemical properties, including infrared spectra, polarographic characteristics, proton magnetic resonance, and the like, have also been studied (7–9,14). Nuclear magnetic resonance (NMR) spectra are particularly useful in characterizing various isomers of vitamin A (5,14). Vitamin A in crystalline form or in oil, if kept under a dry nitrogen atmosphere in a dark cool place, is stable for long periods. In contrast, vitamin A and its analogs are particularly sensitive to oxidation by air in the presence of light, particularly when spread as a thin surface film. In its natural form, whether present in liver or bound to protein in the plasma, vitamin A also is quite stable when stored in the frozen state, preferably at −70°C, in a hermetically closed container in the dark (15). On the other hand, upon extraction from biological materials, care must be taken to prevent both oxidation and isomerization of vitamin A (15).

Commercial firms have been successful in preparing several stabilized forms of vitamin A. For example, retinyl acetate, propionate, or palmitate can be coated in the presence of antioxidants into a gelatin-carbohydrate matrix. The beadlets thus formed, upon being mixed in animal feeds or in human food, retain 90% or more of their activity for at least 6-months if kept under good storage conditions. However, in the presence of high humidity, heat, and oxygen, the loss is considerably greater.

III. ANALYTICAL PROCEDURES

The most common method for analyzing vitamin A and its analogs in pharmaceutical preparations, feedstuffs, and tissues is high-performance liquid chromatography (HPLC) combined with a UV detector, usually set at 325 nm (15). Two major types of chromatographic columns are employed, so-called straight-phase and reverse-phase supports. In the former, hydrophobic compounds are eluted first and more polar compounds later, whereas in the latter the reverse order of elution occurs. Isomers of a given retinoid are best separated on the former (16), whereas different retinyl esters are best resolved on the latter. Carotenoids can also be well resolved by HPLC, usually in conjunction with a UV-visible (UV-Vis) detector set at 450 nm (17–19). Other retinoids are usually measured at their maximal absorption wavelengths.

One of the most sensitive methods for measuring vitamin A is by its intensive greenish yellow fluorescence (Table 1). Because many other natural compounds fluoresce, such as the polyene pigment phytofluene, chromatographic separation by HPLC or by other procedures is usually necessary to ensure adequate specificity in the response. The intense fluorescence of vitamin A has recently been used as well in separating liver cells containing large amounts of retinyl ester (stellate cells) from those containing smaller amounts by means of laser-activated flow cytometry (20).

Mass spectroscopy (MS) is also being increasingly used in the analysis of retinoids and carotenoids (15). The combination of gas chromatography on selected capillary columns with MS has been used to study the in vivo kinetics and the equilibration of deuterated retinol with endogenous reserves of vitamin A in humans (21,22). By using β-carotene richly labeled with carbon 13, isotope ratio mass spectrometry has provided important new information about its metabolism in humans (23,24).

The older conventional methods, such as the colorimetric measurement of the transient blue complex formed at 620 nm when vitamin A is dehydrated in the presence of Lewis acids and the decrease in absorbance at 325 nm when vitamin A in serum extracts is inactivated by UV light, remain of value. Currently, however, these procedures are the methods of choice only in field surveys in which more sophisticated instrumentation is not available (25).

IV. BIOASSAY METHODS

Biological tests have a utility that cannot be replaced by specific chemical or physical methods. The physiological response of a species to a given provitamin, a mixture of provitamins, or a vitamin A preparation can only be assessed by biological procedures. Indeed, the complexities of absorption, metabolism, storage, transport, and uptake by tissues is integrated into a meaningful whole only by biological tests.

Biological methods for vitamin A include the classical growth response tests in vitamin A–deficient rats, liver storage assays in rats and chicks, and the vaginal smear technique. The requisite design and pitfalls of using the classical rat growth test have been well reviewed (26). These classical procedures are rarely used now.

Cell culture systems are now commonly employed for measuring the biological activity, including the toxicity, of retinoids (27–29). A major focus has been the effects of retinoids in inducing the differentiation of transformed cells in vitro (28–30).

V. NUTRITIONAL ASPECTS

A. Foods

Common dietary sources of preformed vitamin A are various dairy products, such as milk, cheese, butter, ice cream, and eggs; liver; other internal organs, such as kidney and heart;

and many fish, such as herring, sardines, and tuna. The very richest sources are liver oils of shark; of marine fish, such as halibut; and of marine mammals, such as polar bear.

With respect to carotenoids, carrots and green leafy vegetables, such as spinach and amaranth, generally contain large amounts. Although tomato contains some vitamin A–active carotenoids, the major pigment is lycopene, which has no nutritional activity. Fruits like papaya and orange have appreciable quantities of carotenoids. The cereal grains generally contain very little vitamin A, particularly when milled (an important consideration in dealing with the problem of vitamin A deficiency among very young children throughout the world). Yellow maize does have vitamin A–active carotenoids, whereas white maize, which is popular in many parts of Africa, does not. The richest source of carotenoids is red palm oil, which contains about 0.5 mg of mixed α- and β-carotene per milliliter. Thus, about 7 mL of red palm oil per day should meet the nutritional needs of a preschool child.

Food composition tables are commonly used to determine the content of vitamin A, including carotenoid sources, in the diet. Although the adequacy of vitamin A and carotenoid intake can usually be estimated by their use, such tables are much less helpful for quantifying mean intakes of vitamin A.

B. Assessment of Vitamin A Status

Vitamin A status can be classified into five categories: deficient, marginal, satisfactory, excessive, and toxic. The deficient and toxic states are characterized by clinical signs, whereas the other three states are not. The most commonly used indicator of clinical deficiency in surveys is the presence of xerophthalmia (25). Of various eye signs that appear, Bitot's spots (XlB) in preschool children have proven to be the most useful. Corneal involvement, although quite specific, is a relatively rare sign, and conjunctival xerosis (XlA) is fairly nonspecific. Of course, the disappearance of any reversible sign in response to vitamin A treatment strengthens its validity. Night blindness, which is difficult to evaluate in young children by direct measurement, can often be satisfactorily assessed by interviewing the mother (25).

Other methods of assessing vitamin A status include dietary, physiological, histological, and biochemical procedures. *Dietary methods* provide useful information about food habits but, as already indicated, are less useful for quantifying mean intakes. The current availability of better values for various carotenoids in foods (31), however, enhances the utility of this approach. Several *physiological techniques* exist for measuring the impaired sensitivity of rod cells in the eye (25). The major histological method is *conjunctival impression cytology*. In vitamin A–depleted individuals, goblet cells in the conjunctiva of the eye tend to disappear, and epithelial cells often assume abnormal shapes (25). However, the subjectivity of the technique and the existence of confounding variables, e.g., trachoma, limit its specificity somewhat. *Biochemical methods* include the measurement of concentrations of retinol and retinol-binding protein (RBP) in plasma and tears (25). Because plasma retinol is maintained homeostatically and is affected by other nutrients and infections, its measurement is most useful when individuals are depleted of the vitamin. Two response tests, the relative dose response (RDR) and modified relative dose response (MRDR) methods, provide more specific information about the vitamin A statuses of individuals and populations from deficiency through marginal status (25). These latter tests are increasingly being used to assess vitamin A status. Finally, isotope dilution techniques, employing vitamin A labeled with deuterium or ^{13}C, are starting to be used to determine total body stores of vitamin A in populations at risk (21–23,25,32).

C. Requirements and Recommendations

A nutritional requirement, whether for animals or humans, must be defined operationally in terms of nutrient-specific indicators. If a different indicator is selected, the value of the requirement may well change. Thus, the presumed requirement will progressively increase as survival, prevention of clinical signs of deficiency, reproductive performance, and longevity are used as end points. As a consequence, several tiers of requirements might be defined, based on the physiological level of satisfactory performance that is desired. Many vitamins, including vitamin A, also show biological *actions*, some beneficial and some adverse at yet higher levels, e.g., the adjuvant effect of large doses of vitamin A in stimulating the immune system. These effects, which are pharmacological in nature, should *not* be considered in setting the nutritional requirement. However, some responses to nutrients that fall between clearly nutritional and clearly pharmacological effects can be difficult to classify.

Strictly speaking, the nutritional requirement for a population, as defined by a given selected indicator, should be expressed as the mean requirement, i.e., when 50% of those in the population show satisfactory values and 50% show unsatisfactory values of a given indicator. However, because the objective of nutritional analysis is the maintenance of a satisfactory status in most individuals in a population, the recommended dietary intake (RDI), a term used worldwide (33), or recommended dietary allowance (RDA), as used in the United States and in a few other countries, has been defined as the average intake over time that theoretically would meet or exceed the nutritional needs of 97.7% (mean + 2 SD) of a selected healthy population group, or, more pragmatically, of "nearly all" members of the group. Although reliable data on the frequency distribution of requirements for many species are not available, a coefficient of variation of 20% seems to apply to several vitamins in humans and other species.

Often, the RDI of a nutrient is set at a level generously above that which provides for any possible physiological need but significantly lower than that which causes toxic effects. In such instances, a "safety factor," usually selected somewhat arbitrarily, is included in the calculation of the desired intake level.

Because all of these procedures have been used in setting recommended intakes of vitamin A, a range of recommended values exists. In selecting an appropriate level for a given application, the basis of various recommendations must consequently be probed.

Three sets of recommended dietary intakes for vitamin A are given in Table 2: the 1988 RDI of the Food and Agriculture Organization and the World Health Organization (FAO/WHO) (34); the 1989 RDAs of the Food and Nutrition Board, National Research Council, U.S. National Academy of Sciences (35); and the 1991 dietary reference values (DRV) for the United Kingdom (36). The age categories cited in the table (37) are similar but not identical in the three recommendations. Three different systems have been used: a single value in the RDA (USA), a two-tier system in the RDI (FAO/WHO), and a three-tier system in the DRV (UK). In a conceptual sense, the higher tier in the FAO/WHO system, termed "safe" intake, the top tier in the UK system, termed "reference nutrient intake" (RNI), and the single RDA values in the U.S. system are roughly equivalent (37). All presume the presence of an adequate total body reserve of vitamin A for most individu-

Table 2 Recommended Dietary Intakes of Vitamin A in µg Retinol Equivalents[a,b]

Group	RDI (FAO/WHO)[c]		RDA (USA)[d]	DRV (UK)[e]		
	Basal	Safe		Lower reference nutrient intake	Estimated average requirement	Reference nutrient intake
Infants						
0–0.5 y	180	350	375	150	250	350
0.5–1 y	180	350	375	150	250	350
Children						
1–2 y	200	400	400	200	300	400
2–6 y	200	400	500	200	300	400
6–10 y	250	400	700	250	350	500
Males						
10–12 y	300	500	1000	250	400	600
12–70+ y	300	600	1000	300	500	700
Females						
10–70+ y	270	500	800	250	400	600
Pregnancy	+100	+100	0			+100
Lactation						
0–6 mo	+180	+350	+500			+350
>6 mo	+180	+350	+400			+350

[a]A µg retinol equivalent is defined as 1 µg all-*trans* retinol, which is considered equal to 6 µg all-*trans* β-carotene or 12 µg of mixed provitamin A carotenoids. In molar terms, 1 µg retinol equivalent = 3.5 nmol retinol = 11.2 nmol β-carotene.
[b]Modified from Ref. 37.
[c]Ref. 34.
[d]Ref. 35.
[e]Ref. 36.

als but also include the requirements of "those few members of the community with particularly high needs" (36).

The lowest tiers in the FAO/WHO system and in the DRV system show similar values but are defined differently. The FAO/WHO defines basal requirements as the "(average daily) amount needed to prevent clinically demonstrable impairment of function" (34). The lower reference nutrient intake (LRNI) is defined as the "(average daily) amount of the nutrient that is enough for only the few people in a group who have low needs" (36).

The estimated average requirement (EAR) in the UK system is the (average daily) amount of a nutrient that meets the needs of 50% of the analyzed group of people. Thus, it is truly a mean requirement. In the UK system, the LRNI is 2 SD below the EAR and the RNI is 2 SD above the EAR. The assigned values for the LRNI in the UK system, which in large part is an *inadequate intake* for most people, and for the basal requirement in the FAO/WHO classification, which is a *minimally adequate intake* for most people, are similar. This confounding fact may in part be explained by the different reference standards for weight that are used. The reference weights for adult men and women, respectively, in the FAO/WHO report are 65 and 55 kg, in the UK report are 74 and 60 kg, and in the U.S. report are 76 and 62 kg. Despite these differences, the RDA in the United States is still very generous relative to those of many other countries (37).

Recommended nutrient intakes are established for healthy individuals. Febrile conditions and lipid malabsorption can markedly increase needs. Genetic defects in the handling of vitamin A also can significantly affect requirements. Homozygous defects have not been identified for vitamin A, however, probably because they are inconsistent with survival. Nonetheless, cases of vitamin A intolerance, in which toxic signs appear in some individuals who ingest moderate amounts of vitamin A, have been reported (38). These instances, though rare, seem to have a genetic basis (37).

Another aspect of nutritional requirements that merits comment is the concept of a single "optimal" intake. Little evidence favors such a view. Rather, most studies indicate that a broad range of satisfactory intakes exists between a low level that prevents deficiency and a high level that causes toxicity. The "satisfactory range" viewpoint is inherent in the establishment of a "safe and adequate range" for several nutrients (35).

The most suitable indicator of vitamin A nutriture, as indicated earlier, is its total body reserve. The status of many other nutrients, such as iron, folic acid, vitamin B_{12}, and vitamin C, can also be related to their total body reserves. The relation of various indicators of nutritional status to the total body reserve and the selection of a total body reserve that meets specified operational criteria of adequacy are essential for the future definition of more scientifically based nutritional requirements and recommendations.

VI. PHYSIOLOGIC RELATIONSHIPS

A. Metabolism

1. Absorption

After foods are ingested, preformed vitamin A of animal tissues and the provitamin carotenoids of vegetables and fruits are released from proteins by the action of pepsin in the stomach and of proteolytic enzymes in the small intestine. In the stomach the free carotenoids and retinyl esters tend to aggregate in fatty globules, which then enter the duodenum. In the presence of bile salts, the globules are broken up into smaller lipid aggregates, which can be more easily digested by pancreatic lipase, retinyl ester hydrolase, cholesteryl ester hydrolase, and the like. The resultant mixed micelles, which contain retinol, the carotenoids, sterols, some phospholipids, mono- and diglycerides, and fatty acids, then diffuse into the glycoprotein layer surrounding the microvillus and make contact with the cell membranes. Various components of the micelles, except for the bile salts, are then readily absorbed into mucosal cells, mainly in the upper half of the intestine (39). A specific transporter exists for vitamin A (40) but evidently not for carotenoids. Nonetheless, carotenoids do interact with each other during absorption (41,42).

The bioavailability and the digestion of vitamin A and carotenoids, needless to say, are affected by the overall nutritional status of the individual and the integrity of the intestinal mucosa. Nutritional factors of importance are protein, fat, vitamin E, zinc, and probably iron. Some types of fiber, e.g., highly methoxylated pectins, markedly reduce carotenoid absorption (43).

Bile salts, which are detergents, promote the rapid cleavage of retinyl and carotenoid esters and assist in the transfer of these lipids into mucosal cells. Interestingly, carotenoid absorption has an absolute requirement for bile salts, independent of its dispersion in a suitable micelle, whereas vitamin A in any properly solubilized form is readily absorbed (24,39).

The overall absorption of dietary vitamin A is approximately 80–90%, with somewhat less efficient absorption at very high doses. The efficiency of absorption of carotenoids from foods is 50–60%, depending on their bioavailability. The absorption efficiency of carotenoids decreases markedly at high intakes or doses (24,39,44).

2. Transport

Retinol in intestinal mucosal cells, whether derived directly from the diet or formed from carotenoids, is largely converted to retinyl esters in these cells. Retinyl esters, together with some unesterified retinol, phospholipids, triglycerides, and apolipoproteins, are incorporated into chylomicra, which are released into the lymph.

During the transport of chylomicra from the lymph into the general circulation, the triglycerides are degraded by lipoprotein lipase of the plasma. The latter is activated by divalent metal ions and requires the presence of albumin as a fatty acid acceptor. After much of the triglyceride is digested and removed from the chylomicron, the resultant chylomicron remnant, which contains retinyl ester, cholesteryl ester, some other lipids, and several apolipoproteins, is taken up primarily by the liver but also to some extent by other tissues (39). Uptake by liver involves interaction between apolipoproteins E and B48 in the chylomicron remnant and high-affinity receptors on the cell surface of parenchymal cells (39).

Whereas dietary retinol is transported to the liver largely as an ester in lipoproteins formed in the intestine, the mobilization of vitamin A from the liver stores and its delivery to peripheral tissues is a highly regulated process. A precursor (pre-RBP) of a specific retinol-binding protein is synthesized in liver parenchymal cells (45). After the removal of a 3500-d polypeptide, the resultant apo-RBP binds all-*trans* retinol, and the complex (holo-RBP), together with transthyretin, is secreted into the plasma. Human RBP, first isolated in the late 1960s, is a single polypeptide chain, has a molecular weight of 21,000, and possesses a single binding site for retinol. Under normal conditions of vitamin A nutriture, approximately 90% of RBP is saturated with retinol. In children the usual level of RBP in plasma is approximately 1–1.5 μmol/L (20–30 μg/mL), which rises at puberty to adult levels of 2–2.5 μmol/L (40–50 μg/mL). In human plasma RBP is largely complexed with transthyretin, a tetramer that also binds thyroid hormones, in a 1:1 molar complex. The association constant (K_a) between transthyretin and RBP is approximately 10^6 M/L. The amino acid sequence and three-dimensional structure of RBP have been determined, and a cDNA clone for it has been identified and studied (45).

Within RBP, retinol resides in a hydrophobic β barrel (46), which protects it from oxidation or destruction during transport (Fig. 5). RBP has been isolated from the plasma of many other species, including rat, monkey, rabbit, cow, dog, and chicken. In all of these species, RBP is similar in size to that found in the human, and in most instances it forms a complex with transthyretin.

The formation of the larger transthyretin–RBP complex may minimize the loss of RBP in the urine during its passage through the kidney (45). In chronic renal disease, the levels of RBP and of plasma retinol are greatly elevated, whereas in severe protein calorie malnutrition, the amounts of both are reduced to approximately 50%. Thus, the steady-state concentrations of both vitamin A and RBP in the plasma are dependent on factors other than vitamin A status alone.

Under normal physiological conditions, the turnover of holo-RBP in the plasma is quite rapid. When associated with transthyretin, its half-life is approximately 11 to 16 h,

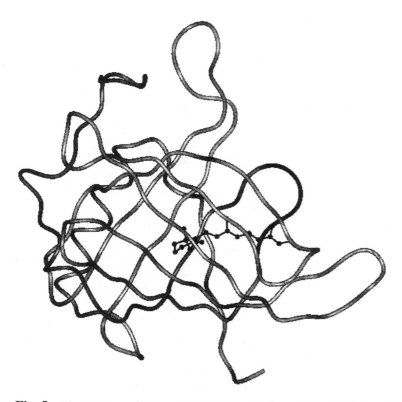

Fig. 5 The structure of human holo-RBP, depicted as a ribbon. The loop that changes conformation between the holo and apo forms of RBP is depicted in black. The N-terminal end is at the upper left, the C-terminal end at the lower right. Retinol is also depicted in black within a eight-stranded β barrel. (Graciously provided by Dr. Marcia Newcomer, Vanderbilt University, Nashville, TN. See also Ref. 46.)

whereas in the apo form it is removed much more rapidly. These half-life values increase, i.e., turnover decreases, about 50% in severe protein calorie malnutrition.

The retinol ligand of holo-RBP in the plasma not only is taken up by peripheral target tissues but also is recycled back to the liver. By use of in vivo kinetic analysis, Green and Green (47) found that an average retinol molecule in the plasma recycles 7–13 times before being irreversibly lost. As another expression of the efficiency of recycling, the turnover rate of plasma retinol is 10 times the irreversible disposal rate of whole-body vitamin A (47). The kidney is the major organ involved in recycling. If its function is adversely affected by disease, infection, and high fever, large amounts of holo-RBP are excreted in the urine (48).

Most of the retinol recycled from tissues back to the liver, on kinetic grounds, is in the form of holo-RBP. However, other transport agents, e.g., retinyl ester on lipoproteins and retinyl β-glucuronide, may play some role in recycling. The key point is that retinol is carefully conserved by the body and is not indiscriminately lost through excretion. Indeed, the irreversible daily loss of vitamin A in the feces and urine of vitamin A–depleted animals, and presumably of humans, is much reduced relative to the rate of loss in vitamin A–sufficient animals (47).

Of all processes relative to the uptake, storage, and transport of vitamin A, the combination of retinol with RBP in the liver seems to be one of the most specific. Thus, the only ligand found under normal physiological conditions is all-*trans* retinol. Other vitamin A analogs will combine with apo-RBP in vitro, however, and 3-dehydroretinol, 15-methylretinol, and 15-dimethylretinol form holo-RBP analogs in vivo (49,50). Indeed, the abilities of specific analogs to saturate RBP in vivo can be closely related to their biological activities (49).

Retinoic acid is primarily transported on plasma albumin and retinyl esters in low-density lipoproteins (39). The β-glucuronides of retinol and retinoic acid, although water-soluble, are bound to plasma albumin and possibly to other plasma proteins in vivo (51).

3. Uptake, Storage, and Release by Cells

The major circulating form of vitamin A in plasma is holo-RBP. Two mechanisms have been explored for its uptake by cells, one of which is dependent on cell surface receptors and the other of which is not (45,52).

Cell surface receptors for holo-RBP have been best characterized from the retinal pigment epithelium of the vertebrate eye. The receptor on the external surface of these cells, i.e., that in contact with plasma, binds holo-RBP strongly, i.e., with a K_a of $1-3 \times 10^7$ M/L (45). The estimated size of the receptor is 63 kd. Transthyretin is not bound to the receptor, nor is an RBP–transthyretin complex necessary for binding. Each cell of the retinal pigment epithelium has about 50,000 cell surface receptors for RBP. Cells of other tissues, such as kidney, lung, and muscle, do not possess this receptor, although other proteins that facilitate retinol uptake may be present on them (45,52). Receptors for holo-RBP may also be present on cells of the liver, skin, placenta, and the barriers between the blood and testes and the blood and brain (52). After complexing with the cell surface receptor, holo-RBP, in all likelihood, is internalized within target cells by receptor-mediated endocytosis. Thereafter, retinol is released into the cytosol, and the apo-RBP is either degraded or secreted from cells in a modified form (45,52). Apo-RBP is bound to surface receptors less tightly than holo-RBP.

Receptor-independent uptake of retinol has been demonstrated in keratinocytes and in some liver cell lines cultured in vitro (45). Retinol can dissociate from RBP at a rate sufficient to explain its direct uptake by lipid membranes (52). In all probability, both mechanisms operate with different cells and under different conditions.

Under normal conditions, more than 90% of the vitamin A in the animal body is stored in the liver, although small quantities are also found in all other tissues. Thus, most of the attention has been given to elucidating the mechanism of its storage and release from the liver.

Endocytosed chylomicron remnants are initially located in low-density endosomes of parenchymal cells (52,53), where retinyl esters are hydrolyzed. Retinol is then translocated, presumably in a complex with CRBP-I (cellular retinol-binding protein I), to the endoplastic reticulum, which is rich in apo-RBP and contains both lecithin:retinol acyltransferase (LRAT) and acyl-CoA:retinol acyltransferase (ARAT). Retinol might continue to circulate as holo-CRBP-I, be transferred to apo-RBP for release into the plasma or to stellate cells, be transferred to stellate cells by some other mechanism, or be esterified to retinyl ester. Retinyl esters, which predominantly consist of palmitate with smaller amounts of stearate, oleate, and linoleate, are then stored in vitamin A–containing globules (VAGs) within the parenchymal cells. When liver reserves of vitamin A are low, parenchymal cells are the major storage site of vitamin A (39,52,53).

When liver reserves are adequate ($\geq$20 μg retinol per gram), approximately 80% of the newly absorbed vitamin A is transferred from parenchymal cells to a specialized type of perisinusoidal cell, termed the stellate cell (fat-storing cell, lipocyte, Ito cell) (52–54). The mechanism of intercellular transport is not clear, although complexes of retinol with CRBP and RBP have been postulated to serve as possible carriers. Within stellate cells, retinol is rapidly esterified to fatty acids in a pattern similar to that found in parenchymal cells. The resultant retinyl esters are stored in vitamin A–containing globules, which contain up to 60% retinyl esters, a significant amount of triglyceride, less phospholipid, and small amounts of cholesterol, cholesteryl esters, and α-tocopherol. Several proteins, including retinyl ester synthase and hydrolase, are associated with these globules. Stellate cells also contain appreciable amounts of CRBP-I and cellular retinoic acid–binding protein (CRABP). Whether stellate cells can synthesize RBP is controversial (52,53). Circulating radiolabeled RBP, however, can be taken up by stellate cells. All other cells of the liver, including Kupffer's cells, endothelial cells, and other nonparenchymal cells, contain very small amounts (<5%) of the total liver reserve of vitamin A.

In the storage of dietary vitamin A, the hydrolysis and formation of retinyl esters is important. Similarly, in the release of vitamin A from the liver as holo-RBP, hydrolysis of retinyl esters is a key step. In regard to the control of these processes, it is interesting that several forms of retinyl ester hydrolase exist in the liver (53) and that their activity is enhanced by a poor vitamin A status and by vitamin E deficiency. On the other hand, the esterifying enzymes are enhanced when large amounts of vitamin A are ingested. Thus, these enzymes should be included among factors that control vitamin A metabolism in vivo. Another important factor may be the ratio of apo-CRBP-I to holo-CRBP-I (55). When apo-CRBP-I predominates, the hydrolysis of vitamin A is favored, but when holo-CRBP-I is the major form, the storage of retinyl esters is enhanced.

Stellate cells are found not only in rat liver but in many other tissues and in many other species (54). Besides storing vitamin A, stellate cells are highly active in synthesizing collagen and other structural proteins. The factors underlying the association of these two important processes in stellate cells have not been elucidated.

As already mentioned, retinol is released from parenchymal cells of the liver as a 1:1 complex of holo-RBP and transthyretin. In mice that cannot make transthyretin, holo-RBP concentrations in the plasma are very low (56). Stellate cells of the liver may also release holo-RBP directly or may transfer retinol to circulating apo-RBP (52). Cells of other tissues, and particularly of kidney, seem to release holo-RBP for transport back to the liver. No structural differences have been noted between RBP synthesized in the liver and that formed in peripheral tissues (45). Retinyl esters are released from the intestine as chylomicra, as already noted, and from the liver as very low-density lipoproteins (VLDLs) (52,53). Although retinoic acid, retinyl β-glucuronide, and retinoyl β-glucuronide are present in plasma in low concentrations, the mode of their release from tissues has not been defined.

4. Transformations

Dietary forms of vitamin A are retinyl esters, derived entirely from animals sources, and provitamin A carotenoids, derived largely from plant foods. Of the total carotenoids present in nature, however, less than 10% serve as precursors of vitamin A. Retinol is released from its dietary esters by the action of pancreatic hydrolases, which act in the presence of bile salts within the intestinal lumen. Within intestinal mucosal cells, retinol combines with CRBP-II (cellular retinol-binding protein II) to give a complex that can undergo either esterification to retinyl ester or oxidation to retinal (39,40,53,55). The major source of the acyl moiety is the usually saturated fatty acid in the α position of phosphatidylcholine, but a transfer from acyl-CoA can also occur (39,40,53,55).

The major reactions of vitamin A metabolism, which are summarized in Fig. 6, are esterification, oxidation at C-15, oxidation at C-4, conjugation, isomerization, other miscellaneous oxidative reactions, and chain cleavage (53,55). Retinol and retinal, as well

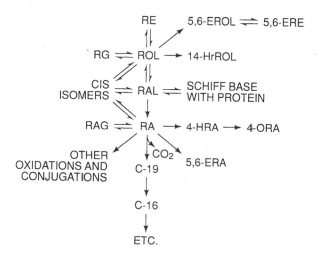

Fig. 6 Major metabolic transformations of vitamin A. ROL, retinol; RE, retinyl ester; RAL, retinal; RA, retinoic acid; RG, retinyl β-glucuronide; RAG, retinoyl β-glucuronide; 5,6-EROL, 5,6-epoxyretinol; 5,6-ERE, 5,6-epoxyretinyl ester; 14-HrROL, 14-hydroxy-4,14-retro retinol; 4-HRA, 4-hydroxyretinoic acid; 4-ORA, 4-oxoretinoic acid; 5,6-ERA, 5,6-epoxyretinoic acid; C-19 and C-16, chain-shortened, oxidized products with the indicated number of carbon atoms. Double arrows indicate reversible reactions; single arrows irreversible changes. (From Ref. 39.)

as other metabolites reversibly converted to them, all possess significant biological activity. Retinoic acid and its glucuronide are active in growth but not in vision or, in most species, in reproduction. Except for 14-hydroxy-4,14-*retro* retinol, more oxidized products, such as 4-hydroxyretinoic acid, 5,6-epoxyretinoic acid, and C-19 metabolites, are largely devoid of biological activity. Retinoyl β-glucuronide, retinyl β-glucuronide, and retinoic acid are normally present in small amounts (3–11 nmol/L, or 1–5 μg/L) in human plasma. Retinoyl β-glucuronide is not hydrolyzed in some cells and slowly hydrolyzed in vivo. Retinoic acid can also be covalently bound to proteins, possibly by means of a coenzyme A intermediate (39,57).

The enzymology of these reactions in most instances is conventional. Retinol is reversibly oxidized to retinal by a nicotinamide adenine dinucleotide (NAD)–dependent pathway in many tissues. Although alcohol dehydrogenase can catalyze this reaction, specific retinol dehydrogenases (retinal reductases) also exist.

Retinal is then irreversibly converted to retinoic acid in many tissues by aldehyde dehydrogenases and oxidases (53). The rate of conversion of retinal to retinoic acid is several fold faster than the oxidation of retinol to retinal. Retinoic acid is inactivated biologically either by hydroxylation at the C-4 position or by epoxidation at the C-5/6 positions. In animal tissues, both of these oxidative reactions are irreversible. 4-Hydroxyretinoic acid can be further oxidized to the 4-keto derivative, followed by a variety of oxidative, chain-cleaving, and conjugative reactions. One of the major conjugated cleavage products is retinotaurine, which appears in the bile in significant amounts (58).

Vitamin A also forms several β-glucuronides by interacting with UDP-glucuronic acid in the presence of glucuronyltransferases. The most interesting of these conjugates are retinyl β-glucuronide and retinoyl β-glucuronide, both of which retain high biological activity (53). Both are synthesized in the intestine, in the liver, and probably in other tissues. In addition, both are endogenous components of human blood and, together with the glucuronides of 4-ketoretinoic acid and of other oxidized metabolites, appear in the bile (53). The rate of formation of β-glucuronides of 9-*cis* and of 13-*cis* retinoic acid by rat liver microsomes in vitro is significantly faster than that of the all-trans isomer (59).

When retinyl acetate or retinyl β-glucuronide is administered to rats, 5,6-epoxyretinyl ester is found in discrete amounts in the liver (60). Thus, epoxidation occurs at the level of retinol as well as of retinoic acid.

β-Carotene and several other provitamin A carotenoids are converted by oxidative cleavage at the 15,15′ double bond to yield two molecules of retinal. The enzyme catalyzing this reaction, 15,15′-carotenoid dioxygenase, is found in the intestine, liver, kidney, and some other tissues. This enzyme will also convert β-apocarotenoids found in plants and in some animal tissues to retinal (53). 9-*cis* β-Carotene is also cleaved by the enzyme to yield 9-*cis* retinal, which can be oxidized to 9-*cis* retinoic acid (61,62), the ligand for RXR receptors. β-Carotene can also be cleaved eccentrically to β-apocarotenals, which can be further degraded oxidatively to retinal (63). Although the relative importance of the two pathways in vivo in mammals has not been determined, cell-free preparations of pig and guinea pig intestines yield retinal as the major, if not sole, product of β-carotene cleavage (64,65).

5. Retinoid-Binding Proteins

Retinoid-binding proteins play key roles in the metabolism and function of their ligands (39,45,53,55,66,67). Major well-characterized binding proteins for retinoids are listed in

Table 3 Properties of Major Retinoid-Binding Proteins[a]

Name	Abbreviation	Molecular weight (kd)	Major ligand
Retinol-binding protein	RBP	21.2	All-*trans* retinol
Cellular retinol-binding protein, type I	CRBP-I	15.7	All-*trans* retinol
Cellular retinol-binding protein, type II	CRBP-II	15.6	All-*trans* retinol
Cellular retinoic acid–binding protein, type I	CRABP-I	15.5	All-*trans* retinoic acid
Cellular retinoic acid–binding protein, type II	CRABP-II	15.0	All-*trans* retinoic acid
Cellular retinaldehyde-binding protein	CRALBP	36.0	11-*cis* retinol and retinal
Interphotoceptor retinol-binding protein	IRBP	135.0	11-*cis* retinal and all-*trans* retinol
Epididymal retinoic acid–binding protein	E-RABP	18.5	All-*trans* and 9-*cis* retinoic acid

[a]Based on Refs. 46, 55, and 68.

Table 3. Several are coligands for enzymatic transformations of retinoids, namely, CRBP-II, CRBP-I, CRABP-I, and possibly CRABP-II (66,67). RBP (as already mentioned), interphotoreceptor retinal-binding protein (IRBP) and epididymal retinoic acid–binding protein (E-RABP) primarily serve transport functions, the first in the plasma, and the second in the eye, and the third in the testes (39,45,68). Cellular retinaldehyde-binding protein (CRALBP) probably also serves a transport function in retinal pigment epithelial cells and possibly in Müller cells of the eye, but it may also have some role during morphogenesis of the eye (68). The ratio of apo to holo forms of some of these proteins, as already mentioned, may regulate metabolic processes (66,67). Apart from these major functions, all binding proteins sequester retinoids in hydrophobic cavities, thereby protecting the ligand from degradation and membrane systems of the cell from unwanted interaction with these amphiphilic ligands. Finally, the RBPs render the hydrophobic ligands water-soluble, much like glucuronidation does. Some redundancy in their functions may exist in that transgenic mice without CRABP-II and/or CRABP-I developed normally in one study (69) but showed polydactyly in another (70).

Other proteins, such as several fatty acid–binding proteins, serum albumin, and β-lactoglobulin, also bind retinoids. These interactions, although relatively nonspecific, might play some physiological role.

6. Excretion

In a quantitative sense, ingested vitamin A is generally metabolized in the following way: (a) 10–20% is not absorbed and, hence, is excreted within 1–2 days into the feces; (b) of the 80–90% absorbed, 20–60% is either conjugated or oxidized to products that are excreted within approximately 1 week in the feces or urine, with a small amount in expired CO_2; and (c) the remainder (30–60%) of the absorbed vitamin A is stored, primarily in the liver. However, when the initial liver reserves are depleted, the relative amount stored in the liver is much lower (39).

Stored vitamin A is metabolized much more slowly in the liver and peripheral tissues to conjugated and oxidized forms of vitamin A, which then are excreted. The half-life for the overall depletion rate in humans is 128–156 days (71). As a general rule, derivatives of vitamin A with an intact carbon chain are excreted in the feces, whereas acidic chain–shortened products tend to be excreted in the urine. In the steady state, approximately equal amounts of metabolites are excreted in the feces and the urine. During periods of severe infection, however, large amounts of holo-RBP, often exceeding the daily requirement for vitamin A, are excreted in the urine (48).

B. Functions

1. Vision

When a photon of light strikes the dark-adapted retina of the eye, 11-*cis* retinal, present as a protonated Schiff base of lysine residue 296 in rhodopsin, is converted to a highly strained transoid form in bathorhodopsin. The more stable all-trans form of bathorhodopsin is then converted to metarhodopsin I, and subsequently by deprotonation to metarhodopsin II (68).

Rhodopsin contains both hydrophilic and hydrophobic regions; has a molecular weight of approximately 38,000 in the cow; is asymmetrical, with a folded length of approximately 70Å; and absorbs maximally at 498–500 nm. Like several other transmembrane proteins, rhodopsin contains seven helical segments that extend back and forth across the disk membrane. Approximately 60% of the total amino acid structure is found in these helical proteins. In addition to loops between helical portions, less structured C-terminal and N-terminal portions extend into the cytosol and the intradisk space, respectively. The chromophore resides in a hydrophobic pocket formed by several transmembrane segments near the cytosolic side of the disk membrane. Residues close to the chromophore include Phe115, Ala117, Glu122, Trp126, Ser127, Trp265, and Tyr268 (68). Rhodopsin contains an acetylated N terminus and two oligosaccharides on asparagines 2 and 15. The complete amino acid sequence of rhodopsin is known.

Three color pigments exist in the cone cells of the human eye, with maximal absorption at 420 nm (blue cones), 534–540 nm (green cones), and 563–570 nm (red cones). The amino acid sequences of these iodopsins are similar, but not identical, to each other and to that of rhodopsin. All adopt a similar conformation within the disk membrane, being composed of seven transmembrane segments. Only seven amino acids differ among the transmembrane segments of the cone pigments. Thus, the differences in absorption maxima must reside in subtle changes in the ambience surrounding the chromophore (68).

Metarhodopsin II, the penultimate conformational state of the light-activated visual pigment, reacts with transducin, a G protein attached to the disk membrane that contains three subunits. In response, the α subunit of transducin binds GTP in place of endogenously associated GDP, thereby activating cGMP phosphodiesterase, which hydrolyzes cGMP to GMP. Because cGMP specifically maintains in an open state the sodium pore in the plasmalemma of the rod cell, a decrease in its concentration causes a marked reduction in the influx of sodium ions into the rod outer segment. The membrane consequently becomes hyperpolarized, which triggers the nerve impulse to other cells of the retina through the synaptic terminal of the rod cell (68). This sequence of events is summarized in Fig. 7.

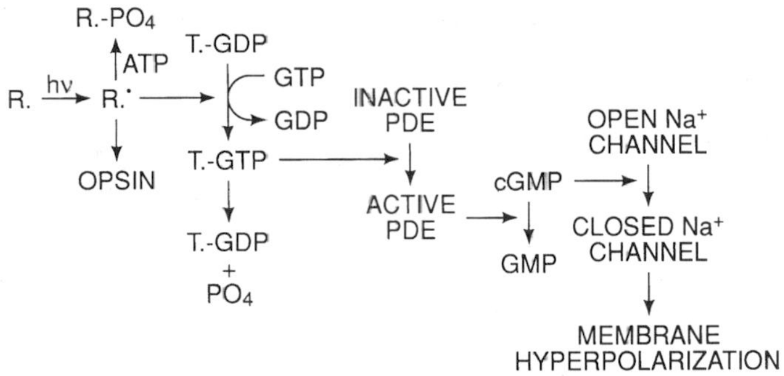

Fig. 7 Possible sequence of steps between the light-induced activation of rhodopsin and hyperpolarization of the rod cell membrane. R., Rhodopsin; R: light-activated metarhodopsin II; R-PO$_4$, phosphorylated rhodopsin; ATP, adenosine triphosphate; GTP, guanosine triphosphate; GDP, guanosine diphosphate; cGMP, cyclic 3′–5′-guanosine monophosphate; GMP, guanosine monophosphate; T., transducin; T.-GDP, a complex of transducin with GDP; T.-GTP, a complex of transducin with GTP; PDE, phosphodiesterase. (From Ref. 39.)

Recovery from this activated state occurs in three ways: (a) The α subunit of transducin, which also shows GTPase activity, hydrolyzes bound GTP to GDP, thereby leading to subunit reassociation to inactive transducin. As a consequence, cGMP phosphodiesterase activity falls and the cGMP level increases back to a normal level. (b) Metarhodopsin II is phosphorylated at sites in the C-terminal portion by ATP, which then reacts with arrestin to form a complex that no longer activates transducin. (c) Metarhodopsin II dissociates through metarhodopsin III to yield all-*trans* retinal and opsin, which also does not activate transducin. These events are also summarized in Fig. 7.

Retinol is very much involved in the process of vision, as shown in Fig. 8 (68). All-*trans* retinol bound to RBP interacts with a cell surface receptor and is internalized into retinal pigment epithelial cells (RPEs) (reaction 8). Thereafter, probably as a CRBP complex, retinol is transacylated from phosphatidylcholine to yield retinyl ester (reaction 3), which in turn undergoes a concerted hydrolysis and isomerization to 11-*cis* retinol (reaction 4) (72). The latter, probably as a CRALBP complex, can either be esterified by transacylation (reaction 3) or be oxidized to 11-*cis* retinal (reaction 5). The latter compound is transferred to IRBP, which ferries it to the rod outer segment in the neural retina. 11-*cis* Retinal, probably in association with a protein but not with CRALBP, then combines with opsin in the disk membranes to form rhodopsin (reaction 7). Another source of both all-*trans* and 11-*cis* retinal in the RPEs is the discarded tips of the constantly regenerating rod outer segments, which are phagocytized by the RPE (not shown in Fig. 8).

Light activation of rhodopsin, as already noted, ultimately gives rise to all-*trans* retinal (reaction 1), which is reduced in the rod outer segment to all-*trans* retinol (reaction 2). The latter is transported by IRBP back to the RPE, where it is stored as all-*trans* retinyl ester (reaction 3). Thus, the cycle is complete.

During continued exposure of the retina to strong light, rhodopsin is largely converted to opsin plus all-*trans* retinal, which tends to accumulate as all-*trans* retinyl ester in the RPE. During dark adaptation, the reverse process occurs. As the rhodopsin concentration increases in the rod outer segment, the retina becomes increasingly sensitive to light of very low intensity.

RETINOID BINDING PROTEINS   REACTIONS   COMPARTMENTS

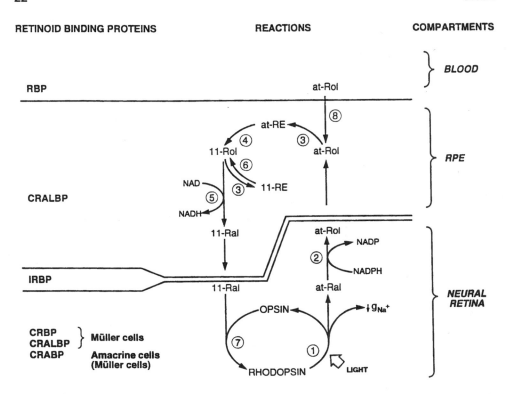

Fig. 8   Reactions of the mammalian visual cycle. Horizontal lines separate cellular and extracellular compartments in the retina involved in retinoid metabolism. Most of the enzymatic reactions of the cycle occur in RPE, including the generation of 11-*cis* retinol. The reactions shown below the double line occur within the photoreceptor cell. Retinoid-binding proteins found within the indicated compartments are shown on the left. CRALBP, cellular retinaldehyde-binding protein; CRBP, cellular retinol-binding protein; IRBP, interphotoreceptor retinoid-binding protein; RBP, retinol-binding protein; at-, all-*trans*-; 11-, 11-*cis*-; Rol, retinol; Ral, retinaldehyde; RE, retinyl ester; g_{Na}, sodium conductance; RPE, retinal pigment epithelium. Reactions shown are: 1, photoisomerization; 2, all-*trans* retinol dehydrogenase; 3, lecithin:retinol acyltransferase (LRAT); 4, retinyl ester isomerohydrolase; 5, 11-*cis* retinol dehydrogenase; 6, retinyl ester hydrolase; 7, regeneration of rhodopsin (nonenzymatic); 8, uptake of plasma retinol. (From Ref. 68.)

Most of the processes involved in the visual cycle have been extensively studied, including the enzymes that are involved (68).

2.   Growth

The earliest assays for vitamin A were based on the growth response of rats fed a purified diet. The fact that maintenance of normal vision and enhancement of growth are two separate properties of the vitamin A molecule was dramatically demonstrated by the observation that retinoic acid stimulated growth but could not maintain vision (73). In nutritional studies, the onset of vitamin A deficiency has often been detected by a so-called growth plateau, which after several days is followed by a rapid loss of weight and, ultimately, death. However, in animals cycled on retinoic acid, i.e., given retinoic acid for 18 days followed by 10 days of deprivation, rats become exquisitely sensitive to the removal of retinoic acid from the diet (74). In these animals loss of appetite occurs within 1–2 days after the withdrawal of retinoic acid, which is closely followed by a depression in growth. Loss of appetite is therefore one of the first symptoms noted in all vitamin A–deficient animals. In humans, however, night blindness and mild xerophthalmia seem to be the earliest signs of clinical deficiency.

Decreased food intake is not due in this case to impaired taste function or to the poor palatability of the deficient diet. Because many factors affect appetite, it has not been possible to define specifically the molecular effect of vitamin A on this process. Inasmuch as distortions in nitrogen metabolism and in amino acid balances within tissues and in the plasma occur concomitantly, the effect of vitamin A deficiency on appetite may be related to these latter abnormalities as well as to disturbances in cell differentiation.

3. Cellular Differentiation

In vitamin A deficiency, mucus-secreting cells are replaced by keratin-producing cells in many tissues of the body. Conversely, the addition of vitamin A to vitamin A–deficient keratinizing cells in tissue culture induces a shift to mucus-producing cells. Retinoids also rapidly induce F-9 teratocarcinoma cells, as well as many other cell lines, to differentiate. In this process, many new proteins appear in the newly differentiated cells. Thus, vitamin A and its analogs, both in vivo and in vitro, markedly influence the way in which cells differentiate (30,39).

The mechanism by which retinoids induce cellular differentiation is becoming clear (Fig. 9). Within tissue cells, all-*trans* retinol, in association with CRBP, can be oxidized to all-*trans* retinoic acid and presumably can also be isomerized to 9-*cis* retinol, which in turn can be oxidized to 9-*cis* retinoic acid. The latter can also arise by isomerization of all-*trans* retinoic acid or by central cleavage of 9-*cis* β-carotene (61,62). All-*trans* or 9-*cis* retinoic acid is transported on CRABP or on other RBPs to the nucleus, where it is tightly bound to one or more of the three (α, β, γ) retinoic acid receptors (RAR) or to one or more of the three (α, β, γ) retinoid X receptors (RXR), respectively (39,75–79).

The RAR and RXR receptors, like other nuclear hormone receptors, possess six protein domains with specific functions. At the N-terminal end, domains A and B serve as physiological activators of the receptor; domain C, which is highly conserved, contains zinc–sulfhydryl interactions (''zinc fingers'') that bind to DNA; domain D is a hinge region that provides the necessary conformation of the receptor; domain E binds the ligand; and domain F, at the C-terminal end, enhances dimerization. All nuclear retinoid receptors contain 410–467 amino acids and have molecular weights of 45–51 kd (39,75–79). RARα and RXRα show very limited homology to each other (only 61% in the DNA binding domain and 27% in the ligand binding domain). Each of the six human retinoid receptors shows different chromosomal locations, developmental expression, and organ localization. Thus, RARα and RXRβ are widespread in their occurrence; RARβ and RXRγ are both present in adult muscle and heart but differ in their embryonic appearance and localization; RARγ is found in adult skin and lung but has a variable embryonic pattern; and RXRα is found in liver, skin, and kidney of both adults and embryos. Thus, each receptor seems to show its own pattern of expression and function. Furthermore, no functional relationship exists between the respective α, β, and γ forms of the two receptors (75–79).

Of the two retinoid receptor families, RXR shows the broadest actions. In the absence of 9-*cis* retinoic acid, RXR forms heterodimers with the vitamin D receptor (VDR), the thyroid receptor (TR), and RAR but only when the latter are charged with their respective ligands, namely, 1α, 25-calcitriol, triiodothyronine, and all-*trans* or 9-*cis* retinoic

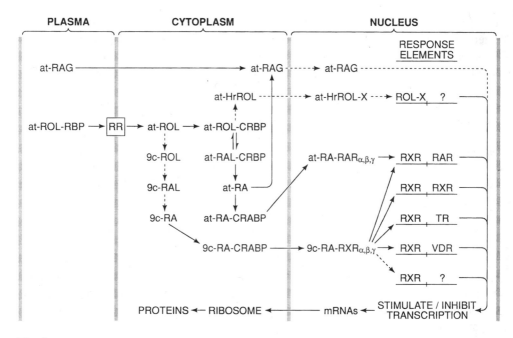

Fig. 9   Roles of retinoids in cellular differentiation at, All-*trans*; 9c, 9-*cis*, ROL, retinol; HrROL, 14-hydroxy-4,14-*retro* retinol; RAL, retinal; RA, retinoic acid; RAG, retinoyl β-glucuronide; RR, cell surface receptor for holo-RBP; RBP, plasma retinol-binding protein; CRBP, cellular retinol-binding protein; CRABP, cellular retinoic acid–binding protein; RARα, β, γ, nuclear retinoic acid receptors, forms α, β, and γ; RXRα, β, γ, nuclear 9-*cis* retinoic acid receptors, forms α, β, γ; TR, triiodothyronine bound to the nuclear thyroid receptor; VDR, calcitriol bound to the nuclear vitamin D receptor; X, unknown nuclear receptors; solid arrows, known transformations or effects; dashed arrows, postulated transformations or effects; ?, unknown occupancy. (From Ref. 39.)

acid (75–79). In the presence of 9-*cis* retinoic acid, RXR will bind its ligand and form a homodimer. Thus, heterodimer and homodimer formation are competitive actions that are dependent on the ratio of all-*trans* retinoic acid to 9-*cis* retinoic acid and the latter's concentration in the nucleus.

The hormone response element (HRE) in DNA for both RXR and RAR is the consensus sequence AGGTCA. For gene activation, a direct repeat, i.e., AGGTCA–other bases–AGGTCA, rather than a palindromic arrangement, is most common. The spacing between the two sequences and the number of tandem repeats largely defines the specificity of the interaction. This observation has given rise to the DR 1-2-3-4-5 rule, in which the numbers refer to the number of intervening deoxynucleotide base pairs between the directly repeating (DR) consensus sequences (75–79). The RXR response element usually employs DR-1 spacing, the RAR response element uses DR-2 or DR-5 spacing, and the VDR and TR response elements use DR-3 and DR-4 spacing, respectively. The sequence of intervening bases often is important as well, but not always. The RXR homodimer is demonstrably formed in the presence of 9-*cis* retinoic acid, whereas the RAR homodimer only seems to form under special circumstances within cells (75–79).

In addition to the interaction with RAR, VDR, and TR, RXR also forms heterodimers with two orphan receptors, the chicken ovalbumin upstream promoter transcription factor (COUP-TF) and the peroxisomal proliferator activated receptor (PPAR). The ligands for these two receptors are not known. Although most heterodimers of RXR, including that with PPAR, activate gene expression, RXR-COUP-TF, either as such or as a COUP-TF homodimer, inhibits the expression mediated by the RXR homodimer of the *CRABP-II* gene. Similarly, RAR, probably by combining with cJun in a nonproductive complex, also inhibits the activation of the AP-1 site, which is important for cell proliferation. The AP-1 site normally is activated by the cJun-cFos heterodimer. Thus, the retinoid receptors can show both activation and suppression of gene expression, depending on the nature of the heterodimers formed. Genes are activated by retinoids as a result of the binding of an appropriate homo- or heterodimer to a hormone response element in DNA, whereas gene expression seems to be suppressed by the competition between a retinoid receptor and some other transcription factor for the latter's activating partner protein (75–79).

Many genes contain response elements for the retinoid receptors, as shown in Table 4 (75–80). Although all of these effects cannot be placed in a cohesive physiological framework, some deserve special mention, namely: (a) the stimulation of both certain cytosolic and nuclear binding proteins for retinoids by retinoic acid via retinoic acid response elements; (b) the stimulation of *Hox a-1* (*Hox 1.6*) and *Hox b-1* (*Hox 2.9*), initiating genes of embryonic development in several cells and species and (c) enhancement of class I alcohol dehydrogenase type 3, which may well induce the synthesis of more retinoic acid from retinol.

Some retinoids may stimulate differentiation by a different pathway; for example, retinoyl β-glucuronide does not bind to CRBP, CRABP, or nuclear RAR but nonetheless is highly active biologically (39,53). Similarly, B lymphocytes differentiate in response to 14-hydroxy-4,14-*retro* retinol but not to all-*trans* retinoic acid (81). Furthermore, the binding to RAR of various acidic retinoids, both natural and synthetic, relates closely to their invoked cellular responses but not to their binding affinities for CRABP, at least in some cellular systems (39).

4. Morphogenesis

Both a deficiency and an excess of vitamin A and of most other retinoids adversely affect embryogenesis (79,82). Studies on the role of vitamin A in embryogenesis were greatly stimulated by the demonstration that an implant containing all-*trans* retinoic acid, when placed in the anterior part of the developing chick limb bud, mimics the activity of the naturally occurring zone of polarizing activity (ZPA) (82,83). Thus, the hypothesis arose

Table 4   Some Proteins Whose Genes Contain Response
Elements for Retinoid Receptors

RARα2	Hox a-1 (Hox 1.6)
RARβ2	Hox b-1 (Hox 2.9)
RARγ2	Phosphoenolpyruvate carboxykinase
CRBP-I	Apolipoprotein A-I
CRBP-II	Ovalbumin
CRABP-II	Complement factor H
Laminin B1	Alcohol dehydrogenase (class I, type 3)
	Acyl coenzyme A oxidase

that all-*trans* retinoic acid might well be one of a presumed host of morphogens that control embryologic development (39).

The initial hypothesis was refined to postulate that a posterior-to-anterior gradient of all-*trans* retinoic acid, but not of retinol, was primarily responsible for pattern formation in the chick limb bud (82,83). In a historical context, pattern formation in the skin was postulated to depend on gradients of vitamin A more than 20 years ago (84). A significant body of experimental data accord with the gradient hypothesis. On the other hand, many other observations do not, including the observation that the gene product of *Sonic hedgehog* induces the polarizing activity of the ZPA, independent of the presence of RA (85). Thus, the gradient hypothesis is not as attractive today as it was when first formulated (80,82,84). Nonetheless, retinoic acid, some of its metabolites, and its receptors clearly play crucial roles in development (79,80,82,83).

Other retinoids also induce limb bud duplication: tetrahydrotetramethylnaphthalenylpropenylbenzoic acid (TTNPB), CD-367, a structurally related compound, and 9-*cis* retinoic acid are much more active than all-*trans* retinoic acid; all-*trans* 3,4-didehydroretinoic acid and Am-580 are as active; and 13-*cis* retinoic acid and many other retinoids are less active (82).

Retinoid receptors and cellular retinoid-binding proteins are expressed in a temporally dependent manner in various parts of the developing embryo. In 9.5- to 10-d mouse embryos, for example, RARα and RARγ are uniformly expressed throughout the mesenchyme, whereas RARβ is restricted to the proximal part of the limb bud. Concomitantly, CRABP-I is expressed uniformly in the mesenchyme as well as in the apical ectodermal ridge. As the embryo develops, the distribution and concentrations of both retinoid-binding proteins and retinoid receptors change.

To define the possible roles of specific retinoid receptors in development, mice lacking one allele of various isoforms of the RAR family, of all RARα isoforms, and of all RARγ isoforms were created (79). Null mice homozygous for single RAR isoforms showed no deformities, whereas mice homozygous for either all RARα or all RARγ isoforms showed an altered phenotype. Although indistinguishable at birth from their wild-type littermates, mice null for RARα were rejected by their mothers, grew slowly, and eventually died, even though no malformations could be detected (79). In contrast, most RARγ null mice showed abnormalities of the axial skeleton but not of the limbs. Null mice for both RARα and RARγ showed more severe abnormalities and died soon after birth (79). The RXRα null mutation was embryolethal. These interesting findings strongly implicate the retinoid receptors in embryogenesis but also indicate that considerable redundancy exists in the functions of specific isoforms. The observation that limbs develop normally in the absence of RARα or RARγ is particularly interesting in view of the marked actions of retinoic acid on that process (79).

Genes that clearly play direct roles in development are the four *Hox* gene clusters (80,82). As already indicated, *Hox a-1* (*Hox 1.6*) contains a retinoic acid response element (RARE) at its 3′ end. Activated *Hox a-1* could set in motion sequential activation of other 5′-located genes in its cluster (*Hox a-2* to *a-13*). *Hox b-1* (*Hox 2.9*) of the *Hox b* family, which is involved in hindbrain development, contains an enhancing RARE at its 3′ end and a suppressing RARE at its 5′ end (86). Genes of the *Hox d* family, which are known to be involved in limb development, are also activated by retinoic acid (80,82). Other genes are also activated or suppressed by treatment with retinoic acid (80). Their interplay in development is slowly being clarified.

Retinoids not only influence the normal physiological development of the embryo but also, in larger amounts, induce fetal abnormalities. Thus, the regulation of retinoic acid metabolism in the developing embryo is a crucial aspect of the overall complex process.

5. The Immune Response

Vitamin A was early termed the "antiinfective" vitamin, based on the increased number of infections noted in vitamin A–deficient animals and humans (81). In vitamin A deficiency, both specific and nonspecific protective mechanisms are impaired: namely, the humoral response to bacterial, parasitic, and viral infections; cell-mediated immunity; mucosal immunity; natural killer cell activity; and phagocytosis (39). Large doses of vitamin A can also serve as an adjuvant. When vitamin A–deficient animals are supplemented with vitamin A, immune responses generally improve. The immune responses to certain antigens in vitamin A–depleted children are also enhanced by vitamin A supplementation (87). However, increased responsiveness in children is less marked than in animals, probably because of the presence of multiple nutritional deficiencies in malnourished populations and a poor, but not acutely deficient, vitamin A status (39).

The primary immune response to protein antigens, such as tetanus toxoid, is markedly reduced in vitamin A deficiency (81). On the other hand, the process of immunological memory, essential for a marked secondary response, does not seem to be adversely affected. In a similar manner, the primary immune response to membrane polysaccharides of bacteria is severely depressed by vitamin A depletion but quickly recovers with vitamin A supplementation (81). In contrast, the immune response to bacterial lipopolysaccharides is unaffected by vitamin A status. Interestingly, the injection of lipopolysaccharide into vitamin A–deficient rats greatly enhances their immune response to tetanus toxoid and to bacterial polysaccharides (81). The nature of this stimulatory effect of lipopolysaccharide has not been clarified. Nonetheless, the observation shows that the antibody-forming process in vitamin A–deficient animals is functionally intact (81).

The humoral response to many viruses, including measles, herpes simplex, and the Newcastle disease virus, is impaired in vitamin A deficiency and improved by repletion. Similarly, the host response to some parasitic infections is reduced by vitamin A depletion (81).

In most instances, except as noted below, retinoic acid was more potent than retinol in restoring the humoral response in vitamin A–depleted animals (88). At least one mechanism of action for retinoic acid in the stimulation of antibody formation is now becoming clear (88). In vitamin A–sufficient animals, an antigen is phagocytized by an antigen-presenting cell (APC), often a macrophage, which metabolizes the antigen to a fragment that is presented on the cell surface bound to class II molecules of the major histocompatibility complex (MHC). T-helper lymphocytes are stimulated by contact with APC to form interleukin-2 (IL-2), which in turn enhances T-cell proliferation as well as B-cell growth and differentiation. Activated B cells (plasma cells) initially produce IgM antibodies but switch to the production of high-affinity IgG antibodies upon maturation (81).

Two types of T-helper cells exist, T_H1 and T_H2. T_H1 cells secrete cytokines that stimulate cell-mediated immunity (CMI) and T_H2 cells secrete cytokines that enhance antibody production (88). Thus, the balance between T_H1 and T_H2 will determine the nature of the response to a given antigen. T_H1 cells produce interferon γ (IFN-γ) and T_H2

cells produce IL-10, IL-5, and IL-4. Cross-regulation exists, inasmuch as IFN-γ inhibits T_H2 cell proliferation and IL-10 and IL-4, via different mechanisms, inhibit both T_H1 cell development and IFN-γ production (88).

In mesenteric lymph node cells of vitamin A–deficient mice, IFN-γ is overproduced and IL-10 and IL-5 are underproduced (88). Thus, the ratio of T_H1 to T_H2 cells is increased, presumably leading to an enhancement of the CMI response and to a decrease in the antibody response (88). These interesting observations explain the decrease in humoral immunity in vitamin A deficiency but do not accord with the general observation that CMI responses are usually depressed in vitamin A deficiency (81). Nonetheless, these valuable studies pinpoint the T-helper cell as a major site of vitamin A action in the immune response.

Much interest is now being shown in α-14-hydroxy-*retro* retinol (HRR). B-Lymphoblastoid cells transformed with Epstein–Barr virus die unless HRR, which is derived from retinol, is present in the medium (81). Retinoic acid is inactive in this system. Retinol, presumably via HRR, also is involved in the proliferation of normal B cells and T cells. Various cytokines can modulate the process but cannot replace HRR in it. HRR has also been identified in many other types of cells. These studies have stimulated much interest for two reason: (a) retinoic acid, which in most instances is presumed to be the active form of vitamin A, is inactive in these systems, and (b) other retro derivatives of vitamin A, such as anhydroretinol and retroretinol, do not enhance animal growth. As yet, however, the enzyme responsible for the conversion of retinol to HRR has not been studied. These observations indicate that another signaling pathway for retinoids, albeit specialized in given cells, may exist.

Phagocytosis, particularly the "oxygen burst" following the ingestion of a foreign body, and IgA secretion are also depressed in vitamin A deficiency (81). The synthesis of goblet cell mucins is reduced as well by vitamin A depletion, both in the intestinal mucosa and in the conjunctiva of the eye. Because protein malnutrition also adversely affects the immune response, children afflicted with both deficiencies are clearly at increased risk of developing severe infections (39).

In vitamin A–sufficient animals, carotenoids also enhance the immune response. Both nutritionally inactive (canthaxanthin) and nutritionally active (β-carotene) carotenoids have similar effects (39). Their mechanism of action is not known.

6. Transmembrane Proton Transfer

The past generalization that vitamin A was found only in the animal kingdom was upset in 1971 by the dramatic discovery of a new retinaldehyde-containing pigment, bacteriorhodopsin, in the membrane of the purple bacterium *Halobacterium halobium* (89). Soon thereafter, its basic function was discovered, namely, that under the influence of light, protons are pumped from the inside to the outside of the bacterial cell. The chemiosmotic gradient thereby established could be used for the active transport of nutrients into the cell and for the formation of ATP and other energy storage compounds.

Bacteriorhodopsin and rhodopsin are similar in many ways: Bacteriorhodopsin contains 248 amino acids and has a molecular weight of 26,000, 70% of which is in the form of α helices. Seven helical segments stretch back and forth across the membrane, each of which contains approximately 23 amino acid residues (90). The two chromophores of bacteriorhodopsin are 13-*cis* and all-*trans* retinaldehyde, both of which are bound to the protein as a Schiff base at Lys216 in approximately a 1:1 ratio at thermal equilibrium. The chromophore lies in the middle of the membrane at a tilted angle of 27 degrees from the plane of the membrane. In the light, the 13-cis isomer is converted to the all-trans form. In the dark-adapted pigment, as in the case of rhodopsin, the Schiff base is protected from reagents like hydroxylamine; however, after exposure to light, the reaction sites are exposed. After the absorption of a photon, a sequence of conformational changes occur in bacteriorhodopsin that are analogous to those seen with rhodopsin.

In the photocycle of bacteriorhodopsin, the Schiff base is deprotonated and then reprotonated. Because only one proton is transferred across the membrane in one photocycle, the Schiff base proton released probably plays a role in proton transfer.

In addition to bacteriorhodopsin, halobacteria contain four other related pigments: halorhodopsin, involved in chloride transport; sensory photosystems SR-I and SR-II, which control phototaxis; and slow-cycling rhodopsin (SCR), the function of which is unclear (91). SR-I (λ_{max} 587 nm) provides a light-attractant signal and SR-II (λ_{max} 373) a light-repellant signal. SR-I forms a molecular complex with a membrane-bound transducer protein, Htrl (92). The phototactic action of the SR-I–Htrl complex does not involve deprotonation and reprotonation. However, SR-I, when separated from Htrl, pumps protons across the membrane just like bacteriorhodopsin. Thus, the phototactic function of SR-I, as contrasted with the proton-pumping, energy-yielding action of bacteriorhodopsin, depends crucially on its interaction with the transducer protein Htrl (92). The actions of SR-I in photosynthetic bacteria and of rhodopsin in mammalian physiology clearly are highly analogous.

In animals, retinoids do not seem to play similar roles, although they can inhibit some processes involving active transport.

7. Gap Junction Communication

All-*trans* retinoic acid and some carotenoids elevate mRNA in cells in vitro for connexin 43, a gap junctional protein involved in intercellular communication (93). Interestingly, the action of canthaxanthin in enhancing connexin 43 synthesis is mediated by its cleavage product, 4-oxoretinoic acid (94). A transcription factor for 4-oxoretinoic acid, however, has not as yet been identified.

VII. PHARMACOLOGY

Vitamin A and various retinoids have been used to treat nutritional inadequacy, some skin disorders, and certain forms of cancer.

A. Nutrition

Oral doses of vitamin A of 200,000 IU_a (60 mg or 2.1 mmol) in oil have been used as a prophylactic measure in many less industrialized countries, but particularly in Asia (95,96). Such doses are usually given one to three times a year to children 2–6 years of age. Smaller doses (25,000–100,000 IU_a) have been used in younger children. Because vitamin A is generally well absorbed and is stored in the liver and other tissues, the procedure has been effective in reducing the incidence of acute vitamin A deficiency in children. Although vitamin A is relatively inexpensive, the logistics of maintaining a public health measure involving the individual dosing of children has not always proved easy.

Lactating women in less industrialized countries are also at risk. Thus, the administration of 200,000 IU_a of vitamin A to a mother soon after the birth of her child is helpful in enhancing her vitamin A reserves as well as in increasing the concentration of vitamin

A in her milk. The only danger is that a lactating woman may again become pregnant and, as a consequence, the dose may have teratogenic effects.

Other intervention strategies for improving the vitamin A status of groups at risk include the fortification of foods with vitamin A, horticultural programs featuring carotenoid-rich foods, and nutrition education. All of these interventions have been applied successfully in various countries. Needless to say, each approach has its own set of advantages and drawbacks (95,96).

In general, the vitamin A status of children in most less industrialized countries is improving. This encouraging positive development may be attributed to a variety of factors: the above-cited intervention strategies, other government-sponsored programs in nutrition and health, improved communication, a better standard of living, and political stability (95,96).

B. Skin Disorders

Various retinoids, and particularly 13-*cis* retinoic acid (Fig. 3a), etretinate (Fig. 3b), and acitretin (Fig. 3c), are used to treat acne, psoriasis, and other skin disorders (97,98). Although highly effective, retinoic acid and etretinate are teratogenic at high doses when given orally, and retinoic acid is a skin irritant when applied topically. Although most efficacious retinoids are also toxic, some conjugated forms, such as retinoyl β-glucuronide and hydroxyphenylretinamide, retain their therapeutic actions with less, if any, toxicity (99). Topical all-*trans* retinoic acid can also reduce wrinkling and hyperpigmentation caused by photo-induced aging (39,97,98).

The mechanism of these effects is probably multifactorial. On the physiological level, retinoic acid can alter epidermal differentiation pathways, induce epidermal hyperplasia, inhibit transglutaminase, and, as a result, reduce the cross-linking of proteins in the cornified envelope, change the pattern of keratins formed by keratinocytes, alter membrane viscosity, stimulate gap junction formation, influence various facets of the immune system (including the inflammatory response), inhibit collagenase formation, and reduce sebum production (97,98).

On the molecular level, skin cells contain the usual cytosolic binding proteins and some nuclear receptors for retinoids (98,100). The distribution of the latter, however, seems unique for skin: RXRs predominate over RARs by a ratio of 5:1. Of the RARs, RARγ represents 87% and RARα 13% of the total. RARβ is not detected. RXRα is the only detected receptor of the RXR family. The major heterodimer present, as expected, is RXRα/RARγ (100). Topically applied all-*trans* retinoic acid induces CRABP-II strongly, but not CRABP-I, in human skin. Thus, all of the elements for the control of differentiation by retinoids are present in skin cells, but the key factors involved in the therapeutic effects of RA on abnormal skin are only being identified slowly (98,101).

C. Cancer

Retinoids have been used both as therapeutic and as chemopreventive agents against a large variety of tumors in both experimental animals (102) and humans (103).

In experimental animals, retinoids have been tested, in most instances with positive results, as chemopreventive agents against chemically induced cancers of the mammary gland, skin, lung, bladder, pancreas, liver, digestive tract, and prostate gland (102). Combination therapy, e.g., chemopreventive use of hydroxyphenylretinamide and tamoxifen against *N*-methyl-*N*-nitrosourea-induced mammary cancers, has often proved more effective, and sometimes synergistically so, than the use of only one agent (102). Retinoids alone show the best chemopreventive effects with tumors that follow discrete promotional stages of carcinogenesis, e.g., mammary and skin tumors.

In humans, trials with retinoids or carotenoids, often together with other nutrients, have been conducted to prevent oral premalignancies, bronchial metaplasia, laryngeal papillomatosis, Barrett's esophagus, actinic keratosis, and cervical dysplasia—all considered to be premalignant lesions (103). Chemoprevention trials against secondary primary tumors of the head and neck, skin, breast, lung, and bladder have also been initiated or conducted, as have several primary prevention trials for epithelial cancer. Therapeutic trials have been conducted with a variety of hematological and epithelial malignancies (103).

The most promising results have been found with all-*trans* retinoic acid for acute promyelocytic leukemia, but drug resistance developed rapidly. Other positive responses were obtained with 13-*cis* retinoic acid for squamous cell cancers of the skin; with 13-*cis* retinoic acid for mycosis fungoides; with 13-*cis* retinoic acid, etretinate or β-carotene for oral leukoplakia; with 13-*cis* retinoic acid for laryngeal papillomatosis; with etretinate for actinic keratosis; with all-*trans* retinoic acid for cervical dysplasia; and with 13-*cis* retinoic acid for secondary primary tumors in patients with head and neck cancer (103). Responses to retinoids were poor or unconvincing in a large number of other cancers. Responses to retinoids tended understandably to be better in diseases that were not too far advanced. The major problem encountered was the toxicity caused by the most efficacious doses (103).

In only one disease, acute promyelocytic leukemia (APL), is the mechanism of action of retinoic acid fairly well defined (30,75,103). In essence, the *RARα* gene and the *PML* gene form a hybrid because of a chromosomal translocation. Presumably higher concentrations of retinoic acid are therefore required to activate the modified RARα transcription factor and to induce cellular differentiation. After several months, resistance to retinoic acid treatment develops, which may be due to an increased rate of retinoic acid metabolism, increased sequestration of retinoic acid by induced CRABP, or reduced permeability of the cell to retinoic acid. Thus, the treatment of APL by retinoic acid, which initially was heralded with enthusiasm, is clearly less effective as long-term therapy than initially hoped.

Cancer is such a complex and variegated group of diseases that generalizations about possible mechanisms of action of retinoic acid have not been very fruitful. Nonetheless, retinoids probably act by stimulating the differentiation of precancerous stem cells, whereas carotenoids probably are involved in a network of antioxidants that include vitamin E, vitamin C, sulfhydryl groups, and a variety of other enzymatic and nonenzymatic processes.

D. Other Diseases

Retinoids and carotenoids have been implicated as protective agents in a variety of other diseases, including heart disease, cortical cataract, and age-related macular degeneration (104,105). Epidemiological and intervention studies with carotenoids are considered later.

VIII. TOXICITY

Three types of toxicity to retinoids exist: acute, chronic, and teratogenic (106–111). Acute toxicity is due to a single dose or a limited number of large doses taken during a short period, whereas chronic toxicity is caused by moderately high doses taken frequently, usually daily, over a span of months or years. Teratogenicity is caused by the ingestion of moderate to high doses during the first trimester of pregnancy in humans and at selected "sensitive" periods during gestation in animals.

A. Acute Vitamin A Toxicity

When a single dose of more than 0.7 mmol of vitamin A (>200 mg, or $>660,000$ IU_a) is ingested by adults or when a dose larger than 0.35 mmol (>100 mg or $>330,000$ IU_a) is ingested by children, nausea, vomiting, headache, increased cerebrospinal fluid pressure, vertigo, blurred (double) vision, muscular incoordination, and (in infants) bulging of the fontanelle may occur (39,106). Some infants can be adversely affected by single doses of only 0.1 mmol. These signs are generally transient and subside within 1 to 2 days. When the dose is extremely large, drowsiness, malaise, inappetence, reduced physical activity, skin exfoliation, itching around the eyes, and recurrent vomiting soon follow. Finally, when lethal doses are given to young monkeys, an excellent model for vitamin A toxicity in the human, the animals have deepening coma, convulsions, and respiratory irregularities, and they finally die of either respiratory failure or convulsions (107). The median lethal dose (LD_{50} value) of vitamin A injected intramuscularly in a water-miscible form in the young monkey is 0.6 mmol (168 mg) retinol per kilogram body weight. Extrapolated to a 3-kg child and a 70-kg adult, the total LD_{50} dose would be 1.8 mmol (500 mg) and 41 mmol (11.8 g), respectively. A newborn child, who mistakenly was given 0.09 mmol (25 mg) daily, or 28 μmol/kg for 11 days, died of vitamin A toxicity (108). The total dose received was 0.31 mmol/kg, 50% of the LD_{50} value for young monkeys. Such enormous amounts of vitamin A are present only in high-potency preparations of vitamin A or in large amounts ($\sim$500 g) of livers particularly rich in vitamin A (>0.035 mmol/g or >10 mg/g) (39).

The LD_{50} values of all-*trans* retinyl ester, 13-*cis* retinoic acid, and all-*trans* retinoic acid in the adult rat are 27.7 mmol/kg body weight, 13.3 mmol/kg body weight, and 6.7 mmol/kg body weight, respectively (109). These LD_{50} values reflect the general observation that the relative toxicities of these retinoids are all-*trans* retinoic acid $>$ 13-*cis* retinoic acid $>$ all-*trans* retinol and its esters (109,110). Retinoids injected intraperitoneally are much more toxic than those administered orally, which in turn are more toxic than those applied topically to the skin. Species differences in toxicity also exist.

B. Chronic Vitamin A Toxicity

Chronic toxicity is induced in humans by the recurrent intake of vitamin A in amounts at least 10 times the RDA, i.e., 13 μmol (3.75 mg retinol equivalents or 12,500 IU_a) for an infant or 35 μmol (10 mg retinol equivalents or 33,300 IU_a) for an adult. A health-food enthusiast who ingested 26 μmol (25,000 IU_a) of vitamin A as a supplement daily plus a similar amount in food showed severe signs of toxicity (39,109,110). Approximately 50 signs of chronic toxicity have been reported, of which the most common are alopecia, ataxia, bone and muscle pain, cheilitis, conjunctivitis, headache, hepatotoxicity, hyperlipemia, hyperostosis, membrane dryness, pruritis, pseudotumor cerebri, various skin disorders, and visual impairment (39,109,110). Particular attention has recently been paid to serious adverse effects on the liver caused by daily doses of 26–52 μmol (25,000–50,000 IU_a) of vitamin A (111,112). When the supplemental intake of vitamin A is eliminated, these signs usually, but not always, disappear over a period of weeks to months (39).

In chronic hypervitaminosis A, holo-RBP in the plasma is not much elevated, whereas retinyl esters are usually increased markedly. Factors that enhance toxicity include alcohol ingestion, low protein intake, viral hepatitis, other diseases of the liver and kidney, and possibly tetracycline use. Elderly individuals may be more sensitive because of a slower rate of storage in the liver and a reduced plasma clearance of administered vitamin A. Tocopherol, taurine, and zinc are protective in tissue culture cells, but they may or may not be effective in vivo (39,110).

Some individuals seem to suffer from vitamin A intolerance, i.e., the appearance of signs of toxicity upon routinely ingesting moderate amounts of vitamin A. This relatively rare condition, which seems to be genetic, mainly affects males (38,39,110).

C. Teratogenicity

Vitamin A and other retinoids are powerful teratogens both in experimental animals and in women (39,109,110,113). A single extremely large dose, exposure for as short as a week on high daily doses (0.1–0.3 mmol, or 30–90 mg), or long-term daily intakes of 26 μmol (25,000 IU_a or 7500 retinol equivalents) of vitamin A during early pregnancy can induce spontaneous abortions or major fetal malformations. Common defects are craniofacial abnormalities, including microcephaly, microtia, and harelip; congenital heart disease; kidney defects; thymic abnormalities; and central nervous system disorders (39,109,110,113).

Permanent learning disabilities have been noted in otherwise normal rat pups whose dams received nonteratogenic doses of vitamin A (114), and similar effects have been noted in children whose mothers received large doses of 13-*cis* retinoic acid (115,116). Most of the children with severely impaired intelligence quotients in this study, however, also had major physical malformations. In general, the doses that induce demonstrable learning disabilities in rats are approximately 40% (range 10–70%) of those that cause terata (114). Many procedures have been employed to measure behavior, and usually the measurement has been conducted within weeks of birth. Whether the observed effects become worse or are ameliorated at a later time is still unclear.

Synergism between vitamin A and other teratogens, such as alcohol and drugs, at nonteratogenic doses of each is probable. Thus, women who are pregnant, or who might become so, should carefully control their intake of vitamin A, both in regard to rich food sources, such as liver, and to vitamin A supplements (39).

D. Relative Toxic Effects of Different Retinoids

Most attention has focused on the relative teratogenic effects of natural and synthetic retinoids that have been, or might be, used therapeutically. As shown in Table 5, a 10,000-fold range exists, with TTNPB as the most teratogenic and retinoyl β-glucuronide as the least (117). The reasons for this wide range of teratogenic activities are only partly understood. All-*trans* retinoyl β-glucuronide, for example, when given in large doses, seems to be absorbed from the intestinal tract less well than all-*trans* retinoic acid, is transferred across the placenta much less well than all-*trans* retinoic acid, is converted to a significant extent to the less toxic 13-cis isomer, is taken up by the embryo less well than all-*trans* retinoic acid, and is inherently much less teratogenic than all-*trans* retinoic acid (118). Indeed, the formation of glucuronides of retinoids may well be a protective metabolic process to prevent toxic effects (113). Isomerization may also serve a protective action, e.g., 9-*cis* retinoic acid is rapidly converted to the much less active 9,13-*cis* retinoic acid

Table 5   Relative Teratogenicity of Retinoids in Hamsters[a]

Retinoid	TD[b]$_{50}$ (mg/kg)
TTNPB	<0.02
Etretinate	6
All-*trans* retinoic acid	11
13-*cis* retinoic acid	22
All-*trans* retinal	23
All-*trans* retinol	23
All-*trans* hydroxyphenylretinamide	139
All-*trans* retinoyl β-glucuronide	>200[c]

[a]Selected data from Refs. 117 and 118.
[b]Teratogenic dose to the dam that induced malformations in 50% of the offspring.
[c]Extrapolated value.

during pregnancy (119,120). Nonetheless, 9,13-*cis* retinoic acid might possibly play some functional role in pregnancy (120).

TTNPB, as a stable aromatic molecule, might be metabolized less rapidly than retinoic acid. Thus, its high activity both physiologically and as a teratogen might be attributed to a lack of its conversion to less active products. Inasmuch as the rate of glucuronidation of TTNPB and of all-*trans* retinoic acid by rat liver microsomes is similar (G. Genchi and J. A. Olson, unpublished observations), some other metabolic step might be affected.

E.   Mechanisms of Toxic Action

Acute toxicity is probably caused by the presence of significant amounts of "free" retinol and retinyl ester in blood and tissues. Retinol bound to RBP is much less cytotoxic than unbound retinol. Similarly, the activity of all-*trans* retinoic acid in cell differentiation is reduced by the presence of large amounts of CRABP in the cells (30). When large doses of retinol are given to monkeys, the amount of retinol in the brain markedly increases (107), consistent with the many signs of central nervous system involvement in acute toxicity.

Chronic toxicity and teratogenicity seem to involve primarily the nuclear retinoid receptors (80,109,113). The teratogenic effects of AM-580 (Fig. 3g), a RARα-specific agonist, in pregnant mice are offset by a sulfur-containing analog of TTNBP that serves as a RARα antagonist (109). As already mentioned, retinoic acid activates via its receptors several families of *Hox* genes that are closely implicated with specific developmental structures. Thus, the concentration of retinoic acid at various locations in the embryo must be very carefully regulated so that a proper spacial and temporal sequence of activations can occur. By flooding the embryo with retinoic acid, regulation will be lost. Thus, in early mouse embryos treated with all-*trans* retinoic acid, *Hox a-1* (*Hox 1.6*) expression at 7.5 d is much more extensive and less localized than in control embryos (109).

Other possible actions of retinoids include effects on cell proliferation, cell differentiation, cell migration, programmed cell death (apoptosis), and membrane integrity. Many of these effects also involve known retinoid receptors, but others do not (109,113).

The structural characteristics of highly teratogenic retinoids comprise the following (117):

1. A polar terminus with an acidic pK_a
2. A lipophilic polyene side chain with good π electron delocalization, e.g., cis isomers are less active than the all-trans isomer
3. A fairly lipophilic ring with no set nature or dimensions opposite the polar terminus, and
4. Conformational restriction (for acidic retinoids only)

Interestingly, the two *least* teratogenic compounds, 4-hydroxyphenylretinamide (Fig. 3d) and retinoyl β-glucuronide (Fig. 1j) are conjugated, more polar molecules.

F. Safe Levels of Vitamin A Intake in Humans

This issue of crucial public health importance can be approached in two ways: (a) by assessing the no-effect level of vitamin A intake experimentally in various populations at risk and (b) by reviewing the formal recommendations made by expert groups. The major groups at risk are infants, young children, pregnant women, and lactating women. The issue is complicated in part by whether or not a given population will clearly benefit as well as be put at-risk by the ingestion of a large dose of vitamin A. In the following discussion, however, only the issue of toxicity will be considered.

Statistical issues are also crucial. It clearly is easier to define an appropriate maximal intake for a population in which an allowable percentage of that group, e.g., 1%, 3%, or 5%, is adversely affected than to specify an intake that presumably will affect nobody. As already mentioned, genetic vitamin A intolerance is known to exist in a handful of persons (38,110,121). However, recommendations for a healthy population certainly should not be based on the adverse reactions of a few hypersensitive individuals. Finally, some relatively nonspecific reactions to moderate doses might not be causally related to the vitamin A in the dose.

Nonetheless, despite these caveats, some useful guidelines can be defined for vitamin A intake. RDIs of vitamin A are completely safe insofar as we know. RDA values for infants, children 1–6 years, and pregnant women are 375, 400–500, and 800 µg retinol equivalents, respectively (35); (Table 2). These values, however, assume that intakes are a mixture of preformed vitamin A and corotenoids. Thus, the estimated amount of *preformed* vitamin A in the diet of infants, children 1–6 years, and pregnant women are approximately 340 µg (1133 IU_a), 300–375 µg (1000–1250 IU_a), and 600 µg (2000 IU_a), respectively.

1. Pregnant Women

Although additional vitamin A is not usually needed by healthy pregnant women, daily multivitamin tablets containing 5000 IU_a (1500 µg) of vitamin A are ingested by large numbers of women. Toxic reactions to their ingestion have not been reported.

Toxic reactions have been noted, however, in women, both pregnant and nonpregnant, ingesting daily doses of $\geq$18,000 IU_a ($\geq$5400 µg) (106,110–113), whereas only sporadic claims, often poorly documented, have been made for toxic effects at lower daily doses. In a study of retinoid embryopathy in pregnant women, the lowest dose of 13-*cis* retinoic acid that was associated with birth defects was 10 mg/day (122). Thus, 1500 µg of vitamin A seems safe for essentially all pregnant women, whereas $\geq$5400 µg probably is not.

Responding to concerns about the teratogenicity of vitamin A, the Teratology Society recommended that women who might become pregnant limit their daily intake of preformed vitamin A to 8000 IU_a (2400 μg) and ingest provitamin A carotenoids as a primary source of dietary vitamin A (123). Furthermore, they recommended that the unit dose of commercially available vitamin A be limited to 5000–8000 IU_a and that the hazards of excessive intakes of vitamin A be indicated on the labels of such products. Similarly, the American Institute of Nutrition, the American Society for Clinical Nutrition, and the American Dietetic Association issued a joint formal statement that supplements of vitamins and minerals were not needed by well-nourished, healthy individuals, including pregnant women, except in some specific instances (124). The Council for Responsible Nutrition, a group sponsored by industry, has also advised that pregnant women, while needing to ensure an adequate intake of vitamin A, should prudently limit their intake of nutritional supplements of vitamin A to 5000–10,000 IU_a (125). Subsequently, they recommended that the unit dosage of retinol in commercial vitamin A preparations be limited to 10,000 IU_a.

In recognition of the fact that a deficiency of vitamin A in the mother can also cause abortion and fetal abnormalities, the International Vitamin A Consultative Group (IVACG) has recommended that the average daily diet of pregnant women should supply 620 μg retinol equivalents, in keeping with FAO/WHO recommendations (34). However, in areas of the world where this level of intake does not occur and little opportunity exists for dietary improvement, or in emergency situations in which food supplies are disrupted, IVACG recommends that daily supplements of 3000 μg retinol equivalents (10,000 IU_a) can be given safely anytime during pregnancy. They do not suggest, by the way, that well-nourished women take supplements.

To summarize, well-nourished healthy women of reproductive potential should include carotenoid-rich fruits and vegetables in their diet. They should also avoid taking supplements of preformed vitamin A during the first trimester of pregnancy, during which increased nutritional demands are small and the risk of fetal abnormalities is high. If supplements of vitamin A are subsequently taken, the daily dose should be carefully limited to 5000–10,000 IU_a.

2. Infants and Young Children

As already indicated, the probable daily amounts of preformed vitamin A in recommended diets of infants and young children are 1133 IU_a and 1250 IU_a, respectively. As a general public health measure, oral doses (200,000 IU_a) of all-*trans* retinyl palmitate in oil have been administered one to three times a year to preschool children, usually 1–6 years of age, in less industrialized countries (95). Side effects, e.g., nausea, vomiting, and bulging of the fontanelle in infants, have usually been reported in <5% of the treated children and have been transient (1–3 days) in nature. Because vitamin A toxicity is a function of weight, younger children tend to be most affected.

Because of international public health interest in combining vitamin A supplementation with the expanded program of immunization (EPI), the effects of dosing infants with vitamin A from 6 weeks to 9 months of age has been explored. When 50,000 IU_a of vitamin A was given orally in oil to Bangladesh infants at 6, 11, and 16 weeks of age, 11% showed transient bulging of the fontanelle. In a subsequent study in which 25,000 IU_a was given orally at approximately the same three times to Bangladesh infants, 8%, corrected for the 2.5% incidence in the placebo group, showed the same effect (126). In this study, nearly all of the affected children had received three doses of vitamin A; none showed any toxicity after one dose. If we assume that the infants' daily intake from breast milk was approximately 100 μg (333 μg IU_a), the calculated mean daily increment from the dose is only an additional 715 IU_a, giving a total of 1048 IU_a, essentially the same as the recommended daily intake. A single oral dose of 50,000 IU_a given to Indonesian infants showed much less toxicity, i.e., 4.5% in the vitamin A–treated group and 2.5% in the control group (127). Furthermore, the cerebral fluid volume, but *not* the pressure, was transiently increased in these infants, and no lasting side effects were noted. Bulging of the fontanelle, in consequence, may be more of a transient physiological response to a dose of this magnitude than an indicator of toxicity.

As yet, the relationship between an acceptably safe dose and the age of infants has not been defined. Quite possibly, infants suffering from inadequate intakes of protein and calories may be more susceptible to vitamin A toxicity than better nourished infants. If so, the dose of vitamin A that is selected for a given country might well be based on anthropometric indices within that country.

Thus, a generally safe single dose for most infants, although currently undefined, probably will fall in the range of 4000–5000 IU_a per kilogram body weight with an interval between doses of $\geq$5 weeks.

IX. CAROTENOIDS

Unlike retinoids, including vitamin A, carotenoids are generally nontoxic. However, individuals who routinely ingest large amounts of carotenoids, either in tomato or carrot juice or in commercial supplements of β-carotene, can develop hypercarotenosis, characterized by a yellowish coloration of the skin and a very high concentration of carotenoids in the plasma. This benign condition, although resembling jaundice, gradually disappears upon correcting the excessive intake of carotenoids. The only known toxic manifestation of carotenoid intake is canthaxanthin retinopathy, which can develop in patients with erythropoietic porphyria and related disorders who are treated with large daily doses (50–100 mg) of canthaxanthin, the 4,4'-diketo derivative of β-carotene, for long periods (128). In most instances, however, these deposits of canthaxanthin disappear slowly upon termination of treatment (128). Canthaxanthin-containing supplements are not currently available in the United States. β-Carotene at similar doses is not known to cause retinopathy. Carotenoids, even when ingested in large amounts, are not known to cause birth defects or hypervitaminosis A, primarily because the efficiency of their absorption from the intestine falls rapidly as the dose increases and because their conversion to vitamin A is not sufficiently rapid to induce toxicity (39).

Quite apart from their function as precursors of vitamin A, carotenoids are distributed widely in mammalian tissues, can quench singlet oxygen, can serve as an antioxidant in tissues (particularly under conditions of low oxygen tension), and can stimulate the immune response (39,104).

Thus, by using provitamin A activity as the nutritional function and singlet oxygen-quenching and antioxidant activity as the biological action, four classes of carotenoids might be defined: those that are both nutritionally and biologically active, such as β-carotene; those that are nutritionally active and biologically inactive, such as 14'-β-apocarotenal; those that are nutritionally inactive but biologically active, such as lycopene and violaxanthin; and those that are both nutritionally and biologically inactive, such as phytoene.

Because over 90% of the 600 characterized carotenoids in nature are not precursors of vitamin A, their biological effects in mammalian physiology, independent of their provitamin A activity, are being followed with interest.

X. POTENTIAL HEALTH BENEFITS OF RETINOIDS AND CAROTENOIDS

A. Retinoids

The requirements for vitamin A and safe levels of intake have already been discussed. A positive response to chemopreventive treatment with large doses of vitamin A has been shown in leukoplakia and actinic keratosis (103). The retinoids most commonly used for therapy and chemoprevention, however, are 13-*cis* retinoic acid, all-*trans* retinoic acid, hydroxyphenylretinamide, etretinate, and acitretin (103). As indicated earlier, these agents have shown promising results in the prevention or treatment of some carcinomas (103). The major drawbacks in their use are their toxicity at highly efficacious doses and, in the case of APL, a rapidly developing resistance to the drug. Of the retinoids listed, all-*trans* retinoic acid and etretinate, because of the latter's slow turnover in the body, are the most toxic and hydroxyphenylretinamide the least. The search consequently continues to identify new retinoids with high efficacy but low toxicity. Retinoyl β-glucuronide, a naturally occurring metabolite of retinoic acid, shows these properties (99,118). It is active in treating acne but has not been tested against other diseases. In epidemiological surveys, the dietary intake of preformed vitamin A and the plasma concentration of retinol are rarely associated with a reduced incidence of chronic diseases.

B. Carotenoids

1. Cancer

One of the most dramatic and consistent observations in epidemiological studies is the inverse association between β-carotene intake and the incidence of *lung cancer* (104,129). These findings have stimulated intervention trials in two high-risk groups: asbestos workers in Tyler, Texas and middle-aged male smokers in Finland. The results of these intervention trials are disappointing. In asbestos workers, no differences in the prevalence of sputum atypia was noted between treated (50 mg β-carotene + 25,000 IU_a retinol every other day) and control groups over a 5-year period. In the Finnish study, the group treated daily with β-carotene (20 mg) showed a significantly higher incidence of lung cancer [relative risk (RR) = 1.18, 95% confidence interval (CI) = 1.03–1.36] and total mortality (RR = 1.08, 95% CI = 1.01–1.16) than did the placebo group. Supplemental β-carotene did not affect the incidence of other major cancers found in this population (129).

The unexpected negative finding in the Finnish study has several possible explanations: (a) Supplemental β-carotene is interfering with the intestinal absorption of other possible chemopreventive nutrients. In this regard, β-carotene inhibits the absorption in humans of lutein, which shows good antioxidant activity (42). In that same vein, α-carotene, which shows chemopreventive properties, might be similarly affected. (b) Supplemental β-carotene may be serving as a pro-oxidant in the well-oxygenated ambient of the lung (130). (c) The population of middle-aged male smokers is not representative of other groups, who might well benefit from a higher intake of carotenoids. (d) A comparison between treated and control subjects that fall only in the lowest quartile of initial plasma carotene values might yield different results. (e) Vitamin C, which is low in the plasma of most Finns, may have played some role in the outcome. (f) Alcohol intake may also play an important role in the outcome.

Several other trials have provided results that support the findings in the Finnish study, namely, the Carotene and Retinal Efficacy Trial (CARET) and the Physician's Health Study (129).

The development of *head and neck cancers*, including those of the oral cavity, pharynx, and larynx, are influenced by many factors, including smoking, other uses of tobacco, alcohol, and diet (129). Serum carotene concentrations, adjusted for smoking, are inversely related to the incidence of these carcinomas. Supplements of β-carotene can markedly reduce leukoplakia, although the lesion returns upon cessation of treatment (129). A chemoprevention trial designed to assess the effect of daily supplements of β-carotene (50 mg) on the recurrence of head and neck cancer is currently under way (129).

The effects of various nutrient combinations on *esophageal cancer* and on *stomach cancer* was evaluated in Linxian, China, where the incidence of esophageal cancer is 100-fold higher than in the United States (129). Of four nutrient treatments, only one, involving supplements of β-carotene, selenium, and α-tocopherol, showed a positive effect: the reductions in total deaths, cancer deaths, esophageal cancer deaths, and gastric cancer deaths were 9% (RR = 0.91, 95% CI = 0.84–0.99), 13% (RR = 0.87, 95% CI = 0.75–1.00), 4% (RR = 0.96, 95% CI = 0.78–1.18), and 21% (RR = 0.79, 95% CI = 0.64–0.99), respectively (129). Although these results support the concept that diet influences cancer incidence, the general nutritional status of the population was poor. Thus, whether the mixed supplement, or one component of it, was protective as a result of generally improved health or of a more specific anticancer effect is not clear (129).

The dietary intake and serum concentrations of carotenoids are often inversely associated with the risk of *colorectal cancer* (129). However, by using adenomas as an indicator, supplemental β-carotene (25 mg/day) was found to be ineffective (RR = 1.01, 95% CI = 0.85–1.20) in preventing the recurrence of this lesion (129).

β-Carotene intake has been associated with an improved survival rate in *breast cancer* patients (129). Whether supplements of carotenoids reduce the incidence of breast cancer in a well-designed clinical trial is not known. In an ongoing trial, the effect of hydroxyphenylretinamide on the recurrence of breast cancer is being explored (103).

The risk of *cervical cancer* has been correlated with the prediagnostic serum levels of α-, β-, and total carotenoids (RR = 2.7–3.1, 95% CI = ≥1.1–≤8.1) (129). On the other hand, invasive cervical cancer among white women in the United States did not relate to any specific food group of the diet or to the use of supplements of vitamins A, C, and E and folic acid (131). However, cervical dysplasia, considered to be a precancerous lesion, did respond to β-carotene supplements (30 mg/day) (129).

The recurrence of skin cancer was not affected by β-carotene supplements (50 mg/day) over a 5-year period (RR 1.05, 95% CI = 0.91–1.22) (129).

Thus, a dichotomy exists. Most of the associations found between diseases in dietary or plasma level studies do not agree with the results of intervention trials. The former tend to show strong significant correlations and the latter, in large part, do not. Possible explanations are as follows: (a) β-Carotene, which is only one of approximately 600 known carotenoids, might not be the most active one, or indeed, might inhibit the absorption of other more chemopreventive carotenoids and other nutrients. (b) Carotenoids might be only one of a group of chemopreventive agents in foods that act synergistically in preventing carcinogenesis. The fact that the relative risk values for colored fruits and vegetables usually are less (more protective) than those for carotenoids or for any other component of the food supports this viewpoint. (c) Carotenoids may serve solely as a useful *marker* for a healthful lifestyle. (d) The preventive action of carotenoids might occur very early in disease progression but be ineffective later. Thus, subjects in identified high-risk groups, who often have had a primary tumor, may be resistant to nutritional supplements. (e) The associations found in observational epidemiology are not causal and can be confounded by a variety of unanticipated and unmeasured factors. In essence, the intervention trials may well be providing more valid answers (129).

2.   Photosensitivity Disorders

Patients with erythropoietic porphyria and similar diseases benefit by ingesting supplements (180 mg/day) of β-carotene. Canthaxanthin, though also protective, is no longer used because of the reversible retinopathy that results (128). Although the concentrations of β-carotene and vitamin A are elevated in the livers of these patients, the side effects of β-carotene ingestion over a period of years are minimal (129).

3.   Cardiovascular Disease

Epidemiologic studies suggest protective effects of carotenoid intake against both coronary events and stroke (104,129). In a European study (WHO/MONICA, i.e., monitoring cardiovascular disease), mortality from ischemic heart disease correlated inversely with serum vitamin E concentrations ($r^2 = 0.63$) but not with β-carotene levels ($r^2 = 0.04$) (132). If the 3 Finnish sites, which were outliers, of the 16 examined were excluded, however, the inverse correlation with β-carotene concentrations improved markedly ($r^2 = 0.50$). A mean serum β-carotene concentration in populations of 0.4 µmol/L or higher was associated with good health in the European studies, whereas a concentration <0.25 µmol/L in populations was related to an increased risk of coronary disease, stroke, and cancer (132). The risk of myocardial infarction was inversely related to adipose β-carotene content in smokers (RR = 2.62, 95% CI = 1.79–3.83) but not in nonsmokers (RR = 1.07) (129). Furthermore, physicians with stable angina or prior coronary revascularizations, who were supplemented with β-carotene for 5 years, showed a 51% reduction in the risk of major coronary events (129). β-Carotene did not show beneficial effects, however, in the total population enrolled in the Physicians Health Study (129). The incidence of cardiovascular deaths in the Finnish lung cancer study also was not affected by β-carotene supplementation.

The overall results, therefore, are somewhat mixed. The most likely mechanism of action of carotenoids, but by no means the only one, is a reduction in the oxidation of low-density lipoproteins, which seem to play a key role in atherogenesis (104). Of various antioxidants studied both in vivo and in vitro, however, β-carotene does not seem to be very protective, if at all. Thus, the relationship among dietary intakes of carotenoids, their plasma and tissue concentrations, and cardiovascular disease remains unclear.

4.   Age-Related Macular Degeneration

The macular of the eye predominantly contains two pigments, lutein and zeaxanthin (133,134). Because these two pigments account for less than 25% of plasma carotenoids, their uptake from plasma and deposition in the macula show specificity. These pigments might consequently play a role in protecting the macula from damage caused by light and particularly by blue light. In a recent study with patients suffering from age-related macular degeneration (ARMD) vs. matched controls, subjects in the highest quintile of carotenoid intake had a 43% lower risk (RR = 0.57, 95% CI = 0.35–0.92) of suffering from ARMD than those in the lowest quintile (129). Of various carotenoids, the intake of spinach and collard greens, rich in these two carotenoids, was most strongly associated with reduced risk. However, not all studies support this finding (105). Nonetheless, higher plasma concentrations of lutein and zeaxanthin as well as β-carotene showed a significant trend toward a lower risk of developing ARMD. These interesting findings are currently being investigated.

5. Senile Cataract

Cataract consists of gradual opacification of the lens with aging, which may in part result from oxidative stress. Carotenoid intake, as well as that of vitamins C and E, has been associated with a reduced risk of cataract (105,129). In the main Linxian, China trial, however, combined supplements of β-carotene, selenium, and α-tocopherol were not associated with a reduction in the incidence of cataracts, and inconclusive results have been reported by others (105,129). Thus, whereas the concept that antioxidant nutrients might prevent oxidative damage to a fairly exposed structure, such as the lens, is highly feasible, the data supporting a protective role of dietary components in the process are mixed.

6. HIV Infection

In HIV infection T-helper (CD4) cells are destroyed, thereby impairing the immune response. In humans as well as in experimental animals, both β-carotene, which is a provitamin A carotenoid, and canthaxanthin, which is not, can enhance the immune response (135). Indeed, in HIV-infected patients, large doses of β-carotene increased the CD4/CD8 ratio, which is usually depressed in HIV infection, and improved the response to vaccines (135). UV light tends both to activate human HIV expression, at least in transgenic mice, and to reduce plasma carotenoid concentrations in humans. In phase II HIV-infected subjects, plasma carotenoid concentrations are reduced by 50%. AIDS patients treated daily with a combination of β-carotene supplementation (120 mg) and whole-body hyperthermia (42°C, 1 h) showed a better and longer lasting response than either treatment separately (136). Thus, carotenoids seem to ameliorate the condition of AIDS patients, probably, at least in part, by enhancing the immune response.

A quite different effect of a carotenoid has also been reported, namely, that halocynthiaxanthin (5,6-epoxy-3,3′-dihydroxy-7′,8′-didehydro-5,6,7,8-tetrahydro-β,β-carotene-8-one) strongly and rather specifically inhibits RNA-dependent DNA polymerase of the HIV virus (135). The use of carotenoids in the treatment of subjects with HIV infections clearly merits further attention.

XI. CONCLUDING REMARKS

Revising a chapter that was written approximately 9 years ago (137) is a valuable but humbling experience. The facts cited earlier have not changed, and many of the concepts have been modified only slightly; however, interests have shifted markedly and a whole new body of information and hypotheses—some quite clear, some conflicting—has arisen. Old observations are viewed in new ways, and a completely new set of research questions are being asked. Thus, in revising the chapter, the addition of a paragraph here and a reference there was just not feasible. As a consequence, the chapter is largely rewritten, and the reference list is in large part new. This new chapter and the previous one are, therefore, complementary to each other.

A major recent advance has been the discovery of the nuclear retinoid receptors and their impact on embryogenesis, cell differentiation, disease, and pharmacology. In truth, the paradigm of interpreting vitamin A actions has markedly changed. A new chemistry has arisen, with the focus on finding compounds that serve as specific agonists and antagonists for given retinoid receptors. The linkage between nutrition and molecular biology, seemingly such diverse fields, has been strengthened by the observation that the RXR receptors for vitamin A interact meaningfully with the thyroid receptor, dependent on iodine for its activity, and with the vitamin D receptor, dependent of course on another fat-soluble vitamin. The hint (not the demonstration) that α-tocopherol may also have specific nuclear effects adds further interest to this linkage.

Diet, admittedly along with a variety of other factors, is known to affect the onset and possibly the severity of major chronic diseases. In the past several years, an explosion of information about specific nutrients that may play roles in these processes has appeared. Despite the great care with which most of these studies have been done, these surveys have inherent constraints. Thus, the findings in large part have tantalized us rather than presented a coherent picture. Carotenoids, primarily together with vitamin E, have played a central role in these surveys.

Although these studies have great potential impact, some biases have arisen: namely, that carotenoids, vitamin A, vitamin E, vitamin C, and selenium all act solely as antioxidants in these processes. Indeed, the term "antioxidant vitamins" has become common parlance. To stimulate a broader, less constrained view of their potential actions, these nutrients, as well as many other naturally occurring compounds, both of endogenous as well as of dietary origin, might better be called "physiological modulators" (138). By so doing, the mechanism of action is not automatically inferred from the outset for whatever beneficial or adverse effect that they might show (138,139).

To keep the reference list within bounds, references are largely made to reviews, which in turn can serve as a guide to the primary research literature. Recent reviews, monographs, and articles of particular interest are cited in references 140–162. I regret the necessary omission of many specific research papers that have enriched our knowledge in this dynamic field.

ACKNOWLEDGMENTS

This study has been supported in part by grants from the National Institutes of Health (DK-39733), the USDA (NRICGP 94-37200-0490 and ISU/CDFIN/CSRS 94-34115-2835), and the W. S. Martin Fund. This is Journal Paper J-16524 of the Iowa Agriculture and Home Economics Experiment Station, Ames, IA (Iowa Project No. 3335). The author is indebted to Ms. Margaret Haaland for outstanding administrative and secretarial assistance.

REFERENCES

1. G. Wolf, Historical note on mode of administration of vitamin A for cure of night blindness, *Am. J. Clin. Nutr. 31*:290–292 (1978).
2. T. Moore, *Vitamin A*, Elsevier, Amsterdam, 1957.
3. Nomenclature policy: Generic descriptors and trivial names for vitamins and related compounds, *J. Nutr. 120*:12–19 (1990).
4. A. E. Asato, A. Kim, M. Denny, and R. S. H. Liu, 7-cis, 9-cis, 11-cis Retinal, all-cis vitamin

A, and 7-cis, 9-cis, 11-cis, 12-fluororetinal. New geometric isomers of vitamin A and carotenoids. *J. Am. Chem. Soc. 105*:2923–2924 (1983).

5. R. S. H. Liu and A. E. Asato, Photochemistry and synthesis of stereoisomers of vitamin A. *Tetrahedron 40*:1931–1969 (1984).

6. C. G. Knudsen, S. C. Carey, and W. H. Okamura, [1,5]-Sigmatropic rearrangement of vinyl allenes: a novel route to geometric isomers of the retinoids possessing 11-cis linkages including 9-cis, 11-cis, 13-cis retinal. *J. Am. Chem. Soc. 102*:6355–6356 (1980).

7. O. Isler (ed.), *Carotenoids*, Birkhäuser Verlag, Basel, 1971.

8. G. Britton, S. Liaaen-Jensen, and H. Pfander (eds.), *Carotenoids*, Vols. 1A and 1B, Birkhäuser Verlag, Basel, 1995.

9. M. I. Dawson and P. D. Hobbs, The synthetic chemistry of retinoids, in *The Retinoids: Biology, Chemistry, and Medicine*, 2nd ed. (M. B. Sporn, A. B. Roberts, and D. S. Goodman, eds.), Raven Press, New York, 1994, pp. 5–178.

10. W. Bollag and A. Matter, From vitamin A to retinoids in experimental and clinical oncology: achievements, failures, and outlook, *Ann. N.Y. Acad. Sci. 359*:9–23 (1981).

11. A. Stephens-Jarnegan, D. A. Miller, and H. F. DeLuca, The growth supporting activity of a retinoidal benzoic acid derivative and 4,4-difluororetinoic acid. *Arch. Biochem. Biophys. 237*:11–16 (1985).

12. B. A. Bernard, J. M. Bernardon, C. Delescluse, B. Martin, M. C. Lenoir, J. Maignan, B. Charpentier, W. R. Pilgrim, U. Reichert, and B. Shroot, Identification of synthetic retinoids with selectivity for human nuclear retinoic acid receptor γ, *Biochem. Biophys. Res. Commun. 186*:977–983 (1992).

13. J. M. Lehmann, L. Jong, A. Fanjul, J. F. Cameron, X. P. Lu, P. Haefner, M. I. Dawson, and M. Pfahl, Retinoids selective for retinoid X receptor response pathways. *Science 258*:1944–1946 (1992).

14. F. Frickel, Chemistry and physical properties of retinoids, *The Retinoids* (B. Sporn, A. B. Roberts, and D. S. Goodman, eds.), Academic Press, Orlando, FL, 1984, pp. 7–145.

15. H. C. Furr, A. B. Barua, and J. A. Olson, Analytical methods, in *The Retinoids: Biology, Chemistry, and Medicine*, 2nd ed. (M. B. Sporn, A. B. Roberts, and D. S. Goodman, eds.), Raven Press, New York, 1994, pp. 179–209.

16. G. M. Landers and J. A. Olson, Rapid simultaneous determination of isomers of retinaldehyde, retinal oxime, and retinol by high-performance liquid chromatography, *J. Chromatogr. 438*:383–392 (1988).

17. L. Packer (ed.), Carotenoids Part A, *Meth. Enzymol. 213*:1–538 (1992).

18. L. Packer (ed.), Carotenoids Part B, *Meth. Enzymol. 214*:1–468 (1993).

19. A. B. Barua, D. Kostic, and J. A. Olson, New simplified procedures for the extraction and simultaneous high-performance liquid chromatographic analysis of retinol, tocopherols and carotenoids in human serum, *J. Chromatogr. 617*:257–264 (1993).

20. Y. Tanaka, R. Hirata, Y. Minato, Y. Hasumura, and J. Takeuchi, Isolation and higher purification of fat-storing cells from rat liver with flow cytometry, in *Cells of the Hepatic Sinusoids*, Vol. 1 (A. Kien, D. L. Knook, and E. Wisse, eds.), The Kupffer Cell Foundation, Rijswijk, The Netherlands, 1986, pp. 473–477.

21. H. C. Furr, O. Amedee-Manesme, A. J. Clifford, H. R. Bergen III, A. D. Jones, D. P. Anderson, and J. A. Olson, Relationship between liver vitamin A concentration determined by isotope dilution assay with tetradeuterated vitamin A and by biopsy in generally healthy adult humans. *Am. J. Clin. Nutr. 49*:713–716 (1989).

22. A. J. Clifford, A. D. Jones, and H. C. Furr. Stable isotope dilution mass spectrometry to assess vitamin A status, *Meth. Enzymol. 189*:94–104 (1990).

23. R. S. Parker, J. E. Swanson, B. Marmor, K. J. Goodman, A. B. Spielman, J. T. Brenna, S. M. Viereck, and W. K. Canfield, Study of β-carotene metabolism in humans using ^{13}C-β-carotene and high precision isotope ratio mass spectrometry, *Ann. N.Y. Acad. Sci. 691*:86–95 (1993).

24. R. S. Parker, Absorption, metabolism and transport of carotenoids, *FASEB J. 10*:542–551 (1996).

25. B. A. Underwood and J. A. Olson (eds.), *A Brief Guide to Current Methods of Assessing Vitamin A Status*, Int. Vitamin A Consult. Group, ILSI Press, Washington, D.C., 1993.

26. P. L. Harris, Bioassay of vitamin A compounds, *Vitam. Horm. (U.S.) 18*:341–370 (1960).

27. M. B. Sporn, N. M. Dunlop, D. L. Newton, and W. R. Henderson, Relationship between structure and activity of retinoids, *Nature 263*:110–113 (1976).

28. A. M. Jetten, Induction of differentiation of embryonal carcinoma cells by retinoids, in *Retinoids and Cell Differentiation* (M. I. Sherman, ed.), CRC Press, Boca Raton, FL, 1986, pp. 105–136.

29. M. Imaizumi and T. R. Breitman, Retinoic acid-induced differentiation of the human promyelocytic leukemia cell line HL-60 and fresh human leukemia cells in primary culture, a model for differentiation-inducing therapy of leukemia, *Eur. J. Haematol. 38*: 289–302 (1987).

30. L. J. Gudas, M. B. Sporn, and A. B. Roberts, Cellular biology and biochemistry of the retinoids, in *The Retinoids: Biology, Chemistry, and Medicine*, 2nd ed. (M. B. Sporn, A. B. Roberts, and DeWitt S. Goodman, eds.), Raven Press, New York, 1994, pp. 443–520.

31. A. R. Mangels, J. M. Holden, G. R. Beecher, M. R. Forman, and E. Lanza, Carotenoid content of fruits and vegetables: an evaluation of analytic data, *J. Am. Diet. Assoc. 93*:284–296 (1993).

32. A. J. Clifford, A. D. Jones, Y. Tondeur, H. C. Furr, H. R. Bergen III, and J. A. Olson, Assessment of vitamin A status of humans by isotope dilution GC/MS, 34th Annual Conference on Mass Spectrometry and Allied Topics, Cincinnati, OH, 1986, pp. 327–328.

33. A. S. Truswell, Recommended dietary intakes around the world, *Nutr. Abstr. Rev. 53*:939–1015, 1075–1119 (1983).

34. Food and Agriculture Organization/World Health Organization, Requirements of vitamin A, iron, folate, and vitamin B_{12}. Report of a joint FAO/WHO Expert Committee. FAO Food and Nutrition Series 23, FAO, Rome, 1989, pp. 1–107.

35. National Research Council, *Recommended Dietary Allowances*, 10th ed., National Academy of Sciences Press, Washington, D.C., 1989, pp. 1–284.

36. Department of Health, Dietary reference values for food energy and nutrients for the United Kingdom, Report No. 41 on Health and Social Subjects, HMSO, London, 1991, pp. 1–210.

37. J. A. Olson, Vitamin A, in *Present Knowledge in Nutrition*, 7th ed. (E.E. Ziegler and L.J. Filer Jr, eds.), ILSI Press, Washington, D.C., 1996, pp. 109–119.

38. J. A. Olson, Upper limits of vitamin A in infant formulas, with some comments on vitamin K, *Am. J. Clin. Nutr. 119*:1820–1824 (1989).

39. J. A. Olson, Vitamin A, retinoids and carotenoids, in *Modern Nutrition in Health and Disease*, 8th ed. (M.E. Shils, J.A. Olson, and M. Shike, eds.), Lea & Febiger, Philadelphia, 1994, pp. 287–307.

40. D. E. Ong, Absorption of vitamin A, in *Vitamin A in Health and Disease* (R. Blomhoff, ed.), Marcel Dekker, New York, 1994, pp. 37–72.

41. W. S. White, M. Stacewicz-Sapuntzakis, J. W. Erdman, Jr., and P. E. Bowen, Pharmacokinetics of β-carotene and canthaxanthin after ingestion of individual and combined doses by human subjects, *J. Am. Coll. Nutr. 13*:665–671 (1994).

42. D. Kostic, W. S. White, and J. A. Olson, Intestinal absorption, serum clearance, and interactions between lutein and β-carotene when administered to human adults in separate or combined oral doses, *Am. J. Clin. Nutr. 62*:604–610 (1995).

43. J. W. Erdman, G. C. Fahey, and C. B. White, Effects of purified dietary fiber sources on β-carotene utilization by the chick, *J. Nutr. 116*:2415–2423 (1986).

44. G. Brubacher and H. Weiser, The vitamin A activity of beta-carotene, *Intern. J. Vitam. Nutr. Res. 55*:5–15 (1985).

45. D. R. Soprano and W. S. Blaner, Plasma retinol-binding protein, in *The Retinoids: Biology,*

Chemistry, and Medicine, 2nd ed. (M.B. Sporn, A.B. Roberts, and D.S. Goodman, eds.), Raven Press, New York, 1994, pp. 257–281.

46. M. E. Newcomer, Retinoid-binding proteins: structural determinants important for function, *FASEB J. 9*:229–239 (1995).

47. M. H. Green and J. B. Green, Dynamics and control of plasma retinol, in *Vitamin A in Health and Disease* (R. Blomhoff, ed.), Marcel Dekker, New York, 1994, pp. 119–133.

48. C. B. Stephenson, J. O. Alvarez, J. Kohatsu, R. Hardmeier, J. I. Kennedy, Jr., and R. B. Gammon, Jr., Vitamin A is excreted in the urine during acute infection, *Am. J. Clin. Nutr. 60*:388–392 (1994).

49. P. Tosukhowong and J. A. Olson, The syntheses, biological activity and metabolism of 15-methyl retinone, 15-methyl retinol and 15-dimethyl retinol in rats, *Biochem. Biophys. Acta. 529*:438–453 (1978).

50. T. C. M. Wilson and G. A. J. Pitt, 3,4-Didehydroretinol (vitamin A_2) has vitamin A activity in the rat without conversion to retinol, *Biochem. Soc. Trans. 14*:950–951 (1986).

51. B. Becker, A. B. Barua, M. Barua, and J. A. Olson, Relative effects of various carbohydrate conjugates of retinoic acid on the differentiation and viability of HL-60 cells, *FASEB J. 7*: A304 (1993).

52. R. Blomhoff, Transport and metabolism of vitamin A, *Nutr. Rev. 52*:13–23 (1994).

53. W. S. Blaner and J. A. Olson, Retinol and retinoic acid metabolism, in *The Retinoids: Biology, Chemistry, and Medicine*, 2nd ed. (M. B. Sporn, A. B. Roberts, D. S. Goodman, eds.), Raven Press, New York, 1994, pp. 229–255.

54. K. Wake, Role of perisinusoidal stellate cells in vitamin A storage, in *Vitamin A in Health and Disease* (R. Blomhoff, ed.), Marcel Dekker, New York, 1994, pp. 73–86.

55. D. E. Ong, M. E. Newcomer, and F. Chytil, Cellular retinoid-binding proteins, in *The Retinoids: Biology, Chemistry, and Medicine*, 2nd ed. (M. B. Sporn, A. B. Roberts, and D. S. Goodman, eds.) Raven Press, New York, 1994, pp. 283–317.

56. S. H. Wei, V. Episkopou, R. Piantedosi, S. Maeda, K. Shimada, M. E. Gottesman, and W. S. Blaner, Studies on the metabolism of retinol and retinol-binding protein in transthyretin-deficient mice produced by homologous recombination, *J. Biol. Chem. 270*:866–870 (1995).

57. N. Takahashi and T. R. Breitman, Retinoylation of proteins in mammalian cells, in *Vitamin A in Health and Disease* (R. Blomhoff, ed.), Marcel Dekker, New York, 1994, pp. 257–273.

58. K. L. Skare, H. K. Schnoes, and H. F. DeLuca, Biliary metabolites of all-trans retinoic acid: isolation and identification of a novel polar metabolite, *Biochemistry 21*:3308–3317 (1982).

59. G. Genchi, W. Wang, A. Barua, W. R. Bidlack, and J. A. Olson, Formation of β-glucuronides and of β-galacturonides of various retinoids catalyzed by induced and noninduced microsomal UDP-glucuronosyltransferases of rat liver. *Biochim. Biophys. Acta 1289*:284–290 (1996).

60. A. B. Barua, R. O. Batres, and J. A. Olson, Synthesis and metabolism of all-trans [11-^{3}H]retinyl β-glucuronide in rats in vivo, *Biochem. J. 252*:415–420 (1988).

61. A. Nagao and J. A. Olson, Enzymatic formation of 9-cis, 13-cis, and all-trans retinals from isomers of β-carotene, *FASEB J. 8*:968–973 (1994).

62. X.-D. Wang, N. I. Krinsky, P. N. Benotti, and R. M. Russell, Biosynthesis of 9-cis-retinoic acid from 9-cis-β-carotene in human intestinal mucosa in vitro, *Arch. Biochem. Biophys. 313*:150–155 (1994).

63. N. I. Krinsky, X.-D. Wang, G. Tang, and R. M. Russell, Mechanism of carotenoid cleavage to retinoids, *Ann. N.Y. Acad. Sci. 691*:167–176 (1993).

64. J. Devery and B. V. Milborrow, β-Carotene-15,15′-dioxygenase (EC 1.13.11.21) isolation reaction mechanism and an improved assay procedure, *Br. J. Nutr. 72*:397–414 (1994).

65. A. Nagao, A. During, C. Hoshino, J. Terao, and J. A. Olson, Stoichiometric conversion of all-trans β-carotene to retinal by pig intestinal extract. *Arch. Biochem. Biophys. 328*:57–63 (1996).

66. J. L. Napoli, Retinoic acid homeostasis: prospective roles of β-carotene, retinol, CRBP, and CRABP, in *Vitamin A in Health and Disease* (R. Blomhoff, ed.), Marcel Dekker, New York, 1994, pp. 135–188.

67. D. E. Ong, Cellular transport and metabolism of vitamin A: roles of the cellular retinoid-binding proteins, *Nutr. Rev. 52*:24–31 (1994).

68. J. C. Saari, Retinoids in photosensitive systems, in *The Retinoids: Biology, Chemistry, and Medicine*, 2nd ed. (M. B. Sporn, A. B. Roberts, and D. S. Goodman, eds.), Raven Press, New York, 1994, pp. 351–385.

69. C. Lampron, C. Rochette-Egly, P. Gorry, P. Dolle, M. Mark, T. Lufkin, M. Lemeur, and P. Chambon, Mice deficient in cellular retinoic acid-binding protein II (CRABP-II) or in both CRABP-I and CRABP-II are essentially normal, *Development 121*:539–548 (1995).

70. D. Fawcett, P. Pasceri, R. Fraser, M. Colbert, J. Rossant, and V. Giguere, Postaxial polydactyly in forelimbs of CRABP-II mutant mice, *Development 121*:671–679 (1995).

71. J. A. Olson, Recommended dietary intakes (RDI) of vitamin A in humans, *Am. J. Clin. Nutr. 45*:704–716 (1987).

72. R. R. Rando, Isomerization reactions of retinoids in the visual system, *Pure Appl. Chem. 66*:989–994 (1994).

73. J. E. Dowling and G. Wald, The role of vitamin A acid, *Vitam. Horm. (U.S.) 18*:515–541 (1960).

74. A. J. Lamb, P. Apiwatanaporn, and J. A. Olson, Induction of rapid synchronous vitamin A deficiency in the rat, *J. Nutr. 104*:1140–1148 (1974).

75. D. J. Mangelsdorf, K. Umesono, and R. M. Evans, The retinoid receptors, in *The Retinoids: Biology, Chemistry, and Medicine*, 2nd ed. (M. B. Sporn, A. B. Roberts, and D. S. Goodman, eds.), Raven Press, New York, 1994, pp. 319–349.

76. D. J. Mangelsdorf, Vitamin A receptors, *Nutr. Rev. 52*:32–44 (1994).

77. S. M. Pemrick, D. A. Lucas, and J. F. Grippo, The retinoid receptors, *Leukemia 8*:1797–1806 (1994).

78. M. Pfahl, R. Apfel, I. Bendik, A. Fanjul, G. Graupner, M.-O. Lee, N. La-Vista, X. P. Lu, J. Piedrafita, M. A. Ortiz, G. Salbert, and X.-K. Zhang, Nuclear retinoid receptors and their mechanism of action, *Vitam. Horm. (U.S.) 49*:327–382 (1994).

79. P. Chambon, The molecular and genetic dissection of the retinoid signaling pathway, *Rec Progr Hormone Res 50*:317–332 (1995).

80. L. J. Gudas, Retinoids and vertebrate development, *J. Biol. Chem. 269*:15399–15402 (1994).

81. A. C. Ross and U. G. Hämmerling, Retinoids and the immune system, in *The Retinoids: Biology, Chemistry, and Medicine*, 2nd ed. (M. B. Sporn, A. B. Roberts, and D. S. Goodman, eds.), Raven Press, New York, 1994, pp. 521–543.

82. C. Hofmann and G. Eichele, Retinoids in development, in *The Retinoids: Biology, Chemistry, and Medicine*, 2nd ed. (M. B. Sporn, A. B. Roberts, and D. S. Goodman, eds.), Raven Press, New York, 1994, pp. 387–441.

83. M. Maden, Vitamin A in embryonic development, *Nutr. Rev. 52*:3–12 (1994).

84. J. A. Olson, The biological role of vitamin A in maintaining epithelial tissues, *Isr. J. Med. Sci. 8*:1170–1178 (1972).

85. Y.-P. Chen, D. Dong, M. Solursh, and M. H. Zile, Direct evidence that retinoic acid is not a morphogen in the chick limb bud, *FASEB J. 9*:A833 (1995).

86. M. Studer, H. Pöpperl, H. Marshall, A. Kuroiwa, and R. Krumlauf, Role of a conserved retinoic acid response element in rhombomere restriction of Hoxb-1, *Science 265*:1728–1732 (1994).

87. R. D. Semba, Vitamin A, immunity, and infection, *Clin. Infect. Dis. 19*:489–499 (1994).

88. M. T. Cantorna, F. E. Nashold, and C. E. Hayes, In vitamin A deficiency multiple mechanisms establish a regulatory T helper cell imbalance with excess Th1 and insufficient Th2 function, *J. Immunol. 152*:1515–1522 (1994).

89. W. Stoeckenius and R. A. Bogomolni, Bacteriorhodopsin and related pigments of halobacteria, *Annu. Rev. Biochem. 51*:587–616 (1982).

90. J. K. Lanyi, Proton translocation mechanism and energetics in the light-driven pump bacteriorhopsin, *Biochim. Biophys. Acta 1183*:241–261 (1990).

91. N. A. Dencher, The five retinal-protein pigments of halobacteria: bacteriorhodopsin, halorhodopsin, P565, P370, and slow-cycling rhodopsin, *Photochem. Photobiol. 38*:753–757 (1983).

92. J. L. Spudich, Protein-protein interaction converts a proton pump into a sensory receptor, *Cell 79*:747–750 (1994).

93. J. S. Bertram, Cancer prevention by carotenoids: mechanistic studies in cultured cells, *Ann. N.Y. Acad. Sci. 691*:177–191 (1993).

94. M. Hanusch, W. Stahl, W. A. Schulz, and H. Sies, Induction of gap junctional communication by 4-oxoretinoic acid generated from its precursor canthaxanthin, *Arch. Biochem. Biophys. 317*:423–428 (1995).

95. B. A. Underwood, Vitamin A in animal and human nutrition, in *The Retinoids*, Vol. 1 (M. B. Sporn, A. B. Roberts, and D. S. Goodman, eds.), Academic Press, New York, 1984, pp. 281–392.

96. B. A. Underwood, Vitamin A in human nutrition: public health considerations, in *The Retinoids: Biology, Chemistry, and Medicine*, 2nd ed. (M.B. Sporn, A.B. Roberts, and D.S. Goodman, eds.), Raven Press, New York, 1994, pp. 211–227.

97. G. L. Peck and J. J. DiGiovanna, Synthetic retinoids in dermatology, in *The Retinoids: Biology, Chemistry, and Medicine*, 2nd ed. (M. B. Sporn, A. B. Roberts, and D. S. Goodman, eds.), Raven Press, New York, 1994, pp. 631–658.

98. A. Vahlquist, Role of retinoids in normal and diseased skin, in *Vitamin A in Health and Disease* (R. Blomhoff, ed.), Marcel Dekker, New York, 1994, pp. 365–424.

99. D. B. Gunning, A. B. Barua, R. A. Lloyd, and J. A. Olson, Retinoyl β-glucuronide: a nontoxic retinoid for the topical treatment of acne, *J. Dermatol. Treat. 5*:181–185 (1994).

100. G. J. Fisher, H. S. Talwar, J.-H. Xiao, S. C. Datta, A. P. Reddy, M.-P. Gaubs, C. Rochette-Egly, P. Chambon, and J. J. Voorhees, Immunological identification and functional quantitation of retinoic acid and retinoid X receptor proteins in human skin, *J. Biol. Chem. 269*: 20629–20635 (1994).

101. C. E. Griffiths and J. J. Voorhees, Human in vivo pharmacology of topical retinoids, *Arch. Dermatol. Res. 287*:53–60 (1994).

102. R. C. Moon, R. G. Mehta, and K. V. N. Rao, Retinoids and cancer in experimental animals, in *The Retinoids: Biology, Chemistry, and Medicine*, 2nd ed. (M. B. Sporn, A. B. Roberts, and D. S. Goodman, eds.), Raven Press, New York, 1994, pp. 573–595.

103. W. K. Hong and L. M. Itri, Retinoids and human cancer, in *The Retinoids: Biology, Chemistry, and Medicine*, 2nd ed. (M. B. Sporn, A. B. Roberts, and D. S. Goodman, eds.), Raven Press, New York, 1994, pp. 597–630.

104. L. M. Canfield, N. I. Krinsky, and J. A. Olson (eds.), Carotenoids in human health, *Ann. N.Y. Acad. Sci. 691*:1–300 (1993).

105. W. Schalch and P. Weber, Vitamins and carotenoids—a promising approach to reducing the risk of coronary heart disease, cancer and eye diseases, in *Free Radicals in Diagnostic Medicine* (D. Armstrong, ed.), Plenum Press, New York, 1994, pp. 335–350.

106. J. C. Bauernfeind, *The Safe Use of Vitamin A*, Int. Vitamin A Consult. Group, Nutrition Foundation, Washington, D.C., 1980, pp. 1–44.

107. M. P. Macapinlac and J. A. Olson, A lethal hypervitaminosis A syndrome in young monkeys following a single intramuscular dose of a water-miscible preparation containing vitamins A, D_2 and E, *Int. J. Vitam. Nutr. Res. 51*:331–341 (1981).

108. M. E. Bush and B. B. Dahms, Fatal hypervitaminosis in a neonate, *Arch. Pathol. Lab. Med. 108*:838–842 (1984).

109. R. B. Armstrong, K. O. Ashenfelter, C. Eckhoff, A. A. Levin, and S. S. Shapiro, General and reproductive toxicology of retinoids, in *The Retinoids: Biology, Chemistry, and Medicine*,

2nd ed. (M. B. Sporn, A. B. Roberts, and D. S. Goodman, eds.), Raven Press, New York, 1994, pp. 545–572.

110. J. N. Hathcock, D. G. Hattan, M. Y. Jenkins, J. T. McDonald, P. R. Sundaresan, and V. L. Wilkening, Evaluation of vitamin A toxicity, *Am. J. Clin. Nutr. 52*:183–202 (1990).

111. A. P. Geubel, C. De Galocsy, N. Alves, J. Rahier, and C. Dive, Liver damage caused by therapeutic vitamin A administration: estimate of dose-related toxicity in 41 cases, *Gastroenterology 100*:1701–1709 (1991).

112. T. E. Kowalski, M. Falestiny, E. Furth, and P. F. Malet, Vitamin A hepatotoxicity: a cautionary note regarding 25,000 IU supplements, *Am. J. Med. 97*:523–528 (1994).

113. H. Nau, I. Chahoud, L. Dencker, E. J. Lammer, and W. J. Scott, Teratogenicity of vitamin A and retinoids, in *Vitamin A in Health and Disease* (R. Blomhoff, ed.), Marcel Dekker, New York, 1994, pp. 615–663.

114. J. Adams, Structure–activity and dose–response relationships in the neural and behavioral teratogenesis of retinoids, *Neurotoxicol. Teratol. 15*:193–202 (1993).

115. J. Adams, Neural and behavioral pathology following prenatal exposure to retinoids, in *Retinoids in Clinical Practice* (G. Koren, ed.), Marcel Dekker, New York, 1993, pp. 111–115.

116. J. Adams and E. J. Lammer, Neurobehavioral teratology of isotretinoin, *Reprod. Toxicol. 7*: 175–177 (1993).

117. C. C. Willhite, Molecular correlates in retinoid pharmacology and toxicology, in *Chemistry and Biology of Synthetic Retinoids* (M. L. Dawson and W. H. Okamura, eds.), CRC Press, Boca Raton, FL, 1990, pp. 539–573.

118. D. B. Gunning, A. B. Barua, and J. A. Olson, Comparative teratogenicity and metabolism of all-trans retinoic acid, all-trans retinoyl β-glucose, and all-trans retinoyl β-glucuronide in pregnant Sprague-Dawley rats, *Teratology 47*:29–36 (1993).

119. G. Tzimas, J. O. Sass, W. Wittfoht, M. M. A. Elmazar, K. Ehlers, and H. Nau, Identification of 9,13-di-cis-retinoic acid as a major plasma metabolite of 9-cis-retinoic acid and limited transfer of 9-cis-retinoic acid and 9-13-di-cis-retinoic acid to the mouse and rat embryos, *Drug Metab. Dispos. 22*:928–936 (1994).

120. R. L. Horst, T. A. Reinhardt, J. P. Goff, B. J. Nonnecke, V. K. Gambhir, P. D. Fiorella, and J. L. Napoli, Identification of 9-cis, 13-cis-retinoic acid as a major circulating retinoid in plasma, *Biochemistry 34*:1203–1209 (1995).

121. T. O. Carpenter, J. M. Pettifor, R. M. Russell, J. Pitha, S. Mobarhan, M. S. Ossip, S. Wainer, and C. S. Anast, Severe hypervitaminosis A in siblings: evidence of variable tolerance to retinol intake, *J. Pediatr. 111*:507–512 (1987).

122. F. W. Rosa, Retinoid embryopathy in humans, in *Retinoids in Clinical Practice* (G. Koren, ed.), Marcel Dekker, New York, 1993, pp. 77–109.

123. Recommendations for vitamin A use during pregnancy: Teratology Society position paper, *Teratology 35*:269–275 (1987).

124. C. W. Callaway, K. W. McNutt, R. S. Rivlin, A. C. Ross, H. H. Sandstead, and A. P. Simopoulos, Statement on vitamin and mineral supplements, The Joint Public Information Committee of the American Institute of Nutrition and the American Society for Clinical Nutrition, *J. Nutr. 117*:1649 (1987).

125. Council for Responsible Nutrition, *Safety of Vitamins and Minerals: A Summary of Findings of Key Reviews*, Washington, D.C., 1986, p. 12.

126. A. H. Baqui, A. de Francisco, S. E. Arifeen, A. K. Siddique, and R. B. Sack, Bulging fontanelle after supplementation with 25,000 IU vitamin A in infancy using EPI contacts, Proc. 16th Int. Vitam. A Consult. Group, 24–28 Oct. 1994, Chiangrai, Thailand, p. 74.

127. J. H. Humphrey, T. Agoestina, G. Taylor, A. Usman, K. P. West, Jr., and A. Sommer, Acute and long term risks and benefits to neonates of 50,000 IU oral vitamin A, Proc. XVI Int. Vitam. A Consult. Group, 24–28 Oct. 1994, Chiangrai, Thailand, p. 74.

128. U. Weber, G. Georz, H. Baseler, and L. Michaelis, Canthaxanthin retinopathy: follow-up of over 6 years, *Klin. Monatsbl. Augenheilkd, 201*:174–177 (1992).

129. S. T. Mayne, β-Carotene, carotenoids and disease prevention in humans, *FASEB J. 10*:690–701 (1996).

130. G. W. Burton, Antioxidant action of carotenoids, *J. Nutr. 119*:109–111 (1989).

131. R. G. Ziegler, L. A. Brinton, R. F. Hamman, H. F. Lehman, R. S. Levine, K. Mallin, S. A. Norman, J. F. Rosenthal, A. C. Trumble, and R. N. Hoover, Diet and the risk of invasive cervical cancer among white women in the United States, *Am. J. Epidemiol. 132*:432–445 (1990).

132. K. F. Gey, U. K. Moser, P. Jordan, H. B. Stähelin, M. Eichholzer, and E. Lüdin, Increased risk of cardiovascular disease at suboptimal plasma concentrations of essential antioxidants: an epidemiological update with special attention to carotene and vitamin C, *Am. J. Clin. Nutr. 57*:787S–797S (1993).

133. R. A. Bone, J. T. Landrum, G. W. Hime, A. Cains, and J. Zamor, Stereochemistry of the human macular carotenoids, *Invest. Ophthalmol. Vis. Sci. 34*:2033–2040 (1993).

134. G. J. Handelman, E. A. Dratz, C. C. Reay, and F. J. G. M. van Kuijk, Carotenoids in the human macula and whole retina, *Invest. Ophthalmol. Vis. Sci. 29*:850–855 (1988).

135. A. Bendich, Recent advances in clinical research involving carotenoids, *Pure Appl. Chem. 66*:1017–1024 (1994).

136. P. Pontiggia, A. B. Santamaria, K. Alonso, and L. Santamaria, Whole body hyperthermia associated with beta-carotene supplementation in patients with AIDS, *Biomed. Pharmacother. 49*:263–265 (1995).

137. J. A. Olson, Vitamin A, in *Handbook of Vitamins*, 2nd ed. (L. Machlin, ed.), Marcel Dekker, New York, 1991, pp. 1–57.

138. J. A. Olson, Benefits and liabilities of vitamin A and carotenoids, *J. Nutr. 126*:1208S–1212S (1996).

139. A. Bendich and J. A. Olson, Biological actions of carotenoids, *FASEB J. 3*:1927–1932 (1989).

140. P. Chambon, J. A. Olson, and A. C. Ross, coordinators, The retinoid revolution, *FASEB J. 10*:939–1107 (1996).

141. A. Sommer and K. P. West, Jr., Vitamin A deficiency, in *Health, Survival and Vision*, Oxford University Press, New York, 1996, pp. 1–438.

142. A. B. Barua, Retinoyl beta-glucuronide: a biologically active form of vitamin A, *Nutr. Rev. 55*:259–267 (1997).

143. R. K. Miller, A. G. Hendricks, J. L. Mills, H. Hummler, and U. W. Wiegand, Periconceptional vitamin A use: how much is teratogenic? *Reprod. Toxicol. 12*:75–88 (1998).

144. J. E. Thurman and A. D. Mooradian, Vitamin supplementation therapy in the elderly, *Drugs Aging 11*:433–449 (1997).

145. R. D. Semba, Overview of the potential role of vitamin A in mother-to-child transmission of HIV-1, *Acta Paediatr. Suppl. 421*:107–112 (1997).

146. R. W. Evans, B. J. Shaten, B. W. Day, and L. H. Kuller, Prospective association between lipid soluble antioxidants and coronary heart disease in men. The Multiple Risk Factor Intervention Trial, *Am. J. Epidemiol. 147*:180–186 (1998).

147. C. E. Orfanos, C. C. Zouboulis, B. Almond-Roesler, et al., Current use and future potential role of retinoids in dermatology, *Drugs 53*:358–388 (1997).

148. A. C. Ross, Vitamin A and retinoids, in *Modern Nutrition in Health and Disease*, 9th ed. (M. E. Shils, J. A. Olson, M. Shike, and A. C. Ross, eds.), Williams and Wilkins, Baltimore, 1999, pp. 305–327.

149. K. D. Lawson, S. J. Middleton, and C. D. Hassall, Olestra, a nonabsorbed, noncaloric replacement for dietary fat: a review, *Drug Metab. Rev. 29*:651–703 (1997).

150. G. Britton, ed., Eleventh international symposium on carotenoids, *Pure Appl. Chem. 69*: 2027–2173 (1997).

151. H. C. Furr, and R. M. Clark, Intestinal absorption and tissue distribution of carotenoids, *J. Nutr. Biochem. 8*:364–377 (1997).

152. R. S. Parker, Bioavailability of carotenoids, *Eur. J. Clin. Nutr. 51(Suppl.)*: S86–S90 (1997).

153. C. A. Rice-Evans, J. Sampson, P. M. Bramley, and D. K. Holloway, Why do we expect carotenoids to be antioxidants in vivo? *Free Radic. Res. 26*:381–398 (1997).

154. J. A. Olson, Carotenoids, in *Modern Nutrition in Health and Disease*, 9th ed. (M. E. Shils, J. A. Olson, M. Shike, and A. C. Ross, eds.), Williams & Wilkins, Baltimore, 1999, pp. 525–541.

155. J. A. Olson, N. Loveridge, G. G. Duthie, and M. J. Shearer, Fat soluble vitamins, in *Human Nutrition and Dietetics*, 10th ed. (J. S. Garrow, W. P. T. James, and A. Ralph, eds.), Churchill Livingstone, Edinburgh, 2000, pp. 211–247.

156. M. I. Dawson, The importance of vitamin A in nutrition, *Curr. Pharm. Des. 6*:311–325 (2000).

157. J. P. Shenai, Vitamin A supplementation in very low birth weight neonates: rational and evidence, *Pediatrics 104*:1369–1374 (1999).

158. A. Coutsoudis, K. Pillay, E. Spooner, L. Kuhn, and H. M. Coovadia, Influence of infant-feeding patterns on early mother-to-child transmission of HIV-1 in Durban, South Africa: a prospective cohort study, *Lancet 354*:471–476 (1999).

159. J. J. Castenmiller and C. E. West, Bioavailability and bioconversion of carotenoids, *Ann. Rev. Nutr. 18*:19–38 (1998).

160. IARC Handbooks of Cancer Prevention, Vol. 2. *Carotenoids*, Intern. Agency for Research on Cancer Press, Lyon, 1998, pp. 1–326.

161. IARC Handbooks of Cancer Prevention, Vol. 3. *Vitamin A*, Intern. Agency for Research on Cancer Press, Lyon, 1999, pp. 1–261.

162. IARC Handbooks of Cancer Prevention, Vol. 4. *Retinoids*, Intern. Agency for Research on Cancer Press, Lyon, 1999, pp. 1–331.

2

Vitamin D

ELAINE D. COLLINS

San José State University, San José, California

ANTHONY W. NORMAN

University of California, Riverside, California

I. INTRODUCTION

Vitamin D designates a group of closely related compounds that possess antirachitic activity. The two most prominent members of this group are ergocalciferol (vitamin D_2) and cholecalciferol (vitamin D_3). Ergocalciferol is derived from a common plant steroid, ergosterol, and is the form that was employed for vitamin D fortification of foods from the 1940s to the 1960s. Cholecalciferol is the form of vitamin D obtained when radiant energy from the sun strikes the skin and converts the precursor 7-dehydrocholesterol into vitamin D_3. Since the body is capable of producing cholecalciferol, vitamin D does not meet the classical definition of a vitamin. It is more accurate to call vitamin D a *prohormone*; thus, vitamin D is metabolized to a biologically active form that functions as a steroid hormone (1–4). However, since vitamin D was first recognized as an essential nutrient, it has historically been classified among the lipid-soluble vitamins. Even today it is thought of by many as a vitamin, although it is now known that there exists a vitamin D endocrine system that generates the steroid hormone $1\alpha,25$-dihydroxyvitamin D_3 [$1\alpha,25(OH)_2D_3$].

Vitamin D functions to maintain calcium homeostasis together with two peptide hormones, calcitonin and parathyroid hormone (PTH). Vitamin D is also important for phosphorus homeostasis (5,6). Calcium and phosphorus are required for a wide variety of biological processes (Table 1). Calcium is necessary for muscle contraction, nerve pulse transmission, blood clotting, and membrane structure. It also serves as a cofactor for such enzymes as lipases and ATPases and is needed for eggshell formation in birds. It is an important intracellular signaling molecule for signal transduction pathways, such as those involving calmodulin and protein kinase C. Phosphorus is an important component of

Table 1   Biological Calcium and Phosphorus

Calcium	Phosphorus
Utilization	
Body content: 70-kg man has 1200 g Ca^{2+}	Body content: 70-kg man has 770 g P
Structural: bone has 95% of body Ca	Structural: Bone has 90% of body P
Plasma $[Ca^{2+}]$ is 2.5 mM, 10 mg %	Plasma $[P_i]$ is 2.3 mM, 2.5–4.3 mg %
Muscle contraction	Intermediary metabolism (phosphorylated in-
Nerve pulse transmission	termediates)
Blood clotting	Genetic information (DNA and RNA)
Membrane structure	Phospholipids
Enzyme cofactors (amylase, trypsinogen,	Enzyme/protein components (phosphohisti-
lipases, ATPases)	dine, phosphoserine)
Eggshell (birds)	Membrane structure
Daily requirements (70-kg man)	
Dietary intake: 700[a]	Dietary intake: 1200[a]
Fecal excretion: 300–600[a,b]	Fecal excretion: 350–370[a,b]
Urinary excretion: 100–400[a,b]	Urinary excretion: 200–600[a,b]

[a]Values in milligrams per day.
[b]Based on the indicated level of dietary intake.

DNA, RNA, membrane lipids, and the intracellular energy–transferring ATP system. The phosphorylation of proteins is important for the regulation of many metabolic pathways. Furthermore, the maintenance of serum calcium and phosphorus levels within narrow limits is important for normal bone mineralization. Any perturbation in these levels results in bone calcium accretion or resorption. Disease states, such as rickets, can develop if the serum ion product is not maintained at a level consistent with that required for normal bone mineralization. Maintaining a homeostatic state for these two elements is of considerable importance to a living organism.

Recently, $1\alpha,25(OH)_2D_3$ has been shown to act on novel target tissues not related to calcium homeostasis. There have been reports characterizing receptors for the hormonal form of vitamin D and activities in such diverse tissues as brain, pancreas, pituitary, skin, muscle, immune cells, and parathyroid (Table 2). These studies suggest that vitamin D status is important for insulin and prolactin secretion, muscle function, immune and stress response, melanin synthesis, and cellular differentiation of skin and blood cells.

There are a number of recent books (7–9) and comprehensive reviews (10–19) that cover many aspects of vitamin D, including its endocrinological aspects.

II.   HISTORY

Rickets, a deficiency disease of vitamin D, appears to have been a problem in ancient times. There is evidence that rickets occurred in Neanderthal man about 50,000 BC (20). The first scientific descriptions of rickets were written by Dr. Daniel Whistler (21) in 1645 and by Professor Francis Glisson (22) in 1650. Rickets became a health problem in northern Europe, England, and the United States during the Industrial Revolution when many people lived in urban areas with air pollution and little sunlight. Prior to the discovery of vitamin D, theories on the causative factors of rickets ranged from heredity to syphilis (2).

Table 2 Distribution of $1,25(OH)_2D_3$ Actions[a]

Tissue Distribution of Nuclear $1,25(OH)_2D_3$ Receptor		
Adipose	Intestine	Pituitary
Adrenal	Kidney	Placenta
Bone	Liver (fetal)	Prostrate
Bone marrow	Lung	Retina
Brain	Muscle, cardiac	Skin
Breast	Muscle, embryonic	Stomach
Cancer cells	Muscle, smooth	Testis
Cartilage	Osteoblast	Thymus
Colon	Ovary	Thyroid
Eggshell gland	Pancreas β cell	Uterus
Epididymus	Parathyroid	Yolk sac (bird)
Hair follicle	Parotid	

Distribution of Nongenomic Responses	
Intestine	Transcaltachia
Osteoblast	Ca^{2+} channel opening
Osteoclast	Ca^{2+} channel opening
Liver	Lipid metabolism
Muscle	A variety

[a]Summary of the tissue location of the nuclear receptor for $1\alpha,25(OH)_2D_3$ (nVDR) (top) and tissues displaying "rapid" or membrane-initiated biological responses (bottom).

Some of the important scientific discoveries leading to the understanding of rickets were dependent on discoveries about bone. As reviewed by Hess (23), the first formal descriptions of bone were made by Marchand (1842), Bibard (1844), and Friedleben (1860). In 1885, Pommer wrote the first description of the pathological process taking place in the rachitic skeleton. In 1849, Trousseau and Lasque recognized that osteomalacia and rickets were different manifestations of the same disorder. In 1886 and 1890, Hirsch and Palm did a quantitative geographical study of the worldwide distribution of rickets and found that the incidence of rickets paralleled the incidence of lack of sunlight (23). This was substantiated in 1919 when Huldschinsky demonstrated that ultraviolet (UV) rays were effective in healing rickets (24).

In the early 1900s, the concept of vitamins was developed and nutrition emerged as an experimental science, allowing for further advances in understanding rickets. In 1919, Sir Edward Mellanby (25,26) was able to experimentally produce rickets in puppies by feeding synthetic diets to over 400 dogs. He further showed that rickets could be prevented by the addition of cod liver oil or butterfat to the feed. He postulated that the nutritional factor preventing rickets was vitamin A since butterfat and cod liver oil were known to contain vitamin A (26). Similar studies were also conducted by McCollum et al. (27).

The distinction between the antixerophthalmic factor, vitamin A, and the antirachitic factor, vitamin D, was made in 1922 when McCollum's laboratory showed that the antirachitic factor in cod liver oil could survive both aeration and heating to 100°C for 14 h, whereas the activity of vitamin A was destroyed by this treatment. McCollum named the new substance vitamin D (28).

Although it was known that UV light and vitamin D are equally effective in preventing and curing rickets, the close interdependence of the two factors was not immediately recognized. Then, in 1923, Goldblatt and Soames (29) discovered that food that was irradiated and fed to rats could cure rickets; food that was not irradiated could not cure rickets. In 1925, Hess and Weinstock demonstrated that a factor with antirachitic activity was produced in the skin upon UV irradiation (30,31). Both groups demonstrated that the antirachitic agent was in the lipid fraction. The action of the light appeared to produce a permanent chemical change in some component of the diet and the skin. They postulated that a provitamin D existed that could be converted to vitamin D by UV light absorption. Much more work ultimately demonstrated that the antirachitic activity resulted from the irradiation of 7-dehydrocholesterol.

The isolation and characterization of vitamin D was now possible. In 1932, the structure of vitamin D_2 was simultaneously determined by Windaus in Germany, who named it vitamin D_2 (32), and by Askew in England, who named it ergocalciferol (33). In 1936, Windaus identified the structure of vitamin D found in cod liver oil, vitamin D_3 (34). Thus, the ''naturally'' occurring vitamin is vitamin D_3, or cholecalciferol. The structure of vitamin D was determined to be that of a steroid or, more correctly, a *seco*-steroid. However, the relationship between its structure and its mode of action was not realized for an additional 30 years.

Vitamin D was believed for many years to be the active agent in preventing rickets. It was assumed that vitamin D was a cofactor for reactions that served to maintain calcium and phosphorus homeostasis. However, when radioisotopes became available, more precise measurements of metabolism could be made. Using radioactive $^{45}Ca^{2+}$, Linquist found that there was a lag period between the administration of vitamin D and the initiation of its biological response (35). Stimulation of intestinal calcium absorption required 36–48 h for a maximal response. Other investigators found delays in bone calcium mobilization and serum calcium level increases after treatment with vitamin D (36–40). The duration of the lag and the magnitude of the response were proportional to the dose of vitamin D used (37).

One explanation for the time lag was that vitamin D had to be further metabolized before it was active. With the development of radioactively labeled vitamin D, it became possible to study the metabolism of vitamin D. Norman et al. were able to detect three metabolites that possessed antirachitic activity (41). One of these metabolites was subsequently identified as the 25-hydroxy derivative of vitamin D_3 [25(OH)D_3] (42). Because 25(OH)D_3 was found to have 1.5 times more activity than vitamin D in curing rickets in the rat, it was thought that this metabolite was the biologically active form of vitamin D (43). However, in 1968, Haussler et al. reported a more polar metabolite that was found in the nuclear fraction of the intestine from chicks given tritiated vitamin D_3 (44). Biological studies demonstrated that this new metabolite was 13–15 times more effective than vitamin D in stimulating intestinal calcium absorption and 5–6 times more effective in elevating serum calcium levels (45). The new metabolite was also as effective as vitamin D in increasing growth rate and bone ash (45). In 1971, the structural identity of this metabolite was reported to be the 1α,25-dihydroxy derivative of vitamin D[1α,25(OH)$_2D_3$] (46–48), the biologically active metabolite of vitamin D.

In 1970, the site of production of 1α,25(OH)$_2D_3$ was demonstrated to be the kidney (49). This discovery, together with the finding that 1α,25(OH)$_2D_3$ is found in the nuclei of intestinal cells, suggested that vitamin D was functioning as a steroid hormone (44,50). Subsequently, a nuclear receptor protein for 1α,25(OH)$_2D_3$ was identified and characterized (50,51). Since the cDNA for the 1α,25(OH)$_2D_3$ nuclear receptor from several species has now been cloned and sequenced (52–55), the relationship between vitamin D and the other steroid hormones has been clearly established (56). The discovery that the biological actions of vitamin D could be explained by the classical model of steroid hormone action marked the beginning of the modern era of vitamin D.

III. CHEMISTRY

A. Structure

As previously mentioned, vitamin D refers to a family of compounds that possess antirachitic activity. Members of the family are derived from the cyclopentanoperhydrophenanthrene ring system (Fig. 1), which is common to other steroids, such as cholesterol. However, vitamin D has only three intact rings; the B ring has undergone fission of the 9,10 carbon bond, resulting in the conjugated triene system of double bonds that is possessed by all D vitamins. The structure of vitamin D_3 is shown in Fig. 1. Naturally occurring mem-

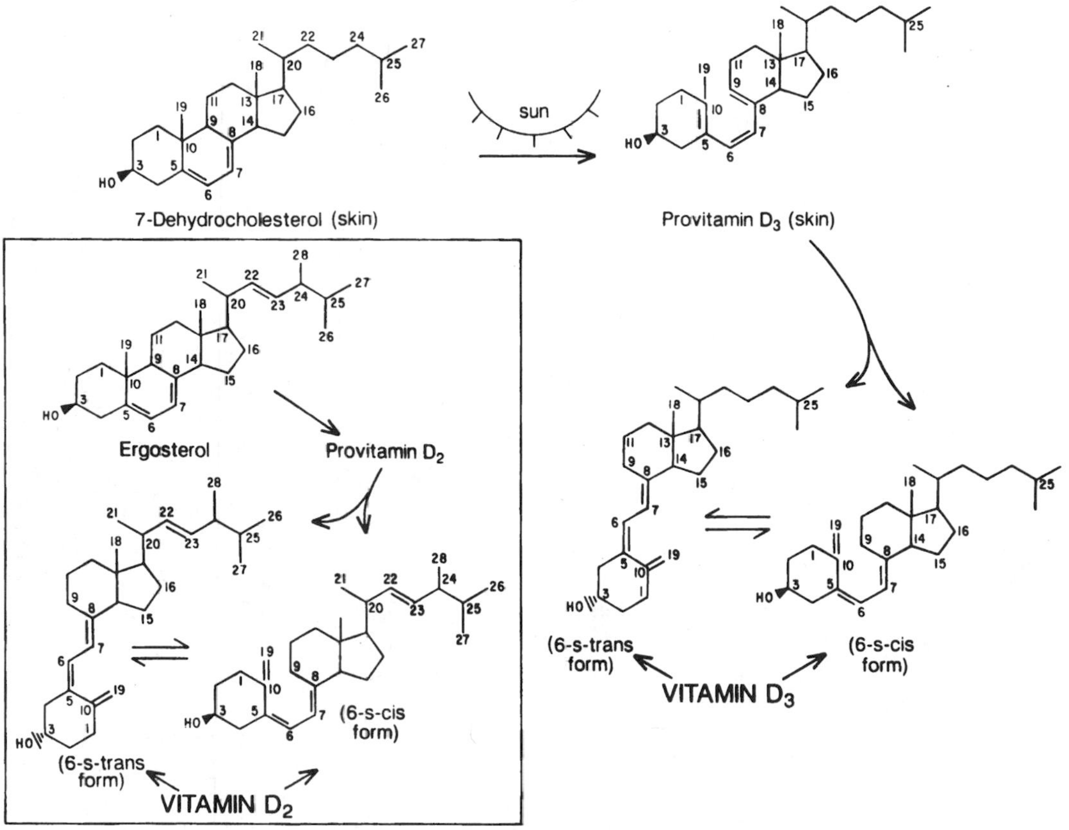

Fig. 1 Chemistry and irradiation pathway for production of vitamin D_3 (a natural process) and vitamin D_2 (a commercial process). In each instance the provitamin, with a $\Delta5,\Delta7$ conjugated double-bond system in the B ring, is converted to the *seco*-B previtamin, with the 9,10 carbon–carbon bond broken. Then the previtamin D thermally isomerizes to the "vitamin" form, which contains a system of three conjugated double bonds. In solution vitamin D is capable of assuming a large number of conformations due to rotation about the 6,7 carbon–carbon bond of the B ring. The 6-s-cis conformer (the steroid-like shape) and the 6-s-trans conformer (the extended shape) are presented for both vitamin D_2 and vitamin D_3.

Table 3 Side Chains of Provitamin D

Provitamin trivial name	Vitamin D produced upon irradiation	Empirical formula (complete steroid)	Side chain Structure
Ergosterol	D_2	$C_{28}H_{44}O$	
7-dehydrocholesterol	D_3	$C_{27}H_{44}O$	
22,23-dihydroergosterol	D_4	$C_{28}H_{46}O$	
7-dehydrositosterol	D_5	$C_{29}H_{48}O$	
7-dehydrostigmasterol	D_6	$C_{29}H_{46}O$	
7-dehydrocampesterol	D_7	$C_{28}H_{46}O$	

bers of the vitamin D family differ from each other only in the structure of their side chains; the side chain structures of the various members of the vitamin D family are given in Table 3.

From the x-ray crystallographic work of Nobel laureate Crowfoot-Hodgkin et al., it is now known that the diene system of vitamin D that extends from C-5 to C-8 is transoid and nearly planar (57,58). However, the C-6 to C-19 diene system is cisoid and not planar. The C-10 to C-19 double bond is twisted out of the plane by 60°. As a result, the A ring exists in one of two possible chair conformations. It is also known that the C and D rings are rigid and that the side chain prefers an extended configuration. In 1974, Okamura et al. reported that vitamin D and its metabolites have a high degree of conformational mobility (59). Using nuclear magnetic resonance (NMR) spectroscopy, they were able to detect that the A ring undergoes rapid interconversion between the two chair conformations, as shown in Fig. 2. This conformational mobility is unique to the vitamin D molecule and is not observed for other steroid hormones. It is a direct consequence of the breakage of the 9,10 carbon bond of the B ring, which serves to "free" the A ring. As a result of this mobility, substituents on the A ring are rapidly and continually alternating between the axial and equatorial positions.

B. Nomenclature

Vitamin D is named according to the new revised rules of the International Union of Pure and Applied Chemists (IUPAC). Since vitamin D is derived from a steroid, the structure retains its numbering from the parent steroid compound. Vitamin D is designated "seco" because its B ring has undergone fission. Asymmetrical centers are named using R, S

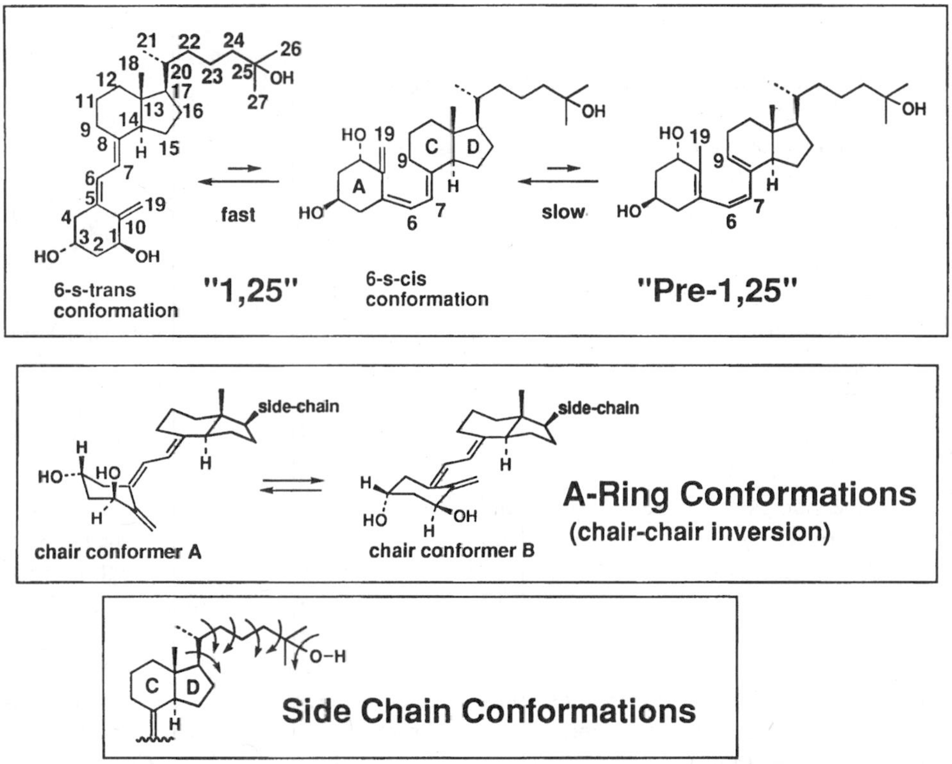

Fig. 2 The dynamic behavior of 1α,25(OH)$_2$D$_3$. The topological features of the hormone 1α,25(OH)$_2$D$_3$ undergo significant changes as a consequence of rapid conformational changes (due to single-bond rotation) or, in one case, as a consequence of a hydrogen shift (resulting in the transformation of 1α,25(OH)$_2$D$_3$ to pre-1α,25(OH)$_2$D$_3$). (top) The dynamic changes occurring within the *seco*-B conjugated triene framework of the hormone (C5, 6, 7, 8, 9, 10, 19). All of the carbon atoms of the 6-s-trans conformer of 1α,25(OH)$_2$D$_3$ are numbered using standard steroid notation for the convenience of the reader. Selected carbon atoms of the 6-s-cis conformer are also numbered as are those of pre-1α,25(OH)$_2$D$_3$. (middle) The rapid chair–chair inversion of the A ring of the secosteroid. (bottom) The dynamic single-bond conformational rotation of the cholesterol-like side chain of the hormone. The C/D *trans*-hydrindane moiety is assumed to serve as a rigid anchor about which the A-ring, *seco*-B triene, and side chain are in dynamic equilibrium.

notation and Cahn's rules of priority. The configuration of the double bonds is notated E, Z; E for "trans," Z for "cis." The formal name for vitamin D$_3$ is 9,10-*seco*(5Z,7E)-5,7,10(19)-cholestatriene-3β-ol and for vitamin D$_2$ is 9,10-*seco*(5Z,7E)-5,7,10(19),21-ergostatetraene-3β-ol.

C. Chemical Properties

1. Vitamin D$_3$(C$_{27}$H$_{44}$O)

Three double bonds
Melting point, 84–85°C
UV absorption maximum at 264–265 nm with a molar extinction coefficient of 18,300 in alcohol or hexane, α$_D$20 +84.8° in acetone

Molecular weight, 384.65
Insoluble in H_2O
Soluble in benzene, chloroform, ethanol, and acetone
Unstable in light
Will undergo oxidation if exposed to air at 24°C for 72 h
Best stored at 0°C

2. Vitamin $D_2(C_{28}H_{44}O)$

Four double bonds
Melting point, 121°C
UV absorption maximum at 265 nm with a molar extinction coefficient of 19,400
   in alcohol or hexane, $\alpha_D 20 + 106°$ in acetone
Same solubility and stability properties as D_3

D. Isolation

Many of the studies that have led to our understanding of the mode of action of vitamin D have involved tissue localization and identification of vitamin D and its various metabolites. Since vitamin D is a steroid, it is isolated from tissue by methods that extract total lipids. The technique most frequently used for this extraction is that of Bligh and Dyer (60).

Over the years, a wide variety of chromatographic techniques have been used to separate vitamin D and its metabolites. These include paper, thin-layer, column, and gas chromatographic methods. Paper and thin-layer chromatography usually require long development times with unsatisfactory resolutions and have limited capacity. Column chromatography, with alumina, Floridin, Celite, silicic acid, and Sephadex LH-20 as supports, has been used to rapidly separate many closely related vitamin D compounds (2). However, none of the above methods are capable of resolving and distinguishing vitamin D_2 from vitamin D_3. Gas chromatography can separate these two compounds, but in the process vitamin D is thermally converted to pyrocalciferol and isopyrocalciferol, resulting in two peaks. High-performance liquid chromatography (HPLC) has become the method of choice for the separation of vitamin D and its metabolites (61,62). This powerful technique is rapid and gives good recovery with high resolution.

E. Synthesis of Vitamin D

1. Photochemical Production

In the 1920s, it was recognized that provitamins D were converted to vitamins D upon treatment with UV radiation (Fig. 1). The primary structural requirement for a provitamin D is a sterol with a C-5 to C-7 diene double-bond system in ring B. The conjugated double-bond system is a chromophore which upon UV irradiation initiates a series of transformations resulting in the production of the vitamin D *seco*-steroid structure. The two most abundant provitamins D are ergosterol (provitamin D_2) and 7-dehydrocholesterol (provitamin D_3).

2. Chemical Synthesis

There are two basic approaches to the synthesis of vitamin D. The first involves the chemical synthesis of a provitamin that can be converted to vitamin D by UV irradiation. The second is a total chemical synthesis.

Since vitamin D is derived from cholesterol, the first synthesis of vitamin D resulted from the first chemical synthesis of cholesterol. Cholesterol was first synthesized by two groups in the 1950s. The first method involves a 20-step conversion of 4-methoxy-2,5-toluquinone to a progesterone derivative, which is then converted in several steps to progesterone, testosterone, cortisone, and cholesterol (63). The other method uses the starting material 1,6-dihydroxynaphthalene. This is converted to the B and C rings of the steroid. A further series of chemical transformations leads to the attachment of the A ring and then the D ring. The final product of the synthesis was epiandrosterone, which could be converted to cholesterol (64). The cholesterol was then converted to 7-dehydrocholesterol and UV-irradiated to give vitamin D. The yield of vitamin D from photochemical conversion is normally 10–20%.

The first pure chemical synthesis of vitamin D, without any photochemical irradiation steps, was accomplished in 1967 (65). This continuing area of investigation allows for the production of many vitamin D metabolites and analogs without the necessity of a photochemical step. Pure chemical synthesis also allows for the synthesis of radioactive vitamin D and metabolites for the study of the metabolism of vitamin D.

Figure 3 summarizes some of the currently used synthetic strategies (14). Method A involves the photochemical ring opening of a 1-hydroxylated side-chain-modified deriv-

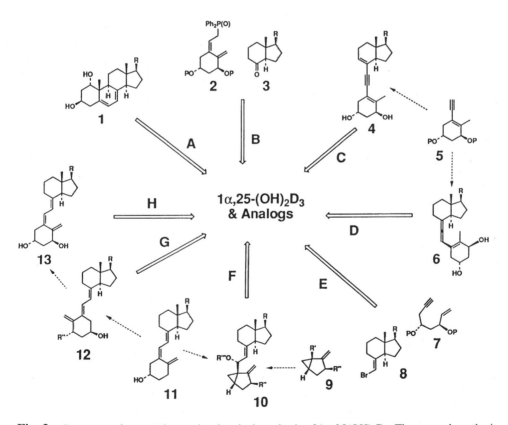

Fig. 3 Summary of approaches to the chemical synthesis of $1\alpha,25(OH)_2D_3$. The general synthetic approaches A–H, which are discussed in the text, represent some of the major synthetic approaches used in recent years to synthesize the hormone $1\alpha,25(OH)_2D_3$ and analogs of $1\alpha,25(OH)_2D_3$.

ative of 7-dehydrocholesterol **1** producing a provitamin that is thermolyzed to vitamin D (66,67). Method B is useful for producing side chain and other analogs. In this method, the phosphine oxide **2** is coupled to a Grundmann's ketone derivative **3**, producing the $1\alpha,25(OH)_2D_3$ skeleton (68,69). In method C, dienynes such as **4** are semihydrogenated to a previtamin structure that undergoes rearrangement to the vitamin D analog (70,71). Method D involves the production of the vinylallene **6** from compound **5** and the subsequent rearrangement with heat- or metal-catalyzed isomerization followed by sensitized photoisomerization (72). Method E starts with an acyclic A ring precursor **7** that is intramolecularly cross-coupled to bromoenyne **8**, resulting in the $1,25(OH)_2D_3$ skeleton (73,74). Method F starts with the tosylate of **11**, which is isomerized to the i-steroid **10**. This structure can be modified at C-1 and then reisomerized under sovolytic conditions to $1\alpha,25(OH)_2D_3$ or analogs (75,76). In method G, vitamin D derivatives **11** are converted to 1-oxygenated 5,6-*trans* vitamin D derivatives **12** (77). Finally, method H involves the direct modification of $1\alpha,25(OH)_2D_3$ or an analog **13** through the use of protecting groups, such as transition metal derivatives, or by other direct chemical transformations on **13** (78). These synthetic approaches have allowed the synthesis of more than 300 analogs of $1\alpha,25(OH)_2D_3$. For an article that gives an extensive review of all the synthetic approaches, see (14).

IV. METABOLISM

The elucidation of the metabolic pathway by which vitamin D is transformed into its biologically active form is one of the most important advances in our understanding of how vitamin D functions and the development of the vitamin D endocrine system. It is now known that both vitamin D_2 and vitamin D_3 must be hydroxylated at the C-1 and C-25 positions (Fig. 4) before they can produce their biological effects. The activation of vitamin D_2 occurs via the same metabolic pathway as does the activation of vitamin D_3, and the biological activities of both vitamin D_2 and vitamin D_3 have been shown to be identical in all animals except birds and the New World monkey. Apparently, these animals have the ability to discriminate against vitamin D_2 (79).

A. Absorption

Vitamin D can be obtained from the diet, in which case it is absorbed in the small intestine with the aid of bile salts (80,81). In rat, baboon, and human, the specific mode of vitamin D absorption is via the lymphatic system and its associated chylomicrons (82,83). It has been reported that only about 50% of a dose of vitamin D is absorbed (83,84). However, considering that sufficient amounts of vitamin D can be produced daily by exposure to sunlight, it is not surprising that the body has not evolved a more efficient mechanism for vitamin D absorption from the diet.

Although the body can obtain vitamin D from the diet, the major source of this prohormone is its production in the skin from 7-dehydrocholesterol. The 7-dehydrocholesterol is located primarily in the malpighian layer of the skin. Upon exposure to UV light, it is photochemically converted to previtamin D, which then isomerizes to vitamin D over a period of several days (85). Once formed, vitamin D is preferentially removed from the skin into the circulatory system by the blood transport protein for vitamin D, the vitamin D–binding protein (DBP).

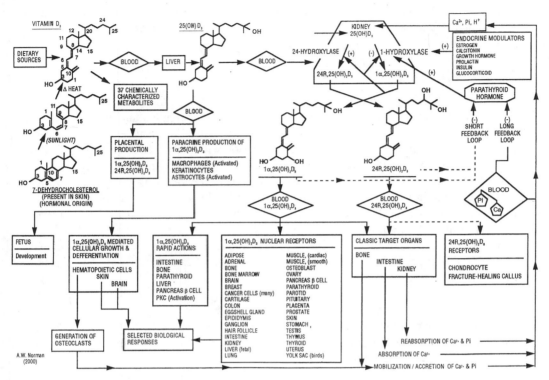

Fig. 4 Overview of the vitamin D endocrine and paracrine system. Target organs and cells for 1α,25(OH)₂D₃ by definition contain receptors for the hormone. Biological effects are generated by both genomic and nongenomic signaling pathways.

B. Transport

The actual site of transfer of vitamin D from the chylomicrons to its specific plasma carrier protein, DBP, is unknown. After an oral dose of radioactive vitamin D, the radioactivity becomes associated with the lipoprotein fraction of the plasma (86). As time passes, there is a progressive shift from this fraction to the γ-globulin fraction (87,88). It has been shown that the electrophoretic mobility of DBP is identical to that of γ₂-globulins and albumins (89,90).

In mammals, vitamin D, 25(OH)D₃, 24R,25(OH)₂D₃, and 1α,25(OH)₂D₃ are transported on the same protein, DBP (91,92). DBP, also known as group-specific protein (Gc protein), is a globulin protein with a molecular weight in humans of 58,000. It is a bifunctional protein, responsible both for the transport of vitamin D and its metabolites as well as functioning as a scavenger for actin which may be inappropriately present in the plasma. DBP possesses a high-affinity binding site for monomeric actin and forms a high molecular weight complex with it (93). DBP also possesses a high-affinity binding site for 25(OH)D₃ and binds other vitamin D metabolites with somewhat lower affinity (92). Sequence analysis of the cDNA for DBP indicates that it shares homology with serum albumin and α-fetoprotein (94). DBP has been proposed to help in the cellular internalization of vitamin D sterols, and levels of DBP influence the concentration of "bound" and 'free' hormone in the plasma (95). The concentration of the free hormone may be important in determining the biological activity of the hormone (95–99). Several review articles on DBP are available (97–99).

C. Storage

Vitamin D is taken up rapidly by the liver. Since it was known that the liver serves as a storage site for retinol, another fat-soluble vitamin, it was thought that the liver also functioned as a storage site for vitamin D. However, it has since been shown that blood has the highest concentration of vitamin D, in comparison with other tissues (100,101). From studies in rats it was concluded that no rat tissue can store vitamin D or its metabolites against a concentration gradient (82). The persistence of vitamin D in animals during periods of vitamin D deprivation may be explained by the slow turnover rate of vitamin D in certain tissues, such as skin and adipose tissue. During times of deprivation, the vitamin D in these tissues is released slowly, thus meeting the vitamin D needs of the animal over a period of time. In contrast, it was found that in pig tissue concentrations of $1\alpha,25(OH)_2D_3$, especially in adipose tissue, are three- to seven-fold higher than plasma levels (102).

Similarly, Mawer et al. carried out studies in humans on the distribution and storage of vitamin D and its metabolites (103). In human tissue, adipose tissue and muscle were found to be major storage sites for vitamin D. Their studies also indicated that adipose tissue serves predominantly as the storage site for vitamin D_3 and that muscle serves as the storage site for $25(OH)D_3$.

D. Metabolism

The parent vitamin D is largely biologically inert; before vitamin D can exhibit any biological activity, it must be metabolized by the body to its active forms. $1\alpha,25(OH)_2D_3$ is the most active metabolite known, but there is evidence that $24R,25(OH)_2D_3$ is required for some of the biological responses attributed to vitamin D (16,104,105). Both of these metabolites are produced *in vivo* following carbon-25 hydroxylation of the parent vitamin D molecule.

1. $25(OH)D_3$

In the liver, vitamin D undergoes its initial transformation, which involves the addition of a hydroxyl group to the 25-carbon. The metabolite thus formed is $25(OH)D_3$, which is the major circulating form of vitamin D. Although there is some evidence that this metabolite can be formed in other tissues, such as intestine and kidney, it is generally accepted that the formation of $25(OH)D_3$ occurs predominantly in the liver.

The production of $25(OH)D_3$ is catalyzed by the enzyme vitamin D_3 25-hydroxylase. The 25-hydroxylase is found in liver microsomes and mitochondria (106–109). It is a P450-like enzyme that is poorly regulated (110). Therefore, circulating levels of $25(OH)D_3$ are a good index of vitamin D status, i.e., they reflect the body content of the parent vitamin D_3 (111,112). Recent studies have suggested that the 25-hydroxylation of vitamin D is partially regulated by $1\alpha,25(OH)_2D_3$. Other studies suggest that 25-hydroxylation is dependent on intracellular calcium levels (113). However, the extent, nature, and physiological significance of any regulatory mechanism of this step in the metabolism of vitamin D remains uncertain. The 25-hydroxylase has been cloned (114) and expressed in yeast cells (115).

2. $1\alpha,25(OH)_2D_3$

From the liver, $25(OH)D_3$ is returned to the circulatory system where it is transported via DBP to the kidney. In the kidney, a second hydroxyl group can be added at the C-1 position. The enzyme responsible for the 1α-hydroxylation of $25(OH)D_3$ is the 25-hydroxyvitamin D_3-1α-hydroxylase (1-hydroxylase) (116).

1-Hydroxylase is located in the mitochondria of the proximal tubules in the kidney. The enzyme belongs to a class of enzymes known as mitochondrial mixed-function oxidases. Mixed-function oxidases use molecular oxygen as the oxygen source instead of water. 1-Hydroxylase is composed of three proteins that are integral components of the mitochondrial membrane: they are renal ferredoxin reductase, renal ferredoxin, and a cytochrome P450.

The most important point of regulation of the vitamin D endocrine system occurs through the stringent control of the activity of the renal 1-hydroxylase (117). In this way, the production of the hormone $1\alpha,25(OH)_2D_3$ can be modulated according to the calcium needs of the organism. Although extrarenal production of $1\alpha,25(OH)_2D_3$ has been demonstrated in placenta (118,119), cultured pulmonary alveolar and bone macrophages (120–122), cultured embryonic calvarial cells (123), and cultured keratinocytes (124,125), which can provide the hormone to adjacent cells in a paracrine fashion, the kidney is considered the primary source of circulating $1\alpha,25(OH)_2D_3$. Several regulatory factors have been identified that modulate 1-hydroxylase activity, but some are functional only in certain species and under certain experimental conditions. The major factors are $1\alpha,25(OH)_2D_3$ itself, PTH, and the serum concentrations of calcium and phosphate (126).

Probably the most important determinant of 1-hydroxylase activity is the vitamin D status of the animal. When circulating levels of $1\alpha,25(OH)_2D_3$ are low, the production of $1\alpha,25(OH)_2D_3$ in the kidney is high, and when circulating levels of $1\alpha,25(OH)_2D_3$ are high, synthesis of $1\alpha,25(OH)_2D_3$ is low (117). The changes of enzyme activity induced by $1\alpha,25(OH)_2D_3$ can be inhibited by cycloheximide and actinomycin D (127), which suggests that $1\alpha,25(OH)_2D_3$ is acting at the level of transcription. Another modulator of renal $1\alpha,25(OH)_2D_3$ production is PTH. PTH is released when plasma calcium levels are low, and in the kidney it stimulates the activity of the 1-hydroxylase and decreases the activity of the 24-hydroxylase. $1\alpha,25(OH)_2D_3$ and $24R,25(OH)_2D_3$ also operate in a feedback loop to modulate and/or reduce the secretion of PTH. Other modulators of renal $1\alpha,25(OH)_2D_3$ production are shown in Fig. 4.

3. $24R,25(OH)_2D_3$

A second dihydroxylated metabolite of vitamin D is produced in the kidney, namely, $24R,25(OH)_2D_3$. Also, virtually all other tissues that have receptors for $1\alpha,25(OH)_2D_3$ can also produce $24R,25(OH)_2D_3$. There is some controversy concerning the possible unique biological actions of $24R,25(OH)_2D_3$. However, there is some evidence that $24R,25(OH)_2D_3$ plays a role in the suppression of PTH secretion (128,129), in the mineralization of bone (130,131), and in fracture healing (132,133). Other studies demonstrated that the combined presence of $24R,25(OH)_2D_3$ and $1\alpha,25(OH)_2D_3$ are required for normal egg production, fertility, and hatchability in chickens (104) and quail (134). From these studies it was apparent that only combination doses of both compounds were capable of eliciting the same response as the parent vitamin D. Thus, it appears that both $1\alpha,25(OH)_2D_3$ and $24R,25(OH)_2D_3$ may be required for some of the known biological responses to vitamin D.

The enzyme responsible for the production of $24R,25(OH)_2D_3$ is the 25-hydroxy-vitamin D_3-24R-hydroxylase (24-hydroxylase). Experimental evidence suggests that this enzyme is also a mixed-function oxidase. The activity of this enzyme is regulated so that when $1\alpha,25(OH)_2D_3$ levels are low, the activity of the 24R-hydroxylase is also low, but when $1\alpha,25(OH)_2D_3$ levels are high, the activity of the 24R-hydroxylase is high. Under normal physiological conditions both $1\alpha,25(OH)_2D_3$ and $24R,25(OH)_2D_3$ are secreted from the kidney and circulated in the plasma of all classes of vertebrates.

In addition to these three metabolites, many other vitamin D_3 metabolites have been chemically characterized, and the existence of others appears likely. The chemical struc-

Fig. 5 Summary of the metabolic transformations of vitamin D_3. Shown here are the structures of all known chemically characterized vitamin D_3 metabolites.

tures of the 37 known metabolites are shown in Fig. 5. Most of these metabolites appear to be intermediates in degradation pathways of $1\alpha,25(OH)_2D_3$. None of these other metabolites have been shown to have biological activity except for the $1\alpha,25(OH)_2D_3$-26,23-lactone. The lactone is produced by the kidney when the plasma levels of $1\alpha,25(OH)_2D_3$ are very high. The metabolite appears to be antagonistic to $1\alpha,25(OH)_2D_3$ because it mediates a decrease in serum calcium levels in the rat. Other experiments suggest that lactone inhibits bone resorption and blocks the resorptive action of $1\alpha,25(OH)_2D_3$ on the bone (135), perhaps functioning as a natural antagonist of $1\alpha,25(OH)_2D_3$ to prevent toxic effects from overproduction of $1\alpha,25(OH)_2D_3$.

E. Catabolism and Excretion

Several pathways exist in humans and animals to further metabolize $1\alpha,25(OH)_2D_3$. These include oxidative cleavage of the side chain, hydroxylation of C-24 to produce $1\alpha,24,25(OH)_3D_3$, formation of 24-oxo-$1\alpha,25(OH)_2D_3$, formation of $1\alpha,25(OH)_2D_3$-26,23-lactone, and formation of $1\alpha,25,26(OH)_3D_3$ (Fig. 5). It is not known which of these pathways are involved in the breakdown or clearance of $1\alpha,25(OH)_2D_3$ in humans.

The catabolic pathway for vitamin D is obscure, but it is known that the excretion of vitamin D and its metabolites occurs primarily in the feces with the aid of bile salts. Very little appears in the urine. Studies in which radioactively labeled $1\alpha,25(OH)_2D_3$ was administered to humans have shown that 60–70% of the $1\alpha,25(OH)_2D_3$ was eliminated in the feces as more polar metabolites, glucuronides, and sulfates of $1\alpha,25(OH)_2D_3$. The half-life of $1\alpha,25(OH)_2D_3$ in plasma has two components. Within 5 min, only half of an administered dose of radioactive $1\alpha,25(OH)_2D_3$ remains in the plasma. A slower component of elimination has a half-life of about 10 h. $1\alpha,25(OH)_2D_3$ is catabolized by a number of pathways that result in its rapid removal from the organism (136).

V. BIOCHEMICAL MODE OF ACTION

The major classical physiological effects of vitamin D are to increase the active absorption of Ca^{2+} from the proximal intestine and to increase the mineralization of bone. This is achieved via two major signal transduction pathways: genomic and nongenomic.

A. Genomic

Vitamin D, through its daughter metabolite $1\alpha,25(OH)_2D_3$, functions in a manner homologous to that of steroid hormones. A model for steroid hormone action is shown in Fig. 6. In the general model, the hormone is produced in an endocrine gland in response to a physiological stimulus and then circulates in the blood, usually bound to a protein carrier (i.e. DBP), to target tissues where the hormone interacts with specific, high-affinity, intracellular receptors. The receptor–hormone complex localizes in the nucleus, undergoes some type of "activation" perhaps involving phosphorylation (137–140), and binds to a hormone response element (HRE) on the DNA to modulate the expression of hormone-sensitive genes. The modulation of gene transcription results in either the induction or repression of specific mRNAs, ultimately resulting in changes in protein expression needed to produce the required biological response. High-affinity receptors for $1\alpha,25(OH)_2D_3$ have been identified in at least 26 target tissues (12,16,141) and more than 50 genes are known to be regulated by $1\alpha,25(OH)_2D_3$ (142).

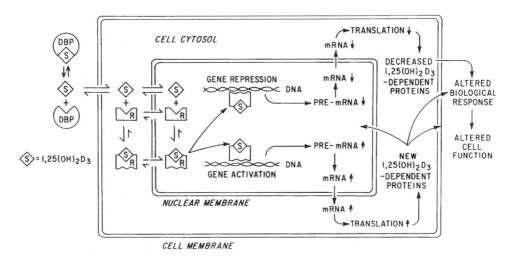

Fig. 6   General model for the mode of action of steroid hormones. Target tissues contain receptors for the steroid which confer on them the ability to modulate gene transcription. S, steroid; R, receptor protein, which may be present inside the cell in either the cytosol or nuclear compartment; SR, steroid–receptor complex; DBP, serum vitamin D–binding protein, which functions to transport the steroid hormone from the endocrine gland to its various target tissues.

Genes that have been shown to be transcriptionally regulated by $1\alpha,25(OH)_2D_3$ are listed in Table 4.

1. Nuclear Receptor

The $1\alpha,25(OH)_2D_3$ receptor was originally discovered in the intestine of vitamin-D deficient chicks (50,51). It has been extensively characterized and the cDNA for the nuclear receptor has been cloned and sequenced (52–55). The $1\alpha,25(OH)_2D_3$ receptor is a DNA-binding protein with a molecular weight of about 50,000 da. It binds $1\alpha,25(OH)_2D_3$ with high affinity with a K_D in the range of $1-50 \times 10^{-10}$ M (143–145). The ligand specificity of the nuclear $1\alpha,25(OH)_2D_3$ receptor is illustrated in Table 5. The $1\alpha,25(OH)_2D_3$ receptor protein belongs to a superfamily of homologous nuclear receptors (52). To date only a single form of the receptor has been identified.

The superfamily of ligand-dependent nuclear receptors includes receptors for glucocorticoids (GR), progesterone (PR), estrogen (ER), aldosterone, androgens, thyroid hormone (T3R), hormonal forms of vitamins A (RAR, RXR) and D (VDR), and several orphan receptors (141,146,147). Comparative studies of these receptors reveal that they have the common structural organization consisting of five domains (148), shown in Fig. 7. The different domains act as distinct modules that can function independently of each other (149–151).

The DNA binding domain, C, is the most conserved domain throughout the family. About 70 amino acids fold into two zinc finger–like motifs. Conserved cysteines coordinate a zinc ion in a tetrahedral arrangement. The first finger, which contains four cysteines and several hydrophobic amino acids, determines the DNA response element specificity. The second zinc finger, which contains five cysteines and many basic amino

Table 4 Genes Regulated by $1\alpha,25(OH)_2D_3$

Gene	Reg.	Evidence	Tissue/cell
α-Tubulin	Down	mRNA	Chick intestine
Aldolase subunit B	Up	mRNA	Chick kidney
Alkaline phosphatase	Up	mRNA	Rat intestine
			Chick intestine
			TE-85 cells
ATP synthase	Up	mRNA	Rat intestine
			Chick intestine
	Down	mRNA	Chick kidney
c-FMS	Up	mRNA	HL-60 cells
c-FOS	Up	mRNA	MG-63 cells
			HL-60 cells
c-KI-RAS	Up	mRNA	BALB-3T3 cells
c-MYB	Down	mRNA	HL-60 cells
c-MYC	Up	mRNA	MG-63
	Down	mRNA	U937 cells
		Transcription	HL-60 cells
			HL-60 cells
Calbindin$_{28K}$	Up	mRNA	Chick intestine
		Transcription	Mouse kidney
			Chick intestine
Calbindin$_{9K}$	Up	mRNA	Mouse kidney
		VDRE	Rat
Carbonic anhydrase	Up	mRNA	Marrow cells
		Transcription	Myelomonocytes
CD-23	Down	mRNA	PBMC
Collagen type I	Down	mRNA/VDRE	Rat
Cytochrome oxidase subunit I	Up	mRNA	Rat intestine
			Chick intestine
	Down	mRNA	Chick kidney
Cytochrome oxidase subunit II	Up	mRNA	Chick intestine
	Down	mRNA	Chick kidney
Cytochrome oxidase subunit III	Up	mRNA	Rat intestine
			Chick intestine
	Down	mRNA	Chick kidney
Cytochrome B	Down	mRNA	Chick kidney
Fatty acid–binding protein	Down	mRNA	Chick intestine
Ferridoxin	Down	mRNA	Chick kidney
Fibronectin	Up	mRNA	MG-63
			TE-85
			HL-60 cells
γ-Interferon	Down	mRNA	T lymphocytes
			PBMC
Glyceraldehyde-3-phosphate dehydrogenase	Up	mRNA	BT-20 cells

Table 4 Continued

Gene	Reg.	Evidence	Tissue/cell
GM-colony-stimulating factor	Down	mRNA	T lymphocytes
Heat shock protein 70	Up	mRNA	PBMC
Histone H4	Down	mRNA/ Ttranscription	HL-60 cells
1-Hydroxyvitamin D-24-hydroxylase	Up	mRNA	Rat kidney
		mRNA/ Transcription	Rat kidney
Integrin$_{\alpha v \beta 3}$	Up	mRNA/ Transcription	Avian osteoclast precursor cells
Interleukin-6	Up	mRNA	U937
Interleukin-4	Up	mRNA	U937 cells
Interleukin-2	Down	mRNA	T lymphocytes
Interleukin-3 receptor	Up	mRNA	MC3T3 cells
Matrix gla protein	Up	mRNA	UMR106-01, ROS 25/1, 25/4 cells
Metallothionein	Up	mRNA	Rat kertinocytes
			Mouse liver/kidney/skin
			Chick kidney
Monocyte-derived neutrophil-activating peptide	Up	mRNA/transcription	HL-60 cells
NADH DH subunit I	Down	mRNA	Chick kidney
NADH DH subunit III	Up	mRNA	Chick intestine
NADH DH subunit IV	Up	mRNA	Chick intestine
Nerve growth factor	Up	mRNA	L-929 cells
Osteocalcin	Up	mRNA	ROS 17/2.8
			ROS 25/1
		VDRE	ROS 17/2.8
			Rat
Osteopontin	Up	mRNA	ROS 17/2.8
		VDRE	ROS 17/2.8
Plasma membrane calcium pump	Up	mRNA	Chick intestine
Pre-pro-PTH	Down	mRNA	Rat
		mRNA/transcription	Bovine parathyroid
Prolactin	Up	mRNA	GH$_4$C$_1$ cells
Protein kinase inhibitor	Down	mRNA	Chick kidney
Protein kinase C	Up	mRNA/transcription	HL-60 cells
PTH	Down	mRNA	Rat parathyroid
PTH-related protein	Down	mRNA/transcription	TT cells
Transferrin receptor	Down	mRNA	PBMC
Tumor necrosis factor α	Up	mRNA	U937 cells
		Transcription	HL-60 cells
VDR	Up	mRNA	Rat intestine
			Rat pituitary
			MG-60 cells

Source: Ref. 142.

Table 5 Ligand Specificity of the Nuclear $1\alpha,25(OH)_2D_3$ Receptor

Ligand	Structural modification	RCI (%)[a]
$1\alpha,25(OH)_2D_3$		100
$1\alpha,25(OH)_2$-24-nor-D_3	Shorten side chain by 1 carbon	67
$1\alpha,25(OH)_2$-3-epi-D_3	Orientation of 3β-OH altered	24
$1\alpha,25(OH)_2$-24a-dihomo-D_3	Lengthen side chain by 2 carbons	24
$1\beta,25(OH)_2D_3$	Orientation of 1α-OH changed	0.8
$1\alpha(OH)D_3$	Lacks 25-OH	0.15
$25(OH)D_3$	Lacks 1α-OH	0.15
$1\alpha,25(OH)_2$-7-dehydrocholesterol	Lacks a broken B ring; is not a *seco* steroid	0.10
Vitamin D_3	Lacks 1α and 25-OH	0.0001

[a]The *Relative Competitive Index* (RCI) is a measure of the ability of a nonradioactive ligand to compete, under in vitro conditions, with radioactive $1\alpha,25(OH)_2D_3$ for binding to the nuclear $1\alpha,25(OH)_2D_3$ receptor (VDR)
Source: Ref. 14.

acids, is also necessary for DNA binding and is involved in receptor dimerization (146,150,152,153).

The next conserved region is the steroid binding domain (region E). This region contains a hydrophobic pocket for ligand binding and also contains signals for several other functions, including dimerization (154–157), nuclear translocation, and hormone-dependent transcriptional activation (149,150,158).

The A/B domain is also known as the immuno- or transactivation domain. This region is poorly conserved in amino acids and in size, and its function has not been clearly defined. The VDR has the smallest A/B domain (25 amino acids) of the known receptors; mineralocorticoid receptor has the largest (603 amino acids). An independent transcriptional activation function is located within the A/B region (146,150,151) that is constitutive in receptor constructs lacking the ligand binding domain (region E). The relative importance of the transcriptional activation by this domain depends on the receptor, the context of the target gene promoter, and the target cell type (159).

Region D is the hinge region between the DNA binding domain and the ligand bonding domain. The hinge region in the VDR contains 156 amino acids and has immunogenic properties. The VDR has the longest hinge region of the known receptors (160). Human GR and PR have hinge regions of 92 and 101 amino acids, respectively.

The VDR belongs to a subgroup of the receptors designated group II, which includes T3R, RAR, RXR, and several orphan receptors. All of the group II receptors can form heterodimers with RXR (161,162), and other heterodimeric interactions have also been

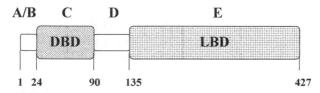

Fig. 7 Schematic representation of the human nuclear VDR. The DNA binding domain (C) and ligand binding domain (E) are boxed.

reported (163). T3R lacking the DNA binding domain can inhibit the transactivation of RAR (156) and VDR (152), but not a chimeric receptor containing the ligand binding domain of GR. The VDR can also form heterodimers with RAR (163,164). The ability to form heterodimers with other receptors allows for enhanced affinity for distinct DNA targets, generating the diverse range of physiological effects.

2. Calbindin D

One of the major effects of $1\alpha,25(OH)_2D_3$ in many of its target tissues is the induction of the calcium binding protein, calbindin D. In the mammalian kidney and brain and in avians, a larger form of the protein is expressed, calbindin D_{28K} (165), whereas in the mammalian intestine and placenta a smaller form is expressed, calbindin D_{9K} (166). The expression of calbindins in various tissues and species appears to be regulated to differing degrees by $1\alpha,25(OH)_2D_3$ (167).

Early experiments showed that actinomycin D and α-amanitin, transcriptional inhibitors, could block the induction of calbindin D_{28K} by $1\alpha,25(OH)_2D_3$ (168). Later experiments showed that $1\alpha,25(OH)_2D_3$ was able to stimulate total RNA synthesis in the chick intestine (169) in addition to specifically inducing the mRNA for calbindin D_{28K} (170). Nuclear transcription assays have shown that transcription of calbindin D_{28K} mRNA is directly induced by $1\alpha,25(OH)_2D_3$ in the chick intestine and is correlated to the level of occupied $1\alpha,25(OH)_2D_3$ receptors (171). The gene for calbindin D_{28K} has now been cloned and sequenced, but there is still much to learn about how $1\alpha,25(OH)_2D_3$ induces this gene (172).

B. Nongenomic Actions

Recent studies (173) suggest that not all of the actions of $1\alpha,25(OH)_2D_3$ can be explained by receptor–hormone interactions with the genome. $1\alpha,25(OH)_2D_3$ can stimulate the intestinal transport of calcium within 4–6 min, i.e., too quickly to involve genome activation. The rapid transport of calcium mediated by $1\alpha,25(OH)_2D_3$ in the intestine has been termed "transcaltachia" ("trans" = across; "cal" = calcium; "tachia" = swiftly). Transcaltachia is not inhibited by actinomycin D but is inhibited by colchicine, an antimicrotubule agent, and by leupeptin, an antagonist of lysosomal cathepsin B. Transcaltachia induced by $1\alpha,25(OH)_2D_3$ in the intestine appears to involve the internalization of calcium in endocytic vesicles at the brush-border membrane, which then fuse with lysosomes and travel along microtubules to the basal lateral membrane where exocytosis occurs. Therefore, some of the actions of $1\alpha,25(OH)_2D_3$ may be mediated at the cell membrane or by extranuclear subcellular components.

Other effects of $1\alpha,25(OH)_2D_3$ that do not appear to be mediated by the nuclear receptor are phosphoinositide breakdown (174), enzymatic activity in osteoblast-derived matrix vesicles (175), certain secretion events in osteoblasts (176), rapid changes in cytosolic Ca^{2+} levels in primary cultures of osteoblasts and osteosarcoma cells (177–179), and increases in cyclic guanosine monophosphate levels in fibroblasts (180). These rapid effects appear to be mediated by a membrane receptor–like protein for $1\alpha,25(OH)_2D_3$ (181); a candidate membrane receptor for $1\alpha,25(OH)_2D_3$ has been proposed (182). Other steroid hormones, i.e., estrogen (183), progesterone (184–187), testosterone (188), glucocorticoids (189,190), corticosteroid (191), and thyroid (192,193), have been shown to have similar membrane effects (181). A model for the nongenomic signal transduction pathway is shown in Fig. 8.

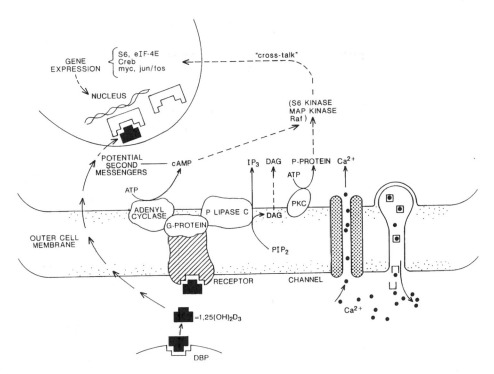

Fig. 8 Model describing the signal transduction pathways associated with the nongenomic response of transcaltachia. The general model of vesicular Ca^{2+} transport includes formation of Ca^{2+}-containing endocytic vesicles at the brush-border membrane, fusion of endocytic vesicle with lysosomes, movement of lysosomes along microtubules, and exocytotic extrusion of Ca^{2+} via fusion of the lysosomes with the basal lateral membrane of the intestinal enterocyte. The binding of $1\alpha,25(OH)_2D_3$ to a membrane receptor results in an increase of several second messengers, including IP3, cAMP, activation of PKC or intracellular Ca^{2+}, which may result in the transient opening of Ca^{2+} channels. The increased Ca^{2+} concentration may then initiate the exocytosis of the lysosomal vesicles.

VI. SPECIFIC FUNCTIONS OF $1\alpha,25(OH)_2D_3$

A. $1\alpha,25(OH)_2D_3$ and Mineral Metabolism

The classical target tissues for $1\alpha,25(OH)_2D_3$ are those tissues that have been found to be directly involved in the regulation of mineral homeostasis. In humans, serum calcium levels are normally maintained between 9.5 and 10.5 mg/100 mL, whereas the phosphorus concentration is between 2.5 and 4.3 mg/100 mL (2). Together with PTH and calcitonin, $1\alpha,25(OH)_2D_3$ maintains serum calcium and phosphate levels by its actions on the intestine, kidney, bone, and parathyroid gland.

In the intestine, one of the best characterized effects of $1\alpha,25(OH)_2D_3$ is the stimulation of intestinal lumen–to–plasma flux of calcium and phosphate (39,194,195). Although extensive evidence exists showing that $1\alpha,25(OH)_2D_3$ interacting with its receptor upregulates calbindin D in a genome-mediated fashion, the relationship between calbindin D and calcium transport is not clear (196). In the vitamin D–deficient state, both mammals and birds have severely decreased intestinal absorption of calcium with no detectable levels of calbindin. There is a linear correlation between the increased cellular levels of calbindin D and calcium transport. When $1\alpha,25(OH)_2D_3$ is given to vitamin D–deficient chicks, the transport of calcium reaches maximal rates at 12–14 h, whereas calbindin D does not reach its maximal levels until 48 h (173). In one study employing immunohisto-chemical techniques, it was demonstrated that the cellular location of calbindin D_{28K} changed with the onset of calcium transport (197).

$1\alpha,25(OH)_2D_3$ treatment also is known to alter the biochemical and morphological characteristics of the intestinal cells (198,199). The size of the villus and the size of the microvilli increase upon $1\alpha,25(OH)_2D_3$ treatment (200). The brush border undergoes no-ticeable alterations of structure and composition of cell surface proteins and lipids, oc-curring in a time frame corresponding to the increase in Ca^{2+} transport mediated by $1\alpha,25(OH)_2D_3$ (201). However, despite extensive work, the exact mechanisms involved in the vitamin D–dependent intestinal absorption of calcium remain unknown (202–204).

The kidney is the major site of synthesis of $1\alpha,25(OH)_2D_3$ and of several other hydroxylated vitamin D derivatives. Probably the most important effect of $1\alpha,25(OH)_2D_3$ on the kidney is the inhibition of $25(OH)D_3$-1α-hydroxylase activity, which results in a decrease in the synthesis of $1\alpha,25(OH)_2D_3$ (205,206). Simultaneously, the activity of the $25(OH)D_3$-24-hydroxylase is stimulated. The actions of vitamin D on calcium and phos-phorus metabolism in the kidney has been controversial, and more research is needed to clearly define the actions of $1\alpha,25(OH)_2D_3$ on the kidney.

Although vitamin D is a powerful antirachitic agent, its primary effect on bone is the stimulation of bone resorption leading to an increase in serum calcium and phosphorus levels (207). With even slight decreases in serum calcium levels, PTH is synthesized, which then stimulates the synthesis of $1\alpha,25(OH)_2D_3$ in the kidney. Both of these hor-mones stimulate bone resorption. Maintaining constant levels of calcium in the blood is crucial, whether calcium is available from the diet or not. Therefore, the ability to release calcium from its largest body store—bone—is vital. Bone is a dynamic tissue that is constantly being remodeled. Under normal physiological conditions, bone formation and bone resorption are tightly balanced (208). The stimulation of bone growth and mineraliza-tion by $1\alpha,25(OH)_2D_3$ appears to be an indirect effect of the provision of minerals for bone matrix incorporation through an increase of intestinal absorption of calcium and phosphorus. In bone, nuclear receptors for $1\alpha,25(OH)_2D_3$ have been detected in normal osteoblasts (209), osteoblast-like osteosarcoma cells, but not in osteoclasts. In addition, $1\alpha,25(OH)_2D_3$ can induce rapid changes in cytosolic Ca^{2+} levels in osteoblast and osteosar-coma cells by opening voltage-gated Ca^{2+} channels via a nongenomic signal transduction pathway (178,179).

Some of the actions of $1\alpha,25(OH)_2D_3$ in bone are related to changes in bone cell differentiation. $1\alpha,25(OH)_2D_3$ is known to affect a number of osteoblast-related functions. For example, $1\alpha,25(OH)_2D_3$ decreases type I collagen production (210), and increases alkaline phosphatase production and the proliferation of cultured osteoblasts (211); $1\alpha,25(OH)_2D_3$ increases the production of osteocalcin (212) and matrix Gla protein (213), and decreases the production of type I collagen by fetal rat calvaria (214).

$1\alpha,25(OH)_2D_3$ also affects the growth and differentiation of osteoclasts and osteo-clast-like cells in vivo in rats (215) and in primate bone marrow cell cultures (216). Since the osteoclast does not have a nuclear VDR, $1\alpha,25(OH)_2D_3$ must affect osteoclasts indi-rectly or by nongenomic mechanisms. There is some evidence that a factor produced by osteoblasts promotes the formation of osteoclasts (217,218). It is possible that the only effect of $1\alpha,25(OH)_2D_3$ on the osteoclasts is to stimulate its generation from progenitor cells.

PTH is an important tropic stimulator of $1\alpha,25(OH)_2D_3$ synthesis by the kidney. High circulating levels of $1\alpha,25(OH)_2D_3$ have been shown to decrease the levels of PTH by two different mechanisms: an indirect mechanism due to the resulting increase in serum calcium levels, which is an inhibitory signal for PTH production, and a direct mechanism involving the interaction of $1\alpha,25(OH)_2D_3$ and its receptor, which directly suppresses the expression of the prepro-PTH gene.

During pregnancy and lactation, large amounts of calcium are needed for the developing fetus and for milk production. Hormonal adjustments in the vitamin D endocrine system are critical to prevent depletion of minerals leading to serious bone damage for the mother. Although receptors for $1\alpha,25(OH)_2D_3$ have been found in placental tissue and in the mammary gland, the role of vitamin D is not clear.

B. Vitamin D in Nonclassical Systems

In the 1970s and 1980s, nuclear receptors for $1\alpha,25(OH)_2D_3$ were discovered in a variety of tissues and cells not directly involved in calcium homeostasis. Thus, the role of the vitamin D endocrine system has expanded to include general effects on cell regulation and differentiation (12,18). Nuclear VDRs are present in muscle, hematolymphopoietic, reproductive, and nervous tissue, as well as in other endocrine tissues and skin. More than 50 proteins are known to be regulated by $1\alpha,25(OH)_2D_3$, including several oncogenes (56,142) (Table 2), which extend by far the classical limits of vitamin D actions on calcium homeostasis. In many of these systems it is not yet clear what the effect of vitamin D is on the tissue or its mode of action.

Skeletal muscle is a target organ for $1\alpha,25(OH)_2D_3$. Clinical studies have shown the presence of muscle weakness or myopathy during metabolic bone diseases related to vitamin D deficiency (19,22,219). These abnormalities can be reversed with vitamin D therapy. Experimental evidence has shown that $1\alpha,25(OH)_2D_3$ has a direct effect on Ca^{2+} transport in cultured myoblasts and skeletal muscle tissue. Furthermore, there is evidence that the action of $1\alpha,25(OH)_2D_3$ on skeletal muscle may be important for the calcium homeostasis of the entire organism because the hormone induces a rapid release of calcium from muscle into the serum of hypocalcemic animals. $1\alpha,25(OH)_2D_3$ receptors have been detected in myoblast cultures, and the changes in calcium uptake have been shown to be RNA- and protein synthesis–dependent, suggesting a genomic mechanism. $1\alpha,25(OH)_2D_3$ has also been shown to be important for cardiac muscle function (220–223).

In the skin, $1\alpha,25(OH)_2D_3$ appears to exert effects on cellular growth and differentiation. Receptors for $1\alpha,25(OH)_2D_3$ have been found in human (224) and mouse skin (225). $1\alpha,25(OH)_2D_3$ inhibits the synthesis of DNA in mouse epidermal cells (225). The hormone induces changes in cultured keratinocytes, which are consistent for terminal differentiation of nonadherent cornified squamous cells (226). Additional experiments have shown that human neonatal foreskin keratinocytes produce $1\alpha,25(OH)_2D_3$ from $25(OH)D_3$ under in vitro conditions (227), suggesting that keratinocyte-derived $1\alpha,25(OH)_2D_3$ may affect epidermal differentiation locally. Psoriasis is a chronic hyperproliferative skin disease. Some forms of psoriasis have been shown to improve significantly when treated topically with calcipotriol, a nonhypercalcemic analog of $1\alpha,25(OH)_2D_3$ (228–230). In mouse skin carcinogenesis, $1\alpha,25(OH)_2D_3$ blocks the production of tumors induced by 12-O-tetradecanoylphorbol-12-acetate (231).

In the pancreas, $1\alpha,25(OH)_2D_3$ has been found to be essential for normal insulin secretion. Experiments with rats have shown that vitamin D increases insulin release from the isolated perfused pancreas, in both the presence and the absence of normal serum calcium levels (232–236). Human patients with vitamin D deficiency, even under conditions of normal calcemia, exhibit impaired insulin secretion but normal glucagon secretion, suggesting that $1\alpha,25(OH)_2D_3$ directly affects β-cell function (237).

Receptors for $1\alpha,25(OH)_2D_3$ have been found in some sections of the brain (238,239). However, the role of $1\alpha,25(OH)_2D_3$ in the brain is not well understood. Both calbindins D have been found in the brain, but neither the expression of calbindin D_{28K} nor that of calbindin D_{9K} appears to be directly modulated by vitamin D (238,239). In the rat, $1\alpha,25(OH)_2D_3$ appears to increase the activity of the choline acetyltransferase in specific regions of the brain (238). Other steroid hormones have also been shown to affect the metabolism of specific brain regions (240,241).

Also, normal, benign, hyperplastic and malignant prostatic epithelial and fibroblastic cells contain receptors for $1\alpha,25(OH)_2D_3$ (242). In hematopoietic tissue, $1\alpha,25(OH)_2D_3$ promotes the differentiation and inhibits proliferation of both malignant and nonmalignant hematopoietic cells. Human promyelocytic leukemia cells, HL-60, have been shown to have receptors for $1\alpha,25(OH)_2D_3$ and to differentiate toward macrophages upon treatment with $1\alpha,25(OH)_2D_3$ (243,244). Other effects of $1\alpha,25(OH)_2D_3$ on the immune system will be discussed in the next section.

C. Immunoregulatory Roles

In 1979, when the VDR was discovered in several neoplastic hematopoietic cell lines as well as in normal human peripheral blood mononuclear cells, monocytes, and activated lymphocytes (245,246), a role for $1\alpha,25(OH)_2D_3$ in immune function was suggested. Since then, $1\alpha,25(OH)_2D_3$ has been shown to affect cells of the immune system in a variety of ways. $1\alpha,25(OH)_2D_3$ reduces the proliferation of HL-60 cells and induces their differentiation to monocytes (243) and macrophages (244,247,248). The actions of $1\alpha,25(OH)_2D_3$ on normal monocytes are controversial, but it appears that the molecule may enhance monocyte function. $1\alpha,25(OH)_2D_3$ appears to reduce levels of HLA-DR and $CD4^+$ class II antigens on monocytes or macrophages with no effect on the expression of class I antigens (249). The enhancement of class II antigen expression is a common feature of autoimmunity and often precedes the onset of autoimmune diseases.

$1\alpha,25(OH)_2D_3$ also promotes the differentiation of leukemic myeloid precursor cells toward cells with the characteristics of macrophages (243). Subsequent experiments have shown that $1\alpha,25(OH)_2D_3$ does not alter the clonal growth of normal myeloid precursors but does induce the formation of macrophage colonies preferentially over the formation of granulocyte colonies (247). In addition, macrophages derived from different tissues can synthesize $1\alpha,25(OH)_2D_3$ when activated by γ-interferon (121). Also, $1\alpha,25(OH)_2D_3$ can suppress immunoglobulin production by activated B lymphocytes (250) and inhibit DNA synthesis and proliferation of both activated B and T lymphocytes (251–253). These findings suggest that a vitamin D paracrine system exists that involves activated macrophages and activated lymphocytes (Fig. 4).

$1\alpha,25(OH)_2D_3$ also affects some functions of T and B lymphocytes and natural killer (NK) cells. In T lymphocytes, the mitogen activation of lymphocyte proliferation is blocked in the presence of $1\alpha,25(OH)_2D_3$ (251,254,255), apparently by interference with cell cycle progression from early G1 to late G1 phase (254). $1\alpha,25(OH)_2D_3$ exhibits a permissive or enhancing effect on T-cell suppressor activity. In an in vitro model of transplant compatibility, the mixed lymphocyte reaction, $1\alpha,25(OH)_2D_3$ significantly enhanced T-cell suppressor activity (256). $1\alpha,25(OH)_2D_3$ also affects the cytotoxicity of NK and T-cytotoxic cells probably by interfering with their generation from precursor cells.

$1\alpha,25(OH)_2D_3$ has been shown to decrease mRNA levels for IL-2 (257), γ-interferon (258) and granulocyte-macrophase colony-stimulating factor (GM-CSF) (259,260). $1\alpha,25(OH)_2D_3$ also attenuates the inducing effect of T-helper cells on IgG synthesis by B cells (261).

Despite the wide range of actions of $1\alpha,25(OH)_2D_3$ on various immune cells, no general immunomodulatory role for $1\alpha,25(OH)_2D_3$ has been defined. In contrast to the serum calcium elevation observed in patients with sarcoidosis, lymphoma, and an anephric patients with end-stage renal disease, no systemic immunosuppressive activity of $1\alpha,25(OH)_2D_3$ has been described in these disease states to date, suggesting that $1\alpha,25(OH)_2D_3$ acts in an autocrine or paracrine fashion to modulate local immune function (261,262). $1\alpha,25(OH)_2D_3$ and cyclosporine, a potent immunosuppressive drug, appear to affect the immune system in a similar fashion. They both affect T lymphocytes during initial activation by antigen, select the generation of T-helper cells by inhibiting lympho-kine production at a genomic level, and inhibit the generation of T-cytotoxic and NK cells. Both are involved in the enhancement of T-suppressor function, a key element in the efficacy of cyclosporine as a drug that reduces allograft tissue rejection (263). $1\alpha,25(OH)_2D_3$ appears to work synergistically with cyclosporine when the two compounds are used in combination (264,265).

The use of nonhypercalcemic $1\alpha,25(OH)_2D_3$ analogs can result in enhanced immuno-suppressive effects without the toxicity risks of $1\alpha,25(OH)_2D_3$. Because of the synergistic effect when $1\alpha,25(OH)_2D_3$ is used with cyclosporine, synthetic $1\alpha,25(OH)_2D_3$ analogs may be used in the treatment of autoimmune diseases (266) or for transplantation (267) in combination with cyclosporine to reduce the toxicity of both compounds.

D. Structures of Important Analogs

In nephrectomized animals, vitamin D compounds cannot be hydroxylated at the C-1 position because the kidney is the site where this hydroxylation occurs. Researchers found that neither vitamin D nor 25(OH)D was able to elicit a significant biological response when administered in physiological doses to nephrectomized animals (48,268). Also, it was noted in the 1940s and 1950s that dihydrotachysterol$_3$ (a 5,6-trans analog of vitamin D_3) was biologically active under circumstances where the parent vitamin D demonstrated little or no biological activity. These findings raised the question of the functional importance of the various structural elements of the vitamin D molecule. Studies using analogs of vitamin D have been used to address this question. The ability of analogs to bind to the nuclear receptor for $1\alpha,25(OH)_2D_3$, to increase intestinal calcium absorption (ICA) and bone calcium mobilization (BCM), and to promote cellular differentiation are then determined. Because of recent advances in new vitamin D syntheses described above and in Fig. 3, analogs have been synthesized with modifications in the A ring, *seco*-B ring, C ring, C/D ring junction, D ring, and/or side chain (14, 269).

The importance of the configuration of the A ring has been studied by synthesizing 5,6-trans analogs. Because of the rotation of the A ring, these analogs cannot undergo 1-hydroxylation and have been found to be only 1/1000 as biologically effective as $1\alpha,25(OH)_2D_3$. The relative significance of the 3β-hydroxyl group has been assessed by preparing analogs such as 3-deoxy-$1\alpha,25(OH)_2D_3$. Although this analog is active in vivo, it is interesting in that it preferentially stimulates intestinal calcium absorption over bone calcium mobilization (270). Of all the analogs synthesized, only a few show such selective biological activity.

The effect of altering the length of the side chain has been studied. The 27-*nor*-25(OH)D$_3$ and 26,27-bis-*nor*-25(OH)D$_3$ are reportedly able to stimulate intestinal calcium absorption and bone calcium mobilization in both normal and anephric rats but are 10–100 times less active than 25(OH)D$_3$ (271). The 24-*nor*-25(OH)D$_3$ was found to have no biological activity (272), although it was able to block the biological response to vitamin D but not to 25(OH)D$_3$ or 1α,25(OH)$_2$D$_3$. This suggests that it might have anti–vitamin D activity.

One of the most interesting side-chain analogs of 1α,25(OH)$_2$D$_3$ is 1α(OH)D$_3$. This metabolite appears to have the same biological activity in the chick as 1α,25(OH)$_2$D$_3$ (273) and is approximately half as active in the rat (274). In an attempt to determine if the biological activity of 1α(OH)D$_3$ is the result of in vivo 25-hydroxylation, the 25-fluoro-1α(OH)D$_3$ derivative was prepared (275). The fluorine on C-25 prevents hydroxylation of this carbon. The fluoro compound was found to be one-fiftieth as active as 1α,25(OH)$_2$D$_3$, suggesting that 1α(OH)D$_3$ has some activity even without 25-hydroxylation.

From such studies, the particular attributes of the structure of 1α,25(OH)$_2$D$_3$ that enables it to elicit its biological responses are being defined. It is now known that the 3β-hydroxy group does not appear to be as important for biological activity as the 1α- or 25-hydroxyl groups; the cis configuration of the A ring is preferred over the trans configuration; and the length of the side chain appears critical, as apparently there is little tolerance for its being shortened or lengthened.

Analogs of 1α,25(OH)$_2$D$_3$ have been used to study the in vivo metabolism and mode of action of vitamin D compounds. There has also been widespread interest in developing 1α,25(OH)$_2$D$_3$ analogs to use as therapeutic agents in the treatment of osteoporosis, renal osteodystrophy, cancer, immunodeficiency syndromes, autoimmune diseases, and some skin disorders. Of particular interest are analogs that separate the calcemic effects from the proliferation and differentiation effects of 1α,25(OH)$_2$D$_3$.

One of the most successful analogs in terms of separating biological activities is a cyclopropyl derivative of 1α,25(OH)$_2$D$_3$, 1α,24S(OH)$_2$-22ene-26,27-dehydrovitamin D$_3$, designated calcipotriol; this analog has weak systemic effects on calcium metabolism but potent effects on cell proliferation and differentiation (276,277). It is rapidly converted to inactive metabolites in vivo (278,279) and is 200-fold less potent than 1α,25(OH)$_2$D$_3$ in causing hypercalciuria and hypercalcemia in rats (276). It is as effective in binding to the nuclear receptor as 1α,25(OH)$_2$D$_3$ and has similar effects on the growth and differentiation of keratinocytes (280,281). It is currently marketed as a topical treatment for psoriasis, a proliferative disorder of the skin (228,282–285).

Another analog that has potential as a therapeutic agent is 22-oxa-1α,25(OH)$_2$D$_3$. This analog has been shown to suppress the secretion of PTH and may be useful in the treatment of secondary hyperparathyroidism (286). It is 10 times more potent in suppressing proliferation and inducing differentiation than 1α,25(OH)$_2$D$_3$, with only 1/50 to 1/100 of the in vitro bone-resorbing activity of 1α,25(OH)$_2$D$_3$ (287).

Still another set of analogs of 1α,25(OH)$_2$D$_3$ with potential therapeutic applications are the compounds with a double bond at C-16 and/or a triple bond at C-23. The best characterized of these compounds is 1α,25(OH)$_2$-16ene-23yne-D$_3$ (288–290). This analog is 300-fold less active in intestinal calcium absorption (ICA) and bone calcium mobilization (BCM) and 10 to 15 times less active in inducing hypercalcemia in vivo in mice than 1α,25(OH)$_2$D$_3$. In three leukemia models, therapy with the analog resulted in a significant increase in survival (288,289). All of the 16-ene and or 23-yne analogs that have been tested are equivalent or more potent than 1α,25(OH)$_2$D$_3$ in the induction of HL-60 cell differentiation and inhibition of clonal proliferation (289,291), and ten to two hundred fold less active in ICA and BCM (291,292).

Fluorinated analogs of $1\alpha,25(OH)_2D_3$ have been especially useful for studying the in vivo metabolism of $1\alpha,25(OH)_2D_3$. Fluorine groups have been substituted for the hydroxyls at C-25, C-1, and C-3 to study the importance of these hydroxylations for the biological activity of $1\alpha,25(OH)_2D_3$. Also, fluorine groups have been substituted for hydrogens at C-23, C-24, and C-26 to facilitate the study of $1\alpha,25(OH)_2D_3$ catabolism. The analog $1\alpha,25(OH)_2$-26,26,26,27,27,27-hexafluoro-D_3 has been shown to be 10 times more potent than $1\alpha,25(OH)_2D_3$ in calcium mobilization, with longer lasting effects due to its slower rate of catabolism and metabolic clearance (293). This analog is also 10 times more potent than $1\alpha,25(OH)_2D_3$ in suppressing proliferation and inducing differentiation of HL-60 cells (247,294,295).

VII. BIOLOGICAL ASSAYS

With the exception of vitamin B_{12}, vitamin D is the most potent of the vitamins (as defined by the amount of vitamin required to elicit a biological response). Consequently, biological samples and animal tissues usually contain very low concentrations of vitamin D. For example, the circulating plasma level of vitamin D_3 in humans is only 10–20 ng/mL, or $2–5 \times 10^{-8}$ M (296). In order to detect such low concentrations of vitamin D, assays that are specific for and sensitive to vitamin D and its biologically active metabolites are required.

A. Rat Line Test

From 1922 to 1958, the only official assay for determination of the vitamin D content of pharmaceutical products or food was the rat line test. The term "official" indicates that the reproducibility and accuracy of the assay are high enough that the results of the test can be accepted legally. This assay, which is capable of detecting 1–12 IU (25–300 ng) of vitamin D, is still widely used today to determine the vitamin D content of many foods, particularly milk (297–299). The rat line test for vitamin D employs recently weaned rachitic rats; these rats are fed a rachitogenic diet for 19–25 days until severe rickets develops. The rats are then divided into groups of 7–10 animals and are fed diets that have been supplemented either with a graded series of known amounts of vitamin D_3 as standards or with the unknown test sample. (Although vitamin D oils can be directly assayed, milk, vitamin tablets, and vitamin D–fortified foods must be saponified and the residue taken up into a suitable oil vehicle prior to assay.) The rats are maintained on their respective diets for 7 days. The animals are sacrificed and their radii and ulnae dissected out and stained with a silver nitrate solution. Silver is deposited in areas of bone where new calcium has been recently deposited. The regions turn dark when exposed to light. Thus, the effects of the unknown sample on calcium deposition in the bone can be determined by visual comparison with the standards. Typical results for the rat line test are shown in Fig. 9.

B. AOAC Chick Assay

Since the rat line test is done in rats, it cannot discriminate between vitamin D_2 and vitamin D_3. In the chick, vitamin D_3 is 10 times more potent than vitamin D_2, so it is important to accurately determine the amount of vitamin D_3 in poultry feeds. The Association of

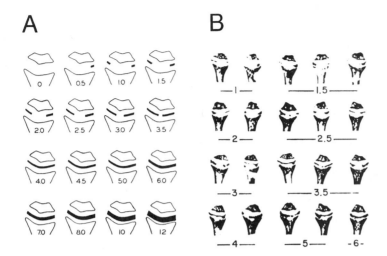

Fig. 9 Rat line test chart is shown in panel A. Photographs of radii sections scored according to the line test chart are shown in panel B.

Official Analytical Chemists (AOAC) chick test was developed to specifically measure vitamin D_3 (300,301).

Groups of 20 newly hatched chicks are placed on vitamin D–deficient diets containing added levels of vitamin D_3 (1–10 IU) or the test substance. After 3 weeks on the diet, the birds are sacrificed and the percentage of bone ash of their tibia is determined. A rachitic bird typically has 25–27% bone ash, whereas a vitamin D–supplemented has 40–45% bone ash. This assay is not used frequently because it is time consuming and expensive.

C. Intestinal Calcium Absorption

Other biological assays have been developed that make use of the ability of vitamin D to stimulate the absorption of calcium across the small intestine. Two basic types of assays measure this phenomenon: those that measure the effect of the test substance on intestinal calcium uptake in vivo (302) and those that employ in vitro methods (39,303). Each is capable of detecting physiological quantities, i.e., 2–50 IU (50–1250 ng; 0.13–3.2 nmol), of vitamin D.

1. In Vivo Technique

The in vivo technique for measuring intestinal calcium absorption uses rachitic chicks that have been raised on a low-calcium (0.6%), rachitogenic diet for 3 weeks. The birds are then given one dose of the test compound orally, intraperitoneally, or intracardially. Twelve to 48 h later, the chicks are anesthetized and 4.0 mg of $^{40}Ca^{2+}$ and approximately 6×10^6 dpm $^{45}Ca^{2+}$ are placed in the duodenal loop. Thirty minutes later, the chicks are killed by decapitation and serum is collected. Aliquots of serum are measured for $^{45}Ca^{2+}$ in a liquid scintillation counter (302).

2. In Vitro Technique

The general design of this technique is the same as that of the in vivo technique because vitamin D activity is measured in terms of intestinal calcium transport. In these assays, a vitamin D standard or test compound is given orally or intraperitoneally 24–48 h before the assay. At the time of the assay, the animals are killed and a 10-cm length of duodenum is removed and turned inside out. A gut sac is formed by tying off the ends of the segment so that the mucosal surface is on the outside and the serosal surface on the inside. The everted intestinal loop is incubated with solutions of $^{45}Ca^{2+}$. The mucosal surface of the intestine actively transports the calcium through the tissue to the serosal side. The ratio of calcium concentration on the serosal vs. the mucosal side of the intestine is a measure of the "active" transport of calcium (39,304,305). In a vitamin D–deficient animal this ratio is 1–2.5; in a vitamin D–dosed animal it can be as high as 6–7. The chick in vivo assay is usually preferred because of the tedious nature of preparing the everted gut sacs. The in vitro technique is used primarily for studies with mammals rather than birds.

D. Bone Calcium Mobilization

Another assay for vitamin D activity that often is performed simultaneously with the chick in vivo intestinal calcium absorption assay is measurement of the vitamin D–mediated elevation of serum calcium levels. If 3-week-old rachitic chicks are raised on a zero-calcium diet for at least 3 days before the assay and then are given a compound containing vitamin D, their serum calcium levels will rise in a highly characteristic manner, proportional to the amount of steroid given (302). Since there is no dietary calcium available, the only calcium source for elevation of serum calcium is bone. By carrying out this assay simultaneously with the intestinal calcium absorption assay, it is possible to measure two different aspects of the animal's response to vitamin D at the same time.

E. Growth Rate

The administration of vitamin D to animals leads to an enhanced rate of whole-body growth. An assay for vitamin D was developed in the chick using the growth-promoting properties of the steroid (45,306). One-day-old chicks are placed on a rachitogenic diet and given standard doses of vitamin D_3 or the test compound three times weekly. The birds are weighed periodically, and their weight is plotted vs. age. In the absence of vitamin D, the rate of growth essentially plateaus by the fourth week, whereas 5–10 IU of vitamin D_3 per day is sufficient to maintain a maximal growth rate in the chick. The disadvantage of this assay is the 3- to 4-week time period needed to accurately determine the growth rate.

F. Radioimmunoassay and Enzyme-Linked Immunosorbent Assay for Calbindin D_{28K}

Additional biological assays utilize the presence of calbindin D_{28K} protein as an indication of vitamin D activity. Calbindin D_{28K} is not present in the intestine of vitamin D–deficient chicks and is only synthesized in response to the administration of vitamin D. Therefore, it is possible to use the presence of calbindin D_{28K} to determine vitamin D activity. A radioimmunoassay (RIA) and an enzyme-linked immunosorbent assay (ELISA), both capable of detecting nanogram quantities of calbindin D_{28K}, have been developed for this purpose (307).

A comparison of the sensitivity and working range of the biological assays for vitamin D is given in Table 6.

Table 6 Comparison of Sensitivity and Working Range of Biological Assays for Vitamin D

Assay	Time required for assay	Minimal level detectable in assay		Usual working range
		ng	nmol	
Rat line test	7 d	12	0.03	25–300 ng
AOAC chick	21 d	50	0.113	50–1250 ng
Intestinal Ca^{2+} absorption				
In vivo				
$^{45}Ca^{2+}$	1 d	125	0.33	0.125–25 g
$^{47}Ca^{2+}$	1 d	125	0.33	0.125–25 g
In vitro				
Everted sacs	1 d	250	0.65	250–1000 ng
Duodenal uptake of $^{45}Ca^{2+}$	1 d	250	0.65	250–1000 ng
Bone Ca^{2+} mobilization				
In vivo	24 h	125	0.32	0.125–25 g
Body growth	21–28 d	50	0.06	50–1250 ng
Immunoassays for calcium-binding protein	1 d	1	0.0025	1 ng

VIII. ANALYTICAL PROCEDURES

Although considerable progress has been made in the development of chemical or physical means to measure vitamin D, these methods at present generally lack the sensitivity and selectivity of the biological assays. Thus, they are not adequate for measuring samples that contain low concentrations of vitamin D. However, these physical and chemical means of vitamin D determination have the advantage of not being as time consuming as the biological assays and so are frequently used on samples known to contain high levels of vitamin D.

A. Ultraviolet Absorption

The first techniques available for quantitation of vitamin D were based on the measurement of the UV absorption at 264 nm. The conjugated triene system of double bonds in the vitamin D *seco*-steroids produces a highly characteristic absorption spectra (Fig. 10). The absorption maxima for vitamin D occurs at 264 nm, and at this wavelength the molar extinction coefficient for both vitamins D_2 and D_3 is 18,300. Thus, the concentration of an unknown solution of vitamin D can be calculated once its absorption at 264 nm is known. Although this technique is both quick and easy, it suffers from the disadvantage that the sample must be scrupulously purified prior to assay in order to remove potential UV-absorbing contaminants.

B. Colorimetric Methods

Several colorimetric methods for the quantitation of vitamin D have been developed over the years. Among these various colorimetric assays is a method based on the isomerization of vitamin D to isotachysterol. This procedure, which employs antimony trichloride, can detect vitamin D in the range of 1–1000 µg. Because it can detect such large amounts

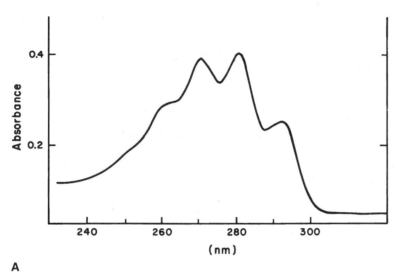

A

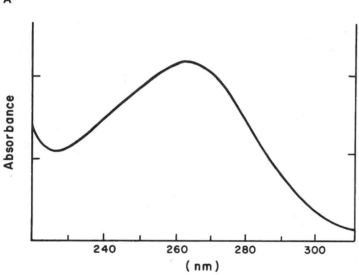

B

Fig. 10 Ultraviolet spectrum of provitamin D and vitamin D. Panel A illustrates the characteristic UV absorption spectrum of provitamin D. The wavelengths of the several absorption maxima are 262, 271, 282, and 293 nm. The molar extinction coefficient at 282 nm is 11,500. Panel B illustrates the characteristic UV spectrum of vitamin D. The molar extinction coefficient at the 265- to 265-nm absorption maxima is 18,300.

of vitamin D, this assay is now used primarily to determine the vitamin D content of pharmaceutical preparations and has become the official U.S. Pharmacopoeia (USP) color-imetric assay for vitamin D_3.

In this assay, three types of tubes are normally prepared: one containing the standard vitamin D or unknown sample plus the color reagent; one containing only the solvent, ethylene chloride; and a third containing ethylene chloride, acetic anyhdride, and the color reagent. The absorbance at 500 nm is measured 45 s after addition of the color reagent.

The concentration of vitamin D in the assay tube is proportional to its absorbance, which is corrected for the solvent blank and for the tube containing the acetic anhydride. This procedure has been found to follow Beer's law for solutions containing 3.25–6.5 nmol vitamin D per milliliter assay solution. The major disadvantage to this assay is that the copresence of vitamin A, which is often present in pharmaceutical samples along with vitamin D, interferes with the assay. Thus, purification procedures that include adsorption chromatography and partition chromatography are required, making this assay procedure rather laborious and time consuming. Another disadvantage is the necessity for careful timing of the reaction because of the short time required for the appearance of maximal intensity of color. However, since this is the only direct chemical method routinely available, it has widespread industrial application (308).

C. Fluorescence Spectroscopy

There have been two reports in the literature describing assays that depend on the reaction of vitamin D with a substance capable of fluorescing (309,310). Both of these procedures are based on the fact that an acetic anhydride–sulfuric acid solution containing vitamin D is capable of fluorescence if the solution is activated by light of the correct wavelength. The lower limits of detectability of this assay are the same as those of the antimony trichloride colorimetric assay, and the same compounds that interfere with the colorimetric assay also interfere with this fluorescence assay. As a result, this assay is not normally used for the analytical determination of vitamin D concentration.

D. Gas Chromatography–Mass Spectrometry

One of the most powerful modern techniques available to steroid chemists for the analytical determination of samples containing mixtures of steroids is mass spectrometry (MS), or mass spectrometry coupled with prior separation by gas chromatography (GC). The gas chromatography–mass spectrometry (GC-MS) technique can be coupled to an on-line computer that collects information on the fragmentation patterns of steroids in the mass spectrometer. In this way a sophisticated quantitative assay can be developed with a sensitivity and selectivity approaching that of RIAs. There have recently appeared several GC-MS procedures applicable to vitamin D *seco*-steroids (311), but they are not yet widely employed.

E. High-Performance Liquid Chromatography

Several papers that describe the separation of vitamin D and its various metabolites by HPLC have appeared (312,313). This separation process has an exceedingly high resolving capability due to the large number of theoretical plates present in a typical column. Of equal importance to this technique is the sensitivity of the detector used for observing the separated compounds. All of the published procedures for the separation of vitamin D by HPLC have used an UV detector, and so their sensitivity is limited to approximately 5 ng. The chief advantage of using HPLC is the reduction in labor and time required to separate vitamin D and its metabolites. In the current official USP method for the determination of vitamin D, two prepurification steps, requiring up to 8 h, are necessary before the colorimetric analysis can be performed (314). However, with HPLC, reproducible separation of closely related compounds can be achieved in less than 1 h. HPLC has great potential once a more sensitive detection method is developed. There are now available several official AOAC procedures for the HPLC determination of vitamin D in a variety of sample types, including vitamin preparations, multivitamin preparations, vitamin D oil concentrates, and fortified milk and milk powder (308).

F. Competitive Binding Assays

Various competition assays that can specifically quantitate the levels of $25(OH)D_3$, $24,25(OH)_2D_3$, or $1\alpha,25(OH)_2D_3$ in a sample are now available. Such assays were developed as a consequence of the discovery of specific vitamin D–binding proteins in the serum and tissues of mammals and birds, along with the availability of high specific activity tritiated $25(OH)D_3$ and $1\alpha,25(OH)_2D_3$. Since these steroid competition assays are sensitive (they are capable of detecting picogram quantities), they are now routinely used to measure vitamin D metabolite levels in plasma.

Two different types of steroid competition assays have been developed for the detection of $1\alpha,25(OH)_2D_3$. The first employs incubation of intestinal mucosal cytosol plus the nuclear chromatin fractions with standardized amounts of tritiated $1\alpha,25(OH)_2D_3$. The $1\alpha,25(OH)_2D_3$ in the sample competes with the tritiated hormone for the binding sites of the $1\alpha,25(OH)_2D_3$ receptors present in the cytosol and chromatin fractions. By measuring the amount of tritiated $1\alpha,25(OH)_2D_3$ bound to the receptor, the amount of $1\alpha,25(OH)_2D_3$ in the sample can be determined. The first such assay developed that could be used to measure $1\alpha,25(OH)_2D_3$ levels in plasma was that of Brumbaugh and Haussler (315). Their technique requires a minimum of 10 mL plasma and involves a laborious three-stage chromatographic procedure. The final $1\alpha,25(OH)_2D_3$ peak is then assayed. Separation of bound from free steroid is achieved by filtration of the incubation media through glass filters. The steroid associated with the chromatin-cytosol receptor is specifically bound to these filters. A similar assay has been described by Procsal et al., except that separation of the bound from free steroid is achieved by high-speed differential centrifugation (316).

The second type of competition assay involves the use of calf thymus cytosol as the source of binding protein. Reinhardt et al. developed a radioreceptor assay for vitamin D_2, vitamin D_3, and their metabolites that does not require HPLC (317). Their technique includes the use of a stable $1\alpha,25(OH)_2D_3$ receptor preparation from calf thymus, nonequilibrium assay conditions, and solid-phase extraction of vitamin D metabolites from serum or plasma samples. This procedure requires 0.2–1.0 mL plasma. $1\alpha,25(OH)_2D_3$ is removed from the plasma on a C18-silica cartridge. The cartridge is first reverse-phase-eluted and then switched to normal-phase elution (318). The $1\alpha,25(OH)_2D_3$ is recovered and incubated with $[^3H]1\alpha,25(OH)_2D_3$ and reconstituted thymus receptor. Separation of receptor-bound hormone from free hormone is achieved by the addition of dextran-coated charcoal. Similar assays have been developed for $25(OH)D_3$ and $24,25(OH)_2D_3$ (146,319,320). This assay has a sensitivity of 0.7 pg.

IX. NUTRITIONAL REQUIREMENTS FOR VITAMIN D

A. Humans

The vitamin D requirement for healthy adults has never been precisely defined. Since vitamin D_3 is produced in the skin upon exposure to sunlight, a human being does not have a requirement for vitamin D when sufficient sunlight is available. However, vitamin D does become an important nutritional factor in the absence of sunlight. In addition to geographical and seasonal factors, UV light from the sun may be blocked by factors such as air pollution and sun screens. In fact, as air pollution became prevalent during the industrial revolution, the incidence of rickets became widespread in industrial cities. It is now known that the rickets epidemic was partly caused by lack of sunlight due to air pollution. Thus, rickets has been called the first air pollution disease. Any condition that blocks sunlight from skin, such as the wearing of clothes, use of sunscreens, living indoors and in cities with tall buildings, or living in geographical regions of the world that do not receive adequate sunlight, can contribute to the inability of the skin to biosynthesize sufficient amounts of vitamin D. Under these conditions, vitamin D becomes a true vitamin in that it must be supplied in the diet on a regular basis.

Since vitamin D can be endogenously produced by the body and since it is retained for long periods of time by vertebrate tissue, it is difficult to determine minimum daily requirements for this substance. The requirement for vitamin D is also dependent on the concentration of calcium and phosphorus in the diet, the physiological stage of development, age, gender, degree of exposure to the sun, and the amount of pigment in the skin.

The current allowance of vitamin D recommended in 1989 by the National Research Council is 300 IU/day [1 IU = 0.025 μg vitamin D (321)] for infants from birth to 6 months in age; 400 IU/day for children, adolescents, and pregnant and lactating women; and 200 IU/day for other adults (322). Because rickets is more prevalent in preschool children, the Food and Agricultural Organization/World Health Organization (FAO/WHO) committee recommended that children receive 400 IU/day until the age of 6 years, after which the recommended daily allowance (RDA) is 100 IU/day.

In the United States, adequate amounts of vitamin D can readily be obtained from the diet and from casual exposure to sunlight. However, in some parts of the world where food is not routinely fortified and sunlight is often limited, obtaining adequate amounts of vitamin D becomes more of a problem. As a result, the incidence of rickets in these countries is higher than in the United States. Rickets was practically eradicated from the United States in the mid-1920s, but it has reappeared in the past two decades. The increase is predominantly due to changes in infant feeding and dietary preferences. Of 27 cases of infant rickets reported within the first year of life, 26 were found in breast-fed infants (323). Of 62 cases involving older children, 56 were from families following strict vegetarian diets that included no meat products, milk products, fish, or eggs.

B. Animals

The task of assessing the minimum daily vitamin D requirement for animals is no easier than it is for humans. Such factors as the dietary calcium/phosphorus ratio, physiological stage of development, gender, amount of fur or hair, color, and perhaps even breed, all affect the daily requirement for vitamin D in animals. Also, some animals, such as chicken and turkey, do not respond as well to vitamin D_2 as to vitamin D_3. As with humans, animals that are maintained in sunlight can produce their own vitamin D, so that dietary supplementation is not really necessary. For animals that are kept indoors or that live in climates where the sunlight is not adequate for vitamin D production, the vitamin D content of food becomes important. Sun-cured hays are fairly good sources of vitamin D, but

Table 7 Vitamin D Requirements of Animals

Animal	Daily requirements (IU)
Chickens, growing	90[a]
Dairy cattle:	
Calves	660[b]
Pregnant, lactating	5000–6000[c]
Dogs:	
Growing puppies	22[d]
Adult maintenance	11[d]
Ducks	100[d]
Monkey, growing	
rhesus	25[d]
Mouse, growing	167[d]
Sheep:	
Lambs	300[e]
Adults	250[e]
Swine:	
Breed sows	550[c]
Lactating sows	1210[c]
Young boars	690[c]
Adult boars	550[c]
Turkeys	400[a]

[a]IU required per pound of feed.
[b]IU required for 100 kg body weight.
[c]IU required per animal.
[d]IU required per kg body weight.
[e]IU required per 45 kg body weight.
Source: Published information by the Committee of Animal
Nutrition, Agricultural Board (National Research Council).

dehydrated hays, green feeds, and seeds are poor sources. A brief list of the RDAs for
animals is given in Table 7.

X. FOOD SOURCES OF VITAMIN D

For the most part, vitamin D is present in unfortified foods in only very small and variable quantities (Table 8). The vitamin D that occurs naturally in unfortified foods is
generally derived from animal products. Saltwater fish, such as herring, salmon, and sar-
dine, contain substantial amounts of vitamin D, and fish liver oils are extremely rich
sources. However, eggs, veal, beef, unfortified milk, and butter supply only small quantities of the vitamin. Plants are extremely poor sources of vitamin D; fruits and nuts contain
no vitamin D, and vegetable oils contain only negligible amounts of the provitamin. As
a consequence, in the United Stated dietary requirements for vitamin D are met by the
artificial fortification of suitable foods. Among these fortified foods are milk, both fresh
and evaporated; margarine and butter; cereals; and chocolate mixes. Milk is fortified to
supply 400 IU vitamin D per quart, and margarine usually contains 2000 IU or more per
pound. A more complete listing of the vitamin D values of food is given by Booher et
al. (324).

Table 8   Vitamin D Content of Unfortified Foods

Food source	Vitamin D (IU/100 g)
Beef steak	13
Beet greens	0.2
Butter	35
Cabbage	0.2
Cheese	12
Cod	85
Cod liver oil	10,000
Corn oil	9
Cream	50
Egg yolk	25
Herring (canned)	330
Herring liver oil	140,000
Liver:	
Beef (raw)	8–40
Calf (raw)	0–15
Pork (raw)	40
Chicken (raw)	50–65
Lamb (raw)	20
Mackerel	120
Milk:	
Cow (100 mL)	0.3–4
Human (100 mL)	0–10
Salmon (canned)	220–440
Sardines (canned)	1500
Shrimp	1150
Spinach	0.2

Source: Refs. 2 and 324.

XI. SIGNS OF VITAMIN D DEFICIENCY

A. Humans

A deficiency of vitamin D results in inadequate intestinal absorption and renal reabsorption of calcium and phosphate. As a consequence, serum calcium and phosphate levels fall and serum alkaline phosphatase activity increases. In response to these low serum calcium levels, hyperparathyroidism occurs. The result of increased levels of PTH, along with whatever $1\alpha,25(OH)_2D_3$ is still present at the onset of the deficiency, is the demineralization of bone. This ultimately leads to rickets in children and osteomalacia in adults. The classical skeletal symptoms associated with rickets, i.e., bowlegs, knock-knees, curvature of the spine, and pelvic and thoracic deformities (Fig. 11), result from the application of normal mechanical stress to demineralized bone. Enlargement of the bones, especially in the knees, wrists, and ankles, and changes in the costochondral junctions also occur. Since in children bone growth is still occurring, rickets can result in epiphysial abnormalities not seen in adult osteomalacia. Rickets also results in inadequate mineralization of tooth

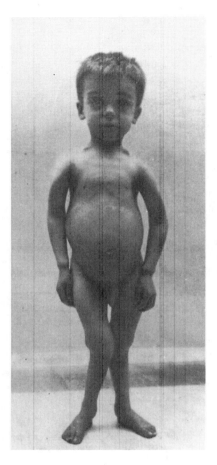

Fig. 11 Classic appearance of rickets in a child.

enamel and dentin. If the disease occurs during the first 6 months of life, convulsions and tetany can occur. Few adults with osteomalacia develop tetany.

Low serum calcium levels in the range of 5–7 mg per 100 mL and high serum alkaline phosphatase activity can be used to diagnose rickets and osteomalacia. Also, a marked reduction in circulating $1\alpha,25(OH)_2D_3$ levels in individuals with osteomalacia or rickets has been reported (325–327). Radiographic changes are also evident and can be used in diagnosis.

B. Animals

The response to vitamin D deficiency in animals closely resembles that in humans. Among the first symptoms of the deficiency is a decline in the plasma concentration of calcium and phosphorus. This is followed by an abnormally low growth rate and the characteristic alteration of bones, including faulty calcification of the bone matrix. As the disease progresses, the forelegs bend sideways and the joints become swollen. In laying birds, the eggs are thin-shelled, egg production declines, and hatchability is markedly reduced (328); classic symptoms of rickets develop, followed by tetany and death.

XII.   HYPERVITAMINOSIS D

Excessive amounts of vitamin D are not available from natural sources. However, vitamin D intoxication is a concern in patients being treated with vitamin D or vitamin D analogs for hypoparathyroidism, vitamin D–resistant rickets, renal osteodystrophy, osteoporosis, psoriasis, some cancers, or in those who are taking supplemental vitamins. Hypervitaminosis D is a serious problem because it can result in irreversible calcification of the heart, lungs, kidneys, and other soft tissues. Therefore, care should be taken to detect early signs of vitamin D intoxication in patients receiving pharmacological doses. Symptoms of intoxication include hypercalcemia, hypercalciuria, anorexia, nausea, vomiting, thirst, polyuria, muscular weakness, joint pains, diffuse demineralization of bones, and disorientation. If allowed to go unchecked, death will eventually occur.

Vitamin D intoxication is thought to occur as a result of high 25(OH)D levels rather than high $1\alpha,25(OH)_2D$ levels (329,330). Patients suffering from hypervitaminosis D have been shown to exhibit a 15-fold increase in plasma 25(OH)D concentration as compared to normal individuals. However, their $1\alpha,25(OH)_2D$ levels are not substantially altered (331). Furthermore, anephric patients can still suffer from hypervitaminosis D even though they are for the most part incapable of producing $1\alpha,25(OH)_2D$. It has also been shown that large concentrations of 25(OH)D can mimic the actions of $1\alpha,25(OH)_2D$ at the level of the receptor (315,329,332,333).

In the early stages of intoxication, the effects are usually reversible. Treatment consists of merely withdrawing vitamin D and perhaps reducing dietary calcium intake until serum calcium levels fall. In more severe cases, treatment with glucocorticoids, which are thought to antagonize some of the actions of vitamin D, may be required to facilitate the correction of hypercalcemia. Since calcitonin can bring about a decline in serum calcium levels, it may also be used in treatment.

XIII.   FACTORS THAT INFLUENCE VITAMIN D STATUS

A.   Disease

In view of the complexities of the vitamin D endocrine system, it is not surprising that many disease states are vitamin D–related. Figure 12 classifies some of the human disease states that are believed to be associated with vitamin D metabolism according to the metabolic step where the disorder occurs.

1.   Intestinal Disorders

The intestine functions as the site of dietary vitamin D absorption and is also a primary target tissue for the hormonally active $1\alpha,25(OH)_2D_3$. Impairment of intestinal absorption of vitamin D can occur in those intestinal disorders that result in the malabsorption of fat. Patients suffering from such disorders as tropical sprue, regional enteritis, and multiple jejunal diverticulosis often develop osteomalacia because of what appears to be a malabsorption of vitamin D from the diet (334). Surgical conditions, such as gastric resection and jejunoileal bypass surgery for obesity, may also impair vitamin D absorption. Also, patients receiving total parenteral nutrition in the treatment of the malnutrition caused by profound gastrointestinal disease often develop bone disease (335).

On the other hand, intestinal response to vitamin D can be affected by certain disease states. Patients suffering from idiopathic hypercalciuria exhibit an increased intestinal

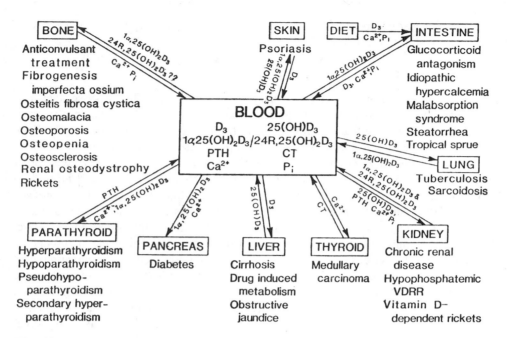

Fig. 12   Human disease states related to vitamin D. PTH, parathyroid hormone; CT, calcitonin; VDRR, vitamin D–resistant rickets; Pi, inorganic phosphate; Ca^{2+}, calcium.

absorption of calcium that may result from an enhanced intestinal sensitivity to $1\alpha,25(OH)_2D_3$ or from an overproduction of $1\alpha,25(OH)_2D_3$. The disease sarcoidosis also results in enhanced sensitivity to vitamin D. Sarcoidosis is characterized by hypercalcemia and hypercalciuria in patients receiving only modest amounts of vitamin D. Experiments have shown that these patients have elevated levels of serum $1\alpha,25(OH)_2D_3$. The excess $1\alpha,25(OH)_2D_3$ is likely of extrarenal origin and therefore not regulated by circulating levels of PTH (336). Other experiments have shown that macrophages from patients with sarcoidosis can produce $1\alpha,25(OH)_2D_3$ (120).

Other disease states that can result in extrarenal production of $1\alpha,25(OH)_2D_3$ are tuberculosis (337), leprosy (338), and some lymphomas (339).

2.   Liver Disorders

The liver plays an important role in the vitamin D endocrine system; not only is it the primary site for the production of 25(OH)D, but it is also the source of bile salts that aid in the intestinal absorption of vitamin D. Furthermore, it is likely that the liver is the site where binding of 25(OH)D by vitamin D-binding protein occurs, and it may even be the site at which this binding protein is synthesized. Hence, malfunctions of the liver can possibly interfere with the absorption, transport, and metabolism of vitamin D. Malabsorption of calcium and the appearance of bone disease have been reported in patients suffering from either primary biliary cirrhosis or prolonged obstructive jaundice. The disappearance of radioactive vitamin D from the plasma of these patients is much slower than in normal humans (340), and their plasma 25(OH)D levels are reduced (341). Although these patients respond poorly to vitamin D treatment, they immediately respond if treated with

$25(OH)D_3$. Thus, it appears that the bone disease experienced by these patients results from their inability to produce 25(OH)D.

3. Renal Disorders

The kidney functions as the endocrine gland for $1\alpha,25(OH)_2D_3$. Thus, disease states that affect the kidney can concomitantly alter the production of this calcium homeostatic hormone. It is well known that patients suffering from renal failure also often suffer from skeletal abnormalities. Termed renal osteodystrophy, these skeletal abnormalities include growth retardation, osteitis fibrosa, osteomalacia, and osteosclerosis. It became apparent with the discovery that under normal conditions $1\alpha,25(OH)_2D_3$ is produced in the kidney that these skeletal abnormalities result from the failure of patients to produce $1\alpha,25(OH)_2D_3$. Support for this theory came from studies on the metabolism of radioactively labeled vitamin D in normal persons vs. patients with chronic renal failure. From these studies, anephric or uremic individuals appeared incapable of producing $1\alpha,25(OH)_2D_3$. Direct evidence for this came from the observation that circulating level of $1\alpha,25(OH)_2D_3$ in the normal subject is in the range of 30–35 pg/mL, whereas in chronic renal failure the levels have been reported as low as 3–6 pg/mL (342,343). However, after a successful renal transplant, $1\alpha,25(OH)_2D_3$ levels return to the normal range. Also, the administration of $1\alpha,25(OH)_2D_3$ to these patients results in the stimulation of intestinal calcium absorption and an elevation of serum calcium levels (344).

4. Parathyroid Disorders

As previously outlined, PTH influences the production of $1\alpha,25(OH)_2D_3$, so that any disease state that affects the secretion of PTH may, in turn, have an effect on the metabolism of vitamin D. Hyperactivity of the parathyroid glands, as in primary hyperparathyroidism, results in the appearance of bone disease resembling osteomalacia. Circulating $1\alpha,25(OH)_2D_3$ levels in these subjects have been reported to be significantly elevated (345), as is their intestinal calcium transport (346). On the other hand, in hypoparathyroidism, hypocalcemia occurs. In these patients, a slight reduction in circulating $1\alpha,25(OH)_2D_3$ levels has been reported (347). When these patients are treated with $1\alpha,25(OH)_2D_3$ their serum levels are increased to normal.

There are several published reviews on the role of vitamin D in disease (18,334,348).

B. Genetics

Vitamin D–resistant rickets (hypophosphatemic rickets) appears to be an X-linked, dominant genetic disorder. Winters et al. presented evidence that this disease is almost always inherited and is usually congenital (349). Males are usually more severely affected by this disease than females. Associated with the disease are skeletal abnormalities, such as rickets or osteomalacia, and a diminished renal tubular reabsorption of phosphate that results in hypophosphatemia. Individuals with this disease do not respond to physiological doses of vitamin D; treatments with $25(OH)D_3$ and $1\alpha,25(OH)_2D_3$ are also ineffective, although an increase in intestinal calcium absorption does occur (350). These patients have also been reported to have normal serum $1\alpha,25(OH)_2D_3$ levels. Thus, it appears that this disorder does not result from an alteration in the metabolism of vitamin D or from an impaired intestinal response to $1\alpha,25(OH)_2D_3$, but rather from a specific defect in renal tubular reabsorption of phosphate.

A genetic defect that interferes with vitamin D metabolism has also been suggested in vitamin D–dependent rickets type I. This ailment differs from rickets in that it appears in children who are receiving adequate amounts of vitamin D and requires pharmacological doses of vitamin D or $25(OH)D_3$ to reverse the harmful effect of disease on bone. However, the disease is responsive to physiological amounts of $1\alpha,25(OH)_2D_3$, which suggests that the defect occurs in the metabolism of $25(OH)D_3$ to $1\alpha,25(OH)_2D_3$. This disease state appears to be the result of an autosomal recessively inherited genetic defect (351). It is not known how this defect affects the metabolism of $25(OH)D_3$.

Vitamin D–dependent rickets type II also has a genetic basis. This ailment is similar to vitamin D–dependent rickets type I except that children do not respond to large doses of vitamin D, $25(OH)D_3$, or $1\alpha,25(OH)_2D_3$. The combination of symptoms, i.e., defective bone mineralization, decreased intestinal calcium absorption, hypocalcemia, and increased serum levels of $1\alpha,25(OH)_2D_3$, suggest end-organ resistance to the action of $1\alpha,25(OH)_2D_3$. Experiments have shown that these children have a single-point mutation in the nuclear receptor for $1\alpha,25(OH)_2D_3$ (352–355).

C. Drugs

Recent evidence suggests that prolonged use of anticonvulsant drugs, such as diphenylhydantoin or phenobarbital, can result in an impaired response to vitamin D; this results in an alteration of calcium metabolism and the appearance of rickets or osteomalacia. Serum $25(OH)D_3$ levels in patients receiving these drugs have been reported to be markedly reduced (356). Also, studies in animals suggest that these drugs stimulate the hepatic microsomal cytochrome P450 enzymes, which could lead to an increased catabolism of $25(OH)D_3$ (357). However, $1\alpha,25(OH)_2D_3$ levels have been shown to be normal or even increased after drug treatment (358). It appears that this drug-induced osteomalacia may not be the result of an effect of the drug on vitamin D metabolism. Studies on rat and chick duodena in organ culture indicate that anticonvulsant drugs may act on the gastrointestinal tract and affect the absorption of calcium (359). Anticonvulsant drugs have also been shown to inhibit calcium reabsorption in organ culture mouse calvaria (360). Further research is needed to determine the mechanism by which these anticonvulsant drugs affect calcium metabolism.

D. Alcohol

Persons suffering from chronic alcoholism exhibit a decrease in plasma $25(OH)D_3$ levels and in intestinal calcium absorption and bone mineral content. This is observed in patients with and without cirrhosis of the liver. Current evidence indicates that the impairment of intestinal calcium absorption is the result of low $25(OH)D_3$ levels (361,362). However, how chronic alcoholism results in low $25(OH)D_3$ levels is at present not understood.

E. Age

The fact that changes in the metabolism of vitamin D may occur with aging has been suggested by the observation that the ability to absorb dietary calcium decreases with age (363). In addition, loss of bone increases in the elderly and an age-related hypoplasia of bone cells occurs. $1\alpha,25(OH)_2D_3$ levels in the plasma decrease with age, possibly due to an age-related reduction of epidermal concentrations of 7-dehydrocholesterol, resulting in less photochemical production of $1\alpha,25(OH)_2D_3$ by the skin. Also, the 1α-hydroxylase

Table 9 Drug Forms of Vitamin D Metabolites

Compound name	Generic name	Commercial name	Pharmaceutical company	Effective daily dose (micrograms)[a]	Approved use
$1\alpha,25(OH)_2D_3$	Calcitriol	Rocaltrol	Hoffmann-La Roche	0.5–1.0	RO, HP, O[b]
$1\alpha,25(OH)_2D_3$	Calcitriol	Calcijex	Abbott	0.5 (i.v.)	HC
$1\alpha,24(OH)_2$-19-nor-D_3	Paricalcitol	Zemplar	Abbott	2.8–7 (eod)	SHP
$1\alpha,24(OH)_2D_3$	Tacalcitol	Bonalfa	Teijin Ltd.—Japan	40–80 (topical)	PP
$1\alpha,24S(OH)_2$-22-ene-24-cyclopropyl-D_3	Calcipotriene	Dovonex	Leo—Denmark	40–80 (topical)	PP
$1\alpha,24S(OH)_2$-22-ene-24-cyclopropyl-D_3	Calcipotriene	Dovonex	Westwood-Squibb	40–80 (topical)	PP
1α-OH-D_3	Alfacalcidol	One-Alfa	Leo—Denmark	1–2	RO, HP, O, VDRR
1α-OH-D_3	Alfacalcidol	Alpha-D_3	Teva-Israel	0.25–1.0	RO, O, HC, HP
1α-OH-D_3	Alfacalcidol	Onealfa	Teijin Ltd.—Japan	0.25–1.0	RO, O
1α-OH-D_3	Alfacalcidol	Onealfa	Chugai—Japan	0.25–1.0	RO, O
1α-OH-D_2	Doxercalciferol	Hectorol	Bone Care	10 3×/WEEK	SHP
$25(OH)D_3$	Calcifediol	Calderol	Organon-USA	50–500	RO
$25(OH)D_3$	Calcifediol	Dedrogyl	Roussel-Uclaf-France	50–500	RO
10,19-dihydrotachysterol$_3$	Dihydrotachysterol$_3$	Hytakerol	Winthrop	200–1000	RO

The key to the approved uses of the vitamin D analogs is as follows: RO, renal osteodystrophy; O, postmenopausal osteoporosis; PP, plaque psoriasis; HC, hypocalcemia (frequently present in patients with renal osteodystrophy who are subjected to hemodialysis); HP, hypoparathyroidism and associated hypocalcemia (which may be encountered in patients with hypoparathyroidism, pseudohypoparathyroidism, or in circumstances of postsurgical hypoparathyroidism); SHP, secondary hyperparathyroidism associated with renal osteodystrophy; VDRR, vitamin D-resistant rickets.

[a]Oral dose unless otherwise indicated; eod = every other day.

[b]The use of Rocaltrol for postmenopausal osteoporosis is approved in Argentina, Australia, Austria, Czech Republic, Colombia, India, Ireland, Italy, Japan, Malaysia, Mexico, New Zealand, Peru, Philippines, South Africa, South Korea, Switzerland, Turkey, and the United Kingdom.

enzyme is less responsive to induction by PTH due to the decrease of glomerular filtration with age (364).

F. Sex

Gray et al. demonstrated that men and women differ in their metabolism of vitamin D in response to various physiological stimuli (365). In women, they observed that dietary phosphate deprivation resulted in a decrease in serum phosphorus levels with a concomitant increase in plasma $1\alpha,25(OH)_2D_3$ concentrations. However, no change of either of these parameters was noted in men. Thus, the mechanism by which men and women respond to dietary phosphate deprivation seems to differ.

XIV. EFFICACY OF PHARMACOLOGICAL DOSES

Several ailments are known to respond to massive doses of vitamin D. For example, the intestinal malabsorption of calcium that results from chronic renal failure and the subsequent development of rickets or osteomalacia can be overcome by administration of 100,000–300,000 IU vitamin D per day (364). Patients suffering from hypoparathyroidism can usually be treated by giving 80,000–100,000 IU vitamin D per day (366). Also, children afflicted with vitamin D–dependent rickets type I can be treated with 10,000–100,000 IU per day (2). The therapeutic effect of such massive doses can be explained by the fact that 25(OH)D in sufficiently high concentrations will mimic the action of $1\alpha,25(OH)_2D_3$ at the receptor. However, as mentioned earlier, the administration of such pharmacological doses of vitamin D to patients over a prolonged period of time carries with it the danger of vitamin D toxicity.

Table 9 shows the drug forms of vitamin D metabolites that are currently available for the treatment of several disease states. Experiments are in progress to develop additional vitamin D analogs that will be useful pharmacological agents without the hypercalcemic side effect of $1\alpha,25(OH)_2D_3$.

XV. RECENT DEVELOPMENTS

Recently, the x-ray crystal structure of the ligand-binding domain (LBD) of VDR (367) and several other nuclear receptors have been determined. These LBDs are composed primarily of 11–13 α-helices that are folded to form three layers. The interior of the LBD is composed primarily of hydrophobic residues designed to bind lipophilic ligands like $1,25(OH)D_3$. Ligand binding to its cognate receptor induces conformational changes in the receptor, which increases its ability to activate gene transcription. The major structural difference between unoccupied and ligand-bound nuclear receptors is the repositioning of the C-terminal helix 12 containing the ligand-dependent transactivation domain, AF-2. Upon ligand binding, helix 12 moves from projecting away from the LBD to a position tightly packed against helix 3 of the LBD (368,369). Repositioning of the AF-2 domain results in the formation of new surfaces, including a hydrophobic cleft (370), needed to interact with other nuclear factors called coactivators. Coactivators are adapter proteins that link transcriptional activators such as nuclear receptors to basal transcription machinery resulting in increased gene transcription (371).

Three families of coactivator proteins are known to interact with nuclear receptors: SRC-1 (372)/NcoA-1 (373), TIF2 (374)/GRIP-1 (375), and ACTR/pCIP (376). These proteins contain a conserved leucine-containing motif (LXXLL), important for interactions between coactivators and nuclear receptors (377). The coactivator complexes have activities associated with them such as histone acetyltransferase activity (378) or methyl transferase activity (379), which aid in the decondensation of chromatin to facilitate binding of RNA polymerase II transcription complex (RNA Pol II) to DNA. Some of the proteins in the coactivator complex can also interact with other proteins that are part of the RNA Pol II core complex. In this way, the coactivators can link upstream activators such as nuclear receptors to RNA Pol II and modulate gene transcription (380). Identifying these intermediary factors involved in transcriptional control of specific target genes is essential for understanding the biological actions of vitamin D and will continue to be an exciting area of research for several years.

XVI. CONCLUSION

Vitamin D, through its hormonally active metabolite $1\alpha,25(OH)_2D_3$, is known to act on bone and intestine to maintain calcium homeostasis. However, the actions of $1\alpha,25(OH)_2D_3$ are much broader than was originally thought. There is evidence that $1\alpha,25(OH)_2D_3$ is involved in the physiology of tissues not related to calcium homeostasis, such as skin, pancreas, pituitary, muscle, and hematopoietic cells. Figure 4 demonstrates the complexity of the vitamin D endocrine system as it is understood today. Although many advances have been made in the past decade, there is still much to learn about the detailed cellular and molecular mode of action of vitamin D and its metabolites.

REFERENCES

1. A. W. Norman, 1,25-(OH)$_2$-D$_3$ as a steroid hormone, in *Vitamin D: Molecular Biology and Clinical Nutrition* (A. W. Norman, ed.), Marcel Dekker, New York, 1980, pp. 197–250.
2. A. W. Norman, *Vitamin D: The Calcium Homeostatic Steroid Hormone*, Academic Press, New York, 1979, pp. 1–490.
3. D. E. M. Lawson (ed.), *Vitamin D*, Academic Press, New York, 1978, pp. 1–433.
4. H. F. DeLuca, Vitamin D metabolism and function, *Arch Intern Med 138*:836 (1978).
5. H. E. Harrison, Vitamin D and calcium and phosphate transport, *Pediatrics 28*:531 (1961).
6. D. B. N. Lee, M. W. Walling, G. V. Silis, and J. W. Coburn, Calcium and inorganic phosphate transport in rat colon, *J Clin Invest 65*:1326 (1980).
7. A. W. Norman, K. Schaefer, H.-G. Grigoleit, and D. V. Herrath (eds.), *Vitamin D: Molecular, Cellular and Clinical Endocrinology*, 7th ed., Walter de Gruyter, Berlin, 1988, pp. 1–1069.
8. A. W. Norman, R. Bouillon, and M. Thomasset (eds.), *Vitamin D: Gene Regulation, Structure Function Analysis and Clinical Application*, 8th ed., Walter de Gruyter, Berlin, 1991, pp. 1–991.
9. A. W. Norman, R. Bouillon, and M. Thomasset (eds.), *Vitamin D, A Pluripotent Steroid Hormone: Structural Studies, Molecular Endocrinology and Clinical Applications*, 9th ed., Walter de Gruyter, Berlin, 1994, pp. 1–973.
10. M. R. Haussler, Vitamin D receptors: nature and function, *Ann Rev Nutr 6*:527 (1986).
11. M. R. Haussler, D. J. Mangelsdorf, B. S. Komm, C. M. Terpenning, K. Yamaoka, E. A. Allegretto, A. R. Baker, J. Shine, D. P. McDonnell, M. Hughes, N. L. Weigel, B. W. O'Malley, and J. W. Pike, Molecular biology of the vitamin D hormone, in *Recent Progress in Hormone Research* (R. O. Greep, ed.), Academic Press, New York, 1988, pp. 263–305.
12. M. R. Walters, Newly identified actions of the vitamin D endocrine system, *Endocr Rev 3*: 719 (1992).

13. D. D. Bikle and S. Pillai, Vitamin D, calcium, and epidermal differentiation, *Endocr Rev* *14*:3 (1993).

14. R. Bouillon, W. H. Okamura, and A. W. Norman, Structure–function relationships in the vitamin D endocrine system, *Endocr Rev 16*:200 (1995).

15. S. Christakos, C. Gabrielides, and W. B. Rhoten, Vitamin D–dependent calcium binding proteins: chemistry, distribution, functional considerations, and molecular biology, *Endocr Rev 10*:3 (1989).

16. A. W. Norman, J. Roth, and L. Orci, The vitamin D endocrine system: steroid metabolism, hormone receptors and biological response (calcium binding proteins), *Endocr Rev 3*:331 (1982).

17. R. E. Weishaar and R. U. Simpson, The involvement of the endocrine system in regulating cardiovascular function: emphasis on vitamin D_3. *Endocr Rev 10*:351 (1989).

18. H. Reichel, H. P. Koeffler, and A. W. Norman, The role of the vitamin D endocrine system in health and disease, *N Engl J Med 320*:980 (1989).

19. R. L. Boland, Role of vitamin D in skeletal-muscle function, *Endocr Rev 7*:434 (1986).

20. R. S. Soleki, *Shanidar: The Humanity of Neanderthal Man*, Knopf, New York, 1971, pp. 1–220.

21. D. Whistler, Morbo puerli Anglorum, quem patrio idiomate indigenae voacant "the rickets," 1645, pp. 1–13.

22. F. Glisson, A treatise of the rickets being a disease common to children, 1660, pp. 1–373.

23. A. F. Hess, *Rickets Including Osteomalacia and Tetany*, Lea & Febiger, Philadelphia, 1929, pp. 1–485.

24. K. Huldschinsky, Heilung von Rachitis durch kuenstliche Hohensonne, *Dtsch Med Wschr 45*:712 (1919).

25. E. Mellanby and M. D. Cantab, Experimental investigation on rickets, *Lancet 196*:407 (1919).

26. E. Mellanby, Experimental rickets, Spec Rep Ser Med Res Council (GB) *SRS-61*:1–78 (1921).

27. E. V. McCollum, N. Simmonds, H. T. Parsons, P. G. Shipley, and E. A. Park, Studies on experimental rickets. I. The production of rachitis and similar disease in the rat by deficient diets, *J Biol Chem 45*:333 (1920).

28. E. V. McCollum, N. Simmonds, J. E. Becker, and P. G. Shipley, Studies on experimental rickets. XXI. An experimental demonstration of the existence of a vitamin which promotes calcium deposition, *J Biol Chem 53*:293 (1922).

29. H. Goldblatt and K. N. Soames, A study of rats on a normal diet irradiated daily by the mercury vapor quartz lamp or kept in darkness, *Biochem J 17*:294 (1923).

30. A. F. Hess and M. Weinstock, The antirachitic value of irradiated cholesterol and phytosterol. II. Further evidence of change in biological activity, *Meth Enzymol 64*:181 (1925).

31. A. F. Hess and M. Weinstock, The antirachitic value of irradiated cholesterol and phytosterol. III. Evidence of chemical change as shown by absorption spectra, *Meth Enzymol 64*:193 (1925).

32. A. Windaus, O. Linsert, A. Luttringhaus, and G. Weidlinch, Uber das krystallistierte Vitamin D_2, *Justus Liebigs Ann Chem 492*:226 (1932).

33. T. C. Angus, F. A. Askew, R. B. Bourdillon, H. M. Bruce, R. Callow, C. Fischmann, L. Philpot, and T. A. Webster, A crystalline antirachitic substance, *Proc R Soc Ser. B 108*:340 (1931).

34. A. Windaus, F. Schenck, and v.f. Werder, Uber das antirachitisch wirksame Bestrahlungs-produkt aus 7-dehydro-cholesterin, *H -S Zeit Physiol Chem 241*:100 (1936).

35. A. Carlsson and B. Lindquist, Comparison of intestinal and skeletal effects of vitamin D in relation to dosage, *Acta Physiol Scand 35*:53 (1955).

36. B. B. Migicovsky, Influence of vitamin D on calcium resorption and accretion, *Can J Biochem Physiol 35*:1267 (1957).

37. A. W. Norman, Actinomycin D effect on lag in vitamin D–mediated calcium absorption in the chick, *Am J Physiol 211*:829 (1966).

38. J. E. Zull, E. Czarnowska-Misztal, and H. F. DeLuca, On the relationship between vitamin D action and actinomycin-sensitive processes, *Proc Natl Acad Sci USA 55*:177 (1966).

39. D. Schachter, D. V. Kimberg, and H. Schenker, Active transport of calcium by intestine: action and bio-assay of vitamin D, *Am J Physiol 200*:1263 (1961).

40. V. W. Thompson and H. F. DeLuca, Vitamin D and phospholipid metabolism, *J Biol Chem 239*:984 (1964).

41. A. W. Norman, J. Lund, and H. F. DeLuca, Biologically active forms of vitamin D_3 in kidney and intestine, *Arch Biochem Biophys 108*:12 (1964).

42. J. W. Blunt, H. F. DeLuca, and H. K. Schnoes, 25-Hydroxycholecalciferol: a biologically active metabolite of vitamin D_3, *Biochemistry 6*:3317 (1968).

43. H. F. DeLuca, 25-Hydroxycholecalciferol, the probable metabolically active form of vitamin D_3: its identification and subcellular site of action, *Arch Intern Med 124*:442 (1969).

44. M. R. Haussler, J. F. Myrtle, and A. W. Norman, The association of a metabolite of vitamin D_3 with intestinal mucosa chromatin, in vivo, *J Biol Chem 243*:4055 (1968).

45. A. W. Norman and R. G. Wong, The biological activity of the vitamin D metabolite 1,25-dihydroxycholecalciferol in chickens and rats, *J Nutr 102*:1709 (1972).

46. D. E. M. Lawson, D. R. Fraser, E. Kodicek, H. R. Morris, and D. H. Williams, Identification of 1,25-dihydroxycholecalciferol, a new kidney hormone controlling calcium metabolism, *Nature 230*:228 (1971).

47. A. W. Norman, J. F. Myrtle, R. J. Midgett, H. G. Nowicki, V. Williams, and G. Popjak, 1,25-Dihydroxycholecalciferol: identification of the proposed active form of vitamin D_3 in the intestine, *Science 173*:51 (1971).

48. M. F. Holick, H. K. Schnoes, H. F. Deluca, T. Suda, and R. J. Cousins, Isolation and identification of 1,25-dihydroxycholecalciferol: a metabolite of vitamin D active in intestine, *Biochemistry 10*:2799 (1971).

49. D. R. Fraser and E. Kodicek, Unique biosynthesis by kidney of a biologically active vitamin D metabolite, *Nature 288*:764 (1970).

50. M. R. Haussler and A. W. Norman, Chromosomal receptor for a vitamin D metabolite, *Proc Natl Acad Sci USA 62*:155 (1969).

51. H. C. Tsai and A. W. Norman, Studies on calciferol metabolism. VIII. Evidence for a cytoplasmic receptor for 1,25-dihydroxyvitamin D_3 in the intestinal mucosa, *J Biol Chem 248*:5967 (1972).

52. A. R. Baker, D. P. McDonnell, M. Hughes, T. M. Crisp, D. J. Mangelsdorf, M. R. Haussler, J. W. Pike, J. Shine, and B. W. O'Malley, Cloning and expression of full-length cDNA encoding human vitamin D receptor, *Proc Natl Acad Sci USA 85*:3294 (1988).

53. H. Goto, K. Chen, J. M. Prahl, and H. F. DeLuca, A single receptor identical with that from intestine/T47D cells mediates the action of 1,25-dihydroxyvitamin D_3 in HL-60 cells, *Biochim Biophys Acta 1132*:103 (1992).

54. J. K. Burmester, N. Maeda, and H. F. DeLuca, Isolation and expression of rat 1,25-dihydroxyvitamin D_3 receptor cDNA, *Proc Natl Acad Sci USA 85*:1005 (1988).

55. M. A. Elaroussi, J. M. Prahl, and H. F. DeLuca, The avian vitamin D receptors: primary structures and their origins, *Proc Natl Acad Sci USA 91*:11596 (1994).

56. P. P. Minghetti and A. W. Norman, 1,25(OH)$_2$-vitamin D_3 receptors: gene regulation and genetic circuitry, *FASEB J 2*:3043 (1988).

57. D. Crowfoot and J. D. Dunitz, Structure of calciferol, *Nature 162*:608 (1948).

58. D. C. Hodgkin, B. M. Rimmer, J. D. Dunitz, and K. N. Trueblood, The crystal structure of a calciferol derivative, *J Chem Soc 947*:4945 (1963).

59. W. H. Okamura, A. W. Norman, and R. M. Wing, Vitamin D: concerning the relationship between molecular topology and biological function, *Proc Natl Acad Sci USA 71*:4194 (1974).

60. E. G. Bligh and W. J. Dyer, An extraction procedure for removing total lipids from tissue, *Can J Biochem 37*:911 (1958).

61. N. Ikekawa and N. Koizumi, Separation of vitamin D_3 metabolites and their analogues by high-pressure liquid chromatography, *J Chromatogr 119*:227 (1976).

62. B. P. Halloran, D. D. Bikle, and J. O. Whitney, Separation of isotopically labeled vitamin-D metabolites by high-performance liquid-chromatography, *J Chromatogr 303*:229 (1984).

63. R. B. Woodward, F. Sondheimer, D. Taub, K. Heusler, and W. M. McLamore, The total synthesis of steroids, *J Am Chem Soc 74*:4223 (1952).

64. H. M. E. Cardwell, J. W. Cornforth, S. R. Duff, H. Holtermann, and R. Robinson, Experiments on the synthesis of substances related to the sterols. Ll. Completion of the syntheses of androgenic hormones and of the cholesterol group of sterols, *Chem Soc London J*:361 (1953).

65. P. R. Bruck, R. D. Clark, R. S. Davidson, W. H. Gunther, P. S. Littlewood, and B. Lythgoe, Calciferol and its relatives. VIII. Ring A intermediates for the synthesis of tachysterol. *J Chem Soc 23*:2529 (1967).

66. D. H. R. Barton, R. H. Hesse, M. M. Pechet, and E. Rizzardo, A convenient synthesis of 1α-hydroxy-vitamin D_3, *J Am Chem Soc 95*:2748 (1973).

67. D. H. R. Barton, R. H. Hesse, M. M. Pechet, and E. Rizzardo, Convenient synthesis of crystalline 1α,25-dihydroxyvitamin D_3, *J Chem Soc Chem Commun* 203 (1974)

68. E. G. Baggiolini, J. A. Iacobelli, B. M. Hennessy, A. D. Batcho, J. F. Sereno, and M. R. Uskokovic, Stereocontrolled total synthesis of 1α,25-dihydroxycholecalciferol and 1α,25-dihydroxyergocalciferol, *J Org Chem 51*:3098 (1986).

69. J. DeSchrijver and P. J. Declerco, A novel synthesis of an A-ring precursor to 1α-hydroxyvitamin D, *Tetrahedron Lett 34*:4369 (1993).

70. R. G. Harrison, B. Lythgoe, and P. W. Wright, Calciferol and its relatives. XVIII. Total synthesis of 1α-hydroxyvitamin D_3, *J Chem Soc Perkin Trans 1*:2654 (1974).

71. S. A. Barrack, R. A. Gibbs, and W. H. Okamura, Potential inhibitors of vitamin D metabolism: an oxa analogue of vitamin D, *J Org Chem 53*:1790 (1988).

72. W. H. Okamura, J. M. Aurrecoechea, R. A. Gibbs, and A. W. Norman, Synthesis and biological activity of 9, 11-dehydrovitamin D_3 analogues: stereoselective preparation of 6b-vitamin D vinylallenes and a concise enynol synthesis for preparing the A-ring, *J Org Chem 54*:4072 (1989).

73. B. M. Trost, J. Dumas, and M. Villa, New strategies for the synthesis of vitamin D metabolites via Pd-catalyzed reactions, *J Am Chem Soc 114*:9836 (1992).

74. K. Nagasawa, Y. Zako, H. Ishihara, and I. Shimizu, Stereoselective synthesis of 1α-hydroxyvitamin D_3 A-ring synthons by palladium-catalyzed cyclization, *Tetrahed Lett 32*:4937 (1991).

75. M. Sheves and Y. Mazur, The vitamin D-3, 5-cyclovitamin D rearrangement, *J Am Chem Soc 97*:6249 (1975).

76. H. E. Paaren, H. F. DeLuca, and H. K. Schnoes, Direct C(1) hydroxylation of vitamin D_3 and related compounds, *J Org Chem 45*:3253 (1980).

77. D. R. Andrews, D. H. R. Barton, K. P. Cheng, J. P. Finet, R. H. Hesse, G. Johnson, and M. M. Pechet, A direct, regioselective and stereoselective 1-alpha-hydroxylation of (5E)-calciferol derivatives, *J Org Chem 51*:1635 (1986).

78. L. Vanmaele, P. J. Declercq, and M. Vandewalle, An efficient synthesis of 1-α,25-dihydroxy vitamin-D_3, *Tetrahedron 41*:141 (1985).

79. A. W. M. Hay and G. Watson, Binding of 25-hydroxy vitamin D_2 to plasma protein in New World monkeys, *Nature 256*:150 (1975).

80. W. Heymann, Metabolism and mode of action of vitamin D. V. Intestinal excretion of vitamin D, *J Biol Chem 122*:257 (1938).

81. W. Heymann, Metabolism and mode of action of vitamin D. IV. Importance of bile in the absorption and excretion of vitamin D, *J Biol Chem 12*:249 (1938).

82. S. J. Rosenstreich, C. Rich, and W. Volwiler, Depostition in and release of vitamin D_3 from body fat: evidence for a storage site in the rat. *J Clin Invest 50*:679 (1971).

83. D. Schachter, J. D. Finelstein, and S. Kowarski, Metabolism of vitamin D. I. Preparation of radioactive vitamin D and its intestinal absorption in the rat, *J Clin Invest 43*:787 (1964).

84. A. W. Norman and H. F. DeLuca, The preparation of H^3-vitamins D_2 and D_3 and their localization in the rat, *Biochemistry 2*:1160 (1963).

85. M. F. Holick, The cutaneous photosynthesis of previtamin D_3: a unique photoendocrine system, *J Invest Dermatol 76*:51 (1981).

86. H. F. DeLuca, Mechanism of action and metabolic fate of vitamin D, *Vitam Horm 25*:315 (1967).

87. P. G. Walsh and J. G. Haddad, Rocket immunoelectrophoresis assay of vitamin D–binding protein (Gc-globulin), *Clin Chem 28*:1781 (1982).

88. D. Dykes, H. Polesky, and E. Cox, Isoelectric focusing of Gc (vitamin D binding globulin) in parentage testing, *Hum Genet 58*:174 (1981).

89. S. P. Daiger, M. S. Schanfield, and L. L. Cavalli-Sforza, Group-specific component (Gc) proteins bind vitamin D and 25-hydroxyvitamin D, *Proc Natl Acad Sci USA 72*:2076 (1975).

90. R. Bouillon, H. van Baelen, W. Rombauts, and P. De Moor, The purification and characterization of the human-serum binding protein for the 25-hydroxycholecalciferol (transcalciferin), *Eur J Biochem 66*:285 (1976).

91. J. Silver and M. Fainaru, Transport of vitamin D sterols in human plasma: effect of excess vitamin D, 25-hydroxyvitamin D and 1,25-dihydroxyvitamin D, *Eur J Clin Invest 9*:433 (1979).

92. J. P. Mallon, D. Matuskewski, and H. Sheppard, Binding specificity of the rat serum vitamin D transport protein, *J Steroid Biochem Molec Biol 13*:409 (1980).

93. H. van Baelen, R. Bouillon, and P. Demoor, Vitamin D–binding protein (Gc-globulin) binds actin, *J Biol Chem 255*:2270 (1980).

94. N. E. Cooke and E. V. David, Serum vitamin D–binding protein is a third member of the albumin and alpha-fetoprotein gene family, *J Clin Invest 76*:2420 (1985).

95. R. Bouillon, F. A. van Assche, H. van Baelen, W. Heyns, and P. De Moor, Influence of the vitamin D–binding protein on the serum concentration of 1,25-dihydroxyvitamin D_3. *J Clin Invest 67*:589 (1981).

96. D. D. Bikle, P. K. Siiteri, E. Ryzen, and J. G. Haddad, Serum protein binding of 1,25-dihydroxyvitamin D: a reevaluation by direct measurement of free metabolite levels, *J Clin Endocrinol Metab 61*:969 (1985).

97. H. van Baelen, K. Allewaert, and R. Bouillon, New aspects of the plasma carrier protein for 25-hydroxycholecalciferol in vertebrates, *Ann NY Acad Sci 538*:60 (1988).

98. R. Bouillon and H. van Baelen, Transport of vitamin D: significance of free and total concentrations of the vitamin D metabolites, *Calcif Tissue Int 33*:451 (1981).

99. N. E. Cooke and J. G. Haddad, Vitamin D binding protein (Gc- globulin). *Endocr Rev 10*: 294 (1989).

100. A. W. M. Hay and G. Watson, The binding of 25-hydroxycholecalciferol and 25-hydroxyergocalciferol to receptor proteins in a New World and an Old World primate, *Comp Biochem Physiol 56B*:131 (1977).

101. J. E. Compston, L. W. L. Horton, and M. F. Laker, Vitamin-D analogues and renal function, *Lancet 1*:386 (1979).

102. J. Rungby, L. Mortensen, K. Jakobsen, A. Brock, and L. Mosekilde, Distribution of hydroxylated vitamin D metabolities [25(OH)D_3 and 1,25(OH)$_2$$D_3$] in domestic pigs: evidence that 1, 25(OH)$_2$$D_3$ is stored outside the blood circulation, *Comp Biochem Physiol 104A*:483 (1993).

103. E. B. Mawer, J. Backhouse, C. A. Holman, G. A. Lumb, and S. W. Stanbury, The distribution and storage of vitamin D and its metabolites in human tissues, *Clin Sci 43*:413 (1972).

104. H. L. Henry and A. W. Norman, Vitamin D: two dihydroxylated metabolites are required for normal chicken egg hatchability, *Science 201*:835 (1978).

105. A. W. Norman, V. L. Leathers, and J. E. Bishop, Studies on the mode of action of calciferol. XLVIII. Normal egg hatchability requires the simultaneous administration to the hen of 1α,25-dihydroxyvitamin D_3 and 24R,25-dihydroxyvitamin D_3, *J Nutr 113*:2505 (1983).

106. K. Saarem, S. Bergseth, H. Oftebro, and J. I. Pedersen, Subcellular localization of vitamin D_3 25-hydroxylase in human liver, *J Biol Chem 259*:10936 (1984).

107. E. Axén, T. Bergman, and K. Wikvall, Microsomal 25-hydroxylation of vitamin D_2 and vitamin D_3 in pig liver, *J Steroid Biochem Mol Biol 51*:97 (1994).

108. M. H. Bhattacharyya and H. F. DeLuca, Subcellular location of rat liver calciferol-25-hydroxylase, *Arch Biochem Biophys 160*:58 (1974).

109. T. Bergman and H. Postlind, Characterization of mitochondrial cytochromes P-450 from pig kidney and liver catalysing 26-hydroxylation of 25-hydroxyvitamin D_3 and C_{27} steroids, *Biochem J 276*:427 (1991).

110. G. Tucker, R. E. Gagnon, and M. R. Haussler, Vitamin D_3-25-hydroxylase: tissue occurrence and apparent lack of regulation, *Arch Biochem Biophys 155*:47 (1973).

111. J. M. Pettifor, F. P. Ross, and J. Wang, Serum levels of 25-hydroxycholecalciferol as a diagnostic aid in vitamin D deficiency states, *South Afr Med J 51*:580 (1977).

112. E. B. Mawer, Clinical implication of measurements of circulating vitamin D metabolites, *J Clin Endocrinol Metab 9*:63 (1980).

113. D. T. Baran and M. L. Milne, 1,25 Dihydroxyvitamin-D increases hepatocyte cytosolic calcium levels: a potential regulator of vitamin-D-25-hydroxylase, *J Clin Invest 77*:1622 (1986).

114. E. Usui, M. Noshiro, and K. Okuda, Molecular cloning of cDNA for vitamin D_3 25-hydroxylase from rat liver mitochondria. *FEBS Lett 262*:135 (1990).

115. M. Akiyoshi-Shibata, E. Usui, T. Sakaki, Y. Yabusaki, M. Noshiro, K. Okuda, and H. Ohkawa, Expression of rat liver vitamin D_3 25-hydroxylase cDNA in *Saccharomyces cerevisiae*, *FEBS Lett 280*:367 (1991).

116. H. L. Henry and A. W. Norman, Studies on calciferol metabolism. IX. Renal 25-hydroxyvitamin D_3-1-hydroxylase. Involvement of cytochrome P-450 and other properties, *J Biol Chem 249*:7529 (1974).

117. H. L. Henry, Vitamin D hydroxylases, *J Cell Biochem 49*:4 (1992).

118. Y. Weisman, A. Harell, S. Edelstein, M. David, Z. Spirer, and A. Golander, 1α,25-Dihydroxyvitamin D_3 and 24,25-dihydroxyvitamin D_3 in vitro synthesis by human decidua and placenta, *Nature 281*:317 (1979).

119. Y. Tanaka, B. P. Halloran, H. K. Schnoes, and H. F. DeLuca, In vitro production of 1,25-dihydroxyvitamin D_3 by rat placental tissue, *Proc Natl Acad Sci USA 76*:5033 (1979).

120. J. S. Adams, F. R. Singer, M. A. Gacad, O. P. Sharma, M. J. Hayes, P. Vouros, and M. F. Holick, Isolation and structural identification of 1,25-dihydroxyvitamin D_3 produced by cultured alveolar macrophages in sarcoidosis, *J Clin Endocrinol Metab 60*:960 (1985).

121. H. Reichel, H. P. Koeffler, and A. W. Norman, Synthesis in vitro of 1,25-dihydroxyvitamin D_3 and 24,25-dihydroxyvitamin D_3 by interferon-gamma-stimulated normal human bone marrow and alveolar macrophages, *J Biol Chem 262*:10931 (1987).

122. H. Reichel, J. E. Bishop, H. P. Koeffler, and A. W. Norman, Evidence for 1,25-dihydroxyvitamin D_3 production by cultured porcine alveolar macrophages, *Mol Cell Endocrinol 75*:163 (1991).

123. J. E. Puzas, R. T. Turner, G. A. Howard, J. S. Brand, and D. J. Baylink, Synthesis of 1,25-dihydroxycholecalciferol and 24,25-dihydroxycholecalciferol by calvarial cells: characterization of the enzyme-systems, *Biochem J 245*:333 (1987).

124. S. Pillai, D. D. Bikle, and P. M. Elias, 1,25-Dihydroxyvitamin D production and receptor binding in human keratinocytes varies with differentiation, *J Biol Chem 263*:5390 (1988).

125. K. Matsumoto, Y. Azuma, M. Kiyoki, H. Okumura, K. Hashimoto, and K. Yoshikawa, Involvement of endogenously produced 1,25-dihydroxyvitamin D_3 in the growth and differentiation of human keratinocytes, *Biochim Biophys Acta 1092*:311 (1991).

126. H. L. Henry, C. Dutta, N. Cunningham, R. Blanchard, R. Penny, C. Tang, G. Marchetto, and S.-Y. Chou, The cellular and molecular regulation of 1,25(OH)$_2$D$_3$ production, *J Steroid Biochem Mol Biol 41*:401 (1992).

127. K. W. Colston, I. M. Evans, T. C. Spelsberg, and I. Macintyre, Feedback regulation of vitamin D metabolism by 1,25-dihydroxycholecalciferol. *Biochem J 164*:83 (1977).

128. H. L. Henry, A. N. Taylor, and A. W. Norman, Response of chick parathyroid glands to the vitamin D metabolites, 1,25-dihydroxycholecalciferol and 24,25-dihydroxycholecalciferol, *J Nutr 107*:1918 (1977).

129. J. M. Canterbury, G. Gavellas, J. J. Bourgoignie, and E. Reise, Metabolic consequences of oral administration of 24,25-dihydroxycholecalciferol to uremic dogs, *J Clin Invest 65*:571 (1980).

130. H. Yamato, R. Okazaki, T. Ishii, E. Ogata, T. Sato, M. Kumegawa, K. Akaogi, N. Taniguchi, and T. Matsumoto, Effect of 24R,25-dihydroxyvitamin D$_3$ on the formation and function of osteoclastic cells, *Calcif Tissue Int 52*:255 (1993).

131. R. A. Evans, Hills E., S. Y. P. Wong, Dunstan C. R., and A. W. Norman, The use of 24,25-dihydroxycholecalciferol alone and in combination with 1,25-dihydroxycholecalciferol in chronic renal failure, in *Vitamin D: Chemical, Biochemical and Clinical Endocrinology of Calcium Metabolism* (A. W. Norman, K. Schaefer, H.-G. Grigoleit, and D. v. Herrath, eds.), Walter de Gruyter, Berlin, 1982, pp. 835–840.

132. C. Lidor, I. Atkin, A. Ornoy, S. Dekel, and S. Edelstein, Healing of rachitic lesions in chicks by 24R,25-dihydroxycholecalciferol administered locally into bone, *J Bone Miner Res 2*:91 (1987).

133. C. Lidor, S. Dekel, and S. Edelstein, The metabolism of vitamin D$_3$ during fracture-healing in chicks, *Endocrinology 120*:389 (1987).

134. A. W. Norman, V. L. Leathers, J. E. Bishop, Kadowaki S., and B. E. Miller, 24R,25-Dihydroxyvitamin D$_3$ has unique receptors (parathyroid gland) and biological responses (egg hatchability), in *Vitamin D: Chemical, Biochemical, and Clinical Endocrinology of Calcium Metabolism* (A. W. Norman, K. Schaefer, H.-G. Grigoleit, and D. Herrath, eds.), Walter de Gruyter, Berlin, 1982, pp. 147–151.

135. S. Ishizuka, T. Ohba, and A. W. Norman, 1α,25(OH)$_2$D$_3$-26,23-Lactone is a major metabolite of 1α,25(OH)$_2$D$_3$ under physiological conditions, in *Vitamin D: Molecular Cellular and Clinical Endocrinology* (A. W. Norman, K. Schaefer, H. G. Grigoleit, and D. v. Herrath, eds.), Walter de Gruyter, Berlin, 1988, pp. 143–144.

136. R. Kumar, The metabolism and mechanism of action of 1,25-dihydroxyvitamin D$_3$, *Kidney Int 30*:793 (1986).

137. E. Orti, J. E. Bodwell, and A. Munck, Phosphorylation of steroid hormone receptors, *Endocr Rev 13*:105 (1992).

138. H. M. Darwish, J. K. Burmester, V. E. Moss, and H. F. DeLuca, Phosphorylation is involved in transcriptional activation by the 1,25-dihydroxyvitamin D$_3$ receptor, *Biochim Biophys Acta 1167*:29 (1993).

139. P. W. Jurutka, J.-C. Hsieh, and M. R. Haussler, Phosphorylation of the human 1,25-dihydroxyvitamin D$_3$ receptor by cAMP-dependent protein kinase, in vitro, and in transfected COS-7 cells, *Biochem Biophys Res Commun 191*:1089 (1993).

140. J.-C. Hsieh, P. W. Jurutka, S. Nakajima, M. A. Galligan, C. A. Haussler, Y. Shimizu, N. Shimizu, G. K. Whitfield, and M. R. Haussler, Phosphorylation of the human vitamin D receptor by protein kinase C: biochemical and functional evaluation of the serine 51 recognition site, *J Biol Chem 268*:15118 (1993).

141. K. E. Lowe, A. C. Maiyar, and A. W. Norman, Vitamin D–mediated gene expression, *Crit Rev Eukar Gene Exp 2*:65 (1992).

142. S. S. Hannah and A. W. Norman, 1,25-Dihydroxyvitamin D$_3$ regulated expression of the eukaryotic genome, *Nutr Rev 52*:376 (1994).

143. W. R. Wecksler and A. W. Norman, A kinetic and equilibrium binding study of 1α,25-

dihydroxyvitamin D_3 with its cytosol receptor from chick intestinal mucosa, *J Biol Chem* *255*:3571 (1980).

144. W. R. Wecksler, F. P. Ross, R. S. Mason, S. Posen, and A. W. Norman, Studies on the mode of action of calciferol. XXIV. Biochemical properties of the 1α, 25-dihydroxyvitamin D_3 cytoplasmic receptors from human and chick parathyroid glands, *Arch Biochem Biophys 201*: 95 (1980).

145. W. R. Wecksler, F. P. Ross, R. S. Mason, and A. W. Norman, Studies on the mode of action of calciferol. XXI. Biochemical properties of the 1α,25-dihydroxyvitamin D_3 cytosol receptors from human and chicken intestinal mucosa, *J Clin Endocrinol Metab 50*:152 (1980).

146. R. M. Evans, The steroid and thyroid hormone receptor superfamily, *Science 240*:889 (1988).

147. M. G. Parker, *Nuclear Hormone Receptors: Molecular Mechanisms, Cellular Functions, Clinical Abnormalities*, Academic Press, London, 1991.

148. A. Krust, S. Green, P. Argos, V. Kumar, P. Walter, J. M. Bornert, and P. Chambon, The chicken oestrogen receptor sequence: homology with v-erbA and the human oestrogen and glucocorticoid receptors, *Embo J 5*:891 (1986).

149. M. Beato, Gene regulation by steroid hormones, *Cell 56*:335 (1989).

150. S. Green and P. Chambon, Nuclear receptors enhance our understanding of transcription regulation, *Trends Genet 4*:309 (1988).

151. J. Ham and M. G. Parker, Regulation of gene expression by nuclear hormone receptors, *Curr Opin Cell Biol 1*:503 (1989).

152. B. M. Forman and H. H. Samuels, Interactions among a subfamily of nuclear hormone receptors: the regulatory zipper model, *Mol Endocrinol 4*:1293 (1990).

153. F. Rastinejad, T. Perlmann, R. M. Evans, and P. B. Sigler, Structural determinants of nuclear receptor assembly on DNA direct repeats, *Nature 375*:203 (1995).

154. S. E. Fawell, J. A. Lees, R. White, and M. G. Parker, Characterization and colocalization of steroid binding and dimerization activities in the mouse estrogen receptor. *Cell 60*:953 (1990).

155. C. K. Glass, S. M. Lipkin, O. V. Devary, and M. G. Rosenfeld, Positive and negative regulation of gene transcription by a retinoic acid–thyroid hormone receptor heterodimer, *Cell 59*: 697 (1989).

156. B. M. Forman, C. Yan, M. Au, J. Casanova, J. Ghysdael, and H. H. Samuels, A domain containing leucine zipper like motifs may mediate novel in vivo interactions between the thyroid hormone and retinoic acid receptors, *Mol Endocrinol 3*:1610 (1989).

157. W. Bourguet, M. Ruff, P. Chambon, H. Gronemeyer, and D. Moras, Crystal structure of the ligand-binding domain of the human nuclear receptor RXR-α, *Nature 375*:377 (1995).

158. D. Picard, B. Khursheed, M. J. Garabedian, M. G. Fortin, S. Lindquist, and K. R. Yamamoto, Reduced levels of hsp90 compromise steroid receptor action in vivo, *Nature 348*:166 (1990).

159. L. Tora, J. White, C. Brou, D. Tasset, N. Webster, E. Scheer, and P. Chambon, The human estrogen receptor has two independent nonacidic transcriptional activation functions, *Cell 59*:477 (1989).

160. D. P. McDonnell, J. W. Pike, and B. W. O'Malley, The vitamin D receptor: a primitive steroid receptor related to thyroid hormone receptor, *J Steroid Biochem 30*:41 (1988).

161. V. C. Yu, C. Delsert, B. Andersen, J. M. Holloway, O. V. Devary, A. M. Näär, S. Y. Kim, J.-M. Boutin, C. K. Glass, and M. G. Rosenfeld, RXRβ: a coregulator that enhances binding of retinoic acid, thyroid hormone, and vitamin D receptors to their cognate response elements, *Cell 67*:1251 (1991).

162. S. A. Kliewer, K. Umesono, D. J. Mangelsdorf, and R. M. Evans, Retinoid X receptor interacts with nuclear receptors in retinoic acid, thyroid hormone and vitamin D_3 signalling, *Nature 355*:446 (1992).

163. C. Carlberg, RXR-independent action of the receptors for thyroid hormone, retinoid acid and vitamin D on inverted palindromes, *Biochem Biophys Res Commun 195*:1345 (1993).

164. M. Schräder, I. Bendik, M. Becker-André, and C. Carlberg, Interaction between retinoic acid and vitamin D signaling pathways, *J Biol Chem 268*:17830 (1993).

165. R. H. Wasserman, R. A. Corradino, and A. N. Taylor, Vitamin D–dependent calcium-binding protein: purification and some properties, *J Biol Chem 243*:3978 (1968).

166. C. Perret, C. Desplan, A. Brehier, and M. Thomasset, Characterization of rat 9-kDa cholecalcin (CaBP) messenger RNA using a complementary DNA: absence of homology with 28-kDa cholecalcin mRNA. *Eur J Biochem 148*:61 (1985).

167. L. Cancela, H. Ishida, J. E. Bishop, and A. W. Norman, Local chromatin changes accompany the expression of the calbindin-D_{28K} gene: tissue specificity and effect of vitamin D activation, *Mol Endocrinol 6*:468 (1992).

168. R. A. Corradino, 1,25-Dihydroxycholecalciferol: inhibition of action in organ-cultured intestine by actinomycin D and α-amanitin, *Nature 243*:41 (1973).

169. H. C. Tsai and A. W. Norman, Studies on the mode of action of calciferol. VI. Effect of 1,25-dihydroxyvitamin D_3 on RNA synthesis in the intestinal mucosa, *Biochem Biophys Res Commun 54*:622 (1973).

170. P. Siebert, W. Hunziker, and A. W. Norman, Studies on the mode of action of calciferol. XLVIII. Cell-free translation analysis of vitamin D–dependent calcium binding protein mRNA activity present in total RNA and polysomal extracts from chick intestine, *Arch Biochem Biophys 219*:286 (1982).

171. G. Theofan, A. P. Nguyen, and A. W. Norman, Regulation of calbindin-D_{28k} gene expression by 1,25-dihydroxyvitamin D_3 is correlated to receptor occupancy, *J Biol Chem 261*:16943 (1986).

172. P. P. Minghetti, L. Cancela, Y. Fujisawa, G. Theofan, and A. W. Norman, Molecular structure of the chicken vitamin D-induced calbindin-D_{28K} gene reveals eleven exons, six Ca^{2+}-binding domains, and numerous promoter regulatory elements. *Mol Endocrinol 2*:355 (1988).

173. I. Nemere and A. W. Norman, Studies on the mode of action of calciferol. LII. Rapid action of 1,25-dihydroxyvitmin D_3 on calcium transport in perfused chick duodenum: effect of inhibitors, *J Bone Miner Res 2*:99 (1987).

174. M. Lieberherr, B. Grosse, P. Duchambon, and T. Drüeke, A functional cell surface type receptor is required for the early action of 1,25-dihydroxyvitamin D_3 on the phosphoinositide metabolism in rat enterocytes, *J Biol Chem 264*:20403 (1989).

175. B. D. Boyan, Z. Schwartz, L. Bonewald, and L. Swain, Localization of $1,25(OH)_2D_3$-responsive alkaline phosphatase in osteoblast-like cells (ROS 17/2.8, MG 63, and MC 3T3) and growth cartilage cells in culture, *J Biol Chem 264*:11879 (1989).

176. M. C. Meikle, Hypercalcaemia of malignancy, *Nature 336*:311 (1988).

177. L. Cancela, I. Nemere, and A. W. Norman, $1\alpha,25(OH)_2$ Vitamin D_3: a steroid hormone capable of producing pleiotropic receptor-mediated biological responses by both genomic and nongenomic mechanisms, *J Steroid Biochem Mol Biol 30*:33 (1988).

178. J. M. Caffrey and M. C. Farach-Carson, Vitamin D_3 metabolites modulate dihydropyridine-sensitive calcium currents in clonal rat osteosarcoma cells, *J Biol Chem 264*:20265 (1989).

179. M. Lieberher, Effects of vitamin-D_3 metabolites on cytosolic free calcium in confluent mouse osteoblasts, *J Biol Chem 262*:13168 (1987).

180. J. Barsony and S. J. Marx, Rapid accumulation of cyclic GMP near activated vitamin D receptors, *Proc Natl Acad Sci USA 88*:1436 (1991).

181. I. Nemere, L.-X. Zhou, and A. W. Norman, Nontranscriptional effects of steroid hormones, *Receptor 3*:277 (1993).

182. I. Nemere, M. C. Dormanen, M. W. Hammond, W. H. Okamura, and A. W. Norman, Identification of a specific binding protein for $1\alpha,25$-dihydroxyvitamin D_3 in basal-lateral membranes of chick intestinal epithelium and relationship to transcaltachia, *J Biol Chem 269*: 23750 (1994).

183. P. Morley, J. F. Whitfield, B. C. Vanderhyden, B. K. Tsang, and J. L. Schwartz, A new,

nongenomic estrogen action: the rapid release of intracellular calcium, *Endocrinology 131*: 1305 (1992).

184. C. Mendoza and J. Tesarik, A plasma-membrane progesterone receptor in human sperm is switched on by increasing intracellular free calcium, *FEBS Lett 330*:57 (1993).

185. C. Aurell Wistrom and S. Meize, Evidence suggesting involvement of a unique human sperm steroid receptor/Cl⁻ channel complex in the progesterone-initiated acrosome reaction, *Dev Biol 159*:679 (1993).

186. P. F. Blackmore, S. J. Beebe, D. R. Danforth, and N. Alexander, Progesterone and 17α-progesterone: novel stimulators of calcium influx in human sperm, *J Biol Chem 265*:1376 (1990).

187. M. D. Majewska and D. B. Vaupel, Steroid control of uterine motility via gamma-aminobutyric acid$_A$ receptors in the rabbit: a novel mechanism, *J Endocrinol 131*:427 (1991).

188. H. Koenig, C.-C. Fan, A. D. Goldstone, C. Y. Lu, and J. J. Trout, Polyamines mediate androgenic stimulation of calcium fluxes and membrane transport in rat myocytes, *Circ Res 64*: 415 (1989).

189. P. Rehberger, M. Rexin, and U. Gehring, Heterotetrameric structure of the human progesterone receptor, *Proc Natl Acad Sci USA 89*:8001 (1992).

190. B. Gametchu, C. S. Watson, and S. Wu, Use of receptor antibodies to demonstrate membrane glucocorticoid receptor in cells from human leukemic patients, *FASEB J 7*:1283 (1993).

191. M. Orchinik, T. F. Murray, and F. L. Moore, A corticosteroid receptor in neuronal membranes, *Science 252*:1848 (1991).

192. T. J. Smith, F. B. Davis, and P. J. Davis, Stereochemical requirements for the modulation by retinoic acid of thyroid hormone activation of Ca^{2+}-ATPase and binding at the human erythrocyte membrane, *Biochem J 284*:583 (1992).

193. J. Segal, Thyroid hormone action at the level of the plasma membrane, *Thyroid 1*:83 (1990).

194. D. Schachter, E. B. Dowdle, and H. Schenker, Active transport of calcium by the small intestine of the rat, *Am J Physiol 198*:263 (1960).

195. H. E. Harrison and H. C. Harrison, Vitamin D and permeability of intestinal mucosa to calcium, *Am J Physiol 208*:370 (1965).

196. V. L. Leathers, Biophysical characterization and functional studies on calbindin-D$_{28K}$: a vitamin D–induced calcium binding protein, Ph.D. dissertation, University of California, Riverside, 1989.

197. I. Nemere, V. L. Leathers, B. S. Thompson, R. A. Luben, and A. W. Norman, Redistribution of calbindin-D$_{28K}$ in chick intestine in response to calcium transport, *Endocrinology 129*: 2972 (1991).

198. J. A. Putkey, A. M. Spielvogel, R. D. Sauerheber, Dunlap C. S., and A. W. Norman, Studies on the mode of action of calciferol. XXXIX. Vitamin D–mediated intestinal calcium transport: effect of essential fatty acid deficiency and spin label studies of enterocyte membrane lipid fluidity, *Biochim Biophys Acta 688*:177 (1982).

199. J. A. Putkey and A. W. Norman, Studies on the mode of action of calciferol. XLV. Vitamin D: its effect on the protein composition and core material structure of the chick intestinal brush border membrane, *J Biol Chem 258*:8971 (1983).

200. A. M. Spielvogel, R. D. Farley, and A. W. Norman, Studies on the mechanism of action of calciferol. V. Turnover time of chick intestinal epithelial cells in relation to the intestinal action of vitamin D, *Exp Cell Res 74*:359 (1972).

201. J. T. McCarthy, S. S. Barham, and R. Kumar, 1,25-Dihydroxyvitamin D$_3$ rapidly alters the morphology of the duodenal mucosa of rachitic chicks: evidence for novel effects of 1,25-dihydroxyvitamin D$_3$, *J Steroid Biochem Mol Biol 21*:253 (1984).

202. R. H. Wasserman, M. E. Brindak, M. M. Buddle, Q. Cai, F. C. Davis, C. S. Fullmer, R. F. Gilmour, Jr., C. Hu, H. M. Mykkanen, and D. N. Tapper, Recent studies on the biological actions of vitamin D on intestinal transport and the electrophysiology of peripheral nerve and cardiac muscle, *Prog Clin Biol Res 332*:99 (1990).

203. I. Nemere and A. W. Norman, Transport of calcium, in *Handbook of Physiology* (M. Field and R. A. Frizzell, eds.), *Am. Physiol. Soc.*, Bethesda, MD, 1991, pp. 337–360.

204. I. Nemere and A. W. Norman, Transcaltachia, vesicular calcium transport, and microtubule-associated calbindin-D28K: emerging views of 1,25-dihydroxyvitamin D_3–mediated intestinal calcium absorption, *Min Elect Metab 16*:109 (1990).

205. M. R. Clements, L. Johnson, and D. R. Fraser, A new mechanism for induced vitamin-D deficiency in calcium deprivation, *Nature 325*:62 (1987).

206. H. L. Henry and A. W. Norman, Vitamin D: metabolism and biological action, *Annu Rev Nutr 4*:493 (1984).

207. J. L. Underwood and H. F. DeLuca, Vitamin-D is not directly necessary for bone-growth and mineralization, *Am J Physiol 246*:E493 (1984).

208. A. W. Norman and S. Hurwitz, The role of the vitamin D endocrine system in avian bone biology, *J Nutr 123*:S30 (1993).

209. M. R. Walters, D. M. Rosen, A. W. Norman, and R. A. Luben, 1,25-Dihydroxyvitamin D receptors in an established bone cell line: correlation with biochemical responses, *J Biol Chem 257*:7481 (1982).

210. J. R. Harrison, D. N. Petersen, A. C. Lichtler, A. T. Mador, D. W. Rowe, and B. E. Kream, 1,25-Dihydroxyvitamin D_3 inhibits transcription of type I collagen genes in the rat osteosarcoma cell line ROS 17/2.8, *Endocrinology 125*:327 (1989).

211. N. Kurihara, K. Ikeda, Y. Hakeda, M. Tsunoi, N. Maeda, and M. Kumegawa, Effect of 1,25-dihydroxyvitamin D_3 on alkaline phosphatase activity and collagen synthesis in osteoblastic cells, clone MC3T3-E1, *Biochem Biophys Res Commun 119*:767 (1984).

212. L. C. Pan and P. A. Price, The effect of transcriptional inhibitors on the bone gamma-carboxyglutamic acid protein response to 1,25-dihydroxyvitamin D_3 in osteosarcoma cells, *J Biol Chem 259*:5844 (1984).

213. J. D. Fraser and P. A. Price, Induction of matrix Gla protein synthesis during prolonged 1,25-dihydroxyvitamin D_3 treatment of osteosarcoma cells. *Calcif Tissue Int 46*:270 (1990).

214. D. W. Rowe and B. E. Kream, Regulation of collagen synthesis in fetal rat calvaria by 1,25-dihydroxyvitamin D_3, *J Biol Chem 257*:8009 (1982).

215. M. E. Holtrop, K. A. Cox, M. B. Clark, M. F. Holick, and C. S. Anast, 1,25-Dihydroxycholecalciferol stimulates osteoclasts in rat bones in the absence of parathyroid hormone, *Endocrinology 108*:2293 (1981).

216. G. D. Roodman, K. J. Ibbotson, B. R. MacDonald, T. J. Kuehl, and G. R. Mundy, 1,25-Dihydroxyvitamin D_3 causes formation of multinucleated cells with several osteoclast characteristics in cultures of primate marrow, *Proc Natl Acad Sci USA 82*:8213 (1985).

217. N. Takahashi, T. Akatsu, N. Udagawa, T. Sasaki, A. Yamaguchi, J. M. Moseley, T. J. Martin, and T. Suda, Osteoblastic cells are involved in osteoclast formation, *Endocrinology 123*: 2600 (1988).

218. E. Abe, Y. Ishimi, N. Takahashi, T. Akatsu, H. Ozawa, H. Yamana, S. Yoshiki, and T. Suda, A differentiation-inducing factor produced by the osteoblastic cell line MC3T3-E1 stimulates bone resorption by promoting osteoclast formation, *J Bone Miner Res 3*:635 (1988).

219. M. Peacock and P. J. Heyburn, Effect of vitamin D metabolites on proximal muscle weakness, *Calcif Tissue Res 24*:78 (1977).

220. M. R. Walters, D. C. Wicker, and P. C. Riggle, 1,25-Dihydroxyvitamin D_3 receptors identified in the rat heart, *J Mol Cell Cardiol 18*:67 (1986).

221. A. Rabie, A. Brehier, S. Intrator, M. C. Clavel, C. O. Parkes, C. Legrand, and M. Thomasset, Thyroid state and cholecalcin (calcium-binding protein) in cerebellum of the developing rat, *Dev Brain Res 29*:253 (1986).

222. J. Merke, W. Hofmann, D. Goldschm., and E. Ritz, Demonstration of $1,25(OH)_2$ vitamin-D_3 receptors and actions in vascular smooth-muscle cells in vitro, *Calcif Tissue Res 41*:112 (1987).

223. R. Boland, A. W. Norman, E. Ritz, and W. Hasselbach, Presence of a 1,25-dihydroxyvitamin

D_3 receptor in chick skeletal muscle myoblasts, *Biochem Biophys Res Commun 128*:305 (1992).

224. W. E. Stumpf, M. Sar, F. A. Reid, Y. Tanaka, and H. F. DeLuca, Target cells for 1,25-dihydroxyvitamin D_3 in intestinal tract, stomach, kidney, skin, pituitary, and parathyroid, *Science 206*:1188 (1979).

225. J. Hosomi, J. Hosoi, E. Abe, T. Suda, and T. Kuroki, Regulation of terminal differentiation of cultured mouse epidermal cells by 1-α,25-dihydroxyvitamin D_3, *Endocrinology 113*:1950 (1983).

226. M. Regnier and M. Darmon, 1,25-Dihydroxyvitamin D_3 stimulates specifically the last steps of epidermal differentiation of cultured human keratinocytes, *Differentiation 47*:173 (1991).

227. D. D. Bikle, M. K. Nemanic, J. O. Whitney, and P. W. Elias, Neonatal human foreskin keratinocytes produce 1,25-dihydroxyvitamin D_3, *Biochemistry 25*:1545 (1986).

228. K. Kragballe, H. I. Beck, and H. Sogaard, Improvement of psoriasis by a topical vitamin D_3 analogue (MC 903) in a double-blind study, *Br J Dermatol 119*:223 (1988).

229. K. Kragballe, Vitamin D analogues in the treatment of psoriasis, *J Cell Biochem 49*:46 (1992).

230. M. F. Holick, Will 1,25-dihydroxyvitamin D_3, MC 903, and their analogues herald a new pharmacologic era for the treatment of psoriasis, *Arch Dermatol 125*:1692 (1989).

231. K. Chida, H. Hashiba, M. Fukushim, T. Suda, and T. Kuroki, Inhibition of tumor promotion in mouse skin by 1 α,25-dihydroxyvitamin D_3, *Cancer Res 45*:5426 (1985).

232. C. Cade and A. W. Norman, Vitamin D_3 improves impaired glucose-tolerance and insulin-secretion in the vitamin D-deficient rat in vivo, *Endocrinology 119*:84 (1986).

233. S. Kadowaki and A. W. Norman, Studies on the mode of action of calciferol. XLIX. Dietary vitamin D is essential for normal insulin secretion from the perfused rat pancreas, *J Clin Invest 73*:759 (1984).

234. C. Cade and A. W. Norman, Rapid normalization/stimulation by 1,25(OH)$_2$-vitamin D_3 of insulin secretion and glucose tolerance in the vitamin D-deficient rat, *Endocrinology 120*:1490 (1987).

235. S. Kadowaki and A. W. Norman, Studies on the mode of action of calciferol. LIX. Time-course study of the insulin secretion after 1,25-dihydroxyvitamin D_3 administration, *Endocrinology 117*:1765 (1985).

236. S. Kadowaki and A. W. Norman, Demonstration that the vitamin D metabolite 1,25(OH)$_2$-vitamin D_3 and not 24R,25(OH)$_2$-vitamin D_3 is essential for normal insulin secretion in the perfused rat pancreas, *Diabetes 34*:315 (1985).

237. O. Gedik and S. Akalin, Effects of vitamin D deficiency and repletion on insulin and glucagon secretion in man, *Diabetologia 29*:142 (1986).

238. J. Sonnenberg, V. N. Luine, L. C. Krey, and S. Christakos, 1,25-Dihydroxyvitamin D_3 treatment results in increased choline acetyltransferase activity in specific brain nuclei, *Endocrinology 118*:1433 (1986).

239. M. Thomasset, C. O. Parkes, and P. Cuisinier-Gleizes, Rat calcium-binding proteins: distribution, development, and vitamin D dependence, *Am J Physiol 243*:E483 (1982).

240. A. L. Morrow, J. R. Pace, R. H. Purdy, and S. M. Paul, Characterization of steroid interactions with gamma-aminobutyric acid receptor–gated chloride ion channels: evidence for multiple steroid recognition sites. *Mol Pharmacol 37*:263 (1990).

241. B. J. Bowers and J. M. Wehner, Biochemical and behavioral effects of steroids on GABA$_A$ receptor function in long- and short-sleep mice, *Brain Res Bull 29*:57 (1992).

242. D. M. Peehl, R. J. Skowronski, G. K. Leung, S. T. Wong, T. A. Stamey, and D. Feldman, Antiproliferative effects of 1,25-dihydroxyvitamin D_3 on primary cultures of human prostatic cells, *Cancer Res 54*:805 (1994).

243. E. Abe, C. Miyaura, H. Sakagami, M. Takeda, K. Konno, T. Yamazaki, S. Yoshiki, and T. Suda, Differentiation of mouse myeloid leukemia cells induced by 1α,25-dihydroxyvitamin D_3, *Proc Natl Acad Sci USA 78*:4990 (1981).

244. D. M. McCarthy, J. F. San Miguel, H. C. Freake, P. M. Green, H. Zola, D. Catovsky, and J. M. Goldman, 1,25-Dihydroxyvitamin D_3 inhibits proliferation of human promyelocytic leukaemia (HL60) cells and induces monocyte-macrophage differentiation in HL60 and normal human bone marrow cells, *Leuk Res 7*:51 (1983).

245. D. M. Provvedini, C. D. Tsoukas, L. J. Deftos, and S. C. Manolagas, 1,25-Dihydroxyvitamin D_3 receptors in human leukocytes. *Science 221*:1181 (1983).

246. A. K. Bhalla, E. P. Amento, T. L. Clemens, M. F. Holick, and S. M. Krane, Specific high-affinity receptors for 1,25-dihydroxyvitamin D_3 in human peripheral blood mononuclear cells: presence in monocytes and induction in T lymphocytes following activation, *J Clin Endocrinol Metab 57*:1308 (1983).

247. H. P. Koeffler, T. Amatruda, N. Ikekawa, and Y. Kobayashi, Induction of macrophage differentiation of human normal and leukemic myeloid stem-cells by 1,25-dihydroxyvitamin D_3 and its fluorinated analogs, *Cancer Res 44*:5624 (1984).

248. S. Murao, M. A. Gemmell, M. F. Callaham, N. L. Anderson, and E. Huberman, Control of macrophage cell differentiation in human promyelocytic HL-60 leukemia cells by 1,25-dihydroxyvitamin D_3 and phorbol-12-myristate-13-acetate, *Cancer Res 43*:4989 (1983).

249. W. F. C. Rigby, M. Waugh, and R. F. Graziano, Regulation of human monocyte HLA-DR and CD4 antigen expression, and antigen presentation by 1,25-dihydroxyvitamin D_3. *Blood 76*:189 (1990).

250. S. Iho, T. Takahashi, F. Kura, H. Sugiyama, and T. Hoshino, The effect of 1,25-dihydroxyvitamin D_3 on in vitro immunoglobulin production in human B cells. *J Immunol 136*:4427 (1986).

251. C. D. Tsoukas, D. M. Provvedini, and S. C. Manolagas, 1,25-Dihydroxyvitamin D_3 a novel immunoregulatory hormone, *Science 224*:1438 (1984).

252. I. Nemere and C. M. Szego, Early actions of parathyroid hormone and 1,25-dihydroxycholecalciferol on isolated epithelial cells from rat intestine. I. Limited lysosomal enzyme release and calcium uptake, *Endocrinology 108*:1450 (1981).

253. W. F. Rigby, S. Denome, and M. W. Fanger, Regulation of lymphokine production and human T lymphocyte activation by 1,25-dihydroxyvitamin D_3: specific inhibition at the level of messenger RNA, *J Clin Invest 79*:1659 (1987).

254. W. F. Rigby, R. J. Nowlle, K. Krause, and M. W. Fanger, The effects of 1,25-dihydroxyvitamin D_3 on human T lymphocyte activation and proliferation: a cell cycle analysis, *J Immunol 135*:2279 (1985).

255. A. K. Bhalla, E. P. Amento, B. Serog, and L. H. Glimcher, 1,25-Dihydroxyvitamin D_3 inhibits antigen-induced T cell activation, *J Immunol 133*:1748 (1984).

256. M. A. Meehan, R. H. Kerman, and J. M. Lemire, 1,25-Dihydroxyvitamin D_3 enhances the generation of nonspecific suppressor cells while inhibiting the induction of cytotoxic cells in a human MLR, *Cell Immunol 140*:400 (1992).

257. W. F. C. Rigby, S. Denome, and M. W. Fanger, Regulation of lymphokine production and human T lymphocyte activation by 1,25-dihydroxyvitamin D_3, *J Clin Invest 79*:1659 (1987).

258. H. Reichel, H. P. Koeffler, A. Tobler, and A. W. Norman, 1α,25-Dihydroxyvitamin D_3 inhibits gamma-interferon synthesis by normal human peripheral blood lymphocytes, *Proc Natl Acad Sci USA 84*:3385 (1987).

259. A. Tobler, J. Gasson, H. Reichel, A. W. Norman, and H. P. Koeffler, Granulocyte macrophage colony simulating factor: sensitive and receptor mediated regulation by 1,25-dihydroxyvitamin D_3 in normal human peripheral blood lymphocytes, *J Clin Invest 79*:1700 (1987).

260. A. Tobler, C. W. Miller, A. W. Norman, and H. P. Koeffler, 1,25-Dihydroxyvitamin D_3 modulates the expression of a lymphokine (granulocyte-macrophage colony-stimulating factor) posttranscriptionally, *J Clin Invest 81*:1819 (1988).

261. J. M. Lemire, Immunomodulatory role of 1,25-dihydroxyvitamin D_3, *J Cell Biochem 49*:26 (1992).

262. M. Hewison, Vitamin D and the immune system, *J Endocrinol 132*:173 (1992).

263. A. D. Hess, P. M. Colobani, and A. Esa, *Kidney Transplantation: Diagnosis and Treatment,* Marcel Dekker, New York, 1986, pp. 353–382.

264. M.-C. Boissier, G. Chiocchia, and C. Fournier, Combination of cyclosporine A and calcitriol in the treatment of adjuvant arthritis, *J Rheumatol 19*:754 (1992).

265. C. Fournier, P. Gepner, M. Sadouk, and J. Charreire, In vivo beneficial effects of cyclosporin A and 1,25-dihydroxyvitamin D₃ on the induction of experimental autoimmune thyroiditis, *Clin Immunol Immunopathol 54*:53 (1990).

266. S. T. Lillevang, J. Rosenkvist, C. B. Andersen, S. Larsen, E. Kemp, and T. Kristensen, Single and combined effects of the vitamin D analogue KH1060 and cyclosporin A on mercuric-chloride-induced autoimmune disease in the BN rat, *Clin Exp Immunol 88*:301 (1992).

267. E. Lewin and K. Olgaard, The in vivo effect of a new, in vitro, extremely potent vitamin D₃ analog KH1060 on the suppression of renal allograft rejection in the rat, *Calcif Tissue Int 54*:150 (1994).

268. R. G. Wong, A. W. Norman, C. R. Reddy, and J. W. Coburn, Biologic effects of 1,25-dihydroxycholecalciferol (a highly active vitamin D metabolite) in acutely uremic rats, *J Clin Invest 51*:1287 (1972).

269. R. A. Gibbs and W. H. Okamura, Studies on vitamin D (calciferol) and its analogs. XXXII. Synthesis of 3-deoxy-1-α,25-dihydroxy-9, 11-dehydrovitamin-D₃–selective formation of 6-β-vitamin-D vinylallenes and their thermal <1,5*-sigmatropic hydrogen shift, *Tetrahedron Lett 28*: 6021 (1987).

270. W. H. Okamura, M. N. Mitra, D. A. Procsal, and A. W. Norman, Studies on vitamin D and its analogs. VIII. 3-Deoxy-1,25-dihydroxy-vitamin D₃, a potent new analog of 1,25-(OH)₂-D₃, *Biochem Biophys Res Commun 65*:24 (1975).

271. M. F. Holick, M. Garabedian, H. K. Schnoes, and H. F. DeLuca, Relationship of 25-hydroxyvitamin D₃ side chain structure to biological activity, *J Biol Chem 260*: 226 (1975).

272. R. L. Johnson, W. H. Okamura, and A. W. Norman, Studies on the mode of action of calciferol. X. 24-Nor-25-hydroxyvitamin D₃, an analog of 25-hydroxyvitamin D₃ having ''anti-vitamin'' activity, *Biochem Biophys Res Commun 67*:797 (1975).

273. M. R. Haussler, J. E. Zerwekh, R. H. Hesse, E. Rizzardo, and M. M. Peche, Biological activity of 1α-hydroxycholecalciferol, a synthetic analog of the hormonal form of vitamin D₃, *Proc Natl Acad Sci USA 70*:2248 (1973).

274. M. F. Holick, P. Kasten-Schraufrogel, T. Tavela, and H. F. DeLuca, Biological activity of 1 α-hydroxyvitamin D₃ in the rat, *Arch Biochem Biophys 166*:63 (1975).

275. J. L. Napoli, M. A. Fivizzani, H. K. Schnoes, and H. F. DeLuca, 1Alpha-hydroxy-25-fluoro-vitamin D₃: a potent analogue of 1alpha, 25-dihydroxyvitamin D₃, *Biochemistry 17*:2387 (1978).

276. L. Binderup and E. Bramm, Effects of a novel vitamin D analogue MC-903 on cell proliferation and differentiation in vitro and on calcium metabolism in vivo. *Biochem Pharmacol 37*: 889 (1988).

277. D. D. Bikle, E. Gee, and S. Pillai, Regulation of keratinocyte growth, differentiation, and vitamin D metabolism by analogs of 1,25-dihydroxyvitamin D, *J Invest Dermatol 101*:713 (1993).

278. H. Sorensen, L. Binderup, M. J. Calverley, L. Hoffmeyer, and N. R. Andersen, In vitro metabolism of calcipotriol (MC 903), a vitamin D analogue. *Biochem Pharmacol 39*:391 (1990).

279. S. Masuda, S. Strugnell, M. J. Calverley, H. L. J. Makin, R. Kremer, and G. Jones, In vitro metabolism of the anti-psoriatic vitamin D analog, calcipotriol, in two cultured human keratinocyte models, *J Biol Chem 269*:4794 (1994).

280. J. W. Pike, C. A. Donaldson, S. L. Marion, and M. R. Haussler, Development of hybridomas secreting monoclonal antibodies to the chicken intestinal 1α-25-dihydroxyvitamin D₃ receptor, *Proc Natl Acad Sci USA 79*:7719 (1982).

281. P. H. Itin, M. R. Pittelkow, and R. Kumar, Effects of vitamin D metabolites on proliferation

and differentiation of cultured human epidermal keratinocytes grown in serum-free or defined culture medium, *Endocrinology 135*:1793 (1994).

282. R. A. El-Azhary, M. S. Peters, M. R. Pittelkow, P. C. Kao, and S. A. Muller, Efficacy of vitamin D_3 derivatives in the treatment of psoriasis vulgaris: a preliminary report, *Mayo Clin Proc 68*:835 (1993).

283. C. Nieboer and C. A. Verburgh, Psoriasis treatment with vitamin D_3 analogue MC 903, *Br J Dermatol 126*:302 (1992).

284. P. C. M. van de Kerkhof and E. M. G. J. De Jong, Topical treatment with the vitamin D_3 analogue MC903 improves pityriasis rubra pilaris: clinical and immunohistochemcial observations, *Br J Dermatol 125*:293 (1991).

285. T. Trydal, J. R. Lillehaug, L. Aksnes, and D. Aarskog, Regulation of cell growth, c-*myc* mRNA, and 1,25-$(OH)_2$ vitamin D_3 receptor in C3H/10T1/2 mouse embryo fibroblasts by calcipotriol and 1,25-$(OH)_2$ vitamin D_3, *Acta Endocrinol 126*:75 (1992).

286. A. J. Brown, C. R. Ritter, J. L. Finch, J. Morrissey, K. J. Martin, E. Murayama, Y. Nishii, and E. Slatopolsky, The noncalcemic analogue of vitamin D, 22-oxacalcitriol, suppresses parathyroid hormone synthesis and secretion. *J Clin Invest 84*:728 (1989).

287. J. Abe, M. Morikawa, K. Miyamoto, S. Kaiho, M. Fukushima, C. Miyaura, E. Abe, T. Suda, and Y. Nishi, Synthetic analogues of vitamin D_3 with an oxygen atom in the side chain skeleton, *FEBS Lett 226*:58 (1987).

288. J. Y. Zhou, A. W. Norman, D. Chen, G. Sun, M. R. Uskokovic, and H. P. Koeffler, 1,25-Dihydroxy-16-ene-23-yne-vitamin D_3 prolongs time of leukemic mice. *Proc Natl Acad Sci USA 87*: 3929 (1990).

289. A. W. Norman, J. Y. Zhou, H. L. Henry, M. R. Uskokovic, and H. P. Koeffler, Structure–function studies on analogs of 1,25-dihydroxyvitamin D_3 differential effects on leukemic cell growth, differentiation, and intestinal calcium absorption, *Cancer Res 50*:6857 (1990).

290. S. J. Jung, Y. Y. Lee, S. Pakkala, S. De Vos, E. Elstner, A. W. Norman, J. Green, M. Uskokovic, and H. P. Koeffler, 1,25$(OH)_2$-16ene-Vitamin D_3 is a potent antileukemic agent with low potential to cause hypercalcemia, *Leuk Res 18*:453 (1994).

291. J. Y. Zhou, A. W. Norman, M. Lubbert, E. D. Collins, M. R. Uskokovic, and H. P. Koeffler, Novel vitamin D analogs that modulate leukemic cell growth and differentiation with little effect on either intestinal calcium absorption or bone calcium mobilization. *Blood 74*:82 (1989).

292. P. M. Wovkulich, A. D. Batcho, E. G. Baggioloni, A. Boris, G. Truitt, and M. R. Uskokovic, Synthesis of 1 α,25S,26-trihydroxy-22-ene-cholecalciferol, a potent inducer of cell differentiation, in *Vitamin D: Chemical, Biochemical and Clinical Update* (A. W. Norman, K. Schaefer, H.-G. Grigoleit, and D. v. Herrath, eds.), Walter de Gruyter, Berlin, 1985, pp. 755–764.

293. Y. Tanaka, H. F. DeLuca, Y. Kobayashi, and N. Ikekawa, 26,26,26,27,27,27-Hexafluoro-1,25-dihydroxyvitamin-D_3: a highly potent, long-lasting analog of 1,25-dihydroxyvitamin-D_3, *Arch Biochem Biophys 229*:348 (1984).

294. M. Inaba, S. Okuno, Y. Nishizawa, K. Yukioko, S. Otani, I. Matsui-Yuasa, S. Morisawa, H. F. DeLuca, and H. Morii, Biological activity of fluorinated vitamin D analogs at C-26 and C-27 on human promyelocytic leukemia cells, HL-60, *Arch Biochem Biophys 258*:421 (1987).

295. K. Yukioka, S. Otani, I. Matsui-Yuasa, H. Goto, S. Morisawa, S. Okuno, M. Inaba, Y. Nishizawa, and H. Morii, Biological activity of 26,26,26,27,27,27-hexafluorinated analogs of vitamin D_3 in inhibiting interleukin-2 production by peripheral blood mononuclear cells stimulated by phytohemagglutinin, *Arch Biochem Biophys 260*:45 (1988).

296. R. Belsey, H. F. DeLuca, and J. T. Potts, Competitive binding assay for vitamin D and 25-OH vitamin D, *J Clin Endocrinol Metab 33*:554 (1971).

297. Vitamin D assay, in *The United States Pharmacopeia*, U.S. Pharmacopeial Convention, 1995, pp. 1756–1760.

298. Association of Official Analytical Chemists, Conduct of the rat line test for measurement of vitamin D activity, in *Official Methods of Analyses*, Washington DC, 1975, pp. 885–888.

299. Vitamins and other nutrients: AOAC official method for determination of vitamin D in milk, vitamin preparations and feed concentrates via the rat bioassay, in *Official Methods of Analysis of AOAC International* (M. J. Deutsch, ed.), AOAC International, Arlington, VA, 1995, pp. 53–57.

300. Determination of vitamin D in vitamin D/AD concentrates, in multivitamin preparations, in fortified milk and milk powder, in mixed feeds, pre mixes and pet foods: liquid chromatographic method, *Official Methods of Analysis of the Association of Official Analytical Chemists* (K. Helrich, ed.), Association of Official Analytical Chemists, Inc., Washington, DC, 1990, pp. 1067–1070.

301. Vitamins and other nutrients: AOAC official method of determination of vitamin D_3 in poultry feed supplements via the chick bioassay, in *Official Methods of Analysis of AOAC International*, AOAC International, Arlington, VA, 1995, pp. 57–59.

302. K. A. Hibberd and A. W. Norman, Comparative biological effects of vitamins D_2 and D_3 and dihydrotachysterol$_2$ and dihydrotachysterol$_3$ in the chick, *Biochem Pharmacol 18*:2347 (1969).

303. E. B. Olson and H. F. DeLuca, 25-Hydroxycholecalciferol: direct effect on calcium transport, *Science 165*:405 (1979).

304. D. Schachter and S. M. Rosen, Active transport of Ca^{45} by the small intestine and its dependence on vitamin D, *Am J Physiol 196*:357 (1959).

305. D. V. Kimberg, D. Schachter, and H. Schenker, Active transport of calcium by intestine: effects of dietary calcium, *Am J Physiol 200*:1256 (1961).

306. H. L. Henry, A. W. Norman, A. N. Taylor, D. L. Hartenbower, and J. W. Coburn, Biological activity of 24,25-dihydroxycholecalciferol in chicks and rats, *J Nutr 106*:724 (1976).

307. B. E. Miller and A. W. Norman, Enzyme-linked immunoabsorbent assay (ELISA) and radio-immunoassay (RIA) for the vitamin D–dependent 28,000 dalton calcium-binding protein, *Meth Enzymol 102*:291 (1983).

308. Vitamins and other nutrients: AOAC official methods for vitamin D in vitamin preparations, in *Official Methods of Analysis of AOAC International* (M. J. Deutsch, ed.), AOAC International, Arlington, VA, 1995, pp. 19–30.

309. C. H. Nield, W. C. Russell, and A. Zimmerli, The spectrophotometric determination of vitamin D_2 and D_3, *J Biol Chem 136*:73 (1940).

310. A. J. Passannante and L. V. Avioli, Studies on the ultraviolet fluorescence of vitamin D and related compounds in acid-alcohol solutions, *Anal Biochem 15*:287 (1966).

311. R. D. Coldwell, C. E. Porteous, D. J. Trafford, and H. L. Makin, Gas chromatography–mass spectrometry and the measurement of vitamin D metabolites in human serum or plasma. *Steroids 49*:155 (1987).

312. K. A. Tartivita, J. P. Sciarello, and B. C. Rudy, High-performance liquid chromatographic analysis of vitamins. I. Quantitation of cholecalciferol or ergocalciferol in presence of photochemical isomers of the provitamin and application to cholecalciferol resins. *J Pharmacol Sci 65*:1024 (1976).

313. G. Jones and H. F. DeLuca, High-pressure liquid chromatography: separation of the metabolites of vitamins D_2 and D_3 on small-particle silica columns, *J Lipid Res 16*:448 (1975).

314. F. J. Mulder, E. De Vries, and B. Borsje, Analysis of fat-soluble vitamins. XXII. High performance liquid chromatographic determination of vitamin D in concentrates: a collaborative study, *J Assoc Off Anal Chem 62*:1031 (1979).

315. P. F. Brumbaugh and M. R. Haussler, 1 α,25-Dihydroxyvitamin D_3 receptor: competitive binding of vitamin D analogs, *Life Sci 13*:1737 (1973).

316. D. A. Procsal, W. H. Okamura, and A. W. Norman, Structural requirements for the interaction of 1 α,25-$(OH)_2$-vitamin D_3 with its chick intestinal receptor system, *J Biol Chem 250*:8382 (1975).

317. T. A. Reinhardt, R. L. Horst, J. W. Orf, and B. W. Hollis, A microassay for 1,25-dihydroxyvitamin D not requiring high performance liquid chromatography: application to clinical studies. *J Clin Endocrinol Metab 58*:91 (1984).

318. B. W. Hollis and T. Kilbo, The assay of circulating 1,25(OH)$_2$D using non-end-capped C-18 silica (C$_{18}$OH): performance and validation, in *Vitamin D: Molecular, Cellular and Clinical Endocrinology* (A. W. Norman, K. Schaefer, H.-G. Grigoleit, and D. v. Herrath, eds.), Walter de Gruyter, Berlin, 1988, pp. 710–719.

319. N. Horiuchi, T. Shinki, S. Suda, N. Takahashi, S. Yamada, H. Takayama, and T. Suda, A rapid and sensitive in vitro assay of 25-hydroxyvitamin D$_3$-1α-hydroxylase and 24-hydroxylase using rat kidney homogenates, *Biochem Biophys Res Commun 121*:174 (1984).

320. T. A. Reinhardt and R. L. Horst, Simplified assays for the determination of 25(OH)D, 24,25(OH)$_2$D and 1,25(OH)$_2$D, *Vitamin D: Molecular, Cellular and Clinical Endocrinology* (A. W. Norman, K. Schaefer, H.-G. Grigoleit, and D. v. Herrath, eds.), Walter de Gruyter, Berlin, 1988, pp. 720–726.

321. Report of the subcommittee on fat-soluble vitamins, *WHO Bull 3*:7–11 (1950).

322. Subcommittee on the Tenth Edition of the RDAs, Food and Nutrition Board, Commission on Life Sciences, National Research Council, *Recommended Dietary Allowances*, 10th ed., National Academy Press, Washington, DC, 1989, pp. 1–285.

323. S. K. Bhowmick, K. R. Johnson, and K. R. Rettig, Rickets caused by vitamin D deficiency in breast-fed infants in the southern United States, *Am J Dis Child 145*:127 (1991).

324. L. E. Booher, E. R. Hartzler, and E. M. Hewston, A compilation of the vitamin values of foods in relation to processing and other variants, U.S. Department of Agriculture Circular 638: (1942).

325. F. R. Greer and S. Marshall, Bone mineral content, serum vitamin D metabolite concentrations, and ultraviolet B light exposure in infants fed human milk with and without vitamin D$_2$ supplements, *J Pediatr 114*:204 (1989).

326. P. Wilton, Cod-liver oil, vitamin D and the fight against rickets, *Can Med Assoc J 152*:1516 (1995).

327. P. Lips, F. C. van Ginkel, M. J. M. Jongen, F. Rubertus, W. J. F. van der Vijgh, and J. C. Netelenbos, Determinants of vitamin D status in patients with hip fracture and in elderly control subjects, *Am J Clin Nutr 46*:1005 (1987).

328. T. Suda, H. F. DeLuca, H. K. Schnoes, G. Ponchon, Y. Tanaka, and M. F. Holick, 21,25-Dihydroxycholecalciferol: a metabolite of vitamin D$_3$ perferentially active on bone, *Biochemistry 9*:2917 (1970).

329. S. J. Counts, D. J. Baylink, F. Shen, D. J. Sherrard, and R. O. Hickman, Vitamin D intoxication in an anephric child, *Ann Intern Med 82*:196 (1975).

330. J. M. Pettifor, D. D. Bikle, M. Cavaleros, D. Zachen, M. C. Kamdar, and F. P. Ross, Serum levels of free 1,25-dihydroxyvitamin D in vitamin D toxicity, *Ann Intern Med 122*:511 (1995).

331. M. R. Hughes, D. J. Baylink, P. G. Jones, and M. R. Haussler, Radioligand receptor assay for 25-hydroxyvitamin D$_2$/D$_3$ and 1 α,25-dihydroxyvitamin D$_2$/D$_3$, *J Clin Invest 58*:61 (1976).

332. J. M. Gertner and M. Domenech, 25-Hydroxyvitamin D levels in patients treated with high-dosage ergo- and cholecalciferol, *Clin Pathol 30*:144 (1977).

333. R. L. Morrissey, R. M. Cohn, R. N. Empson, H. L. Greene, O. D. Taunton, and Z. Z. Ziporin, Relative toxicity and metabolic effects of cholecalciferol and 25-hydroxycholecalciferol in chicks, *J Nutr 107*:1027 (1977).

334. J. W. Coburn and N. Brautbar, Disease states in man related to vitamin D, in *Vitamin D: Molecular Biology and Clinical Nutrition* (A. W. Norman, ed.), Marcel Dekker, New York, 1980, pp. 515–577.

335. G. L. Klein, C. M. Targoff, M. E. Ament, D. J. Sherrard, R. Bluestone, J. H. Young, A. W. Norman, and J. W. Coburn, Bone disease associated with total parenteral nutrition, *Lancet 15*:1041 (1980).

336. J. K. Maesaka, V. Batuman, N. C. Pablo, and S. Shakamuri, Elevated 1,25-dihydroxyvitamin D levels: occurrence with sarcoidosis with end-stage renal disease. *Arch Intern Med 142*: 1206 (1982).

337. S. Epstein, P. H. Stern, N. H. Bell, I. Dowdeswell, and R. T. Turner, Evidence for abnormal regulation of circulating 1α,25-dihydroxyvitamin D in patients with pulmonary tuberculosis and normal calcium metabolism, *Calcif Tissue Int 36*:541 (1984).

338. V. N. Hoffman and O. M. Korzenio, Leprosy, hypercalcemia, and elevated serum calcitriol levels, *Ann Intern Med 105*:890 (1986).

339. A. Mudde, H., H. van den Berg, P. G. Boshuis, F. C. Breedveld, H. M. Markusse, P. M. Kluin, O. L. Bijvoet, and S. E. Papapoulos, Ectopic production of 1,25-dihydroxyvitamin D by B-cell lymphoma as a cause of hypercalcemia, *Cancer 59*:1543 (1987).

340. L. V. Avioli, S. W. Lee, J. E. McDonald, J. Lurid, and H. F. DeLuca, Metabolism of vitamin D_3-^{3}H in human subjects: distribution in blood, bile, feces, and urine. *J Clin Invest 46*:983 (1967).

341. J. G. Haddad and K. J. Chyu, Competitive protein-binding radioassay for 25-hydroxycholecalciferol, *J Clin Endocrinol Metab 33*:992 (1971).

342. C. Christiansen, M. S. Christiansen, F. Melsen, P. Rodbro, and H. F. DeLuca, Mineral metabolism in chronic renal failure with special reference to serum concentrations of $1,25(OH)_2D$ and $24,25(OH)_2D$, *Clin Nephrol 15*:18 (1981).

343. E. B. Mawer, C. M. Taylor, J. Backhouse, G. A. Lumb, and S. W. Stanbury, Failure of formation of 1,25-dihydroxycholecalciferol in chronic renal insufficiency, *Lancet 1*:626 (1973).

344. A. S. Brickman, J. W. Coburn, A. W. Norman, and S. G. Massry, Short-term effects of 1,25-dihydroxycholecalciferol on disordered calcium metabolism of renal failure, *Am J Med 57*: 28 (1974).

345. M. R. Haussler, D. J. Baylink, M. R. Hughes, P. F. Brumbaugh, J. E. Wergedal, F. H. Shen, R. L. Nielsen, S. J. Counts, K. M. Bursac, and T. A. McCain, The assay of 1 alpha,25-dihydroxyvitamin D_3: physiologic and pathologic modulation of circulating hormone levels, *Clin Endocrinol 5*:151S (1976).

346. A. S. Brickman, J. Jowsey, D. J. Sherrard, G. Friedman, F. R. Singer, D. J. Baylink, N. Maloney, S. G. Massry, A. W. Norman, and J. W. Coburn, Therapy with 1,25-dihydroxyvitamin D_3 in the management of renal osteodystrophy, in *Vitamin D and Problems Related to Uremic Bone Disease* (A. W. Norman, K. Schaefer, H. G. Grigoleit, D. v. Herrath, and E. Ritz, eds.), Walter de Gruyter, Berlin, 1975, pp. 241–247.

347. B. Lund, O. H. Sorensen, B. Lund, J. E. Bishop, and A. W. Norman, Vitamin D metabolism in hypoparathyroidism, *J Clin Endocrinol Metab 51*:606 (1980).

348. H. Reichel and A. W. Norman, Systemic effects of vitamin D, *Annu Rev Med 40*:71 (1989).

349. R. W. Winters, J. B. Graham, T. F. Williams, V. W. McFalls, and C. H. Burnett, A genetic study of familial hypophosphatemia and vitamin D resistant rickets with a review of the literature, *Medicine 37*:97 (1958).

350. A. S. Brickman, J. W. Coburn, K. Kurokawa, J. E. Bethune, H. E. Harrison, and A. W. Norman, Actions of 1,25-dihydroxycholecalciferol in patients with hypophosphatemic vitamin D–resistant rickets, *N Engl J Med 289*:495 (1973).

351. C. Arnaud, R. Maijer, T. Reade, C. R. Scriver, and D. T. Whelan, Vitamin D dependency: an inherited postnatal syndrome with secondary hyperparathyroidism, *Pediatrics 46*:871 (1970).

352. H. H. Ritchie, M. R. Hughes, E. T. Thompson, P. J. Malloy, Z. Hochberg, D. Feldman, J. W. Pike, and B. W. O'Malley, An ochre mutation in the vitamin D receptor gene causes hereditary 1,25-dihydroxyvitamin D_3–resistant rickets in three families, *Proc Natl Acad Sci USA 86*:9783 (1989).

353. T. Sone, R. A. Scott, M. R. Hughes, P. J. Malloy, D. Feldman, B. W. O'Malley, and J. W. Pike, Mutant vitamin D receptors which confer hereditary resistance to 1,25-dihydroxyvitamin D_3 in humans are transcriptionally inactive in vitro, *J Biol Chem 264*:20230 (1989).

354. A. R. Rut, M. Hewison, K. Kristjansson, B. Luisi, M. R. Hughes, and J. L. H. O'Riordan, Two mutations causing vitamin D resistant rickets: modelling on the basis of steroid hormone receptor DNA-binding domain crystal structures, *Clin Endocrinol 41*:581 (1994).

355. K. Kristjansson, A. R. Rut, M. Hewison, J. L. H. O'Riordan, and M. R. Hughes, Two mutations in the hormone binding domain of the vitamin D receptor cause tissue resistance to 1,25 dihydroxyvitamin D_3, *J Clin Invest 92*:12 (1993).

356. T. J. Hahn, B. A. Hendin, C. R. Scharp, and J. G. Haddad, Effect of chronic anticonvulsant therapy on serum 25-hydroxycalciferol levels in adults, *N Engl J Med 287*:900 (1972).

357. A. W. Norman, J. D. Bayless, and H. C. Tsai, Biologic effects of short term phenobarbital treatment on the response to vitamin D and its metabolites in the chick, *Biochem Pharmacol 25*:163 (1976).

358. W. Jubiz, M. R. Haussler, T. A. McCain, and K. G. Tolman, Plasma 1,25-dihydroxyvitamin D levels in patients receiving anticonvulsant drugs, *J Clin Endocrinol Metab 44*:617 (1977).

359. R. A. Corradino, Diphenylhydantoin: direct inhibition of the vitamin D_3–mediated calcium absorptive mechanism in organ-cultured duodenum. *J Clin Invest 74*:1451 (1976).

360. M. V. Jenkins, M. Harris, and M. R. Wills, The effect of phenytoin on parathyroid extract and 25-hydroxycholecalciferol-induced bone resorption: adenosine 3,5 cyclic monophosphate production, *Calcif Tissue Res 16*:163 (1974).

361. S. A. Mobarhan, R. M. Russell, R. R. Recker, D. B. Posner, F. L. Iber, and P. Miller, Metabolic bone-disease in alcoholic cirrhosis: a comparison of the effect of vitamin-D_2, 25-hydroxyvitamin-D, or supportive treatment, *Hepatology 4*:266 (1984).

362. J. M. Barragry, R. G. Long, M. W. France, M. R. Wills, B. J. Boucher, and S. Sherlock, Intestinal absorption of cholecalciferol in alcoholic liver disease and primary biliary cirrhosis, *Gut 20*:559 (1979).

363. J. R. Bullamore, R. Wilkinson, J. C. Gallagher, B. E. Nordin, and D. H. Marshall, Effect of age on calcium absorption, *Lancet 2*:535 (1970).

364. R. D. Lindeman, J. Tobin, and N. W. Shock, Longitudinal studies on the rate of decline in renal function with age, *J Am Geriatr Soc 33*:278 (1985).

365. R. W. Gray, D. R. Wilz, A. E. Caldas, and J. Lemann, The importance of phosphate in regulating plasma 1,25-$(OH)_2$-vitamin D levels in humans: studies in healthy subjects, in calcium-stone formers and in patients with primary hyperparathyroidism, *J Clin Endocrinol Metab 45*:299 (1977).

366. A. W. Ireland, J. S. Clubb, F. C. Neale, S. Posen, and T. S. Reeve, The calciferol requirements of patients with surgical hypoparathyroidism, *Ann Intern Med 69*:81 (1968).

367. N. Rochel, J. M. Wurtz, A. Mitschler, B. Klaholz, and D. Moras, The crystal structure of the nuclear receptor for vitamin D bound to its natural ligand, *Mol Cell 5*:175–179 (2000).

368. J.-P. Renaud, N. Rochel, M. Ruff, V. Vivat, P. Chambon, H. Gronemeyer, and D. Moras, Crystal structure of the RAR-y ligand-binding domain bound to *all-trans* retinoic acid, *Nature 378*:681–689 (1995).

369. W. Bourguet, M. Ruff, P. Chambon, H. Gronemeyer, and D. Moras, Crystal structure of the ligand-binding domain of the human nuclear receptor RXR-a, *Nature 375*:377–382 (1995).

370. W. Feng, R. C. J. Ribeiro, R. L. Wagner, H. Nguyen, J. W. Apriletti, R. J. Fletterick, J. D. Baxter, P. J. Kushner, and B. L. West, Hormone-dependent coactivator binding to a hydrophobic cleft on nuclear receptors, *Science 280*:1747–1749 (1998).

371. D. Moras and H. Gronemeyer, The nuclear receptor ligand-binding domain: structure and function, *Curr Opin Cell Biol 10*:384–391 (1998).

372. S. A. Onate, V. Boonyaratanakornkit, T. E. Spencer, S. Y. Tsai, M. J. Tsai, D. P. Edwards, and B. W. O'Malley, The steroid receptor coactivator-1 contains multiple receptor interacting and activation domains that cooperatively enhance the activation function 1 (AF1) and AF2 domains of steroid receptors, *J Biol Chem 273*:12101–12108 (1998).

373. R. M. Lavinsky, K. Jepsen, T. Heinzel, J. Torchia, T. M. Mullen, R. Schiff, A. L. Del-Rio,

M. Ricote, S. Ngo, J. Gemsch, S. G. Hilsenbeck, C. K. Osborne, C. K. Glass, M. G. Rosenfeld, and D. W. Rose, Diverse signaling pathways modulate nuclear receptor recruitment of N-CoR and SMRT complexes, *Proc Natl Acad Sci USA 95*:2920–2925 (1998).

374. J. J. Voegel, M. J. S. Heine, C. Zechel, P. Chambon, and H. Gronemeyer, TIF2, a 160 kDa transcriptional mediator for the ligand-dependent activation function AF-2 of nuclear receptors, *EMBO J 15*:3667–3675 (1996).

375. H. Hong, K. Kohli, M. J. Garabedian, and M. R. Stallcup, GRIP1, a transcriptional coactivator for the AF-2 transactivation domain of steroid, thyroid, retinoid, and vitamin D receptors, *Mol Cell Biol 17*:2735–2744 (1997).

376. J. Torchia, D. W. Rose, J. Inostroza, Y. Kamei, S. Westin, C. K. Glass, and M. G. Rosenfeld, The transcriptional co-activator p/CIP binds CBP and mediates nuclear-receptor function, *Nature 387*:677–684 (1997).

377. M. Mannervik, Y. Nibu, H. Zhang, and M. Levine, Transcriptional coregulators in development, *Science 284*:606–609 (1999).

378. V. V. Ogryzko, R. L. Schiltz, V. Russanova, B. H. Howard, and Y. Nakatani, The transcriptional coactivators p300 and CBP are histone acetyltransferases, *Cell 87*:953–959 (1996).

379. D. Chen, H. Ma, H. Hong, S. S. Koh, S-M. Huang, B. T. Schurter, D. W. Aswad, and M. R. Stallcup, Regulation of transcription by a protein methyltransferase, *Science 284*:2174–2177 (1999).

380. R. E. Kingston, A shared but complex bridge, *Nature 399*:199–200 (1999).

3

Vitamin K

JOHN W. SUTTIE

University of Wisconsin, Madison, Wisconsin

I. HISTORY

The discovery of vitamin K was the result of a series of experiments by Henrik Dam on the possible essential role of cholesterol in the diet of the chick. Dam (1) noted that chicks ingesting diets that had been extracted with nonpolar solvents to remove the sterols developed subdural or muscular hemorrhages and that blood taken from these animals clotted slowly. The disease was subsequently observed by McFarlene et al. (2), who described a clotting defect seen when chicks were fed ether-extracted fish or meat meal, and also by Holst and Halbrook (3). Studies in a number of laboratories soon demonstrated that this disease could not be cured by the administration of any of the known vitamins or other known physiologically active lipids. Dam continued to study the distribution and lipid solubility of the active component in vegetable and animal sources and in 1935 proposed (4,5) that the antihemorrhagic vitamin of the chick was a new fat-soluble vitamin, which he called vitamin K. Not only was K the first letter of the alphabet that was not used to described an existing or postulated vitamin activity at that time, but it was also the first letter of the German word *Koagulation*. Dam's reported discovery of a new vitamin was followed by an independent report of Almquist and Stokstad (6,7), describing their success in curing the hemorrhagic disease with ether extracts of alfalfa and clearly pointing out that microbial action in fish meal and bran preparations could lead to the development of antihemorrhagic activity.

The only plasma proteins involved in blood coagulation that were clearly defined at that time were prothrombin and fibrinogen, and Dam et al. (8) succeeded in preparing a crude plasma prothrombin fraction and demonstrating that its activity was decreased when it was obtained from vitamin K–deficient chick plasma. At about the same period of time, the hemorrhagic condition resulting from obstructive jaundice or biliary problems was shown to be due to poor utilization of vitamin K by these patients, and the bleeding episodes were attributed to a lack of plasma prothrombin. The prothrombin assays used at that time were not specific for prothrombin, and it was widely believed that the defect in the plasma of animals fed vitamin K–deficient diets was due solely to a lack of prothrombin. A real understanding of the various factors involved in regulating the generation of thrombin from prothrombin did not begin until the mid-1950s, and during the next 10 years, factors VII, IX, and X were discovered and shown to be dependent on vitamin K for synthesis.

A number of groups were involved in the attempts to isolate and characterize this new vitamin, and Dam's collaboration with Karrer of the University of Zurich resulted in the isolation of the vitamin from alfalfa as a yellow oil. Subsequent studies soon established that the active principle was a quinone and vitamin K_1 was characterized as 2-methyl-3-phytyl-1,4-naphthoquinone (9) and synthesized by Doisy's group in St. Louis. Their identification was confirmed by independent synthesis of this compound by Karrer et al. (10), Almquist and Klose (11), and Fieser (12). The Doisy group also isolated a form of the vitamin from putrified fish meal, which in contrast to the oil isolated from alfalfa was a crystalline product. Subsequent studies demonstrated that this compound called vitamin K_2, contained an unsaturated side chain at the 3-position of the naphthoquinone ring. Early investigators recognized that sources of the vitamin, such as putrified fish meal, contained a number of different vitamins of the K_2 series with differing chain-length polyprenyl groups at the 3-position. Early observations suggested that the alkylated forms of vitamin K could be formed in animal tissues from the parent compound, menadione. This was not definitely established until Martius and Esser (13) demonstrated that they could isolate a radioactive polyprenylated form of the vitamin from tissues of rats fed radioactive menadione. Much of the early history of the discovery of vitamin K has been reviewed by Almquist (14), and he and others (15–17) have reviewed the literature in this field shortly after discovery of the vitamin.

II. CHEMISTRY

A. Isolation

Vitamin K can be isolated from biological material by standard methods used to obtain physiologically active lipids. The isolation is always complicated by the small amount of desired product in the initial extracts. Initial extractions are usually made with the use of some type of dehydrating conditions, such as chloroform-methanol, or by first grinding the wet tissue with anhydrous sodium sulfate and then extracting it with acetone followed by hexane or ether. Large samples (kilogram quantities) of tissues can be extracted with acetone alone, and this extract is partitioned between water and hexane to obtain the crude vitamin. Small samples, such as in vitro incubation mixtures or buffered subcellular fractions, can be effectively extracted by shaking the aqueous suspension with a mixture of isopropanol and hexane. The phases can be separated by centrifugation and the upper layer analyzed directly. Methods for the efficient extraction of vitamin K from various food matrices have been developed (18) and will be useful in developing accurate food composition data.

Crude nonpolar solvent extracts of tissues contain large amounts of contaminating lipid in addition to the desired vitamin. Further purification and identification of vitamin K in this extract can be facilitated by a preliminary fractionation of the crude lipid extract on hydrated silicic acid (19). A number of the forms of the vitamin can be separated from each other and from other lipids by reversed-phase partition chromatography, as described

by Matschiner and Taggart (20). These general procedures appear to extract the majority
of vitamin K from tissues. Following separation of the total vitamin K fraction from much
of the contaminated lipid, the various forms of the vitamin can be separated by the procedures described in Sec. III.

B.   Structure and Nomenclature

The nomenclature of compounds possessing vitamin K activity has been modified a number of times since the discovery of the vitamin. The nomenclature in general use at the
present time is that most recently adopted by the IUPAC-IUB Subcommittee on Nomenclature of Quinones (21). The term *vitamin K* is used as a generic descriptor of 2-
methyl-1,4-naphthoquinone (I) and all derivatives of this compound that exhibit an
antihemorrhagic activity in animals fed a vitamin K–deficient diet. The compound 2-
methyl-3-phytyl-1,4-naphthoquinone (II) is generally called vitamin K_1, but is preferably
called phylloquinone. The USP nomenclature for phylloquinone is phytonadione. The
compound, first isolated from putrified fish meal and called at that time vitamin K_2, is
one of a series of vitamin K compounds with unsaturated side chains, called multiprenyl-
menaquinones, that are found in animal tissues and bacteria. This particular menaquinone
(2-methyl-3-farnesylgeranylgeranyl-1,4-naphthoquinone) had 7 isoprenoid units, or 35
carbons in the side chain; it was once called vitamin K_2 (35) but now is called menaqui-
none-7 (MK-7) (III). Vitamins of the menaquinone series with up to 13 prenyl groups
have been identified, as well as several partially saturated members of this series. The
parent compound of the vitamin K series, 2-methyl-1,4-naphthoquinone, has often been
called vitamin K_3 but is more commonly and correctly designated as menadione.

I   II

III

C.   Structures of Important Analogs, Commercial Forms, and Antagonists

1.   Analogs and Their Biological Activity

Following the discovery of vitamin K, a number of related compounds were synthesized
in various laboratories and their biological activity compared with that of the isolated
forms. A large number of compounds were synthesized by the Fieser group (22), and data
on the biological activity of these and other compounds have been reviewed and summarized elsewhere (23,24). The data from some of the early studies are somewhat difficult
to compare because of variations in methods of assay, but a number of generalities were
apparent rather early. Although there were early suggestions that menadione (I) might be

functioning as a vitamin, it is now usually assumed that the compound is alkylated to a biologically active menaquinone either by intestinal microorganisms or by tissue alkylating enzymes. The range of compounds that can be utilized by animals (or intestinal bacteria) is wide, and compounds such as 2-methyl-4-amino-1-naphthol or 2-methyl-1-naphthol have biological activity similar to that of menadione when fed to animals. Compounds such as the diphosphate, the disulfate, the diacetate, and the dibenzoate of reduced vitamin K series have been prepared and have been shown to have full biological activity. Most early studies compared the activities of various compounds with that of menadione. The 2-methyl group is usually considered essential for activity, and alterations at this position, such as the 2-ethyl derivative (IV), result in inactive compounds. This is not due to the inability of the 2-ethyl derivative to be alkylated, as 2-ethyl-3-phytyl-1,4-naphthoquionone is also inactive. There is some evidence from feeding experiments that the methyl group may not be absolutely essential. 2-Phytyl-1,4-naphthoquionone (V) has biological activity, but the available data do not establish that V functions as a biologically active desmethyl-phylloquinone rather than being methylated to form phylloquinone.

Studies with substituted 2-methyl-1,4-naphthoquinones have revealed that polyisoprenoid side chains are the most effective substituents at the 3-position. The biological activity of phylloquinone, 2-methyl-3-phytyl-1,4-naphthoquinone (II), is reduced by saturation of the double bond to form 2-methyl-3-(β,γ-dihydrophytyl)-1,4-naphthoquinone (VI). This compound is, however, considerably more active than 2-methyl-3-octadecyl-1,4-naphthoquinone (VII), which has an unbranched alkyl side chain of similar size. Natural phylloquinone is the trans isomer, and although there has been some confusion in the past, Matschiner and Bell (25) have shown that the cis isomer of phylloquinone (VIII) is essentially inactive. The naphthoquinone nucleus cannot be altered appreciably, as methylation to form 2,6-dimethyl-3-phytyl-1,4-naphthoquinone (IX) results in loss of activity, and the benzoquinone most closely corresponding to phylloquinone, 2,3,5-trimethyl-6-phytyl-1,4-benzoquinone, has been reported (22) to have no activity.

IV   V

VI   VII

VIII   IX

The activity of various structural analogs of vitamin K in whole-animal assay systems is, of course, a summation of the relative absorption, transport, metabolism, and effectiveness of this compound at the active site as compared with that of the reference compound. Much of the data on biological activity of various compounds were obtained by the use of an 18-h oral dose curative test utilizing vitamin K–deficient chickens. This type of assay allows metabolic alterations of the administered form to an active form of the vitamin, and significant activity of a compound in this assay may result from bioconversion to an active form. Activity varies with length of the isoprenoid side chain, and isoprenalogs with three to five isoprenoid groups have maximum activity (24) when administered orally. The lack of effectiveness of higher isoprenalogs in this type of assay may be due to the relatively poor absorptions of these compounds. Matschiner and Taggart (26) have shown that when the intracardial injection of vitamin K in deficient rats is used as a criterion, the very high molecular weight isoprenalogs of the menaquinone series are the most active; maximum activity was observed with MK-9 (Table 1). Structure–function relationships of vitamin K analogs have also been studied utilizing in vitro assays of the vitamin K–dependent γ-glutamylcarboxylase, and these will be discussed in Sec. VII.

2. Commercial Form of Vitamin K

Only a few forms of the vitamin are commercially important. The major use of vitamin K in the animal industry is in poultry diets. Chicks are very sensitive to vitamin K restriction, and antibiotics that decrease intestinal vitamin synthesis are often added to poultry diets. Supplementation is therefore required to ensure an adequate supply. Phylloquinone is too expensive for this purpose, and different forms of menadione have been used. Menadione itself possesses high biological activity in a deficient chick, but its effectiveness depends on the presence of lipids in the diet to promote absorption. There are also problems of its stability in feed products, and because of this, water-soluble forms are used. Menadione forms a water-soluble sodium bisulfite addition product, menadione sodium bisulfite (MSB) (X), which has been used commercially but which is also somewhat unstable in mixed feeds. In the presence of excess sodium bisulfite, MSB crystallizes as a complex with an additional mole of sodium bisulfite; this complex, known as menadione

Table 1 Effect of Route of Administration on Biological Activity[a]

Number of C atoms in side chain	Phylloquinone series	Menaquinone series	
	Oral (chick)	Oral (chick)	Intracardial (rat)
10	10	15	<2
15	30	40	—
20	100	100	13
25	80	>120	15
30	50	100	170
35	—	70	1700
40	—	68	—
45	—	60	2500
50	—	25	1700

[a] Data are expressed on a molar basis with phylloquinone assigned a value of 100.
Source: From Refs. 24 and 26.

sodium bisulfite complex (MSBC) (XI), has increased stability and it is widely used in the poultry industry. A third water-soluble compound is a salt formed by the addition of dimethylpyridionol to MSB; it is called menadione pyridionol bisulfite (MPB) (XII). Comparisons of the relative biopotency of these compounds have often been made on the basis of the weight of the salts rather than on the basis of menadione content, and this has caused some confusion in assessing their value in animal feeds. However, a number of studies (27–29) have indicated that MPB is somewhat more effective in chick rations than in MSBC. This form of the vitamin has also been demonstrated to be effective in swine rations (30).

The clinical use of vitamin K is largely limited to two forms. A water-soluble form of menadione, menadiol sodium diphosphate, which is sold as Kappadione or Synkayvite, is still used in some circumstances, but the danger of hyperbilirubinemia associated with menadione usage (see Sec. XI) had led to the use of phylloquinone as the desired form of the vitamin. Phylloquinone (USP phytonadione) is sold as AquaMEPHYTON, Konakion, Mephyton, and Mono-Kay.

3. Antagonists of Vitamin Action

The history of the discovery of the first antagonists of vitamin K, the coumarin derivatives, has been documented and discussed by Link (31). A hemorrhagic disease of cattle, traced to the consumption of improperly cured sweet clover hay, was described in Canada and the U.S. Midwest in the 1920s. If serious hemorrhages did not develop, animals could be aided by transfusion with whole blood from healthy animals, and by the early 1930s it was established that the cause of the prolonged clotting times was a decrease in the concentration of prothrombin in the blood. The compound present in spoiled sweet clover that was responsible for this disease had been studied by a number of investigators but was finally isolated and characterized as 3,3′-methylbis-(4-hydroxycoumarin) by Link's group (32,33) during the period from 1933 to 1941 and was called dicumarol (XIII). Dicumarol was successfully used as a clinical agent for anticoagulant therapy in some early studies, and a large number of substituted 4-hydroxycoumarins were synthesized both in Link's laboratory and elsewhere. The most successful of these, both clinically for long-term lowering of the vitamin K–dependent clotting factors and subsequently as a rodenticide, have been warfarin, 3-(α-acetonylbenzyl)-4-hydroxycoumarin (XIV), or its sodium salt; phenprocoumon, 3-(1-phenylpropyl)-4-hydroxycoumarin (XV); and ethyl biscovmacetate, 3,3′-carboxymethylenebis-(4-hydroxycoumarin) ethyl ester (XVI). The various drugs that have been used differ in the degree to which they are absorbed from the intestine, in their plasma half-life, and presumably in their effectiveness as a vitamin K antagonist in the active site. Because of this, their clinical use differs. Much of the information on the structure–activity relationships of the 4-hydroxycoumarins has been reviewed by Renk and Stoll (34). These drugs are synthetic compounds, and although the clinically used compound is the racemic mixture, studies of the two optical isomers of warfarin have shown that they differ both in their effectiveness as anticoagulants and in the influence of other drugs on their metabolism. The clinical use of these compounds and many of their pharmacodynamic interactions have been reviewed by O'Reilly (35).

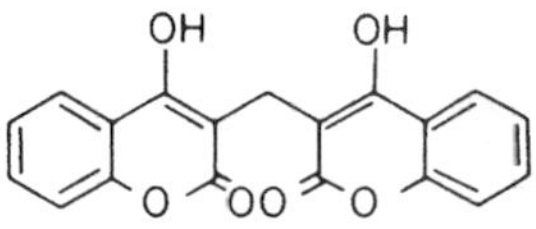

XIII 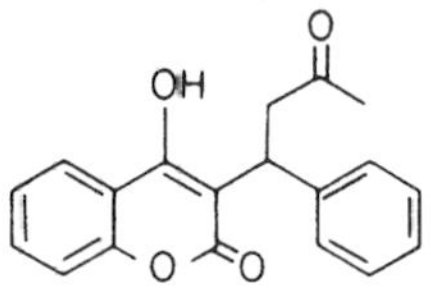   XIV

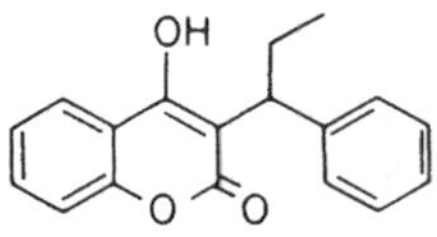

XV 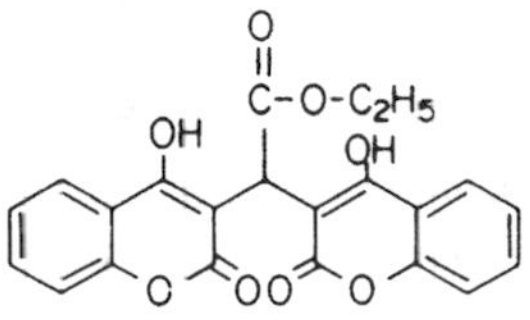   XVI

Warfarin is widely used as a rodenticide, and concern has been expressed in recent years because of the identification of anticoagulant-resistant rat populations. These were first observed in northern Europe (39) and subsequently in the United States (37). Resistance is now a significant problem in both North America (38) and Europe (39), and concern over the spread of resistance has led to the synthesis of more effective coumarin derivatives. Two of the most promising appear to be 3-(3-*p*-diphenyl-1,2,3,4-tetrahydronaphth-1-yl)-4-hydroxycoumarin, difenacoum (XVII) and 3-(3-[4′-bromobiphenyl-4-yl]-1,2,3,4-tetrahydronaphth-1-yl)-4 hydroxycoumarin, bromodifenacoum (XVIII). The genetics of resistance and the differences between various resistant strains are now better understood (40), and it appears that the problems can be brought under control.

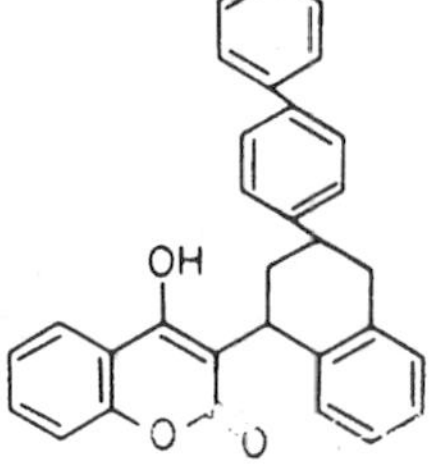

XVII 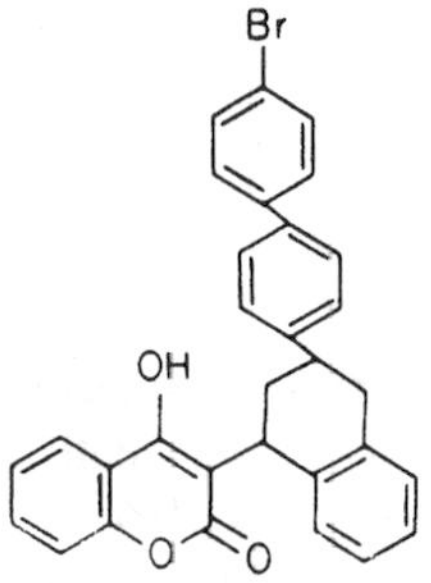   XVIII

A second class of chemical compounds having anticoagulant activity that can be reversed by vitamin K administration (41) are the 2-substituted 1,3-indandiones. Many of these compounds also have been synthesized, and two of the more commonly used members of the series have been 2-phenyl-1,3-indandione (XIX) and 2-pivalyl-1, 3-indandione (XX). These compounds have had some commercial use as rodenticides, but because of the potential for hepatic toxicity (35), they are no longer used clinically. Studies on the mechanism of action of these compounds have not been as extensive as those on the 4-hydroxycoumarins, but the observations that warfarin-resistant rats are also resistant to the indandiones and their effects on vitamin K metabolism (42) suggest that the mechanism of action of the indandiones is similar to that of the 4-hydroxycoumarins.

XIX   **XX**

During the course of a series of investigations into the structural requirements for vitamin K activity, it was shown (43) that replacement of the 2-methyl group of phylloquinones by a chlorine atom to form 2-chloro-3-phytyl-1,4-naphthoquinone (XXI) or by a bromine atom to form 2-bromo-3-phytyl-1,4-naphthoquinone resulted in compounds that were potent antagonists of vitamin K. The most active of these two compounds is the chloro derivative (commonly called chloro-K. Lowenthal (44) has shown that, in contrast to the coumarin and indandione derivatives, chloro-K acts like a true competitive inhibitor of the vitamin at its active site(s). Because its mechanism of action is distinctly different from that of the commonly used anticoagulants, chloro-K has been used to probe the mechanism of action of vitamin K, and, as it is an effective anticoagulant in coumarin anticoagulant–resistant rats (45), it has been suggested as a possible rodenticide. Lowenthal's studies of possible agonists and antagonists of vitamin K indicated (46) that, in contrast to earlier reports (22), some of the *para*-benzoquinones do have biological activity. The benzoquinone analog of vitamin K_1, 2,5,6-trimethyl-3-phytyl-1,4-benzoquinone, was found to have weak vitamin K activity, and 2-chloro-5,6-dimethyl-3-phytyl-1,4-benzoquinone (XXII) is an antagonist of the vitamin. When these compounds were modified to contain shorter isoprenoid side chains at the 3-position, they were neither agonists nor antagonists.

XXI

Two compounds that appear to be rather unrelated to either vitamin K or the coumarins and that have anticoagulant activity have recently been described. Marshall (47) has shown that 2,3,5,6-tetrachloro-4-pyridinol (XXIII) has anticoagulant activity, and on the basis of its action in warfarin-resistant rats (42), it would appear that it is functioning as a direct antagonist of the vitamin, as does chloro-K. A second series of compounds even less structurally related to the vitamin are the 6-substituted imidazole-[4-5-*b*]-pyrimidines. These compounds were described by Bang et al. (48) as antagonists of the vitamin, and the action of 6-chloro-2-trifluoromethylimidazo-[4,5-*b*]-pyrimidine (XXIV) in warfarin-resistant rats (48,49) suggests that they function in the same way as a coumarin or indandione type of compound. As is the case with potentially active forms of the vitamin, studies of vitamin K antagonists have more recently been studied utilizing in vitro assays and will be discussed in Sec. VII.

XXII

XXIII   **XXIV**

A hypoprothrombinemia can also be produced in some species by feeding animals sulfa drugs and antibiotics. There is little evidence that these compounds are doing anything other than decreasing intestinal synthesis of the vitamin by altering the intestinal flora. They should, therefore, not be considered antagonists of the vitamin.

D. Synthesis of Vitamin K

The methods used in the synthesis of vitamin K have remained essentially those originally described by Doisy's group (50), Almquist and Klose (11,51), and Fieser (12,52) in 1939. Those procedures involved the condensation of phytol or its bromide with menadiol or its salt to form the reduced addition compound, which is then oxidized to the quinone. Purification of the desired product from unreacted reagents and side products occurred either at the quinol stage or after oxidation. These reactions have been reviewed in considerable detail, as have methods to produce the specific menaquinones rather than phylloquinoone (53,54). The major side reactions in this general scheme is the formation of the cis rather than the trans isomer at the Δ^2 position and alkylation at the 2- rather than the 3-position to form the 2-methyl-2-phytyl derivative. The use of monoesters of menadiol and newer acid catalysts for the condensation step (55) is the basis for the general method of industrial preparation used at the present time. Naruta and Maruyama (56) have described a new method for the synthesis of compounds of the vitamin K series based on the coupling of polyprenyltrimethyltins to menadione. This method is a regio- and stereocontrolled synthesis that gives a high yield of the desired product. It is likely that this method may have particular utility in the synthesis of radiolabeled vitamin K for metabolic studies, as the purification of the desired product appears to be somewhat simpler than with the synthesis currently in use.

E. Physical and Chemical Properties

Compounds with vitamin K activity are substituted 1,4-naphthoquinones and, therefore, have the general chemical properties expected of all quinones. The chemistry of quinoids

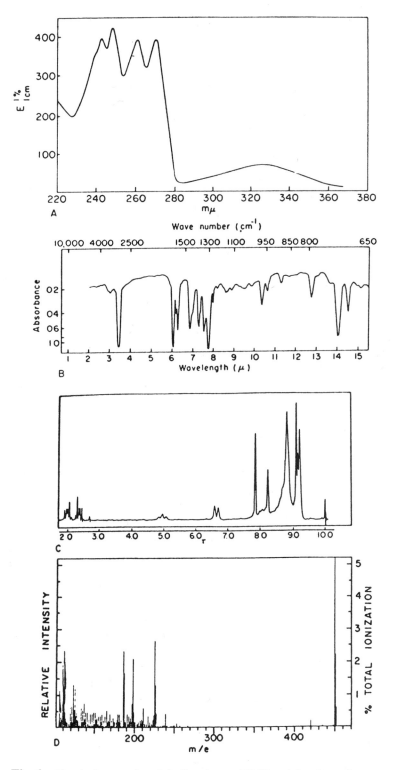

Fig. 1 Physical properties of phylloquinone. (A) Ultraviolet absorption spectra in petroleum ether. (B) Infrared absorption spectra. (C) Nuclear magnetic resonance spectra in CDCl₃ at 60 Mc. (D) Mass fragmentation spectrum of the parent molecular ion is seen at m/e 450.

has been reviewed in a book edited by Patai (57), and much of the data on the spectral and other physical characteristics of phylloquinone and the menaquinones have been summarized by Sommer and Kofler (58) and Dunphy and Brodie (59). The oxidized form of the K vitamins exhibits an ultraviolet (UV) spectrum that is characteristic of the naphthoquinone nucleus, with four distinct peaks between 240 and 280 nm and a less sharp absorption at around 320–330 nm. The extinction coefficient ($E_{1cm}^{1\%}$) decreases with chain length and has been used as a means of determining the length of the side chain. The molar extinction value ϵ for both phylloquinone and the various menaquinones is about 19,000. The absorption spectrum changes drastically upon reduction to the hydroquinone, with an enhancement of the 245-nm peak and disappearance of the 270-nm peak. These compounds also exhibit characteristic infrared and nuclear magnetic resonance (NMR) absorption spectra that are again largely those of the naphthoquinone ring. NMR analysis of phylloquinone has been used to establish firmly that natural phylloquinone is the trans isomer and can be used to establish the cis–trans ratio in synthetic mixtures of the vitamin. Mass spectroscopy has been useful in determining the length of the side chain and the degree of saturation of vitamins of the menaquinone series isolated from natural sources. The UV, infrared, NMR and mass fragmentation spectra of phylloquinone are shown in Fig. 1. Phylloquinone is an oil at room temperature; the various menaquinones can easily be crystallized from organic solvents and have melting points from 35 to 60°C, depending on the length of the isoprenoid chain.

III. ANALYTICAL PROCEDURES

Vitamin K can be analyzed by a variety of color reactions or by direct spectroscopy (58–60). Chemical reactivity of the vitamin is a function of the naphthoquinone nucleus, and as other quinones also react with many of the colorimetric assays, they lack specificity for the vitamin. The number of interfering substances present in crude extracts is such that a significant amount of separation is required before UV absorption spectra can be used to quantitate the vitamin in anything other than reagent grade preparations. These methods are therefore of little value in the determination of the small amount of vitamin present in natural sources. Early quantitation determinations depended on a biological assay, and analytical methods suitable for the small amounts of vitamin K present in tissues and most food sources have been available only recently. The separation of the extensive mixtures of menaquinones in bacteria and animal sources was first achieved with various thin-layer or paper chromatographic systems (53,58,59,61). All separations involving concentrated extracts of vitamin K should be carried out in subdued light to minimize UV decomposition of the vitamin. Compounds with vitamin K activity are also sensitive to alkali, but they are relatively stable to an oxidizing atmosphere and to heat and can be vacuum-distilled with little decomposition.

Interest in the quantitation of vitamin K in serum and animal tissues has led to an increasing emphasis on the use of high-performance liquid chromatography (HPLC) as an analytical tool to investigate vitamin K metabolism. This method was first demonstrated to be applicable to the separation of vitamin K by Williams et al. (62). The general development of HPLC techniques for the quantitation of phylloquinone and menaquinones in biological materials has been reviewed (63), and specific methods will be discussed in Secs. V and VI.

IV. BIOASSAY PROCEDURES

The classical assay for the vitamin K content of an unknown source was based on determination of the whole-blood clotting time of the chick (64) and at the present time is largely of historical interest. A small amount of material to be assayed was placed in the crop of vitamin K–deficient chicks, and the response at 20 h was compared with that of known amounts of vitamin K. When larger amounts of material were available, they were fed for 2 weeks and the degree of vitamin sufficiency compared with known amounts of the vitamin. Utilization of more sensitive plasma clotting factor assays (65,66) improves sensitivity, as does better standardization of the degree of deficiency (67,68). Some investigators have modified the degree of hypoprothrombinemia of the test chicks by anticoagulant administration (29,65,66), but this modification makes the assay somewhat insensitive to menadione and other unalkylated forms of the vitamin.

Because of the ease in producing a vitamin K deficiency in this species, chicks have most often been used in biological assays; however, rats have also been used (26). All oral bioassay procedures are complicated by the effects of different rates and extents of absorption of the desired nutrients from the various products being assayed. They have been superseded by HPLC techniques and will be used in the future only to compare the biological activity of different forms of the vitamin or to investigate nutrient interactions that modify biological activity.

V. CONTENT IN FOOD

Satisfactory tables of the vitamin K content of various commonly consumed foods have not been made available until recently. Many of the values commonly quoted have apparently been recalculated in an unspecified way from data obtained by a chick bioassay (69,70) that was not intended to be more than qualitative and should not in any sense be used to give absolute values. Tables in various texts and reviews (71–73) may also contain data from this source, as well as considerable amounts of unpublished data. A major problem has been the lack of suitable bioassays or chemical assays for the vitamin. The range of values reported for the same foods has therefore been wide and is illustrated in Table 2. The significant variation observed suggests that reporting a single value for the vitamin K content of different foods gives an erroneous impression of the confidence that should be given such a value.

Current methodology utilizes HPLC analysis of lipid extracts, and has been reported (18) to have a within-sample coefficient of variation for different foods in the range of 7–14% and a between-sample coefficient of variation of 9–45%. Although green leafy vegetables have been known for some time to be the major source of vitamin K in the diet, it is now apparent that cooking oils, particularly soybean oil and rapeseed oil (79), are major contributors. Interest in hemorrhagic diseases of the newborn has led to a determination of phylloquinone in cow (80) and human milk (80–82) and in infant formulas (80). Human milk contains only 1–2 ng of phylloquinone per milliliter, which is somewhat less than that found in cow's milk. Infant formulas are currently supplemented with vitamin K, providing a much higher intake than that provided by breast milk.

The data in Table 3 are taken from a survey of literature (83), which considered most of the reported HPLC derived values for various food items and from analyses of the FDA total diet study (83a). In general, green and/or leafy vegetables are the best sources of the vitamin, with cooking oils being the next major sources. Due to the low

Table 2 Determination of Vitamin K by Bioassay and Chemical Methods

Food	Vitamin K (µg per 100 g fresh)				
	Bioassay		Chemical assay		
	Ia	IIb	IIIc	IVd	Ve
Beans, green	14	22	46	—	—
Broccoli	20	65	147	230	—
Cabbage	95	37	110	110	61
Cauliflower	136	<10	27	—	—
Carrots	20	—	5	—	—
Peas	19	—	39	50	—
Potatoes, white	4	—	<1	—	—
Spinach	177	130	415	260	231
Tomato, ripe	11	—	6	—	—

[a] Doisy (75), chick bioassay of fresh food with phylloquinone as a standard.
[b] Richardson et al. (74), chick bioassay of frozen food with menadione as a standard.
[c] Shearer et al. (76), HPLC assay for phylloquinone.
[d] Kodaka et al. (77), HPLC assay for phylloquinone.
[e] Seifert (78), gas chromatography assay of phylloquinone; data have been recalculated on a dry weight basis.

Table 3 Vitamin K Content of Ordinary Foods[a]

µg Phylloquinone/100 g of edible portion					
Vegetables		Nuts, Oils, Seeds		Fruits	
Kale	817	Soybean oil	193	Avocado	40
Parsley	540	Rapeseed oil	141	Grapes	3
Spinach	400	Olive oil	49	Cantaloupe	1
Endive	231	Walnut oil	15	Bananas	0.5
Green onions	207	Safflower oil	11	Apples	0.1
Broccoli	205	Sunflower oil	9	Oranges	0.1
Brussels sprouts	177	Corn oil	3		
Cabbage	147	Dry soybeans	47	Meat & Dairy	
Lettuce	122	Dry kidney beans	19	Ground beef	0.5
Green beans	47	Sesame seeds	8	Chicken	0.1
Peas	36	Dry navy beans	2	Pork	<0.1
Cucumbers	19	Raw peanuts	0.2	Turkey	<0.1
Tomatoes	6			Tuna	<0.1
Carrots	5	Grains		Butter	7
Cauliflower	5	Bread	3	Cheddar cheese	3
Beets	3	Oat meal	3	3.5% Milk	0.3
Onions	2	White rice	1	Yogurt	0.3
Potatoes	0.8	Wheat flour	0.6	Skim milk	<0.1
Sweet corn	0.5	Dry spaghetti	0.2	Mayonnaise	81
Mushrooms	<0.1	Shredded wheat	0.7	Egg yolk	2
		Corn flakes	<0.1	Egg white	<0.1

[a] Values are taken from a provisional table (83) and are median values from a compilation of reported assays. Only HPLC assays are included in the database.

dietary requirement for vitamin K, a few servings of foods with high vitamin K content contribute a significant amount toward satisfying this requirement, and an uncomplicated dietary deficiency is unlikely. Utilizing current food composition data and 14-day food diaries obtained from 2000 households, the mean adult phylloquinone intake in the United States has been determined to be 80 μg for men and 73 μg for women (83b). Dihydrophylloquinone formed during hydrogenation of cooking oils has lower biological activity and is present in the U.S. diet at about 20% of the amount of phylloquinone. Because most dietary vitamin K is present in a few foods, it is possible (84) to construct palatable research diets that meet the recommended dietary allowances for other nutrients but that have a phylloquinone content of less than 10 μg/day.

The vitamin K content of individual food items is probably more variable than that of many other nutrients. Significant changes in the vitamin K content of plants during growth and maturity and at various geographical locations have been reported (85). The effect of food processing and cooking on vitamin K content has not been carefully considered. Richardson et al. (74) have investigated the effect of heat processing (canning) or of sterilization by ionizing radiation on the vitamin K content of six vegetables. The vitamin K content of these foods, as determined by chick bioassay, did not differ consistently from that of the fresh frozen food, nor did storage of any of the products for 15 months have a significant effect. Phylloquinone in cooking oils has been shown to be relatively stable to heat but rapidly destroyed by both daylight and fluorescent light (79). Other than being affected by light, vitamin K present in food is most likely relatively stable; however, only limited data are available.

VI. METABOLISM

A. Absorption

The absorption of nonpolar lipids, such as vitamin K, into the lymphatic system depends on their incorporation into mixed micelles, and optimal formation of these micellar structures requires the presence of both bile and pancreatic juice (86,87). Normal subjects have been reported (88) to excrete less than 20% of a radiolabeled 1-mg dose of phylloquinone in the feces, but as much as 70–80% of the ingested phylloquinone was excreted unaltered in the feces of patients with impaired fat absorption caused by obstructive jaundice, pancreatic insufficiency, or adult celiac disease. Much of the fecal radioactivity found in studies such as these is due to polar biliary excretion products rather than unmetabolized phylloquinone.

Very little is known about the absorption of vitamin K from different food sources. Vitamin K in spinach (89) was found to be absorbed 13.3% as well as from a commercial detergent-solubilized preparation (Konakion) of phylloquinone when 25 g of butter was included with the meal, but only 1.4% as well in the absence of butter. A second study (89a) also indicates the phylloquinone in food sources is only 15–20% as bioavailable as added phylloquinone. These limited data suggest that the absorption of vitamin K is probably low but may be highly dependent on food sources and total dietary composition.

The potential for absorption of the large-bowel menaquinone pool has been reviewed (90,91). Early, rather nonphysiological, experiments (92) demonstrated that MK-9 disappeared from the lumen of the isolated rat ileum or large intestine but did not document its appearance in plasma or lymph. Ichihashi et al. (93) have shown that in the presence of bile MK-9 is absorbed via the lymphatic pathway from rat jejunum but that in the absence of bile no uptake of MK-9 from the colon to lymph or blood occurred within 6 h. The oral administration of 1 mg mixed long-chain menaquinones to anticoagulated human subjects has been shown (94) to effectively decrease the extent of the acquired hypoprothrombinemia. This demonstrates that the human digestive tract can absorb these more hydrophobic forms of the vitamin when they are presented to the small intestine but does not address their absorption from the large bowel. In addition, a small but nutritionally significant portion of the intestinal content of the vitamin is located not in the large bowel but in a region where bile acid–mediated absorption could occur (95).

Menadione is widely used in poultry, swine, and laboratory animal diets as a source of vitamin K. It can be absorbed from both the small intestine and the colon by a passive process (96,97), and after absorption it can be alkylated to MK-4, a biologically active form of the vitamin.

B. Transport

The lymphatic system has been demonstrated to be the major route of transport of absorbed phylloquinone from the intestine. There is no evidence that the vitamin in the lymph is in any way modified. Phylloquinone is transported in chylomicrons and other low-density lipoprotein particles (91), and its clearance is affected by apolipoprotein E polymorphism (98). Hyperlipidemic patients have elevated phylloquinone concentrations, and this increase is associated with the triglyceride-rich very low-density lipoprotein (VLDL) fraction of the lipoprotein pool (91,98). The clearance of an injected dose of radioactive phylloquinone from plasma has been investigated in humans (99,100) and shown to consist of a two-phase exponential decline in radioactivity. The first phase had a half-life of 20–30 min and the second a half-life of 120–165 min. Although the body pool of vitamin K cannot be calculated from such data, it can be calculated (99) that the total body pool of vitamin K is replaced approximately every 2.5 h.

C. Plasma Concentrations

The resolution provided by the development of HPLC system (63) and the increased sensitivity provided by newer methods of detection have made it possible to obtain a quantitative measure of the amount of vitamin K in serum or plasma. The clinical significance of these measurements is not yet fully established, and little information on the relationship between dietary intake and serum levels is available. Measurements of endogenous serum phylloquinone concentrations (0.5–2 ng/mL) require a preliminary semipreparative column to rid the sample of contaminating lipids followed by an analytical column. The chief alterations and improvement in methodology in recent years have been associated with the use of different methods of detection. Early methods utilized UV detectors, which lack sensitivity, and electrochemical detection or fluorescence detection of the vitamin following chemical or electrochemical reduction have replaced this methodology. These techniques have been recently reviewed (101,102).

Initial reports of plasma or serum phylloquinone concentrations were probably too high, and it now appears (103,104) that normal fasting values are in the region of 0.5 ng/mL (1.1 nmol/L). This is a lower concentration than is found for the other fat-soluble vitamins. Although concentrations of individual menaquinones were at one time thought too low to be measured, this does not appear to be the case. The combined plasma concentration of MK-7 and MK-8 in three studies (105–107) are in the same range as the concentration of phylloquinone (108), and substantial amounts of plasma MK-6 have been reported by other investigators (109,110). The significance of these observations is not yet known. Circulating concentrations of phylloquinone vary with dietary intake, and in animal (111,112) or human (113–115) studies where vitamin K intake has been restricted or in patients with decreased food intake (116) there is a rapid decrease in plasma vitamin K concentrations.

D. Tissue Deposition and Storage

Early studies of the distribution of dietary vitamin K in tissues were hampered by the low specific radioactivity of the vitamin available. These studies, which utilized milligram quantities of vitamin K injected into rats, indicated that phylloquinone was specifically concentrated and retained in the liver but that menadione was poorly retained in this organ. Menadione was found to be widely distributed in all tissues and to be very rapidly excreted. The metabolism of more physiological doses of the vitamin has also been studied, and significant differences in the distribution of phylloquinone and menadione have been seen (117,118). About 50% of a 10-μg dose of phylloquinone administered to a rat was found in liver at 3 h, but only 2% of a 2-μg dose of menadiol diphosphate was found in the liver at that time. The half-life of phylloquinone in rat liver is in the range of 10–15 h (111,112,119), so hepatic concentrations are greatly dependent on recent intake. Vitamin K distribution has also been studied (120) by the technique of whole-body radiography following administration of radioactive menadione, phylloquinone, or MK-4. Essentially equal body distribution of phylloquinone and MK-4 were found at 24 h. Radioactive menadione was spread over the whole body much faster than the other two compounds, but the amount retained in the tissues was low. Whole-body radiography also confirmed that vitamin K is concentrated by organs other than liver, and the highest activity was seen in the adrenal glands, lungs, bone marrow, kidneys, and lymph nodes.

Early studies of vitamin K distribution in the liver (121) depended on bioassay of isolated fractions of normal liver and indicated that the vitamin was distributed in all subcellular organelles. When 0.02 or 3μg of phylloquinone was injected (122) into vitamin K–deficient rats, more than 50% of the liver radioactivity was recovered in the microsomal fraction and substantial amounts were found in the mitochondria and cellular debris fractions. The specific activity (picomoles vitamin K per milligram protein) of injected radioactive phylloquinone has been studied (118), and only the mitochondrial and microsomal fractions had a specific activity that was enriched over that of the entire homogenate, with the highest activity in the microsomal fraction. The less biologically active cis isomer of phylloquinone was found to be concentrated in liver microsomes (25), but the mitochondrial fraction had the highest specific activity. Nyquist et al. (123) found the highest specific activity (dpm/mg protein) of radioactive phylloquinone to be in the Golgi and smooth microsomal membrane fractions. Only limited data on the distribution of menaquinones are available, and MK-9 has been reported (124) to be preferentially localized in a mitochondrial rather than a microsomal subcellular fraction. Factors influencing intracellular distribution of the vitamin are not well understood, and only preliminary evidence of an intracellular vitamin K–binding protein that might facilitate intraorganelle movement has been presented (125). The vitamin does appear to be lost more rapidly from the cytosol than from membrane fractions as a deficiency develops (126).

Because of the small amounts of vitamin K in animal tissues, it has been difficult to determine which of the vitamers are present in tissue from different species. Only limited data are available, and they have been reviewed by Matschiner (127,128). These data, obtained largely by thin-layer chromatography, indicate that phylloquinone is found in the liver of those species ingesting plant material and that, in addition to this, menaquinones containing 6–13 prenyl units in the alkyl chain are found in the liver of most species. Some of these long-chain vitamers appear to be partially saturated. More recently, analysis of a limited number of human liver specimens has shown that phylloquinone represents only about 10% of the total vitamin K pool and that a broad mixture (Table 4) of menaquinones is present. The predominant forms appear to be MK-7, MK-8, MK-10, and MK-11. Kayata et al. (131) have reported that the hepatic menaquinone content of five 24-month-old infants was approximately sixfold higher than that of three infants less than 2 weeks of age, and another study (132) failed to find menaquinones in neonatal livers. Although the long-chain menaquinones are potential sources of vitamin K activity in liver, the extent to which they are utilized is not known. A recent study (124) has demonstrated that the utilization of MK-9 as a substrate for the vitamin K–dependent carboxylase is only about 20% as extensive as phylloquinone when the two compounds are present in the liver in equal concentrations. Recent data have also suggested that MK-4 may play a unique role in satisfying the vitamin K requirement of some species or tissues. Most analyses of liver from various species have not detected significant amounts of MK-4. However, chicken liver has been shown (133,134) to contain more MK-4 than phylloquinone, and some nonhepatic tissues of the rat have been shown (135) to contain much more MK-4 than phylloquinone. The significance of these observations and the possibility that MK-4 may play a unique role in some aspect of vitamin K nutrition is not yet known.

E. Metabolism

1. Alkylation of Menadione

Animals cannot synthesize the naphthoquinone ring, and this portion of the vitamin must be furnished in the diet. Bacteria and plants synthesize this aromatic ring system from shikimic acid, and these pathways have been reviewed (136). It appears that bacterial synthesis of menaquinones does not usually proceed through free menadione as an intermediate, but rather 1,4-dihydroxy-2-naphthoic acid is prenylated, decarboxylated, and then methylated to form the menaquinones. The transformation of menadione to MK-4 in animal tissues was first observed by Martius and Esser (13). Radioactive menadione was used (137) to establish that menadione could be converted to a more lipophilic compound that, on the basis of limited characterization, appeared to be MK-4. It was then demonstrated (138) that menadione could be converted to MK-4 by an in vitro incubation of rat or chick liver homogenates with geranylgeranyl pyrophosphate. The activity in chick liver was much higher than that in the rat. These studies have been extended (139), and it has been demonstrated that other isoprenoid pyrophosphates could serve as alkyl donors for menaquinone synthesis. Early studies (140) also demonstrated that administered phylloquinone or other alkylated forms could be converted to MK-4. It was originally believed that the dealkylation and subsequent realkylation with a geranylgeranyl side chain occurred in the liver, but subsequently it was concluded that phylloquinone was not converted to MK-4 unless it was administered orally. This suggested that intestinal bacterial action

Table 4 Phylloquionone and Menaquinone Content of Human Liver

K[b]						Vitamin K (pmol/g liver)[a]							
	MK-5	MK-6	MK-7	MK-8	MK-9	MK-10	MK-11	MK-12	MK-13	Ref.
22 ± 5	12 ± 18	12 ± 13	57 ± 59	95 ± 157	2 ± 4	67 ± 71	90 ± 74	15 ± 13	5 ± 6	(129)
18 ± 4	NR	NR	122 ± 61	11 ± 2	4 ± 2	96 ± 16	94 ± 26	21 ± 6	8 ± 3	(130)
28 ± 4	NR	NR	34 ± 12	9 ± 2	2 ± 1	75 ± 10	99 ± 15	14 ± 2	5 ± 1	(116)

[a] Values are mean ± SEM for 6 or 7 subjects in each study. Values from Refs. 129 and 130 have been recalculated from data presented as ng/g liver. NR, not reported.
[b] Phylloquinone.

was required for the dealkylation step. More recent studies (133–135) have demonstrated that the phylloquinone-to-MK-4 conversion is very extensive in some tissues and it has been demonstrated that gut bacteria are not needed for this conversion (135a,135b).

2. Metabolic Degradation and Excretion

Menadione metabolism has been studied in both whole animals (142,143) and isolated perfused livers (135). The phosphate, sulfate, and glucuronide of menadiol have been identified in urine and bile, and studies with hepatectomized rats (145) have suggested that extrahepatic metabolism is significant. Early studies of phylloquinone metabolism (146) demonstrated that the major route of excretion was in the feces and that very little unmetabolized phylloquinone was present. Wiss and Gloor (147) observed that the side chains of phylloquinone and MK-4 were shortened by the rat to seven carbon atoms, yielding a carboxylic acid group at the end that cyclized to form a γ-lactone. This lactone was excreted in the urine, presumably as a glucuronic acid conjugate. The metabolism of radioactive phylloquinone has now been studied in humans by Shearer's group (88). They found that about 20% of an injected dose of either 1 mg or 45 μg of vitamin K was excreted in the urine in 3 days, and that 40–50% was excreted in the feces via the bile. Two different aglycones of phylloquinone were tentatively identified as the 5- and 7-carbon side-chain carboxylic acid derivatives, respectively. It was concluded that the γ-lactone previously identified was an artifact formed by the acidic extraction conditions used in previous studies.

More recent studies of vitamin K metabolism have been conducted following the discovery of the significance of the 2,3-epoxide of the vitamin (XXV). This metabolite, commonly called vitamin K oxide, was discovered by Matschiner et al. (148), who were investigating an observation (122) that warfarin treatment caused a buildup of liver phylloquinone. This increase in radioactive vitamin was shown to be due to the presence of a significant amount of a metabolite more polar than phylloquinone that was isolated and characterized as phylloquinone 2,3-epoxide. Further studies of this compound (149) revealed that about 10% of the vitamin K in the liver of a normal rat is present as the epoxide and that this can become the predominant form of the vitamin following treatment with coumarin anticoagulants.

XXV

As might be expected from the effects of coumarin anticoagulants on tissue metabolism of vitamin K, anticoagulant treatment has a profound effect on vitamin K excretion. Disappearance of radiolabeled vitamin K from the plasma is not significantly altered by coumarin anticoagulant administration, but the ratio of plasma vitamin K epoxide to vitamin K does increase drastically under these conditions. Warfarin administration also greatly increases urinary excretion and decreases fecal excretion of phylloquinone (150) The distribution of the various urinary metabolites of phylloquinone is also substantially altered by warfarin administration. The amounts of the two major urinary glucuronides that have tentatively been identified as the conjugates of 2-methyl-3-(5'-carboxy-3'-methyl-2'pentenl)-1,4-naphthoquinone (XXVI) and 2-methyl-3-(3'-carboxy-3'-methylpropyl)-

1,4-naphthoquinone (XXVII) are substantially decreased, and three new metabolites appear (151).

          XXVI   **XXVII**

It is likely that there are a number of excretion products of vitamin K that have not yet been identified. Matschiner (127) has fed double-labeled phylloquinone—a mixture of [6,7-^{3}H]phylloquinone and [^{14}C]phylloquinone-(phytyl-U) with a ^{3}H/^{14}C ratio of 66—to vitamin K–deficient rats and recovered radioactivity in the urine with a ^{3}H/^{14}C ratio of 264. This ratio is higher than would be expected for the excretion of a metabolite with seven carbons remaining in the side chain and is more nearly that expected from the excretion of the metabolite with five carbons remaining in the side chain. These data suggest that some more extensively degraded metabolites of phylloquinone are also formed. An exchange of the phytyl side chain of phylloquinone for the tetraisoprenoid chain of MK-4 before degradative metabolism would also be expected to contribute to an increase in the ^{3}H/^{14}C ratio, and current data suggest that this exchange may occur. It is doubtful, however, that this exchange is extensive enough to account for the increased ratio of naphthoquinone ring atoms to side-chain atoms that was observed in this study.

The available data relating to vitamin K metabolism therefore suggest that menadione is rapidly metabolized and excreted, and that only a relatively minor portion of this synthetic form of the vitamin gets converted to biologically active MK-4. The degradative metabolism of phylloquinone and the menaquinones is much slower. The major products appear to have been identified, but there may be a number of urinary and biliary products not yet characterized. The ratio of the various metabolites formed is drastically altered by anticoagulant administration. However, there is no evidence to indicate that the isoprenoid forms of the vitamin must be subjected to any metabolic transformation before they serve as a cofactor for the vitamin K–dependent carboxylases.

VII. BIOCHEMICAL FUNCTION

A. Vitamin K–Dependent Clotting Factors

Soon after Dam's discovery of a hemorrhagic condition in chicks that could be cured by certain plant extracts, it was demonstrated that the plasma of these chicks contained a decreased concentration of prothrombin. At this time, the complex series of reactions involved in the conversion of circulating fibrinogen to a fibrin clot were poorly understood. A basis for a clear understanding of the role of the multitude of factors involved in coagulation came with the realization that many of the proteins involved could be looked at as zymogens that could be activated by a specific protease and that these modified proteins could in turn activate still other zymogens. This led to the development of a generalized "cascade" or "waterfall" theory of blood coagulation, which has now been shown (152) to be an oversimplification. As shown in Fig. 2, the key step is the activation of prothrombin (factor II) to thrombin by the activated form of factor X (factor Xa). Factor X can be

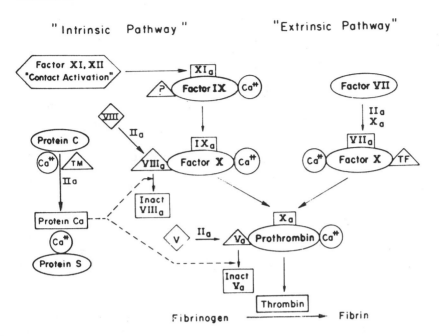

Fig. 2   Involvement of vitamin K–dependent clotting factors in coagulation. The vitamin K–dependent proteins (ellipses) circulate as inactive forms of serine proteases until converted to their active (subscript a) forms. These conversions occur in stages where an active protease, a substrate, and a protein cofactor (triangles) form a Ca^{2+}-mediated associated with a phospholipid surface. The protein cofactors V and VII are activated by thrombin (II_a) to achieve their full activity. The clotting system is traditionally divided into two pathways: the extrinsic pathway, which involves a tissue factor (TF) in addition to blood components, and an intrinsic pathway, which involves components present in the blood. Protein C is activated by II_a in the presence of an endothelial cell protein called thrombomodulin (TM). Protein C is not a procoagulant but rather functions in a complex with protein S to inactivate V_a and VIII_a.

activated by two pathways involving the vitamin K–dependent proteins factor VII and factor IX. These activations are carried out in a Ca^{2+}/phospholipid-dependent manner, and the enzymology of these conversions is now well understood. The four proteins involved in thrombin generation were collectively called the "vitamin K–dependent clotting factors" for 25 years before protein C and protein S were discovered. These two proteins have an anticoagulant role as they are able to inactivate the accessory proteins, factor Va and factor VIIIa (153). A seventh vitamin K–dependent protein, termed protein Z, has been described but its function is not known.

There was a period of about 40 years between the discovery of vitamin K and the elucidation of the metabolic function of the vitamin. Theories of the participation of vitamin K in oxidative phosphorylation somehow being related to the synthesis of specific proteins could not be substantiated, nor could theories of the vitamin-controlled production of specific proteins at a transcriptional level be proven. Involvement of an intracellular precursor in the biosynthesis of prothrombin was first clearly stated by Hemker et al. (154) who postulated that an abnormal clotting time in anticoagulant-treated patients was due to a circulating inactive form of plasma prothrombin. Direct evidence of the presence of a liver precursor protein was obtained when Shah and Suttie (155) demonstrated that the prothrombin produced when hypoprothrombinemic rats were given vitamin K and cycloheximide was not radiolabeled if radioactive amino acids were administered at the same time as the vitamin. These observations were consistent with the presence of a hepatic precursor protein pool in the hypoprothrombinemic rat that was rapidly being synthesized and that could be converted to prothrombin in a step that did not require protein synthesis. This hypothesis was strengthened by direct observations (156) that the plasma of patients treated with coumarin anticoagulants contained a protein that was antigenically similar to prothrombin but lacked biological activity. A similar protein was first demonstrated in bovine plasma by Stenflo (157), but it appears (158) to be present in low concentrations or altogether absent in other species.

Studies of this "abnormal" prothrombin (159) demonstrated that it contained normal thrombin, had the same molecular weight and amino acid composition, but did not absorb to insoluble barium salts as did normal prothrombin. This difference, and the altered calcium-dependent electrophoretic and immunochemical properties, suggested a difference in calcium binding properties of these two proteins that was subsequently demonstrated by direct calcium binding measurements. The critical difference in the two proteins was the inability of the abnormal protein to bind to calcium ions, which are needed for the phospholipid-stimulated activation of prothrombin by factor X (160). These studies of the abnormal prothrombin clearly implicated the calcium binding region of prothrombin as the vitamin K–dependent region. Acidic, Ca^{2+}-binding peptides could be isolated from a tryptic digest of the fragment 1 region of normal bovine prothrombin but could not be obtained when similar isolation procedures were applied to preparations of abnormal prothrombin. The nature of the vitamin K–dependent modification was elucidated when Stenflo et al. (161) succeeded in isolating an acidic tetrapeptide (residues 6–9 of prothrombin) and demonstrating that the glutamic acid residues of this peptide were modified so that they were present as γ-carboxyglutamic acid (3-amino-1,1,3-propanetricarboxylic acid) residues (Fig. 3). Nelsestuen et al. (162) independently characterized γ-carboxyglutamic acid (Gla) from a dipeptide (residues 33 and 34 of prothrombin), and these characterizations of the modified glutamic acid residues in prothrombin were confirmed by Magnusson

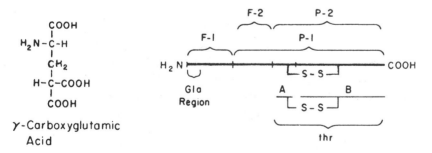

Fig. 3 Structure of γ-carboxyglutamic acid (Gla) and a diagramatic representation of the prothrombin molecule. Specific proteolysis of prothrombin by factor X_a and thrombin will cleave prothrombin into the specific large peptides shown: fragment 1 (F-1), fragment 2 (F-2), prethrombin 1 (P-1), prethrombin 2 (P-2), and thrombin (thr). For details of the activation of prothrombin to thrombin and for sequences of vitamin K–dependent proteins, see Ref. (146). The Gla residues in bovine prothrombin are located at residues 7, 8, 15, 17, 20, 21, 26, 27, 30, and 33, and they occupy homologous positions in the other vitamin K–dependent plasma proteins.

et al. (163), who demonstrated that all 10 Glu residues in the first 33 residues of prothrombin are modified in this fashion.

B. Vitamin K–Dependent Carboxylase

After the vitamin K–dependent step in prothrombin synthesis was shown to be the formation of γ-carboxyglutamic acid residues, Esmon et al. (164) demonstrated that the addition of vitamin K and $H^{14}CO_3$ to vitamin K–deficient rat liver microsomal preparations resulted in the fixation of CO_2 into microsomal proteins. It was possible to isolate radioactive prothrombin from this incubation mixture and show that essentially all of the incorporated radioactivity was present as γ-carboxyglutamic acid residues in the fragment 1 region of prothrombin. These observations would appear to offer final proof of the biochemical role of vitamin K as a cofactor for this microsomal glutamyl carboxylase (Fig. 4).

This enzyme activity was soon shown to be active when solubilized in a number of detergents, and it was demonstrated (165) that the pentapeptide Phe-Leu-Glu-Glu-Val would serve as a substrate for this enzyme. Most subsequent studies have utilized this or a similar peptide substrate, rather than the endogenous microsomal substrates, to study enzyme activity. The rough microsomal fraction of liver is highly enriched in carboxylase activity, and lower but significant levels are found in smooth microsomes. Mitochondria, nuclei, and cytosol have negligible activities. The data obtained from protease sensitivity studies are consistent with the hypothesis that the carboxylation event occurs on the lumen side of the rough endoplasmic reticulum (166). Details of the properties of the enzyme can be found in a number of reviews (167–173). This carboxylation reaction does not require ATP, and the available data are consistent with the view that the energy to drive this carboxylation reaction is derived from the reoxidation of the reduced form of vitamin K. This unique carboxylase requires O_2, and studies ruled out the involvement of biotin in the system. These findings and a direct study of the CO_2/HCO_3^- requirement indicate that carbon dioxide rather than HCO_3^- is the active species in the reaction. The vitamin K antagonist 2-chloro-3-phytyl-1,4-naphthoquinone is an effective inhibitor of the carboxylase, and the reduced form of this analog has been shown to be competitive with the reduced vitamin site. Polychlorinated phenols are strong inhibitors, and substitution of a trifluoromethyl group, a hydroxymethyl group, or a methoxymethyl group at the 2-position also results in inhibitory compounds (174).

A review of studies of the substrate specificity at the vitamin site of the carboxylase suggests that the only important structural features of this substrate in a detergent-solubilized system are a 2-methyl-1,4-naphthoquinone substituted at the 3-position with a somewhat hydrophobic group (174). Methyl substitution of the benzenoid ring has little effect or increases the apparent K_m for the reduced vitamin. A large number of low molecular

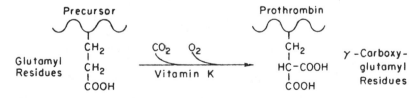

Fig. 4   The vitamin K–dependent carboxylation reaction.

weight peptide substrates of the enzyme have been synthesized, and their assay has failed to reveal any unique sequence needed as a signal for carboxylation. In general, peptides with Glu-Glu sequences are better substrates than those with single Glu residues. Gln, D-Glu, or Homo-Glu residues have been demonstrated to be noncarboxylated residues, while Asp residues are poorly carboxylated. Why only the first of the two adjacent Glu residues in these synthetic, nonphysiological substrates is carboxylated by the enzyme is not yet apparent. A consensus sequence within the Gla region that may be important for efficient carboxylation has been identified (175), but its significance is not known. The manner by which the enzyme carboxylates its normal physiological protein substrates is not known. Data obtained by characterizing the partially γ-carboxylated forms of prothrombin secreted by a dicoumarol-treated cow would suggest that carboxylation may begin at the most amino-terminal of the 10 potential Gla sites in prothrombin and move toward the more carboxy terminal sites (176). However, an in vitro study utilizing des-γ-carboxyosteocalcin has shown carboxylation preferentially occurring at the most carboxy terminal of the three potential Gla sites (177), and this aspect of the carboxylation reaction is not understood. The stereochemistry of the reactions at the glutamyl residue has also been determined (178). The incorporation of *threo-* or *erythro-*γ-fluoroglutamate into a peptide substrate results in a stereospecific hydrogen abstraction that corresponds to the elimination of the 4-pro-S-hydrogen of the glutamyl residues. Studies with tritium-labeled substrates have also established that the hydrogen exchange catalyzed by the enzyme in the absence of CO_2 proceeds with a stereospecific abstraction of the same 4-pro-S-hydrogen of the glutamyl residue that is eliminated in the carboxylation reaction. The stereochemistry associated with the addition of CO_2 to the active intermediate to form Gla has been determined by the demonstration (179) that carboxylation of 4-S-fluoroglutamate proceeds with inversion of configuration.

The substrate Phe-Leu-Glu-Glu-Leu, tritiated at the γ carbon of each Glu residue, was used (180) to demonstrate that the enzyme catalyzed a vitamin KH_2-dependent and O_2-dependent, but CO_2-independent release of tritium from this substrate, thus establishing the role of the vitamin in removing the γ hydrogen of the Glu substrate. The 2,3-epoxide of vitamin K is a coproduct of Gla formation, and at saturating concentrations of CO_2 there is an apparent equivalent stoichiometry between epoxide formation and Gla formation (181). At lower CO_2 concentrations a large excess of vitamin K epoxide is produced. How epoxide formation is coupled to γ-hydrogen abstraction has not yet been firmly established, but presumably involves an oxygenated intermediate that is on the pathway to epoxide formation. This enzyme will catalyze a vitamin KH_2- and oxygen-dependent exchange of ^3H from ^3H_2O into the γ position of a Glu residue of a peptide substrate (182,183). Exchange of ^3H from water with the γ-carbon hydrogen is decreased as the concentration of HCO_3^- in the media is increased. Studies utilizing γ-^3H-labeled Glu substrates have also demonstrated a close association between epoxide formation, Gla formation, and γ-C—H bond cleavage. The efficiency of the carboxylation reaction, Gla/γ-C—H bond cleaved, is independent of Glu substrate concentration, and the data suggest that this ratio approaches unity at high CO_2 concentrations (184). The available data are consistent with the model shown in Fig. 5, which indicates that the role of vitamin K is to abstract the γ-methyl hydrogen to leave a carbanion.

A major gap in an understanding of the mechanism of action of this enzyme has been the assumption that abstraction of the γ-methyl hydrogen requires a strong base (presumably formed from the vitamin) and the lack of evidence for such an intermediate. A series of reports by Dowd and co-workers (185,186) has suggested that an initial attack

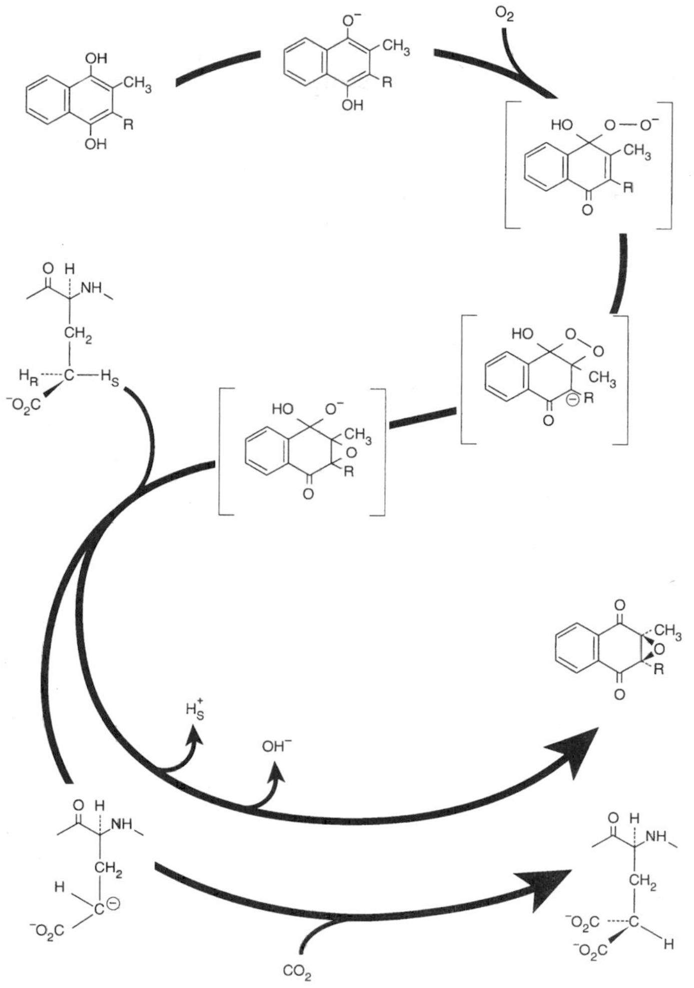

Fig. 5 Proposed mechanism of action for the vitamin K–dependent carboxylase enzyme. The available data support the interaction of O_2 with the reduced form of vitamin K to form an oxygenated intermediate that is sufficiently basic to abstract the γ-hydrogen of the glutamyl residue. The products of this reaction would be vitamin K epoxide and a glutamyl carbanion. The bracketed peroxy, dioxetane, and alkoxide intermediates have not been identified in the enzyme-catalyzed reaction but are postulated based on model organic reactions.

of O_2 at the carbonyl carbon adjacent to the methyl group results in the formation of a dioxetane ring, which generates an alkoxide intermediate. This intermediate is hypothesized to be the strong base that abstracts the γ-methyl hydrogen. This pathway leads to the possibility that a second atom of molecular oxygen can be incorporated into the carbonyl group of the epoxide product, and this activity can be followed by utilization of either $^{18}O_2$ or vitamin K with ^{18}O incorporated into the carbonyl oxygens in the reaction. This partial dioxygenase activity of the carboxylase has been verified by a second group (187), and although the general scheme (188) shown in Fig. 5 is consistent with all of the available data, the mechanism remains a hypothesis at this time.

Normal functioning of the vitamin K–dependent carboxylase poses an interesting question in terms of enzyme–substrate recognition. This microsomal enzyme recognizes a small fraction of the total hepatic secretory protein pool and then carboxylates 9–12 Glu sites in the first 45 residues of these proteins. Cloning of the vitamin K–dependent proteins has revealed that the primary gene products contain a very homologous "propeptide" between the amino terminus of the mature protein and the signal peptide (189). This region appears from both early in vitro and in vivo lines of evidence (190) to be a "docking" or "recognition site" for the enzyme. This domain of the carboxylase substrates has also been shown (191) to be a modulator of the activity of the enzyme by decreasing the apparent K_m of the Glu site substrate. The structural features of the propeptide domain that are important in stimulating the activity of the enzyme have been demonstrated (192) to be the same as those that have been shown to be involved in targeting these proteins for carboxylation. This propeptide domain is undoubtedly of major importance in directing the efficient carboxylation of the multiple Glu sites in these substrates.

Significant progress toward a detailed understanding of the properties of the vitamin K–dependent carboxylase were limited for years by the lack of a purified enzyme. A preparation that has 500–1000 times the specific activity of microsomes and can be routinely prepared in 2 days has been available for a number of years (193). The first report of purification of the enzyme to homogeneity (194) was of a 77-kd protein that was subsequently acknowledged (195) to represent the purification of the endoplasmic reticulum heat shock protein BiP or GRP78. Two microsomal proteins of 94 kd (196,197) and 98 kd (198) have been purified and claimed to be the carboxylase. The subsequent expression of the 94-kd protein in an insect cell line that lacks any endogenous carboxylase activity utilizing a baculovirus expression system (199) has established that this protein carries the activity previously associated with crude preparations of the carboxylase. The role of the 98-kd protein in the overall carboxylase mechanism, if any, is not known.

Although recombinant carboxylase is available, little information on the structure of this enzyme is available. The affinity of the propeptide region of various vitamin K–dependent proteins for the carboxylase has been studied (199a), as has the influence of the Glu substrate on the epoxidase activity (199b) of the enzyme. The fate of proteins that are not carboxylated has also been investigated, and, at least in the case of prothrombin, it has been shown (199c) that in the rat, but not in the human or bovine, lack of carboxylation leads to degradation in the endoplasmic reticulum rather than secretion.

The substrates for the vitamin K–dependent carboxylase contain multiple potential Gla residues, and the available data (199d, 199e) suggest that carboxylation is an ordered process leading to complete carboxylation once initiated. The enzyme has been shown (199f) to be a substrate for itself, and this modification may be involved in control of the activity of the enzyme.

C. Vitamin K–Dependent Proteins in Skeletal Tissues

Hauschka et al. (200) first reported the presence of Gla in the EDTA-soluble proteins of chick bone and demonstrated that it was located in an abundant low molecular weight protein that would bind tightly to $BaSO_4$. Price et al. (201) independently discovered a low molecular weight Gla-containing protein in bovine bone that bound tightly to hydroxyapatite and that inhibited hydroxyapatite formation from saturated solutions of calcium phosphate. This protein was soon sequenced and shown to contain three Gla residues in a 49-residue sequence (MW 5700) that shows no apparent homology to the vitamin K–

dependent plasma proteins. This protein has been called either "osteocalcin" or bone Gla protein (BGP), and the terms are used interchangeably by most investigators. A second protein, matrix Gla protein (MGP), that has been found in bone in cartilage is structurally related to bone Gla protein (BGP) (202) and is also expressed in numerous other tissues (203). Protein S, a vitamin K–dependent protein with an antithrombotic role in hemostasis, has also been shown (204) to be present in bone matrix. A number of reviews of the properties and metabolic role of these skeletal tissue components are available (205–208).

Clear evidence for the physiological role of these vitamin K–dependent bone proteins has been difficult to obtain. Early reports tended to assume some role of osteocalcin in bone mineralization, but no evidence to support this role has been obtained. Price and Williamson (209) developed a vitamin K–deficient rat model in which animals were maintained by administration of a ratio of warfarin to vitamin K that prevented hemorrhage but resulted in bone osteocalcin levels only 2% of normal. No defect in bone size, morphology, or mineralization was observed in these animals, and calcium homeostasis was not impaired. Subsequent studies (210) demonstrated that when rats were maintained on this protocol for 8 months rather than 2 months, a mineralization disorder characterized by complete fusion of proximal tibia growth plate and cessation of longitudinal growth was observed. Whether this response is due to a defect in osteocalcin, MGP, or protein S synthesis is not yet known. These data do, however, suggest that a skeletal vitamin K–dependent protein is involved in regulating the deposition of bone mineral.

Osteocalcin is synthesized in bone rather than accumulating there after synthesis in a different organ. However, small amounts do circulate and can be measured by radioimmunoassay. Circulating osteocalcin is four- to fivefold higher in young children than in adults and reaches the adult level of 5–7 ng/mL at puberty. Levels are probably slightly higher in males, and a slight increase with advanced age has been found. In males, it has been shown that plasma osteocalcin increases in Paget's disease, bone metastasis, renal osteodystrophy, hyperparathyroidism, and osteopenia. In general, plasma osteocalcin increased markedly in Paget's disease and is also elevated in other diseases characterized by increased resorption and increased bone formation (211,212).

A number of reports have suggested that vitamin K status might influence skeletal health. Low concentrations of circulating vitamin K have been reported in patients with bone fractures, and under-γ-carboxylated circulating osteocalcin has been associated with low bone density and fracture rate (208,212,214). A large multicenter study (214a) has identified increased under-γ-carboxylated osteocalcin as a predictor of hip fracture, and low phylloquinone intake has been correlated with hip fractures (214b). If vitamin K status is related to osteoporosis incidence, the anticoagulant-treated population should be at risk. However, only marginal associations between warfarin treatment and lower bone mineral density were found in a recent meta-analysis of nine studies (214c). If there is a relationship between bone health and vitamin K status, it is unlikely to be mediated through the rather small alterations in the amount of biologically active osteocalcin that have been observed in these reports. Both osteocalcin and MGP have been cloned, and the phenotype of mice lacking the osteocalcin gene was increased bone mass and stronger bones (214d) and of MGP an increase in aortic calcification (214e).

D. Other Vitamin K–Dependent Proteins

Proteins containing Gla residues have been found in mineralized tissues other than bone and in pathologically calcified tissues (215–217). These proteins have not been extensively characterized, and their relationship to osteocalcin is not clear. A Gla-containing protein has been purified from kidney (218) and shown to be distinct from osteocalcin, but its function has not been established. The reported presence of a Gla-containing protein in liver mitochondria (219) has been questioned (220) but later confirmed by a third investigator (221). Spermatozoa (222) and urine (223) have also been reported as sources of specific Gla-containing proteins. Specific proteins have not been characterized following these reports; however, two novel proline-rich proteins have been cloned (223a). The role of these proteins has not been elucidated, but a vitamin K–dependent protein encoded by a growth arrest–specific gene (gas-b) (224) with considerable homology to protein S appears to be important in cellular development. These proteins are not restricted to higher vertebrates, and proteins containing Gla residues have been found in elasmobranch species (225), snake venoms (226), and murine gastropods (227).

These reports of purified or nearly purified proteins containing Gla residues make it abundantly clear that the action of vitamin K–dependent carboxylase is not limited to the hepatic modification of a few mammalian plasma proteins. With the exception of muscle, most mammalian tissues and organs contain significant levels of carboxylase activity. In considering the physiological role of vitamin K–dependent proteins without a known function, the widespread use of vitamin K antagonists as clinical anticoagulants should be considered. Patients are routinely given sufficient amounts of coumarin anticoagulants to depress vitamin K–dependent clotting factor levels of 15–35% of normal. It might, therefore, have been expected that a number of problems unrelated to clotting factor synthesis would have been observed in these patients. However, widespread clinical problems associated with an effect on synthesis of other vitamin K–dependent proteins during routine anticoagulant therapy has not been observed. Vitamin K metabolism in these other tissues may not be as sensitive to coumarin anticoagulants as that in liver, or the normal level of these proteins may be in excess of that needed for their physiological function. It may also be that these proteins are part of a secondary backup system to a metabolic pathway that is warfarin-insensitive.

E. Metabolic Effects of 4-Hydroxycoumarins

The oral anticoagulants are not only effective antithrombotic clinical drugs (228), but they have also been widely used as rodenticides. Vitamin K and warfarin or other 4-hydroxycoumarins exhibit a general in vivo agonist/antagonist relationship, and early investigators recognized that the relationship was not what would be expected for the competition of two compounds for a single active site. When the vitamin K–dependent carboxylase was discovered, it was shown that the enzyme was not particularly sensitive to inhibition by these drugs and that in contrast to the situation in animals in vitro inhibition of the carboxylase enzyme by warfarin could not be reversed by high concentrations of vitamin. The microsomal associated activities that have been identified as being involved in the metabolic interconversions of the liver vitamin K pool are shown in Fig. 6. In addition to the carboxylase/epoxidase, they include a vitamin K epoxide reductase and two or more vitamin K quinone reductases. The microsomal epoxide reductase requires a dithiol rather than a reduced pyridine nucleotide for activity and is commonly studied in vitro with dithiothreitol as a reductant. This membrane-associated activity has been difficult to purify but has now been reported (228a) to be composed of a microsomal epoxide hydrase and a second protein. The thioredoxin/thioredoxin reductase systems can

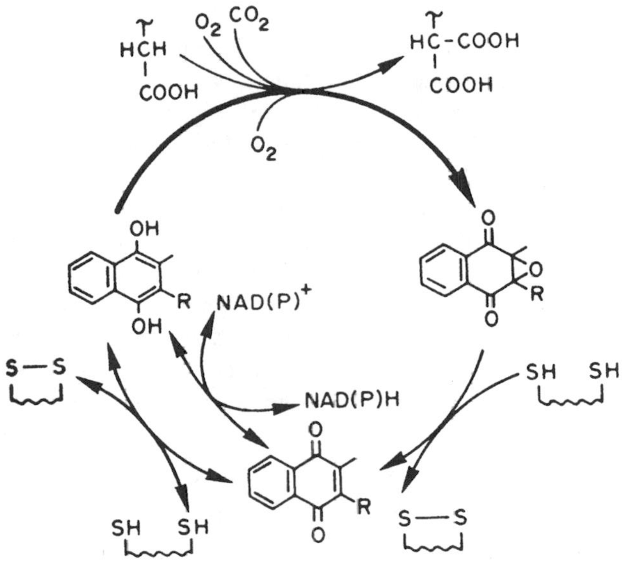

Fig. 6 Vitamin K–related activities in rat liver microsomes. Vitamin K epoxide formed in the carboxylation reaction is reduced by a warfarin-sensitive pathway "vitamin K epoxide reductase" that is driven by dithiothreitol (DTT) as a reducing agent in in vitro studies. The quinone form of the vitamin can be reduced to the hydroquinone either by a warfarin-sensitive DTT-driven quinone reductase or by one or more nicotinamide nucleotide–linked quinone reductases which are less sensitive to warfarin. Warfarin and other coumarin anticoagulants do not have a significant effect on the carboxylase/epoxidase activity.

drive the reactions in vitro, and it is possible that this is the physiologically relevant reductase (229,230).

A real understanding of the mechanism of action of warfarin began when Matschiner and colleagues (231,232) demonstrated that the 2,3-epoxide of vitamin K was a normal metabolite in rat liver and that the ratio of vitamin K epoxide to vitamin K was increased by warfarin administration. Although it was originally thought that these high levels of vitamin K epoxide inhibited the carboxylase, this was shown to be unlikely (233). The current general theory is that inhibition of vitamin K epoxide reduction by warfarin prevents efficient recycling of the vitamin to its enzymatically active form and therefore limits the action of the carboxylase.

A strain of wild rats that were resistant to the action of the common 4-hydroxycoumarin anticoagulants were used to establish this theory. The vitamin K epoxide reductase preparation obtained from livers of the warfarin-resistant rats were relatively insensitive to inhibition by warfarin (234,235). These preparations were, however, strongly inhibited by a second 4-hydroxycoumarin, difenacoum, which had been developed as an effective rodenticide for control of the warfarin-resistant rat population. These data appeared to provide the final proof that the inhibition of the epoxide reductase by warfarin was related to its anticoagulant action. The dithiol-dependent vitamin K quinone reductase has also been reported to be warfarin-sensitive, and it has been shown that this activity, like that of the epoxide reductase, is less sensitive to warfarin when assayed in tissues of the warfarin-resistant rat (236,237). It is, therefore, likely that the effects of the 4-hydroxycoumarin anticoagulants involve not only the reduction of vitamin K epoxide to the quinone but also the reduction of the quinone to the hydroquinone (238). The NADH-dependent quinone reductases are less sensitive to warfarin inhibition and constitute a pathway for vitamin K quinone reduction in the anticoagulant-treated animal. The presence of this pathway explains the ability of administered vitamin K to counteract the hemorrhagic condition resulting from a massive dose of warfarin (239).

The alteration of two enzyme activities by what has been assumed to be a single mutation responsible for the development of warfarin resistance in the wild rat population raises interesting questions in terms of the structural relationship of the proteins involved. Whether or not the two microsomal dithiol-dependent activities, vitamin K epoxide reductase and vitamin K quinone reductase, are catalyzed by the same active site, two active sites on the same protein, or on two subunits of the same protein is unclear at present. A significant number of data do, however, support the theory that the same enzyme catalyzes both activities (240).

VIII. DEFICIENCY SIGNS AND METHODS OF NUTRITIONAL ASSESSMENT

The classical method used to define an inadequate intake of vitamin K was to measure the plasma concentration of one of the vitamin K–dependent clotting factors, prothrombin (factor II), factor VII, factor IX, or factor X. The various tests of clotting function used in clinical practice, which are based on the activity of these factors, have been summarized by Denson and Biggs (241).

Whole-blood clotting times, which were used in early work, are notoriously inaccurate, variable, and insensitive and should not be used. Tests used at present measure the time it takes recalcified citrated or oxalated plasma to form a fibrin clot. The standard "one-stage prothrombin time" assay measures clotting time after the addition of calcium and a lung or brain extract (thromboplastin) preparation to furnish phospholipids and tissue factor. Variations of this assay have been developed, and commercial reagent kits are available. Because of the presence of tissue extract, factor IX is bypassed, and the assay responds to the level of prothrombin and factors VII and X. Of these, factor VII has the shortest half-life, and its concentration decreases earliest when vitamin K action is impaired. It is likely, therefore, that these one-stage prothrombin assays often measure the level of factor VII rather than prothrombin. Specific assays for factors VII and X are also available but are seldom used in studies of vitamin K sufficiency. The classic assay for prothrombin is a "two-stage" assay in which thrombin is generated from prothrombin in one tube and a sample of this added to fibrinogen in a second tube to measure thrombin concentration. This is a much more tedious procedure; although capable of giving an accurate measurement of the amount of prothrombin in a plasma sample, it has seldom been used for routine assay of vitamin K status. A number of snake venom preparations liberate thrombin from prothromin and have been used (158,242,243) to develop one-stage clotting assays for prothrombin. The enzymes in these preparations do not require that prothrombin be present in a calcium-dependent phospholipid complex for activation, and they will therefore activate the descarboxyprothrombin formed in vitamin K–deficient animals. For this reason they cannot be used to monitor a vitamin K deficiency.

All of these methods depend on the observation of the formation of a fibrin clot as an end point to assay and are not amenable to automation. The vitamin K–dependent clotting factors are serine proteases, and in recent years there has been considerable interest in the development of chromogenic substrates for the assay of these proteins, particularly factor X and prothrombin. These assays, when utilized to assay prothrombin activity, actually measure the concentration of thrombin that has been generated from prothrombin by various methods (242). Because they can be readily adapted to an automated analysis, these substrates are receiving a great deal of attention as clinical tools to follow anticoagulant therapy but have not yet been utilized in any attempt to monitor vitamin K adequacy.

Human vitamin K deficiency results in the secretion into the plasma of partially carboxylated species of vitamin K–dependent proteins. Because they lack the full complement of γ-carboxyglutamic acid residues, their calcium binding affinity is altered and they can be separated from normal prothrombin by alterations in their ability to bind to barium salts or by electrophoresis. Antibodies that are specific for these ''abnormal'' prothrombins have been developed and can also be used to detect a vitamin K deficiency. These assays or similar methods used to detect the concentration of under-γ-carboxylated osteocalcin have greatly increased the sensitivity with which a vitamin K deficiency can be detected (244). It is also likely that vitamin K status is reflected in alterations of circulating levels of the vitamin. The extremely low concentration of vitamin K in plasma made these measurements very difficult at one time, but satisfactory HPLC methods for the determination of plasma or serum phylloquinone have now been developed (see Sec. VI.C). The amount of vitamin K found in ''normal'' plasma appears to be about 0.5 ng/mL, and limited information on the response of circulating vitamin K to changes in dietary vitamin K is currently available.

IX. NUTRITIONAL REQUIREMENT

A. Animals

The establishment of a dietary vitamin K requirement for various species has been difficult. The challenge in demonstrating dietary requirement in many species presumably comes from the varying degrees to which they utilize the large amount of vitamin K synthesized by intestinal bacteria and the degree to which different species practice coprophagy. A spontaneous deficiency of vitamin K was first noted in chicks, and poultry are much more likely to develop symptoms of a dietary deficiency than any other species. This has usually been assumed to be due to the rapid transit rate of material through the relatively short intestinal tract of the chick or to limited synthesis of menaquinones in this species. A more recent study (133) suggests that limited recycling of vitamin K because of low epoxide reductase activity may be the cause of the increased requirement.

Ruminal microorganisms synthesize large amounts of vitamin K, and ruminants do not appear to need a source of vitamin in the diet. Deficiencies have, however, been produced in most monogastric species. Estimations of vitamin K requirements by different workers are difficult to compare because of the different forms of the vitamin that were used and different methods that were employed to establish the requirement. Some studies have utilized a curative assay in which animals were first made hypoprothrombinemic by feeding a vitamin K–deficient diet. They were then either injected with the vitamin and prothrombin levels assayed after a few hours or a day, or fed diets containing various amounts of the vitamin for a number of days and the response in clotting factor synthesis noted. Preventive or prophylactic assays have also been used and an attempt made to determine the minimum concentration of the vitamin that must be present in a diet to maintain normal clotting factor levels.

Phylloquinone has often been used in experimental nutrition studies, whereas other forms of vitamin K are usually used in practical rations. Menadione is usually considered to be from 20% to 40% as effective as phylloquinone on a molar basis, but this depends a great deal on the type of assay which is used. It is rather ineffective in a curative assay, where the rate of its alkylation to a menaquinone is probably the rate-limiting factor, but often shows activity nearly equal to phylloquinone in a long-term preventive assay. Practical nutritionists have often preferred to utilize a water-soluble form of menadione, such as menadione sodium bisulfite complex (MSBC). This compound appears to be about as active on a molar basis as phylloquinone in poultry rations, and at least in this species, the activity of menadione, MSBC, and phylloquinone are roughly equal on a weight basis.

Detailed discussions of the vitamin K requirements of various species are available in articles by Scott (245), Doisy and Matschiner (246), and Griminger (67). The data indicate that the requirement for most species falls in a range of 2–200 µg vitamin K per kilogram body weight per day. The data in Table 5, which have been adopted from a table presented by Griminger (67), give an indication of the magnitude of the requirement for various species. It should be remembered that this requirement can be altered by age, sex, or strain, and that any condition influencing lipid absorption or conditions altering intestinal flora will have an influence of these values (see Sec. X). A considerably higher level of dietary vitamin K has been recommended for most laboratory animals by the National Research Council (247). Recommendations for most species are in the range of 3000 µg/kg of diet, but the rat requirement has been set at 50 µg/kg.

Although this level is sufficient in most cases, it does not prevent all signs of deficiency (111), and the American Institute of Nutrition (248) has now recommended that purified diets for laboratory rodents should have 750 µg of phylloquinone added to each kilogram of diet.

B. Humans

The requirement of the adult human for vitamin K is extremely low, and there seems little possibility of a simple dietary deficiency developing in the absence of complicating factors. Until recently, the low requirement and the relatively high levels of vitamin K found in most diets had prevented an accurate assessment of the requirement. Frick et al. (249)

Table 5 Vitamin K Requirements of Various Species[a]

Species	Daily intake (µg/kg per day)	Dietary concentration (µg/kg diet)
Dog	1.25	60
Pig	5	50
Rhesus monkey	2	60
Rat, male	11–16	100–150
Chicken	80–120	530
Turkey poult	180–270	1200

[a] Data have been summarized from a more extensive table (Ref. 67) and are presented as the amount of vitamin needed to prevent the development of a deficiency. No correction for differences in potency of equal weights of different forms of the vitamin has been made.

studied the vitamin K requirement of starved intravenously fed debilitated patients given antibiotics to decrease intestinal vitamin K synthesis. They determined that 0.1 µg/kg per day was not sufficient to maintain normal prothrombin levels and that 1.5 µg/kg per day was sufficient to prevent any decreases in clotting factor synthesis. Their data indicate that the requirement was of the order of 1 µg/kg per day. Two other studies (75,250) with very limited numbers of subjects successfully decreased vitamin K intake to the extent that clotting factor activities were lowered and also suggested that the vitamin requirement of the human is in the range of 0.5–1.0 µg vitamin K per kilogram per day.

A major problem in determining a dietary requirement for vitamin K has been the relative insensitivity of the commonly used prothrombin time measurement (244). A more recent study modifying the vitamin K intake of young adults by restriction of foods with a high phylloquinone content (251) has resulted in two symptoms of a mild deficiency: increased circulating under-γ-carboxylated prothrombin and decreased Gla excretion. These responses were reversed by additional dietary vitamin, and the data obtained are consistent with a requirement in the range previously suggested. A more carefully controlled metabolic ward study (113) found alterations in the same two sensitive measures of deficiency in subjects consuming about 10 µg phylloquinone per day, and increased circulating under-γ-carboxylated prothrombin has been observed in a second metabolic ward controlled study (252). Based largely on these studies, the RDA for vitamin K has been set at 1 µg phylloquinone per kilogram body weight for adults, and 5 µg phylloquinone per day for infants from birth to 6 months and 10 µg phylloquinone per day from 6 to 12 months. The 1 µg/kg body weight was also applied to children.

This requirement may be subject to change as more data become available. Because of the close relationship of plasma phylloquinone concentration to recent intakes, this measurement lacks utility for assessing vitamin K status. Assessment of adequacy by use of the somewhat insensitive one-stage prothrombin time has meant that a large decrease in vitamin K–dependent clotting factor synthesis was needed to produce an apparent deficiency. More sensitive clotting assays and the ability to immunochemically detect circulating des-γ-carboxyprothrombin (253) provide an opportunity to monitor much milder forms of vitamin K deficiency. There are a number of reports that vitamin K status may be important in maintaining skeletal health (214), and in recent studies (254,255,255a) the extent of under-γ-carboxylation of circulating osteocalcin has been shown to be a very sensitive criterion of vitamin K sufficiency. It is likely that this method will become increasingly important in future studies directed at defining the human requirement for vitamin K.

X. FACTORS INFLUENCING VITAMIN K STATUS AND POSSIBLE GROUPS AT RISK

A. Adult Human Considerations

The human population normally consumes a diet containing a great excess of vitamin K, but a vitamin K–responsible hypoprothrombinemia can sometimes be a clinically significant human problem. O'Reilly (35) has reviewed the potential problem areas and has pointed out the basic factors needed to prevent a vitamin K deficiency: (a) a normal diet containing the vitamin (b) the presence of bile in the intestine, (c) a normal absorptive surface in the small intestine, and (d) a normal liver. Cases of an acquired vitamin K deficiency do, therefore, occur in the adult population and, though relatively rare, present a significant problem for some individuals. It has usually been assumed that a general deficiency is not possible, but Hazell and Bloch (256) have observed that a relatively high percentage of an older adult hospital-admitted population has a hypoprothrombinemia that responds to administration of oral vitamin K. The basis for this apparent increase in vitamin K requirement was not determined and was probably multicausal. It has, however, been shown (257) that it is much easier to develop a vitamin K deficiency in older than in young rats. Whether this is related to anything other than an increased intake of nutrients/unit of body weight in the younger animals has not been determined.

B. Hemorrhagic Disease of the Newborn

A vitamin K–responsive hemorrhagic disease of the human newborn has been a long recognized syndrome. The classical disease is seen in infants at the age of 2–5 days, but ''late hemorrhagic disease'' is a condition also seen in infants older than 1 month (258). The newborn infant has low levels of the vitamin K–dependent clotting factors (259), and these normally increase with time. Infants are at risk because the gut of the newborn is relatively sterile and provides little vitamin K, and human breast milk is low in vitamin K. Serious problems appear to be associated almost exclusively with breast-fed infants. Because of the possibility of hemorrhagic episodes, the American Academy of Pediatrics recommends that phylloquinone be administered parenterally to all newborn infants at a dose of 0.5–1.0 mg. Commercial infant formulas are routinely fortified with vitamin K.

Improvement in methods for detecting mild vitamin K deficiencies (260) has led to a renewed interest in vitamin K nutrition in the infant, and the importance of the low vitamin K intake of breast-fed infants to the development of signs of deficiency is now more clearly understood. A critical factor in determining the vitamin K status of the breast-fed infant has been shown to be milk intake (261,262). There is currently increased interest in oral rather than parenteral administration of vitamin K to infants, and efforts are being made to develop satisfactory products.

C. Effects of Drugs

Early studies (see Ref. 257) of vitamin K action and requirement established the need to prevent coprophagy to produce a vitamin K deficiency in the rat. It is easily demonstrated that the vitamin K requirement of the rat is greatly increased under germ-free conditions (263) and that feeding sulfa drugs increases the vitamin K requirement of the chick (264). The importance of menaquinones in satisfying the human requirement for vitamin K is unclear (108). Numerous cases of vitamin K–responsive hemorrhagic events in patients receiving antibiotics have been reviewed by Savage and Lindenbaum (265). These have been assumed to be due to decreased menaquinone utilization by these patients, but it is possible that many cases may represent low dietary intake alone and that the presumed adverse drug interaction is not always related to the hypoprothrombinemia. The second- and third-generation cephalosporins have been implicated in a large number of hypothrombinemic episodes; although it has been suggested that these drugs have a direct effect on the vitamin K–dependent carboxylase (266), it is more likely that they are exerting a weak coumarin-like response (267,268).

A series of reports have demonstrated (269,270) that dietary butylated hydroxytoluene causes a hemorrhagic condition in rats that can be cured by vitamin K supplementation. The mechanism by which the effect is mediated has not yet been clarified. Phenobarbital and diphenylhydantoin administration to mothers has been reported (271) to produce a vitamin K–responsive hemorrhage in the newborn, and clofibrate appears to alter coumarin responsiveness (272) through an effect on vitamin K utilization of availability. The nature of these effects has not been extensively investigated.

The widely used coumarin anticoagulants (see Secs. II.C and VII.E) effectively antagonize the action of vitamin K and also influence metabolism of the vitamin. Clinically, the effect of this vitamin K antagonism has been thought to be limited to their effect on clotting factor synthesis. Oral anticoagulant therapy during the first trimester is known to be associated with development of the fetal "warfarin syndrome" (273). Whether this outcome is related to the influence of warfarin on the action of osteocalcin, matrix Gla protein, protein S, or an unknown vitamin K–dependent protein is not known. It is also possible that this adverse drug effect is related not to the action of warfarin as a vitamin K antagonist but to an unrelated action of the drug.

D. Influence of Hormones

Early studies of vitamin K requirements indicated that female rats had higher plasma prothrombin concentrations and a lower vitamin K requirement. Plasma prothrombin levels are also higher in pregnant rats, and this increased concentration of prothrombin in female and pregnant rats results from an effect on rate of synthesis rather than rate of prothrombin degradation (274). Castration of both sexes unifies that vitamin K response, and in the castrated rat, prothrombin concentrations can be increased with estrogens and decreased with androgens (275). There are some indications (126) from studies utilizing [^{3}H]vitamin K_1 that the estrogen effects are related to the amount of vitamin needed in the liver to maintain normal levels of prothrombin. The available evidence suggests that the influence of estrogens on rate of synthesis is reflected in a higher rate of synthesis and accumulation of prothrombin precursors in the microsomes (276,277).

The effect of other hormones on vitamin K metabolism or action is less clear. Nishino (278) has shown that hypophysectomy prevents the estrogen stimulation of prothrombin synthesis in castrated females, and that prolactin injections protect intact males from the development of hypoprothrombinemia. It has also been shown that hypothyroidism in humans results in a decrease in both the rate of synthesis and destruction of the vitamin K–dependent clotting factors (279). It is likely that the effects of these hormones are related to rates of synthesis of the proteins involved rather than to any effect on vitamin K metabolism. But it is possible that there are also effects on the vitamin itself.

E. Other Dietary and Disease State Factors

Early studies of vitamin K functions (280) established that the inclusion of mineral oil in diets prevented its absorption, and mineral oil has often been used in vitamin K–deficient diets. High dietary vitamin A has also been recognized for some time to adversely influence vitamin K action (246). Whether this is a general effect on nonpolar lipid absorption or a specific vitamin K antagonism is not clear, but it can be observed at relatively low dietary levels of retinol acetate and retinoic acid (257). Dietary oxidized squalene also has a potent hemorrhagic effect (257), and a d-α-tocopherol hydroquinone administration has been shown to produce a vitamin K–responsive hemorrhagic syndrome in the pregnant rat (281). High vitamin E consumption may be of some clinical significance, as the addition of vitamin K to the diet of a patient on coumarin anticoagulant therapy has been shown to result in a hemorrhagic episode (282). It is possible that high vitamin E intakes may exacerbate a borderline vitamin K deficiency. There are indications that α-tocopherol quinone, rather than α-tocopherol (283), may be the causative agent, and this may explain some of the differences in the reported results.

Insufficient assimilation of vitamin K can occur in adults on protracted antibiotic treatment or those receiving long-term parenteral hyperaliminentation without vitamin K supplementation. Malabsorption of vitamin K has also occurred as a result of obstructive jaundice, biliary fistula, pancreatic insufficiency, steatorrhea, or chronic diarrhea. Specific references to observations of vitamin K–responsive hemorrhagic conditions in these various diseases states can be found in Refs. 35 and 265.

XI. EFFICACY AND HAZARDS OF PHARMACOLOGICAL DOSES OF VITAMIN K

No hazards attribute to the long-term ingestion of elevated amounts of the natural forms of vitamin K have been reported (284,285). For treatment of prolonged clotting times when hemorrhage is not a problem, vitamin K can be given orally or parenterally. If given orally to patients with impaired biliary function, bile salts should also be administered. Vitamin K_1 is available as the pure compound or as an aqueous colloidal solution that can be given intramuscularly or intravenously. Some adverse reactions have been noted following intervenous administration, and unless a severe hemorrhagic episode is present, intramuscular infection is the recommended route of therapy. Effective therapy requires synthesis of normal clotting factors, and a couple of hours may be necessary before a substantial decrease in clotting times is apparent.

The relative safety of phylloquinone and, presumably, menaquinones does not hold for menadione or its water-soluble derivatives. These compounds can be safely used at low levels to prevent the development of a deficiency but should not be used as a pharmacological treatment for a hemorrhagic condition. Although once prescribed for treatment of the hemorrhagic disease of the newborn, these compounds are known to react with free sulfhydryl groups of various tissues and to cause hemolytic anemia, hyperbilirubinemia, and kernicterus. This marked increase in conjugated bilirubin is extremely toxic to the neonatal brain and has caused death in some instances (284).

REFERENCES

1. H. Dam, Cholesterinstoffwechsel in Hühnereiern and Hünchen. *Biochem. Z. 215*:475–492 (1929).
2. W. D. McFarlane, W. R. Graham, and F. Richardson, The fat-soluble vitamin requirements of the chick. I. The vitamin A and vitamin D content of fish meal and meat meal. *Biochem. J. 25*:358–366 (1931).
3. W. F. Holst and E. R. Halbrook, A ''scurvy-like'' disease in chicks. *Science 77*:354 (1933).
4. H. Dam. The antihaemorrhagic vitamin of the chick. *Biochem. J. 29*:1273–1285 (1935).
5. H. Dam. The antihaemorrhagic vitamin of the chick. Occurrence and chemical nature. *Nature 135*:652–653 (1935).
6. H. J. Almquist and E. L. R. Stokstad, Dietary haemorrhagic disease in chicks. *Nature 136*: 31 (1935).
7. H. J. Almquist and E. L. R. Stokstad, Hemorrhagic chick disease of dietary origin. *J. Biol. Chem. 111*:105–113 (1935).
8. H. Dam, F. Schønheyder, and E. Tage-Hansen, Studies on the mode of action of vitamin K. *Biochem. J. 30*:1075–1079 (1936).
9. D. W. MacCorquodale, L. C. Cheney, S. B. Binkley, W. F. Holcomb, R. W. McKee, S. A.

Thayer, and E. A. Doisy, The constitution and synthesis of vitamin K. *J. Biol. Chem. 131*: 357–370 (1939).

10. P. Karrer, A. Geiger, R. Legler, R. Rüegger, and H. Salomon, Über die isolierung des α-Phyllochinones (vitamin K aus Alfalfa) sowie über dessen Entdeckungsgeschechter. *Helv. Chim. Acta 22*:1464–1470 (1939).

11. H. J. Almquist and A. A. Klose, Synthetic and natural antihemorrhagic compounds. *J. Am. Chem. Soc. 61*:2557–2558 (1939).

12. L. F. Fieser, Synthesis of 2-methyl-3-phytyl-1,4-naphthoquinone. *J. Am. Chem. Soc. 61*: 2559–2561 (1939).

13. C. Martius and H. O. Esser. Über die Konstitution des in Tierkorper aus Methyl Naphthochinongelbildenten K-Vitamines. *Biochem Z. 331*:1–9 (1958).

14. H. J. Almquist, The early history of vitamin K. *Am. J. Clin. Nutr. 28*:656–659 (1975).

15. H. J. Almquist, Vitamin K. *Physiol. Rev. 21*:194–216 (1941).

16. H. Dam, Vitamin K, its chemistry and physiology. *Adv. Enzymol. 2*:285–324 (1942).

17. E. A. Doisy, S. B. Binkley, and S. A. Thayer, Vitamin K. *Chem. Rev. 28*:477–517 (1941).

18. S. L. Booth, K. W. Davidson, and J. A. Sadowski, Evaluation of an HPLC method for the determination of phylloquinone (vitamin K_1) in various food matrices. *J. Agric. Food Chem. 42*:295–300 (1994).

19. J. T. Matschiner, W. V. Taggart, and J. M. Amelotti. The vitamin K content of beef liver, detection of a new form of vitamin K. *Biochemistry 6*:1243–1248 (1967).

20. J. T. Matschiner and W. V. Taggart, Separation of vitamin K and associated lipids by reversed-phase partition column chromatography. *Anal. Biochem. 18*:88–93 (1967).

21. IUPAC-IUB, Commission on Biochemical Nomenclature. Nomenclature of quinones with isoprenoid side chains. *Eur. J. Biochem. 53*:15–18 (1975).

22. L. F. Fieser, M. Tishler, and W. L. Sampson, Vitamin K activity and structure. *J. Biol. Chem. 137*:659–692 (1941).

23. P. Griminger, Biological activity of the various vitamin K forms. *Vitam Hormones 24*:605–618 (1966).

24. F. Weber and O. Wiss, Vitamin K group: active compounds and antagonists, in *The Vitamins*, Vol. 3 (W. H. Sebrell and R. S. Harris, eds.). Academic Press, New York, 1971, pp. 457–466.

25. J. T. Matschiner and R. G. Bell, Metabolism and vitamin K activity of cis phylloquinone in rats. *J. Nutr. 102*:625–630 (1972).

26. J. T. Matschiner and W. V. Taggart, Bioassay of vitamin K by intracardial injection in deficient adult male rats. *J. Nutr. 94*:57–59 (1968).

27. P. Griminger, Relative vitamin K potency of two water-soluble menadione analogues. *Poultry Sci. 44*:211–213 (1965).

28. P. N. Dua and E. J. Day, Vitamin K activity of menadione dimethylpyrimidinol bisulfite in chicks. *Poultry Sci. 45*:94–96 (1966).

29. O. W. Charles, B. C. Dilworth, and E. J. Day, Chick bioassay of vitamin K compounds using dicumarol and pivalyl as anticoagulants. *Poultry Sci. 47*:754–760 (1968).

30. R. W. Seerley, O. W. Charles, H. C. McCampbell, and S. P. Bertsch, Efficacy of menadione dimethylpyrimidinol bisulfite as a source of vitamin K in swine diets. *J. Anim. Sci. 42*:599–607 (1976).

31. K. P. Link, The discovery of dicumarol and its sequels. *Circulation 19*:97–107 (1959).

32. H. A. Campbell and K. P. Link. Studies on the hemorrhagic sweet clover disease. IV. The isolation and crystallization of the hemorrhagic agent. *J. Biol. Chem. 138*:21–33 (1941).

33. M. A. Stahmann, C. F. Huebner, and K. P. Link, Studies on the hemorrhagic sweet clover disease. V. Identification and synthesis of the hemorrhagic agent. *J. Biol. Chem. 138*:513–527 (1941).

34. E. Renk and W. G. Stoll, Orale Antikoagulantien. *Prog. Drug. Res. 11*:226–355 (1968).

35. R. A. O'Reilly, Vitamin K and the oral anticoagulant drugs. *Annu. Rev. Med. 27*:245–261 (1976).

36. C. M. Boyle, Case of apparent resistance of *Rattus norvegicus* Berkenhout to anticoagulant poisons. *Nature 188*:517 (1960).

37. W. B. Jackson and D. Kaukeinen, Resistance of wild norway rats in North Carolina to warfarin rodenticide. *Science 176*:1343–1344 (1972).

38. W. B. Jackson, A. D. Ashton, and K. Delventhal, Overview of anticoagulant rodenticide usage and resistance, in *Current Advances in Vitamin K Research* (J. W. Suttie, ed.). Elsevier, Science Publisher, New York, 1988, pp. 381–388.

39. M. Lund, Detection and monitoring of resistance to anticoagulant rodenticies in populations of brown rats (*Rattus norvegicus*) in Denmark, in *Current Advances in Vitamin K Research* (J. W. Suttie, ed.). Elsevier, New York, 1988, pp. 399–405.

40. J. H. Greaves and P. B. Cullen-Ayres, Genetics of Difenacoum resistance in the rat, in *Current Advances in Vitamin K Research* (J. W. Suttie, ed.). Elsevier, New York, 1988, pp. 389–397.

41. H. Kobat, E. F. Stohlman, and M. I. Smith, Hypoprothrombinemia induced by administration of indandione derivatives. *J. Pharmacol. Exp. Ther. 80*:160–170 (1944).

42. P. Ren, R. E. Laliberte, and R. G. Bell, Effects of warfarin, phenylindanedione, tetrachloropyridinol, and chloro-vitamin K_1 on prothrombin synthesis and vitamin K metabolism in normal and warfarin-resistant rats. *Mol. Pharmacol. 10*:373–380 (1974).

43. J. Lowenthal, J. A. MacFarlene, and K. M. McDonald, The inhibition of the antidotal activity of vitamin K_1 against coumarin anticoagulant drugs by its chloro analogue. *Experientia 16*: 428–429 (1960).

44. J. Lowenthal, Vitamin K analogs and mechanisms of action of vitamin K, in *The Fat-Soluble Vitamins* (H. F. DeLuca and J. W. Suttie, eds.). University of Wisconsin Press, Madison, 1970, pp. 431–446.

45. J. W. Suttie, Anticoagulant-resistant rats: possible control by the use of the chloro analog of vitamin K. *Science 180*:741–743 (1973).

46. J. Lowenthal and J. A. MacFarlene, Vitamin K-like and antivitamin K activity of substituted *para*-benzoquinones. *J. Pharmacol. Exp. Ther. 147*:130–138 (1965).

47. M. N. Marshall, Potency and coagulation factor effects of 2,3,5,6-tetrachloropyridinol compared to warfarin and its antagonism by vitamin K. *Proc. Soc. Exp. Biol. Med. 139*:806–810 (1972).

48. N. U. Bang, G. O. P. O'Doherty, and R. D. Barton, Selective suppression of vitamin K–dependent procoagulant synthesis by compounds structurally unrelated to vitamin K. *Clin. Res. 23*:521A (1975).

49. P. A. Friedman and A. E. Griep, *In vitro* inhibition of vitamin K-dependent carboxylation by tetrachloropyridinol and the imidazopyridines. *Biochemistry 19*:3381–3386 (1980).

50. S. B. Binkley, L. C. Cheney, W. F. Holcomb, R. W. McKee, S. A. Thayer, D. W. MacCorquodale, and E. A. Doisy, The constitution and synthesis of vitamin K_1. *J. Am. Chem. Soc. 61*:2558–2559 (1939).

51. H. J. Almquist, Vitamin K: discovery, identification, synthesis, functions. *Fed. Proc. 28*: 2687–2689 (1979).

52. L. F. Fieser, Identity of synthetic 2-methyl-3-phytyl-1,4-napththoquinone and vitamin K_1. *J. Am. Chem. Soc. 61*:2561 (1939).

53. H. Mayer and O. Isler, Vitamin K group—chemistry, in *The Vitamins*, 2nd ed., Vol. 3 (W. H. Sebrell and ???

54. H. Mayer and O. Isler, Synthesis of vitamins K, in *Methods in Enzymology*, Vol. 18C (D. B. McCormick and L. D. Wright, eds.). Academic Press, New York, 1971, pp. 491–547.

55. H. Mayer and O. Isler, Vitamin K group—industrial preparation, in *The Vitamins*, 2nd ed., Vol. 3 (W. H. Sebrell and R. S. Harris, eds.). Academic Press, New York, 1971, pp. 444–445.

56. Y. Naruta and K. Maruyama, Regio- and stereocontrolled polyprenylation of quinones. A new synthetic method of vitamin K series. *Chem. Lett.*:991–884 (1979).

57. S. Patai, *The Chemistry of the Quinoid Compounds*, Parts 1 and 2. John Wiley and Sons, New York, 1974.

58. P. Sommer and M. Kofler, Physiochemical properties and methods of analysis of phyllo-quinones, menaquinones, ubiquinones, phostoquinones, menadione, and related compounds. *Vitamins and Hormones* 24:349–399 (1966).

59. P. J. Dunphy and A. F. Brodie, The structure and function of quinones in respiratory metabolism, in *Methods in Enzymology*, Vol. 18C (D. B. McCormick and L. D Wright, eds.). Academic Press, New York, 1971, pp. 407–461.

60. H. Mayer and O. Isler, Isolation of vitamin K, in *Methods in Enzymology*, Vol. 18C (D. B. McCormick and L. D. Wright, eds.). Academic Press, New York, 1971, pp. 469–491.

61. J. T. Matschiner and J. M. Amelotti, Characterization of vitamin K from bovine liver *J. Liver Res.* 9:176–179 (1968).

62. R. C. Williams, J. A. Schmidt, and R. A. Henry, Quantitative analysis of fat-soluble vitamins by high-speed liquid chromatography. *J. Chromatogr. Sci.* 10:494–501 (1972).

63. M. J. Shearer, High-performance liquid chromatography of K vitamins and their antagonists, in *Advances in Chromatography*, Vol. 21 (J. C. Giddings, E. Grushka, J. Cazes, and P. R. Brown, eds.). Marcel Dekker, New York, 1983, pp. 243–301.

64. H. Dam and E. Søndergaard, The determination of vitamin K, in *The Vitamins*, 2nd ed., Vol. 6 (P. Gyorgy and W. N. Pearson, eds.). Academic Press, New York, 1967, pp. 245–260.

65. P. Griminger and O. Donis, Potency of vitamin K_1 and two analogs in counteracting the effects of dicumarol and sulfaquinoxaline in the chick. *J. Nutr.* 70:361–368 (1960).

66. J. T. Matschiner and E. A. Doisy, Jr., Bioassay of vitamin K in chicks. *J. Nutr.* 90:97–100 (1966).

67. P. Griminger, Nutritional requirements for vitamin K-animal studies, in *Symposium Proceedings on the Biochemistry, Assay, and Nutritional Value of Vitamin K and Related Compounds.* Assn. Vitamin Chemists, Chicago, 1971, pp. 39–59.

68. H. Weiser and A. W. Kormann, Biopotency of vitamin K. *Int. J. Vitamin Nutr. Res.* 53:143–155 (1983).

69. H. Dam and F. Schønheyder, The occurrence and chemical nature of vitamin K. *Biochem. J.* 30:897–901 (1936).

70. H. Dam and J. Glavind. Vitamin K in the plant. *Biochem. J.* 32:485–487 (1938).

71. R. E. Olson, Vitamin K, in *Modern Nutrition in Health and Disease* (M. E. Shils and V. R. Young, eds.). Lea & Febiger, Philadelphia, 1988, pp. 328–339.

72. S. W. Souci, W. Fachmann, and H. Kraut, *Food Composition and Nutrition Tables, 1986/1987*, 3rd ed., Wisenschaftliche Verlagsgesellschaft, Stuttgart, 1986.

73. J. A. T. Pennington, *Bowes and Church's Food Values of Portions Commonly Used.* J. B. Lippincott, Philadelphia, 1989.

74. L. R. Richardson, S. Wilkes, and S. J. Ritchey, Comparative vitamin K activity of frozen, irradiated and heat-processed foods. *J. Nutr.* 73:369–373 (1961).

75. E. A. Doisy, Vitamin K in human nutrition, in *Symposium Proceedings on the Biochemistry, Assay and Nutritional Value of Vitamin K and Related Compounds.* Assn. Vitamin Chemists, Chicago, 1971, pp. 79–92.

76. M. J. Shearer, V. Allan, Y. Haroon, and P. Barkhan, Nutritional aspects of vitamin K in the human, in *Vitamin K Metabolism and Vitamin K–Dependent Proteins* (J. W. Suttie, ed.). University Park Press, Baltimore, 1980, pp. 317–327.

77. K. Kodaka, T. Ujiie, T. Ueno, and M. Saito, Contents of vitamin K_1 and chlorophyll in green vegetables. *J. Jap. Soc. Nutr. Food Sci.* 39:124–126 (1986).

78. R. M. Seifert, Analysis of vitamin K_1 in some green leafy vegetables by gas chromatography. *J. Agric. Food Chem.* 27:1301–1304 (1979).

79. G. Ferland and J. A. Sadowski, Vitamin K_1 (phylloquinone) content of edible oils: effects of heating and light exposure. *J. Agric. Food Chem.* 40:1869–1873 (1992).

80. Y. Haroon, M. J. Shearer, S. Rahim, W. G. Gunn, G. McEnery, and P. Barkhan, The content

of phylloquinone (vitamin K_1) in human milk, cows' milk and infant formula foods determined by high-performance liquid chromatography. *J. Nutr. 112*:1105–1117 (1982).

81. R. v. Kries, M. Shearer, P. T. McCarthy, M. Haug, G. Harzer, and U. Gobel, Vitamin K_1 content of maternal milk: influence of the stage of lactation, lipid composition, and vitamin K_1 supplements given to the mother. *Pediatr. Res. 22*:513–517 (1987).

82. L. M. Canfield, J. M. Hopkinson, A. F. Lima, G. S. Martin, K. Sugimoto, J. Burri, L. Clark, and D. L. McGee, Quantitation of vitamin K in human milk. *Lipids 25*:06–411 (1990).

83. S. L. Booth, J. A. Sadowski, J. L. Weihrauch, and G. Ferland, Vitamin K_1 (phylloquinone) content of foods: a provisional table. *J. Food Comp. Anal. 6*:109–120 (1993).

83a. S. L. Booth, J. A. Sadowski, and J. A. T. Pennington, Phylloquinone (vitamin K_1) content of foods in the U.S. Food and Drug Administration's total diet study. *J. Agric. Food Chem. 43*:1574–1579 (1995).

83b. S. L. Booth, R. Webb, and J. C. Peters, Assessment of phylloquinone and dihydrophylloquinone dietary intakes among a nationally representative sample of U.S. consumers using 14-day food diaries. *J. Am. Diet. Assoc. 99*:1072–1076 (1999).

84. G. Ferland, D. L. MacDonald, and J. A. Sadowski, Development of a diet low in vitamin K-1 (phylloquinone). *J. Am. Dietet. Assoc. 92*:593–597 (1992).

85. G. Ferland and J. A. Sadowski, Vitamin K_1 (phylloquinone) content of green vegetables: effects of plant maturtion and geographical growth location. *J. Agric. Food Chem. 40*:1874–1877 (1992).

86. D. Hollander, Intestinal absorption of vitamins A, E, D, and K. *J. Lab. Clin. Med. 97*:449–462 (1981).

87. R. Blomstrand and L. Forsgren, Vitamin K_1-^{3}H in man. Its intestinal absorption and transport in the thoracic duct lymph. *Int. Z. Vitamin Forschung 38*:46–64 (1968).

88. M. J. Shearer, A. McBurney, and P. Barkhan, Studies on the absorption and metabolism of phylloquinone (vitamin K_1) in man. *Vitamins and Hormones 32*:513–542 (1974).

89. K-S. G. Jie, B. L. M. G. Gijsbers, and C. Vermeer, in *The Role of Vitamin K-Dependent Proteins in Tissue Calcification*, Ph.D. thesis, University of Limburg, Maastricht, The Netherlands, 1995, pp. 77–83.

89a. A. K. Garber, N. C. Binkley, D. C. Krueger, and J. W. Suttie, Comparison of phylloquinone bioavailability from food sources or a supplement in human subjects. *J. Nutr. 129*:1201–1203 (1999).

90. J. J. Lipsky, Subject review: nutritional sources of vitamin K. *Mayo Clin. Proc. 69*:462–466 (1994).

91. M. J. Shearer, Vitamin K metabolism and nutriture. *Blood Rev. 6*:92–104 (1992).

92. D. Hollander, E. Rim, and P. E. Ruble, Jr., Vitamin K_2 colonic and ileal in vivo absorption: Bile, fatty acids, and pH effects on transport. *Am. J. Physiol. 233*:E124–E129 (1977).

93. T. Ichihashi, Y. Takagishi, K. Uchida, and H. Yamada, Colonic absorption of menaquinone-4 and menaquinone-9 in rats. *J. Nutr. 122*:506–512 (1992).

94. J. M. Conly and K. E. Stein, The absorption and bioactivity of bacterially synthesized menaquinones. *Clin. Invest. Med. 16*:45–57 (1993).

95. J. M. Conly and K. Stein, Quantitative and qualitative measurements of K vitamins in human intestinal contents. *Am. J. Gastroenterol. 87*:311–316 (1992).

96. D. Hollander and T. C. Truscott, Colonic absorption of vitamin K-3. *J. Lab. Clin. Med. 83*: 648–656 (1974).

97. D. Hollander and T. C. Truscott, Mechanism and site of vitamin K-3 small intestinal transport. *Am. J. Physiol. 226*:1526–1522 (1974).

98. J. Saupe, M. J. Shearer, and M. Kohlmeier, Phylloquinone transport and its influence on γ-carboxyglutamate residues of osteocalcin in patients on maintenance hemodialysis. *Am. J. Clin. Nutr. 58*:204–208 (1993).

99. T. D. Bjornsson, P. J. Meffin, S. E. Swezey, and T. F. Blaschke, Disposition and turnover

of vitamin K₁ in man, in *Vitamin K Metabolism and Vitamin K–Dependent Proteins* (J. W. Suttie, ed.). University Park Press, Baltimore, 1980, pp. 328–332.

100. M. J. Shearer, C. N. Mallinson, G. R. Webster, and P. Barkhan, Clearance from plasma and excretion in urine, faeces and bile of an intravenous dose of tritiated vitamin K₁ in man. *Br. J. Haematol. 22*:579–588 (1972).

101. J. P. Hart, The exploitation of the electrochemical properties of K vitamins for their sensitive measurement in tissues, in *Vitamin K and Vitamin K–Dependent Proteins: Analytical, Physiological, and Clinical Aspects* (M. J. Shearer and M. J. Seghatchian, eds.). CRC Press, Boca Raton, 1993, pp. 27–54.

102. Y. Haroon, Reaction detection methods for K vitamins and their 2′3′-epoxy metabolite in liquid chromatography, in *Vitamin K and Vitamin K–Dependent Proteins: Analytical, Physiological, and Clinical Aspects* (M. J. Shearer and M. J. Seghatchian, eds.). CRC Press, Boca Raton, 1993, pp. 55–90.

103. J. A. Sadowski, S. J. Hood, G. E. Dallal, and P. J. Garry, Phylloquinone in plasma from elderly and young adults: factors influencing its concentration. *Am. J. Clin. Nutr. 50*:100–108 (1989).

104. M. J. Shearer, P. T. McCarthy, O. E. Crampton, and M. B. Mattock, The assessment of human vitamin K status for tissue measurements, in *Current Advances in Vitamin K Research* (J. W. Suttie, ed.). Elsevier, New York, 1988, pp. 437–452.

105. S. J. Hodges, M. J. Pilkington, M. J. Shearer, L. Bitensky, and J. Chayen, Age-related changes in the circulating levels of congeners of vitamin K₂, menaquinone-7 and menaquinone-8. *Clin. Sci. 78*:63–66 (1990).

106. S. J. Hodges, M. J. Pilkington, T. C. B. Stamp, A. Catterall, M. J. Shearer, L. Bitensky, and J. Chayen, Depressed levels of circulating menaquinones in patients with osteoporotic fractures of the spine and femoral neck. *Bone 12*:387–389 (1991).

107. S. J. Hodges, K. Akesson, P. Vergnaud, K. Obrant, and P. D. Delmas, Circulating levels of vitamin K₁ and K₂ decreased in elderly women with hip fracture. *J. Bone Miner. Res. 8*:1241–1245 (1993).

108. J. W. Suttie, The importance of menaquinones in human nutrition. *Annu. Rev. Nutr. 15*:399–417 (1995).

109. M. Shino, Determination of endogenous vitamin K (phylloquinone and menaquinone-*n*) in plasma by high-performance liquid chromatography using platinum oxide catalyst reduction and fluorescence detection. *Analyst 113*:393–397 (1988).

110. K. Hirauchi, T. Sakano, T. Nagaoka, A. Morimoto, and S. Masuda, Measurement of K vitamins in biological materials by high-performance liquid chromatography with fluorometric detection, in *Proceedings of the 7th Symposium on Analytical Chemistry of Biological Substances*, Pharmaceutical Soc. Japan, 1985, pp. 43–46.

111. C. G. Kindberg, and J. W. Suttie, Effect of various intakes of phylloquinone on signs of vitamin K deficiency and serum and liver phylloquinone concentrations in the rat. *J. Nutr. 119*:175–180 (1989).

112. B. H. Will and J. W. Suttie, Comparative metabolism of phylloquinone and menaquinone-9 in rat liver. *J. Nutr. 122*:953–958 (1992).

113. P. M. Allison, L. L. Mummah-Schendel, C. G. Kindberg, C. S. Harms, N. U. Bang, and J. W. Suttie, Effects of a vitamin K-deficient diet and antibiotics in normal human volunteers. *J. Lab. Clin. Med. 110*:180–188 (1987).

114. J. W. Suttie, L. L. Mummah-Schendel, D. V. Shah, B. J. Lyle, and J. L. Greger, Vitamin K deficiency from dietary vitamin K restriction in humans. *Am. J. Clin. Nutr. 47*:475–480 (1988).

115. G. Ferland, J. A. Sadowski, and M. E. O'Brien, Dietary induced subclinical vitamin K deficiency in normal human subjects. *J. Clin. Invest. 91*:1761–1768 (1993).

116. Y. Usui, H. Tanimura, N. Nishimura, N. Kobayashi, T. Okanoue, and K. Ozawa, Vitamin

K concentrations in the plasma and liver of surgical patients. *Am. J. Clin. Nutr. 51*:846–852 (1990).

117. M. J. Thierry and J. W. Suttie, Distribution and metabolism of menadiol diphosphate in the rat. *J. Nutr. 97*:512–516 (1969).

118. M. J. Thierry and J. W. Suttie, Effect of warfarin and the chloro analog of vitamin K on phylloquinone metabolism. *Arch. Biochem. Biophys. 147*:430–435 (1971).

119. M. J. Thierry, M. A. Hermodson, and J. W. Suttie, Vitamin K and warfarin distribution and metabolism in the warfarin-resistant rat. *Am. J. Physiol. 219*:854–859 (1970).

120. T. Konishi, S. Baba, and H. Sone, Whole-body autoradiographic study of vitamin K distribution in rat. *Chem. Pharm. Bull. 21*:220–224 (1973).

121. J. P. Green, E. Søndergaard, and H. Dam, Intracellular distribution of vitamin K in beef liver. *Biochim. Biophys. Acta 19*:182–183 (1956).

122. R. G. Bell and J. T. Matschiner, Intracellular distribution of vitamin K in the rat. *Biochim. Acta 184*:597–603 (1969).

123. S. E. Nyquist, J. T. Matschiner, and D. J. James Morre, Distribution of vitamin K among rat liver cell fractions. *Biochim. Biophys. Acta 244*:645–649 (1971).

124. C. K. Reedstrom and J. W. Suttie, Comparative distribution, metabolism, and utilization of phylloquinone and menaquinone-9 in rat liver. *Proc. Soc. Exp. Biol. Med. 209*:403–409 (1995).

125. C. E. Kight, C. K. Reedstrom, and J. W. Suttie, Identification, isolation, and partial purification of a cytosolic binding protein for vitamin K from rat liver. *FASEB J 9*:A725 (1995).

126. T. E. Knauer, C. M. Siegfried, and J. T. Matschiner, Vitamin K requirement and the concentration of vitamin K in rat liver. *J. Nutr. 106*:1747–1756 (1976).

127. J. T. Matschiner, Occurrence and biopotency of various forms of vitamin K, in *The Fat-Soluble Vitamins* (H. F. DeLuca and J. W. Suttie, eds.). University of Wisconsin Press, Madison, 1970, pp. 377–397.

128. J. T. Matschiner, Isolation and identification of vitamin K from animal tissue, in *Symposium Proceedings on the Biochemistry, Assay, and Nutritional Value of Vitamin K and Related Compounds*. Assn. Vitam. Chem., Chicago, 1971, pp. 21–37.

129. K. Uchida and T. Komeno, Relationships between dietary and intestinal vitamin K, clotting factor levels, plasma vitamin K and urinary Gla, in *Current Advances in Vitamin K Research* (J. W. Suttie, ed.). Elsevier, New York, 1988, pp. 477–492.

130. Y. Usui, N. Nishimura, N. Kobayashi, T. Okanoue, M. Kimoto, and K. Ozawa, Measurement of vitamin K in human liver by gradient elution high-performance liquid chromatography using platinum-black catalyst reduction and fluorimetric detection. *J. Chromatogr. 489*:291–301 (1989).

131. S. Kayata, C. Kindberg, F. R. Greer, and J. W. Suttie, Vitamin K_1 and K_2 in infant human liver. *J. Pediatr. Gastroenterol. Nutr. 8*:304–307 (1989).

132. M. J. Shearer, P. T. McCarthy, O. E. Crampton, and M. B. Mattock, The assessment of human vitamin K status from tissue measurements, in *Current Advances in Vitamin K Research* (J. W. Suttie, ed.). Elsevier, New York, 1988, pp. 437–452.

133. B. H. Will, Y. Usui, and J. W. Suttie, Comparative metabolism and requirement of vitamin K in chicks and rats. *J. Nutr. 122*:2354–2360 (1992).

134. M. Guillaumont, H. Weiser, L. Sann, B. Vignal, M. Ledercq, and A. Frederich, Hepatic concentration of vitamin K active compounds after application of phylloquinone to chickens on a vitamin K deficient or adequate diet. *Int. J. Vitam. Nutr. Res. 62*:15–20 (1992).

135. H. H. W. Thijssen and M. J. Drittij-Reijnders, Vitamin K distribution in rat tissues: dietary phylloquinone is a source of tissue menaquinone-4. *Br. J. Nutr. 72*:415–425 (1994).

135a. R. T. Davidson, A. L. Foley, J. A. Engelke, and J. W. Suttie, Conversion of dietary phylloquinone to tissue menaquinone-4 in rats is not dependent on gut bacteria. *J. Nutr. 128*:220–223 (1998).

135b. J. E. Ronden, M. J. Drittij-Reijnders, C. Vermeer, and H. H. W. Thijssen, Intestinal flora is not an intermediate in the phylloquinone-menaquinone-4 conversion in the rat. *Biochem. Biophys. Acta 1379*:69–75 (1998).

136. R. Bentley and R. Meganathan, Biosynthesis of vitamin K (menaquinone) in bacteria. *Microbiol. Rev. 46*:241–280 (1982).

137. C. Martius, Chemistry and function of vitamin K, in *Blood Clotting Enzymology* (W. H. Seegers, ed.). Academic Press, New York, 1967, pp. 551–575.

138. C. Martius, The metabolic relationships between the different K vitamins and the synthesis of the ubiquinones. *Am. J. Clin. Nutr. 9*:97–103 (1961).

139. G. H. Dialameh, K. G. Yekundi, and R. E. Olson, Enzymatic alkylation of menaquinone-O to menaquinones by microsomes from chick liver. *Biochim. Biophys. Acta 223*:332–338 (1970).

140. C. Martius, Recent investigations on the chemistry and function of vitamin K, in *Ciba Fdn. Symposium on Quinones in Electron Transport* (G. E. W. Wolstenholme and C. M. O'Connor, eds.). Little, Brown, Boston, 1961, pp. 312–326.

141. M. Billeter, W. Bollinger, and C. Martius, Untersuchungen über die Umwandlung von verfutterten K-Vitaminen durch Austausch der Seitenkette und die Rolle der Darmbakterien hierbei. *Biochem. Z. 340*:290–303 (1964).

142. F. C. G. Hoskin, J. W. T. Spinks, and L. B. Jaques, Urinary excretion products of menadione (vitamin K_3). *Can. J. Biochem. Physiol. 32*:240–250 (1954).

143. K. T. Hart, Study of hydrolysis of urinary metabolites of 2-methyl-1,4-naphthoquinone. *Proc. Soc. Exp. Biol. Med. 97*:848–851 (1958).

144. R. Losito, C. A. Owen, Jr., and E. V. Flock, Metabolism of [C^{14}] menadione. *Biochemistry 6*:62–68 (1967).

145. R. Losito, C. A. Owen, Jr., and E. V. Flock, Metabolic studies of vitamin K_1-^{14}C and menadione-^{14}C in the normal and hepatectomized rats. *Thromb. Diath. Haemorrhag. 19*:383–388 (1968).

146. J. D. Taylor, G. J. Miller, L. B. Jaques, and J. W. T. Spinks, The distribution of administered vitamin K_1-^{14}C in rats. *Can. J. Biochem. Physiol. 34*:1143–1152 (1956).

147. O. Wiss and H. Gloor, Absorption, distribution, storage and metabolites of vitamin K and related quinones. *Vitam Hormones 24*:575–586 (1966).

148. J. T. Matschiner, R. G. Bell, J. M. Amelotti, and T. E. Knauer, Isolation and characterization of a new metabolite of phylloquinone in the rat. *Biochim. Biophys. Acta 201*:309–315 (1970).

149. R. G. Bell, J. A. Sadowski, and J. T. Matschiner, Mechanism of action of warfarin. Warfarin and metabolism of vitamin K_1. *Biochemistry 11*:1959–1961 (1972).

150. M. J. Shearer, A. McBurney, and P. Barkhan, Effect of warfarin anticoagulation on vitamin-K_1 metabolism in man. *Br. J. Haematol. 24*:471–479 (1973).

151. M. J. Shearer, A. McBurney, A. M. Breckenridge, and P. Barkhan, Effect of warfarin on the metabolism of phylloquinone (vitamin K_1): dose–response relationships in man. *Clin. Sci. Mol. Med. 52*:621–630 (1977).

152. E. W. Davie, K. Fujikawa, and W. Kisiel, The coagulation cascade: initiation, maintenance, and regulation. *Biochemistry 30*:10363–10370 (1991).

153. L. H. Clouse and P. C. Comp, The regulation of hemostasis: the protein C system. *N. Engl. J. Med. 314*:1298–1304 (1986).

154. H. C. Hemker, J. J. Veltkamp, A. Hensen, and E. A. Loeliger, Nature of prothrombin biosynthesis: preprothrombinaemia in vitamin K-deficiency. *Nature 200*:589–590 (1963).

155. D. V. Shah and J. W. Suttie, Mechanism of action of vitamin K: evidence for the conversion of a precursor protein to prothrombin in the rat. *Proc. Natl. Acad. Sci. USA 68*:1653–1657 (1971).

156. P. O. Ganrot and J. E. Nilehn, Plasma prothrombin during treatment with dicumarol. II. Demonstration of an abnormal prothrombin fraction. *Scand. J. Clin Lab. Invest. 22*:23–28 (1968).

157. J. Stenflo, Dicumarol-induced prothrombin in bovine plasma. *Acta Chem. Scand. 24*:3762–3763 (1970).

158. T. L. Carlisle, D. V. Shah, R. Schlegel, and J. W. Suttie, Plasma abnormal prothrombin and microsomal prothrombin precursor in various species. *Proc. Soc. Exp. Biol. Med. 148*:140–144 (1975).

159. J. Stenflo and J. W. Suttie, Vitamin K-dependent formation of γ-carboxyglutamic acid. *Annu. Rev. Biochem. 46*:157–172 (1977).

160. C. T. Esmon, J. W. Suttie, and C. M. Jackson, The functional significance of vitamin K action. Difference in phospholipid binding between normal and abnormal prothrombin. *J. Biol. Chem. 250*:4095–4099 (1975).

161. J. Stenflo, P. Fernlund, W. Egan, and P. Roepstorff, Vitamin K dependent modifications of glutamic acid residues in prothrombin. *Proc. Natl. Acad. Sci. USA 71*:2730–2733 (1974).

162. G. L. Nelsestuen, T. H. Zytkovicz, and J. B. Howard, The mode of action of vitamin K. Identification of γ-carboxyglutamic acid as a component of prothrombin. *J. Biol. Chem. 249*:6347–6350 (1974).

163. S. Magnusson, L. Sottrup-Jensen, T. E. Petersen, H. R. Morris, and A. Dell, Primary structure of the vitamin K–dependent part of prothrombin. *FEBS Lett. 44*:189–193 (1974).

164. C. T. Esmon, J. A. Sadowski, and J. W. Suttie, A new carboxylation reaction. The vitamin K-dependent incorporation of $H^{14}CO_3^-$ into prothrombin. *J. Biol. Chem. 250*:4744–4748 (1975).

165. J. W. Suttie, J. M. Hageman, S. R. Lehrman, and D. H. Rich, Vitamin K–dependent carboxylase: development of a peptide substrate. *J. Biol. Chem. 251*:5827–5830 (1976).

166. T. L. Carlisle and J. W. Suttie, Vitamin K dependent carboxylase: subcellular location of the carboxylase and enzymes involved in vitamin K metabolism in rat liver. *Biochemistry 19*:1161–1167 (1980).

167. B. C. Johnson, Post-translational carboxylation of preprothrombin. *Mol. Cell. Biochem. 38*:77–121 (1981).

168. R. E. Olson, The function and metabolism of vitamin K. *Annu. Rev. Nutr. 4*:281–337 (1984).

169. J. W. Suttie, Vitamin K–dependent carboxylase. *Annu. Rev. Biochem. 54*:459–477 (1985).

170. J. W. Suttie. Vitamin K–dependent carboxylation of glutamyl residues in proteins. *BioFactors 1*:55–60 (1988).

171. J. W. Suttie, Synthesis of vitamin K–dependent proteins. *FASEB J. 7*:445–452 (1993).

172. C. Vermeer and M. A. G. de Boer-van den Berg, Vitamin K–dependent carboxylase. *Haematologia 18*:71–97 (1985).

173. C. Vermeer. γ-Carboxyglutamate-containing proteins and the vitamin K–dependent carboxylase. *Biochem. J. 266*:625–636 (1990).

174. A. Y. Cheung, G. M. Wood, S. Funakawa, C. P. Grossman, and J. W. Suttie, Vitamin K–dependent carboxylase: substrates, products, and inhibitors, in *Current Advances in Vitamin K Research* (J. W. Suttie, ed.). Elsevier, New York, 1988, pp. 3–16.

175. P. A. Price, J. D. Fraser, and G. Metz-Virca, Molecular cloning of matrix Gla protein: implications for substrate recognition by the vitamin K–dependent γ-carboxylase. *Proc. Natl. Acad. Sci. USA 84*:8335–8339 (1987).

176. D. J. Liska and J. W. Suttie, Location of γ-carboxyglutamyl residues in partially carboxylated prothrombin preparations. *Biochemistry 27*:8636–8641 (1988).

177. M. E. Benton, P. A. Price, and J. W. Suttie, Multi-site-specificity of the vitamin K–dependent carboxylase: *in vitro* carboxylation of des-γ-carboxylated bone Gla protein and des-γ-carboxylated pro bone Gla protein. *Biochemistry 34*:9541–9551 (1995).

178. R. Azerad, P. Decottignies-Le Marechal, C. Ducrocq, A. Righini-Tapie, A. Vidal-Cros, S. Bory, J. Dubois, M. Gaudry, and A. Marquet, The vitamin K–dependent carboxylation of peptide substrates: stereochemical features and mechanistic studies with substrate analogues, in *Current Advances in Vitamin K Research* (J. W. Suttie, ed.). Elsevier, New York, 1988, pp. 17–23.

179. J. Dubois, C. Dugave, C. Foures, M. Kaminsky, J-C. Tabet, S. Bory, M. Gaudry, and A.

Marquet, Vitamin K dependent carboxylation: determination of the stereochemical course using 4-fluoroglutamyl-containing substrate. *Biochemistry 30*:10506–10512 (1991).

180. P. A. Friedman, M. A. Shia, P. M. Gallop, and A. E. Griep, Vitamin K–dependent γ-carbon–hydrogen bond cleavage and the non-mandatory concurrent carboxylation of peptide bound glutamic acid residues. *Proc. Natl. Acad. Sci. USA 76*:3126–3129 (1979).

181. A. E. Larson, P. A. Friedman, and J. W. Suttie, Vitamin K–dependent carboxylase: stoichiometry of carboxylation and vitamin K 2,3-epoxide formation. *J. Biol. Chem. 256*:11032–11035 (1981).

182. J. J. McTigue and J. W. Suttie, Vitamin K-dependent carboxylase: demonstration of a vitamin K- and O_2-dependent exchange of ^3H from ^3H_2O into glutamic acid residues. *J. Biol. Chem. 258*:12129–12131 (1983).

183. D. L. Anton and P. A. Friedman, Fate of the activated γ-carbon–hydrogen bond in the uncoupled vitamin K–dependent γ-glutamyl carboxylation reaction. *J. Biol. Chem. 258*:14084–14087 (1983).

184. G. M. Wood and J. W. Suttie, Vitamin K-dependent carboxylase. Stoichiometry of vitamin K epoxide formation, γ-carboxyglutamyl formation, and γ-glutamyl-^3H cleavage. *J. Biol. Chem. 263*:3234–3239 (1988).

185. P. Dowd, S. W. Ham, and S. J. Geib, Mechanism of action of vitamin K. *J. Am. Chem. Soc. 113*:7734–7743 (1991).

186. P. Dowd, S-W. Ham, and R. Hershline, Role of oxygen in the vitamin K–dependent carboxylation reaction: incorporation of a second atom of ^{18}O from molecular oxygen-$^{18}O_2$ into vitamin K oxide during carboxylase activity. *J. Am. Chem. Soc. 114*:7613–7617 (1992).

187. A. Kuliopulos, B. R. Hubbard, Z. Lam, I. J. Koski, B. Furie, B. C. Furie, and C. T. Walsh, Dioxygen transfer during vitamin K–dependent carboxylase catalysis. *Biochemistry 31*: 7722–7728 (1992).

188. P. Dowd, S. W. Ham, S. Naganathan, and R. Hershline, The mechanism of action of vitamin K. *Annu. Rev. Nutr. 15*:419–440 (1995).

189. A. Ichinose and E. W. Davie, The blood coagulation factors: their cDNAs, genes, and expression, in *Hemostasis and Thrombosis: Basic Principles and Clinical Practice*, 3rd ed. (R. W. Colman, J. Hirsh, V. J. Marder, and E. W. Salzman, eds.). J. B. Lippincott, Philadelphia, 1994, pp. 19–54.

190. B. Furie and B. C. Furie, Molecular and cellular biology of blood coagulation. *N. Engl. J. Med. 326*:800–806 (1992).

191. J. E. Knobloch and J. W. Suttie, Vitamin K–dependent carboxylase. Control of enzyme activity by the "propeptide" region of factor X. *J. Biol. Chem. 262*:15334–15337 (1987).

192. A. Cheung, J. W. Suttie, and M. Bernatowicz, Vitamin K–dependent carboxylase: structural requirements for peptide activation. *Biochim. Biophys. Acta 1039*:90–93 (1990).

193. M. C. Harbeck, A. Y. Cheung, and J. W. Suttie, Vitamin K–dependent carboxylase: partial purification of the enzyme by antibody affinity techniques. *Thromb. Res. 56*:317–323 (1989).

194. B. R. Hubbard, M. M. W. Ulrich, M. Jacobs, C. Vermeer, C. Walsh, B. Furie, and B. C. Furie, Vitamin K–dependent carboxylase: affinity purification from bovine liver by using a synthetic propeptide containing the γ-carboxylation recognition site. *Proc. Natl. Acad. Sci. USA 86*:6893–6897 (1989).

195. A. Kuliopulos, C. E. Cieurzo, B. Furie, B. C. Furie, and C. T. Walsh, *N*-Bromoacetyl peptide substrate affinity labeling of vitamin K–dependent carboxylase. *Biochemistry 31*:9436–9444 (1992).

196. S-M. Wu, D. P. Morris, and D. W. Stafford, Identification and purification to near homogeneity of the vitamin K–dependent carboxylase. *Proc. Natl. Acad. Sci. USA 88*:2236–2240 (1991).

197. S-M. Wu, W-F. Cheung, D. Frazier, and D. Stafford, Cloning and expression of the cDNA for human γ-glutamyl carboxylase. *Science 254*:1634–1636 (1991).

198. K. L. Berkner, M. Harbeck, S. Lingenfelter, C. Bailey, C. M. Sanders-Hinck, and J. W.

Suttie, Purification and identification of bovine liver γ-carboxylase. *Proc. Natl. Acad. Sci. USA 89*:6242–6246 (1992).

199.   D. A. Roth, A. Rehemtulla, R. J. Kaufman, C. T. Walsh, B. Furie, and B. C. Furie, Expression of bovine vitamin K–dependent carboxylase activity in baculovirus infected cells. *Proc. Natl. Acad. Sci. USA 90*:8372–8376 (1993).

199a.   B. Furie, B. A. Bouchard, and B. C. Furie, Vitamin K–dependent biosynthesis of γ-carboxy-glutamic acid. *Blood 93*:1798–1808 (1999).

199b.   T. B. Stanley, S. M. Wu, R. J. Houben, V. P. Mutucumarana, and D. W. Stafford, Role of the propeptide and γ-glutamic acid domain of factor IX for in vitro carboxylation by the vitamin K–dependent carboxylase. *Biochemistry 37*:13262–13268 (1998).

199c.   W. Wu, J. D. Bancroft, and J. W. Suttie, Structural features of the kringle domain determine the intracellular degradation of under-γ-carboxylated prothrombin: studies of chimeric rat/human prothrombin. *Proc. Natl. Acad. Sci. USA 94*:13654–13660 (1997).

199d.   M. E. Benton, P. A. Price, and J. W. Suttie, Multi-site-specificity of the vitamin K–dependent carboxylase: in vitro carboxylation of des-γ-carboxylated bone Gla protein and Des-γ-carbox-ylated pro bone Gla protein. *Biochemistry 34*:9541–9551 (1995).

199e.   D. P. Morris, R. D. Stevens, D. J. Wright, and D. W. Stafford, Processive post-translational modification. Vitamin K–dependent carboxylation of a peptide substrate. *J. Biol. Chem. 270*: 30491–30498 (1995).

199f.   K. L. Berkner and B. N. Pudota, Vitamin K–dependent carboxylation of the carboxylase. *Proc. Natl. Acad. Sci. USA 95*:466–472 (1998).

200.   P. V. Hauschka, J. B. Lian, and P. M. Gallop, Direct identification of the calcium-binding amino acid γ-carboxyglutamate, in mineralized tissue. *Proc. Natl. Acad. Sci. USA 72*:3925–3929 (1975).

201.   P. A. Price, A. S. Otsuka, J. W. Poser, J. Kristaponis, and N. Raman, Characterization of a γ-carboxyglutamic acid-containing protein from bone. *Proc. Natl. Acad. Sci. USA 73*:1447–1451 (1976).

202.   P. A. Price and M. K. Williamson, Primary structure of bovine matrix Gla protein, a new vitamin K–dependent bone protein. *J. Biol. Chem. 260*:14971–14975 (1985).

203.   J. D. Fraser and P. A. Price, Lung, heart, and kidney express high levels of mRNA for the vitamin K–dependent matrix Gla protein. *J. Biol. Chem. 263*:11033–11036 (1988).

204.   C. Maillard, M. Berruyer, C. M. Serre, M. Dechavanne, and P. D. Delmas, Protein S, a vitamin K–dependent protein, is a bone matrix component synthesized and secreted by osteo-blasts. *Endocrinology 130*:1599–1604 (1992).

205.   J. B. Lian, P. V. Hauschka, and P. M. Gallop, Properties and biosynthesis of a vitamin K–dependent calcium binding protein in bone. *Fed. Proc. 37*:2615–2620 (1978).

206.   P. A. Price, Vitamin K–dependent formation of bone Gla protein (osteocalcin) and its func-tion. *Vitam Hormones 42*:65–108 (1985).

207.   P. A. Price, Role of vitamin K–dependent proteins in bone metabolism. *Annu. Rev. Nutr. 8*:565–583 (1988).

208.   C. Vermeer, K-S. G. Jie, and M. H. J. Knapen, Role of vitamin K in bone metabolism. *Annu. Rev. Nutr. 15*:1–22 (1995).

209.   P. A. Price and M. K. Williamson, Effects of warfarin on bone. Studies on the vitamin K–dependent protein of rat bone. *J. Biol. Chem. 256*:12754–12759 (1981).

210.   P. A. Price, M. K. Williamson, T. Haba, R. B. Dell, and W. S. S. Jee, Excessive mineralization with growth plate closure in rats on chronic warfarin treatment. *Proc. Natl. Acad. Sci. USA 79*:7734–7738 (1982).

211.   P. A. Price, J. G. Parthemore, and L. J. Deftos, New biochemical marker for bone metabolism. Measurement by radioimmunoassay of bone Gla protein in the plasma of normal subjects and patients with bone disease. *J. Clin. Invest. 66*:878–883 (1980).

212.   P. D. Delmas, L. Malaval, M. E. Arlot, and P. J. Meunier, Serum bone Gla-protein compared to bone histomorphometry in endocrine diseases. *Bone 6*:339–341 (1985).

213. P. A. Price, Vitamin K nutrition and postmenopausal osteoporosis (editorial). *J. Clin. Invest.* *91*:1268 (1993).

214. N. C. Binkley and J. W. Suttie, Vitamin K nutrition and osteoporosis *J. Nutr. 125*:1812–1821 (1995).

214a. P. Vergnaud, P. Garnero, P. J. Meunier, G. Breart, K. Kamihagi, and P. D. Delmas, Undercarboxylated osteocalcin measured with a specific immunoassay predicts hip fractures in elderly women: the EPIDOS study. *J. Clin. Endocrinol. Metab. 82*:719–724 (1997).

214b. D. Feskanich, P. Weber, W. C. Willett, H. Rockett, S. L. Booth, and G. A. Colditz, Vitamin K intake and hip fractures in women: a prospective study. *Am. J. Clin. Nutr. 69*:74–79 (1999).

214c. P. J. Caraballo, S. E. Gabriel, M. R. Castro, E. J. Atkinson, and L. J. Melton, Changes in bone density after exposure to oral anticoagulants: a meta-analysis. *Osteoporosis Int. 9*:441–448 (1999).

214d. P. Ducy, C. Desbois, B. Boyce, G. Pinero, B. Story, C. Dunstan, E. Smith, J. Bonadio, S. Goldstein, C. Gundberg, A. Bradley, and G. Karsenty, Increased bone formation in osteocalcin-deficient mice. *Nature 382*:448–452 (1996).

214e. G. Luo, P. Ducy, M. D. McKee, G. J. Pinero, E. Loyer, R. R. Behringer, G. Karsenty, Spontaneous calcification of arteries and cartilage in mice lacking matrix GLA protein. *Nature 386*: 78–81 (1997).

215. J. B. Lian, R. J. Levy, J. T. Levy, and P. A. Friedman, Other vitamin K dependent proteins, in *Calcium-Binding Proteins: Structure and Function* (F. L. Siegel, E. Carafoli, R. H. Kretsinger, D. H. MacLennan, and R. H. Wasserman, eds.). Elsevier, New York, 1980, pp. 449–460.

216. J. A. Helpern, S. J. McGee, and J. M. Riddle, Observations suggesting a possible link between gamma-carboxyglutamic acid and porcine bioprosthetic valve calcification. *Henry Ford Hosp. Med. J. 30*:152–155 (1982).

217. L. J. M. van Haarlem, B. A. M. Soute, H. C. Hemker, and C. Vermeer, Characterization of Gla-containing proteins from calcified human atherosclerotic plaques, in *Current Advances in Vitamin K Research* (J. W. Suttie, ed.). Elsevier, New York, 1988, pp. 287–292.

218. A. E. Griep and P. A. Friedman, Purification of a protein containing γ-carboxyglutamic acid from bovine kidney, in *Vitamin K Metabolism and Vitamin K-Dependent Proteins* (J. W. Suttie, ed.). University Park Press, Baltimore, 1980, pp. 307–310.

219. A. Gardemann and G. F. Domagk, The occurrence of γ-carboxyglutamate in a protein isolated from ox liver mitochondria. *Arch. Biochem. Biophys. 220*: 347–353 (1983).

220. D. M. Smalley and P. C. Preusch, Analysis of γ-carboxyglutamic acid by reverse phase HPLC of its phenylthiocarbamyl derivative. *Anal. Biochem. 172*:241–247 (1988).

221. X-J. Wen and M. Thierry-Palmer, Identification of a vitamin K–dependent protein in rat liver mitochondria. *FASEB J. 9*:A725, 1995.

222. B. A. M. Soute, W. Muller-Ester, M. A. G. de Boer-van den Berg, M. Ulrich, and C. Vermeer, Discovery of a γ-carboxyglutamic acid-containing protein in human spermatozoa. *FEBS Lett. 190*:137–141 (1985).

223. Nakagawa, Y., V. Abram, F. J. Kezdy, E. T. Kaiser, and F. L. Coe, Purification and characterization of the principal inhibitor of calcium oxalate monohydrate crystal growth in human urine. *J. Biol. Chem. 258*:12594–12600 (1983).

223a. J. D. Kulman, J. E. Harris, B. A. Haldeman, and E. W. Davie, Primary structure and tissue distribution of two novel proline-rich γ-carboxyglutamic acid proteins. *Proc. Natl. Acad. Sci. USA 94*:9058–9062 (1997).

224. G. Manfioletti, C. Bancolini, G. Avanzi, and C. Schneider, The protein encoded by a growth arrest–specific gene (gas6) is a new member of the vitamin K-dependent proteins related to protein S, a negative coregulator in the blood coagulation cascade. *Mol. Cell. Biol. 13*:4976–4985 (1993).

225. P. V. Hauschka, E. A. Mullen, G. Hintsch, and S. Jazwinski, Abundant occurrence of γ-carboxyglutamic acid (Gla)-containing peptides in the marine gastropod family *Conidae*, in

Current Advances in Vitamin K Research (J. W. Suttie, ed.). Elsevier, New York, 1988, pp. 237–243.

226. G. Tans, J. W. P. Govers-Riemslag, J. L. M. L. van Rijn, and J. Rosing, Purification and properties of a prothrombin activator from the venom of *Notechis scutatus scutatus. J. Biol. Chem. 260*:9366–9370 (1985).

227. B. M. Olivera, W. R. Gray, R. Zeikus, J. M. McIntosh, J. Varga, J. R. Victoria de Santos, and L. J. Cruz, Peptide neurotoxins from fish-hunting cone snails. *Science 230*:1338–1343 (1985).

228. J. Hirsh, J. S. Ginsberg, and V. J. Marder, Anticoagulant therapy with coumarin agents, in *Hemostasis and Thrombosis: Basic Principles and Clinical Practice*, 3rd ed. (R. W. Colman, J. Hirsh, V. J. Marder, and E. W. Salzman, eds.). J. B. Lippincott, Philadelphia, 1994, pp. 1567–1583.

228a. D. Cain, S. M. Hutson, and R. Wallin, Assembly of the warfarin-sensitive vitamin K 2,3-epoxide reductase enzyme complex in the endoplasmic reticulum membrane. *J. Biol. Chem. 272*:29068–29075 (1997).

229. L. J. M. van Haarlem, B. A. M. Soute, and C. Vermeer, Vitamin K–dependent carboxylase. Possible role for thioredoxin in the reduction of vitamin K metabolites in liver. *FEBS Lett. 222*:353 (1987).

230. R. B. Silverman and D. L. Nandi, Reduced thioredoxin: a possible physiological cofactor for vitamin K epoxide reductase. Further support for an active site disulfide. *Biochem. Biophys. Res. Commun. 155*:1248 (1988).

231. J. T. Matschiner, R. G. Bell, J. M. Amelotti, and T. E. Knauer, Isolation and characterization of a new metabolite of phylloquinone in the rat. *Biochim. Biophys. Acta 201*:309 (1970).

232. R. G. Bell and J. T. Matschiner, Warfarin and the inhibition of vitamin K activity by an oxide metabolite. *Nature 237*:32 (1972).

233. J. A. Sadowski and J. W. Suttie, Mechanism of action of coumarins. Significance of vitamin K epoxide. *Biochemistry 13*:3696–3699 (1974).

234. A. Zimmermann and J. T. Matschiner, Biochemical basis of hereditary resistance to warfarin in the rat. *Biochem. Pharmacol. 23*:1033 (1974).

235. D. S. Whitlon, J. A. Sadowski, and J. W. Suttie, Mechanism of coumarin action: significance of vitamin K epoxide reductase inhibition. *Biochemistry 17*:1371 (1978).

236. M. J. Fasco and L. M. Principe, R-and S-warfarin inhibition of vitamin K and vitamin K 2,3-epoxide reductase activities in the rat. *J. Biol. Chem. 257*:4894 (1982).

237. M. J. Fasco, E. F. Hildebrandt, and J. W. Suttie, Evidence that warfarin anticoagulant action involves two distinct reductase activities. *J. Biol. Chem. 257*:11210 (1982).

238. E. F. Hildebrandt and J. W. Suttie, Mechanism of coumarin action: sensitivity of vitamin K metabolizing enzymes of normal and warfarin resistant rat liver. *Biochemistry 21*:2406 (1982).

239. R. Wallin, S. D. Patrick, and L. F. Martin, Vitamin K_1 reduction in human liver. *Biochem. J. 260*:879 (1989).

240. S. L. Gardill and J. W. Suttie, Vitamin K epoxide and quinone reductase activities: evidence for reduction by a common enzyme. *Biochem. Pharmacol. 40*:1055 (1990).

241. K. W. E. Denson and R. Biggs, Laboratory diagnosis, tests of clotting function and their standardization, in *Human Blood Coagulation, Haemostasis and Thrombosis* (R. Biggs, ed.). Blackwell Scientific Publishers, 1972, pp. 278–332.

242. B. R. J. Kirchhof, C. Vermeer, and H. C. Hemker, The determination of prothrombin using synthetic chromogenic substrates; choice of a suitable activator. *Thromb. Res. 13*:219–232 (1978).

243. K. W. E. Denson, R. Borrett, and R. Biggs, The specific assay of prothrombin using the Taipan snake venom. *Br. J. Haematol. 21*:219–226 (1971).

244. J. W. Suttie, Vitamin K and human nutrition. *J. Am. Dietet. Assoc. 92*:585–590 (1992).

245. M. L. Scott, Vitamin K in animal nutrition. *Vitam Hormones 24*:633–647 (1966).

246. E. A. Doisy and J. T. Matschiner, Biochemistry of vitamin K, in *Fat-Soluble Vitamins* (R. A. Morton, ed.). Pergamon Press, Oxford, 1970, pp. 293–331.

247. National Academy of Sciences. *Nutritional Requirements of Laboratory Animals*, 3rd ed., Washington, D.C. 1978.

248. P. G. Reeves, F. H. Nielsen, and G. C. Fahey, Jr., AIN-93 purified diets for laboratory rodents: final report of the American Institute of Nutrition *ad Hoc* Writing Committee on the Reformulation of the AIN-76A Rodent Diet. *J. Nutr. 123*:1939–1951 (1993).

249. P. G. Frick, G. Riedler, and H. Brogli, Dose response and minimal daily requirement for vitamin K in man. *J. Appl. Physiol. 23*:387–389 (1967).

250. R. A. O'Reilly, Vitamin K in hereditary resistance to oral anticoagulant drugs. *Am. J. Physiol. 221*:1327–1330 (1971).

251. J. W. Suttie, L. L. Mummah-Schendel, D. V. Shah, B. J. Lyle, and J. L. Greger, Development of human vitamin K deficiency by dietary vitamin K restriction. *Am. J. Clin. Nutr. 47*:475–480 (1988).

252. G. Ferland, J. A. Sadowski, and M. E. O'Brien, Dietary induced subclinical vitamin K deficiency in normal human subjects. *J. Clin. Invest. 91*:1761–1768 (1993).

253. R. A. Blanchard, B. C. Furie, M. Jorgensen, S. F. Kruger, and B. Furie, Acquired vitamin K-dependent carboxylation deficiency in liver disease. *N. Engl. J. Med. 305*:242–248 (1981).

254. K-S. G. Jie, K. Hamulyak, B. L. M. G. Gijsbers, F. J. M. E. Roumen, and C. Vermeer, Serum osteocalcin as a marker for vitamin K-status in pregnant women and their newborn babies. *Thromb. Haemost. 68*:388–391 (1992).

255. M. H. J. Knapen, K. Hamulyak, and C. Vermeer, The effect of vitamin K supplementation on circulating osteocalcin (bone Gla protein) and urinary calcium excretion. *Ann. Intern. Med. 111*:1001–1005 (1989).

255a. S. L. Booth, M. E. O'Brien-Morse, G. E. Dallal, K. W. Davidson, and C. M. Gundberg, Response of vitamin K status to different intakes and sources of phylloquinone-rich foods: comparison of younger and older adults. *Am. J. Clin Nutr. 70*:368–377 (1999).

256. K. Hazell and K. H. Baloch, Vitamin K deficiency in the elderly. *Gerontol. Clin. 12*:10–17 (1970).

257. E. A. Doisy, Jr., Nutritional hypoprothrombinemia and metabolism of vitamin K. *Fed. Proc. 20*:989–994 (1961).

258. P. A. Lane and W. E. Hathaway, Vitamin K in infancy. *J. Pediatr. 106*:351–359 (1985).

259. M. Andrew, B. Paes, R. Milner, M. Johnston, L. Mitchell, D. M. Tollefsen, and P. Powers, Development of the human coagulation system in the full-term infant. *Blood 70*:165–172 (1987).

260. R. v. Kries, F. R. Greer, and J. W. Suttie, Assessment of vitamin K status of the newborn infant. *J. Pediatr. Gastroenterol. Nutr. 16*:231–238 (1993).

261. Kries, R. v., A. Becker, and U. Gobel, Vitamin K in the newborn: influence of nutritional factors on acarboxy-prothrombin detectability and factor II and VII clotting activity. *Eur. J. Pediatr. 146*:123–127 (1987).

262. K. Motohara, I. Matsukane, F. Endo, Y. Kiyota, and I. Matsuda, Relationship of milk intake and vitamin K supplementation to vitamin K status in newborns. *Pediatrics 84*:90–93 (1989).

263. B. E. Gustafsson, F. S. Daft, E. G. McDaniel, and J. C. Smith, Effects of vitamin K-active compounds and intestinal microorganisms in vitamin K–deficient germfree rats. *J. Nutr. 78*: 461–468 (1962).

264. T. S. Nelson and L. C. Norris, Studies on the vitamin K requirement of the chick. II. Effect of sulfaquinoxaline on the quantitative requirements of the chick for vitamin K_1, menadione and menadione sodium bisulfite. *J. Nutr. 73*:135–142 (1961).

265. D. Savage and J. Lindenbaum, Clinical and experimental human vitamin K deficiency, in *Nutrition in Hematology* (J. Lindenbaum, ed.). Churchill Livingstone, New York, 1983, pp. 271–320.

266. J. J. Lipsky, Mechanism of the inhibition of the γ-carboxylation of glutamic acid by *N*-methylthiotetrazole-containing antibiotics. *Proc. Natl. Acad. Sci. USA 81*:2893–2897 (1984).

267. H. Bechtold, K. Andrassy, E. Jahnchen, J. Koderisch, H. Koderisch, L. S. Weilemann, H-G. Sonntag, and E. Ritz, Evidence for impaired hepatic vitamin K_1 metabolism in patients treated with N-methyl-thiotetrazole cephalosporins. *Thromb. Haemostas.* (Stuttg). *51*:358–361 (1984).

268. K. A. Creedon and J. W. Suttie, Effect of N-methyl-thiotetrazole on vitamin K epoxide reductase. *Thromb. Res. 44*:147–153 (1986).

269. H. Suzuki, T. Nakao, and K, Hiraga, Vitamin K deficiency in male rats fed diets containing butylated hydroxytoluene (BHT). *Toxicol. Appl. Pharmacol. 50*:261–266 (1979).

270. O. Takahashi and K. Hiraga, Preventive effects of phylloquinone on hemorrhagic death induced by butylated hydroxytoluene in male rats. *J. Nutr. 109*:453–457 (1979).

271. K. R. Mountain, A. S. Gallus, and J. Hirsch, Neonatal coagulation defection due to anticonvulsant drug treatment in pregnancy. *Lancet: 1*:265 (1970).

272. A. S. Rogen and J. C. Ferguson, Clinical observation on patients treated with atronid and anticoagulants. *J. Atheroscler. Res. 3*:671–676 (1963).

273. J. G. Hall, R. M. Pauli, and K. M. Wilson, Maternal and fetal sequelae on anticoagulation during pregnancy. *Am. J. Med. 68*:122–140 (1980).

274. J. T. Matschiner and R. G. Bell, Effect of sex and sex hormones on plasma prothrombin and vitamin K deficiency. *Proc. Soc. Exp. Biol. Med. 144*:316–320 (1973).

275. J. T. Matschiner and A. K. Willingham, Influence of sex hormones on vitamin K deficiency and epoxidation of vitamin K in the rat. *J. Nutr. 104*:660–665 (1974).

276. D. W. Jolly, B. M. Kadis, and T. E. Nelson, Jr., Estrogen and prothrombin synthesis. The prothrombinogenic action of estrogen. *Biochem. Biophys. Res. Commun. 74*:41–49 (1977).

277. C. M. Siegfried, G. R. Knauer, and J. T. Matschiner, Evidence for increased formation of preprothrombin and the noninvolvement of vitamin K–dependent reactions in sex-linked hyperprothrombinemia in the rat. *Arch. Biochem. Biophys. 194*:486–495 (1979).

278. Y. Nishino, Hormonal control of prothrombin synthesis in rat liver microsomes, with special reference to the role of estradiol, testosterone and prolactin. *Arch. Toxicol. Suppl. 2*:397–402 (1979).

279. A. T. van Oosterom, P. Kerkhoven, and J. J. Veltkamp, Metabolism of the coagulation factors of the prothrombin complex in hypothyroidism in man. *Thrombos. Haemostas.* (Stuttg.) *41*: 273–285 (1979).

280. M. C. Elliott, B. Isaacs, and A. C. Ivy, Production of ''prothrombin deficiency'' and response to vitamins A, D and K. *Proc. Soc. Exp. Biol. Med. 43*:240–245 (1940).

281. G. H. Rao and K. E. Mason, Antisterility and antivitamin K activity of *d*-α-tocopheryl hydroquionone in the vitamin E–deficient female rat. *J. Nutr. 105*:495–498 (1975).

282. J. J. Corrigan, Jr., and F. I. Marcus, Coagulopathy associated with vitamin E ingestion. *JAMA 230*:1300–1301 (1974).

283. L. Uotila, Inhibition of vitamin K-dependent carboxylase by vitamin E and its derivatives, in *Current Advances in Vitamin K Research* (J. W. Suttie, ed.). Elsevier, New York, 1988, pp. 59–64.

284. C. A. Owen, Vitamin K group. XI. Pharmacology and toxicology, in *The Vitamins*, Vol. 3, 2nd ed. (W. H. Sebrell and R. S. Harris, eds.). Academic Press, New York, 1971, pp. 492–509.

285. National Research Council, *Vitamin Tolerance of Animals*, National Academy Press, Washington, D.C., 1987.

4

Vitamin E

CHING K. CHOW

University of Kentucky, Lexington, Kentucky

I. INTRODUCTION

Vitamin E was discovered by Evans and Bishop more than 75 years ago as a lipid-soluble substance in lettuce and wheat, necessary for the prevention of fetal death and resorption in rats fed a rancid lard diet (1). This substance was designated as vitamin E following the recognition of vitamin D (2,3). The term "tocopherol" is used after the Greek words "tokos" (childbirth), "phero" (to bring forth), and "ol" (alcohol). α-Tocopherol was isolated from wheat germ oil in 1936 (4). It was first synthesized by Karrer (5), and its structure was determined by Fernholz (6) in 1938. The antioxidant properties of tocopherols were first reported by Olcott and Emerson in 1937 (7).

In addition to fetal resorption in rats, a number of species-dependent deficiency symptoms of vitamin E, such as liver necrosis in rats and pigs, erythrocyte hemolysis in rats and chicken, and white muscle disease in calves, sheep, mice, and mink, were reported during the 1940s and 1950s (8–10). However, due to the lack of a definite clinical syndrome attributable to vitamin E deficiency and the difficulty of inducing vitamin E deficiency in human adults, the need for or use of this vitamin had been questioned. In the late 1960s, the essentiality of vitamin E for humans was recognized in connection with studies on premature infants in which hemolytic anemia was associated with vitamin E deficiency (9,11). Subsequent studies have shown that neurologic abnormalities do occur in association with malabsorption syndromes of various etiologies (9,11,12). In recent years, the involvement of free radicals in the pathogenesis of degenerative diseases and the possible prevention or slow-down of the disease process by antioxidants have promoted a renewed and expanded interest in vitamin E.

II. CHEMISTRY AND ANTIOXIDANT PROPERTIES

Vitamin E is the term suggested for all tocol and tocotrienol derivatives qualitatively exhibiting the biological activity of α-tocopherol. The term "tocopherols" is the generic description for all mono-, di-, and trimethyltocols and tocotrienols, and is not synonymous with the term "vitamin E." All eight naturally occurring tocopherol compounds isolated from plant sources have a 6-chromanol ring (head) and a phytyl side chain (tail) (8–10,13,14). There are four tocopherols and four tocotrienols that occur naturally, differing in the number and position of methyl groups on the phenolic ring (Fig. 1). Tocotrienols have a structure similar to that of tocopherols, except that the side chain contains three isolated double bonds at the 3′, 7′, and 11′ positions.

The tocopherol molecule has three chiral centers (2, 4′, and 8′) in its phytyl tail. Since a total of eight stereoisomeric forms may exist, tocopherols of unspecified configuration are more accurately called methyl-substituted tocols (Table 1). The term "tocol" is the trivial designation for 2-methyl-2-(4′, 8′, 12′-trimethyltridecylchroman-6-ol. All naturally occurring tocopherols (α-, β-, γ-, and δ-) have the same molecular configuration

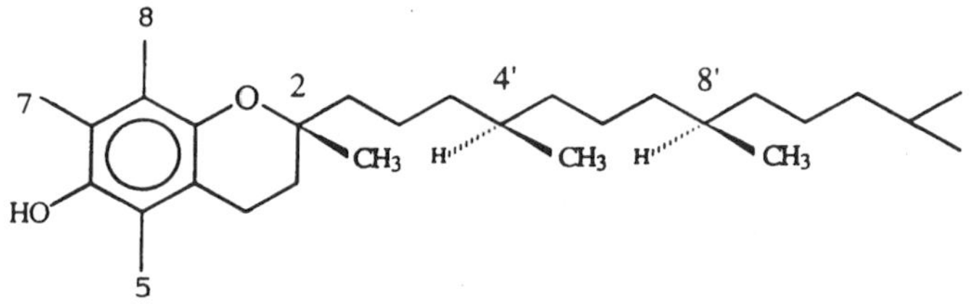

Toco Structure

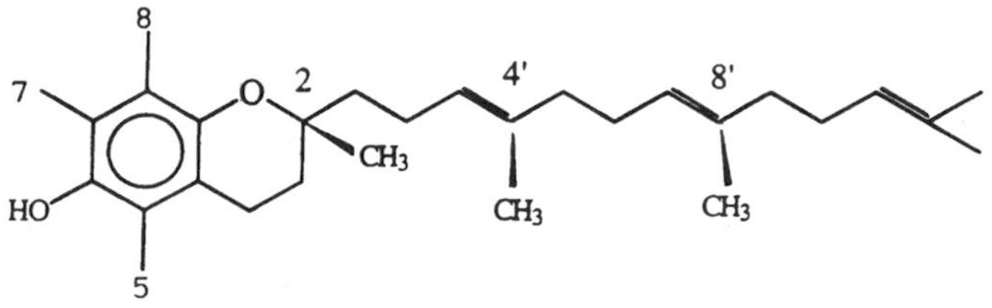

Tocotrienol Structure

Position of methyls	Toco structure	Tocotrienol structure
5,7,8	α-Tocopherol (α-T)	α-Tocotrienol (α-T-3)
5,8	ß-Tocopherol (ß-T)	ß-Tocotrienol (ß-T-3)
7,8	γ-Tocopherol (γ-T)	γ-Tocotrienol (γ-T-3)
8	δ-Tocopherol (δ-T)	δ-Tocotrienol (δ-T-3)

Fig. 1 Structural formula of tocopherols.

Table 1 Vitamin E Compounds

Vitamin	Trivial name	Chemical name
α-Tocopherol	5,7,8-Trimethyltocol	2,5,7,8-Tetramethyl-2-(4′,8′,12′-trimethyltridecyl)-6-chromanol
β-Tocopherol	5,8-Dimethyltocol	2,5,8-Trimethyl-2-(4′,8′,12′-trimethyltridecyl)-6-chromanol
γ-Tocopherol	7,8-Dimethyltocol	2,7,8-Trimethyl-2-(4′,8′,12′-trimethyltridecyl)-6-chromanol
δ-Tocopherol	8-Monomethyltocol	2,8-Dimethyl-2-(4′,8′,12′-trimethyltridecyl)-6-chromanol
α-Tocotrienol	5,7,8-Trimethyl tocotrienol	2,5,7,8-Tetramethyl-2-(4′,8′,12′-trimethyltridecyl)-6-chromanol
β-Tocotrienol	5,8-Dimethyl tocotrienol	2,5,8-Trimethyl-2-(4′,8′,12′-trimethyltridecyl)-6-chromanol
γ-Tocotrienol	7,8-Dimethyl tocotrienol	2,7,8-Trimethyl-2-(4′,8′,12′-trimethyltridecyl)-6-chromanol
δ-Tocotrienol	8-Monomethyl tocotrienol	2,8-Dimthyl-2-(4′,8′,12′-trimethyltridecyl)-6-chromanol

[*RRR*, 2D, 4′D, 8′D, *d*-tocopherol or (+)-tocopherols] in their phytyl groups. Since tocotrienols have only one chiral center at position 2, they can only have 2D and 2L stereoisomers. However, the double bonds at positions 3′ and 7′ of the phytyl tail allow for the existence of four cis/trans geometrical isomers (a total of eight isomers, theoretically) per tocotrienol (13,14).

In addition to naturally occurring isomers, several types of synthetic vitamin E, mainly in ester forms (e.g., α-tocopheryl acetate and tocopheryl succinate), are available commercially. The ester form is less susceptible to oxidation and is therefore more suitable for food and pharmaceutical applications than the free form. To distinguish it from the synthetic one, the naturally occurring stereoisomer of α-tocopherol, formerly known as *d*-α-tocopherol, has been designated as *RRR*-α-tocopherol. The totally synthetic α-tocopherol, previously known as *dl*-α-tocopherol or 2DL, 4′DL, 8′DL-tocopherol, which consists of eight stereoisomers [2D, 4′D, 8′D (*RRR*), 2L, 4′D, 8′D (*SRR*), 2D, 4′D, 8′L (*RRS*), 2L, 4′D,′8′L (*SRS*), 2D, 4′L, 8′D (*RSR*), 2L, 4′L, 8′D (*SSR*), 2D, 4′L, 8′L (*RSS*), and 2L, 4′L, 8′L (*SSS*)] (13,14), has been designated as all-*rac* α-tocopherol. All-*rac* α-tocopherol is usually synthesized by the condensation of trimethylhydroquinone with isophytol. The majority of tocopherol present in plants is not the α form. Therefore, *RRR*-α-tocopherol, which is referred as natural α-tocopherol, is often obtained following methylation of tocopherol mixtures isolated from vegetable oils.

Tocopherols are pale yellow or yellow-brown viscous liquids at room temperature. At high purity they are almost odorless and colorless. Free tocopherols are practically insoluble in water but are soluble in oils, fats, acetone, ethanol, chloroform, and other organic solvents. The boiling point of tocopherols at 0.1 mm Hg pressure is 200–220°C, their molecular weight ranges from 396.6 to 430.69, and the molar absorbance ($E^{1\%}$; 1 cm) at 292–298 nm ranges from 75.8 to 91.4 (Table 2) (14a). Tocopherols can be oxidized by atmospheric oxygen, and the oxidation is accelerated by heat, light, alkali, and metal ions. On the other hand, tocopherols are rather stable to heat and alkali when oxygen is absent.

Table 2 Molecular Weight and UV Absorption Maxima and Molar
Absorbance of Vitamin E Compounds

	Mol w	λ_{max} (nm)	$E^{\%}_{1\,cm}$
α-Tocopherol	430.7	292.0	75.8
β-Tocopherol	416.7	296.0	89.4
γ-Tocopherol	416.7	298.0	91.4
δ-Tocopherol	402.6	298.0	87.3
α-Tocotrienol	424.6	292.5	91.0
β-Tocotrienol	410.6	294.0	87.3
γ-Tocotrienol	410.6	296.0	90.5
δ-Tocotrienol	396.6	297.0	88.1

Source: Adapted from Ref. 14a.

Generally, the first step of tocopherol oxidation is the formation of resonance-stabilized chromanoxyl (chroman-6-oxyl) radical, due to the donation of the phenolic hydrogen to a lipid peroxyl radical (15,16). The chromanoxyl radicals are very reactive toward alkyl and alkylperoxy radicals. Depending on the severity of the oxidation conditions and the presence of other chemicals, different tocopherol oxidation products are obtained. For example, oxidation of α-tocopherol in polar solvents (water or alcohol) leads to the formation of 5-formyl and hemiacetal derivatives of 5-hydroxymethylene chromanol (13,17). In lipophilic solvents, the chromanoxy radicals tend to react mainly via radical–radical coupling reactions and form dimers or trimers (17,18). In the presence of peroxy radicals, α-tocopherol is primarily oxidized to 8α-peroxy-substituted tocopherone, which is then degraded to form α-tocopherylquinone, plus various tocopherone and quinone epoxides and spiro-dimer and trimer (19–21). Relatively little is known about the products of and mechanisms for the antioxidant reactions of tocotrienols. Tocotrienols, however, are expected to yield the same type of radicals and dimeric products as their corresponding tocopherols. Two dimers of γ-tocotrienol [5-(γ-tocotrienyloxy)-γ-tocotrienol and 5-(γ-tocotrienyl)-γ-tocotrienol] are the principal oxidation products of γ-tocotrienol-rich Hevea latex (22).

It has long been recognized that tocopherols possess antioxidant activity (7), and the biological activity of vitamin E is mainly attributed to their ability to donate their phenolic hydrogen to lipid free radicals. Tocopherols, as well as tocotrienols, can react with peroxyl radicals more rapidly than can polyunsaturated fatty acids and, therefore, are very effective free-radical chain-breaking antioxidants. Antioxidant activity of tocopherols is determined by their chemical reactivity with molecular oxygen, superoxide radicals, peroxyl radicals, or other radicals, or by their ability to inhibit autoxidation of fats and oils. The chemical structures of the tocopherols and tocotrienols support a hydrogen-donating power in the order α > β > γ > δ (23). The presence of electron-releasing substituents *ortho* and/or *para* to the hydroxy function increased the lectern density of the active centers, facilitating the homolytic fission of the O-H bond, increasing the stability of the phenoxy radical, and improving the reactivity with peroxy radicals (23). Thus, α-tocopherol is structurally expected to be a more potent hydrogen donor than β-, γ-, and δ-tocopherols because these lack one or two ortho-methyl group. The order of α > β >

$\gamma > \delta$ is generally found in the relative antioxidant activity of the tocopherols in vivo (see Sec. V). The relative antioxidant activity of tocopherols in vitro, however, varies considerably depending on the experimental conditions and the assessment method employed. γ-tocopherol and some other isomers, for example, have sometimes been shown to exhibit a higher antioxidant activity than the α form in a number of in vitro test systems (24). Also, tocotrienols are more effective than tocopherols in preventing oxidation under certain conditions (25). In vitro antioxidant activities of the tocopherols depend not only on their absolute chemical reactivities toward hydroperoxy and other free radicals, but also on such factors as tocopherol concentrations, temperature, light, type of substrate and solvent, and other chemical species, which may act as pro-oxidants and synergists, in the test system (13,23).

Unlike free tocopherols, tocopheryl esters are much more stable to oxidation, and do not function as antioxidants in vitro.

III. ABSORPTION, TRANSPORT AND METABOLISM

A. Intestinal Absorption

Similar to lipid components, intestinal absorption of vitamin E depends on pancreatic function, biliary secretion, micelle formation, and transport across intestinal membrane (26,27). Tocopherol is first emulsified and solubilized within bile salt micelles and transported across the water layer to come in contact with the absorption brush-border membrane of the enterocyte. Tocopheryl esters, such as tocopheryl palmitate, tocopheryl acetate, and tocopheryl succinate, are hydrolyzed to free tocopherol during the absorptive process. Under physiological conditions the hydrolysis occurs in gut lumen, although a mucosal esterase has also been found in the endoplasmic reticulum of the enterocytes (28). Free tocopherol is absorbed by a passive diffusion process from small intestine to the enterocyte (29). The absorption process is nonsaturable, non-carrier-mediated, and does not require energy. The area at the junction between the upper and middle thirds of the small intestine appears to be the region for tocopherol uptake by enterocytes (29,30). Within the enterocyte, tocopherol is incorporated into chylomicrons and secreted into the intracellular spaces and lymphatics, and thus into the bloodstream. This transfer process does not appear to require a specific protein.

The efficacy of tocopherol absorption varies considerably depending on the conditions and methods for its estimation. Generally, the rate of absorption decreases as the amount taken increases. Medium-chain triglycerides have been shown to enhance the absorption process, whereas long-chain polyunsaturated fatty acids reduce the absorption of α-tocopherol (31,32). The adverse effect of dietary long-chain polyunsaturated fatty acids on vitamin E absorption may partly be due to an increase in oxidation of tocopherol during digestion.

In humans, the majority of α-tocopherol is absorbed via the lymph ducts, and a small portion is absorbed by the portal vein (26,27). The absorption of tocopherols other than the α form is not as well understood. However, studies have shown that there are no major differences in the rate of intestinal absorption between α- and non-α-tocopherols, such as γ-tocopherol, and that excess of the α form does not reduce the absorption of γ-tocopherols (33,34). Thus, a higher level of α-tocopherols than other tocopherols in human plasma and tissues is the result of factors other than absorption.

B. Plasma Transport

Due to its hydrophobicity vitamin E requires special transport mechanisms in the aqueous milieu of the plasma, body fluid, and cells. Similar to dietary lipids, tocopherols absorbed in the small intestine are incorporated into triglyceride-rich chylomicrons, secreted into intestinal lymph, and then delivered to the liver. There is no discrimination between various forms of vitamin E during chylomicron secretion (35,36).

Following delivery to the liver from small intestine, tocopherol is repackaged with very low-density lipoproteins (VLDLs) and secreted into plasma. At this stage, *RRR*-α-tocopherol is preferentially secreted with VLDL over *SRR*-α-tocopherol and *RRR*-γ-tocopherol (35–37). During the conversion of VLDLs to low-density lipoproteins (LDLs) in the circulation, a portion of tocopherol is transferred to LDLs. Exchange of tocopherol also occurs between LDLs and high-density lipoproteins (HDLs) (38). Therefore, tocopherol is distributed in all lipoproteins (39,40). In humans, the highest concentration of vitamin E is found in LDL and HDL. Vitamin E can also be exchanged between lipoproteins and liposomes, and between plasma and red blood cells. While no specific plasma/serum carrier protein for vitamin E has been reported, a phospholipid transfer protein that accelerates exchange/transfer of α-tocopherol between lipoproteins and cells has been identified in human plasma (41). This protein may play a role in determining the concentration of tocopherols present in the circulation and in tissues.

Normal plasma vitamin E concentration in humans ranged from 11 to 37 μmol/L (5 to 16 mg/L), 1.6 to 5 μmol α-tocopherol/mmol lipid (0.8 to 1.7 mg/g), or 2.5 to 8 μmol α-tocopherol/mmol cholesterol (2.8 to 8 mg/g). Approximately 10–20% of total tocopherols in human plasma is the γ form. In malabsorption states, such as cystic fibrosis, plasma lipids, lipoproteins, and vitamin E concentration are frequently reduced concurrently (10–12,26).

C. Tissue Uptake

The mechanisms involved in the uptake of tocopherols by tissue remain unclear. Lipoprotein lipase bound to the surface of the endothelial lining of capillary walls catabolizes the triglycerides in the core of chylomicrons and forms chylomicron remnants (26,27). The latter is then taken up by the parenchymal cells in the liver via the remnant receptors on the surface of hepatocytes (42,43). Along with the free fatty acids, some vitamin E is also taken up by peripheral tissues during catabolism of lipoproteins by lipoprotein lipase. Tocopherol is taken up by tissues via several mechanisms. This includes lipoprotein receptor (apolipoprotein B/E)–dependent and receptor-independent pathways, independent transport and cotransport of α-tocopherol and LDL, and uptake from a number of lipoproteins (44). The action of lipoprotein lipase is an important process prior to tissue uptake of tocopherols.

Most tissues, including liver, skeletal muscle, and adipose tissue, have the capacity to accumulate α-tocopherol (45–47). In adipose tissue, tocopherol is located mainly in the bulk lipid droplet and its turnover is slow. The adrenal gland has the highest concentration of α-tocopherol, although lung and spleen also contain relatively high concentrations (47,48). α-Tocopherol is primarily taken up and located in parenchymal cells (49). The transfer of tocopherol between parenchymal and nonparenchymal cells takes place after uptake in parenchymal cells. Parenchymal, but not nonparenchymal, cells have a large storage capacity for surplus α-tocopherol and are more resistant to tocopherol depletion than nonparenchymal cells. Light mitochondria has the highest concentration of α-tocopherol, whereas the concentration is low in cytosol (49,50). The majority of tocopherol is localized in the membranes. For example, approximately three-fourths of mitochondrial α-tocopherol is found in the outer membrane, and one-fourth is associated with the inner membrane (49). Also, essentially all tocopherol in the red cells is found in the membranes (51).

D. Hepatic Secretion

Vitamin E taken up by the liver is stored in the parenchymal cells or secreted into the bloodstream within nascent VLDLs. Following the action of lipoprotein lipase, some vitamin E in the VLDLs may end up in LDLs as these lipoproteins are transformed in plasma. Vitamin E in LDLs may again be taken up by liver via the LDLs (apolipoprotein B/E) receptor or by non-receptor-mediated uptake (26,42,43). Some vitamin E in association with chylomicrons and VLDL may also be transferred to peripheral cells and HDLs during lipolysis by lipoprotein lipase. Secreted tocopherols are either rapidly returned from plasma to the liver during the course of lipid metabolism or excreted via bile. The secretory pathway via nascent VLDLs from liver seems to be critical in maintaining tocopherol concentrations in plasma. Studies have shown that discrimination between the stereoisomers of tocopherol occurs during hepatic secretion of nascent VLDLs (36). *RRR*-α-Tocopherol is preferentially incorporated over other tocopherols into VLDLs, which are then secreted into plasma. This is evidenced by the much higher levels of *RRR*-α-tocopherol than either *SRR*-α- or *RRR*-γ-tocopherol found in the nascent VLDLs and plasma of cynomolgus monkeys after the same amount of each was given (37). The preference of a specific hepatic tocopherol-binding protein to *RRR*-α-tocopherol appears to be responsible for this discrepancy.

E. Tocopherol-Binding Proteins

A cystosolic tocopherol-binding protein has been shown to selectively facilitate the incorporation of *RRR*-α-tocopherols into nascent VLDLs. By preferentially binding the α form over the other forms of tocopherols, the binding protein determines which tocopherols are returned to the liver from plasma via HDLs and LDLs (52–55). The discrimination between the tocopherol isomers and homologues appears to be due to the preference of the tocopherol-binding protein to recognize specific structural features: a fully methylated aromatic ring, a saturated phytyl side chain, and a stereochemical RRR configuration of the methyl group's branching of the side chain (14,36). There is a chiral discrimination between different isomers of α-tocopherol, with preference retention of the *RRR* stereoisomer over the SRR form (35,36). By regulating the binding and/or transfer of tocopherol, the transfer protein plays a key role in determining plasma concentration and biological activity of tocopherols (55). The process results in the preferential enrichment of LDL and HDL with α-tocopherol in plasma. The critical role of tocopherol-binding protein in regulating plasma tocopherol concentration has been indicated in patients with familial isolated vitamin E deficiency (56,57). Those patients have clear signs of vitamin E deficiency (extremely low plasma vitamin E and neurological abnormalities) but have no fat malabsorption or lipoprotein abnormalities. Absence of the transfer protein in those patients, which impairs secretion of tocopherol into hepatic lipoproteins (VLDLs), has been postulated to be responsible for their low plasma vitamin E status (26,55). The human α-tocopherol transfer protein gene was shown to be located at chromosome 8q13 (57a). Studies of the gene structure and mutations of α-tocopherol transfer protein in patients with familial vitamin E deficiency have confirmed that mutations of the gene for α-tocopherol transfer protein are responsible for isolated vitamin E deficiency (58,58a). The degree of functionality of the mutant α-tocopherol transfer protein seems to be associated with the degree of severity of the neurological damage and age of onset (59).

A tocopherol/binding protein with a molecular weight of 30–36 kd has been purified from rat liver cytosol (54,55,60,61). The protein binds α-tocopherol and enhances its transfer between membranes. The purified protein has two isoforms with isoelectric points at 5.0 and 5.1, respectively (61). The binding protein exhibits a structural homology with the cellular retinaldehyde-binding protein present only in visual tissues (62). A tocopherol-binding protein with a molecular weight of 36.6 kd has also been identified and purified from human liver cytosol (63). An apparently different α-tocopherol binding protein with a molecular mass of 14.2 kd has been isolated and purified from rabbit heart (64) and rat liver and heart (65). The binding of this smaller binding protein to α-tocopherol is rapid, reversible, and saturable, and neither γ- nor δ-tocopherol can replace the binding of α-tocopherol. It also markedly stimulates the transfer of tocopherol from liposomes to mitochondria. The binding protein is different from the cytosolic fatty acid–binding protein and may be involved in intracellular transport and metabolism of α-tocopherol (66). The relationship between those two binding proteins, if any, is not known.

F.   Metabolism

The major route of excretion of absorbed tocopherol is fecal elimination. Excess α-tocopherol and other forms of tocopherols are excreted first into the bile and then into the feces. After exerting its antioxidant activity tocopherol is first converted to tocopheryl chromanoxy radical. The chromanoxy radical is readily reverted to tocopherol, and the process can be facilitated by such reducing agents as glutathione and ascorbate and/or a yet to be identified enzymic system. Tocopheryl chromanoxy radicals can also form dimer or trimer, or be further oxidized to form tocopheryl quinone (Fig. 2). A small amount of α-tocopheryl quinone is found in liver (67). Studies on the metabolic fate of C^{14}-labeled α-tocopheryl quinone and α-tocopheryl hydroquinone have showed that there is no conversion to α-tocopherol in vivo (68). α-Tocopheryl quinone, however, can be reduced likely by an enzymic process to α-tocopheryl hydroquinone, and then further metabolized to α-tocopheronic acid. Higher activity of NADPH-dependent tocopheryl quinone reductase is found in the mitochondria and microsomes than in the cytosol of rat hepatocytes (69). Also, NADPH-cytochrome P450 reductase is capable of catalyzing tocopheryl hydroquinone formation from tocopheryl quinone (69). Tocopheronic acid is subsequently conjugated with glucuronic acid or other compounds, and excreted in urine (68). The urinary metabolite, α-tocopheronic acid, was first isolated by Simon and co-workers (70) after administration of large doses of α-tocopherol to rabbits and humans. A portion of bound or conjugated α-tocopheryl hydroquinone is secreted in the bile and eliminated in the feces (68). A small portion of dimer and trimer of α-tocopherol has also been found in rat liver (71,72).

Results obtained form recent studies suggest that 2,5,7,8-tetramethyl-1(2′-carboxyethyl)-6-hydroxy chroman, instead of α-tocopheronic acid, may be the major urinary metabolite of α-tocopherol (72a). The urinary metabolite appears to be formed in liver directly from a side chain degradation of α-tocopherol without oxidative splitting of the chroman ring (Fig. 2). The compound can be oxidized readily to form α-tocopheronic

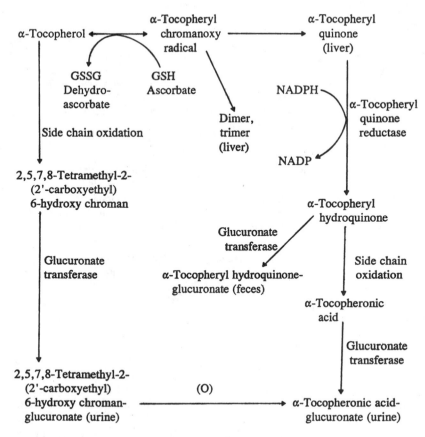

Fig. 2 Metabolic fate of α-tocopherol. GSH represents reduced glutathione; GSSG, oxidized glutathione; NADPH and NADP, reduced and oxidized nicotinamide dinucleotide phosphate, respectively.

acid (or tocopheronolactone). Similarly, 2,7,8-trimethyl-2-(2′-carboxyethyl)-6-hydroxy chroman and 2,8-dimethyl-2(2′-carboxyethyl)-6-hydroxy chroman have been identified as the principal urinary metabolites for γ- and δ-tocopherols, respectively (72b,72c).

IV. DEFICIENCY SYMPTOMS AND TOXICITY

A. Deficiency Symptoms

A number of species-dependent and tissue-specific symptoms of vitamin E deficiency have been reported (Table 3). The development and severity of certain vitamin E deficiency symptoms are associated with the status of other nutrients, including polyunsaturated fatty acids, selenium, and sulfur amino acids (8,9). For example, the most common sign of deficiency is necrotizing myopathy, which occurs in almost all species in the skeletal muscle, as well as in some heart and smooth muscles. However, in lambs and calves, myopathy primarily results from selenium deficiency, and in rabbits and guinea pigs, severe debilitating myopathy develops when fed a low vitamin E diet alone. On the other

Table 3 Pathology of Vitamin E Deficiency

Condition	Animal	Tissues affected
Reproductive failure:		
Embronic degeneration	Female: rat, hen, turkey, ewe	Vascular system of embryo
	Male: rat, guinea pig, hamster, dog, cock	Male gonads
Liver, blood, brain, capillaries:		
Liver necrosis	Rat, pig	Liver
Erythrocyte destruction	Rat, chick	Blood (hemolysis)
Anemia	Monkey	Bone marrow
Blood protein loss	Chick, turkey	Serum albumin
Encephalomalacia	Chick	Cerebellum (Purkinje cells)
Exudative diathesis	Chick, turkey	Vascular system
Depigmentation	Rat	Incisors
Kidney degeneration	Rat, monkey, mink	Kidney tubular epithelium
Steatitis	Mink, pig, chick	Depot fat
Eosinophilia	Rat	Blood, small intestine, liver, stomach, and skeletal muscle
Nutritional myopathies:		
Nutritional muscular dystrophy	Rabbit, guinea pig, rat, monkey, duck, turkey	Skeletal muscle
Stiff lamb	Lamb, kid	Skeletal muscle
White-muscle disease	Calf, sheep, mouse, mink	Skeletal and heart muscle
Myopathy of gizzard and heart	Turkey, poult	Gizzard, heart, and skeletal muscle
Neuromuscular disorder	Rat, monkey, humans	Skeletal muscle, spinal cord

hand, rats manifest a relatively benign myopathy when fed a vitamin E–deficient diet, and chickens develop no myopathy unless the diet is depleted of both vitamin E and sulfur amino acids. In the chicken, encephalomalacia, a disorder characterized by localized hemorrhage and necrosis in the cerebellum, is prevented by vitamin E but not by selenium. However, in chicken, exudative diathesis (a disorder characterized by increased capillary permeability) can be prevented by either vitamin E or selenium (8,9). Similarly, eosinophilic enteritis and eosinophilia in rats is preventable by vitamin E or selenium (73). The cause of the species-dependent and tissue-specific vitamin E deficiency symptoms remains to be delineated.

In humans, lower plasma/serum levels of vitamin E (<0.5 mg/dL) are associated with a shorter life span of the red cells and increased susceptibility to hemolytic stress. Most of the vitamin E deficiency states in humans are associated with fat malabsorption disorders (9,11,12,74). Low serum vitamin E levels have been observed in patients with a variety of fat malabsorption conditions (Table 4). Recent studies of children and adults with diseases that cause fat/vitamin E malabsorption, such as abetalipoproteinemia, chronic cholestatic hepatobiliary disorder, and cystic fibrosis, as well as patients with the primary form of vitamin E deficiency, familial isolated vitamin E deficiency syndrome, have conclusively demonstrated that neurological dysfunction is associated with vitamin E deficiency (12,26,56).

Table 4   Tocopherol Levels in Human Vitamin E Deficiency States

Condition	Mean plasma/ serum α-tocopherol (mg/100 mL)	Range
Premature birth	0.25	0–0.56
Cystic fibrosis	0.24	0–0.97
Malabsorption syndromes other than cystic fibrosis:		
Biliary atresia	0.11	—
Biliary cirrhosis	0	—
Biliary obstruction	0.08	0–0.14
Celiac disease	0.20	0–0.35
Chronic pancreatitis	0.40	0.17–0.79
Gastrectomy	0.36	0.10–0.80
Intestinal lymphangiectasia	0.28	—
Intestinal resection	0.31	—
Nontropical sprue	0.25	0.12–0.32
Regional enteritis plus intestinal resection	0.27	0.15–0.39
Tropical sprue	0.28	—
Ulcerative colitis	0.24	0.17–0.36
Whipple's disease	0.30	0.14–0.45
Protein-calorie malnutrition	0.48	0–0.91
Normal	1.05	7.7–15.5

Source: After Ref. (11.)

B. Toxicity

Both acute and chronic studies with several species of animals have shown that high doses of vitamin E are relatively nontoxic. However, extremely high levels of vitamin E (16,000–64,000 IU/kg diet) have been shown to result in decreased pigmentation and a waxy, feather like appearance (75), and abnormal mineralization and clotting abnormalities (76) in the chick. Also, feeding pelicans with a high dose of vitamin E (5000 IU/kg diet) results in vitamin K deficiency (77). These findings suggest that interference with absorption or metabolic interaction with vitamins D and K is associated with extreme dietary levels of vitamin E. In humans, supplementation of 800 IU vitamin E per day or more for several years has consistently shown no adverse effects (78). Although increased post-surgery bleeding and one individual case of breeding in patient receiving anticoagulant therapy has been reported, a recent study (78a) has shown that moderate to large doses of vitamin E (up to 1200 IU daily) can be safely used in patients receiving warfarin.

V. BIOPOTENCY AND BIOLOGICAL FUNCTIONS

A. Biopotency

The utilization or function of tocopherols in biological tissues is governed not solely by their chemical reactivities but by the biokinetics of their distribution, transport, retention, and localization as well. Tocopherols differ widely in their antioxidant and biological activities. A vitamin E activity of *RRR*-α-tocopherol higher than that of the other forms suggests that the RRR configuration of the phytyl tail must be optimal for maximum biopotency. The biological activity of tocopherol is assessed in terms of its relative ability to prevent such deficiency-related symptoms as fetal resorption-gestation and erythrocyte hemolysis (Table 5), and, more recently, by a curative myopathy test in rats (8,9,79,79a). The curative myopathy test is based on the ability of tocopherol to suppress the increase of pyruvate kinase activity in the plasma of vitamin E–deficient rats. Increased plasma pyruvate kinase activity resulting from myodegeneration is a sensitive and specific indicator for assessing the extent of vitamin E deficiency in rats (80). The increase of plasma pyruvate kinase activity in rats with vitamin E deficiency is related to the nutritional status of selenium (81). Markedly increases in plasma pyruvate kinase activity is also found in dystrophic chickens and hamsters (82). Similarly, the pyruvate kinase activity in the serum of patients with neuromuscular disorders is approximately 20- to 40-fold that of the control values (83–85).

The biological activity of various forms of tocopherol is expressed as units of activity in relation to that of *all-rec*-α-tocopheryl acetate. The relative values in international units (IU/mg) are: *all-rec*-α-tocopheryl acetate, 1.00; *all-rec*-α-tocopherol, 1.10; *RRR*-α-tocopheryl acetate, 1.36; *RRR*-α-tocopherol, 1.49; *all-rec*-α-tocopheryl succinate, 0.89; and *RRR*-α-tocopheryl succinate, 1.21.

B. Biological Functions

Although many biochemical abnormalities are associated with vitamin E deficiency, the mechanism by which vitamin E prevents various metabolic and pathological lesions has not yet been elucidated. Vitamin E is the major lipid-soluble chain-breaking antioxidant found in plasma, red cells, and tissues (14,86,87), and it plays an essential role in maintaining the integrity of biological membranes. Among the biological functions proposed for vitamin E, prevention against free-radical–initiated lipid peroxidation tissue damage is the most important one accepted by most investigators. The antioxidant role of vitamin E in vivo is supported by the findings that synthetic antioxidants can prevent or lessen certain vitamin E–deficient symptoms, and that increased production of peroxidation prod-

Table 5   Vitamin E Activity of Tocopherols[a]

Structure	Resorption-gestation (%)	Hemolysis (%)
RRR-α-tocopherol	100	100
RRR-β-tocopherol	25–50	15–27
RRR-τ-tocopherol	8–19	3–20
RRR-δ-tocopherol	0.1–3	0.3–2
RRR-α-tocotrienol	21–50	17–25
RRR-β-tocotrienol	4–5	1–5
RRR-τ-tocotrienol	—	—
RRR-δ-tocotrienol	—	—
RRR-α-tocopheryl acetate	91	—
All-rec-α-tocopherol[b]	74	—
All-rec-α-tocopheryl acetate[b]	67	—

[a] Bioassay methods.
[b] A mixture of eight steroisomers.

ucts, such as malondialdehyde, ethane, and pentane, are found in vitamin E–depleted animals (9,87). The suggestion that vitamin E may play a structural role in the control of membrane permeability and stability (88) does not conflict with this view.

Several non-antioxidant functions of vitamin E have also been suggested. Vitamin E, for example, has been reported to regulate the de novo synthesis of xanthine oxidase (89), modulate the activities of microsomal enzymes (90,91), and down-regulate protein kinase activity and cell proliferation (92,92a). It is interesting to note that *RRR*-β-tocopherol prevents the inhibition of protein kinase activity and cell growth of smooth-muscle cells caused by *RRR*-α-tocopherol (93). Also, vitamin E may regulate immune response or cell-mediated immunity by modulating the generation of prostaglandins and lipid peroxidation products, and the metabolism of arachidonic acid (94–96). Additionally, vitamin E has been shown to be essential for optimally developing and maintaining the integrity and function of nervous mechanism and skeletal muscle (12,97). Recently, vitamin E has been shown to be capable of down-regulate mitochondrial generation of superoxide and hydrogen peroxide (97a,97b). By reducing mitochondrial generation of superoxide and related ROS, dietary vitamin E not only attenuates oxidative damage but also modulates the expression and activation of signal transduction pathways and other redox-sensitive biological modifiers, and thereby may prevent or delay the onset of degenerative tissue changes.

VI. VITAMIN E AND FREE-RADICAL–INDUCED PEROXIDATIVE DAMAGE

A. Free-Radical–Induced Peroxidative Damage

The harmful effects resulting from inhalation of high concentration of oxygen have been attributed to the formation of reactive oxygen species rather than molecular oxygen per se. Superoxide radicals, nitric oxide, lipid alkoxyl and peroxyl radicals are the most significant reactive oxygen species generated in systems living in aerobic environments. Among them, peroxy radical derived from polyunsaturated fatty acids has special significance due to its involvement in lipid peroxidation, the most common indicator of free-radical processes in living systems (14).

The process of lipid peroxidation (or autoxidation) can be divided into three phases: initiation, propagation, and termination. In the initiation phase (reaction 1), carbon-centered lipid radicals (R$^{\cdot}$) can be produced by proton abstraction from, or addition to, a polyunsaturated fatty acid (RH) when a free-radical initiator (I*) is present. The initiation reaction is generally very slow and is dependent on the type of initiator employed. However, the reaction can be catalyzed by heat, light, trace metals, and/or certain enzymes (e.g., lipoxygenases). During the propagation phrase, the lipid radical reacts readily with available molecular oxygen to form a peroxy radical (ROO$^{\cdot}$) at a very high rate (reaction 2). The peroxyl radical formed can react with another polyunsaturated fatty acid (R′H), at a slow rate, to form a hydroperoxide (ROOH) and a new carbon-centered radical (reaction 3):

$$RH \xrightarrow{\text{I*}} R^{\cdot} \tag{1}$$

$$R^{\cdot} + O_2 \rightarrow ROO^{\cdot} \tag{2}$$

$$ROO^{\cdot} + R'H \rightarrow ROOH + R'' \tag{3}$$

The propagative process can continue until all polyunsaturated fat is consumed or the chain reaction is broken (termination phase). Free radicals can also be scavenged and the chain reaction terminated by self-quenching to form dimers (reaction 4) or by the action of antioxidant (AH) (reaction 5).

$$R^{\cdot} \text{ (or ROO}^{\cdot}) + R^{\prime\prime} \rightarrow R\text{-}R^{\prime} \text{ (or ROOR}^{\prime}) \tag{4}$$

$$R^{\cdot} \text{ (or ROO}^{\cdot}) + AH \rightarrow RH \text{ (or ROOH)} + A^{\cdot} \tag{5}$$

Of the free-radical process inhibitors, tocopherols are among the most effective chain breakers available. This is because tocopherols can react more rapidly with peroxy radicals than do polyunsaturated fatty acid (14). Since lipid peroxy radicals can react with tocopherols, several orders of magnitude faster ($K_a = 10^4–10^9 \text{ M}^{-1} \text{ s}^{-1}$) than their reaction with acyl lipids ($K_a = 10–60 \text{ M}^{-1} \text{ s}^{-1}$) (14,23), one tocopherol molecule can protect up to $10^3–10^8$ polyunsaturated fatty acid molecules at low peroxide levels. Similarly, a small ratio of α-tocopherol to polyunsaturated fatty acid molecules (1:1500) in the red cell membrane is sufficient to interrupt the free-radical chain reactions (98).

Tocopherols can also inhibit the oxidations induced by the electronically excited singlet oxygen. The relative effectiveness of α-, β-, γ-, and δ-tocopherols as singlet oxygen quenchers is 100:55:26:10, respectively (99,100). In addition, tocopherols can react with hydroxyl, perhydroxy, and superoxide radicals (101,102), as well as nitric oxide (103,104), and form tocopheryl quinones, epoxytocopherones, tocored, and/or other oxidation products.

The antioxidant property of various tocopherols is dependent on their stoichiometric factors, the rate of their reaction with peroxy radicals, and their concentrations. Under certain conditions, tocopherol may exert pro-oxidant property. The pro-oxidant effect of α-tocopherol reported in in vitro studies appears to be related to its tocopheroxy radicals (23). When the concentration of tocopheroxy radical is high, the radicals may react reversibly with unperoxidized lipids and lipid hydroperoxides by chain transfer and generate alkyl and peroxy radicals, respectively.

Reactive oxygen species may react with cellular components, with resultant degradation and/or inactivation of essential cellular constituents (105–109). The products derived from the reaction with reactive oxygen species may be more or less reactive or harmful. Cell membranes, which contain a relative high proportion of polyunsaturated fatty acids, are more susceptible to free-radical–induced lipid peroxidation. The process of free-radical–induced lipid peroxidation has been implicated as a critical initiating event leading to cell injury or organ degeneration. There are a large variety of conditions that are capable of initiating or enhancing oxidative stress within the cellular environment. These conditions include inadequate intake of antioxidants, excess intake of prooxidants, exposure to noxious chemical or physical agents, strenuous physical activities, injuries and wounds, and certain hereditary disorders (87).

B. Antioxidant Systems

In view of the potential adverse effects of oxygen and its reactive intermediates, it is important that a number of enzymatic and nonenzymatic antioxidant systems are present in the cell. Under normal conditions, the overall antioxidant activity of the cell is able to control or prevent most of the adverse effects of oxygen and its reactive intermediates. However, when the antioxidant potential is weakened or oxidative stress is greatly increased, irreversible damage to the cell may occur. The susceptibility of a given organ or

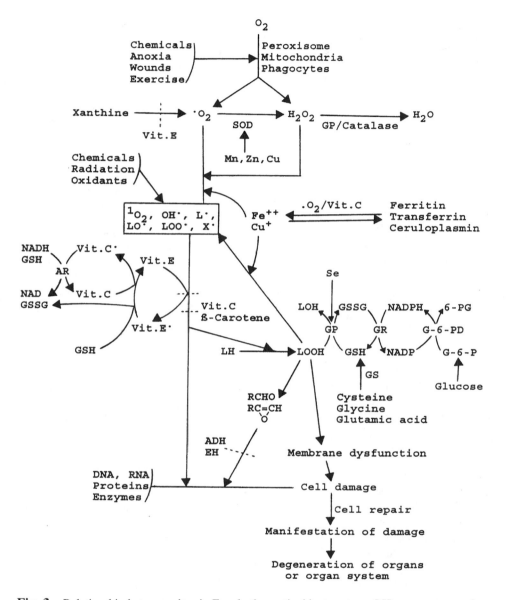

Fig. 3 Relationship between vitamin E and other antioxidant systems. LH represents membrane or polyunsaturated lipids; LOOH, lipid hydroperoxides, LOH, hydroxy acid; LOO', peroxy radical; L', alkyl radical; LO', alkoxy radical; OH', hydroxyl radical; 'O$_2$, superoxide radical; X', other radicals; ^{1}O$_2$, singlet oxygen; H$_2$O$_2$, hydrogen peroxide; RCHO, aldehydes; $\underset{O}{RC=CH}$, epoxides; Vit.E', vitamin E radical; Vit.C', ascorbate radical or dehydroascorbate; 6-PG, 6-phosphogluconate; GSH, reduced glutathione; GSSG, oxidized glutathione; NADH and NAD, reduced and oxidized nicotinamide adenine dinucleotide, respectively; NADPH and NADP, reduced and oxidized nicotinamide dinucleotide phosphate, respectively; Se selenium; SOD, superoxide dismutase; GP, GSH peroxidase or phospholipid hydroperoxide GSH peroxidase; GR, GSSG reductase; G-6-PD, glucose-6-phosphate dehydrogenase; ADH, aldehyde dehydrogenase; EH, epoxide hydrase; AR, ascorbate or dehydroascorbate reductase.

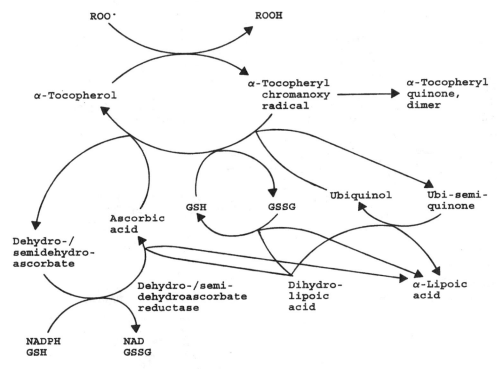

Fig. 4   Possible regeneration pathways of α-tocopherol. LOOH represents lipid hydroperoxides; LOO·, peroxy radical; GSH, reduced glutathione; GSSG, oxidized glutathione; NADH and NAD, reduced and oxidized nicotinamide adenine dinucleotide, respectively; NADPH and NADP, reduced and oxidized nicotinamide dinucleotide phosphate, respectively.

organ system to oxidative damage is determined by the overall balance between the extent of oxidative stress and antioxidant capability.

Major cellular antioxidant mechanisms include (a) direct interaction with oxidants or oxidizing agents by ascorbic acid, glutathione (GSH), and other reducing agents; (b) scavenging of free radicals and singlet oxygen by vitamin E, ascorbic acid, carotenoids, superoxide dismutase, and other scavengers; (c) reduction of hydroperoxides by glutathione peroxidases and catalase; (d) binding or removal of transition metals by ferritin, transferrin, ceruloplasmin, albumin, and other chelators; (e) separation or prevention of reactive oxygen species and other factors from reaching the specific site of action or reacting with essential cellular components by membrane barriers; and (f) repair of resulting damage by dietary nutrients and metabolic activities (Fig. 3). While various antioxidant systems act at different stages and cellular localization, vitamin E occupies a unique position in the overall antioxidant defense (87,110,111). The essential role of vitamin E in maintaining the red cell integrity, for example, is supported by the findings that ascorbic acid and GSH promote oxidative damage to the vitamin E–deficient red cells, while protecting vitamin E–supplemented red cells (111). The contradictory role of these reducing agents appears to be the result of (a) their participation in vitamin E regeneration when certain vitamin E levels are maintained and (b) their involvement in free-radical generation by virtue of their maintaining transition metal ions in a reduced state (112,113) when vitamin E is low or depleted. Approximately 1 part of α-tocopherol is capable of protecting 1000 parts of lipid molecule in biological membranes (98,114). The efficacy of tocopherol is augmented by the presence of other antioxidant systems and the factors involved in its regeneration (Fig. 4).

VII. FUNCTIONAL INTERRELATIONSHIP WITH OTHER NUTRIENTS

It has long been recognized that the function of vitamin E is closely interrelated with other dietary nutrients. In addition to dietary lipids, vitamin E is functionally related to the status of polyunsaturated fatty acids, selenium, vitamin C, iron, β-carotene, and sulfur-containing amino acids.

A. Lipids

Cell membranes contain a high proportion of polyunsaturated fatty acids, and vitamin E is found predominantly in the membranes. The dietary requirement for vitamin E is related to the degree of unsaturation of the fatty acids in tissue lipids, which can be altered by dietary lipids (115,116). The intake of tocopherols generally parallels the intake of polyunsaturated fatty acids. Foods that contained high polyunsaturated fatty acids generally are rich sources of tocopherols. A typical U.S. mixed diet has a ratio of approximately 0.5 mg α-tocopherol equivalent per gram of polyunsaturated fatty acids (114).

B. Selenium

The functional interrelationships between vitamin E and selenium have long been recognized. Selenium has been shown to prevent or reduce the severity of several symptoms of vitamin E deficiency, including necrotic liver degeneration and eosinophilic enteritis in rats and exudative diathesis in chicks (8,9,73). Selenium deficiency in domestic animals is frequently associated with an increased vitamin E requirement, and the situation is often aggravated by the oxidation of vitamin E in feed during storage.

The finding that selenium is an integral part of the enzymes GSH peroxidase (117) and phospholipid hydroperoxide GSH peroxidase (118) provides a reasonable explanation of the metabolic interrelationship between vitamin E and selenium. Selenium may complement the antioxidant function of vitamin E by reducing lipid hydroperoxide (ROOH) to corresponding hydroxy acid (ROH) via the action of GSH peroxidases (reaction 6). This reduction prevents the decomposition of hydroperoxides to form free radicals that may initiate further peroxidation and may thus reduce the requirement of vitamin E.

$$\text{ROOH} + 2\,\text{GSH} \xrightarrow{\text{GSH peroxidase}} \text{ROH} + \text{GSSG} + \text{H}_2\text{O} \qquad (6)$$

C. Vitamin C (Ascorbic Acid)

In addition to being an important water-soluble free-radical scavenger and reducing agent, vitamin C may complement the function of vitamin E and reduce vitamin E requirement. This effect is partly attributable to the involvement of vitamin C in the regeneration or restoration of vitamin E (Fig. 4) after exerting its antioxidant function (119–122). However, unambiguous experimental evidence for the sparing effect of vitamin C on the requirement of vitamin E has yet to be provided (123). In addition to ascorbic acid, GSH, in association with an enzyme or enzyme system(s), seem to be involved in the regeneration of vitamin E (122,124). Also, a GSH-dependent dehydroascorbate reductase and an

NADH-semidehydroascorbate reductase appear to be involved in the regeneration or resto-ration of vitamin C (125). Furthermore, dihydrolipoic acid, NADH-cytochrome $b5$, and ubiquinol have been shown to play a role in vitamin E regeneration (127–129), and dihy-drolipoic acid is also involved in the regeneration of ubiquinol, ascorbate, GSH, and thiore-doxin (129). While the nature of the vitamin E regenerative process in vivo is not entirely clear, the process does provide a rational explanation for the fact that it is very difficult to deplete adult animals or human subjects of the vitamin (115).

D. β-Carotene

Of the more than 500 carotenoids that have been identified, about 50 possess some vitamin A activity (129). In the leafy green and yellow-orange vegetables, the predominant provita-min A activity is due to a hydrocarbon carotene, β-carotene, which makes up over 50% of the total carotenoids. Other carotenoids, such as lycopene and lutein, are found in the circulation but cannot be converted to retinol (130). In addition to being a precursor of vitamin A, β-carotene is an effective quencher of singlet oxygen and free-radical scavenger (131–133). As a more effective scavenger of singlet oxygen, β-carotene may complement vitamin E in protecting cellular components against oxidative damage. β-Carotene may also modulate membrane properties important to cell–cell signaling or possibly serve as an intracellular precursor to retinoic acid (134). Vitamin E has been shown to enhance the lymphatic transport of β-carotene and its conversion to vitamin A in the ferret (135). On the other hand, excessive intake of β-carotene has been reported to decrease the plasma level of vitamin E, and excess vitamin E lowers the levels of β-carotene (136–138).

E. Iron

Separation and removal of iron and other transition metal ions is of critical importance in preventing iron-catalyzed generation of hydroxyl radicals from hydrogen peroxide (re-action 8). Hydrogen peroxide is readily formed from superoxide following the action of superoxide dismutase (SOD) (reaction 7).

$$\overset{\cdot}{O_2^-} + \overset{\cdot}{O_2^-} + 2H^+ \xrightarrow{\text{SOD}} H_2O_2 + O_2 \tag{7}$$

$$H_2O_2 + Fe^{2+} \longrightarrow OH^- + OH^{\cdot} + Fe^{3+} \tag{8}$$

Transition metal ions, particularly Fe^{2+} and Cu^+, can exacerbate membrane damage by catalyzing the decomposition of lipid hydroperoxides to yield more free radicals (reac-tion 9) and aggravate lipid peroxidation membrane damage (112,113):

$$ROOH \longrightarrow R^{\cdot}, RO^{\cdot}, ROO^{\cdot}, \text{etc.} \tag{9}$$

Dietary iron exists mainly in the ferric state but is absorbed in the ferrous form. Iron absorbed from the diet or released from ferritin can be sequestered by transferrin, lactoferrin, citrate, ATP, and other phosphate esters. However, excess iron absorption occurs in certain pathological conditions. The genetic disease idiopathic hemochromatosis, for example, causes a condition similar to iron overload, and the effects of iron overload can be lessened by increased vitamin E intake (139).

F. Sulfur-Containing Amino Acids

It has long been recognized that muscular dystrophy, the condition that occurs in most animal species fed a vitamin E–deficient diet, appears in the chick only when fed a diet that contains low-sulfur amino acids (8,9). This interrelationship of vitamin E with sulfur amino acids appears to be due to a requirement for sulfur amino acids in the synthesis of GSH (140). GSH is needed for the activity of GSH peroxidase and for the restoration/regeneration of vitamin E (Fig. 4). Oxidized glutathione (GSSG), in turn, is reduced by GSSG reductase utilizing NADPH generated in the pentose shunt pathway (reaction 10).

$$\text{NADPH} + \text{GSSG} \xrightarrow{\text{GSSG reductase}} 2\text{GSH} + \text{NADP} \qquad (10)$$

VIII. FOOD CONTENTS AND SOURCES

Tocopherols exist mainly in the free-alcohol form and are widely distributed in a variety of plant life. A notable exception is latex lipids, in which the vast majority of tocopherols (mainly γ-tocotrienol) are present as esters of fatty acids (10,141). Vegetable oils are the primary dietary sources of tocopherols in the United States.

While α-tocopherol predominates in most animal species and is significantly more active biologically than any other form of tocopherol, γ-tocopherol is the major form consumed in the United States. γ-Tocopherol in soybean oil, corn oil, and other vegetable oils, accounts for over half of the estimated total tocopherol intake (Table 6). The significant sources of vitamin E in U.S. diets include regular margarines, regular mayonnaise and salad dressings, vitamin E–fortified breakfast cereals, vegetable shortenings, peanut butter, eggs, cooking oils, potato chips, whole milk, tomato products, and apples (142). Sheppard et al. (143) have summarized the tocopherol content and vitamin E activity in various food products based on recent information.

IX. REQUIREMENTS AND ASSESSMENTS

A. Requirements

In 1968, vitamin E was officially recognized as an essential nutrient, and 30 IU was recommended for men by the Food and Nutrition Board. Reassessment of dietary fat and vitamin E content led in 1974 to a revised recommendation of 15 IU. In the 1980 recommended dietary allowance (RDA), efforts were made to take into account the dietary contribution of non-α-tocopherols, and the requirement was expressed as tocopherol equivalent (TE). One TE is equal to 1 mg *RRR*-α-tocopherol, and the other isomers were converted to TE by multiplying the 1 mg of each isomer by the relative activity factor (e.g., β-tocopherol, 0.5; γ-tocopherol, 0.1; α-tocotrienol, 0.3). The tenth RDA expressed in both TE and IU is shown in Table 7, and the values ranged from 4.5 IU (three α-tocopherol equivalents or 3 mg *d*-tocopherol) in infants, 6.0–10.4 in children (10 years or under), 14.9 in males (>11 years), 11.9 in females (>11 years), 14.9 in pregnant women, 16.4 to 17.9 in lactating women (144). The normal intake of vitamin E in U.S. diets ranges between 4 and 33 TE daily in adults not taking vitamin E supplements, with average values between 11 and 13 TE. This level of intake results in average plasma/serum levels in adults of approximately 9.5 ug α-tocopherol/mL (9).

Table 6 Tocopherol Content of Major Fats and Oils Used in Food Products in the United States

Type of fats and oils	% Total fats/oils[a]	Total tocopherol (mg/100 g)	Individual tocopherol[b] (% of total)								
			α-T	α-T-3	β-T	β-T-3	γ-T	γ-T-3	δ-T	δ-T-3
Animal fats:											
Lard	13.3	0.6–1.3	>90	5	—	—	—	—	—	—
Butter	8.2	1.0–5.0	>90	—	—	—	<5	—	—	—
Tallow	4.9	1.5–2.4	>90	—	—	—	<10	—	—	—
Vegetable oils:											
Soybean	53.4	56–160	4–18	—	—	—	58–69	—	—	—
Cotton seed	8.9	30–81	51–67	—	—	—	33–49	—	—	—
Corn	3.8	53–162	11–24	—	—	—	76–89	—	—	—
Coconut	3.1	1–4	14–67	<14	—	<3	—	<53	<17	—
Peanut	1.4	20–32	48–61	—	—	—	39–52	—	—	—
Palm	1.0	33–73	28–50	16–19	—	4	—	34–39	—	—
Palm kernel	0.7	0	—	—	—	—	—	—	—	—
Safflower	0.7	25–49	80–94	—	—	—	6–20	—	—	—
Olive	0.6	5–15	65–85	—	—	—	15–35	—	—	—

[a] 53.4 pounds per capita.
[b] α-T, β-T, γ-T, and δ-T are α-, β-, γ-, and δ-tocopherol, respectively; α-T-3, β-T-3, γ-T-3, and δ-T-3 are the corresponding tocotrienols.
After Ref. 10.

Table 7 U.S. Recommended Daily Dietary Allowance for Vitamin E

	Age (years)	International units	Tocopherol equivalents
Infants	0–0.5	4.5	3.0
Children	0.5–1.0	6.0	4.0
	1–3	8.9	6.0
	4–10	10.4	7.0
Males	≥11	14.9	10.0
Females	≥11	11.9	8.0
	Pregnant	14.9	10.0
	Lactating, first 6 mo.	17.9	12.0
	Lactating, second 6 mo.	16.4	11.0

Vitamin E deficiency is rarely seen in the population of the United States. When it occurs, it is usually a result of lipoprotein deficiencies or lipid malabsorption syndromes (11,12,26). A high intake of polyunsaturated fat may increase the vitamin E requirement. Increased levels have also been recommended to provide protection against oxidants and oxidizing agents in the environment. In addition, recent information suggests that a higher intake of vitamin E is associated with enhanced immune response and reduced risk of certain chronic diseases, particularly coronary heart disease (see Sec. X).

B. Assessments

The nutritional status of vitamin E is assessed based on the concentration of tocopherols in body store and its physiological functions. Currently, several indices, such as tocopherol concentrations in plasma/serum erythrocytes, platelets, or tissues; degree of erythrocyte hemolysis, and amounts of lipid peroxidation products (e.g., ethane, pentane, and malondi-aldehyde) generated, have been used to assess the nutritional status of vitamin E in humans. Measurement of tocopherol concentrations in plasma/serum is the most commonly used approach. The tocopherol concentration in erythrocytes is an indicator of longer term status for vitamin E than that in plasma/serum. Platelets are more sensitive for measurement of dose response to dietary vitamin E when compared with plasma, erythrocytes, or lymphocytes (145). Also, platelet tocopherol concentrations are independent of serum lipid levels (146), an important advantage relative to serum or plasma tocopherol concentrations. However, none of the above accurately reflects both dietary intake and body stores of vitamin E. Measurement of α-tocopherol tissue store, such as adipose tissue, is a reliable way to assess long-term vitamin E status in humans. However, tissue biopsy is highly invasive and therefore impractical for application to the general public. Measurement of tocopherol concentrations in the relatively noninvasive and easily collectible skin and buccal mucosal cells has shown that the nutritional status of vitamin E, and the concentrations of α- and γ-tocopherol are significantly correlated with that of the plasma (147). The significant plasma–tissue relationships suggest that plasma tocopherol concentrations, in lieu of their target tissue concentrations, can be used as a reliable indicator of vitamin E status.

Long-term intake of vitamin E is difficult to assess accurately because diets vary substantially over time. Also, since tocopherols are widely distributed in foods and products made from them, their consumption is difficult to quantify based on a dietary history. Potential sources of error in the use of interviews for obtaining food/nutrient intake information are as follows: (a) only recent intake is estimated, whereas earlier dietary intake may be very different; (b) a limited number of foods were analyzed; and (c) errors may have occurred in subject estimation of food intake frequency and size of portion eaten.

The accuracy of vitamin E intake information obtained through the use of frequency questionnaires also depends on the reliability of information in the food composition data used. The tocopherol content of foods is highly variable, depending genetic, seasonal, processing, storage, and other factors. Also, the tocopherol content varies greatly among foods, and the tocopherol contents of various foods were obtained using colorimetry procedure rather than HPLC. In addition, food processing, storage, and culinary practices can significantly influence the destruction of tocopherols present in the food. Furthermore, the possible interaction of vitamin E and other nutrients, such as polyunsaturated lipids, β-carotene, selenium, sulfur amino acids, and ascorbic acid, further complicates evaluation of vitamin E status.

X. HEALTH EFFECTS

A. Immune Response

The immune response generally declines with age, and suboptimal immune responses in the elderly may be responsible for the increased risk of infections and immune-mediated diseases (148). Therefore, maintaining proper immune function is important for the elderly. Several indices of immune response, including responses on delayed-type hypersensitivity skin tests, antibody production, lymphocyte proliferation, cytokine production, and counts of the specific subgroups of white blood cells, are influenced by the status of essential nutrients, including vitamin E. Vitamin E supplementation is associated with enhanced production of the cytokine interleukin-2 (IL-2), enhanced lymphocyte proliferation, and decreased production of the immunosuppressive prostaglandin E_2 (149). The antioxidant property is likely to play a role in the immunostimulatory effects of vitamin E for the immune response. For example, vitamin E supplementation (60, 200, or 800 IU/day) for a month or longer resulted in significant increases in delayed-type hypersensitivity skin test response, and T-cell subpopulations and proliferative responses, as well as IL-2 activity in the healthy, free-living elderly (150,151,151a). These findings suggest that current dietary vitamin E intake levels of 10–15 IU/day are insufficient to attain optimal immune responses in the general population. However, the 800 IU/day supplement group showed lower immune responses on some tests than the 200 IU/day group, and was comparable with the 60 IU/day or the placebo group, suggesting that more is not necessarily better (151a,151b).

B. Cardiovascular Disease

Cardiovascular disease is the leading cause of morbidity and mortality in industrialized countries. Several dietary factors, including decreased vitamin E status, have been implicated in the incidence of coronary heart disease. In recent years, several large-scale observational and experimental studies have shown that higher intake or serum/plasma levels of α-tocopherol are associated with a decreased risk of cardiovascular diseases. For example, in a study that included 87,245 female nurses and 39,910 male health professionals, the age-adjusted relative risk of coronary heart disease was found to be significantly lowered in the highest quintile of vitamin E intake compared with the lowest quintile (132,153). The apparent benefit was attributable mainly to subjects consuming 100 IU vitamin E or more daily for 2 years or longer. In another study, which included 21,809 postmenopausal women, higher food-derived vitamin E intake was associated with decreased death rate due to coronary heart disease (154). Results obtained from intervention studies generally corroborated with the epidemiological data. In the Cambridge Heart Antioxidant Study (155), for example, patients with angiographically proven coronary atherosclerosis receiving 400 or 800 IU of vitamin E for 510 days had a significant reduction of the rate of major cardiovascular events (47%) and of nonfatal myocardial infarction (77%), in comparison with those receiving placebo. However, cardiovascular deaths were not significantly different between groups.

While the information available suggests that increased intake of vitamin E is associated with reduced risk of cardiovascular disease, the mechanism by which vitamin E reduces the risk of cardiovascular disease is not clear. It is possible that the ability of vitamin E to prevent LDL oxidation (156,157) and/or to reduce platelet adhesion (158,159) may be responsible. Evidence indicates that enhanced uptake of oxidatively modified LDL by macrophages through a specific receptor gives rise to lipid-laden foam cells, one of the earliest stages in the development of atherosclerosis (160–162). Oxidized LDL is toxic to cells and may be responsible for damage to the endothelial layer and destruction of smooth-muscle cells (160,162).

C. Cancer

Carcinogenesis is a multiple process, and a large number of chemical and physical agents can affect the development of cancer in numerous ways. Vitamin E may protect against cancer development by reacting directly with mutagens/carcinogens, altering metabolic activation, enhancing the immune system, inhibiting cell proliferation, or other mechanisms. A large number of observational and experimental studies have been performed to determine the relationship between vitamin E status and cancer risk (163,164). Also, a number of well-designed large-scale chemoprevention trials for antioxidant micronutrients have been conducted in recent years (165–167). However, only a few of those intervention studies are designed specifically to examine the efficacy of vitamin E. Most of the studies dealing with head/neck, lung, and colorectal cancer suggest a protective effect of vitamin E against cancer risk, though reports dealing with the association between vitamin E status and risk of breast, bladder, and cervical cancer are less consistent (168). Nevertheless, the majority of studies suggest that low vitamin E status is associated with increased risk of certain cancer.

D. Neurodegenerative Diseases

Neurodegenerative diseases, such as dementia of Alzheimer type and Parkinson's disease, are characterized neurochemically by a transmitter-specific loss of neurons, which progresses and extends to several neuronal systems over the course of the disease. While the pathophysiology of the heterogenous disorder is largely unknown, free radical-induced oxidative damage may play a role in the cellular events leading to the degeneration of neurons in the brain. In patients with Alzheimer disease, for example, amyloid beta-peptide accumulates in plaques of the brain. The compound is neurotoxic by a mechanism involving induction of reactive oxygen species. The concentration of α-tocopherol is lower in cerebrospinal fluid from patients with Alzheimer type dementia that the control (169). Also, the concentrations of basal and iron-ascorbate stimulated lipid peroxidation products are higher in the inferior temporal cortex of Alzheimer subjects than the controls (170). These reports suggest that the brain of patients with Alzheimer disease is associated with inadequate antioxidant status and/or increased oxidative stress. In a double-blind, multicenter trial, Sano et al. (171) evaluated the effect of α-tocopherol (2,000 IU daily) on primary outcome in 341 patients with Alzheimer's disease of moderate severity over a 2 year period and found that treatment with α-tocopherol was beneficial in delaying the primary outcome of disease progression (death, severe dementia, loss of the ability to perform basic activities of daily living or the need for institutionalization). The degree of functionality of the mutant α-tocopherol transfer protein in patients with familial vitamin E deficiency are associated with the degree of severity of the neurological damage and age of onset as well as with plasma vitamin E concentration (59). High doses of vitamin E can prevent or mitigate the neurological course of this disease. Plasma vitamin E concentrations in patients increase when they are treated with large doses of vitamin E, presumably because of the direct transfer of tocopherol from chylomicrons to other circulating lipoproteins.

XI. SUMMARY AND CONCLUSION

The essentiality of vitamin E for humans had been questioned after the vitamin was identified as the fat-soluble substance necessary for reproduction in rats more than 75 years ago. This is because a number of tissue-specific deficiency symptoms can be easily produced in experimental animals but not in humans. The essentiality of vitamin E for humans was only recognized in the late 1960s. Subsequent studies of patients with certain neurological disorders have clearly demonstrated the need of the vitamin in the development and maintenance of the integrity and function of the nervous mechanism and skeletal muscle.

Recent studies also provide a much better understanding of the mechanisms of tocopherol transport and selective retention of certain forms or isomers of tocopherols. The finding that tocopherol can be regenerated or restored in association with GSH, ascorbic acid, and other systems provides a feasible explanation as to why it is difficult to deplete vitamin E in human adults. Information is available that supports the view that the function of vitamin E is interrelated with other antioxidant systems and that ascorbic acid, β-carotene, sulfur amino acids, and selenium complement the antioxidant function of vitamin E.

In recent years, public interest in vitamin E has increased drastically. This is mainly promoted by experimental findings which suggest that vitamin E may protect against oxidative stress resulting from exposure to environmental agents, and that increased intake of the vitamin is associated with a reduced risk of cardiovascular disease, cancer, and other degenerative diseases, as well as with enhanced immune response. However, more research is needed to establish more definitively the relationship between vitamin E intake and the risk of degenerative diseases, and to elucidate further the mechanisms by which vitamin E is involved in altering disease processes. In addition, the optimal requirement of various population groups for vitamin E against oxidative stress has yet to be established.

REFERENCES

1. H. M. Evans and K. S. Bishop, On the existence of a hitherto unrecognized dietary factor essential for reproduction. *Science 56*:650 (1922).
2. B. Sure, Dietary requirements for reproduction. II. The existence of a specific vitamin for reproduction. *J. Biol. Chem. 58*:693 (1924).
3. H. M. Evans and G. O. Burr, The anti-sterility vitamin fat soluble E. *Proc. Natl. Acad. Sci. USA 11*:334 (1925).
4. H. M. Evans, O. H. Emerson, and G. A. Emerson, The isolation from wheat germ oil of an alcohol, alpha tocopherol, having the properties of vitamin E. *J. Biol. Chem. 113*:319 (1936).
5. P. von Karrer, H. Fritzsch, B. H. Ringier, and H. Salmon, Alpha-tocopherol. *Helv. Chim. Acta 21*:520 (1938).
6. E. Fernholz, On the constitution of alpha-tocopherol. *J. Am. Chem. Soc. 60*:700 (1938).
7. H. S. Olcott and O. H. Emerson, Antioxidant properties of the tocopherols. *J. Am. Chem. Soc. 59*:1008 (1937).
8. M. L. Scott, Studies on vitamin E and related factors in nutrition and metabolism, in *The Fat-Soluble Vitamins* (H. F. DeLuca and J. W. Suttie, eds.), University of Wisconsin Press, Madison, 1969, pp. 355–368.
9. L. J. Machlin, Vitamin E, in *Handbook of Vitamins*, 2nd ed. (L. J. Machlin, ed.), Second ed., Marcel Dekker, New York, 1991, pp. 99–144.
10. C. K. Chow, Vitamin E and blood. *World Rev. Nutr. Dietet. 45*:133 (1985).
11. J. G. Bieri and P. M. Farrell, Vitamin E. *Vitam. Hormones 34*:31 (1976).
12. R. J. Sokol, Vitamin E deficiency and neurological disease. *Annu. Rev. Nutr. 8*:351 (1988).
13. A. Kamal-Eldin and L.-A. Appelqvist, The chemistry and antioxidant properties of tocopherols and tocotrienols. *Lipids 31*:671 (1996).
14. G. W. Burton and M. G. Traber. Vitamin E: antioxidant activity, biokinetics, and bioavailability. *Annu. Rev. Nutr. 10*:357 (1990).
14a. T. S. Shin and J. S. Godber, Isolation of four tocopherols and four tocotrienols from a variety of natural sources by semi-preparative high performance liquid chromatography. *J. Chromatogr. 678*:48 (1994).
15. T. Doba, G. W. Burton, and K. U. Ingold, EPR spectra of some α-tocopherol model compounds. Polar and conformational effects and their relation to antioxidant activities. *J. Am. Chem. Soc. 105*:6505 (1983).
16. M. Matsuo, S. Matsumoto, and T. Ozawa, Electron spin resonance spectra and hyperfine coupling constants of the tocopheroxy and 2,2,5,7,8-pentamethylchroman-6-oxyl radicals derived from vitamin E and its model and deuterated model compounds. *Org. Magnet. Reson. 21*:261 (1983).
17. C. Suarna, M. Baca, and P. T. Southwell-Keely, Oxidation of the α-tocopherol model compound 2,2,5,7,8-pentamethyl-6-chromanol in the presence of alcohol. *Lipids 27*:447 (1992).
18. J. L. G. Nillson, D. G. Doyle, and K. Folkers, New tocopherol dimers. *Acta Chem. Scand. 22*:200 (1968).
18a. D. R. Nelan and C. D. Robeson, The oxidation product from α-tocopherol and potassium ferricyanide and its reaction with ascorbic and hydrochloric acids. *J. Am. Chem. Soc. 84*:2963 (1962).
19. R. Yamaguchi, T. Matsui, K. Kato, and K. Ueno, Reaction of α-tocopherol with 2,2'-azobis (2,4-dimethyl-valeronitrile) in benzene. *Agric. Biol. Chem. 53*:3257 (1989).

20. O. Igarashi, M. Hagino, and C. Inagaki, Decomposition of α-tocopherol spirodimer by alkaline saponification. *J. Nutr. Sci. Vitaminol. 19*:469 (1973).

21. D. C. Liebler, K. L. Kaysen, and T. A. Kennedy, Redox cycles of vitamin E: hydrolysis and ascorbic acid dependent reduction of 8a-(alkyldioxy) tocopherones. *Biochemistry 28*:9772 (1989).

22. C. K. Chow and H. H. Draper, Isolation of γ-tocotrienol dimers from Hevea latex. *Biochemistry 9*:445 (1970).

23. J. Pokprny, Major factors affecting the autoxidation of lipids, in *Autoxidation of Unsaturated Lipids* (H. W. S., Chan, ed.), Academic Press, London, 1987, pp. 141–206.

24. E. A. Serbinova, V. E. Kagan, D. Han, and L. Packer, Free radical recycling and intramembrane mobility in the antioxidant properties of alpha-tocopherol and alpha-tocotrienol. *Free Rad. Biol. Med. 10*:263 (1991).

25. H. S. Olcott and J. Van der Veen, Comparative antioxidant activities of toco and its methyl derivatives. *Lipids 3*:331 (1968).

26. H. J. Kayden and M. G. Traber, Absorption, lipoprotein transport, and regulation of plasma concentrations of vitamin E in humans. *J. Lipid Res. 34*:343 (1993).

27. C. A. Drevon, Absorption, transport and metabolism of vitamin E. *Free Rad. Res. Commun. 14*:229 (1991).

28. P. M. Mathias, J. T. Harries, T. J. Peters, and D. P. R. Muller, Studies on the in vivo absorption of micellar solution of tocopherol and tocopheryl acetate in the rat. *J. Lipid Res. 22*:829 (1981).

29. D. Hollander, E. Rim, and K. S. Muralidhara, Mechanism and site of small intestinal absorption of α-tocopherol in the rat. *Gastroenterology 68*:1492 (1975).

30. D. Hollander, Intestinal absorption of vitamins A, E, D, and K. *J. Lab. Clin. Med. 97*:449 (1981).

31. M. K. Horwitt, Interrelationship between vitamin E and polyunsaturated fatty acids in adult men. *Vitam. Hormones 20*:541 (1962).

32. A. Bjorneboe, G.-E. Soyland, A. Bjorneboe, G. Rajka, and C. A. Drevon, effect of dietary supplementation with eicosapentaenoic acid in the treatment of atopic dermatitis. *Br. J. Dermatol. 117*:463 (1987).

33. I. R. Peake, H. G. Windmueller, and J. G. Bieri, A comparison of the intestinal absorption, lymph and plasma transport, and tissue uptake of α- and γ-tocopherols in the rat. *Biochim. Biophys. Acta 889*:310 (1972).

34. M. G. Traber, H. J. Kayden, J. B. Green, and M. H. Green, Absorption of water-miscible forms of vitamin E in a patient with cholestasis and in thoracic duct-cannulated rats. *Am. J. Clin. Nutr. 44*:914 (1986).

35. M. G. Traber and H. J. Kayden, Preferential incorporation of α-tocopherol vs γ-tocopherol in human lipoproteins. *Am. J. Clin. Nutr. 49*:517 (1989).

36. M. G. Traber, G. W. Burton, K. U. Ingold, and H. J. Kayden, RRR- and SRR-α-tocopherols are secreted without discrimination in human chylomicrons, but RRR-α-tocopherol is preferentially secreted in very low density lipoproteins. *J. Lipid Res. 31*:675 (1990).

37. M. Traber, L. Rudel, G. Burton, L. Hughes, K. Ingold, and H. Kayden, Nascent VLDL from liver perfusions of cynomolgus monkeys are preferentially enriched in RRR- compared with SRR-α-tocopherol. Studies using deuterated tocopherols. *J. Lipid Res. 31*:687 (1990).

38. M. G. Traber, J. C. Lane, N. Lagmay, and H. J. Kayden, Studies on the transfer of tocopherol between lipoproteins. *Biochim. Biophys. Acta 793*:387 (1984).

39. W. A. Behrens, J. N. Thompson, and R. Madere, Distribution of α-tocopherol in human plasma lipoproteins. *Am. J. Clin. Nutr. 35*:691 (1982).

40. E. C. McMormick, D. G. Cornwell, and J. B. Brown, Studies on the distribution of tocopherol in human serum lipoproteins. *J. Lipid Res. 1*:221 (1960).

41. G. M. Kostner, K. Oettl, M. Jauhiainen, C. Ehnholm, and H. Esterbauer, Human plasma

phospholipid transfer protein accelerates exchange/transfer of alpha-tocopherol between lipoproteins and cells. *Biochem. J. 305*:659 (1995).

42. C. H. Floren and A. Nilson, Binding, interiorization and degradation of cholesteryl ester–labeled chylomicron-remnant particles by rat hepatocyte monolayers. *Biochem. J. 168*:483 (1977).

43. B. C. Sherrill, T. L. Innerarity, and R. W. Mahley, Rapid hepatic clearance of the canine lipoproteins containing only the E apoprotein by a high affinity receptor. *J. Biol. Chem. 255*: 1804 (1980).

44. W. Cohn, M. A. Goss-Sampson, H. Grun, and D. P. Muller, Plasma clearance and net uptake of alpha-tocopherol and low-density lipoprotein by tissue in WHHL and control rabbits. *Biochem. J. 287*:247 (1992).

45. L. J. Machlin and E. Gabriel, Kinetics of tissue α-tocopherol uptake and depletion following administration of high levels of vitamin E. *Ann. NY Acad. Sci. 393*:48 (1982).

46. M. G. Traber and H. J. Kayden, Tocopherol distribution and intracellular localization in human adipose tissue. *Am. J. Clin. Nutr. 46*:488 (1987).

47. A. Bjorneboe, G.-E. Aa. Bjorneboe, E. Bodd, B. F. Hagen, N. Kveseth, and C. A. Drevon, Transport and distribution of α-tocopherol in lymph, serum and liver cells in rats. *Biochim. Biophys. Acta 889*:310 (1986).

48. P. J. Hornsby and J. F. Crivello, The role of lipid peroxidation and biological antioxidants in the function of adrenal cortex. *Mol. Cell Endocrinol. 30*:123 (1983).

49. A. Bjorneboe, M. S. Nenseter, B. F. Hagen, G.-E. A. Bjorneboe, K. Prydz, and C. A. Drevon, Effect of dietary deficiency and supplementation with all-rac-α-tocopherol in hepatic content in rat. *J. Nutr. 121*:1208 (1991).

50. B. F. Hagen, A. Bjorneboe, G.-E., Aa. Bjorneboe, and C. A. Drevon, Effect of chronic ethanol consumption on the content of α-tocopherol in subcellular fractions of rat liver. *Alcoholism (NY) 13*:246 (1989).

51. C. K. Chow, Distribution of tocopherols in human plasma and red blood cells. *Am. J. Clin. Nutr. 28*:756 (1975).

52. W. A. Behrens and R. Madere, Transfer of α-tocopherol to microsomes mediated by a partially purified liver α-tocopherol binding protein. *Nutr. Res. 2*:611 (1982).

53. N. Kaplowitz, H. Yoshida, J. Kuhlenkamp, B. Slitsky, I. Ren, and A. Stolz, Tocopherol binding proteins of hepatic cytosol. *Ann. NY Acad. Sci. 570*:85 (1989).

54. G. L. Catignani and J. G. Bieri, Rat liver α-tocopherol binding protein. *Biochim. Biophys. Acta 497*:349 (1977).

55. M. G. Traber, Determinations of plasma vitamin E concentrations. *Free Rad. Biol. Med. 16*: 229 (1994).

56. M. G. Traber, R. J. Sokol, A. Kohlschutter, and H. J. Kayden, Impaired discrimination between stereoisomers of α-tocopherol in patients with familial isolated vitamin E deficiency. *J. Lipid Res. 34*:201 (1993).

57. A. Kohlschutter, C. Hubner, W. Jansen, and S. G. Lindner, A treatable familial neuromyopathy with vitamin E deficiency, normal absorption, and evidence of increased consumption of vitamin E. *J. Inher. Metab. Dis. 11*:149 (1988).

57a. M. Arita, Y. Sato, A. Miyata, T. Tanabe, E. Takahashi, H.J. Kayden, H. Arai, and K. Inoue. Human α-tocopherol transfer protein: cDNA cloning, expression and chromosomal localization. *Biochem. J. 306*: 437 (1995).

58. A. Hentati, H. X. Deng, W. Y. Hung, M. Nayer, M. G. Ahmed, X. He, R. Tim, D. A. Stumpf, and T. Siddique, Human alpha-tocopherol transfer protein: gene structure and mutations in familial vitamin E deficiency. *Ann. Neurol. 39*:295 (1996).

58a. K. Quahchi, M. Arita, H.J. Kayden, F. Hentati, M.B. Hamida, R. Sokol, H. Arai, K. Inoue, J.L. Mandel, and M. Koenig. Ataxia with isolated vitamin E deficiency is caused by mutations in the α-tocopherol transfer protein. *Nature Genet. 9*: 141 (1995).

59. T. Gotoda, M. Arita, H. Arai, K. Inoue, T. Yokota, Y. Fukuo, Y. Yazaki, and N. Yamada,

Adult-onset spinocerebellar dysfunction caused by a mutation in the gene for the alpha-tocopherol-transfer protein. *N. Engl. J. Med. 333*:1313 (1995).

60.   H. Yoshida, M. Yusin, I. Ren, J. Kuhlenkamp, T. Hirano, A. Soltz, and N. Kaplowitz, Identification, purification and immunochemical characterization of a tocopherol-binding protein in rat liver cytosol. *J. Lipid Res. 33*:343 (1992).

61.   Y. Sato, K. Hagiwara, H. Arai, and K. Inoue, Purification and characterization of the α-tocopherol transfer protein from rat liver. *FEBS Lett. 288*:41 (1991).

62.   Y. Sato, A. Arai, A. Miyata, S. tokita, K. Yamamoto, and T. Tanabe, Primary structure of alpha-tocopherol transfer protein from rat liver. Homology with cellular retinaldehyde-binding protein. *J. Biol. Chem. 268*:17705 (1993).

63.   Kuhlenkamp, J., M. Ronk, M. Yusin, A. Stolz, and N. Kaplowitz, Identification and purification of a human liver cytosolic tocopherol binding protein. *Protein Expression Purif. 4*:382 (1993).

64.   A. K. Dutta-Roy, M. J. Gordon, D. J. Leishman, B. J. Paterson, G. G. Duthie, and W. P. James, Purification and partial characterization of an alpha-tocopherol-binding protein from rabbit heart cytosol. *Mol. Cell. Biochem. 123*:139 (1993).

65.   A. K. Dutta-Roy, D. J. Leishman, M. J. Gordon, F. M. Campbell, and G. G. Duthie, Identification of a low molecular weight mass (14.2 kDa) α-tocopherol-binding protein in the cytosol of rat liver and heart. *Biochem. Biophys. Res. Commun. 196*:1108 (1993).

66.   M. J. Gordon, F. M. Campbell, G. G. Duthie, and A. K. Dutta-Roy, Characterization of a novel alpha-tocopherol-binding protein from bovine heart cystol. *Arch. Biochem. Biophys. 318*:140 (1995).

67.   A. S. Csallany, H. H. Draper, and S. N. Shah, Conversion of d-α-tocopherol-C^{14} to tocopheryl-p-quinone in vivo. *Arch. Biochem. Biophys. 98*:142 (1962).

68.   C. K. Chow, H. H. Draper, A. S. Csallany, and M. Chiu, The metabolism of C^{14}-α-tocopheryl quinone and C^{14}-α-tocopheryl hydroquinone. *Lipids 2*:390 (1967).

69.   T. Hayashi, A. Kanetoshi, M. Nakamura, M. Tamura, and H. Shirahama, Reduction of alpha-tocopherolquinone to alpha-tocopherolhydroquinone in rat hepatocytes. *Biochem. Pharmacol. 44*:489 (1992).

70.   E. J. Simon, A. Eisengart, L. Sundheim, and A. T. Milhorat, The metabolism of vitamin E. II. Purification and characterization of urinary metabolites of α-tocopherol. *J. Biol. Chem. 221*:807 (1956).

71.   A. S. Csallany and H. H. Draper, Dimerization of α-tocopherol in vivo. *Arch. Biochem. Biophys. 100*:335 (1963).

72.   H. H. Draper, A. H. Csallany, and M. Chiu, Isolation of a trimer of α-tocopherol from mammalian liver. *Lipids 2*:47 (1967).

72a. N. Schltz, M. Leist, M. Petrzika, B. Gassmann, and R. Breigelius-Flohe. Novel urinary metabolite of α-tocopherol, 2,5,7,8-tetramethyl-2(2′-carboxyethyl)-6-hydroxyxhroman, as an indicator of an adequate vitamin E supply? *Am. J. Clin. Nutr. 62*: 1527S (1995).

72b. W.J. Wechter, D. Kantoci, E.D. Murry, Jr., D.C. D'Amico, M.E. Jung, and W.-H. Wang. A new endogenous natriuretic factor: LLU-α. *Proc. Natl. Acad. Sci. USA. 93*: 6002 (1996).

72c. S. Chiku, K. Hamamura, and T. Nakamura. Novel urinary metabolite of δ-tocopherol in rats. *J. Lipid Res. 25*: 40 (1984).

73.   C. B. Hong and C. K. Chow, Induction of eosinophilic enteritis and eosinophils in rats by vitamin E and selenium deficiency. *Exp. Mol. Pathol. 48*:182 (1988).

74.   J. Harries and D. Muller, Absorption of vitamin E in children with biliary obstruction. *Gut 12*:579 (1971).

75.   C. F. Nockels, D. L. Menge, and E. W. Kienholz, Effect of excessive dietary vitamin E in the chick. *Poult. Sci. 55*:649 (1976).

76.   B. E. March, E. Wong, L. Seier, L. Sim, and L, Biely, Hypervitaminosis E in the chick. *J. Nutr. 103*:371 (1973).

77. D. K. Nichols, M. J. Wolff, L. G. Philips, Jr., and R. J. Montali, Coagulopathy in pink-backed pelicans (*Pelecanus rufescens*) associated with hypervitaminosis E. *J. Zoo Wildl. Med. 20*:57 (1989).

78. A. Bendich and L. J. Machlin, The safety of oral intake of vitamin E: data from clinical studies from 1986 to 1991, in *Vitamin E in Health and Disease* (L. Packer and J. Fuchs), Marcel Dekker, New York, 1993, pp. 411–416.

78a. J.M. Kim and R.H. White. Effect of vitamin E on the anticoagulant response to warfarin. *Am. J. Cardiol. 77*: 545–546 (1996).

79. L. J. Machlin, E. Gabriel, and M. Brin, Biopotency of α-tocopherol by curative myopathy bioassay in the rat. *J. Nutr. 112*:1437 (1982).

79a. H. Weisser and M. Vecchi, Stereoisomers of α-tocopheryl acetate. II. Biopotencies of eight stereoisomers, individually or in mixtures, as determined by rat resportion-gestation test. *Int. J. Nutr. Res. 52*:351 (1982).

80. C. K. Chow, Increased activity of pyruvate kinase in plasma of vitamin E–deficient rats. *J. Nutr. 105*:1221 (1975).

81. C. K. Chow, Effect of dietary vitamin E and selenium on rats: pyruvate kinase, glutathione peroxidase and oxidative damage. *Nutr. Res. 10*:183 (1990).

82. P. K. S. Liu, E. A. Barnard, and P. J. Barnard, Blood plasma pyruvate kinase as a marker of muscular dystrophy. Properties in dystrophic chickens and hamsters, *Exp. Neurol. 67*:581 (1980).

83. M. C. Alberts and F. J. Samaha, Serum pyruvate kinase in muscle disease and carrier states. *Neurology 24*:462 (1974).

84. I. M. Weinstock, J. R. Behrendt, H. E. Wiltshire, Jr., J. Keleman, and S. Louis, Pyruvate kinase diagnostic value in neuromuscular disease. *Clin. Chim. Acta 80*:415 91977).

85. M. Zata, L. J. Shapiro, D. S. Campion, E. Oda, and M. M. Kaback. Serum pyruvate kinase (PK) and creatinine phosphokinase (CPK) in progressive muscular dystrophies. *J. Neurol. Soc. 36*:349 (1978).

86. A. L. Tappel, Vitamin E and free radical peroxidation of lipids. *Annu. NY Acad. Sci. 203*: 12 (1972).

87. C. K. Chow, Vitamin E and oxidative stress. *Free Rad. Biol. Med. 11*:215 (1991).

88. J. A. Lucy, Functional and structural aspects of biological membranes: a suggested structural role of vitamin E in the control of membrane permeability and stability. *Ann. New York Acad. Sci. 203*:4 (1972).

89. G. L. Catignani, F. Chytil, and W. J. Darby, Vitamin E deficiency: immunochemical evidence for increased accumulation of liver xanthine oxidase. *Proc. Natl. Acad. Sci. USA 71*:1966 (1974).

90. C. K. Chow and C. Gairola, Influence of dietary vitamin E and selenium on metabolic activation of chemicals to mutagens. *J. Agric. Food Chem. 32*:443 (1984).

91. J. Chen, M. P. Goetchius, T. C. Campbell, and G. F. Combs, Jr. Effects of dietary selenium and vitamin E on hepatic mixed function oxidase activities and in vivo covalent binding of aflatoxin B_1 in rats. *J. Nutr. 112*:324 (1982).

92. C. W. Mahoney and A. Azzi, Vitamin E inhibits protein kinase C activity. *Biochem. Biophys. Res. Commun. 154*:694 (1988).

92a. D. Boscoboinik, A. Szewczyk, C. Hensey, and A. Azzi, Inhibition of cell proliferation by alpha-tocopherol. Role of protein kinase C. *J. Biol. Chem. 226*:6188 (1991).

93. A. Tasinato, D. Boscoboinik, G. M., Bartoli, P. Maroni, and A. Azzi, d-Alpha-tocopherol inhibition of vascular smooth muscle cell proliferation occurs at physiological concentrations, correlates with protein kinase c inhibition, and is independent of its antioxidant properties. *Proc. Nat. Acad. Sci. USA 92*:12190 (1995).

94. S. N. Meydani, Micronutrients and immune function in the elderly. *Ann. NY Acad. Sci. 587*: 197 (1990).

95. C. E. Douglas, A. C. Chan, and P. C. Choy, Vitamin E inhibits platelet phospholipase A_2—

protection of membranes from lipid peroxidative damage. *Biochim. Biophys. Acta* *876*:639 (1986).

96.   E. J. Goetzel, Vitamin E modulates the lipoxygenation of arachidonic acid in leukocytes. *Nature* *288*:193 (1981).

97.   C. J. MacEvilly and D. P. R. Muller, Lipid peroxidation in neural tissues and fractions from vitamin E–deficient rats. *Free Rad. Biol. Med. 20*:639 (1996).

97a. C.K. Chow, W. Ibrahim, W. Wei, and A.C. Chan. Vitamin E regulates mitochondrial hydrogen peroxide generation. *Free Rad. Biol. Med. 27*: 580 (1999).

97b. A. Lass and R.S. Sohal. Effect of coenzyme Q(10) and alpha-tocopherol content of mitochondria on the production of superoxide anion radicals. *FASEB J. 14*: 87 (2000).

98.   A. T. Diplock and J. Lucy, The biochemical modes of action of vitamin E and selenium: a hypothesis. *FEBS Lett. 29*:205 (1973).

99.   G. W. Grams and K. Eskins, Dye-sensitized photoxidation of tocopherols: correlation between singlet oxygen reactivity and vitamin E activity. *Biochemistry 11*:606 (1972).

100. C. S. Foote, Quenching of singlet oxygen, in *Singlet oxygen* (H. H. Wasserman and R. W. Muray, eds.), Academic Press, New York, 1979, pp. 139–171.

101. K. Fukuzawa and J. M. Gebicki, Oxidation of α-tocopherol in micelles and liposomes by hydroxyl, perhydroxyl and superoxide free radicals. *Arch. Biochem. Biophys. 226*:242 (1992).

102. M. Nishikimi, H. Yamada, and K. Yagi. Oxidation by superoxide of tocopherols dispersed in aqueous media with deoxycholate. *Biochim. Biophys. Acta 627*:101 (1980).

103. D. Lathia and A. Blum, Role of vitamin E as nitrite scavenger and N-nitrosamine inhibitor: a review. *Int. J. Vitam. Nutr. Res. 59*:430 (1989).

104. R. V. Cooney, A. A. Franke, P. J. Harwood, V. Hatch-Pigott, L. J. Custer, and L. J. Mordan, γ-Tocopherol detoxification of nitrogen dioxide: superiority to α-tocopherol. *Proc. Natl. Acad. Sci. USA 90*:1771 (1993).

105. B. Saltman, Oxidative stress: a radical view. *Semin. Hematol. 26*:249 (1989).

106. B. Halliwell, Oxidants and human disease: some new concepts. *FASEB J. 1*:358 (1987).

107. B. A. Freeman and J. D. Crapo, Biology of disease: free radicals and tissue injury. *Lab. Invest. 47*:412 (1982).

108. B. P. Yu, Cellular defense against damage from reactive oxygen species. *Physiol. Rev. 74*: 139 (1994).

109. H. Sies, Biochemistry of oxidative stress. *Angew. Chem. Int. Ed. Engl. 25*:1058 (1986).

110. C. K. Chow, Nutritional influence in cellular antioxidant defense systems. *Am. J. Clin. Nutr. 32*:1066 (1979).

111. J. Wang, C.-J. Huang, and C. K. Chow, Red cell vitamin E and oxidative damage: a dual role of reducing agents. *Free Rad. Res. 24*:291 (1996).

112. S. D. Aust and L. A. Morehouse, Role of metals in oxygen radical reactions. *Free Rad. Biol. Med. 1*:3 (1986).

113. B. Halliwell and J. M. C. Gutterige, Oxygen radicals and iron in relation to biology and medicine: some problems and concepts. *Arch. Biochem. Biophys. 246*:501 (1986).

114. J. G. Bieri and R. P. Evarts, Tocopherols and fatty acids in American diets. The recommended allowance for vitamin E. *J. Am. Dietet. Assoc. 62*:147 (1993).

115. M. K. Horwitt, Interrelationships between vitamin E and polyunsaturated fatty acids in adult men. *Vitam Hormones 20*:541 (1962).

116. M. K. Horwitt, Data supporting supplementation of human with vitamin E. *J. Nutr. 121*:424 (1991).

117. J. T. Rotruck, A. T. Pope, H. E. Ganther, A. B. Swanson, D. G. Hofeman, and W. G. Hoekstra, Selenium: biochemical role as a component of glutathione peroxidase. *Science 179*:588 (1972).

118. F. Ursini, M. Maiorino, and C. Gregolin, The selenoenzyme phospholipid hydroperoxide glutathione peroxidase. *Biochim. Biophys. Acta 839*:62 (1985).

119. E. Niki, J. Tsuchiya, R. Tanimura, and Y. Kamiya, Regeneration of vitamin E from alpha-chromanoxy radical by glutathione and vitamin C. *Chem. Lett.* 789 (1982).

120. J. E. Packer, T. F. Slater, and R. L. Wilson. Direct observation of a free radical interaction between vitamin E and vitamin C. *Nature 278*:737 (1979).

121. A. C. Chan, K. Tran, T. Raynor, P. R. Ganz, and C. K. Chow, Regeneration of vitamin E in human platelets. *J. Biol. Chem. 266*:17290 (1991).

122. H. Wefers, and H. Sies, The protection by ascorbate and glutathione against microsomal lipid peroxidation is dependent on vitamin E. *Eur. J. Biochem. 174*:353 (1988).

123. G. W. Burton, V. Wronska, L. Stone, D. O. Foster, and K. U. Ingold. Biokinetics of dietary RRR-α-tocopherol in the male guinea pig at the three dietary levels of vitamin C and two levels of vitamin E. Evidence that vitamin C does not ''spare'' vitamin E in vivo. *Lipids 25*:199 (1990).

124. C. C. Reddy, R. W. Scholz, C. E. Thomas, and E. J. Massaro, Vitamin E dependent reduced glutathione inhibition of rat liver microsomal lipid peroxidation. *Life Sci. 31*:571 (1982).

125. E. Diliberto, Jr., G. Dean, C. Carter, and P. L. Allen, Tissue, subcellular and submitochondrial distributions of semidehydroascorbate reductase: possible role of semidehydroascorbate reductase in cofactor regeneration. *J. Neurochem. 39*:563 (1982).

126. V. E. Kagan, E. A. Serbinova, T. Forte, G. Scita, and L. Packer, Recycling of vitamin E in human low density lipoproteins. *J. Lipid Res. 33*:385 (1992).

127. A. Constantunescu, D. Han, and L. Packer, Vitamin E recycling in human erythrocyte membranes. *J. Biol. Chem. 268*:10906 (1993).

128. L. Packer, H. Tritschler, and K. Wessel, Neuroprotection by the metabolic antioxidant α-lipoic acid, *Free Rad. Biol. Med. 22*:359 (1997).

129. A. Bendich and J. A. Olson, Biological actions of carotenoids. *FASEB J. 3*:1927 (1989).

130. M. S. Micozzi, Foods, micronutrients, and reduction of human cancer, in *Nutrition and Cancer Prevention: Investigating the Role of Micronutrients* (T. E. Moon and M. S. Micozzi, eds.), Marcel Dekker, New York, 1989, pp. 213–241.

131. G. W. Burton, Antioxidant action of carotenoids. *J. Nutr. 119*:109 (1989).

132. G. W. Burton and K. U. Ingold, β-Carotene: an unusal type of lipid antioxidant. *Science 224*:569 (1984).

133. T. A. Kennedy and D. C. Liebler, Peroxyl radical scavenging by beta-carotene in lipid bilayers. Effect of oxygen partial pressure. *J. Biol. Chem. 267*:4658 (1992).

134. J. L. Napoli and K. R. Race, The biosynthesis of retinoic acid from retinol by rat tissues in vitro. *Arch. Biochem. Biophys. 255*:95 (1987).

135. X. D. Wang, R. P. Marini, X. Rebuterne, J. G. Fox, N. I. Krinsky, and R. M. Russell, Vitamin E enhances the lymphatic transport of beta-carotene and its conversion to vitamin A in the ferret. *Gastroenterology 108*:719 (1995).

136. S. R. Blakely, E. Grundel, M. Y. Jenkins, and G. V. Mitchell, Alterations in β-carotene and vitamin E stress in rats fed β-carotene and excess vitamin A. *Nutr. Rev. 10*:1035 (1990).

137. L. Arnrich and V. A. Virtamo, Interactions of fat-soluble vitamins in hypervitaminosis. *Ann. NY Acad. Sci. 355*:109 (1980).

138. W. C. Willet, M. J. Stampfer, B. A. Underwood, J. O. Taylor, and C. H. Hennekins, Vitamins A, E and carotene: effects of supplementation on their plasma levels. *Am. J. Clin. Nutr. 38*:559 (1983).

139. B. R. Bacon and R. B. Britton, The pathology of hepatic iron overload: a free radical–mediated process? *Hepatology 11*:127 (1990).

140. R. A. Leedle and S. D. Aust, the effect of glutathione on the vitamin E requirement for inhibition of liver lipid peroxidation. *Lipids 25*:241 (1990).

141. C. K. Chow, H. H. Draper, and C. S. Csallany, Method for the assay of free and esterified tocopherols. *Anal. Biochem. 32*:81 (1969).

142. U.S. Department of Agriculture, Human Nutrition Information Service. Nationwide Food

Consumption Survey, 1987–88. Individual intakes, 3 days. Computer tape PB 90-504044. Springfield, VA. National Technical Information Service, 1990.

143.   A. J. Sheppard and J. A. T. Pennington, Analysis and distribution of vitamin E in vegetable oils and foods. In "Vitamin E in Health and Disease" (L. Packer and J. Fuchs, eds.), Marcer Dekker, New York, 1993, pp. 9–31.

144.   Food and Nutrition Board. Recommended dietary allowances, 10th ed. Washington, D.C.: National Academy of Sciences, 1989.

145.   J. Lehmann, Comparative sensitivities of tocopherol levels of platelets, red blood cells, and plasma for estimating vitamin E nutritional ststus in the *rat. Am. J. Clin. Nutr. 34*:2104 (1981).

146.   G. K., Vatassery, A. M. Krezowski and J. H. Eckfeldt. Vitamin E concentrations in human blood plasma and platelets. *Am. J. Clin. Nutr. 37*:1020 (1983).

147.   Y.-M. Peng, Y.-S. Peng, Y. Lin, T. Moon, D. J. Roe and C. Ritenbaugh. Concentrations and plasma-tissue-diet relationships of carotenoids, retinoids and tocopherols in humans. *Nutr. Cancer 23*:233 (1995).

148.   R. K. Chandra, Nutrition and immunity in the elderly: clinical significance. *Nutr. Rev. 53*: S80 (1995).

149.   S. N., Meydani, M. Meydani, M., L. C. Rall, F. Morrow and J. B. Blumberg. Assessment of the safety of high-dose, short-term supplementation with vitamin E in healthy orer adults. *Am. J. Clin. Nutr. 60*:704 (1994).

150.   S. N., Meydani, M. P. Barklund, S. Liu, R. A. Miller, J. G. Cannon, F. D. Morrow, R. Rocklin and J. B. Blumberg. Vitamin E supplementation enhances cell-mediated immunity in healthy elderly subjects. *Am. J. Clin. Nutr. 52*:557 (1990).

151.   S. N. Meydani, L. Leka and R. Loszewski. Long-term vitamin E supplementation enhances immune response in healthy elderly. *FASEB J. 8*:A272 (1994).

151a. S.N. Meydani, M. Meydani, J.B. Blumberg, L.S. Lynette, G. Siber, R. Loszewski, C. Thompson, M.C. Pedrosa, R.D. Diamond, and B.D. Stollar. Vitamin E supplementation and in vivo immune response in healthy elderly subjects. *J. Am. Med. Assoc. 277*: 1380 (1997).

152.   M. J. Stampfer, C. H. Henneckens, J. E. Mason, G. A. Colditz, B. Rosner and W. C. Willett. Vitamin E consumption and the risk of coronary disease in women. *New England J. Med. 328*:1444 (1993).

152a. R.K. Chandra. Graying of the immune system. Can nutrient supplements improve immunity on the elderly? *J. Am. Med. Assoc. 277*: 1398 (1997).

153.   M. J. Stampfer, C. H. Henneckens, A. Ascherio, E. Giovannucci, G. A. Colditz, B. Rosner and W. C. Willett. Vitamin E consumption and the risk of coronary disease in men. *New England J. Med. 328*:1450 (1993).

154.   L. H. Kushi, A. R. Folson, R. J. prineas, P. J. Mink, Y. Wu, R. M. and Bostick. Dietary antioxidant vitamin intake and coronary disease in postmenopausal women. *New England J. Med. 334*:1156 (1996).

155.   N. G. Stephens, A. Parsons, P. M. Schofield, F. kelly, K. Cheeseman and M. J. Mitchison. Randonmized controlled trial of vitamin E in patients with coronary disease: Cambridge Heart Antioxidant Study (CHAOS). *Lancet 347*:781 (1996).

156.   H. Esterbauer, M. Dieber-Rotheneder, M. Striegel and G. Waeg. Role of vitamin E in preventing the oxidation of low density lipoprotein. *Am. J. Clin. Nutr. 53*:314S (1991).

157.   I. Jialal, C. J. Fuller and B. A. Huet. the effect if alpha-tocopherol supplementation on LDL oxidation: a dose-response study. *Arterioscler. Thromb. Vascular Biol. 15*:190 (1995).

158.   J. Jandak, M. Steiner and P. D. Richardson. Tocopherol, an effective inhibitor of platelet adhesion. *Blood 73*:141 (1989).

159.   M. Steiner, Vitamin E: more than an antioxidant. *Clin. Cardiol. 16* 16S (1993).

160.   H. Esterbauer, M. Dieber-Rotheneder, G. Waeg, G. Striegel and G. Jurgens. Biochemical, structural and functional properties of oxidized low-density lipoprotein. *Chem. Res. Toxicol. 3*:77 (1990).

161. H. Esterbauer, G. Jurgen, O. Quehenberger and E. Koller. Autoixdation of human low density lipoprotein: loss of polyunsaturated fatty acids and vitamin E and generation of aldehydes. *J. Lipid Res. 28*:495 (1987).

162. D. Steinberg, S. Parthasasathy, T. E. Carew, L. C. Khoo and J.-L. Witztum. Beyond cholesterol-modifications of low-density lipoprotein that increase its atherogenicity. *New England J. Med. 320*:915 (1989).

163. H. Sies, W. Stahl and A. R. Sundquist. Antioxidant functions of vitamins: vitamins E and C, beta-carotene, and other carotenoids. *Ann. New York Acad. Sci. 669*:7 (1992).

164. L. Packer, Protective role of vitamin E in biological systems. *Am. J. Clin. Nutr. 53*:1050S (1991).

165. E. R. Greenberg, J. A. Baron, T. D. Tostesom, D. H. Freeman, G. J. Beck, T. A. Colacchino, J. A. Coller, H. D. Frank, R. W. Haile, J. S. Mandel, D. W. Nierenberg, R. Rothstein, D. C. Snover, M. M. Stevens, and R. U. van Stolk, A clinical trial of antioxidant vitamins to prevent corolectoral adenoma. *N Engl. J. Med. 331*:141 (1994).

166. The Alpha-Tocopherol, Beta-Carotene Cancer prevention Group. The effect of vitamin E and beta-carotene on the incidence of lung cancer and other cancer in male smokers. *N. Engl. J. Med. 330*:1029 (1994).

167. W. J. Blot, J.-Y. Li, R. R. Taylor, W. Guo, S. Dawsey, G.-O. Wang, C. S. Yang, S.-F. Zheng, M. Gail, G.-Y. Li, Y. Yu, B.-Q. Liu, J. Tangrea, Y.-H. Sun, F. Lie, J. F. Fraumeni, Y.-H. Zhang, and B. Li, Nutrition intervention trials in Linxian, China: supplementation with specific vitamin/mineral combinations, cancer incidence, and disease-specific mortality in the general population. *J. Natl. Cancer Inst. 85*:1483 (1993).

168. C. K. Chow, Vitamin E and cancer, in *Nutrition and Disease* (K. K. Carroll and D. Kritchevsky, eds.), AOCS Press, Champaign, 1994, pp. 173–233.

169. S.W. Barger, D. Horster, K. Furukawa, Y. Goodman, J. Krieglstein, and M.P. Mattson. Tumor necrosis factors alpha and beta protect neurons against amyloid beta-peptide toxicity: evidence for involvement of a kappa B-binding factor and attenuation of peroxide and Ca2+ accumulation. *Proc. Natl. Acad. Sci. USA 92*: 9328 (1995).

170. H. Tohgi, T. Abe, M. Nakanishi, F. Hamato, K. Sasaki, and S. Takahashi. Concentrations of alpha-tocopherol and its quinone derivative in cerebrospinal fluid from patients with vascular dementia of the Binswanger type and Alzheimer type dementia. *Neurosci. Lett. 174*: 73 (1994).

171. M. Sano, C. Ernesto, R.G. Thomas, M.R. Klauber, K. Schafer, M. Grundman, P. Woodbury, J. Growdon, C.W. Cotman, E. Pfeiffer, L.S. Schneider, and L.J. Thal. A controlled trial of selegiline, alpha-tocopherol of both as treatment for Alzheimer's disease. *N. Engl. J. Med. 336*: 1216 (1997).

5

Bioorganic Mechanisms Important to Coenzyme Functions

DONALD B. McCORMICK

Emory University, Atlanta, Georgia

I. INTRODUCTION

A. Definition

With the use (and misuse) of terms over significant periods of time, there is sufficient uncertainty regarding a specific word that it is best to avoid confusion by defining it. This is the case for the word "coenzyme," which together with "cofactor" and "prosthetic group" have to some become ambiguous biochemical jargon (1). As defined succinctly, a coenzyme is a natural, organic molecule that functions in a catalytic enzyme system (2). Coenzymes bind specifically within protein apoenzymes to constitute catalytically competent holoenzymes. Hence, coenzymes are organic cofactors that augment the diversity of reactions that otherwise would be limited to chemical properties, principally simple acid–base catalysis, of side-chain substituents from amino acid residues within enzymes. With tight binding, coenzymes may be referred to as prosthetic groups; with loose binding, they may be called cosubstrates.

The focus of this chapter will be on the molecular means or mechanisms by which coenzymes participate as the loci of bond making and breaking steps in holoenzyme-catalyzed reactions. Most coenzymes are more complex metabolic derivatives of water-soluble vitamins. Since the subject of this volume is vitamins, coverage of coenzyme mechanisms will bear emphasis on those coenzymes derived from such vitamins covered in other regards in subsequent chapters. References are sparsely used for typical or recent material that otherwise has largely become the domain of mechanistically inclined textbooks of biochemistry (3,4), especially when set for the more advanced student or professional (5).

B. Groupings

Attempts to place each coenzyme in a singular mechanistic group are sometimes fraught with difficulty because some can arguably fit into more than one category. For example, the lipoyl residues of transacylases undergo oxidation–reduction as well as participate in acyl transfers, but it is the latter that distinguishes the biological function. In at least a few instances, the tetrahydro forms of pterin coenzymes are oxidized to the dihydro level during operations of the enzymic systems, e.g., 5,10-methylene tetrahydrofolate in thymidylate synthase and tetrahydrobiopterin in phenylalanine hydroxylase. More conventionally, folyl coenzymes are pulled together in a broad mechanistic view under one-carbon transfers.

The six groups that are subdivisions of this chapter reasonably imply the mechanistic and functional connections of principal coenzymes. These are oxidation–reduction reactions, generation of leaving group potential, acyl activation and transfer, carboxylations, one-carbon transfers, and rearrangements on vicinal carbons. These groupings follow a conventional pattern of biochemical recognition of what major purposes are served by vitamin-derived coenzymes.

II. OXIDATION–REDUCTION REACTIONS

A. Nicotinamide Coenzymes

Nicotinamide adenine dinucleotide (NAD) and its phosphate (NADP) are the two natural pyridine nucleotide coenzymes derived from niacin, as discussed in Chapter 6. Both contain an $N(1)$-substituted pyridine-3-carboxamide (nicotinamide) that is essential to function in redox reactions with a potential near -0.32 V. The nicotinamide coenzymes function in numerous oxidoreductase systems, usually of the dehydrogenase/reductase type, which include such diverse reactions as the conversion of alcohols (often sugars and polyols) to aldehydes or ketones, hemiacetals to lactones, aldehydes to acids, and certain amino acids to keto acids (6). The general reaction stoichiometry is written as:

$$\text{Substrate} + \text{NAD(P)}^+ \rightleftharpoons \text{product} + \text{NAD(P)H} + \text{H}^+$$

The common stereochemical mechanism of operation of nicotinamide coenzymes when associated with enzymes involves the stereospecific abstraction of a hydride ion (H^-) from substrate, with para addition to one or the other side of C-4 in the pyridine ring of the coenzyme, as shown in Fig. 1. The second hydrogen of the substrate group oxidized is concomitantly removed, typically from an electronegative atom (e.g., N, O, or S), as a proton (H^+) that ultimately exchanges as a hydronium ion. The sidedness of the pyridine ring is such that the para hydrogen up from the plane, when the carboxyamide function is near (or to the right) and the pyridine N is on the left (or at the bottom), is

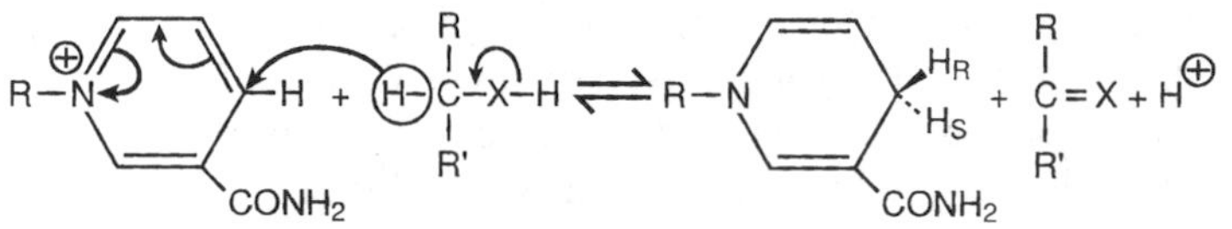

Fig. 1 The stereospecific hydride ion transfer to and from nicotinamide coenzymes generating prochiral R and S forms.

prochiral R because it is oriented toward the *re* face (A side). The hydrogen down from the plane of the ring is then prochiral S and oriented toward the *si* face (B side). There are numerous examples of nicotinamide coenzyme–dependent enzymes that are stereospecific as regards addition/removal of hydride ion from the prochiral R (A) or S (B) position relative to the pyridine ring. These include, for A sidedness, NAD-dependent alcohol dehydrogenase and NADP-dependent cytoplasmic isocitrate dehydrogenase; for B sidedness, NAD-dependent D-glyceraldehyde-3-phosphate dehydrogenase and NADP-dependent D-glucose-6-phosphate dehydrogenase (7). In general, A-side dehydrogenases bind a conformation of the coenzyme in which the nicotinamide ring has an anti orientation with respect to the ribosyl ring, i.e., the carboxamide groups points away; B-side dehydrogenases usually have syn conformation. This is probably due to stabilization of the dihydronicotinamide in the boat conformation attributable to orbital overlap between the lone *n* pair of electrons on the pyridine nitrogen and the antiboding σ orbital of the ribosyl C-O moiety, as depicted in Fig. 2. The pseudoaxial hydrogens, either pro-R with anti or pro-S with syn orientation, are then more easily transferred because of orbital overlap in the transition state.

Nicotinamide coenzymes also participate in other (nonredox) biological reactions that involve ADP ribosylations; however, these are properly considered as substrate rather than coenzymic functions. The larger nucleotide-like structure of nicotinamide coenzymes is important in regard to coenzymic roles primarily to facilitate recognition and specific binding by the protein apoenzymes with which they interact.

B. Flavocoenzymes

Flavin mononucleotide (FMN), flavin adenine dinucleotide (FAD), and covalently linked flavocoenzymes (usually 8α-substituted FAD) are the natural coenzymes derived from riboflavin (vitamin B$_2$), which is discussed in Chapter 7. All contain an isoalloxazine ring system with a redox potential near −0.2 V when free in solution, but the potential is subject to considerable variation when bound within functional flavoproteins. This propensity for being able to shift potential gives flavoenzymes a wider operating range for oxidation–reduction reactions than is the case for enzymes dependent on nicotinamide coenzymes.

Flavoproteins are variably able to catalyze both one- and two-electron redox reactions. The ring portion involving nitrogens 1 and 5 and carbon 4a reflects sequential addition or loss of electrons and hydrogen ions in one-electron processes and addition or loss of other transitory adducts with two-electron processes. There are nine chemically discernible redox forms, i.e., three levels of oxidation–reduction and three species for acid, neutral, and base conditions. Of these only five have biological relevance because

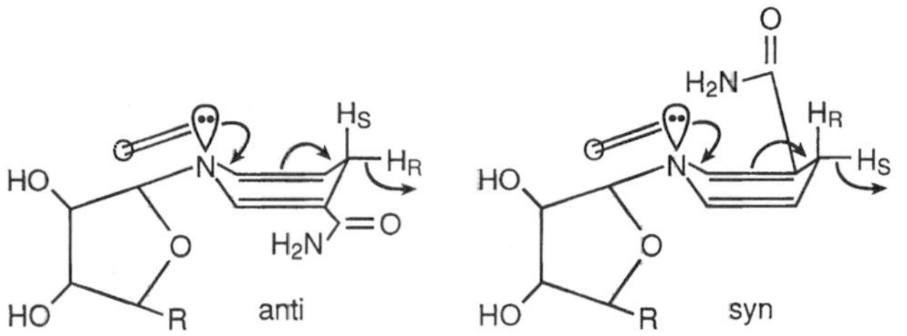

Fig. 2 The syn and anti conformations of bound nicotinamide coenzymes that putatively transfer pro-(S) and pro-(R) hydrogens, respectively.

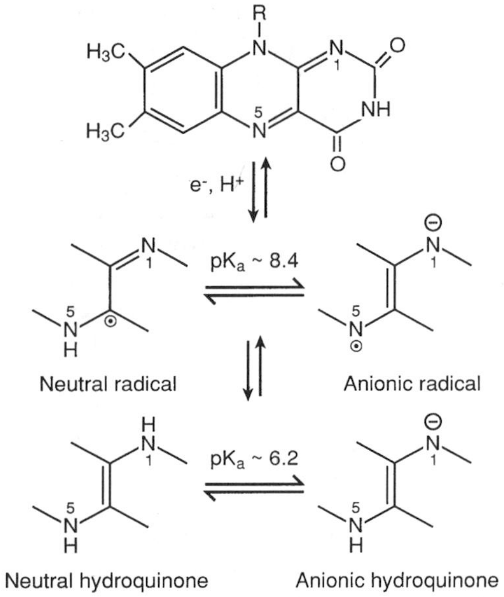

Fig. 3 Biologically important redox states of flavocoenzymes with pK_a values for interconversions of the free species.

of pH considerations (2). These biological forms are summarized in Fig. 3. It should also be noted that free flavin semiquinones (radicals) are quite unstable, and the free flavin hydroquinones react very rapidly with molecular oxygen to become reoxidized. This latter is kinetically dissimilar to reduced nicotinamide coenzymes, which reoxidize more slowly with O_2. Binding to specific enzymes markedly affects the kinetic stability of the half-reduced and reduced forms of flavocoenzymes and again allows for diverse reactions under biological situations. An example of a flavoprotein undergoing a one-electron transfer is with the microsomal NADPH–cytochrome P450 reductase. This enzyme contains both FAD and FMN, which latter cycles between a neutral semiquinone and fully reduced form (8). The FAD in the electron-transferring flavoprotein, which mediates electron flow from fatty acyl coenzyme A (CoA) to the mitochondrial electron transport chain, cycles between oxidized quinone and anionic semiquinone. In addition, a single-step, two-electron transfer from substrate can occur in the nucleophilic reactions shown in Fig. 4. Such cases as hydride ion transfer from reduced nicotinamide coenzymes or the carbanion generated by

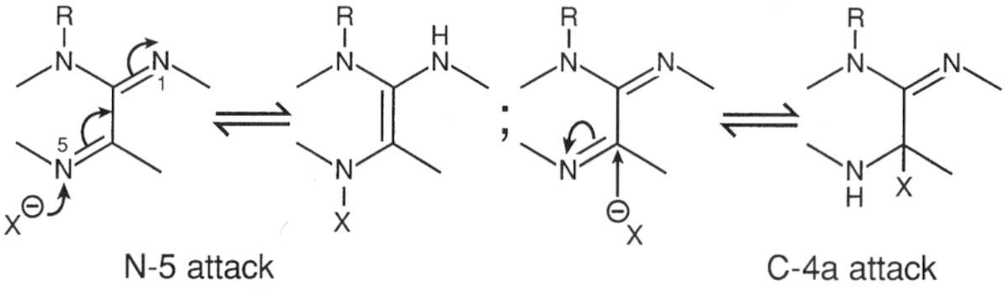

Fig. 4 Reaction types encountered with flavoquinone coenzymes and natural nucleophiles to generate N-5 and C-4a adducts as intermediates.

base abstraction of a substrate proton may lead to attack at the flavin N-5 positions; some nucleophiles, such as the hydrogen peroxide anion, are added on the C-4a position with its frontier orbital.

As regards stereochemistry, the transfer of hydrogen on and off N-5 can take place to or from one or the other face of the isoalloxazine ring system. When visualized with the benzenoid ring to the left and side chain at top (as in the top of Fig. 3), orientation is to the *si* face with *re* on the opposite side. A fair number of flavoproteins have now been categorized on this steric basis (7). Examples include the FAD-dependent glutathione reductase from human erythrocytes, which uses NADPH as substrate and is *re* side interacting, and the FMN-dependent spinach glycolate oxidase, which is *si* side interacting. It should be noted that reduced flavins free in solution have bent or "butterfly" conformations, since the dihydroisoalloxazine has a 144° angle between benzenoid and pyrimidinoid protions. The orientations taken when bound to enzymes, however, may vary and can influence the redox potential.

There are flavoprotein-catalyzed dehydrogenations that are both nicotinamide coenzyme–dependent and independent, reactions with sulfur-containing compounds, hydroxylations, oxidative decarboxylations, dioxygenations, and reduction of O_2 to hydrogen peroxide. The diversity of these systems is covered in periodic symposia on flavins and flavoproteins (9). The intrinsic ability of flavins to be variably potentiated as redox carriers upon differential binding to proteins, to participate in both one- and two-electron transfers, and in reduced (1,5-dihydro) form to react rapidly with oxygen permits wide scope in their operation.

III. GENERATION OF LEAVING GROUP POTENTIAL

A. Thiamine Pyrophosphate

Thiamine pyrophosphate (TPP) is the coenzyme derived from thiamine (vitamin B_1), which is discussed in Chapter 8. Though the substituted pyrimidyl portion shares a role in apoenzymic recognition and binding, the thiazole moiety not only is important in that regard but is involved with substrates to provide a transitory matrix that provides good leaving group potential (2,10). Specifically, TPP as a Mg^{2+} ternary complex within enzymes can react as an ylid, resonance-stabilized carbanion that attacks carbonyl functions as illustrated in Fig. 5.

In all cases, the bond to be labilized (from R' to the C of the original carbonyl carbon) must be oriented for maximal σ-π orbital overlap with the periplanar system extending to the electron-deficient quaternary nitrogen of the thiazole ring. There are two general types of biological reactions in which TPP functions in so-called active aldehyde transfers. First, in decarboxylation of α-keto acids, the condensation of the thiazole moiety of TPP with the α-carbonyl carbon on the acid leads to loss of CO_2 and production of a resonance-stabilized carbanion. Protonation and release of aldehyde occur in fermentative organisms such as yeast, which have only the TPP-dependent decarboxylase, but reaction of the α-hydroxalkyl-TPP with lipoyl residues and ultimate conversion to acyl-CoA occurs in higher eukaryotes, including humans with multienzymic dehydrogenase complexes (see Sec. IV.8). The other general reaction involving TPP is the transformation of α-ketols (ketose phosphates). Specialized phosphoketolases in certain bacteria and higher plants can split ketose phosphates, e.g., D-xylulose-5-phosphate, to simpler, released products, e.g., D-glyceraldehyde-3-phosphate and acetyl phosphate. However, the reaction of impor-

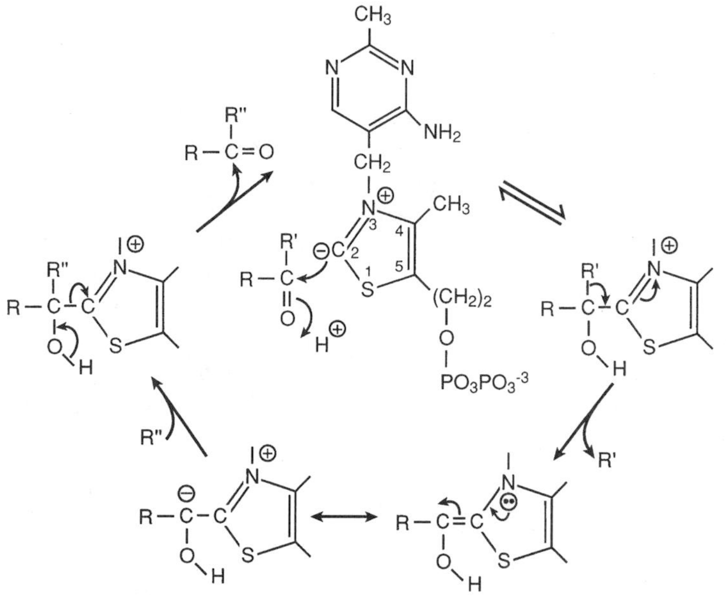

Fig. 5 Function of the thiazole moiety of thiamine pyrophosphate. In α-keto acid decarboxyl-ations, R′ is a carboxylate lost as CO_2 and R″ is a proton generating an aldehyde. In ketolations (both trans- and phospho-), R and R′ are part of a ketose phosphate generating a bound glycoalde-hyde intermediate and R″ is an aldose phosphate or inorganic phosphate generating a different ketose phosphate or acetyl phosphate, respectively.

tance to humans and most animals is a transketolation. Transketolase is a TPP-dependent enzyme found in the cytosol of many tissues, especially liver and blood cells, where princi-pal carbohydrate pathways exist. This enzyme catalyzes the reversible transfer of a gly-coaldehyde moiety (α,β-dihydroxyethyl-TPP) from the first two carbons of a donor ketose phosphate, i.e., D-xylulose-5-phosphate, to the aldehyde carbon of an aldose phosphate, i.e., D-ribose-5-phosphate, of the pentose phosphate pathway wherein D-sedoheptulose-7-phosphate and D-glyceraldehyde-3-phosphate participate as the other substrate–product pair.

B. Pyridoxal-5′-Phosphate

Two of the three natural forms of vitamin B_6, which is discussed in Chapter 10, can be phosphorylated to yield directly functional coenzymes, i.e., pyridoxal-5′-phosphate (PLP) and pyridoxamine-5′-phosphate (PMP). PLP is the predominant and more diversely func-tional coenzymic form, although PMP interconverts as coenzyme during transaminations (2–5,11). At physiological pH, the dianionic phosphates of these coenzymes exist as zwit-terionic *meta*-phenolate pyridinium compounds. Both natural and synthetic carbonyl re-agents (e.g., hydrazines and hydroxylamines) form Schiff bases with the 4-formyl function of PLP, thereby removing the coenzyme and inhibiting PLP-dependent reactions.

PLP functions in numerous reactions that embrace the metabolism of proteins, carbo-hydrates, and lipids. Especially diverse are PLP-dependent enzymes that are involved in amino acid metabolism. By virtue of the ability of PLP to condense its 4-formyl substituent with an amine, usually the α-amino group of an amino acid, to form an azomethine (Schiff base) linkage, a conjugated double-bond system extending from the α carbon of the amine (amino acid) to the pyridinium nitrogen in PLP results in reduced electron density around the α carbon. This potentially weakens each of the bonds from the amine (amino acid) carbon to the adjoined functions (hydrogen, carboxyl, or side chain). A given apoenzyme then locks in a particular configuration of the coenzyme–substrate compound such that maximal overlap of the bond to be broken occurs with the resonant, coplanar, electron-withdrawing system of the coenzyme complex. These events are depicted in Fig. 6.

Selection of the σ bond to be cleaved (R'-C) is achieved by stabilizing a conformation of the external aldimine in which the leaving group is orthogonal to the plane of the ring system, thus ensuring maximal orbital overlap with the π system in the transition state. The type of stereochemistry involved is characteristic of a given enzyme system. Both *re* and *si* faces of the PLP complex can orient toward a reactive function, such as a base, contributed by enzyme protein. This is the case, for instance, with aspartate aminotransferase whereby a base function abstracts a proton from the α position and transfers it to the same side of the π system (syn transfer), adding it to the *si* face of the azomethine group in the ketimine. In this case, the hydrogen is incorporated in a pro-S position at carbon 4' (the methylene of PMP).

Aminotransferases effect rupture of the α-hydrogen bond of an amino acid with ultimate formation of an α-keto acid and PMP; this reversible reaction proceeds by a double-displacement mechanism and provides an interface between amino acid metabolism and that for ketogenic and glucogenic reactions. Amino acid decarboxylases catalyze breakage of the α-carboxyl bond and lead to irreversible formation of amines, including several that are functional in nervous tissue (e.g., epinephrine, norepinephrine, serotonin, and γ-aminobutyrate). The biosynthesis of heme depends on the early formation of δ-aminolevulinate from PLP-dependent condensation of glycine and succinyl-CoA followed by decarboxylation. There are many examples of enzymes, such as cysteine desulfhydrase and serine hydroxymethyltransferase, that affect the loss or transfer of amino acid side chains. PLP is the essential coenzyme for phosphorylase that catalyzes phosphorolysis of the α-1,4 linkages of glycogen. An important role in lipid metabolism is the PLP-dependent condensation of L-serine with palmitoyl-CoA to form 3-dehydrosphinganine, a precursor of sphingolipids. The diversity of PLP functions is covered in periodic symposia

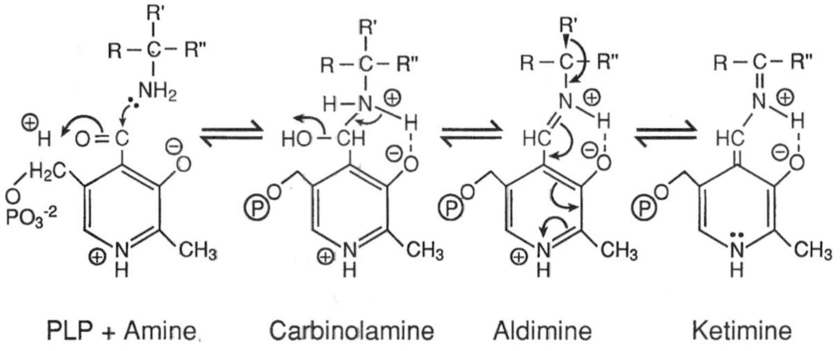

PLP + Amine   Carbinolamine   Aldimine   Ketimine

Fig. 6 Operation of pyridoxal-5'-phosphate with a biological primary amine. Loss of R' shown from the aldimine can be generalized for side-chain cleavages and eliminations, loss of CO_2 in decarboxylations, or loss of a proton in aminotransferations and racemizations.

that now include other carbonyl compounds as cofactors (pyruvyl enzymes, quinoproteins, etc.) (12).

IV. ACYL ACTIVATION AND TRANSFER

A. Phosphopantetheine Coenzymes

4′-Phosphopantetheine is a phosphorylated amide of β-mercaptoethylamine and the vitamin pantothenate, which is discussed in Chapter 9. The phosphopantetheinyl moiety serves as a functional component within the structure of CoA and as a prosthetic group covalently attached to a seryl residue of acyl carrier protein (ACP). Because of a thiol (sulfhydryl) terminus with a pK_a near 9, phosphopantetheine and its coenzymic forms are readily oxidized to the catalytically inactive disulfides.

Esterification of the thiol function to many carboxylic acids is the prelude to numerous acyl group transfers and enolizations (2–5). In classic terminology, the thiol ester enhances both "head" and "tail" activations of acyl compounds, as illustrated in Fig. 7. In head activation, the carbonyl function is attacked by a nucleophile and releases the original thiol. The carbonyl carbon of a thiol ester is more positively polarized than would be an oxy ester counterpart. Hence, the acyl moiety is relatively activated for transfer. The phosphopantetheine terminus of CoA and ACP both function in acyl transfers, the latter only within the fatty acid synthase complex. In tail activation, there is greater tendency for thio than oxy esters to undergo enolization by permitting removal of an α hydrogen as a proton. The enolate is stabilized by delocalization of its negative charge between the α carbon and the acyl oxygen, which makes it thermodynamically accessible as an intermediate. Since the developing charge is also stabilized in the transition state that precedes the enolate, it is also kinetically accessible. This process leads to facile condensation whereby nucleophilic addition of the carbanion-like carbon of the enolate to a neutral activated acyl group (another thiol ester) is a favored process. CoA is a good leaving group from the tetrahedral intermediate. From the above, it follows that thio esters are more like a ketone than an oxy ester. The degree of resonance electron delocalization from the overlap of sulfur p orbitals with the acyl π bond is less than with oxygen in oxy esters.

The myriad acyl thio esters of CoA are central to the metabolism of numerous compounds, especially lipids and the penultimate catabolites of carbohydrates and ketogenic

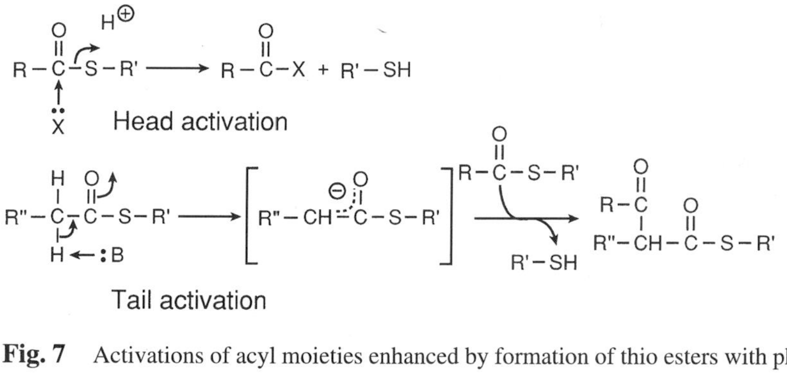

Fig. 7 Activations of acyl moieties enhanced by formation of thio esters with phosphopantetheine coenzymes.

amino acids. For example, acetyl-CoA, which is formed during metabolism of carbohydrates, fats, and some amino acids, can acetylate compounds such as choline and hexosamines to produce essential biochemicals. It can also condense with other metabolites, such as oxalacetate, to supply citrate, and it can lead to formation of cholesterol. The reactive sulfhydryl termini of ACP provide exchange points for acetyl-CoA and malonyl-CoA. The ACP-S-malonyl thio ester can chain-elongate during fatty acid biosynthesis in a synthase complex.

B. Lipoic Acid

α-Lipoic (thioctic) acid is not a vitamin for the human and other animals because it can be biosynthesized from longer chain, essential fatty acids. However, it is indispensable in its coenzymic role, which interfaces with some of the functions of TPP and CoA (2–4). The natural d isomer of 6,8-dithiooctanoic (1,2-dithiolane-3-pentanoic) acid occurs in amide linkage to the ϵ-amino group of lysyl residues within transacylases that are core protein subunits of α-keto acid dehydrogenase complexes of some prokaryotes and all eukaryotes. In such transacylases, the functional dithiolane ring is on an extended flexible arm. The lipoyl group mediates the transfer of the acyl group from an α-hydroxyalkyl-TPP to CoA in a cyclic system that transiently generates the dihydrolipoyl residue, as shown in Fig. 8. Hence, the lipoyl/dihydrolipoyl pair in this cycle serves a dual role of electron transfer and acyl group vector by coupling the two processes.

    The three α-keto acid dehydrogenase complexes of mammalian mitochondria are for pyruvate and α-ketoglutarate of the Krebs citric acid cycle and for the branched-chain α-keto acids from some amino acids. All involve participation of lipoyl moieties within core transacylase subunits surrounded by subunits of TPP-dependent α-keto acid decarboxylase, which generate α-hydroxyalkyl-TPP, and are further associated with subunits of FAD-dependent dihydrolipoyl (lipoamide) dehydrogenase. The number of subunits and their packing varies among cases. For instance, a single particle of *E. coli* pyruvate dehy-

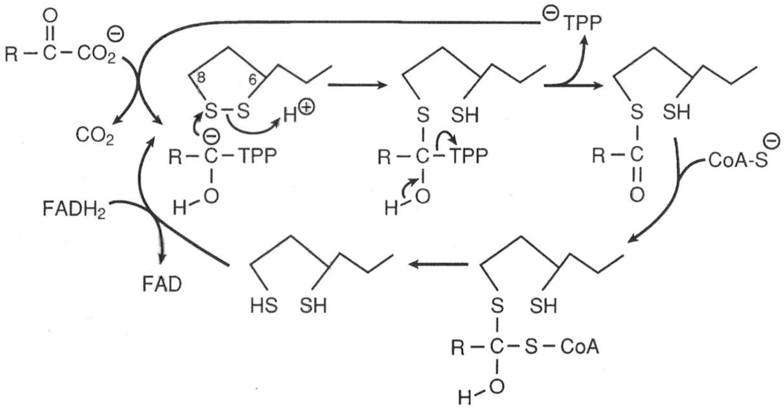

Fig. 8 Function of the lipoyl moiety within enzymes involved in transacylations following α-keto acid decarboxylations (see Fig. 5). In multienzyme dehydrogenases, there is transfer of an acyl moiety from an α-hydroxyalkyl thiamine pyrophosphate to the lipoyl group of a transacylase core and thence to CoA. This results in formation of the dihydrolipoyl group, which is cyclically reoxidized by the FAD-dependent dihydrolipoyl dehydrogenase.

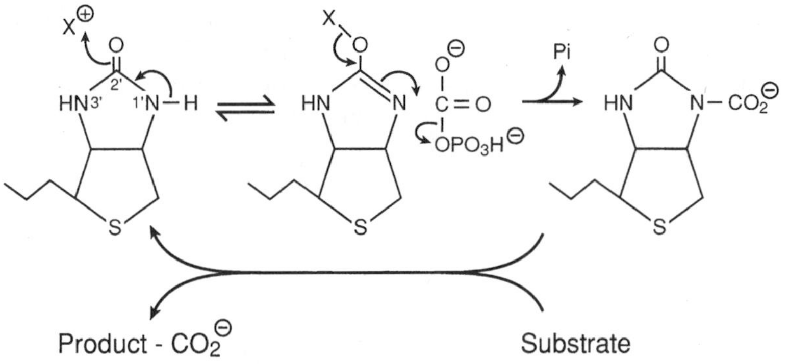

Fig. 9 Function of the biotinyl moiety within enzymes involved in carboxylations with the intermediacy of the putative carbonyl phosphate.

drogenase consists of at least 24 chains of each decarboxylase and transacetylase plus 12 chains of the flavoprotein.

V. CARBOXYLATIONS

A. Biotin

The characteristics of the vitamin biotin are discussed in Chapter 11. The coenzymic form of this vitamin occurs only as the vitamin with its valeric acid side chain amide-linked to the ε-amino group of specific lysyl residues in carboxylase and transcarboxylase (2–5). The length of this flexible arm (~14 Å) is similar to that encountered with the lipoyl attachments in transacylases. Biotin-dependent carboxylases operate by a common mechanism illustrated in Fig. 9. This involves tautomerization of the ureido 1′-N to enhance its nucleophilicity. Though the two ureido nitrogens are essentially isoelectronic in the imidazoline portion of biotin, the steric crowding of the thiolane side chain near the 3′-N essentially prevents chemical additions to position 3′. Phosphorylation of bicarbonate by ATP to form carbonyl phosphate provides an electrophilic mixed-acid anhydride. This latter can then react at the nucleophilically enhanced 1′-N to generate reactive N(1′)-carboxybiotinyl enzyme. This in turn can exchange the carboxylate function with a reactive center in a substrate, typically at a carbon with incipient carbanion character (13).

There are nine known biotin-dependent enzymes: six carboxylases, two decarboxylases, and a transcarboxylase. Of these, only four carboxylases have been found in tissues of humans and other mammals. Acetyl-CoA carboxylase is a cytosolic enzyme that catalyzes formation of malonyl-CoA for fatty acid biosynthesis. Pyruvate carboxylase is a mitochondrial enzyme that forms oxalacetate for citrate formation. Propionyl-CoA carboxylase forms the D isomer of methylmalonyl-CoA on a pathway toward succinyl-CoA in the Krebs cycle. Finally, β-methylcrotonyl-CoA carboxylase forms β-methylglutaconyl-CoA in the catabolic pathway from L-leucine.

VI. ONE-CARBON TRANSFER: TETRAHYDROFOLYL COENZYMES

Among natural compounds with a pteridine nucleus, those most commonly encountered are derivatives of 2-amino-4-hydroxypteridines, which are trivially named pterins. Humans and other mammals normally require only folic acid as a vitamin of the pterion type. This is discussed in Chapter 12. Although a number of pterins when reduced to the 5,6,7,8-tetrahydro level function as coenzymes, the most generally utilized are poly-γ-glutamates of tetrahydrofolate (THF) (2,14). Tetrahydropteroylglutamates and forms of its natural derivatives responsible for vectoring one-carbon units in different enzymic reactions are shown in Fig. 10. All of these bear the substituent for transfer at nitrogen 5 or 10 or are bridged between these basic centers. The number of glutamate residues varies, usually from one to seven, but a few to several glutamyls optimize binding of tetrahydrofolyl coenzymes to most enzymes requiring their function. Less broadly functional, but essential for some coenzymic roles of pterins, is tetrahydrobiopterin, which cycles with its quinoid 7,8-dihydro form during O_2-dependent hydroxylation of such aromatic amino acids as in the conversion of phenylalanine to tyrosine. Also, the recently elucidated molybdopterin functions in some Mo/Fe flavoproteins, such as xanthine dehydrogenase. Pterion coenzymes, most of them at the tetrahydro level, are sensitive to oxidation.

The interconversions of folate with the initial coenzymic relatives, the tetrahydrofolyl polyglutamates, involve dihydrofolate reductase, necessary for reducing the vitamin level compound through 7,8-dihydro to 5,6,7,8-tetrahydro levels. This enzyme is the target of such inhibitory drugs as aminopterin and amethopterin (methotrexate). A similar dihydropterin reductase catalyzes reduction of dihydro- to tetrahydrobiopterin. Tetrahydrofolate is trapped intracellularly and extended to polyglutamate forms that operate with THF-dependent systems. In some cases (e.g., thymidylate synthetase), there is a redox change in tetrahydro to dihydro coenzyme, which is recycled by the NADPH-dependent reductase.

There are different redox levels for the one-carbon fragment carried by THF systems. With formate, phosphorylation by ATP leads to formyl phosphate, a reactive mixed-acid anhydride. This can react with nitrogens at either position 5 or 10 in THF to form the formyl compounds; upon cyclization, the 5,10-methenyl-THF results. With formaldehyde, which becomes an electrophilic cation when protonated ($^+CH_2OH$), the reaction shown in Fig. 11 ensues. This reaction is initiated at the most basic N-5 ($pK_a \sim 4.8$) of THF to generate an N-hydroxymethyl intermediate. The high electron charge (high basicity) and large free valence (high polarizability) confer a low activation energy for electrophilic substitution at N-5. With loss of a hydroxyl and quaternization of this nitrogen, the one-carbon unit becomes canonically equivalent to a carbocation, which can react either inter- or intramolecularly. In the former instance, reaction with a nucleophile can lead to another

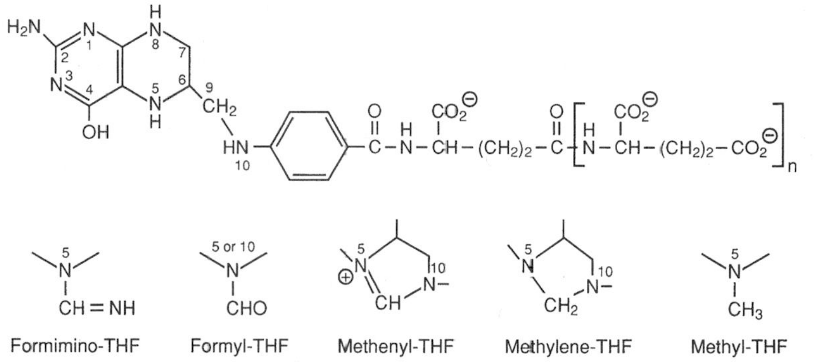

Formimino-THF   Formyl-THF   Methenyl-THF   Methylene-THF   Methyl-THF

Fig. 10 Structures with numbering for tetrahydropteroyl-L-glutamates, including the biological derivatives for one-carbon transfers.

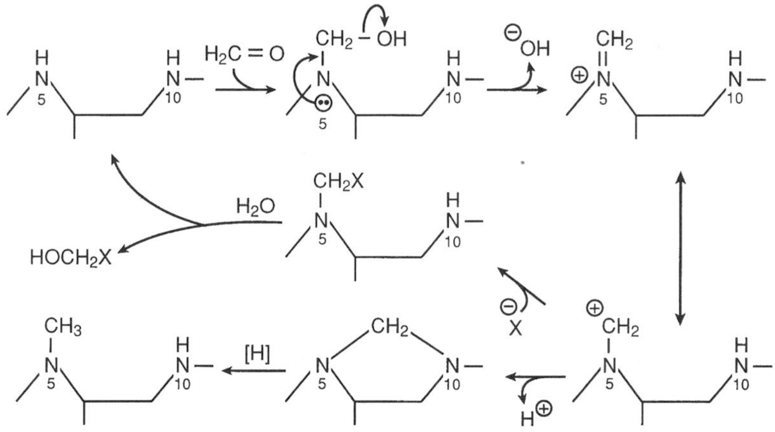

Fig. 11 Reaction of formaldehyde with the basic N-5 of tetrahydrofolyl coenzyme to generate an N-hydroxymethyl intermediate that leads to a mesomeric carbocation. This latter can react intermolecularly in a Mannich-like reaction or intramolecularly to form 5,10-methylene tetrahydrofolate.

hydroxymethyl compound; in the latter, 5,10-methylene-THF is formed. For the methyl level, reduction of the methylene-THF occurs.

An overview of some of the major interconnections among the one-carbon-bearing THF coenzymes and their metabolic origins and roles include a fair range of reaction types. As mentioned above, there is generation and utilization of formate. The important de novo biosynthesis of purine includes two steps wherein glycinamide ribonucleotide and 5-amino-4-imidazole carboxamide ribonucleotide are transformylated by 5,10-methenyl-THF and 10-formyl-THF, respectively. In pyrimidine nucleotide biosynthesis, deoxyuridylate and 5,10-methylene THF form thymidylate and dihydrofolyl coenzyme in a mechanism whereby the tetrahydro coenzyme is oxidized to the dihydro level, as shown in Fig. 12. There are conversions of some amino acids, namely, N-formimino-L-glutamate (from histidine catabolism) with THF to L-glutamate and 5,10-methenyl-THF (via 5-

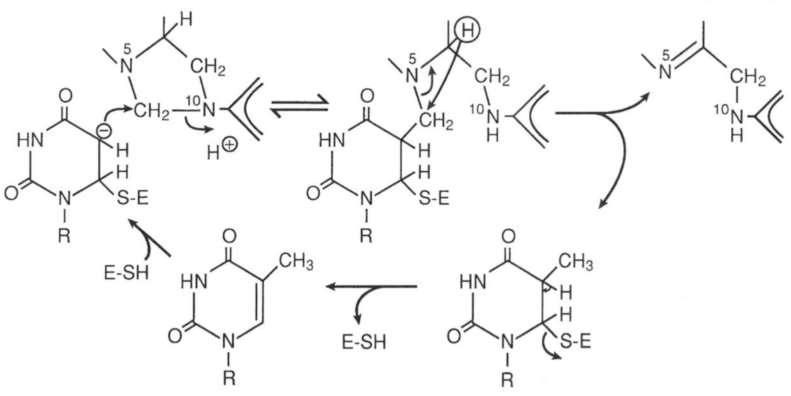

Fig. 12 The 1,3 hydrogen shift and redox nature of 5,10-methylene–tetrahydrofolyl coenzyme within thymidylate synthase.

formimino-THF), L-serine with THF to glycine and 5,10-methylene-THF, and L-homo-cysteine with 5-methyl-THF to L-methionine and regenerated THF.

VII. REARRANGEMENTS ON VICINAL CARBONS: B_{12} COENZYMES

Though there are several biologically active forms derived from vitamin B_{12}, discussed in Chapter 13, only a couple warrant attention as coenzymes in higher eukaryotes (2,15). The coenzyme B_{12} (COB_{12}) known to function in most organisms, including humans, is 5′-deoxyadenosylcobalamin. A second important coenzyme form is methylcobalamin (methyl-B_{12}), in which the methyl group replaces the deoxyadenosyl moiety of CoB_{12}. Some prokaryotes utilize other bases (e.g., adenine) in this position originally occupied by the cobalt-coordinated cyanide anion in cyanocobalamin, the initially isolated form of vitamin B_{12}.

The metabolic interconversion of vitamin B_{12} as the naturally occurring hydroxoco-balamin (B_{12a}) with other vitamin and coenzyme level forms involves sequential reduction of B_{12a} to the paramagnetic or radical B_{12r} and further to the very reactive B_{12s}. The latter reacts in enzyme-catalyzed nucleophilic displacements of tripolyphosphate from ATP to generate CoB_{12} or of THF from 5-methyl-THF to generate methyl-B_{12}.

Seemingly all CoB_{12}-dependent reactions react through a radical mechanism, and all but one (*Lactobacillus leichmanii* ribonucleotide reductase) involve a rearrangement of a vicinal group (X) and a hydrogen atom. This general mechanism is illustrated in Fig. 13. For the CoB_{12}-dependent mammalian enzyme, L-methylmalonyl-CoA mutase, X is the CoA-S-CO- group, which moves with retention of configuration from the carboxyl-bearing carbon of L(*R*)-methylmalonyl-CoA to the carbon β to the carboxyl group in succinyl-CoA. This reaction is essential for funneling propionate to the tricarboxylic acid cycle. Without CoB_{12} (from vitamin B_{12}), more methylmalonate is excreted, but also the CoA

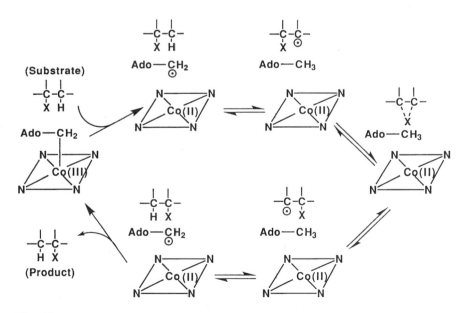

Fig. 13   Radical intermediates in vicinal rearrangements catalyzed by enzymes utilizing 5′-deoxy-adenosyl-B_{12}.

ester competes with malonyl-CoA in normal fatty acid elongation to form instead abnormal, branched-chain fatty acids. Methyl-B_{12} is necessary in the transmethylase-catalyzed formation of L-methionine and regeneration of THF. Without this role, there would not only be no biosynthesis of the essential amino acid, but increased exogenous supply of folate would be necessary to replenish THF, which would not otherwise be recovered from its 5-methyl derivative.

REFERENCES

1. O. H. Hasim and N. Azila Adnan, Coenzyme, cofactor and prosthetic group—ambiguous biochemical jargon, *Biochem. Ed. 22*:93 (1994).
2. D. B. McCormick, Coenzymes, biochemistry, in *Encyclopedia of Human Biology*, Vol. 2 (R. Dulbecco, ed.-in-chief), Academic Press, San Diego, 1991, pp. 527–545.
3. D. E. Metzler, Coenzymes—nature's special reagents, in *Biochemistry: The Chemical Reactions of Living Cells*, Academic Press, New York, 1977, pp. 428–516.
4. P. A. Frey, Vitamins, coenzymes, and metal cofactors, in *Biochemistry* (G. Zubay, ed.), Wm. C. Brown, Dubuque, 1993, pp. 278–304.
5. C. T. Walsh, *Enzymatic Reaction Mechanisms*, Freeman, San Francisco, 1977.
6. D. Dolphin, R. Poulson, and O. Avamovic (eds.), *Pyridine Nucleotide Coenzymes, Parts A and B*, Wiley-Interscience, New York, 1987.
7. D. J. Creighton and N. S. R. K. Murthy, Stereochemistry of enzyme-catalyzed reactions at carbon, *The Enzymes*, Vol. 19 (D. S. Sigmon and P. D. Boyer, eds.), Academic Press, San Diego, 1990, pp. 323–421.
8. A. H. Merrill, Jr., J. D. Lambeth, D. E. Edmondson, and D. B. McCormick, Formation and mode of action of flavoproteins, *Annu. Rev. Nutr. 1*:281–317 (1981).
9. B. Curti, S. Ronchi, and G. Zanetti (eds.), *Flavins and Flavoproteins*, Walter de Gruyter, Berlin, 1991.
10. H. Z. Sable and C. J. Gubler (eds.), Thiamin. Twenty years of progress, *Ann. N.Y. Acad. Sci., 378*, (1982).
11. D. Dolphin, R. Poulson, and O. Avamovic (eds.), *Pyridoxal Phosphate*, Parts A and B, Wiley-Interscience, New York, 1987.
12. T. Fukui, H. Kagamiyama, K. Soda, and H. Wada (eds.), *Enzymes Dependent on Pyridoxal Phosphate and Other Carbonyl Compounds as Cofactors*, Pergamon Press, Oxford, 1991.
13. J. R. Knowles, The mechanism of biotin-dependent enzymes, *Annu. Rev. Biochem. 58*:195, 1989.
14. R. L. Blakley and T. J. Benkovic (Vol. 1) and R. L. Blakley and V. M. Whitehead (Vol. 3), *Folate and Pterins*, John Wiley and Sons, New York, 1986.
15. D. Dolphin, (ed.), B_{12}, Vols. 1 and 2, Wiley-Interscience, New York, 1982.

6

Niacin

JAMES B. KIRKLAND

University of Guelph, Guelph, Ontario, Canada

JEAN M. RAWLING

University of Calgary, Calgary, Alberta, Canada

I. HISTORICAL PERSPECTIVE

The identification of niacin as a vitamin resulted from an urgent need to cure pellagra, which ravaged low socioeconomic groups of the southeastern United States in the early twentieth century and various European populations for the previous two centuries. The history of pellagra and the progression of early research findings in niacin nutrition are extensively documented by many writers, including Harris (1) and Carpenter (2). The first reports of the condition were recorded in the early 1700s in Spain. Corn had been introduced to Europe from the Americas and quickly became a staple food, as it could produce more calories per acre than wheat or rye.

In 1735, the Spanish physician Casal became the first to describe the strange new disease, which he termed *mal de la rosa* (disease of the rose). He documented the skin lesions of the disease quite carefully, and the characteristic rash around the neck of pellagrins is still referred to as ''Casal's necklace.'' The disease spread geographically with the cultivation of corn and became known as *pelle agra* (rough skin) by the Italian peasantry. Due to the widespread incidence of pellagra in eighth-century Europe and the dedication of entire hospitals for victims of the disease, there was a painstaking documentation of its symptoms (1,2). These include diarrhea and neurological disturbances (dementia) (3), as well as the sun-sensitive dermatitis, which, along with the eventual death of the patient, are often referred to as the ''4 D's'' of pellagra. The disease was recognized in populations in Egypt, South Africa, and India in the late 1800s and early 1900s. It reached epidemic proportions in the United States in the first half of the 1900s, producing at least

250,000 cases and 7,000 deaths per year for several decades in the southern states alone (4). Serious outbreaks continue to occur in some developing countries (5). An improved standard of living and supplementation of some breakfast cereals with niacin have limited the disease in North America and Europe, but cases of pellagra still occur, although they are likely underreported due to a lack of familiarity of modern physicians with this disease. It is also likely that subclinical deficiencies of niacin exist in developed countries; 15% of women surveyed in Malmo, Sweden, had blood nucleotide pools that indicated a sub-optimal niacin intake (6).

Until the first half of the twentieth century, the etiology of pellagra was unknown. Early theories suggested that it is a type of leprosy, that it results from a toxin in moldy corn, or that it is an infectious disease communicated through an insect vector. Gradually, an association between pellagra and corn consumption became apparent, even in the absence of mold contamination. This was confirmed by Joseph Goldberger (7), who determined that a pellagra-preventative (''P-P'') factor, missing from corn, was necessary to prevent and cure pellagra in humans. In 1937, Elvehjem and colleagues discovered that nicotinic acid could cure black tongue in dogs, an early animal model for pellagra (8). Paradoxically, chemical analysis of corn revealed that it is not especially low in nicotinic acid content. However, Krehl et al. (9) induced niacin deficiency in rats by feeding the animals a corn-based diet and then alleviated the symptoms with the addition of casein, a protein source rich in tryptophan. There had been suggestions that tryptophan deficiency caused pellagra, and in 1921, a pellagrous patient was treated successfully with tryptophan supplementation (10). It was eventually realized that tryptophan can be used, with low efficiency, as a substrate for the synthesis of nicotinamide adenine dinucleotide (NAD) (11). In 1951, Carpenter's group found that niacin in corn is biologically unavailable and can be released only following prolonged exposure to extremes in pH (12). These findings eventually led to a better understanding of the contribution of a corn-based diet to the development of pellagra; corn-based diets are low in tryptophan, and the preformed nicotinic acid is tightly bound to a protein that prevents its absorption. The release of niacin from this protein at extremes of pH explained the good health of native Americans, who used corn as a dietary staple: in all cases these societies processed their corn with alkali prior to consumption. Corn was in use throughout North and South America for thousands of years as a domesticated crop, and methods of preparation ranged from treatment of corn with ashes from the fireplace, to the well-known use of lime water (water and calcium hydroxide) in the making of tortillas. Had the explorers brought native American cooking techniques along with corn to sixteenth-century Europe, pellagra epidemics might never have occurred.

Pellagra may be difficult to identify, as the symptoms of dermatitis, diarrhea, and dementia occur in an unpredictable order, and it is uncommon to find all three aspects until the disease is very advanced. The earliest sign of deficiency is often inflammation in the oral cavity, which progresses to include the esophagus and eventually the whole digestive tract, associated with a severe diarrhea (1). Burning sensations often discourage food consumption, and the patient progresses to a state of marasmus; the cause of death in most cases is probably a result of the effects of general malnutrition (poor disease resistance) and diarrhea (1). Because of the sensitivity to sunlight, and possibly due to dietary variation, the incidence of pellagra is seasonal. The dermatitis can become severe rapidly, with exposed skin showing hyperpigmentation, bullous lesions, and desquamation (Fig. 1). The skin may heal during fall and winter, leaving pink scars, only to revert to open sores the following summer. The other dramatic symptom of pellagra is dementia,

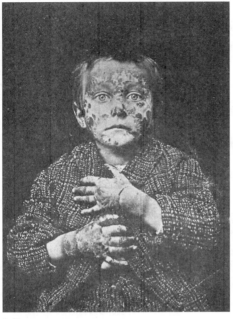

(a)

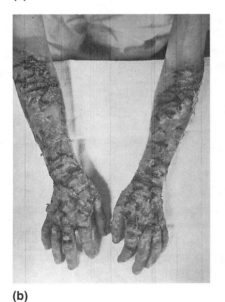

(b)

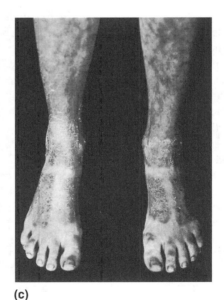

(c)

Fig. 1   (a) An Austrian child with pellagra, showing dermatitis on the exposed skin of the face and hands. Note the unaffected skin on the wrists where the cuffs of the coat are turned up. (b) Severe pellagrous lesions on the arms of a 32-year-old woman. They did not respond to riboflavin supplementation but cleared up with the use of Valentine's whole liver extract, a good source of niacin. (c) Dermatitis on the legs and feet of a typical pellagra patient, showing the pattern of protection from the sun afforded by sandals.
[Fig. 1(a) is reproduced with permission from *Pellagra: History, Distribution, Diagnosis, Prognosis, Treatment, Etiology*, by Stewart R. Roberts, C.V. Mosby Company, 1914. Fig. 1 (b) and 1 (c) are reproduced with permission from *Clinical Pellagra*, by Seale Harris, published by C.V. Mosby, St. Louis, 1941.]

which usually goes beyond the type of depression associated with general malnutrition. Neurological changes in pellagra patients begin peripherally, with signs such as muscle weakness, twitching and burning feelings in the extremities, and altered gait (13). Early psychological changes include depression and apprehension, but these progress to more severe changes, such as vertigo, loss of memory, deep depression, paranoia and delirium, hallucinations, and violent behavior (14). This type of dementia is very similar to schizophrenia (15). There are numerous examples of the pellagrous insane committing murder (16), although it was more common for these patients to turn to suicide. Interestingly, many insane pellagrins are drawn to water, and their most common form of suicide is reported to be drowning (16). While there are pathological changes in the spinal cord in advanced pellagra (17), and some of the motor disturbances are permanent (18), there is a striking recovery of psychological function when insane pellagra patients are treated with nicotinic acid, with a disappearance of symptoms in 1–2 days (14). These observations suggest that a compound derived from niacin is involved in neural signaling pathways.

The first biochemical role established for niacin was its involvement in redox reactions and energy metabolism. For many years, however, the symptoms of pellagra remained mysterious when viewed from this perspective. Other nutrients involved in energy metabolism do cause similar deficiency symptoms. Riboflavin-containing coenzymes are often coupled with nicotinamide-dependent reactions in the transfer of electrons, but riboflavin deficiency is not very similar to pellagra. Iron is intimately involved in electron transport and ATP production, and it functions closely with nicotinamide cofactors in energy metabolism. Iron deficiency and pellagra both cause general weakness, which could be associated with disrupted energy metabolism, but the pellagra patient displays several unique and dramatic clinical characteristics. Whereas other nutrient deficiencies cause dermatitis, only in niacin deficiency is this condition induced by exposure to sunlight. Although thiamine deficiency also causes changes in energy metabolism and neural function, it thought to achieve these effects through separate biochemical functions, and this is likely the case with niacin as well.

Recently, insights into the function of nicotinamide coenzymes in metabolism have improved our appreciation of the biochemical changes that may underlie the 4 D's of pellagra. In 1967, poly(ADP-ribosyl)ation was identified and recognized as a posttranslational modification of nuclear proteins. Poly(ADP-ribose) synthesis makes use of NAD^+ as substrate, rather than as an electron-transporting intermediate. Poly(ADP-ribose) formation has been shown to be important in DNA repair processes, and new knowledge in this area may soon provide an explanation for the sun-sensitive dermatitis of pellagra. Mono(ADP-ribosyl)ation, which also uses NAD^+ as a substrate, was characterized, beginning in the mid-1960s, as a mechanism of action for many bacterial toxins. Mono(ADP-ribosyl)ation is now thought to be important in the endogenous regulation of many aspects of signal transduction and membrane trafficking in eukaryotic cells, but these studies are in their infancy. In 1989, cyclic ADP-ribose was identified as another product of NAD metabolism and was shown to have the ability to regulate cellular calcium homeostasis, a central process in neural transmission. Interestingly, the enzymes that make and degrade cyclic ADP-ribose and the proteins that bind this second messenger are present in the brain in relatively large quantities compared with other tissues. Even more recently, the same enzymes that make cyclic ADP-ribose have been found to produce nicotinic acid adenine dinucleotide phosphate (NAADP), using NADP as a substrate. NAADP appears to also have distinct functions in the regulation of intracellular calcium stores. The studies of mono(ADP-ribosyl)ation reactions and cyclic ADP-ribose and NAADP function are in their early stages, but they may soon provide explanations for the dementia of the pellagra patient and the changes in intestinal cell function that lead to diarrhea.

II. CHEMISTRY

A. Nicotinic Acid and Nicotinamide

The term niacin is accepted as a broad descriptor of vitamers having the biological activity associated with nicotinamide, including nicotinamide, nicotinic acid, and a variety of pyridine nucleotide structures. In the past, niacin has been used to specifically refer to nicotinic acid (pyridine-3-carboxylic acid; Fig. 2A), but for the purposes of this discussion, "niacin" is used in reference to all forms with vitamin activity, while "nicotinic acid" refers to pyridine-3-carboxylic acid. Nicotinic acid is a white crystalline solid, stable in air at normal room temperature. It is moderately soluble in water and alcohol, but insoluble in ether. An aqueous solution has a maximum ultraviolet (UV) absorbance at 263 nm.

Like nicotinic acid, nicotinamide (niacinamide; pyridine-3-carboxamide) (see structure, Fig. 2B) is a white crystalline substance with a maximal UV absorbance at 263 nm. In contrast to nicotinic acid, nicotinamide is highly soluble in water and is soluble in ether—characteristics that allow separation of the two vitamers.

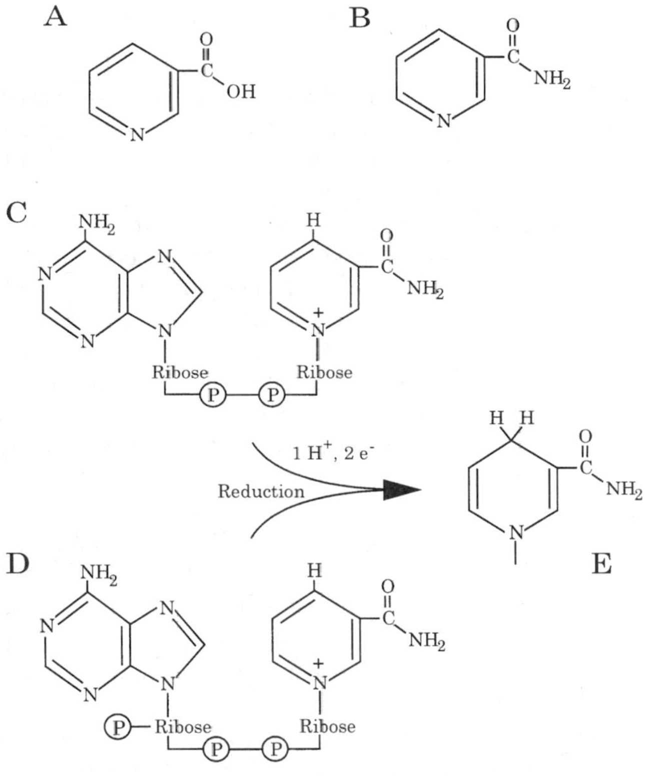

Fig. 2   Chemical structures of niacin compounds. (A) Nicotinic acid; (B) nicotinamide; (C) NAD$^+$; (D) NADP$^+$; (E) site of reduction.

B. Niacin Coenzymes

The biologically active forms of niacin compounds are the NAD and NADP coenzymes (Fig. 2C, D). It is the C-4 position on the pyridine ring of the nicotinamide moiety that participates in oxidation and reduction reactions. Due to the electronegativity of the amide group and the nitrogen at position 1 on this ring, hydride ions can readily reduce the oxidized C-4 position. This is the basis for the enzymatic hydrogen transfer reactions that are ubiquitous among organisms. With respect to the non-redox functions of NAD, the glycosidic linkage between nicotinamide and ribose is a high-energy bond, and cleavage of this bond drives all types of ADP-ribose transfer reactions in the forward direction.

The oxidized and reduced forms of the coenzymes are designated NAD^+ or $NADP^+$ and NADH or NADPH, respectively. The designations NAD and NADP are used to describe the total pools. This is often necessary if the method of quantification does not distinguish between oxidized and reduced forms, or if a general statement about the nucleotide pool is made. The total pool of all four forms may be referred to as NAD(P). Both NAD and NADP are white powders that are freely soluble in water and poorly soluble in ether. Both compounds have strong UV absorption at 340 nm in their reduced forms, with a weaker absorption at 260 nm when oxidized or reduced. The absorbance at 340 nm is often used to monitor the oxidation or reduction of these cofactors in enzyme assays.

III. FOOD CONTENT, DIETARY REQUIREMENTS AND ASSESSMENT OF STATUS

A. Quantification

The traditional analysis for nicotinic acid entails cleavage of the pyridine ring with cyanogen bromide (the Koenig reaction), followed by reaction with an aromatic amine to yield a colored product that can be assayed spectrophotometrically [19]. Because hydrolysis of the nicotinamide pyridine ring with cyanogen bromide is not quantitative, it is advisable to first deamidate nicotinamide enzymatically to form nicotinic acid before preparing samples for the Koenig reaction. Microbiological assay of nicotinic acid and nicotinamide is possible using *Tetrahymena pyriformis* or *Lactobacillus plantarum* [20]. Nicotinic acid can also be specifically determined microbiologically with *Leuconostoc mesenteroides*; this microorganism is an obligate user of nicotinic acid rather than of nicotinamide [21]. Fluorometric measurement of nicotinamide is very sensitive. In this assay, nicotinamide is converted to N'-methylnicotinamide, which is then reacted with a carbonyl compound, commonly acetone or acetophenone, thus producing a fluorescent adduct on ring position 4 [22].

These traditional methods for analysis of nicotinic acid and nicotinamide are gradually being replaced by a variety of new techniques, including gas chromatography and mass spectroscopy [23] or high-performance liquid chromatography (HPLC) [24]. Because of the importance of these vitamers as pharmacological agents, more research on pharmacokinetics following supraphysiological doses is warranted. Specific measurement of nicotinic acid, nicotinamide, and their metabolites will require the use of these more advanced analytical techniques.

Measurement of pyridine nucleotides is easier than measurement of the vitamin precursors. Oxidized forms (NAD^+, $NADP^+$) are extracted by acid, usually 1 N perchlorate. This causes destruction of the reduced forms (NADH, NADPH). The reduced nucleotides are extracted by base, which causes destruction of the oxidized forms [25]. The extracted nucleotides are generally quantified by enzyme cycling techniques that recognize oxidized or reduced nucleotides but are specific to either NAD or NADP (25). The oxidized forms can also be measured by HPLC techniques that provide additional data on ATP, ADP, and AMP levels (26).

B. Food Content

The niacin content of human foods is usually expressed as niacin equivalents (NE), which are equivalent to niacin content (mg) + one-sixtieth tryptophan content (mg). This relationship may not be accurate if the intake of tryptophan is low (see Sec. IV.A). Furthermore, the conversion of tryptophan appears to be improved when the intake of niacin is low (2). Niacin in plant products is mainly in the form of nicotinic acid. Animal products will initially contain mainly NAD and NADP coenzymes, but little is known about breakdown of these nucleotides during the aging of meats. Under anerobic conditions, NAD^+ will be converted to NADH, which may protect it from the wide variety of NAD-catabolizing enzymes in the cell. Analysis of food products for performed niacin is probably best accomplished by microbiological methods that respond to nicotinic acid, nicotinamide, and coenzyme forms of the vitamin. Because plant products contain less tryptophan than animal products and because the nicotinic acid may be largely bound in unavailable forms, some grain products such as breakfast cereals are supplemented with nicotinic acid. Such supplementation, together with the widespread occurrence of niacin and tryptophan in a mixed diet, has greatly diminished the number of clinically obvious cases of pellagra in developed countries. Subclinical niacin deficiency may still be common in developed countries (6,23), and clinical pellagra still occurs in association with alcoholism (27). Outbreaks of pellagra continue in some areas of the world where populations rely on corn (or maize) as a staple (5). The degree of nutrient deficiency in a population is always based on the current perception of optimal intake or function. As our understanding of niacin function evolves, we should be open to revising these end points and reevaluating niacin status.

C. Dietary Requirements and Assessment of Status

Because of the important roles of NAD and NADP in energy metabolism, the niacin requirement is related to energy intake. The recommended daily allowance (RDA) is set at 6.6 NE/1000 kcal/day in the United States. The new Canadian RDAs for niacin range from 2 mg niacin/day for infants to 18 NE/d during pregnancy, following a similar relationship with energy requirement. There is also a minimum intake recommended for those on low-energy diets, and this may reflect the roles of NAD beyond the typical redox reactions of energy metabolism. Niacin status has traditionally been tested by measuring the urinary excretion of various niacin metabolites or the urinary ratio of N-methyl-2-pyridone-5-carboxamide to N-methylnicotinamide (28). The 2-pyridone form decreases to a greater extent in response to a low dietary intake, and a ratio of less than 1.0 is indicative of niacin deficiency. More recently, it has been found that the NAD pool in red blood cells decreases rapidly during niacin deficiency in men, whereas the NADP pool is quite stable (29). This has led to the suggested use of NAD/NADP, referred to as the niacin number, as an easily obtained index of niacin deficiency in humans. Studies using animal models have also shown that blood NAD pools deplete more rapidly and to a greater extent than those of tissues such as liver, heart, or kidney (30). This suggests that a portion of blood NAD may represent a labile storage pool used to support other tissues in the early stages of deficiency.

IV. PHYSIOLOGY

A. Pathways of Synthesis

Although many microorganisms and plants can synthesize the pyridine ring of NAD de novo from aspartic acid and dihydroxyacetone phosphate (31), animals do not have this ability. Nicotinic acid, nicotinamide, pyridine nucleotides and tryptophan represent the dietary sources from which the pyridine ring structure can be derived by most mammalian species. Animals may also practice caprophagy to take advantage of colonic synthesis of niacin by microflora.

In 1958, Preiss and Handler (32) proposed a pathway for the conversion of nicotinic acid to NAD in yeast and erythrocytes (shown in reactions 9, 5, and 6 of Fig. 3). Initially, it was believed that nicotinamide was also metabolized through the Preiss-Handler pathway, following the conversion of nicotinamide to nicotinic acid by nicotinamide deamidase (reaction 8, Fig. 3). However, it was soon demonstrated by Dietrich that nicotinamide reacts first with phosphoribosyl pyrophosphate and then ATP to produce NAD directly (reactions 10 and 11, Fig. 3) (33). Subsequent studies supported this route by showing that inhibition of NAD synthetase, the final enzyme in the Preiss-Handler scheme (reaction 6, Fig. 3), does not affect the incorporation of nicotinamide into NAD (34). It was also found that deamidation of nicotinamide to nicotinic acid only occurs at supraphysiological concentrations of nicotinamide (35), which is not surprising, as the K_m of nicotinamide deamidase for nicotinamide is above physiological nicotinamide levels. It has been concluded, therefore, that the Dietrich pathway is the major route by which physiological concentrations of nicotinamide supply NAD in mammalian cells (36). It should be noted, however, that the predominant pathways of nicotinic acid and nicotinamide utilization change when the vitamers are consumed in pharmacological quantities, such as those used in the treatment of cardiovascular disease or insulin-dependent diabetes mellitus (IDDM). This will be discussed later in this chapter (Sec. VII). Interestingly, there is a common enzyme in the two routes of NAD synthesis. Nicotinamide mononucleotide (NMN) and nicotinic acid mononucleotide (NAMN) adenylyltransferase activities reside within a common protein in the nucleus; this enzyme catalyzes the last step in the conversion of nicotinamide and the second to last step in the conversion of nicotinic acid to NAD (37).

Quinolinic acid, a tryptophan metabolite, can react with phosphoribosyl pyrophosphate to produce nicotinic acid mononucleotide. This reaction occurs only in the kidney and in the liver in mammals, due to the localization of quinolinate phosphoribosyltransferase in these tissues (38). NAD is then synthesized from nicotinic acid mononucleotide via the Preiss-Handler pathway. A number of reactions are required to convert tryptophan to quinolinic acid. Most of the α-amino-β-carboxymuconic-ε-semialdehyde, an intermediate in this pathway, is catabolized to acetyl CoA and CO_2 (39). There is, however, a rate-limiting step in this degradation sequence, which, when overwhelmed, allows some of the semialdehyde to transform spontaneously into quinolinic acid (reaction 3, Fig. 3) and ultimately to produce NAD (40). Therefore, the production of NAD from tryptophan is favored by a low activity of one enzyme (picolinic carboxylase) and a high activity of another (quinolinate phosphoribosyltransferase), leading to a wide range in the efficiency of tryptophan utilization among species.

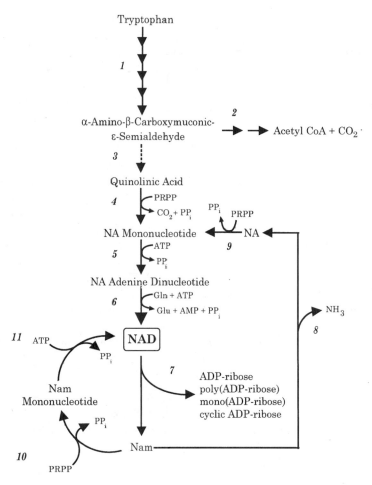

Fig. 3 Pathways of NAD⁺ synthesis in mammals. Reactions 9, 5, and 6 comprise the Preiss-Handler pathway, while reactions 10 and 11 form the Dietrich pathway. The following enzymes correspond to the numbered reactions: (1) 5 step conversion; (2) picolinic carboxylase; (3) spontaneous chemical reaction; (4) quinolinic acid phosphoribosyltransferase; (5) NAMN adenylyltransferase (enzymes 5 and 11 may be identical proteins); (6) NAD synthetase; (7) NAD glycohydrolases, various ADP-ribosylation reactions; (8) nicotinamide deamidase; (9) nicotinic acid phosphoribosyltransferase; (10) nicotinamide phosphoribosyltransferase; (11) NMN adenylyltransferase.

In nutritionally replete humans, there is thought to be a 60 mg/1 mg ratio between tryptophan supply and niacin formation, although individual variation is significant (41). Because of this relationship, dietary niacin content is described in niacin equivalents (1 NE = mg niacin + $^1/_{60}$ mg tryptophan). More recent work has suggested that humans may not utilize tryptophan for niacin synthesis when tryptophan levels are limiting in the diet (29). In these experiments, young men were placed on a diet containing 6 NE/day, and their blood NAD⁺ levels decreased by 70% over 5 weeks. Addition of 240 mg/day of tryptophan to this diet had no effect on blood NAD, although it prevented the decrease in plasma tryptophan that resulted from consumption of the unsupplemented diet. This tryptophan represented, in theory, an additional 4 NE/day, but it appears that protein turnover takes precedence over niacin synthesis when tryptophan levels are low. NE calculations for diets with marginal niacin status may be inaccurate because of this relationship. At the same time, tryptophan supplements have been reported to cure pellagra and a genetic defect in tryptophan absorption, known as Hartnup's disease, also causes pellagra-like signs and symptoms. While the efficiency of conversion of tryptophan to NAD is thought to be increased by niacin deficiency (2), it is also apparent that at low tryptophan intakes, the need for protein metabolism dominates over conversion to NAD (29).

Species vary in their ability to synthesize niacin from tryptophan. Cats, adapted to a diet high in amino acids, have approximately 50 times greater picolinic carboxylase activity than humans (40), and they demonstrate extremely poor utilization of tryptophan as a precursor for NAD (42). Rats, on the other hand, are more efficient than humans in their use of tryptophan for NAD synthesis (ratio of about 33 mg tryptophan: 1 mg niacin) (43). Mice are resistant to niacin deficiency, even on diets containing as little as 6% casein, with 7% gelatin, a protein source lacking tryptophan (unpublished data). Guinea pigs are very susceptible and will eventually die on a niacin-free diet containing 20% casein, demonstrating a very poor conversion of tryptophan to niacin (44). The capacity for NAD generation from tryptophan also depends on adequate status of vitamin B_6 and minerals, including copper, iron, and magnesium, all of which are necessary for reactions in this pathway. When using experimental animals to model niacin deficiency in humans, it is important to remember the tremendous variability between species in the utilization of tryptophan.

B. Absorption

Both nicotinamide and nicotinic acid can be absorbed through the stomach lining, but absorption in the small intestine is more rapid. For intact nucleotides, pyrophosphatase activity in the upper small intestine metabolizes NAD to yield nicotinamide mononucleotide, which is then quickly hydrolyzed to form nicotinamide riboside and eventually free nicotinamide.

Absorption of both nicotinic acid and nicotinamide at low concentrations appears to be via sodium-dependent facilitated diffusion (39) or by carrier-mediated transport making use of proton cotransporters and anion antiporters (45). Higher concentrations of both forms appear to be absorbed by passive diffusion. Once absorbed from the lumen into the enterocyte, nicotinamide may be converted to NAD or released into the portal circulation. Although some nicotinic acid moves into the blood in its native form, the bulk of the nicotinic acid taken up by the enterocyte is converted to NAD via the Preiss-Handler pathway (39). As required, NAD glycohydrolases in the enterocytes release nicotinamide from NAD into the plasma as the principal circulating form of niacin. The kinetics of nicotinamide and nicotinic acid transport and metabolism are influenced significantly by pharmacological intakes (see Sec. VII).

C. Distribution and Metabolism

Niacin compounds entering the portal circulation are either internalized by erythrocytes or transported to the liver. Erythrocytes take up nicotinic acid by facilitated diffusion, using an anionic transport protein that accounts for 30% of the membrane protein content (39). In spite of this, erythrocytes appear to favor nicotinamide as a precursor for NAD synthesis (46).

The liver is the central processing organ for niacin. Aside from its role in the conversion of tryptophan to NAD, it receives nicotinamide and some nicotinic acid via the portal circulation, as well as nicotinamide released from other extrahepatic tissues. In the liver, nicotinic acid and nicotinamide are metabolized to NAD or to yield compounds for urinary excretion, depending on the niacin status of the organism.

The liver has some capacity for NAD storage. Because hepatic regulation of nicotinamide phosphoribosyltransferse by ATP and NAD (positive and negative, respectively) is less effective than in many other tissues, liver NAD concentrations increase significantly following dietary nicotinamide administration (39). In rats, nicotinic acid and nicotinamide, at 1000 mg/kg diet, increase blood and liver NAD^+ to a similar and modest extent (about 50%) (47). NAD glycohydrolases are thought to use the storage pool of hepatic NAD to produce nicotinamide (48), which is released for replenishment of extrahepatic tissues. The function of NAD glycohydrolases is controversial, however, and difficult to distinguish from cyclic ADP-ribose formation and catabolism (see later sections).

The liver plays an important role in the preparation of niacin for urinary excretion, producing a variety of methylation and hydroxylation products of both nicotinic acid and nicotinamide. In humans, nicotinamide is primarily methylated to produce N'-methylnicotinamide, whereas nicotinic acid is conjugated with glycine to form nicotinuric acid. Increasing levels of the untransformed vitamers can be found in the urine as the level of niacin ingestion increases (39).

V. BIOCHEMICAL FUNCTIONS

A. NAD Cofactors in Redox Reactions

Oxidation and reduction of the C-4 position on the pyridine ring of NAD coenzymes is the basis for hydrogen transfer reactions important in oxidative phosphorylation and biosynthetic reactions. NAD^+ is reduced to NADH in glycolytic reactions, oxidative decarboxylation of pyruvate, oxidation of acetate in the tricarboxylic acid (TCA) cycle, oxidation of alcohol, β-oxidation of fatty acids, and a large number of other cellular oxidation reactions. The electrons derived from these oxidation reactions are transferred to the electron transport chain through the oxidation of NADH. The energy resulting from these transfers is used to generate ATP. In contrast to the central role that NAD(H) plays in energy expenditure, NADP(H) is essential for the biosynthetic reactions involved in energy storage. NADPH is produced from reduction of $NADP^+$ in reactions of the pentose phosphate pathway and during the malate/pyruvate shuttle across the mitochondrial membrane. NADPH then acts as a reducing agent for fatty acid production, cholesterol synthesis, and manufacture of deoxyribonucleotides. NADPH is also notable for its involvement in glutathione regeneration.

The extra phosphate group carried on NADP permits the cell to separate the oxidation and reduction activities of the nicotinamide cofactors by allowing most cellular enzymes to be specific for one of these coenzyme species. Because of this specificity, the majority of cellular NADP is maintained in the reduced state by the pentose phosphate pathway and reduction reactions are favored by mass action. Conversely, NAD is predominantly oxidized, improving the oxidant capabilities of this cofactor under cellular conditions. The role of NAD coenzymes in redox reactions has been the function classically associated with niacin since Warburg and Christian showed that nicotinamide was a component of NADP in 1935 and, in the following year, demonstrated the presence of nicotinamide in NAD (49). It is probable that early researchers attributed the pathology associated with niacin deficiency to disruptions in redox cycling, since this was the only function of niacin known at that time. However, the distinctive clinical signs of pellagra may be better explained in relation to the functions of NAD described in the following sections.

B. NAD$^+$ as a Substrate

Although the hydride transfer chemistry of the pyridine nucleotides was initially described in the 1930s (49), the high-energy glycosidic linkage between nicotinamide and ADP-ribose received little attention before the discovery of mono- and poly(ADP-ribosyl)ation reactions in the 1960s. The energy provided by breaking this bond allows the addition of ADP-ribose to a variety of nucleophilic acceptors. These include glutamate side chains and hydroxyl groups of ribose for poly(ADP-ribose) synthesis, a variety of amino acid side chains in mono(ADP-ribosyl)ation reactions, and an internal ribose linkage for cyclic ADP-ribose synthesis (Fig. 4). The catabolic activity of NAD glycohydrolase (if this reaction does exist in isolation) also represents an ADP-ribose transfer, in which water acts as the nucleophilic acceptor.

C. ADP-Ribose Cyclization and NAADP Synthesis

In 1987, a metabolite of NAD$^+$ was found to cause intracellular calcium mobilization in sea urchin eggs (50). The molecule was similar to the potent second messenger inositol-1,4,5-triphosphate (IP$_3$) in its capacity to promote this response. It acted, however, on different cellular pools of calcium. Eventually, the compound was identified as cyclic (ADP-ribose) (51). Cells have two types of channels, the IP$_3$ and the ryanodine receptors, which regulate the release of calcium from internal stores such as the sarcoplasmic or endoplasmic reticulum and mitochondria. While the IP$_3$ receptors have been characterized, the ryanodine receptors have been identified by their nonphysiological stimulation by the plant alkaloid, ryanodine. In the last few years, evidence has accumulated that cyclic ADP-ribose is an endogenous second messenger that controls the ryanodine receptors (52) and that the release of calcium is due to the binding of cyclic ADP-ribose to a small protean, which then dissociates from the ryanodine receptor, allowing calcium transport (53).

Calcium concentrations are about 10,000-fold higher outside cells than within the cytoplasm. Transient increases in cytoplasmic calcium are critical to most types of cell signaling, including nerve transmission and muscle contraction. When a stimulus causes a small amount of calcium to be released, there is a positive feedback leading to a much larger flow of calcium into the cytoplasm. Cyclic ADP-ribose and ryanodine receptors appear to play an important role in this calcium-induced calcium release (CICR) (52).

ADP-ribosyl cyclase and cyclic ADP-ribose hydrolase are the enzyme activities responsible for synthesis and degradation of cyclic ADP-ribose, respectively. NAD glycohydrolase, isolated from canine spleen, actually has both cyclase and hydrolase activities, leading to an overall reaction that appears to be NAD$^+$ glycohydrolysis (54). The leukocyte cell surface antigen CD38 also has these capacities (55). It is possible that other enzymes previously identified as NAD glycohydrolases are similar in their abilities to synthesize and catabolize cyclic ADP-ribose.

At the moment, it is unclear whether cyclic ADP-ribose works as a cofactor to enhance the sensitivity of ryanodine receptors to calcium (as does caffeine) or if it is a true second messenger that responds directly to an extracellular stimulus (56). In rat pancreatic β cells, glucose stimulates the formation of cyclic ADP-ribose, which is thought

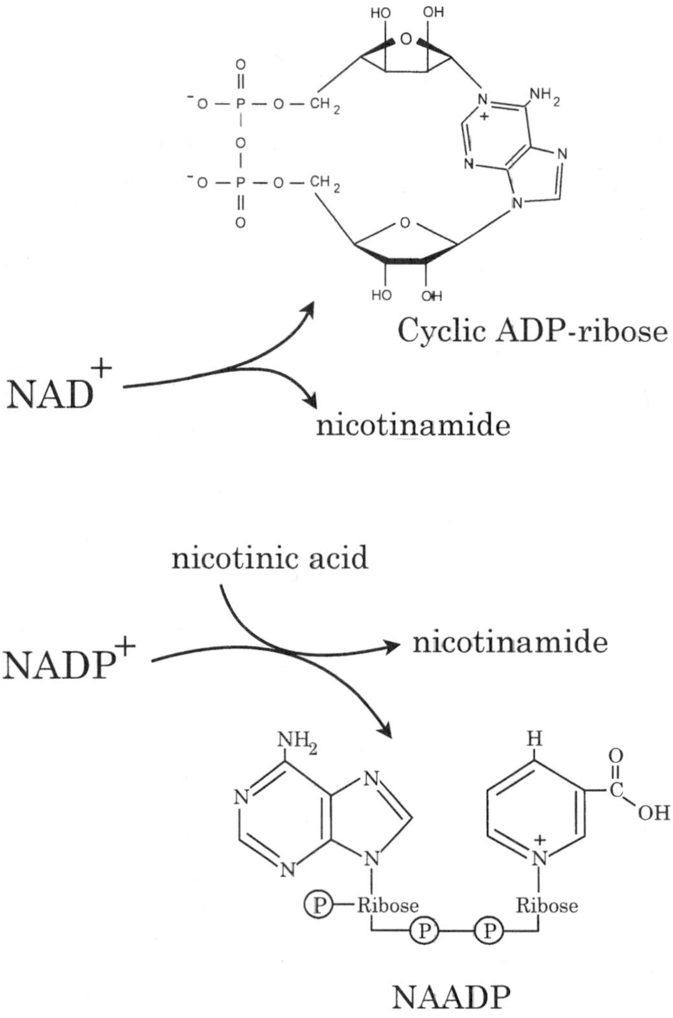

Fig. 4 Structures and origin of cyclic ADP-ribose and NAADP.

to induce insulin secretion via CICR (57). In some systems, there is redundancy between the function of cyclic ADP-ribose and IP_3; in sea urchin eggs, for example, the actions of both must be inhibited to prevent fertilization (52).

In 1995, a metabolite of NADP was also found to cause release of intracellular calcium stores; it was identified as nicotinic acid adenine dinucleotide phosphate (NAADP). Surprisingly, this metabolite is formed by the same enzymes that make cyclic ADP-ribose, although the reaction mechanism seems quite different. NAADP is synthesized by exchanging the nicotinamide on NADP with nicotinic acid, forming one of the most potent calcium-releasing agents known (58). This metabolite is effective at nanomolar concentrations and appears to be physiologically important in various calcium release systems, including sea urchin eggs, brain microsomes, and pancreatic cell insulin release (58). Interestingly, NAADP appears to regulate intracellular calcium pools that are distinct from those controlled by IP_3 or cyclic ADP-ribose.

There are many questions to be answered concerning the roles of cyclic ADP-ribose and NAADP in cell signaling and how they might be affected by niacin deficiency. ADP-ribosyl cyclase activities are widespread; CD38 has been identified in the plasma membrane of a wide variety of mammalian tissues, and a soluble cyclase has been characterized from the testis of *Aplysia* and of dog (59). Another cyclase enzyme from canine spleen is localized to the endoplasmic reticulum (54). CD38 presents a unusual picture in that the catalytic domain is extracellular. This ectoenzyme may obtain NAD^+ or $NADP^+$ from extracellular fluids and release cyclic ADP-ribose or NAADP to interact with extracellular receptors. Cyclic ADP-ribose or NAADP may also enter the cell to interact with intracellular binding proteins. Conversely, CD38 could be internalized under appropriate conditions and function as an intracellular source of cyclic ADP-ribose, drawing on cytosolic NAD^+ or $NADP^+$ pools. To understand the effect of niacin deficiency on these processes, we need to determine the subcellular localization of the active forms of these enzymes and their affinity for nucleotide substrates. Later in this chapter we will try to integrate these functions with other NAD^+-utilizing reactions during niacin deficiency.

The tissue distribution of cyclic ADP-ribose– and NAADP-metabolizing enzymes is also interesting. Levels of CD38 and cyclic ADP-ribose hydrolase activities are particularly high in brain tissue (59,60), and preliminary data show that the small proteins that bind cyclic ADP-ribose are also high in brain tissue (61). Researchers may find over the next few years that neural functions mediated by cyclic ADP-ribose and NAADP are impaired during niacin deficiency and lead to the unusual motor disruptions and dementia of pellagra.

D. NAD$^+$ Glycohydrolysis

NAD^+ glycohydrolase enzymes catalyze the donation of the ADP-ribose moiety of NAD^+ to water, resulting in free ADP-ribose and nicotinamide. There have been many reports of this activity in mammalian plasma membranes and in the cytosol (62). However, difficulty in isolating the enzymes has prevented clear identification of their form and metabolic function. The function of these enzymes has never been clearly understood but may become clearer with the discovery that cyclic ADP-ribose formation and hydrolysis are often catalyzed by the same enzyme, leading to the misleading appearance of simple NAD hydrolysis. It is possible that CD38 and other cyclase/hydrolase complexes are identical to the elusive membrane-bound NAD glycohydrolase (54). To further complicate this issue, poly(ADP-ribose) polymerase (63), bacterial ADP-ribosylating toxins (64), and endogenous mono(ADP-ribosyl) transferases (65) all have the ability to use water as an acceptor of the ADP-ribose moiety, in lieu of a protein, thus leading to glycohydrolase-like activity. If true NAD glycohydrolase activities exist, their function may be related to the control of intracellular NAD levels and to the availability of nicotinamide for export from tissues such as small intestine and liver (62).

E.   Mono(ADP-ribosyl)ation

Mono(ADP-ribosyl)ation is the transfer of a single ADP-ribose moiety, derived from NAD^+, to an amino acid residue on an acceptor protein (66,67). Of these enzyme systems, described in bacteriophages, bacteria, and eukaryotes, bacterial toxins are the best characterized. Various GTP-binding proteins (G proteins) are the targets for *Cholera*, *Pertussis*, *Diphtheria* and *Pseudomonas* toxins. Mono(ADP-ribosyl)ation of $G_{\alpha s}$ by *Cholera* toxin results in the stimulation of adenylate cyclase activity, leading to disruption of ion transport in intestinal epithelial cells and the diarrhea characteristic of this disease. *Pertussis* toxin catalyzes the cysteine-specific ADP-ribosylation of several other forms of G proteins, also leading to an uncoupling of their activities from the associated receptors. *Diphtheria* toxin and *Pseudomonas* exotoxin A ADP-ribosylate a posttranslationally modified histidine residue, called diphthamide, on polypeptide elongation factor 2 (EF-2), also a G protein. This single alteration in EF-2 structure renders the enzyme nonfunctional and halts protein synthesis. *Clostridium* toxins ADP-ribosylate various actin monomers to prevent actin polymerization.

The profound metabolic changes elicited by the bacterial toxins indicate that mono (ADP-ribosyl)ation is a powerful modulator of protein function. Mammalian cells are believed to contain a variety of endogenous mono-ADP-ribosyltransferases, but most of these data are based on the labeling of proteins in the presence of radioactive NAD, and the experiments are usually conducted under nonphysiological conditions. Several transferase activities have been examined in more detail, including those that modify EF-2, grp78, rho, and a more general family of arginine-specific transferases. It appears that the list of endogenous transferases will continue to grow as methodology allows researchers to identify their protein substrates. Several of the endogenous monotransferase enzymes regulate the function of G proteins by changing their interaction with GTP, but this is not a universal mechanism. There is also a broad spectrum of enzyme chemistry, with five or more amino acid side chains acting as ADP-ribose acceptors. Some of the better characterized cellular transferases are discussed in separate sections, below.

The best characterized of these activities, with respect to enzyme purification, are the NAD:arginine ADP-ribosyltransferases and the opposing ADP-ribosylarginine hydrolases. Several transferases and hydrolases have been purified and characterized from different species and tissues (67,68). Different transferase proteins are found in nuclear, cytosolic, and plasma membrane fractions. The enzymes have a fairly broad substrate specificity in vitro, making it difficult to identify the physiologically important protein acceptors. Some of the possible acceptor proteins include histones, cytosolic actin, and Ca^{2+} ATPase (69), but the metabolic roles of these changes in intact cells or tissues are uncertain. All of the acceptors are modified on arginine side chains, and most of the enzymes will ADP-ribosylate free arginine in vitro.

The plasma membrane form of NAD:arginine ADP-ribosyltransferase is generating the most interest. This enzyme is anchored, via glycosylphosphatidylinositol (GPI), to the outer surface of the plasma membrane (67). The GPI-anchored form of the enzyme has been identified in several cell types. A recent report has shown that extracellular NAD^+ causes a depression in cytotoxic T-cell proliferation and that this response is dependent on the extracellular GPI-anchored transferase. The ecto activity appears to lead to the mono(ADP-ribosyl)ation of an intracellular protein (70). This curious chain of events is similar to the ecto activity of the CD38 family of ADP-ribosyl cyclase enzymes, and it raises further questions about the role of extracellular NAD pools and how they might respond to niacin deficiency.

EF-2 is a G protein that is essential to the activity of the peptide elongation cycle. It has a specific histidine residue that is modified posttranslationally to a unique diphthamide structure. No other proteins are known to carry this amino acid residue. Furthermore, the formation of diphthamide on EF-2 occurs in all eukaryotic cells, although it is not necessary for the elongating activity of EF-2. Diphtheria toxin ADP-ribosylates this diphthamide, rendering EF-2 inactive and leading to the clinical manifestations of diphtheria infection. With the idea that humans did not evolve diphthamide synthesis on EF-2 as a site for attack by diphtheria toxin, investigators have searched for an endogenous regulatory system that uses a similar mechanism. Because of an inability to accurately measure the ADP- ribosylation of EF-2, progress has been slow. It appears that EF-2 is modified by an endogenous transferase when cultured cells are growth-arrested through serum deprivation (71). Interestingly, this may be an automodification, as the transferase activity copurifies with EF-2.

Rho is a small molecular weight G protein that is involved in cell division and differentiation through interactions with the cytoskeletal system. The *Clostridium botulinum* C3 toxin ADP-ribosylates rho on a specific asparagine side chain, leading to disruption of the actin cytoskeleton. An endogenous transferase activity has been identified in brain cytosol that appears to be specific to rho as an acceptor and modifies the same asparagine residue (72). No physiological conditions have been identified that cause regulation of rho ADP-ribosylation.

One of the most interesting proposed mechanisms of cellular regulation using mono-ADP-ribosylation involves the 78-kDa glucose-regulated-protein (GRP78). GRP78 is a molecular chaperone that aids in the correct folding of secreted proteins in the lumen of the endoplasmic reticulum. GRP78 may also bind incorrectly glycosylated proteins and prevent their secretion. When cultured hepatoma cells are deprived of glucose or an essential amino acid, GRP78 is mono(ADP-ribosyl)ated by a noncharacterized endogenous transferase. This modification is freely reversible, suggesting the presence of a hydrolase enzyme. The ADP-ribosylated form of GRP78 appears to be inactive in its chaperone functions, and the authors suggest that this is a rapid mechanism to decrease the rate of protein secretion during times of nutritional stress (73). Of interest, GRP78 expression is increased in cultured cells grown in niacin-deficient medium (74).

Cysteine-bound ADP-ribose residues have been found in the membrane fraction of liver tissue (75). An enzyme purified from red blood cell membranes has the ability to ADP-ribosylate a cysteine residue of $G_{i\alpha}$. These observations suggest that an endogenous transferase similar to *Pertussis* toxin participates in the regulation of adenylate cyclase activity.

There are potential artifacts in the study of mono(ADP-ribosyl)ation reactions. For some time it was believed that nitric oxide induced an endogenous mono(ADP-ribosyl)ation of glyceraldehyde-3-phosphate dehydrogenase (76). Research eventually showed that this was actually a nonenzymatic reaction combining an S-nitrosylation with a subsequent addition of the whole NAD molecule to the active site (77). Free ADP-ribose also reacts nonenzymatically with a variety of amino acid side chains, creating the illusion of mono(ADP-ribosyl)ation (78). Careful determination of the enzymatic nature of proposed mono(ADP-ribosyl)ation reactions will be necessary to avoid artifacts in the future. At the same time, it may turn out that nonenzymatic reactions of NAD or ADP-ribose with proteins are physiologically relevant and that these types of reactions could be affected by dietary niacin status.

F. Poly(ADP-ribosyl)ation

The presence of a novel adenylate-containing compound in liver nuclear extracts was described by Mandel's group in 1966 (79). They identified it as a polymer of ADP-ribose and later confirmed that NAD was the substrate and that nicotinamide was released in the reaction. The structure, known as poly(ADP-ribose), was found to be covalently attached

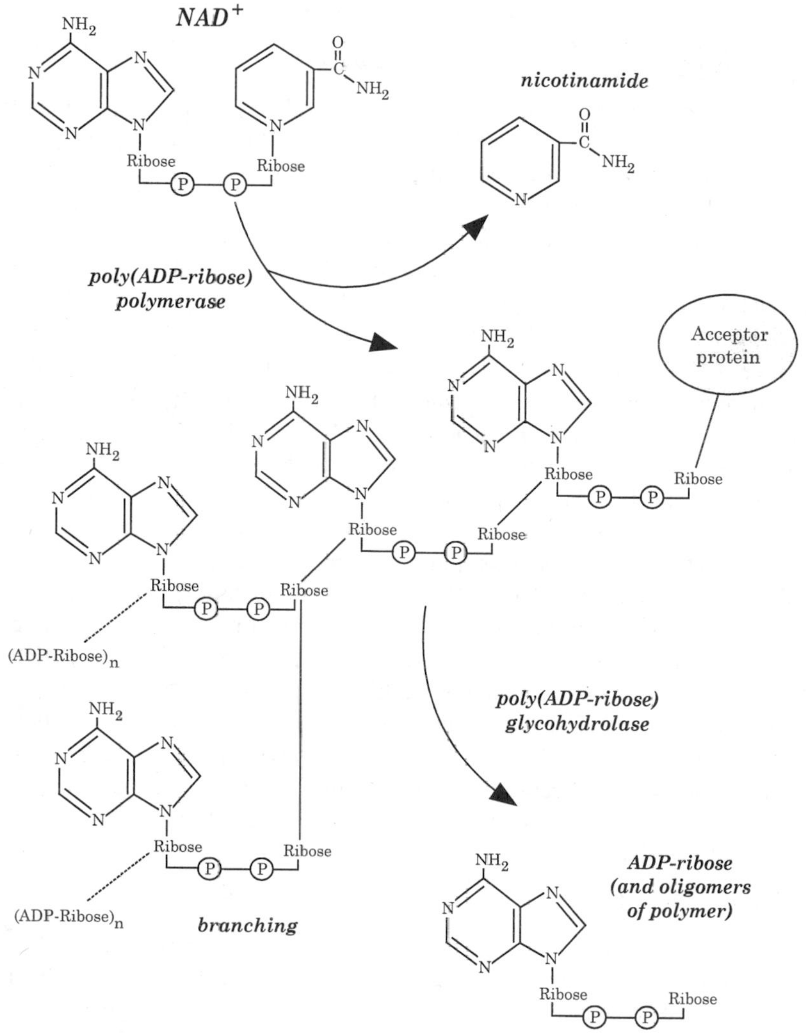

Fig. 5   Synthesis and degradation of poly(ADP-ribose).

to nuclear proteins. [See Figure 5 for an illustration of poly(ADP-ribose) structure and synthesis.]

1. Poly(ADP-ribose) Synthesis

Poly(ADP-ribose) polymerase (PARP; EC 2.4.2.30) was the first enzyme identified that had the ability to synthesize poly(ADP-ribose). PARP contains three functional domains (80). A 42-kDa DNA-binding domain, containing two zinc fingers, is located at the amino terminus. Together, these zinc fingers allow the enzyme to bind specifically to strand breaks in DNA and signal the catalytic portion of the protein to initiate poly(ADP-ribose) synthesis (81). The 55-kDa carboxy terminus region contains the NAD^+ binding and catalytic sites. Although more than 30 nuclear proteins may act as acceptors, most of the poly(ADP-ribose) is synthesized on PARP itself (referred to as automodification). The middle section of the amino acid sequence of PARP includes a 16-kDa region that has been identified as an automodification domain. Here there are 15 sites on which PARP synthesizes poly(ADP-ribose), although more sites for automodification have been found outside of this area of the protein (82). The synthesis of poly(ADP-ribose) on PARP itself is critical in the regulation of its interactions with DNA.

The presence of a DNA strand break is detected by the zinc finger close to the amino terminus, while the second finger is required for catalytic activation (81). Activated PARP initiates poly(ADP-ribose) synthesis by covalently linking the ADP-ribose portion of an NAD^+ molecule to a glutamate or aspartate residue on an acceptor protein, with the release of nicotinamide. A linear sequence of ADP-ribose units is synthesized on the initial protein-bound monomer. At intervals of 40–50 residues, branch points are created on the parent polymer chain and subsequently serve as sites for elongation (83). These initiation, elongation, and branching reactions are all carried out by the catalytic domain at the carboxy terminus of PARP. Automodification of PARP is now thought to occur by two PARP molecules working together as a dimer (84). As PARP becomes more poly(ADP-ribosyl)ated, it takes on an increasingly negative charge because of the accumulation of phosphate groups. This creates electrostatic repulsion between automodified PARP and DNA, causing PARP to dissociate from the DNA nick and stop catalytic activity (85).

It is important to note that inhibition of PARP activity by treatment with competitive inhibitors or removal of NAD from in vitro systems is very different from removing PARP from the system. Inactive PARP binds to strand breaks, preventing access by repair enzymes and possibly impeding the signals that the cell uses to regulate DNA replication and the cell cycle in response to DNA damage. It is not known whether niacin deficiency in vivo causes a similar situation to occur. This depends upon the degree of depletion of NAD and the response of other aspects of poly(ADP-ribose) metabolism, including the rate of degradation of poly(ADP-ribose).

2. Poly(ADP-ribose) Degradation

It has been suggested that two different enzymes are required to degrade poly(ADP-ribose). Poly(ADP-ribose) glycohydrolase cleaves the poly(ADP-ribose) in a combined endo-/exoglycosidic fashion to release free ADP-ribose units and some free oligomers (86). ADP-ribosyl protein lyase appears to release the final ADP-ribose residue from the acceptor proteins, but this activity is poorly characterized. Free ADP-ribose is rapidly degraded to AMP and ribose-phosphate by pyrophosphatase activity; pyrophosphatase could be considered a third enzyme involved in poly(ADP-ribose) catabolism (87). It is not certain why free ADP-ribose is catabolized so rapidly, although it may be to encourage the forward activity of glycohydrolase or to limit the nonenzymatic glycation of proteins, as discussed previously (78). The activity of the glycohydrolase is modified by the nature of the poly(ADP-ribose), with a lower K_m and a higher V_{max} for longer poly(ADP-ribose) chains (88). On short to medium-length chains, PARP and glycohydrolase will compete for the free end of the poly(ADP-ribose) chain. As the ADP-ribose polymer chains elongate, PARP activity decreases (resulting from automodification) and glycohydrolase affinity increases, favoring catabolism.

3. Metabolic Roles of Poly(ADP-ribosyl)ation

The single-strand breaks required to activate PARP are produced in vivo through the processes of DNA replication, transcription, and repair (89). The role played by poly(ADP-ribose) metabolism in these processes has been best characterized in DNA repair. Durkacz et al. (90) demonstrated that poly(ADP-ribosyl)ation was required for cells in culture to repair DNA damage caused by the carcinogen dimethyl sulfate. Exposure of the cells to 3-aminobenzamide (3-AB), a PARP inhibitor, and depletion of cellular NAD^+ (through the use of nicotinamide-free culture medium) resulted in the same inability to recover from DNA damage. This landmark study stimulated a wide variety of research to determine the roles of poly(ADP-ribose) metabolism in the DNA repair process and provided new perspectives on the possible effects of niacin deficiency in the whole animal.

4. Mechanisms of Action

Poly(ADP-ribose) contributes negative charge to acceptor proteins, and the number of ADP-ribose units determines the magnitude of this negative charge. The cell controls poly(ADP-ribose) chain length through the actions of PARP and poly(ADP-ribose) glyco-hydrolase. It would appear that the function of poly(ADP-ribose) is dependent on this anionic nature, which encourages electrostatic repulsion from other polyanions, such as DNA, and attraction to cations, such as basic DNA-binding proteins.

Quantitatively, much of the poly(ADP-ribose) in whole cells and tissues is associated with PARP and histones. The extranucleosomal histone, H1, and histone H2B are modified, leading to localized disruption of the DNA–protein interactions within and among nucleosomes (89). In a process referred to as "histone shuttling," the automodification of PARP may also contribute to chromatin relaxation by drawing nearby histones away from the DNA (85). As a result of these modifications, poly(ADP-ribosylation) leads to a localized relaxation of nucleosomal structure. The exposed DNA is presumably available for interactions with other DNA-binding proteins such as helicases, topoisomerases, polymerases, and ligases involved in replication or repair (89).

A number of studies have also demonstrated poly(ADP-ribose) synthesis on enzymes involved in DNA repair (89). Poly(ADP-ribosyl)ation of most enzymes, including DNA topoisomerases I and II and DNA polymerases α and β, results in inhibition of their activities (89). In addition, the poly(ADP-ribosyl)ation of Ca^{2+}/Mg^{2+}-dependent endonuclease has been shown to suppress the rapid, nonspecific DNA-degrading activity of this enzyme (91), and this may be important in the regulation of apoptosis. In contrast, DNA ligase, the enzyme required to anneal the strand ends once excision repair is complete, appears to be stimulated by poly(ADP-ribosyl)ation (92). Noncovalent association of PARP with DNA polymerase-α has been found to stimulate the activity of DNA polymerase-α (93).

Many approaches have been used to determine the role of poly(ADP-ribose) metabolism in DNA repair, replication, and transcription. In some cases, investigators have created defined in vitro models, in which PARP can be added or removed, and its activity can be controlled by the removal of NAD or the addition of inhibitors. However, these systems lack the complexity of chromatin structure and interactions with the nuclear matrix that exist in the whole cell. Lindhal and co-workers have used a simplified in vitro repair system to show that the combined addition of PARP and NAD does not increase the rate of excision repair. However, in the absence of NAD, PARP addition effectively blocks repair because of its inability to automodify and leave the strand to allow for repair (94). These authors have suggested that PARP is not involved in excision repair directly but may play a role in preventing nonhomologous recombination events between two sites of damage. In support of this, chemical inhibition of PARP in cells has been shown by others to increase homologous recombination and sister chromatid exchanges (95). At the same time, it is apparent that the in vitro models do not contain the complexity of chromatin structure, including nucleosomal organization, DNA supercoiling, and nuclear matrix interactions, seen in intact cells. In attempts to extend these findings to more physiological models, cells expressing PARP antisense mRNA to deplete PARP polypeptide (96) and cells selected for very low PARP expression (97) have been studied. While both of these approaches cause an increased susceptibility to DNA-damaging agents, they lead to conflicting conclusions regarding whether there is a direct role for PARP in excision repair vs. a secondary role in the regulation of recombination events. An excellent way to determine the role of a protein in the whole animal is to delete the gene using recombinant techniques and, if possible, create a fully homozygous animal lacking any functional expression of the gene. This is usually done in mice, and the modified strains are referred to as ''knockouts.'' In the past 3 years a series of publications characterizing two new PARP knockout mouse models has appeared (98). Up to 1 year of life, PARP-null mice appear to be healthy and fertile. In one model, growth of pups is slower and litter size is smaller, but PARP does not appear to be essential for normal development. However, x-ray treated thymocytes from the animals are delayed in their recovery, compared with thymocytes from controls, and there are increases in sister chromatid exchange frequency. Both PARP knockout models show extreme sensitivity to radiation injury, and the one that has been tested is very sensitive to the acute toxicity of a chemical carcinogen that causes DNA alkylation. It will be important to examine the incidence of spontaneous carcinogenesis in these mice in long-term experiments, as well as determining their response to low levels of various carcinogens.

PARP-null mice created an unexpected breakthrough in this field through the finding that they still made small amounts of poly(ADP-ribose), leading to the discovery of four new enzymes that have PARP activity. These are distinct gene products that have eluded discovery due to their small quantities and the small amount of poly(ADP-ribose) that they synthesize. A larger (142 kDa) enzyme, called tankyrase, associates with telomeres (99) and may help regulate the activity of telomerase or inhibit recombination events at chromosome tips. Several PARP homologues are approximately 60 kDa in size, lack both the traditional DNA-binding domain and the automodification zone, but appear to be activated by DNA damage (100–102). The 193-kDa subunit of mammalian vaults is also known to have the ability to synthesize poly(ADP-ribose) (103). Vaults are cytoplasmic ribonucleoprotein complexes of unknown function, but the identification of a cytoplasmic PARP activity explains some earlier observations regarding the existence of cytoplasmic poly(ADP-ribose).

The effects of PARP removal are different from those obtained using inhibitors of PARP activity. It is probably much better for cells to not express PARP at all than to express inactive forms or to express PARP under conditions in which substrate is limiting. Niacin deficiency may fall in this later category, as it has the potential to cause inactive PARP molecules to accumulate on DNA strand breaks, inhibiting their repair. The probability of mutations occurring would further increase if this binding also impaired the signaling events that control cell cycling in the presence of DNA damage. There is evidence that severe niacin deprivation of cells can prevent the DNA damage–induced accumulation of p53, and the removal of PARP protein has a similar effect in this model (104). Other experiments have suggested that p53 is a substrate for poly(ADP-ribosyl)ation and that PARP and p53 may form a complex that binds to certain DNA sequences (98). Increased p53 levels help to remove the cell from active cycling and either promote efficient DNA repair or induce apoptosis. Further DNA replication in the presence of DNA damage favors the accumulation of mutations that lead to the neoplastic phenotype. Similarly, if apoptosis of severely damaged cells is prevented, there is a strong tendency for neoplastic transformation. The responses of the two PARP knockout models differ with respect to p53 induction.

5. DNA Replication and Transcription, Cellular Differentiation

Various roles have been suggested for poly(ADP-ribose) metabolism in other cellular functions, but the experimental results are not as conclusive as those for DNA repair. The amounts of PARP and poly(ADP-ribose) vary during the cell cycle, but inhibition of PARP activity only changes DNA replication or cell division in some experimental models. PARP could be functioning in the complex of proteins at the replication fork through protein–protein interactions, without a requirement for poly(ADP-ribose) synthesis. As mentioned earlier, PARP is known to have protein–protein interactions with DNA polymerase α. If these roles do exist, there must be redundancy of function, given that the PARP knockout mouse grows and develops to adulthood normally (98). Tankyrase, for example, is an alternate PARP that associates with telomerase and could be important in regulation of telomerase activity following DNA replication (99).

PARP appears to associate with areas of DNA that are undergoing active transcription. PARP may modify members of the transcription complex, such as RNA polymerase, or may stabilize cruciform structures in the DNA, but inhibition or removal of PARP from cells does not appear to have dramatic effects on their overall capacity for transcription of RNA (98). PARP itself or poly(ADP-ribose) formation may regulate specific genes; PARP interacts directly with the AP-2 family of transcription factors (89) and also synthesizes poly(ADP-ribose) on transcription factors that bind to AP-2 (105).

There are some dramatic effects in the area of cellular differentiation, especially in cells of myeloid origin. PARP protein is absent from circulating neutrophils; this is the only known example of a nucleated animal cell that does not contain this enzyme (106). Acute promyelocytic leukemia cells provide a model for in vitro differentiation to either neutrophils or monocytes. In this model, neutrophilic differentiation is also associated with a disappearance of PARP protein (106), and when PARP is artificially expressed, this pathway of differentiation is blocked (107). PARP also appears to play a causative role in differentiation in other models using transformed cells in culture (82), but the normal development of the PARP knockout mouse suggests that redundancy exists in the whole animal. For all of these functions, it is important to remember the difference between a loss of catalytic activity that could be caused by niacin deficiency and the removal of PARP protein in the knockout model.

6. Relationship of Poly(ADP-Ribose) Metabolism to Carcinogenesis In Vivo

Given that poly(ADP-ribose) metabolism plays an important role in DNA repair in cultured cells, one would expect PARP activity to be important in the in vivo process of carcinogenesis. There are studies using competitive inhibitors that support this conclusion. Concurrent treatment with 3-aminobenzamide (3-AB), a competitive inhibitor of PARP, caused a 10-fold increase in the formation of altered hepatic foci (precancerous lesions) in rats treated with diethylnitrosamine (108).

It is important to note, for much of the following discussion on PARP inhibition and carcinogenesis, that there are two very different mechanisms in which PARP may be involved in cell death following DNA damage. In the first, low to moderate levels of DNA damage cause a gradual accumulation of genetic changes that initially prevent the cell from dividing and may eventually kill a portion of the cells through the loss of function of essential genes or induction of apoptosis. If one or more mutations occur in genes involved in cell division, this may also lead to the initiation of carcinogenesis. In a second mechanism, large quantities of DNA damage activate PARP to such an extent that cellular NAD is decreased to levels that do not support cellular functions, such as ATP synthesis, and the cell dies quite rapidly. This type of cell death also tends to follow the pattern of apoptosis and has been portrayed by some as the central mechanism for induction of apoptosis via DNA damage. This is not an accurate generalization. Berger proposed the "suicidal NAD depletion" model as a mechanism to encourage cell death when DNA damage was extreme (109). Under these conditions, the risk of carcinogenesis would be high if cells survive. Inhibition of PARP increases cell injury in the first model because of an inhibition of DNA repair processes. In the second model, however, PARP inhibition can save cells from death by NAD depletion, at least in the short term. This can also lead to an increase in carcinogenesis, due to the survival of cells with significant levels of DNA damage.

An in vivo example of this second model is the chemical induction of insulin-dependent diabetes mellitus (IDDM) in animals. Alloxan and streptozotocin are DNA-damaging agents, working via hydroxyl radical formation and base alkylation, respectively, and are fairly selective pancreatic β-cell toxins. They are commonly used to induce IDDM in animal models. In these models, alloxan and streptozotocin induce poly(ADP-ribose) synthesis to the point of NAD depletion and cell death. Treatment with PARP inhibitors, including nicotinamide and 3-AB, results in conservation of pancreatic β cells and maintenance of insulin secretion in rats treated with alloxan and streptozotocin (110). However, the incidence of pancreatic β-cell tumors in these animals is essentially 100% later in life (111). Nicotinamide is a relatively weak PARP inhibitor ($K_i = 10$ mM) (112), and it is rapidly converted to NAD as it enters the cell, so it is unlikely that it accumulates to a level at which it acts effectively as a PARP inhibitor in whole-animal models. Nicotinamide probably functions under these conditions as an NAD precursor (47) and by preventing extreme NAD depletion in damaged β cells allows cells to survive and eventually express a neoplastic genotype.

It is interesting to consider whether suicidal NAD depletion (as a direct response to PARP activation) is relevant to human carcinogenesis. The extreme level of DNA damage required to initiate this response suggests that it is not a major factor in chronic disease states, but there are some interesting exceptions. The concentrated autoimmune attack that initiates IDDM in humans appears to act by this mechanism (see Sec. VII for further discussion). In addition, experiments with PARP knockout mice have shown that PARP activation plays a major role in ischemia-reperfusion injury in the brain (98). The sudden oxidant stress that occurs when blood flow returns to a hypoxic region of the brain causes a PARP-dependent depletion of NAD, which leads to neuronal cell death. This model may provide important clues to the effective treatment of stroke victims. However, most human cancers result from a gradual accumulation of DNA damage under conditions in which poly(ADP-ribose) synthesis rates are low and NAD depletion is modest. Even very aggressive models of chemically induced carcinogenesis cause a high incidence of initiation without severe depletion of NAD (113). If models of cell injury that cause suicidal depletion of NAD are used to examine the effect of niacin deficiency, dramatic results can be obtained (114). This is different from the situation in which a modest amount of

DNA damage causes initiation of apoptosis and NAD depletion occurs as apoptosis progresses. Cell culture models have a great contribution to make in understanding the relationship between niacin status and cellular defense; many parameters can be controlled and/or measured accurately, including NAD levels, degree of DNA damage, progression of apoptosis, and transformation to a neoplastic phenotype. However, researchers do need to relate their in vitro models to the degree of NAD depletion and levels of DNA damage that are possible in the whole animal.

7. Dietary Niacin Deficiency, Poly(ADP-ribose) Metabolism, and Carcinogenesis

The native population of the Transkei region in South Africa has a high risk for esophageal cancer (115). A maize-based, low-protein diet is staple for these people, and pellagra is common. Esophageal ulcerations and esophagitis, common in pellagrins, have been associated with development of carcinoma of the esophagus. Van Rensburg et al. found a greater than fivefold increase in the risk of esophageal cancer in Zulu men who ate maize daily (116). Frequent consumption of maize by natives of the Henan province of China is also associated with increased esophageal cancer risk (117). Consumption of maize in northeastern Italy was associated with increased risk of oral, pharyngeal, and esophageal cancers, especially with heavy consumption of alcohol (118). The frequency of esophageal cancer appears to increase when maize replaces sorghum as a basic dietary component (115,119). Low niacin intake in an American population, in western New York state, was also associated with increased oral cancer risk, when controlled for smoking and alcohol consumption (120), which appeared to be the initiating factors in the disease. This is interesting in that the patients were not deficient in niacin to the degree of showing clinical symptoms of pellagra, and suggests that subclinical niacin deficiency may increase cancer risk.

The Linxian province of northern China also has a very high rate of esophageal and gastric cancers. Recently, a series of nutritional intervention trials was conducted in this area (121,122). During 5 years of intervention, a combined supplement of β-carotene, vitamin E, and selenium decreased total mortality, total cancer mortality, and stomach cancer mortality. A parallel group receiving riboflavin and niacin did not show any benefits. It is important to note that this population has traditionally been dependent on rice as a staple rather than maize, and niacin deficiency is less common than thiamine or selenium deficiency. There is an important need to design an intervention trial in an area dependent on maize as a staple, with a significant incidence of existing niacin deficiency in the population.

A limited number of animal experiments have been conducted to test whether niacin deficiency plays a causal role in the process of carcinogenesis. Miller and Burns studied the interaction of niacin deficiency, protein-energy malnutrition, and renal carcinogenesis in rats (123). The diets were low in tryptophan and total protein (25% of recommended intake), with nicotinamide at zero, requirement, or 10 times requirement levels. The diets did not affect tumor size or number, although death due to renal tumor burden may have been accelerated in the deficient animals. Unfortunately, pyridine nucleotides in the liver and kidney were not decreased by the deficient diet, and it is difficult to determine if niacin deficiency had any impact in this experiment. This experiment highlights the difficulties in working with a nutrient that can be synthesized from an amino acid, requiring, in most species, an imbalanced amino acid diet to create the vitamin deficiency. Parameters must be controlled carefully to get the desired nutritional status.

Van Rensburg et al. (124) developed an animal model of malnutrition and esophageal cancer, induced by N-nitrosomethylbenzylamine (a DNA-alkylating agent) in cornfed rats. Addition of 20 mg of nicotinic acid per kilogram of basal diet resulted in reduction of tumor incidence, size, and progression compared with that seen in rats fed the basal diet alone. NAD and poly(ADP-ribose) levels were not measured in this study. Another study has shown that lymphocytes from niacin-deficient rats were more susceptible to oxygen radical–induced DNA damage (125).

We have developed a model of niacin deficiency in rats that maintains a positive growth rate. The diet is based on components with minimal niacin content and a mixture of casein (7%) and gelatin (6%) as sources of protein, designed to limit tryptophan content. Rats fed this diet develop clinical signs of deficiency, including dermatitis, diarrhea, and ataxia, and also have decreased hepatic NAD^+ and poly(ADP-ribose) levels (30). However, when treated with DEN, hepatic poly(ADP-ribose) accumulation was not affected by niacin deficiency, and there was no long-term effect of diet on the development of preneoplastic altered hepatic foci (126). It is apparent that hepatic NAD^+ levels, which decreased from about 900 to about 600 µM in deficient rats, were still adequate to support the activity of PARP, which has a K_m in the range of 20–80 µM (127).

It is likely that, during niacin deficiency, NAD depletion is not uniform in different organs and tissues. The symptoms of pellagra demonstrate tissue specificity, like most nutrient deficiencies. One factor that can cause more rapid nutrient depletion is cell turnover. The bone marrow has the most rapid rate of cell turnover in the body, with a doubling time of about 12 hours, and these cells leave to other sites in the body, exporting nutrient resources with them. With this in mind, we recently examined the effect of niacin status on NAD and poly(ADP-ribose) metabolism in the bone marrow of rats. We found that NAD depletion in the bone marrow is more dramatic than in any other tissue that we have measured (20% of control) and that basal poly(ADP-ribose) content is decreased in the niacin-deficient marrow to almost undetectable levels. The most common problem with DNA damage to the bone marrow occurs during chemotherapy, so we started to treat niacin-deficient and control rats with nitrosourea drugs that are known to cause bone marrow suppression and induce leukemias in the long term. In this model of the side effects of chemotherapy drugs, niacin deficiency increased the severity of acute bone marrow suppression (anemia, leukopenia) (128) and increased the rate of development of nitrosourea-induced leukemias (129). These results may be important, because a large percentage of cancer patients appear to be niacin deficient (130), and bone marrow suppression and secondary leukemias are probably the two biggest problems in cancer therapy, beyond curing the original disease.

Many more experiments have been conducted on the effect of pharmacological supplementation of nicotinic acid or nicotinamide on various types of carcinogenesis. A wide range of doses of nicotinamide has been used, in the presence and absence of exogenous carcinogens, as reviewed by Bryan (131). Nicotinamide alone does not appear to present risk as a carcinogen, although there may be some risk associated with its use in the prevention of IDDM (further discussion in Sec. VII). When used in conjunction with carcinogenic foods or compounds in animal experiments, nicotinamide has a confusing array of effects, suggesting varied mechanisms of action. When given with diethylnitrosamine, nicotinamide did not affect liver carcinogenesis, but it did increase kidney neoplasms. Following streptozotocin treatment, nicotinamide decreased adenoma formation but increased pancreatic islet cell tumors (111,131). Conversely, nicotinamide has shown significant protective effects against bladder and intestinal cancers when provided in a diet containing bracken fern.

There is much less information on the effects of large doses of nicotinic acid on carcinogenesis. Surprisingly, the best information is probably from human studies, derived from the long-term use of this compound in the treatment of hypercholesterolemia. Nine years after a 6-year period of nicotinic acid use to treat hypercholesterolemic patients, there was a significant decrease in mortality in this group, but this did not appear to be due to a decrease in cancer incidence (132).

We have recently studied animals supplemented with nicotinamide or nicotinic acid and treated with DEN (47). While both supplements increased liver NAD^+ to a modest extent in the absence of carcinogen treatment, only nicotinamide significantly increased basal poly(ADP-ribose) levels (before DEN), whereas only nicotinic acid increased poly (ADP-ribose) levels following DEN. Neither supplement affected the development of DEN-induced preneoplastic altered hepatic foci. In contrast, pharmacological supplementation of nicotinamide in mouse diets caused a dramatic increase in skin NAD^+ and provided significant protection against skin cancer induced by UV radiation (see Sec. V.F.9, below). We have also shown that pharmacological supplementation of both nicotinic acid and nicotinamide cause large increases in bone marrow NAD^+ and poly(ADP-ribose) in rats, and decrease the long term development of leukemia (unpublished data). As with deficiency, it is not surprising that niacin supplementation affects cancer susceptibility in some tissues, like the marrow and skin, and not others, like the liver. We need to continue to build our knowledge of whole animal models to appreciate the complexities of niacin metabolism and allow accurate recommendations for human populations.

Although some of the foregoing responses may have been due to changes in the functions associated with NAD^+ utilization, there are also a variety of pharmacological actions that are not related to NAD^+ synthesis. These may be responsible for the inconsistency of the responses, and some of the potential mechanisms will be discussed in Sec. VII.

8. Niacin Status and Oxidant Lung Injury

Niacin supplementation can decrease the degree of lung injury and fibrosis from a variety of causes, including exposure to lipopolysaccharide, cyclophosphamide, and bleomycin. The best defined model uses bleomycin, an antibiotic chemotherapy drug that intercalates with DNA and induces damage through the local production of oxygen radicals (133). This is a very severe stress, which causes NAD and ATP depletion, perhaps leading to suicidal NAD depletion, as hypothesized by Berger (109). Protection by niacin may be functioning through the maintenance of NAD levels, but nicotinamide is more effective in increasing NAD, while nicotinic acid is more potent in the reduction of lung pathology (134). The protection of lung tissue with these supplements may allow much safer use of bleomycin as a chemotherapy agent, although it is not known whether they will also protect tumor tissue.

Hyperoxia is another popular model of oxidant stress in the lung and has clinical relevance to the care of premature infants and patients with adult respiratory distress syndrome. Hyperoxia has been shown to induce poly(ADP-ribose) synthesis in the lung, and, although poly(ADP-ribose) synthesis is decreased by niacin deficiency, the deficient state does not increase the severity of lung damage (135). Consistent with this finding, pharmacological niacin supplementation is also ineffective in decreasing the severity of hyperoxic lung damage (136). Why is this response different from that observed following bleomycin treatment? Bleomycin induces a more sudden and severe stress to pulmonary cells, and it targets DNA specifically as an intracellular target, leading to depletion of NAD and ATP. Hyperoxic damage occurs gradually over a period of 5 or more days (137) and does not cause NAD depletion (135). The majority of oxygen radicals emanate from the mitochondria and endoplasmic reticulum, and oxidative damage is distributed among the intracellular compartments. Perhaps of greatest interest in this model, hyperoxia actually increases lung NAD content in the niacin-deficient animal to almost that of niacin-replete controls. This result strongly suggests that enzymes or transport systems involved in NAD turnover are regulated in response to certain aspects of cellular damage, perhaps via the oxidant stress response described for the induction of *fos* and *jun* (138). The degree of oxidant stress and the time course of the stress, which differ between hyperoxia and bleomycin toxicity, are likely to be important in determining the ability of cells or tissues to adapt.

9.   Niacin Status and Skin Injury

Because of the sun sensitivity displayed by pellagrins, there has been a long-standing interest in the potential of niacin to improve skin health (139). Unfortunately, very little is known about the effect of niacin deficiency on the susceptibility to UV light or chemical carcinogens. Rainbow trout are more prone to UV light–induced skin damage when niacin-deficient (140), but little is known about other nonhuman species, or whether this finding correlates with a change in ADP-ribose metabolism.

On the other hand, many studies have been conducted using pharmacological doses of various forms of the vitamin. Nicotinic acid acts as a vasodilator in the skin, leading to an increase in blood flow through the microvasculature. Interestingly, this occurs with both oral and topical use and appears, in both cases, to be caused by changes in prostaglandin production (141,142). Over the years, a wide variety of treatment regimens and different forms of nicotinic acid have been used to treat various skin disorders (139). The effects on blood flow occur only at supraphysiological levels of the vitamin and are probably not caused by modulation of NAD pools. However, nicotinic acid supplementation at 1–10 g/kg of diet was shown to decrease UV-induced skin cancers in mice (143). The decrease was linear through the supplementation range, as was the increase in skin NAD content. This work shows that very large dietary doses of niacin may continue to influence cancer susceptibility through modulation of NAD pools, although the authors also showed that niacin supplementation improved immune surveillance of tumor cells.

Nicotinamide supplementation has also been shown to protect the skin from DNA-damaging agents in some animal models. Large doses of nicotinamide, given intraperitoneally, decreased the skin damage caused by sulfur mustard (144). Used as a chemical warfare agent in the First World War, this is a DNA-alkylating agent: that causes edema, necrosis, and microvesical formation in the exposed skin, associated with NAD depletion. Interestingly, nicotinamide prevents the depletion of NAD and subsequent pathological changes, without protecting against the earlier pathological changes that precede NAD depletion.

VI.   COMPETITION FOR NAD$^+$ DURING NIACIN DEFICIENCY

It seems obvious that the most critical cellular functions of the niacin-containing nucleotides are those of electron transport and energy metabolism. A loss in the capacity to deliver reducing equivalents to the electron transport chain would be similar to poisoning the cell with cyanide or suffocating from a lack of oxygen. It makes sense, then, that these functions will be strongly protected when NAD levels start to deplete during niacin deprivation. Cultured cells, in the absence of DNA damage, can grow and divide with less than 5% of control NAD levels (90,104,145), leaving us with a variety of questions to be answered. How do the other pathways of NAD utilization, including poly-, mono-, and cyclic ADP-ribose formation, compete for these limiting substrate pools? What is the nature of this competition at the cellular level with respect to compartmentalization between the nucleus, cytoplasm, and mitochondria? What is the role of extracellular NAD^+ in the function of mono(ADP-ribosyl)transferase and ADP-ribosyl cyclase enzymes on the outer surface of the cell? How are nicotinic acid and nicotinamide distributed among tissues during deficiency, and does this contribute to the distinctive signs and symptoms of pellagra? How do these interactions lead to the specific metabolic lesions that cause the sun-sensitive dermatitis, diarrhea, and dementia?

Possible mechanisms for unequal utilization of NAD at the subcellular level include(a) variation in the affinity of enzymes for NAD^+ (K_m) and (b) compartmentalization. K_m values are used to describe the affinity of an enzyme for its substrate and are defined as the concentration of substrate required to support 50% of the maximal activity. A lower K_m indicates a higher affinity and suggests that an enzyme will compete effectively with enzymes having a higher K_m as NAD^+ concentrations fall during deficiency. Some caution in interpretation is required; enzyme kinetics may change during purification, especially for membrane-bound proteins.

The K_m of PARP for NAD^+ is thought to be between 20 and 80 µM (127). In certain cultured cells, the ability to synthesize poly(ADP-ribose) decreases when cellular NAD content drops to less than half of control levels (145), showing that the synthesis of poly (ADP-ribose) is one of the most sensitive pathways of NAD utilization. This is similar to the proportionate decrease in NAD^+ during niacin deficiency in many tissues in vivo, but tissues vary in their absolute concentrations of NAD^+, and direct extrapolation to intact tissues could be inaccurate. For example, in rats, liver NAD^+ decreases by close to 50% during deficiency but is still at about 500 µM absolute concentration (30).

It has been stated that poly(ADP-ribosyl)ation is the aspect of NAD^+ utilization that is most sensitive to niacin deficiency because of a much higher K_m of PARP for NAD^+. Do the K_m values of other NAD^+-utilizing enzymes suggest that the sensitivity of poly (ADP-ribose) metabolism may be unique? There are scores of dehydrogenase enzymes that use NAD^+ as an electron acceptor, producing NADH for utilization in the electron transport chain. Glyceraldehyde phosphate dehydrogenase is a cytosolic enzyme that is critical to the flow of substrates through glycolysis. It uses NAD^+ as an oxidant and has a K_m for this cofactor of 13 µM (146), smaller than that of PARP, indicating a higher affinity. Other cytosolic enzymes may have slightly higher affinities for NAD^+ than PARP, including alcohol and aldehyde dehydrogenases (17–110 µM and 16 µM, respectively) (146). In the mitochondria, isocitrate is oxidatively decarboxylated in the TCA cycle by a dehydrogenase with a K_m for NAD^+ of 78 µM (146), which is similar to PARP. However, the mitochondrial form of malate dehydrogenase is also critical to the flow of substrate through the TCA cycle, and its K_m for NAD^+ has been reported as 540 µM (146). If these data are correct, the TCA cycle appears to require compartmentalization of NAD^+ during niacin deficiency, and the role of the mitochondria in this regard will be discussed below.

With respect to cyclic ADP-ribose synthesis, the purified microsomal cyclase from canine spleen has a K_m of 10 µM for NAD^+ (54). With access to cytosolic NAD pools, this enzyme should maintain its catalytic activity during niacin deficiency. The CD38 cyclase is reported to have a K_m of 15 μM for NAD^+ (147). While this is a relatively high affinity for substrate, the curious aspect of this enzyme is that it faces the exterior of the cell. Does it have a requirement for extracellular NAD or access to intracellular pools? Cultured kidney epithelial cells synthesize cyclic ADP-ribose, but they require permeabilization to use NAD^+ in the medium and require over 500 μM for a half-maximal response (148). There are many questions to be answered in this area of research, but the potential for cyclic ADP-ribose metabolism to be affected by niacin deficiency is worth considering.

Mono(ADP-ribosyl)transferases are a very diverse group. The only published data on affinity for NAD^+ refers to the arginine specific transferases. In a family of transferases from turkey erythrocytes, two cytosolic enzymes have K_m values of 7 and 36 μM, while a transferase from the membrane fraction has a K_m of 15 μM (67). These enzymes would appear to compete with PARP under conditions of limiting NAD^+ pools, but a transferase from chicken liver nuclei has a K_m of between 200 and 500 μM, and a transferase from mammalian skeletal and heart muscle displays a K_m of 560 μM (67). It appears that some mono(ADP-ribosyl)transferases may be quite sensitive to niacin deficiency. The resulting changes in cell signaling might not appear as problems in cell culture models but could nevertheless present significant problems in the whole organism. As discussed earlier, there is a mono(ADP-ribosyl)transferase anchored to the outer surface of the plasma membrane, which, like CD38, appears to require extracellular NAD^+. This enzyme appears to cause the ADP-ribosylation of an intracellular protein, leading to a depression in T-cell proliferation. Although no attempt has been made to determine the K_m of this enzyme for NAD^+, levels of NAD^+ as low 1 μM in the culture medium are effective in decreasing cell proliferation (70).

It becomes apparent that the physical partitioning of NAD within the cell is a key factor in the availability of the molecule for various metabolic functions. The cytoplasmic pool provides substrate for soluble enzymes, as well as for those on the endoplasmic reticulum and on the inside of the plasma membrane. These would support a host of redox reactions and the activity of a variety of poorly defined mono(ADP-ribosyl) transferases and ADP-ribosylcyclases. The mitochondria isolate a pool of NAD^+ that is predominantly involved in electron transport, although mono(ADP-ribosyl)ation reactions have been reported in this organelle. Nuclear NAD^+ is probably used mainly for poly(ADP-ribosyl) ation reactions, but mono(ADP-ribosyl)transferases also are located here. The least studied pool is extracellular NAD^+, which appears to play a role in some ADP-ribosylcyclase and mono(ADP-ribosyl)transferase activities. How distinct are these pools, and how do they respond to the progression of niacin deficiency?

Because of the presence of nuclear pores, it is unlikely that nuclear and cytosolic NAD^+ concentrations would differ to any great extent. However, the final step in the synthesis of NAD^+ from nicotinamide is catalyzed by a nuclear enzyme (Fig. 3) (36). The same enzyme catalyzes the second to last step in the conversion of nicotinic acid to NAD^+, but the last enzyme in this pathway is in the cytosol. This means that all of the NAD^+ synthesized in the cell from newly arrived nicotinamide, or from nicotinamide released by any of the ADP-ribosylation reactions in the cell, will be available first to nuclear reactions, of which poly(ADP-ribosyl)ation is likely to predominate. Nicotinic acid will lead to the production of cytosolic NAD^+, which may favor different patterns of utilization.

The mitochondria are well equipped to regulate NAD levels. The inner mitochondrial membrane is essentially impermeable to all forms of NAD(P). Reducing equivalents in the form of NADH must be transformed via shuttle mechanisms to enter the mitochondria for ATP production. How does the mitochondrion produce or obtain NAD, and what levels does it maintain? Some researchers believe that mitochondria synthesize NAD (149) whereas others suggest that slow, high-affinity carriers bring the necessary NAD from the cytosol (150,151). The net requirement is probably modest, as most of the reactions identified in this organelle do not degrade the cofactor. The important question concerns the ability of mitochondria to concentrate NAD^+, and it appears that they have potent mechanisms to accomplish this. NAD^+ levels in hepatocyte mitochondria appear to be about 10-fold higher than in the cytoplasm, with absolute concentrations in the neighborhood of 5 mM (152). These NAD^+ concentrations would support enzymes with relatively low affinities for NAD^+, such as malate dehydrogenase (146). With this ability to concentrate NAD^+, the mitochondrial pool could be very well protected during niacin deficiency.

The plasma pool of NAD^+ is poorly characterized. Levels of noncellular NAD^+ in blood samples are extremely low (23), and there is no information on the response of this pool to dietary niacin intake. Further research will be required to determine if this source of NAD^+ has any physiological role.

In addition to the competition for NAD^+ at the cellular level, organs and tissues vary in their ability to conserve NAD pools or compete for precursors during the progression of niacin deficiency (30). Various tissues also start with different levels of NAD^+, which may act as reserves during deficiency. Blood $NADP^+$ is more stable than NAD^+ during niacin deficiency (29), but it may change in other tissues, and the impact of niacin deficiency on NAADP metabolism is not known. These are some of the concepts that must be appreciated as we progress toward a better understanding of the biochemical basis of the pathologies of pellagra, a disease whose clinical symptoms remain unexplained at the molecular level.

VII. PHARMACOLOGY AND TOXICOLOGY

Levels of niacin in excess of the RNI have been used in attempts to treat Hartnup's disease, carcinoid syndrome, poor glucose tolerance, atherosclerosis, schizophrenia, hyperlipidemia, IDDM, and a variety of skin disorders. In some countries, during the shortages of proper medical supplies caused by World War II, nicotinic acid became a popular drug because the dramatic flushing reaction that it caused in the skin was interpreted as a sign of the potency of the treatment (153). Currently, nicotinic acid and nicotinamide are used mainly in the prevention of cardiovascular disease and IDDM, respectively. Very few nutrients are prescribed medicinally in North America for pharmacological purposes that are mechanistically distinct from their known nutrient functions. Both nicotinic acid and nicotinamide fall into this category, and the pharmacological effects of these two vitamers appear to be surprisingly unrelated.

Large oral doses of the two vitamers may be absorbed and distributed quite differently than when provided at levels found in a normal diet. However, there is very little information, especially from humans, on the pharmacokinetics of these compounds. In normal diets, nicotinic acid is obtained mainly from plant products, while nicotinamide and preformed nucleotides are derived mainly from animal-based food products. The intestine and liver are active in the conversion of nicotinic acid to NAD and subsequently to nicotinamide for release into the bloodstream, causing normal plasma levels of nicotinamide to be greater than those of nicotinic acid (23). Large oral doses of nicotinic acid will overcome this regulation, but the concentrations that reach peripheral tissues in patients are poorly defined (153). There are two reasons for this: first, quantification of specific vitamin forms is laborious, and, second, nicotinic acid was approved as a drug under a "grandfather'' clause, which excused it from the rigorous testing required by potential new drugs on the market.

In the case of nicotinamide, which is the main circulating form of the vitamin under normal conditions, large doses can have a depressing effect on relative availability. This is caused by the induction of deamidase enzymes in gut microflora and also by attaining concentrations in the K_m range of similar enzymes in the liver, leading to formation of nicotinic acid (154). However, pharmacokinetic studies of nicotinamide in humans show that plasma levels are significantly elevated by large oral doses (155). More work is needed in defining the pharmacokinetics of both forms of niacin, including an assessment of their effect on NAD pools and their turnover in various tissues.

A. Nicotinic Acid

1. Nicotinic Acid and Hyperlipidemia

Historically, nicotinic acid has been administered to patients with a variety of disorders, often more for the dramatic skin reaction than proven curative powers. Its most successful use is for the treatment of hyperlipidemia (156). At high doses, nicotinic acid causes several changes in lipid and lipoprotein metabolism, including inhibition of lipolysis in adipose tissue (157), inhibition of the synthesis and secretion of very low-density lipoprotein (VLDL) by the liver (158), lowering of serum lipoprotein(a) levels (156), and increasing serum levels of high-density lipoprotein (HDL) (159).

These mechanisms of action of nicotinic acid appear to be unrelated to the formation of pyridine nucleotides or to the actions of nicotinamide. Decreased lipolysis in adipose tissue is due to an inhibition of adenylate cyclase activity (157). The resulting drop in cAMP levels leads to the decreased mobilization of fatty acids. It is not known whether nicotinic acid binds to a cellular receptor or interacts directly with adenylate cyclase. The decrease in fatty acid release from adipose tissue is at least partially responsible for the drop in VLDL formation by the liver as well as for the subsequent drop in LDL levels, although there may also be direct effects on liver lipid metabolism. Studies with radioactive acetate suggest that nicotinic acid inhibits cholesterol synthesis at the level of 3-hydroxy-3-methylglutaryl coenzyme A (HMG-CoA) reductase (156), and this may also play a role in the lowering of VLDL and LDL levels. Unlike many treatments for hyperlipidemias, nicotinic acid also increases circulating levels of HDL (159), the beneficial lipoprotein that removes cholesterol from vascular tissue, but the mechanisms involved are uncertain. The skin flush of the face and upper trunk is a very rapid response to nicotinic acid, and it appears to be caused by the local formation of prostaglandins (142), although the underlying mechanism is unclear.

2. Nicotinic Acid Toxicity

There are a few drawbacks to using high levels of oral nicotinic acid. As mentioned above, the short-term side effects may include vasodilation, burning or stinging sensations in the face and hands, nausea, vomiting, and diarrhea. In the longer term, there may be varying degrees of hyperpigmentation of the skin, abnormal glucose tolerance, hyperuricemia, peptic ulcers, hepatomegaly, and jaundice (156). The chronic doses of time-release nicotinic acid have been reported to cause more hepatotoxicity (160), but this is controversial. It should be noted that all drugs used in the treatment of hyperlipidemia have some side effects, many of which can be managed through changes in dose. Interestingly, nicotinic acid use for 6 years by patients with cardiovascular disease led to a decrease in all-cause mortality measured 8 years after the drug use was discontinued (132). Frequently, the choice of treatment is financially based, with the cost of nicotinic acid treatment being a fraction of that of the newer medications.

B. Nicotinamide

In the past, nicotinamide has been used in the treatment of schizophrenia (161), but more effective drugs have replaced it in this field. It is now being tested as a chemotherapy agent; in this application, nicotinamide potentiates the cytotoxic effects of chemotherapy and radiation treatment against tumor cells (162), an action that appears to be due to increased blood flow and oxygenation of tumor tissue (163). However, most of the current interest in nicotinamide involves its potential use in the prevention or delay of onset of IDDM (156,164).

1. Nicotinamide and IDDM

Interest in this area started with the finding that nicotinamide could prevent diabetes induced by the β-cell toxins alloxan and streptozotocin (110). It was soon shown that the β cells were killed by an interesting mechanism; severe DNA damage led to excessive activation of PARP, which depleted cellular NAD levels to the point that ATP synthesis could not be maintained. The researchers concluded that nicotinamide was preventing diabetes by inhibiting PARP activity and preventing the depletion of NAD pools. Since nicotinamide is not a very high-affinity inhibitor of PARP (112), and cellular levels tend to stay low due to active conversion to NAD, it seems likely that protection in this model was due to the use of nicotinamide as a precursor for NAD synthesis. We have shown that large oral doses of nicotinamide increase NAD^+ and poly(ADP-ribose) levels in the liver (47) and more recently, in the bone marrow (unpublished results). It is unlikely that oral nicotinamide could act as a PARP inhibitor in the pancreas, under similar conditions. Interestingly, all of the animals protected from chemically induced diabetes by nicotinamide developed insulin producing β-cell tumors (111), a form of cancer that is particularly lethal in humans (165). It is not surprising that cells rescued from NAD depletion due to extreme DNA damage, either through inhibition of PARP or through pharmacological support of NAD pools, would be at risk for neoplastic growth due to the survival of cells with significant levels of DNA damage.

One concern in promoting the clinical use of nicotinamide in the prevention of IDDM in humans is that there will also be a long-term risk of pancreatic cancer if survival of β cells can be maintained in the face of autoimmune attack. In spite of the theoretical difficulties and the potential dangers of inhibiting PARP in the face of genotoxic damage to the pancreas, many investigators in the diabetes field are comfortable with the theory of PARP inhibition (166). Fortunately, there are other aspects of the response to nicotinamide that paint a more optimistic picture, although not a very clear one.

In humans, the onset of IDDM occurs spontaneously by immune recognition of β-cell antigens. This is associated with leukocyte infiltration and the presence of anti–islet cell antibodies in the serum. The spontaneous occurrence of IDDM is similar in the non-obese diabetic (NOD) mouse. When nicotinamide is given to weanling NOD mice, the onset of diabetic symptoms is prevented or delayed (167). The important distinction is whether nicotinamide prevents β-cell death by protecting against autoimmune attack or by maintaining the β cell following immune attack. There appear to be several ways in which nicotinamide can prevent the initial damage to the cells of the pancreas. In some experimental models, nicotinamide treatment diminishes some aspects of the immune response itself, including the ability of monocytes to attack cells labeled with antibodies (168), the rate of movement of lymphocytes into tissues (169), and the production of nitric oxide in response to inflammatory cytokines (170). Other studies have shown that nicotinamide acts as a free-radical scavenger to protect the β-cell targets even after the immune response is directed against them (171,172). These mechanisms may provide an explanation for the prevention of IDDM by nicotinamide and lead to more effective protocols in the future.

In a very different experimental model, nicotinamide was found to decrease the severity of diabetes in response to partial pancreatectomy. This appeared to be due to a stimulation of β-cell proliferation, leading to an increase in the size of islets (173). The β cells normally have a low capacity for cell regeneration, and it has been suggested that this predisposes to the development of human diabetes (110). When fetal pancreas is pretreated with nicotinamide, there is an acceleration in the reversal of diabetes after transplantation of the islet cells into diabetic nude mice (174). This is an important model for future directions in the treatment of fully developed IDDM, and these experiments show that nicotinamide has a positive influence on β-cell regeneration and differentiation.

Nicotinamide may have an effect on mono(ADP-ribosyl)ation reactions, either by acting as an inhibitor or by enhancing substrate (NAD^+) levels. In addition, there is a new feeling that cyclic ADP-ribose could have an important role in immune regulation. Cyclase enzymes like CD38 are important proteins in leukocytes, and drugs that interfere with cyclic ADP-ribose regulation of calcium release are immune suppresors (166).

2. Clinical Trials

Encouraged by the type of data summarized above, a number of experiments have been done with human subjects. In the majority of these studies, patients were recruited in an early stage of clinically apparent diabetes. Since these subjects retain a varying degree of β-cell function, it is not surprising that the results have been inconsistent. However, a number of treatment protocols have been successful in inducing remission in some patients (175) and increasing the residual level of plasma insulin for up to 2 years after diagnosis (176).

The animal models show that nicotinamide treatment should start before the disease process is in an advanced stage. To do this in human populations, researchers must identify the susceptible population. Blood levels of islet cell antibodies, human leukocyte antigen, and family history are used as predictors of IDDM for subject recruitment. Several intervention trials starting in the early stages of the disease have produced interesting results. In a 2-year study, 13 of 14 high-risk children treated with oral nicotinamide remained disease-free, while all of the children in the control group became diabetic (177). In another study, nicotinamide treatment for 8 months protected plasma insulin in high-risk children, in spite of the fact that anti–islet cell antibody levels were not decreased (178). This suggests that the autoimmune response is affected downstream of autoantibody production, or, alternatively, that β-cell defenses are improved by nicotinamide treatment.

These results have encouraged the organization of several larger studies, including the European-Canadian Nicotinamide Diabetes Intervention Trial (ENDIT), which will involve the screening of about 30,000 people for subjects that are anti–islet cell antibody–positive with normal glucose tolerance (164). These subjects will be treated with nicotinamide or placebo for 5 years, and the results should provide significant information on the prevention of diabetes by nicotinamide, its mechanism of action, and even the long-term risk of carcinogenesis.

3. Nicotinamide Toxicity

The levels of nicotinamide that are used in the treatment of IDDM (about 3 g/day) have not been reported to cause any adverse side effects on an acute basis. Larger doses (about 10 g/day) have been known to cause liver injury (parenchymal cell injury, portal fibrosis, cholestasis) (179). Chronic intake of nicotinamide can also induce a methyl-group deficiency state due to the methylation reactions involved in excretion (180,181), and physicians recommending nicotinamide therapy should ensure that the subjects have an adequate intake of methyl donors such as choline and methionine. Methyl donor deficiency also appears to increase the risk of carcinogenesis, and more work from this perspective is needed to define the safe use of nicotinamide. The most serious potential side effect of nicotinamide use would be the induction of pancreatic tumors in patients at risk for IDDM, and it will take many years to assess this. There is comfort from the fact that studies of the NOD mouse have not reported pancreatic tumors during long-term nicotinamide treatment.

VIII. SUMMARY

Niacin deficiency has the potential to alter redox reactions, poly and mono(ADP-ribose) synthesis, and the formation of cyclic ADP-ribose and NAADP. During niacin deficiency, the metabolic changes that lead to the dramatic signs and symptoms of pellagra will likely be tissue-specific and reflect subcellular competition for NAD pools. The effect of chronic niacin undernutrition on human health, especially the process of carcinogenesis, appears to be an exciting area that deserves more attention. With a rapidly broadening perspective on the biochemical roles for niacin in metabolism, identification of optimal niacin nutriture should be possible in the coming decade.

Supplementation of nicotinic acid and nicotinamide above the dietary requirement may affect some of the same processes, but these compounds have distinctive pharmacological properties, some of which may be unrelated to their currently defined nutrient functions. Future research in models of niacin deficiency and supplementation may lead us to reevaluate the accepted metabolic roles of niacin and create new guidelines for niacin intake.

ACKNOWLEDGMENTS

The authors thank Dr. William Bettger, Dr. William Woodward, Dr. Elaine Jacobson, and Dr. Myron Jacobson for critical reading of this manuscript. Dr. Doug Lanska helped us to find historical references and early photographs. The Cancer Research Society (Montreal, Canada) has supported research on niacin status and ADP-ribose metabolism, which made the writing of this chapter possible.

REFERENCES

1. S Harris. Clinical Pellagra. St. Louis: C.V. Mosby, 1941.
2. KJ Carpenter. Pellagra. Stroudsburg: Hutchinson Ross, 1981.

3.   SR Roberts. Pellagra, History, Distribution, Diagnosis, Prognosis, Treatment, Etiology. St. Louis: C.V. Mosby, 1914.

4.   AJ Bollet. Politics and pellagra: the epidemic of pellagra in the U.S. in the early twentieth century. Yale J Biol Med 65:211–221, 1992.

5.   P Malfait, A Moren, JC Dillon, A Brodel, G Begkoyian, MG Etchegorry, G Malenga, P Hakewill. An outbreak of pellagra related to changes in dietary niacin among Mozambican refugees in Malawi. Int J Epidemiol 22:504–511, 1993.

6.   EL Jacobson. Niacin deficiency and cancer in women. J Am Coll Nutr 12:412–416, 1993.

7.   J Goldberger. The relation of diet to pellagra. JAMA 78:1676–1680, 1922.

8.   CA Elvehjem, RJ Madden, FM Strong, DM Woolley. Relation of nicotinic acid and nicotinic acid amide to canine black tongue. J Am Chem Soc 59:1767–1768, 1937.

9.   WA Krehl, LJ Teply, CA Elvehjem. Corn as an etiological factor in the production of a nicotinic acid deficiency in the rat. Science 101:283, 1945.

10. WH Sebrell. History of pellagra. Fed Proc 40:1520–1522, 1981.

11. C Heidelberger, EP Abraham, S Lepkovsky. Concerning the mechanism of the mammalian conversion of tryptophan into nicotinic acid. J Biol Chem 176:1461–1462, 1948.

12. J Laguna, KJ Carpenter. Raw versus processed corn in niacin-deficient diets. J Nutr 45:21–28, 1951.

13. Buniva. Observations in pellagra: it would not appear to be contagious. In: KJ Carpenter, ed. Pellagra. Stroudsburg: Hutchinson Ross, 1981, pp 11–12.

14. TD Spies, WB Bean, WF Ashe. Recent advances in the treatment of pellagra and associated deficiencies. In: KJ Carpenter, ed. Pellagra. Stroudsburg: Hutchinson Ross, 1981, pp 213–225.

15. A Hoffer. Pellagra and schizophrenia. Psychosomatics 11:522–525, 1970.

16. S Harris. Clinical Pellagra. St. Louis: C.V. Mosby, 1941, p 366.

17. S Harris. Clinical Pellagra. St. Louis: C.V. Mosby, 1941, p 192.

18. S Harris. Clinical Pellagra. St. Louis: C.V. Mosby, 1941, p 295.

19. M Freed. Methods of Vitamin Assay. New York: The Association of Vitamin Chemists Inc., 1966, p 169.

20. W Friedrich. Vitamins. New York: Walter de Gruyter, 1988, pp 473–542.

21. TR Guilarte, K Pravlik. Radiometric-microbiologic assay of niacin using *Kloeckera brevis*: analysis of human blood and food. J Nutr 113:2587–2594, 1983.

22. HW Huff, WA Perlzweig. The fluorescent condensation product of N^1-methylnicotinamide and acetone: II. A sensitive method for the determination of N^1-methylnicotinamide. J Biol Chem 167:157–167, 1947.

23. EL Jacobson, AJ Dame, JS Pyrek, MK Jacobson. Evaluating the role of niacin in human carcinogenesis. Biochimie 77:394–398, 1995.

24. J Stein, A Hahn, G Rehner. High-performance liquid chromatographic determination of nicotinic acid and nicotinamide in biological samples applying post-column derivatization resulting in bathmochrome absorption shifts. J Chromatogr B Biomed Appl 665:71–78, 1995.

25. R Roskoski. Determination of pyridine nucleotides by fluorescence and other optical techniques. In: D Dolphin, R Poulson, O Avramovic, eds. Pyridine Nucleotide Coenzymes: Chemical, Biochemical and Medical Aspects. New York: John Wiley and Sons, 1987, pp 173–188.

26. DP Jones. Determination of pyridine dinucleotides in cell extracts by high-performance liquid chromatography. J Chromatogr 225:446–449, 1981.

27. JL Spivak, DL Jackson. Pellagra: an analysis of 18 patients and a review of the literature. Johns Hopkins Med J 140:295–309, 1977.

28. SE Sauberlich. Nutritional aspects of pyridine nucleotides. In: D Dolphin, R Poulson, O Avramovic, eds. Pyridine Nucleotide Coenzymes: Chemical, Biochemical and Medical Aspects. New York: John Wiley and Sons, 1987, pp 599–626.

29. CS Fu, ME Swendseid, RA Jacob, RW McKee. Biochemical markers for assessment of niacin status in young men: levels of erythrocyte niacin coenzymes and plasma tryptophan. J Nutr 119:1949–1955, 1989.

30. JM Rawling, TM Jackson, ER Driscoll, JB Kirkland. Dietary niacin deficiency lowers tissue poly(ADP-ribose) and NAD^+ concentrations in Fischer-344 rats. J Nutr 124:1597–1603, 1994.

31. JW Foster, AG Moat. Nicotinamide adenine dinucleotide biosynthesis and pyridine nucleotide cycle metabolism in microbial systems. Microbiol Rev 44:83–105, 1980.

32. J Preiss, P Handler. Biosynthesis of diphosphopyridine nucleotide I. Identification of Intermediates. J Biol Chem 233:488–500, 1958.

33. LS Dietrich, L Fuller, IL Yero, L Martinez. Nicotinamide mononucleotide pyrophosphorylase activity in animal tissues. J Biol Chem 241:188–191, 1966.

34. H Grunicke, HJ Keller, M Liersch, A Benaguid. New aspects of the mechanism and regulation of pyridine nucleotide metabolism. Adv Enzyme Regul 12:397–418, 1974.

35. U Delabar, M Siess. Synthesis and degradation of NAD in guinea pig cardiac muscle: II. Studies about the different biosynthetic pathways and the corresponding intermediates. Basic Res Cardiol 74:571–593, 1979.

36. BM Olivera, AM Ferro. Pyridine nucleotide metabolism and ADP-ribosylation. In: Hayashi O, Ueda K., eds., ADP-ribosylation Reactions: Biology and Medicine. New York: Academic Press, 1982, pp 19–40.

37. S Sestini, C Ricci, V Micheli, G Pompucci. Nicotinamide mononucleotide adenylyltransferase activity in human erythrocytes. Arch Biochem Biophys 302:206–211, 1993.

38. AG Moat, JW Foster. Biosynthesis and salvage pathways of pyridine nucleotides. In: D Dolphin, M Powanda, R Poulson, eds. Pyridine Nucleotide Coenzymes: Chemical, Biochemical and Medical Aspects. New York: John Wiley and Sons, 1982, Part B, pp 1–24.

39. LM Henderson. Niacin. Annu Rev Nutr 3:289–307, 1983.

40. M Ikeda, H Tsuji, S Nakamura, A Ichiyama, Y Nishizuka, O Hayaishi. Studies on the biosynthesis of nicotinamide adenine dinucleotide. II. A role of picolinic carboxylase in the biosynthesis of nicotinamide adenine dinucleotide from tryptophan in mammals. J Biol Chem 240: 1395–1401, 1965.

41. MK Horwitt, AE Harper, LM Henderson. Niacin-tryptophan relationships for evaluating niacin equivalents. Am J Clin Nutr 34:423–427, 1981.

42. AC Da Silva, R Fried, RC De Angelis. Domestic cat as laboratory animal for experimental nutrition studies; niacin requirements and tryptophan metabolism. J Nutr 46:399–409, 1952.

43. LV Hankes, LM Henderson, WL Brickson, CA Elvehjem. Effect of amino acids on the growth of rats on niacin-tryptophan deficient rations. J Biol Chem 174:873–881, 1948.

44. ME Reid. Nutritional studies with the guinea pig. VII. Niacin. J Nutr 75:279–286, 1961.

45. H Takanaga, H Maeda, H Yabuuchi, I Tamai, H Higashida, A Tsuji. Nicotinic acid transport mediated by pH-dependent anion antiporter and proton cotransporter in rabbit intestinal brush-border membrane. J Pharm Pharmacol 48:1073–1077, 1996.

46. V Micheli, HA Simmonds, S Sestini, C Ricci. Importance of nicotinamide as an NAD precursor in the human erythrocyte. Arch Biochem Biophys 283:40–45, 1990.

47. TM Jackson, JM Rawling, BD Roebuck, JB Kirkland. Large supplements of nicotinic acid and nicotinamide increase tissue NAD^+ and poly(ADP-ribose) levels but do not affect diethylnitrosamine- induced altered hepatic foci in Fischer-344 rats. J Nutr 125:1455–1461, 1995.

48. C Bernofsky. Physiology aspects of pyridine nucleotide regulation in mammals. Mol Cell Biochem 33:135–143, 1980.

49. NO Kaplan. History of the pyridine nucleotides. In: D Dolphin, R Poulson, O Avramovic, eds. Pyridine Nucleotide Coenzymes: Chemical, Biochemical and Medical Aspects. New York: John Wiley and Sons, 1987, Part A, pp 1–20.

50. DL Clapper, TF Walseth, PJ Dargie, HC Lee. Pyridine nucleotide metabolites stimulate cal-

cium release from sea urchin egg microsomes desensitized to inositol trisphosphate. J Biol Chem 262:9561–9568, 1987.

51. H Kim, EL Jacobson, MK Jacobson. Position of cyclization in cyclic ADP-ribose. Biochem Biophys Res Commun 194:1143–1147, 1993.
52. HC Lee. Cyclic ADP-ribose: a calcium mobilizing metabolite of NAD$^+$. Mol Cell Biochem 138:229–235, 1994.
53. N Noguchi, S Takasawa, K Nata, A Tohgo, I Kato, F Ikehata, H Yonekura, H Okamoto. Cyclic ADP-ribose binds to FK506-binding protein 12.6 to release Ca^{2+} from islet microsomes. J Biol Chem 272:3133–3136, 1997.
54. H Kim, EL Jacobson, MK Jacobson. Synthesis and degradation of cyclic ADP-ribose by NAD glycohydrolases. Science 261:1330–1333, 1993.
55. E Zocchi, L Franco, L Guida, U Benatti, A Bargellesi, F Malavasi, HC Lee, A De Flora. A single protein immunologically identified as CD38 displays NAD$^+$ glycohydrolase, ADP-ribosyl cyclase and cyclic ADP-ribose hydrolase activities at the outer surface of human erythrocytes. Biochem Biophys Res Commun 196:1459–1465, 1993.
56. A Galione. Cyclic ADP-ribose, the ADP-ribosyl cyclase pathway and calcium signalling. Mol Cell Endocrinol 98:125–131, 1994.
57. S Takasawa, K Nata, H Yonekura, H Okamoto. Cyclic ADP-ribose in insulin secretion from pancreatic beta cells [see comments]. Science 259:370–373, 1993.
58. HC Lee. NAADP: An emerging calcium signaling molecule. J Membr Biol 173:1–8, 2000.
59. HC Lee, R Graeff, TF Walseth. Cyclic ADP-ribose and its metabolic enzymes. Biochimie 77:345–355, 1995.
60. M Mizuguchi, N Otsuka, M Sato, Y Ishii, S Kon, M Yamada, H Nishina, T Katada, K Ikeda. Neuronal localization of CD38 antigen in the human brain. Brain Res 697:235–240, 1995.
61. SH Snyder, DM Sabatini, MM Lai, JP Steiner, GS Hamilton, PD Suzdak. Neural actions of immunophilin ligands. Trends Pharmacol Sci 19:21–26, 1998.
62. J Yamauchi, S Tanuma. Occurrence of an NAD$^+$ glycohydrolase in bovine brain cytosol. Arch Biochem Biophys 308:327–329, 1994.
63. Y Desmarais, L Menard, J Lagueux, GG Poirier. Enzymological properties of poly(ADP-ribose)polymerase: characterization of automodification sites and NADase activity. Biochem Biophys Acta 1078:179–186, 1991.
64. R Antoine, C Locht. The NAD-glycohydrolase activity of the pertussis toxin S1 subunit. Involvement of the catalytic HIS-35 residue. J Biol Chem 269:6450–6457, 1994.
65. RE West, Jr., J Moss. Amino acid specific ADP-ribosylation: specific NAD: arginine mono-ADP-ribosyltransferases associated with turkey erythrocyte nuclei and plasma membranes. Biochemistry 25:8057–8062, 1986.
66. RV Considine, LL Simpson Cellular and molecular actions of binary toxins possessing ADP-ribosyltransferase activity. Toxicon 29:913–936, 1991.
67. A Zolkiewska, IJ Okazaki, J Moss. Vertebrate mono-ADP-ribosyltransferases. Mol Cell Biochem 138:107–112, 1994.
68. T Takada, IJ Okazaki, J Moss. ADP-ribosylarginine hydrolases. Mol Cell Biochem 138:119–122, 1994.
69. M Tsuchiya, M Shimoyama. Target protein for eucaryotic arginine-specific ADP-ribosyltransferase. Mol Cell Biochem 138:113–118, 1994.
70. J Wang, E Nemoto, AY Kots, HR Kaslow, G Dennert. Regulation of cytotoxic T cells by ecto-nicotinamide adenine dinucleotide (NAD) correlates with cell surface GPI-anchored/arginine ADP-ribosyltransferase. J Immunol 153:4048–4058, 1994.
71. WJ Iglewski. Cellular ADP-ribosylation of elongation factor 2. Mol Cell Biochem 138:131–133, 1994.
72. T Maehama, N Sekine, H Nishina, K Takahashi, T Katada. Characterization of botulinum C3-catalyzed ADP-ribosylation of rho proteins and identification of mammalian C3-like ADP-ribosyltransferase. Mol Cell Biochem 138:135–140, 1994.

73. BE Ledford, GH Leno. ADP-ribosylation of the molecular chaperone GRP78/BiP. Mol Cell Biochem 138:141–148, 1994.

74. S Chatterjee, MF Cheng, SJ Berger, NA Berger. Induction of M(r) 78,000 glucose-regulated stress protein in poly(adenosine diphosphate-ribose) polyme. Cancer Res 54:440–4411, 1994.

75. LJ McDonald, J Moss. Enzymatic and nonenzymatic ADP-ribosylation of cysteine. Mol Cell Biochem 138:221–226, 1994.

76. J Zhang, SH Snyder. Nitric oxide stimulates auto-ADP-ribosylation of glyceraldehyde-3-phosphate dehydrogenase. Proc Natl Acad Sci USA 89:9382–9385, 1992.

77. LJ McDonald, J Moss. Nitric oxide and NAD-dependent protein modification. Mol Cell Biochem 138:201–206, 1994.

78. EL Jacobson, D Cervantes-Laurean, MK Jacobson. Glycation of proteins by ADP-ribose. Mol Cell Biochem 138:207–212, 1994.

79. P Chambon, JD Weill, J Doly, MT Strosser, P Mandel. On the formation of a novel adenylic compound by enzymatic extracts of liver nuclei. Biochem Biophys Res Commun 25:638–643, 1966.

80. I Kameshita, Z Matsuda, T Taniguchi, Y Shizuta. Poly(ADP-ribose) synthetase. Separation and identification of three proteolytic fragments as the substrate-binding domain, the DNA-binding domain, and the automodification domain. J Biol Chem 259:4770–4776, 1984.

81. M Ikejima, S Noguchi, R Yamashita, T Ogura, T Sugimura, DM Gill, M Miwa. The zinc fingers of human poly(ADP-ribose) polymerase are differentially required for the recognition of DNA breaks and nicks and the consequent enzyme activation. Other structures recognize intact DNA. J Biol Chem 265:21907–21913, 1990.

82. D Lautier, J Lagueux, J Thibodeau, L Menard, GG Poirier. Molecular and biochemical features of poly(ADP-ribose) metabolism. Mol Cell Biochem 122:171–193, 1993.

83. CC Kiehlbauch, N Aboul-Ela, EL Jacobson, DP Ringer, MK Jacobson. High resolution fractionation and characterization of ADP-ribose polymers. Anal Biochem 208:26–34, 1993.

84. H Mendoza-Alvarez, R Alvarez-Gonzalez Poly(ADP-ribose) polymerase is a catalytic dimer and the automodification reaction is intermolecular. J Biol Chem 268:22575–22580, 1993.

85. FR Althaus, L Hofferer, HE Kleczkowska, M Malanga, H Naegeli, P Panzeter, C Realini. Histone shuttle driven by the automodification cycle of poly(ADP-ribose)polymerase. Environ Mol Mutagen 22:278–282, 1993.

86. S Desnoyers, GM Shah, G Brochu, JC Hoflack, A Verreault, GG Poirier. Biochemical properties and function of poly(ADP-ribose) glycohydrolase. Biochimie 77:433–438, 1995.

87. A Miro, MJ Costas, M Garcia-Diaz, MT Hernandez, JC Cameselle. A specific, low Km ADP-ribose pyrophosphatase from rat liver. FEBS Lett 244:123–126, 1989.

88. K Hatakeyama, Y Nemoto, K Ueda, O Hyashi. Poly(ADP-ribose) glycohydrolase and ADP-ribosyl group turnover. In: ADP-Ribose Transfer Reactions. Berlin: Springer-Verlag, 1989, p 47.

89. D D'Amours, S Desnoyers, I D'Silva, GG Poirier. Poly(ADP-ribosyl)ation reactions in the regulation of nuclear functions. Biochem J 342 (Pt 2):249–268, 1999.

90. BW Durkacz, O Omidiji, DA Gray, S Shall. (ADP-ribose)n participates in DNA excision repair. Nature 283:593–596, 1980.

91. MF Denisenko, VA Soldatenkov, LN Belovskaya, IV Filippovich. Is the NAD-poly (ADP-ribose) polymerase system the trigger in radiation-induced death of mouse thymocytes? Int J Radiat Biol 56:277–285, 1989.

92. Y Ohashi, K Ueda, M Kawaichi, O Hayaishi. Activation of DNA ligase by poly(ADP-ribose) in chromatin. Proc Natl Acad Sci USA 80:3604–3607, 1983.

93. CM Simbulan, M Suzuki, S Izuta, T Sakurai, E Savoysky, K Kojima, K Miyahara, Y Shizuta, S Yoshida. Poly(ADP-ribose) polymerase stimulates DNA polymerase alpha by physical association. J Biol Chem 268:93–99, 1993.

94. MS Satoh, T Lindahl. Role of poly(ADP-ribose) formation in DNA repair. Nature 356:356–358, 1992.

95. TJ Jorgensen, JC Leonard, PJ Thraves, A Dritschilo. Baseline sister chromatid exchange in human cell lines with different levels of poly(ADP-ribose) polymerase. Radiat Res 127:107–110, 1991.

96. T Stevnsner, R Ding, M Smulson, VA Bohr. Inhibition of gene-specific repair of alkylation damage in cells depleted of poly(ADP-ribose) polymerase. Nucleic Acids Res 22:4620–4624, 1994.

97. S Chatterjee, NA Berger. Growth-phase-dependent response to DNA damage in poly(ADP-ribose) polymerase deficient cell lines: basis for a new hypothesis describing the role of poly(ADP-ribose) polymerase in DNA replication and repair. Mol Cell Biochem 138:61–69, 1994.

98. Y Le Rhun, JB Kirkland, GM Shah. Cellular responses to DNA damage in the absence of poly(ADP-ribose) polymerase. Biochem Biophys Res Commun 245:1–10, 1998.

99. S Smith, I Giriat, A Schmitt, T de Lange. Tankyrase, a poly(ADP-ribose) polymerase at human telomeres [see comments]. Science 282:1484–1487, 1998.

100. M Johansson. A human poly(ADP-ribose) polymerase gene family (ADPRTL): cDNA cloning of two novel poly(ADP-ribose) polymerase homologues. Genomics 57:442–445, 1999.

101. H Berghammer, M Ebner, R Marksteiner, B Auer. pADPRT-2: a novel mammalian polymerizing(ADP-ribosyl)transferase gene related to truncated pADPRT homologues in plants and Caenorhabditis elegans. FEBS Lett 449:259–263, 1999.

102. JC Ame, V Rolli, V Schreiber, C Niedergang, F Apiou, P Decker, S Muller, T Hoger, J Menissier-de Murcia, G de Murcia. PARP-2, A novel mammalian DNA damage-dependent poly(ADP-ribose) polymerase. J Biol Chem 274:17860–17868, 1999.

103. VA Kickhoefer, AC Siva, NL Kedersha, EM Inman, C Ruland, M Streuli, LH Rome. The 193-kD vault protein, VPARP, is a novel poly(ADP-ribose) polymerase. J Cell Biol 146:917–928, 1999.

104. CM Whitacre, H Hashimoto, ML Tsai, S Chatterjee, SJ Berger, NA Berger. Involvement of NAD-poly(ADP-ribose) metabolism in p53 regulation and its consequences. Cancer Res 55:3697–3701, 1995.

105. JM Rawling, R Alvarez-Gonzalez. TFIIF, a basal eukaryotic transcription factor, is a substrate for poly(ADP-ribosyl)ation. Biochem J 324 (Pt 1):249–253, 1997.

106. M Bhatia, JB Kirkland, KA Meckling-Gill. Modulation of poly(ADP-ribose) polymerase during neutrophilic and monocytic differentiation of promyelocytic (NB4) and myelocytic (HL-60) leukaemia cells. Biochem J 308 (Pt 1):131–137, 1995.

107. M Bhatia, JB Kirkland, KA Meckling-Gill. Overexpression of poly(ADP-ribose) polymerase promotes cell cycle arrest and inhibits neutrophilic differentiation of NB4 acute promyelocytic leukemia cells. Cell Growth Differ 7:91–100, 1996.

108. S Takahashi, D Nakae, Y Yokose, Y Emi, A Denda, S Mikami, T Ohnishi, Y Konishi. Enhancement of DEN initiation of liver carcinogenesis by inhibitors of NAD^+ADP ribosyl transferase in rats. Carcinogenesis 5:901–906, 1984.

109. NA Berger. Poly(ADP-ribose) in the cellular response to DNA damage. Radiat Res 101:4–15, 1985.

110. H Okamoto. The role of poly(ADP-ribose) synthetase in the development of insulin-dependent diabetes and islet B-cell regeneration. Biomed Biochim Acta 44:15–20, 1985.

111. T Yamagami, A Miwa, S Takasawa, H Yamamoto, H Okamoto. Induction of rat pancreatic B-cell tumors by the combined administration of streptozotocin or alloxan and poly(adenosine diphosphate ribose) synthetase inhibitors. Cancer Res 45:1845–1849, 1985.

112. PW Rankin, EL Jacobson, RC Benjamin, J Moss, MK Jacobson. Quantitative studies of inhibitors of ADP-ribosylation in vitro and in vivo. J Biol Chem 264:4312–4317, 1989.

113. JM Rawling, ER Driscoll, GG Poirier, JB Kirkland. Diethylnitrosamine administration in

vivo increases hepatic poly(ADP-ribose) levels in rats: results of a modified technique for poly(ADP-ribose) measurement. Carcinogenesis 14:2513–2516, 1993.

114. JR Wright, Jr., J Mendola, PE Lacy. Effect of niacin/nicotinamide deficiency on the diabetogenic effect of streptozotocin. Experientia 44:38–40, 1988.

115. GP Warwick, JS Harington. Some aspects of the epidemiology and etiology of esophageal cancer with particular emphasis on the Transkei, South Africa. Adv Cancer Res 17:121–131, 1973.

116. SJ Van Rensburg, ES Bradshaw, D Bradshaw, EF Rose. Oesophageal cancer in Zulu men, South Africa: a case-control study. Br J Cancer 51:399–405, 1985.

117. J Wahrendorf, J Chang-Claude, QS Liang, YG Rei, N Munoz, M Crespi, R Raedsch, D Thurnham, P Correa. Precursor lesions of oesophageal cancer in young people in a high-risk population in China [see comments]. Lancet 2:1239–1241, 1989.

118. S Franceschi, E Bidoli, AE Baron, C La Vecchia. Maize and risk of cancers of the oral cavity, pharynx, and esophagus in northeastern Italy [see comments]. J Natl Cancer Inst 82:1407–1411, 1990.

119. SJ Van Rensburg. Epidemiologic and dietary evidence for a specific nutritional predisposition to esophageal cancer. J Natl Cancer Inst 67:243–251, 1981.

120. JR Marshall, S Graham, BP Haughey, D Shedd, R O'Shea, J Brasure, GS Wilkinson, D West. Smoking, alcohol, dentition and diet in the epidemiology of oral cancer. Eur J Cancer B Oral Oncol 28B:9–15, 1992.

121. WJ Blot, JY Li, PR Taylor, W Guo, S Dawsey, GQ Wang, CS Yang, SF Zheng, M Gail, GY Li. Nutrition intervention trials in Linxian, China: supplementation with specific vitamin/mineral combinations, cancer incidence, and disease-specific mortality in the general population [see comments]. J Natl Cancer Inst 85:1483–1492, 1993.

122. GQ Wang, SM Dawsey, JY Li, PR Taylor, B Li, WJ Blot, WM Weinstein, FS Liu, KJ Lewin, H Wang. Effects of vitamin/mineral supplementation on the prevalence of histological dysplasia and early cancer of the esophagus and stomach: results from the General Population Trial in Linxian, China. Cancer Epidemiol Biomarkers Prev 3:161–166, 1994.

123. EG Miller, H Burns, Jr. N-nitrosodimethylamine carcinogenesis in nicotinamide-deficient rats. Cancer Res 44:1478–1482, 1984.

124. SJ Van Rensburg, JM Hall, PS Gathercole. Inhibition of esophageal carcinogenesis in corn-fed rats by riboflavin, nicotinic acid, selenium, molybdenum, zinc, and magnesium. Nutr Cancer 8:163–170, 1986.

125. JZ Zhang, SM Henning, ME Swendseid. Poly(ADP-ribose) polymerase activity and DNA strand breaks are affected in tissues of niacin-deficient rats. J Nutr 123:1349–1355, 1993.

126. JM Rawling, TM Jackson, BD Roebuck, GG Poirier, JB Kirkland. The effect of niacin deficiency on diethylnitrosamine-induced hepatic poly(ADP-ribose) levels and altered hepatic foci in the Fischer-344 rat. Nutr Cancer 24:111–119, 1995.

127. K Ueda, M Kawaichi, O Hayshi. Poly(ADP-ribose) Synthetase. In: O Hayshi, K Ueda, eds. ADP-ribosylation Reactions, Biology and Medicine. New York: Academic Press, 1982, pp 118–155.

128. AC Boyonoski, LM Gallacher, MM ApSimon, RM Jacobs, GM Shah, GG Poirier, JB Kirkland. Niacin deficiency in rats increases the severity of ethylnitrosourea-induced anemia and leukopenia. J Nutr 130:1102–1107, 2000.

129. AC Boyonoski, LM Gallacher, MM ApSimon, RM Jacobs, GM Shah, GG Poirier, JB Kirkland. Niacin deficiency increases the sensitivity of rats to the short and long term effects of ethylnitrosourea treatment. Mol Cell Biochem 193:83–87, 1999.

130. RI Inculet, JA Norton, GE Nichoalds, MM Maher, DE White, MF Brennan. Water-soluble vitamins in cancer patients on parenteral nutrition: a prospective study. JPEN J Parenter Enteral Nutr 11:243–249, 1987.

131. GT Bryan. The influence of niacin and nicotinamide on in vivo carcinogenesis. In: LA Poirier,

PM Newberne, MW Pariza, eds. Advances in Experimental Medicine and Biology. Vol. 206—Essential Nutrients in Carcinogenesis. New York: Plenum Press, 1986, pp 331–338.

132. PL Canner, KG Berge, NK Wenger, J Stamler, L Friedman, RJ Prineas, W Friedewald. Fifteen year mortality in Coronary Drug Project patients: long-term benefit with niacin. J Am Coll Cardiol 8:1245–1255, 1986.

133. SN Giri, R Blaisdell, RB Rucker, Q Wang, DM Hyde. Amelioration of bleomycin-induced lung fibrosis in hamsters by dietary supplementation with taurine and niacin: biochemical mechanisms. Environ Health Perspect 102 (suppl 10):137–147, 1994.

134. A Nagai, H Matsumiya, M Hayashi, S Yasui, H Okamoto, K Konno. Effects of nicotinamide and niacin on bleomycin-induced acute injury and subsequent fibrosis in hamster lungs. Exp Lung Res 20:263–281, 1994.

135. JM Rawling, MM ApSimon, JB Kirkland. Lung poly(ADP-ribose) and NAD^+ concentrations during hyperoxia and niacin deficiency in the Fischer-344 rat. Free Radic Biol Med 20:865–871, 1996.

136. SG Jenkinson, RA Lawrence, DL Butler. Inability of niacin to protect from in vivo hyperoxia or in vitro microsomal lipid peroxidation. J Toxicol Clin Toxicol 19:975–985, 1982.

137. JD Crapo. Morphologic changes in pulmonary oxygen toxicity. Annu Rev Physiol 48:721–731, 1986.

138. S Bergelson, R Pinkus, V Daniel. Intracellular glutathione levels regulate Fos/Jun induction and activation of glutathione S-transferase gene expression. Cancer Res 54:36–40, 1994.

139. JK Wilkin, TB Bentley, DL Latour, EW Rosenberg. Nicotinic acid treatment of skin disorders. Br J Dermatol 100:471–472, 1979.

140. HA Poston, MJ Wolfe. Niacin requirement for optimum growth, feed conversion and protection of rainbow trout, Salmo gairdneri, from ultraviolet-B radiation. J Fish Dis 8:451–460, 1985.

141. JD Morrow, JA Awad, JA Oates, LJ Roberts. Identification of skin as a major site of prostaglandin D2 release following oral administration of niacin in humans. J Invest Dermatol 98:812–815, 1992.

142. JK Wilkin, G Fortner, LA Reinhardt, OV Flowers, SJ Kilpatrick, WC Streeter. Prostaglandins and nicotinate-provoked increase in cutaneous blood flow. Clin Pharmacol Ther 38:273–277, 1985.

143. HL Gensler, T Williams, AC Huang, EL Jacobson. Oral niacin prevents photocarcinogenesis and photoimmunosuppression in mice. Nutr Cancer 34:36–41, 1999.

144. JJ Yourick, JS Dawson, CD Benton, ME Craig, LW Mitcheltree. Pathogenesis of 2,2'-dichlorodiethyl sulfide in hairless guinea pigs. Toxicology 84:185–197, 1993.

145. EL Jacobson, V Nunbhakdi-Craig, DG Smith, HY Chen, BL Wasson, MK Jacobson. ADP-ribose polymer metabolism: implications for human nutrition. In: GG Poirier, P Moreau, eds. ADP-Ribosylation Reactions. New York: Springer-Verlag, 1992, pp 153–162.

146. TE Barnum ed. Enzyme Handbook. New York: Springer-Verlag, 1969.

147. RM Graeff, TF Walseth, K Fryxell, WD Branton, HC Lee. Enzymatic synthesis and characterizations of cyclic GDP-ribose. A procedure for distinguishing enzymes with ADP-ribosyl cyclase activity. J Biol Chem 269:30260–30267, 1994.

148. KW Beers, EN Chini, HC Lee, TP Dousa. Metabolism of cyclic ADP-ribose in opossum kidney renal epithelial cells. Am J Physiol 268:C741–C746, 1995.

149. H Grunicke, HJ Keller, B Puschendorf, A Benaguid. Biosynthesis of nicotinamide adenine dinucleotide in mitochondria. Eur J Biochem 53:41–45, 1975.

150. JL Purvis, JM Lowenstein. The relationship between intra- and extramitochondrial pyridine nucleotides. J Biol Chem 236:2794–2803, 1961.

151. A Behr, H Taguchi, RK Gholson. Apparent pyridine nucleotide synthesis in mitochondria: an artifact of NMN and NAD glycohydrolase activity? Biochem Biophys Res Commun 101:767–774, 1981.

152. ME Tischler, D Friedrichs, K Coll, JR Williamson. Pyridine nucleotide distributions and

enzyme mass action ratios in hepatocytes from fed and starved rats. Arch Biochem Biophys 184:222–236, 1977.

153. M Weiner, J van Eys. Nicotinic Acid: Nutrient-Cofactor-Drug. New York: Marcel Dekker, 1983.

154. DA Bender, BI Magboul, D Wynick. Probable mechanisms of regulation of the utilization of dietary tryptophan, nicotinamide and nicotinic acid as precursors of nicotinamide nucleotides in the rat. Br J Nutr 48:119–127, 1982.

155. A Petley, B Macklin, AG Renwick, TJ Wilkin. The pharmacokinetics of nicotinamide in humans and rodents. Diabetes 44:152–155, 1995.

156. JR DiPalma, WS Thayer. Use of niacin as a drug. Annu Rev Nutr 11:169–187, 1991.

157. C Marcus, T Sonnenfeld, B Karpe, P Bolme, P Arner. Inhibition of lipolysis by agents acting via adenylate cyclase in fat cells from infants and adults. Pediatr Res 26:255–259, 1989.

158. SM Grundy, HY Mok, L Zech, M Berman. Influence of nicotinic acid on metabolism of cholesterol and triglycerides in man. J Lipid Res 22:24–36, 1981.

159. JD Alderman, RC Pasternak, FM Sacks, HS Smith, ES Monrad, W Grossman. Effect of a modified, well-tolerated niacin regimen on serum total cholesterol, high density lipoprotein cholesterol and the cholesterol to high density lipoprotein ratio. Am J Cardiol 64:725–729, 1989.

160. JI Rader, RJ Calvert, JN Hathcock. Hepatic toxicity of unmodified and time-release preparations of niacin. Am J Med 92:77–81, 1992.

161. A Hoffer. Megavitamin B-3 therapy for schizophrenia. Can Psychiatr Assoc J 16:499–504, 1971.

162. MR Horsman. Nicotinamide and other benzamide analogs as agents for overcoming hypoxic cell radiation resistance in tumours. A review. Acta Oncol 34:571–587, 1995.

163. DJ Chaplin, MR Horsman, MJ Trotter. Effect of nicotinamide on the microregional heterogeneity of oxygen delivery within a murine tumor. J Natl Cancer Inst 82:672–676, 1990.

164. MT Behme. Nicotinamide and diabetes prevention. Nutr Rev 53:137–139, 1995.

165. JA Norton, JL Doppman, RT Jensen. Cancer of the endocrine system. In: VT DeVita, S Hellman, SA Rosenberg, eds. Cancer: Principles and Practice of Oncology. Philadelphia: J.B. Lippincott Co., 1989, pp 1324–1331.

166. H Kolb, V Burkart. Nicotinamide in type 1 diabetes. Mechanism of action revisited. Diabetes Care 22 (suppl 2):B16–B20, 1999.

167. S Reddy, NJ Bibby, RB Elliott. Early nicotinamide treatment in the NOD mouse: effects on diabetes and insulitis suppression and autoantibody levels. Diabetes Res 15:95–102, 1990.

168. H Nakajima, K Yamada, T Hanafusa, H Fujino-Kurihara, J Miyagawa, A Miyazaki, R Saitoh, Y Minami, N Kono, K Nonaka. Elevated antibody-dependent cell-mediated cytotoxicity and its inhibition by nicotinamide in the diabetic NOD mouse. Immunol Lett 12:91–94, 1986.

169. K Yamada, T Hanafusa, H Fujino-Kurihara, A Miyazaki, H Nakajima, J Miyagawa, N Kono, K Nonaka, S Tarui. Nicotinamide prevents lymphocytic infiltration in submandibular glands but not the appearance of anti-salivary duct antibodies in non-obese diabetic (NOD) mice. Res Commun Chem Pathol Pharmacol 50:83–91, 1985.

170. DL Eizirik, S Sandler, N Welsh, K Bendtzen, C Hellerstrom. Nicotinamide decreases nitric oxide production and partially protects human pancreatic islets against the suppressive effects of combinations of cytokines. Autoimmunity 19:193–198, 1994.

171. W Sumoski, H Baquerizo, A Rabinovitch. Oxygen free radical scavengers protect rat islet cells from damage by cytokines. Diabetologia 32:792–796, 1989.

172. B Kallmann, V Burkart, KD Kroncke, V Kolb-Bachofen, H Kolb. Toxicity of chemically generated nitric oxide towards pancreatic islet cells can be prevented by nicotinamide. Life Sci 51:671–678, 1992.

173. Y Yonemura, T Takashima, K Miwa, I Miyazaki, H Yamamoto, H Okamoto. Amelioration of diabetes mellitus in partially depancreatized rats by poly(ADP-ribose) synthetase inhibitors. Evidence of islet B-cell regeneration. Diabetes 33:401–404, 1984.

174.   O Korsgren, A Andersson, S Sandler. Pretreatment of fetal porcine pancreas in culture with
       nicotinamide accelerates reversal of diabetes after transplantation to nude mice. Surgery 113:
       205–214, 1993.
175.   P Vague, R Picq, M Bernal, V Lassmann-Vague, B Vialettes. Effect of nicotinamide treat-
       ment on the residual insulin secretion in type 1 (insulin-dependent) diabetic patients. Diabeto-
       logia 32:316–321, 1989.
176.   P Pozzilli, N Visalli, R Buzzetti, MG Baroni, ML Boccuni, E Fioriti, A Signore, C Mesturino,
       L Valente, MG Cavallo. Adjuvant therapy in recent onset type 1 diabetes at diagnosis and
       insulin requirement after 2 years. Diabete Metab 21:47–49, 1995.
177.   RB Elliott, HP Chase. Prevention or delay of type 1 (insulin-dependent) diabetes mellitus
       in children using nicotinamide. Diabetologia 34:362–365, 1991.
178.   R Manna, A Migliore, LS Martin, E Ferrara, E Ponte, G Marietti, F Scuderi, G Cristiano,
       G Ghirlanda, G Gambassi. Nicotinamide treatment in subjects at high risk of developing
       IDDM improves insulin secretion. Br J Clin Pract 46:177–179, 1992.
179.   SL Winter, JL Boyer. Hepatic toxicity from large doses of vitamin B3 (nicotinamide). N
       Engl J Med 289:1180–1182, 1973.
180.   YA Kang-Lee, RW McKee, SM Wright, ME Swendseid, DJ Jenden, RS Jope. Metabolic
       effects of nicotinamide administration in rats. J Nutr 113:215–221, 1983.
181.   MM ApSimon, JM Rawling, JB Kirkland. Nicotinamide megadosing increases hepatic poly
       (ADP-ribose) levels in choline-deficient rats. J Nutr 125:1826–1832, 1995.

7

Riboflavin (Vitamin B$_2$)

RICHARD S. RIVLIN and JOHN THOMAS PINTO

Weill Medical College of Cornell University and Memorial Sloan-Kettering Cancer Center, New York, New York

I. INTRODUCTION AND HISTORY OF DISCOVERY

In the pioneering studies by McCollum and Kennedy (1) in the early part of the twentieth century, water-soluble tissue extracts were found to be effective in the prevention of the deficiency state of pellagra in experimental animals. As studies progressed, it became evident that there were at least two distinct fractions of these extracts, one of which was heat-labile and the other heat-stable.

Further studies of this heat-stable fraction showed that it was a complex containing a yellow growth factor. This factor had fluorescent properties and was later purified and named riboflavin (B$_2$) (2). Other components of this fraction were later identified as niacin, which was the true antipellagra compound, and vitamin B$_6$, which was particularly effective in preventing dermatitis in animals.

The physiological role of the yellow growth factor remained obscure until the landmark discovery by Warburg and Christian (3) in 1932 of "yellow enzyme" or "old yellow enzyme." This protein was found to be composed of an apoenzyme and a yellow cofactor serving as a coenzyme. This coenzyme was shown subsequently to contain an isoalloxazine ring (4) and a phosphate group (5).

Synthesis of riboflavin was accomplished by Kuhn et al. (6) and Karrer et al. (7). The flavin coenzyme riboflavin-5'-phosphate (flavin mononucleotide, FMN), was identified in 1937 by Theorell (8). In 1938, Warburg and Christian (9) clarified the structure of flavin adenine dinucleotide (FAD), formed from FMN. While many enzymes utilize FMN and FAD as cofactors, flavins bound covalently to specific tissue flavoproteins also have been found to have major biological significance, as reviewed by McCormick (10).

II. CHEMISTRY

Riboflavin is defined chemically as 7,8-dimethyl-10-(1′-D-ribityl)isoalloxazine. The planar isoalloxazine ring provides the basic structure not only for riboflavin (vitamin B_2) but for the naturally occurring phosphorylated coenzymes that are derived from riboflavin (Fig. 1). These coenzymes include FMN, FAD, and flavin coenzymes linked covalently to specific tissue proteins, generally at the 8-α methyl position of the isoalloxazine ring (10).

Among the key mammalian enzymes with covalently bound FAD are included sarcosine and succinic dehydrogenases, which are located in the matrix and inner mitochondrial membrane, respectively; monoamine oxidase in the outer mitochondrial membrane; and L-gulonolactone oxidase, found particularly in liver and kidney microsomes of those mammalian species capable of synthesizing ascorbic acid from its precursors (11).

The sequence of events in the synthesis of the flavin coenzymes from riboflavin and its control by thyroid hormones are shown in Fig. 2. Thyroid hormones regulate the activities of the flavin biosynthetic enzymes (12), the synthesis of the flavoprotein apoenzymes,

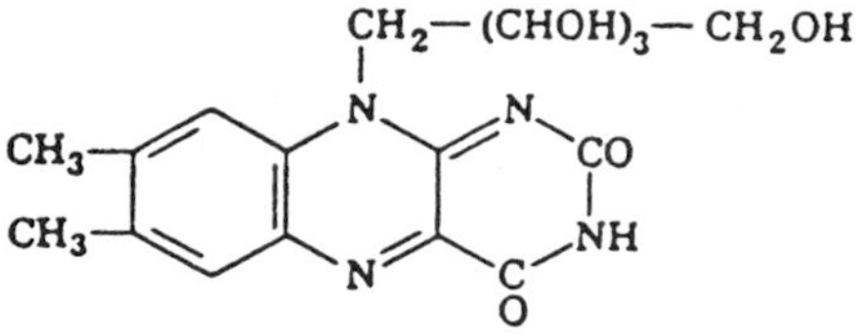

Riboflavin

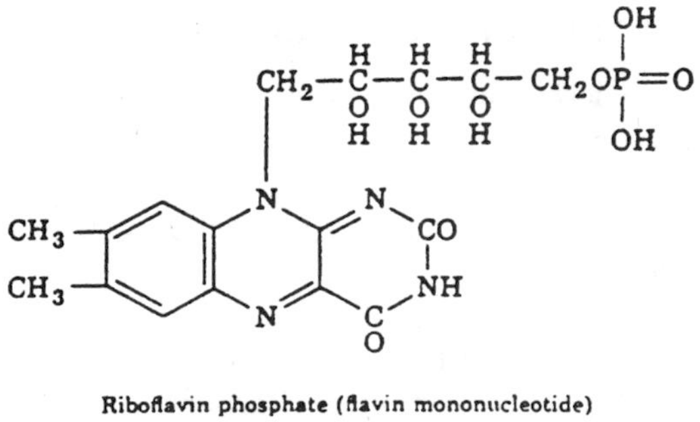

Riboflavin phosphate (flavin mononucleotide)

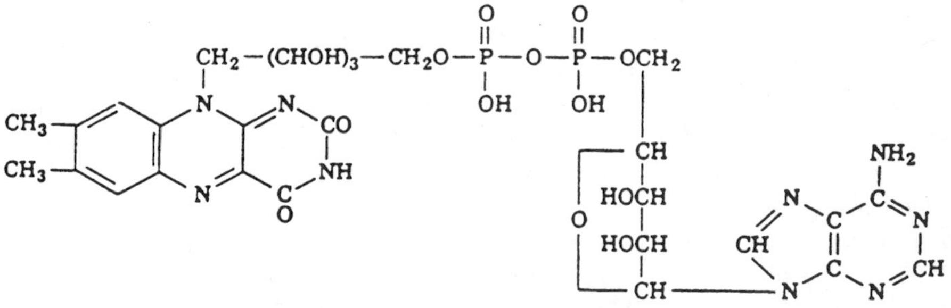

Flavin adenine dinucleotide (FAD)

Fig. 1 Structural formulas of riboflavin, flavin mononucleotide (riboflavin-5′-phosphate, FMN) and flavin adenine dinucleotide (FAD).

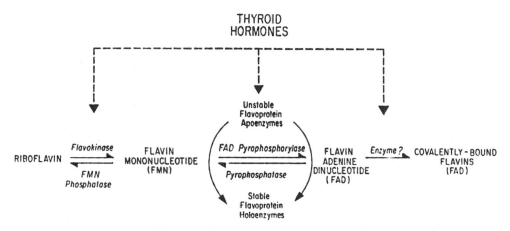

Fig. 2 Metabolic pathway of conversion of riboflavin into FMN, FAD, and covalently bound flavin, together with its control by thyroid hormones (12).

and the formation of covalently bound flavins (13). The first biosynthetic enzyme, flavo-kinase, catalyzes the initial phosphorylation from ATP of riboflavin to FMN. A fraction of FMN is directly utilized in this form as a coenzyme. The largest fraction of FMN, however, combines with a second molecule of ATP to form FAD, the predominant tissue flavin, in a reaction catalyzed by FAD synthetase, also called FAD pyrophosphorylase. The covalent attachment of flavins to specific tissue proteins occurs after FAD has been synthesized. A sequence of phosphatases returns FAD to FMN and FMN, in turn, to riboflavin (12). Most flavoproteins utilize FAD rather than FMN as coenzyme for a wide variety of metabolic reactions. Microsomal NADPH–cytochrome P450 reductase is highly unusual in containing FMN and FAD in equimolar ratios.

Riboflavin is yellow and has a high degree of natural fluorescence when excited by UV light, a property that can be utilized conveniently in its assay. There are a number of variations in structure in the naturally occurring flavins. Riboflavin and its coenzymes are sensitive to alkali and to acid, particularly in the presence of UV light. Under alkaline conditions, riboflavin is photodegraded to yield lumiflavin (7,8,10-trimethylisoalloxazine), which is inactive biologically. Riboflavin is photodegraded to lumichrome (7,8-dimethyl-alloxazine) under acidic conditions, a product that is also biologically inactive (10). Thus, an important physical property of riboflavin and its derivatives is their sensitivity to UV light, resulting in rapid inactivation. Therefore, phototherapy of neonatal jaundice and of certain skin disorders may promote systemic riboflavin deficiency. The structure–function relationships of the various biologically active flavins have been comprehensively reviewed (14).

III. RIBOFLAVIN DEFICIENCY AND FOOD-RELATED ISSUES

A. Riboflavin Deficiency

Clinical riboflavin deficiency is not recognizable at the bedside by any unique or character-istic physical features. The typical glossitis, angular stomatitis, and dermatitis observed is not specific for riboflavin deficiency and may be due to other vitamin deficiencies as well. In fact, when dietary deficiency of riboflavin occurs, it is almost invariably in association with multiple nutrient deficits (15).

With the onset of riboflavin deficiency, one of the adaptations that occurs is a fall in the hepatic free riboflavin pool to nearly undetectable levels, with a relative sparing of the pools of FMN and FAD that are needed to fulfill critical metabolic functions (16). Another adaptation to riboflavin deficiency in its early stages is an increase in the de novo synthesis of reduced glutathione from its amino acid precursors, in response to the diminished conversion of oxidized glutathione back to reduced glutathione. This may represent a compensatory reaction resulting from depressed activity of glutathione reductase, a key FAD-requiring enzyme (Fig. 3).

An emerging concept is that dietary inadequacy is not the only cause of riboflavin deficiency and that certain endocrine abnormalities, such as adrenal and thyroid hormone insufficiency, drugs, and diseases, may interfere significantly with vitamin utilization (17,18). Psychotropic drugs, such as chlorpromazine, antidepressants [including imipramine and amitriptyline (19)], cancer chemotherapeutic drugs (e.g., adriamycin), and some antimalarial agents [e.g., quinacrine (20)], impair riboflavin utilization by inhibiting the conversion of this vitamin into its active coenzyme derivatives. Figure 4 shows the structural similarities among riboflavin, imipramine, chlorpromazine, and amitriptyline. Alco-

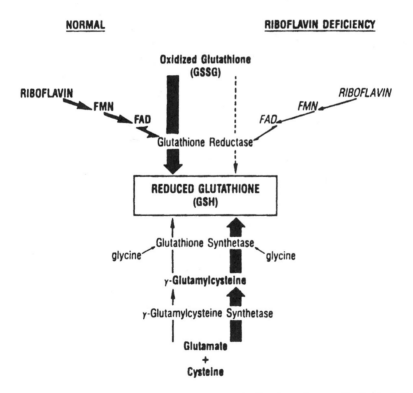

Fig. 3 Diagrammatic representation of metabolic adaptations to riboflavin deficiency. Diminished conversion of oxidized glutathione to reduced glutathione occurs as a result of decreased activity of glutathione reductase, an enzyme that utilizes FAD as a coenzyme. Levels of reduced glutathione are maintained by enhanced de novo synthesis from precursors (31,32).

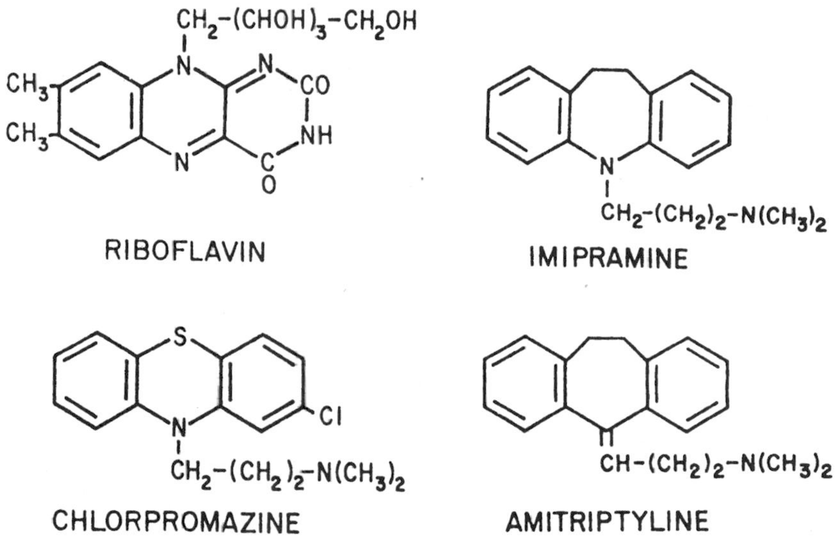

Fig. 4 Structural similarities among riboflavin, imipramine, chlorpromazine, and amitriptyline.

hol appears to cause riboflavin deficiency by inhibiting both its digestion from dietary sources and its intestinal absorption (21).

In approaching riboflavin deficiency, as well as other nutrient deficiencies, it may be useful to think in terms of risk factors. Thus, the consequences of a poor diet may be intensified if the patient is also abusing alcohol, is using certain drugs for prolonged periods, is extremely elderly, or has malabsorption or an underlying illness affecting vitamin metabolism.

In experimental animals, hepatic architecture is markedly disrupted in riboflavin deficiency. Mitochondria in riboflavin-deficient mice increase greatly in size, and cristae increase in both number and size (22). These structural abnormalities may disturb energy metabolism by interfering with the electron transport chain and metabolism of fatty acids. Villi decrease in number in the rat small intestine; villus length increases, as does the rate of transit of developing enterocytes along the villus (23). These findings of structural abnormalities together with accelerated rate of intestine cell turnover (24) may help to explain why dietary riboflavin deficiency leads to both decreased iron absorption and increased iron loss from the intestine.

There are many other effects of riboflavin deficiency on intermediary metabolism, particularly on lipid, protein, and vitamin metabolism. Of particular relevance to vitamin metabolism is the fact that the conversion of vitamin B$_6$ to its coenzyme derivative, pyridoxal 5'-phosphate, may be impaired (25). Riboflavin deficiency has been studied in many animal species and has several vital effects, foremost of which is failure to grow. Additional effects include loss of hair, skin disturbances, degenerative changes in the nervous system, and impaired reproduction. Congenital malformations occur in the offspring of female rats that are riboflavin-deficient. The conjunctiva becomes inflamed, the cornea is vascularized and eventually opaque, and cataract may result (26).

Changes in the skin consist of scaliness and incrustation of red-brown material consistent with changes in lipid metabolism. Alopecia may develop, lips become red and

Table 1 Top Sources of Riboflavin and Their Caloric Content

Top sources	Riboflavin (mg/100 g)	Energy (kcal/100 g)	Top food sources	Riboflavin (mg/100 g)	Energy (kcal/100 g)
Yeast baker's dry (active)	5.41	282	Cheese, pasteurized, process American	3.53	375
Liver, lamb, broiled	5.11	261	Liver, chicken, simmered	1.75	165
Yeast, torula	5.06	277	Corn flakes, w/added nutrients	1.40	380
Kidneys, beef, braised	4.58	252	Almonds, shelled	0.93	598
Liver, hog, fried in margarine	4.36	241	Cheese, natural, Roquefort	0.59	369
Yeast, brewer's, debittered	4.28	283	Eggs, chicken, fried	0.54	210
Liver, beef or calf, fried	4.18	242	Beef, tenderloin steak, broiled	0.46	224
Brewer's yeast, tablet form	4.04	—	Mushrooms, raw	0.46	28
Cheese, pasteurized, process American	3.53	375	Cheese, natural Swiss (American)	0.40	372
Turkey, giblets, cooked (some gizzard fat), simmered	2.72	233	Wheat flour, all-purpose, enriched	0.40	365
Kidneys, lamb, raw	2.42	105	Turnip greens, raw	0.39	28
Kidneys, calf, raw	2.40	113	Cheese, natural Cheddar	0.38	402
Eggs, chicken, dried, white powder	2.32	372	Wheat bran	0.35	353
Whey, sweet, dry	2.21	354	Soybean flour	0.35	333
Eggs, chicken, dried, white flakes	2.16	351	Bacon, cured, cooked, drained, sliced medium	0.34	575
Liver, turkey, simmered	2.09	174	Pork, loin, lean, broiled	0.33	391
Whey, acid dry	2.06	339	Lamb, leg, good or choice, separable lean roasted	0.30	186
Heart, hog, braised	1.89	195	Corn meal, degermed, enriched	0.26	362
Milk, cow's dry, skim, solids, instant	1.78	353	Chicken, dark meat w/o skin, fried	0.25	220
Liver, chicken, simmered	1.75	165	Bread, white, enriched	0.24	270
Liver, beef or calf, fried	4.18	242	Milk, cow's, whole, 3.7%	0.17	66

Figures are given in terms of the riboflavin and calorie amounts in 100 g (approximately 3.5 oz) of the items as usually consumed. Portion size and moisture content will differ among food items. Further details can be found in Table F-36 of Ref. 29.
Source: Ref. 29.

swollen, and filiform papillae on the tongue deteriorate. During late deficiency anemia develops. Fatty degeneration of the liver occurs. Important metabolic changes occur, so that deficient rats require 15–20% more energy than control animals to maintain the same body weight. Thus, in all species studied, riboflavin deficiency causes profound structural and functional changes in an ordered sequence. Early changes are very readily reversible. Later anatomical changes, such as formation of cataract, are largely irreversible despite treatment with riboflavin.

In humans, as noted above, the clinical features of human riboflavin deficiency do not have absolute specificity. Early symptoms may include weakness, fatigue, mouth pain and tenderness, burning and itching of the eyes, and possibly personality changes. More advanced deficiency may give rise to cheilosis, angular stomatitis, dermatitis, corneal vascularization, anemia, and brain dysfunction. Thus, the syndrome of dietary riboflavin deficiency in humans has many similarities to that in animals, with one notable exception. The spectrum of congenital malformations observed in rodents with maternal riboflavin deficiency has not been clearly identified in humans (26).

B. Food-Related Issues

In the United States today, the most significant dietary sources of riboflavin are meat and meat products, including poultry and fish, and milk and dairy products, such as eggs and cheese. In developing countries, plant sources contribute most of the dietary riboflavin intake. Green vegetables, such as broccoli, collard greens, and turnip greens, are reasonably good sources of riboflavin. Natural grain products tend to be relatively low in riboflavin, but fortification and enrichment of grains and cereals has led to a great increase in riboflavin intake from these food items.

The food sources of riboflavin are similar to those of other B vitamins. Therefore, it is not surprising that if a given individual's diet has inadequate amounts of riboflavin, it will very likely be inadequate in other vitamins as well. A primary deficiency of dietary riboflavin has wide implications for other vitamins, as flavin coenzymes are involved in the metabolism of folic acid, pyridoxine, vitamin K, niacin, and vitamin D (27).

Several factors in food preparation and processing may influence the amount of riboflavin that is actually bioavailable from dietary sources. Appreciable amounts of riboflavin may be lost with exposure to UV light, particularly during cooking and processing. Prolonged storage of milk in clear bottles or containers may result in flavin degradation (28). Fortunately, most milk is no longer sold in clear bottles. There has been some controversy as to whether opaque plastic containers provide greater protection than do cartons, particularly when milk is stored on a grocery shelf exposed to continuous fluorescent lighting.

It is highly likely that large amounts of riboflavin are lost during the sun-drying of fruits and vegetables. The precise magnitude of the loss is not known but varies with the duration of exposure. The practice of adding sodium bicarbonate as baking soda to green vegetables to make them appear fresh can result in accelerated photodegradation of riboflavin. The riboflavin content of common food items is shown in Table 1 (29).

IV. PHYSIOLOGY

A. Absorption, Transport, Storage, Turnover, and Excretion

Since dietary sources of riboflavin are largely in the form of coenzyme derivatives, these molecules must be hydrolyzed prior to absorption. Very little dietary riboflavin is found as free riboflavin from sources in nature. Under ordinary circumstances the main sources of free riboflavin are commercial multivitamin preparations.

The absorptive process for flavins occurs in the upper gastrointestinal tract by specialized transport involving a dephosphorylation-rephosphorylation mechanism, rather than by passive diffusion. This process is sodium-dependent and involves an ATPase active transport system that can be saturated (30). It has been estimated that under normal conditions the upper limit of intestinal absorption of riboflavin at any one time is approximately 25 mg (31). This amount represents approximately 15 times the RDA. Therefore, the common practice of some megavitamin enthusiasts to consume massive doses of multivitamins has little benefit with respect to riboflavin, as the additional amounts would be passed in the stool. Dietary covalent-bound flavins are largely inaccessible as nutritional sources. In experimental animals, the uptake of riboflavin from the intestine is increased in dietary riboflavin deficiency (32).

A number of physiological factors influence the rate of intestinal absorption of riboflavin (17). Diets high in psyllium gum appear to decrease the rate of riboflavin absorption, whereas wheat bran has no detectable effect. The time from oral administration to peak urinary excretion of riboflavin is prolonged by the antacids aluminum hydroxide and magnesium hydroxide. Total urinary excretion is unchanged by these drugs, however, and their major effect appears to be delaying the rate of intestinal absorption rather than inhibiting net absorption. Alcohol intake interferes with both the digestion of food flavins into riboflavin and the direct intestinal absorption of the vitamin (21). This observation suggests that the initial rehabilitation of malnourished alcoholic patients may be accomplished more efficiently with vitamin supplements containing riboflavin rather than with food sources comprising predominantly phosphorylated flavin derivatives.

There is evidence that the magnitude of intestinal absorption of riboflavin is increased by the presence of food. This effect of food may be due to decreasing the rates of gastric emptying and intestinal transit, thereby permitting more prolonged contact of dietary riboflavin with the absorptive surface of the intestinal mucosal cells. In general, delaying the rate of gastric emptying tends to increase the intestinal absorption of riboflavin. Bile salts also increase the absorption of riboflavin (17).

A number of metals and drugs form chelates or complexes with riboflavin and riboflavin-5′-phosphate that may affect their bioavailability (30). Among the agents in this category are the metals copper, zinc and iron; the drugs caffeine, theophylline, and saccharin; and the vitamins nicotinamide and ascorbic acid; as well as tryptophan and urea. The clinical significance of this binding is not known with certainty in most instances and deserves further study.

In human blood, the transport of flavins involves loose binding to albumin and tight binding to a number of globulins. The major binding of riboflavin and its phosphorylated derivatives in serum is to several classes of immunoglobulins, i.e., IgA, IgG, and IgM (30).

Pregnancy induces the formation of flavin-specific binding proteins. It has been known for many years that there is an avian riboflavin carrier protein, genetically controlled, that determines the amount of riboflavin in chicken eggs (33). Absence of the protein in autosomal recessive hens results in massive riboflavinuria because there is no mechanism for retaining and binding the vitamin in serum. Eggs become riboflavin-deficient, and embryonic death occurs between the tenth and fourteenth day of incubation. Administration of antiserum to the chicken riboflavin-carrier protein leads to termination of the pregnancy.

A new dimension to concepts of plasma protein binding of riboflavin in mammals was provided by the demonstration that riboflavin-binding proteins can also be found in serum from pregnant cows, monkeys, and humans. A comprehensive review of riboflavin-binding proteins covers the nature of the binding proteins in various species and provides evidence that, as in birds, these proteins are crucial for successful mammalian reproduction (10). The pregnancy-specific binding proteins may help transport riboflavin to the fetus.

Serum riboflavin-binding proteins also appear to influence placental transfer and fetal/maternal distribution of riboflavin. There are differential rates of uptake of riboflavin at the maternal and fetal surfaces of the human placenta (34). Riboflavin-binding proteins regulate the activity of flavokinase, the first biosynthetic enzyme in the riboflavin-to-FAD pathway (16).

The urinary excretion of flavins occurs predominantly in the form of riboflavin; FMN and FAD are not found in urine. McCormick et al. have identified a large number of flavins and their derivatives in human urine. Besides the 60–70% of urinary flavins contributed by riboflavin itself, other derivatives include 7-hydroxymethylriboflavin (10–15%), 8α-sulfonylriboflavin (5–10%), 8-hydroxymethylriboflavin (4–7%), riboflavinyl peptide ester (5%), and 10-hydroxyethylflavin (1–3%), representing largely metabolites from covalently bound flavoproteins and intestinal riboflavin degradation by microorganisms. Traces of lumiflavin and other derivatives have also been found. These findings were described fully by McCormick (10).

Ingestion of boric acid greatly increases the urinary excretion of riboflavin (17). This agent when consumed forms a complex with the side chain of riboflavin and other molecules having polyhydroxyl groups, such as glucose and ascorbic acid. In rodents, riboflavin treatment greatly ameliorates the toxicity of administered boric acid. This treatment should also be effective in humans with accidental exposure of boric acid, although in practice it may be difficult to provide adequate amounts of riboflavin due to its low solubility and limited absorptive capacity from the intestinal tract.

Urinary excretion of riboflavin in rats is also greatly increased by chlorpromazine (19). Levels are twice those of age- and sex-matched pair-fed control rats. In addition, chlorpromazine accelerates the urinary excretion of riboflavin during dietary deficiency. Urinary concentrations of riboflavin are increased within 6 h of treatment with this drug.

V. SPECIFIC FUNCTIONS

The major function of riboflavin, as noted above, is to serve as the precursor of the flavin coenzymes, FMN and FAD, and of covalently bound flavins. These coenzymes are widely distributed in intermediary metabolism and catalyze numerous oxidation–reduction reactions. Because FAD is part of the respiratory chain, riboflavin is central to energy production. Other major functions of riboflavin include drug and steroid metabolism, in conjunction with the cytochrome P450 enzymes, and lipid metabolism. The redox functions of flavin coenzymes include both one-electron transfers and two-electron transfers from substrate to the flavin coenzyme (10).

Flavoproteins catalyze dehydrogenation reactions as well as hydroxylations, oxidative decarboxylations, dioxygenations, and reductions of oxygen to hydrogen peroxide. Thus, many different kinds of oxidative and reductive reactions are catalyzed by flavoproteins.

A. Antioxidant Activity

In the wake of contemporary interest in the dietary antioxidants, one vitamin that is often neglected as a member of this category is riboflavin. Riboflavin does not have significant inherent antioxidant action, but powerful antioxidant activity is derived from its role as a precursor to FMN and FAD. A major protective role against lipid peroxides is provided by the glutathione redox cycle (35). Glutathione peroxidase breaks down reactive lipid peroxides. This enzyme requires reduced glutathione, which in turn is regenerated from its oxidized form (GSSG) by the FAD-containing enzyme glutathione reductase. Thus, riboflavin nutrition should be critical for regulating the rate of inactivation of lipid peroxides. Diminished glutathione reductase activity should be expected to lead to diminished concentrations of reduced glutathione that serve as substrate for glutathione peroxidase and glutathione S-transferase, and therefore would limit the rate of degradation of lipid peroxides and xenobiotic substances (36).

Furthermore, the reducing equivalents provided by NADPH, the other substrate required by glutathione reductase, are primarily generated by an enzyme of the pentose monophosphate shunt, glucose-6-phosphate dehydrogenase. Taniguchi and Hara (37), as well as our laboratory (38), have found that the activity of glucose-6-phosphate dehydrogenase is significantly diminished during riboflavin deficiency. This observation provides an additional mechanism to explain the diminished glutathione reductase activity in vivo during riboflavin deficiency and the eventual decrease in antioxidant activity.

There have been a number of reports (39,40) in the literature indicating that riboflavin deficiency is associated with compromised oxidant defense and furthermore that supplementation of riboflavin and its active analogs improves oxidant status. Investigators have shown that riboflavin deficiency is associated with increased hepatic lipid peroxidation and that riboflavin supplementation limits this process (37–40). In our laboratory, we have shown that feeding a riboflavin-deficient diet to rats increases basal as well as stimulated lipid peroxidation (36).

B. Riboflavin and Malaria

Recent reports provide increasing evidence that riboflavin deficiency is protective against malaria both in experimental animals and in humans (41,42). With dietary riboflavin deficiency, parasitemia is decreased dramatically, and symptomatology of infection may be diminished. In a study with human infants suffering from malaria, normal riboflavin nutritional status was associated with high levels of parasitemia (43,44). In a similar fashion, supplementation of iron and vitamins, which included riboflavin, to children resulted in increased malarial parasitemia (45).

Further evidence for a beneficial role of riboflavin deficiency in malaria is provided by studies utilizing specific antagonists of riboflavin, e.g., galactoflavin and 10-(4′-chlorophenyl)-3-methylflavin (46). These flavin analogs, as well as newer isoalloxazines derivatives (47), are glutathione reductase inhibitors and possess clear antimalarial efficacy (46). The exact mechanism by means of which riboflavin deficiency appears to inhibit malarial parasitemia is not yet established. One possibility relates to effects on the redox status of erythrocytes, which is an important determinant of growth of malaria parasites. Protection from malaria is afforded by several oxidant drugs, vitamin E deficiency, and certain genetic abnormalities compromising oxidative defense (35).

It is well known that malaria parasites (*Plasmodium berghei*) are highly susceptible to activated oxygen species. Parasites are relatively more susceptible than erythrocytes to the damaging effects of lipid peroxidation (35). We have hypothesized that the requirement of the parasites for riboflavin should be higher than that of the host cells and therefore that marginal riboflavin deficiency should be selectively detrimental to parasites. Support for this hypothesis is that the uptake of riboflavin and its conversion to FMN and FAD are significantly higher in parasitized than in unparasitized erythrocytes and furthermore that the rate of uptake of riboflavin is proportional (48) to the degree of parasitemia. These results strongly suggest that parasites have a higher requirement for riboflavin than do host erythrocytes.

C. Riboflavin and Homocysteine

A subject of great contemporary interest is the possible role of homocysteine in the pathogenesis of vascular disease, including cardiovascular, cerebrovascular, and peripheral vascular disorders (49). Blood levels of folic acid sensitively determine serum homocysteine concentrations (50). As shown in Fig. 5, N-5-methyltetrahydrofolate is a cosubstrate with homocysteine in its inactivation by conversion to methionine. Methylcobalamin is also a coenzyme in this enzymatic reaction. Vitamin B$_6$ is widely recognized for its importance in the inactivation of homocysteine by serving as coenzyme of two degradative enzymes, cystathionine β-synthase and cystathioninase.

However, it is not commonly appreciated that riboflavin also has a role in homocysteine metabolism, as the flavin coenzyme, FAD, is required by methyltetrahydrofolate reductase, the enzyme responsible for converting N-5,10-methylenetetrahydrofolate into N-5-methyltetrahydrofolate. Thus, the efficient utilization of dietary folic acid requires adequate riboflavin nutrition. Furthermore, a mutation leading to a heat-sensitive form of methylenetetrahydrofolate reductase has recently been identified (51). Its role in vascular disease has not been defined. Further research is required to determine whether the serum levels of homocysteine and the prevalence of vascular disease can be correlated directly

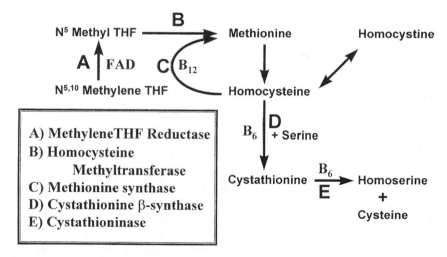

Fig. 5 Diagrammatic representation of homocysteine metabolism, showing the roles of folic acid, B$_{12}$, B$_6$, and riboflavin. FAD is the coenzyme for methylenetetrahydrofolate reductase.

with indices of riboflavin nutrition, and whether effects of marginal as well as overt deficiency of riboflavin are clinically significant with respect to vascular disease.

D. Fat Metabolism

The important role of riboflavin in fat metabolism has been highlighted by recent demonstrations that in certain rare inborn errors administration of riboflavin may be therapeutic. In acyl-CoA dehydrogenase deficiency, infants present with recurrent hypoglycemia and lipid storage myopathy and increased urinary excretion of organic acids. Clinical improvement has occurred rapidly after riboflavin supplementation (52,53). Three varieties of the disorder occur, all of which involve flavoproteins of various types. Five patients with a mitochondrial disorder associated with NADH dehydrogenase deficiency were improved by riboflavin treatment (54).

VI. PHARMACOLOGY/TOXICOLOGY/CARCINOGENESIS INTERRELATIONSHIPS

There is general agreement that dietary riboflavin intake at many times the RDA is without demonstrable toxicity (10,55–57). Because riboflavin absorption is limited to a maximum of about 25 mg at any one time (10), the consumption of megadoses of this vitamin would not be expected to increase the total amount absorbed. Furthermore, classical animal investigations showed an apparent upper limit to tissue storage of flavins that cannot be exceeded under ordinary circumstances (58). The tissue storage capacity for flavins is probably limited by the availability of proteins capable of providing binding sites. Thus, protective mechanisms prevent tissue accumulation of excessive amounts of the vitamin. Because riboflavin has very low solubility, even intravenous administration of the vitamin would not introduce large amounts into the body. FMN is more water-soluble than riboflavin but is not ordinarily available for clinical use.

Nevertheless, the photosensitizing properties of riboflavin raise the possibility of some potential risks. Phototherapy in vitro leads to degradation of DNA and increases in lipid peroxidation, which may have implications for carcinogenesis and other disorders. Irradiation of rat erythrocytes in the presence of FMN increases potassium loss (59). Topical administration of riboflavin to the skin may increase melanin synthesis by stimulation of free-radical formation. Riboflavin forms an adduct with tryptophan and accelerates the photo-oxidation of this amino acid (60). Further research is needed to explore the full implications of the photosensitizing capabilities of riboflavin and its phosphorylated derivatives.

The photosensitization of vinca alkaloids by riboflavin may distort the results of testing of the efficacy of cytotoxic drugs if the studies are carried out in the presence of visible light, as is usually done (61). This property of riboflavin needs to be considered in drug evaluations, inasmuch as cell death will occur even without the addition of the drug being analyzed.

Riboflavin is capable of reacting with chromate (VI), forming a complex and then increasing the DNA breaks due to a chromium-induced free-radical mechanism (62). Treatment of mouse FM3A cells greatly increases mutation frequency and the extent of cellular DNA damage in the presence of light (63). There appears to be increasing evidence that in the presence of visible light, riboflavin and its degradative product, lumiflavin, may enhance mutagenicity (64).

On the other hand, recent studies (65) confirm earlier reports (66) that riboflavin deficiency may enhance carcinogenesis by increasing activation of carcinogens, particularly nitrosamines. Riboflavin may possibly provide protection against damage to DNA caused by certain carcinogens through its action as a coenzyme with a variety of cytochrome P450 enzymes.

It is important to establish the role of riboflavin as a dietary factor capable of preventing carcinogenesis while at the same time determining the full implication of the photosensitizing actions of riboflavin on mutagenesis and carcinogenesis. There are reports raising the possibility that deficient riboflavin nutritional status, together with shortages of other vitamins, may possibly enhance development of precancerous lesions of the esophagus in China (67,68) and in Russia (69).

VII. REQUIREMENTS AND ASSESSMENT

There are a variety of methods available for analysis of riboflavin and its derivatives. Bioassays measure the growth effect of vitamins but lack the precision of more sensitive analytical procedures (70). Fluorometric procedures take advantage of the inherent fluorescent properties of flavins (71). Some degree of purification of the urine or tissues may be required before analysis as there is often significant interference by other natural substances that leads to quenching of fluorescence and methodological artefacts. A procedure has been developed for measuring riboflavin by competitive protein binding that is applicable to studies in human urine (72). Riboflavin binds specifically to the avian egg white riboflavin-binding protein (33) and thereby provides the basis for quantitative analysis (73). Other procedures are also in use based on binding to specific apoenzymes, such as D-amino acid oxidase. Currently, procedures using high-pressure liquid chromatography (HPLC) have been widely applied as they have precision and can be utilized for analysis of riboflavin in pure form as well as in biological fluids and tissues (74). HPLC is the method most widely employed for determination of flavins in the blood and other tissues.

In clinical studies that involve individual patients as well as population groups, the status of riboflavin nutrition is generally evaluated by determining the urinary excretion of riboflavin (75) and the erythrocyte glutathione reductase activity coefficient (EGRAC) (75). Urinary riboflavin determinations are made in the basal state, in random samples, in 24-h collections, or after a riboflavin load test. Normal urinary excretion of riboflavin is approximately 120 μg/g creatinine per 24 h or higher. It is useful to express urinary excretion in terms of creatinine to verify the completeness of the collection and to relate excretion to this biological parameter. Expressed in terms of the total amount, riboflavin excretion in the normal adult is about 1.5–2.5 mg/day, which is very close to the recommended dietary allowance of the National Academy of Sciences.

In deficient adult individuals, urinary riboflavin excretion is reduced to about 40 μg/g creatinine per 24 hr. Thus, deficient individuals have reduced urinary excretion, reflecting diminished dietary intake and depleted body stores. Normal urinary excretion is reduced with age, may be reduced by physical activity (as discussed below), and is stimulated by elevated body temperature, treatment with certain drugs, and various stressful conditions associated with negative nitrogen balance (76). Interpretation of urinary riboflavin excretion must be made with these factors in mind.

Another potential drawback to utilizing urinary riboflavin excretion as an assessment of nutritional status of this vitamin is that the amount excreted reflects recent intake very sensitively. Thus, if an individual has been depleted for a long time but consumes a food item high in riboflavin, the urinary excretion determined a few hours later may not be in the deficient range, but is likely to be normal or even elevated.

It is for this reason that attention has been directed to the development of assessment techniques that more accurately reflect long-term riboflavin status. The method most widely employed that largely meets these needs is assay of EGRAC as noted above. The principle of the method is that the degree of saturation of the apoenzyme with its coenzyme, FAD, should reflect the body stores of FAD. In deficient individuals, relative unsaturation of the apoenzyme with FAD leads to decreased basal activity of the enzyme. Therefore, the addition of FAD to the enzyme contained in a fresh erythrocyte hemolysate from deficient individuals will increase activity in vitro to a greater extent than that observed in a preparation from well-nourished individuals in whom the apoenzyme is more saturated with FAD.

The EGRAC is the ratio of enzyme activity with to that without addition of FAD in vitro. In general, most studies indicate that an activity coefficient of 1.20 or less indicates adequate riboflavin status, 1.2–1.4 borderline-to-low status, and greater than 1.4 a clear riboflavin deficiency (75,76).

It must be kept in mind that a number of physiological variables influence the results of this determination. In the inherited disorder of glucose-6-phosphate dehydrogenase deficiency, associated with hemolytic anemia, the apoenzyme has a higher affinity for FAD than that of the normal erythrocyte that will affect the measured EGRAC. Thyroid function affects glutathion reductase activity, the coefficient being elevated in hypothyroidism and decreased in hyperthyroidism (77), reflecting the fact that hypothyroidism has many biochemical features in common with those of riboflavin deficiency (17).

The latest RDAs issued by the Food and Nutrition Board (55) call for adult males aged 19–50 years to consume 1.7 mg/day and 51 years of age or older, 1.4 mg. Adult females from 19 to 50 years of age should consume 1.3 mg/day and from age 51 years or older 1.2 mg. It is recommended that intake be increased to 1.6 mg/day during pregnancy, to 1.8 mg/day early in lactation, and to 1.7 mg/day later in lactation. Infants should consume 0.4–0.5 mg/day and children 0.8–1.2 mg/day, depending on age.

There has been some concern as to whether these figures are applicable to other population groups around the world. Chinese tend to excrete very little riboflavin, and their requirement may be lower than that of Americans (78). Adults in Guatemala appear to have similar requirements in individuals older than 60 compared to those 51 years or younger (79). This finding may not necessarily be relevant to populations of other countries. The requirements of various national groups require further study. Environmental factors, protein-calorie intake, physical activity, and other factors may have an impact on riboflavin status. More research is needed on the requirements of the extremely old, who form an increasingly large proportion of the population. They are also the population group that consumes the largest number of prescribed and over-the-counter medications.

A point of interest is whether riboflavin requirements are increased in individuals who exercise compared to those who are sedentary. In women aged 50–67 who exercised vigorously for 20–25 min/day, 6 days a week, both a decrease in riboflavin excretion and a rise in the EGRAC was noted, findings consistent with a marginal riboflavin-deficient state, as shown in Table 2 (80). Supplementation with riboflavin did not, however, improve exercise performance. These investigators observed compromised riboflavin status as well in young women exercising vigorously (81). Similar observations of reduced urinary riboflavin excretion and elevated EGRAC were made in young Indian males who exercised actively (82).

Table 2 Group Means ($n = 7$) for EGRAC and Urinary Riboflavin Excretion During Both Exercise and Nonexercise Periods[a]

Group	Nonexercise	Exercise
Low-riboflavin:		
EGRAC	1.224 ± 0.079^b	1.283 ± 0.067^c
Urinary riboflavin (mg/day)	0.17 ± 0.11	0.14 ± 0.10^d
High-riboflavin:		
EGRAC	1.070 ± 0.031	1.109 ± 0.045^c
Urinary riboflavin (mg/day)	0.66 ± 0.49	0.46 ± 0.21^d

[a] $x \pm$ SD. EGRAC, erythrocyte glutathione reductase activity coefficient.
[b] Significantly greater than high-riboflavin group, $p < 0.0005$.
[c] Significantly different from nonexercise: $p < 0.0001$.
[d] Significantly different from nonexercise: $p < 0.01$.
Source: Ref. 62.

To determine whether the status of riboflavin nutrition influences metabolic responses to exercise, blood lactate levels were determined in a group of physically active college students from Finland before and after the exercise period. A number of the students were initially in a state of marginal riboflavin deficiency. Following supplementation with vitamins, including riboflavin, that produced improvement in the elevated EGRAC, the blood lactate levels were unaffected and were related only to the degree of exercise (83).

Thus, to date, while exercise clearly produces biochemical abnormalities in riboflavin metabolism, it has not been shown that these abnormalities lead to impaired performance, nor has it been shown that riboflavin supplementation leads to improved exercise performance.

ACKNOWLEDGMENTS

Research was supported in part by the Clinical Nutrition Research Unit Grant 1-PO1-CA-29502 from the National Institutes of Health, and by grants from the Stella and Charles Guttman Foundation, the Sunny and Abe Rosenberg Foundation, the Rosenstiel Foundation, the Isadore Rosenfeld Heart Foundation, and the Frank J. Scallon Foundation and an industrial agreement with Wakunaga of America Co., Ltd.

REFERENCES

1. E. V. McCollum and C. Kennedy, The dietary factors operating in the production of polyneuritis. *J. Biol. Chem. 24*:491, 1916.
2. A. D. Emmett and G. O. Luros, Water soluble vitamins. I. Are the antineuritic and the growth-promoting water soluble B vitamines the same? *J. Biol. Chem. 43*:265, 1920.
3. O. Warburg and W. Christian, Uber ein neues Oxydationsferment und sein Absorptionsspektrum. *Biochem. Z. 254*:438, 1932.
4. K. G. Stern and E. R. Holiday, Zur Konstitution des Photo-flavins; Versuche in der Alloxazine-Reihe. *Ber. Dtsch. Chem. Gessellsch. 67*:1104, 1934.
5. H. Theorell, Reindarstellung (kristallisation) des gelben Atmungsfermentes und die reversible Spaltung desselben. *Biochem. Z. 272*:155, 1934.

6. R. Kuhn, K. D. Reinemund, H. Kaltschmitt, R. Strobele, and H. Trischmann, Synthetisches 6,7-Dimethyl-9-d-riboflavin. *Naturwissenschaften 23*:260, 1935.

7. P. Karrer, K. Scopp, and F. Benz, Synthesis of flavins IV. *Helv. Chim. Acta 18*:426, 1935.

8. H. Theorell, Die Freie Eiweisskomponente des gelben Ferments und ihre Kupplung mit Lactoflavinphosphorsaure. *Biochem. Z. 290*:293, 1937.

9. O. Warburg and W. Christian, Co-Ferment der d-Aminosaure-Deaminase. *Biochem. Z. 295*: 261, 1938.

10. D. B. McCormick, Riboflavin, in *Modern Nutrition in Health and Disease*, 8th ed. (M. E. Shils, J. A. Olson, and M. Shike, eds.), Lea and Febiger, Philadelphia, 1994, p. 366.

11. K. Yagi, Y. Nakagawa, O. Suzuki, and N. Ohishi, Incorporation of riboflavin into covalently-bound flavins in rat liver. *J. Biochem. 79*:841, 1976.

12. R. Rivlin, Medical progress: riboflavin metabolism. *N. Engl. J. Med. 283*:463, 1970.

13. J. T. Pinto and R. S. Rivlin, Regulation of formation of covalently bound flavins in liver and cerebrum by thyroid hormones. *Arch. Biochem. Biophys. 194*:313, 1979.

14. A. H. Merrill, Jr., J. D. Lambeth, D. E. Edmondson, and D. B. McCormick, Formation and mode of action of flavoproteins. *Annu. Rev. Nutr. 1*:281, 1981.

15. R. S. Rivlin, Vitamin deficiency, in *Conns' Current Therapy* (R. E. Rakel, ed.), W. B. Saunders, Philadelphia, 1994, p. 551.

16. S. Fass and R. S. Rivlin, Regulation of riboflavin-metabolizing enzymes in riboflavin deficiency. *Am. J. Physiol. 217*:988, 1969.

17. R. S. Rivlin, Medical aspects of vitamin B2, in *Chemistry and Biochemistry of Flavins* (F. Muller, ed.), CRC Press, Boca Raton, FL, 1990, p. 201.

18. J. A. Cimino, S. Jhangiani, E. Schwartz, and J. M. Cooperman, Riboflavin metabolism in the hypothyroid human adult. *Proc. Soc. Exp. Biol. Med. 184*:121, 1987.

19. J. T. Pinto, Y. P. Huang, and R. S. Rivlin, Inhibition of riboflavin metabolism in rat tissues by chlorpromazine, imipramine and amitriptyline. *J. Clin. Invest. 67*:1500, 1981.

20. P. Dutta, J. T. Pinto, and R. S. Rivlin, Antimalarial effects of riboflavin deficiency. *Lancet 2*:1040, 1985.

21. J. T. Pinto, J. P. Huang, and R. S. Rivlin, Mechanisms underlying the differential effects of ethanol upon the bioavailability of riboflavin and flavin adenine dinucleotide. *J. Clin. Invest. 79*:1343, 1987.

22. B. Tandler, R. A. Erlandson, and E. L. Wynder, Riboflavin and mouse hepatic cell structure and function. I. Ultrastructural alterations in simple deficiency. *Am. J. Pathol. 52*:69, 1968.

23. E. A. Williams, R. D. E. Rumsey, and H. J. Powers, Cytokinetic and structural responses of the rat small intestine to riboflavin depletion. *Br. J. Nutr. 75*:315, 1996.

24. H. J. Powers, L. T. Weaver, S. Austin, and J. K. Beresford, A proposed intestinal mechanism for the effect of riboflavin deficiency on iron loss in the rat. *Br. J. Nutr. 69*:553, 1993.

25. D. B. McCormick, Two interconnected B vitamins: riboflavin and pyridoxine. *Physiol. Rev. 69*:1170, 1989.

26. G. A. Goldsmith, Riboflavin deficiency, in *Riboflavin* (R. Rivlin, ed.), Plenum Press, New York, 1975, p. 221.

27. R. S. Rivlin, Disorders of vitamin metabolism: deficiencies, metabolic abnormalities and excesses, in *Cecil Textbook of Medicine*, 19th ed. (J. H. Wyngaarden, L. H. Smith, Jr., J. C. Bennett, and F. Plum, eds.), W. B. Saunders, Philadelphia, 1991, p. 1170.

28. R. L. Wanner, Effects of commercial processing of milk and milk products on their nutrient content, in R. S. Harris and H. V. Loesecke (eds.), *The Nutritional Evaluation of Food Processing*, John Wiley and Sons, New York, 1960, p. 173.

29. A. H. Ensminger, M. E. Ensminger, J. E. Konlande, and J. R. K. Robson, *Food and Nutrition Encyclopedia*, CRC Press, Boca Raton, FL, 1994, p. 1927.

30. D. B. McCormick, Riboflavin, in *Present Knowledge in Nutrition*, 6th ed. (M. L. Brown, ed.), International Life Sciences Institute, Washington, DC, 1990, p. 146.

31. J. Zempleni, J. R. Galloway, and D. B. McCormick, Pharmacokinetics of orally and intravenously administered riboflavin in healthy humans. *Am. J. Clin. Nutr. 63*:54, 1996.

32. H. M. Said and R. Mohammadkhani, Uptake of riboflavin across the brush border membrane of rat intestine: regulation by dietary vitamin levels. *Gastroenterology 105*:1294, 1993.

33. C. O. Clagett, Genetic control of the riboflavin carrier protein. *Fed. Proc. 30*:127, 1971.

34. J. Dancis, J. Lehanka, and M. Levitz, Placental transport of riboflavin: differential rates of uptake at the maternal and fetal surfaces of the perfused human placenta. *J. Obstet. Gynecol. 158*:204, 1988.

35. P. Dutta, Disturbances in glutathione metabolism and resistance to malaria: current understanding and new concepts. *J. Soc. Pharm. Chem. 2*:11, 1993.

36. R. S. Rivlin and P. Dutta, Vitamin B2 (riboflavin). Relevance to malaria and antioxidant activity. *Nutr. Today 30*:62, 1995.

37. M. Taniguchi and T. Hara, Effects of riboflavin and selenium deficiencies on glutathione and related enzyme activities with respect to lipid peroxide content of rat livers. *J. Nutr. Vitamin 29*:283, 1983.

38. P. Dutta, J. Seirafi, D. Halpin, J. T. Pinto, and R. S. Rivlin, Acute ethanol exposure alters hepatic glutathione metabolism in riboflavin deficiency. *Alcohol 12*:43, 1995.

39. T. Miyazawa, K. Tsuchiya, and T. Kaneda, Riboflavin tetrabutyrate: an antioxidative synergist of alpha-tocopherol as estimated by hepatic chemiluminescence. *Nutr. Rep. Int. 29*:157, 1984.

40. T. Miyazawa, C. Sato, and T. Kaneda, Antioxidative effects of α-tocopherol and riboflavin-butyrate in rats dosed with methyl linoleate hydroperoxide. *Agric. Biol. Chem. 47*:1577, 1983.

41. P. Kaikai and D. I. Thurnham, The influence of riboflavin deficiency on *Plasmodium berghei* infections in rats. *Trans. R. Soc. Trop. Med. Hyg. 77*:680, 1983.

42. B. S. Das, D. B. Das, R. N. Satpathy, J. K. Patnaik, and T. K. Bose, Riboflavin deficiency and severity of malaria. *Eur. J. Clin. Nutr. 42*:277, 1988.

43. D. I. Thurnham, S. J. Oppenheimer, and R. Bull, Riboflavin status and malaria in infants in Papua New Guinea. *Trans. R. Soc. Trop. Med. Hyg. 77*:423, 1983.

44. S. J. Oppenheimer, R. Bull, and D. I. Thurnham, Riboflavin deficiency in Madang infants. *Papua N. Guinea Med. J. 26*:17, 1983.

45. C. J. Bates, H. J. Powers, and W. H. Lamb, Antimalarial effects of riboflavin deficiency in Madang infants. *Papua N. Guinea Med. J. 26*:17, 1983.

46. K. Becker, R. I. Christopherson, W. B. Cowden, N. H. Hunt, and R. H. Schirmer, Flavin analogs with antimalarial activity as glutathione reductase inhibitors. *Biochem. Pharmacol. 39*:59, 1990.

47. A. Schonleben-Janas, P. Kirsch, P. R. E. Mittl, R. H. Schirmer, and R. L. Krauth-Siegel, Inhibition of human glutathione reductase by 10-arylisoalloxazines: crystallographic, kinetic and electrochemical studies. *J. Med. Chem. 39*:1549, 1996.

48. P. Dutta, Enhanced uptake and metabolism of riboflavin in erythrocytes infected with *Plasmodium falciparium. J. Protozool. 38*:479, 1991.

49. I. A. Graham, L. E. Daly, H. M. Refsum, et al., Plasma homocysteine as a risk factor for vascular disease. The European concerted project. *JAMA 277*:1775, 1997.

50. C. J. Boushey, S. A. A. Beresford, G. S. Omenn, and A. G. Motulsky, A quantitative assessment of plasma homocysteine as a risk factor for vascular disease. *JAMA 274*:1049, 1995.

51. P. Frosst, H. J. Blom, R. Milos, et al., A candidate genetic risk factor for vascular disease: a common mutation in methylenetetrahydrofolate reductase. *Nature Genet. 10*:111, 1995.

52. G. Uziel, B. Garavaglia, E. Ciceri, I. Moroni, and M. Rimoldi, Riboflavin-responsive glutaric aciduria type II presenting as a leukodystrophy. *Pediatr. Neurol. 13*:333, 1995.

53. P. L. J. A. Bernsen, F. J. M. Gabreels, W. Ruitenbeek, and H. L. Hamburger, Treatment of complex I deficiency with riboflavin. *J. Neurol. Sci. 118*:181, 1993.

54. U. A. Walker and E. Byrne, The therapy of respiratory chain encephalomyopathy: a critical review of the past and present perspective. *Acta Neurol. Scand. 92*:273, 1995.

55. Food and Nutrition Board, National Research Council. *Recommended Dietary Allowances*, 10th ed., National Academy Press, Washington, DC, 1989, p. 132.
56. R. S. Rivlin, Effect of nutrient toxicities (excess) in animals and man: riboflavin, in *Handbook of Nutrition and Foods* (M. Recheigl, ed.), CRC Press, Boca Raton, FL, 1979, p. 25.
57. J. M. Cooperman and R. Lopez, Riboflavin, in *Handbook of Vitamins: Nutritional, Biochemical, and Clinical Aspects* (L. J. Machlin, ed.), Marcel Dekker, New York, 1984, p. 299.
58. H. B. Burch, O. H. Lowry, A. M. Padilla, and A. M. Combs, Effects of riboflavin deficiency and realimentation on flavin enzymes of tissues. *J. Biol. Chem. 223*:29, 1956.
59. F. S. Ghazy, T. Kimura, S. Muranishi, and H. Sezaki, The photodynamic action of riboflavin on erythrocytes. *Life Sci. 21*:1703, 1977.
60. M. Salim-Hanna, A. M. Edwards, and E. Silva, Obtention of a photo-induced adduct between a vitamin and an essential amino acid. Binding of riboflavin to tryptophan. *Int. J. Vitam. Nutr. Res. 57*:155, 1987.
61. C. Granzow, M. Kopun, and T. Krober, Riboflavin-mediated photosensitization of vinca alkaloids distorts drug sensitivity assays. *Cancer Res. 55*:4837, 1995.
62. M. Sugiyama, K. Tsuzuki, X. Lin, and M. Costa, Potentiation of sodium chromate (VI)–induced chromosomal aberrations and mutation by vitamin B2 in Chinese hamster V79 cells. *Mutat. Res. 283*:211, 1992.
63. T. Bessho, K. Tano, S. Nishimura, and H. Kasai, Induction of mutations in mouse FM3A cells by treatment with riboflavin plus visible light and its possible relation with formation of 8-hydroxyguanine (7,8-dihydro-8-oxoguanine) in DNA. *Carcinogenesis 14*:1069, 1993.
64. H. Kale, P. Harikumar, S. B. Kulkarni, P. M. Nair, and M. S. Netrawali, Assessment of the genotoxic potential of riboflavin and lumiflavin. B. Effect of light. *Mutat. Res. 298*:17, 1992.
65. R. P. Webster, M. D. Gawde, and R. K. Bhattacharya, Modulation of carcinogen-induced DNA damage and repair enzyme activity by dietary riboflavin. *Cancer Lett. 98*:129, 1996.
66. R. S. Rivlin, Riboflavin and cancer: a review. *Cancer Res. 33*:1997, 1973.
67. N. Munoz, J. Wahrendorf, L. J. Bang, M. Crespi, and A. Grassi. Vitamin intervention on precancerous lesions of the esophagus in a high-risk population in China. *Ann. N.Y. Acad. Sci. 534*:618, 1988.
68. J. Wahrendorf, N. Munoz, J. B. Lu, D. I. Thurnham, M. Crespi, and F. X. Bosch. Blood retinol and zinc riboflavin status in relation to precancerous lesions of the esophagus: findings from a vitamin intervention trial in the People's Republic of China. *Cancer Res. 48*:2280, 1988.
69. D. G. Zaridze, J. U. Bukin, Y. N. Orlov, et al., Relationship between esophageal mucosa pathology and vitamin deficit in population with high frequency of esophageal cancer. *Vop. Onkol. 35*:939, 1989.
70. H. Baker and O. Frank, Analysis of riboflavin and its derivatives in biologic fluids and tissues, in *Riboflavin* (R. S. Rivlin, ed.), Plenum Press, New York, 1975, p. 49.
71. O. A. Bessey, O. H. Lowry, and R. H. Love, Fluorometric measure of the nucleotides of riboflavin and their concentration in tissues, *J. Biol. Chem. 180*:755, 1949.
72. A. G. Fazekas, C. E. Menendez, and R. S. Rivlin, A competitive protein binding assay for urinary riboflavin, *Biochem. Med. 9*:167, 1974.
73. M. J. Kim, H. J. Kim, J. M. Kim, B. Kim, S. H. Han, and G. S. Cha, Homogeneous assays for riboflavin mediated by the interaction between enzyme-biotin and avidin-riboflavin conjugates. *Anal. Biochem. 231*:400, 1995.
74. J. L. Chastain and D. B. McCormick, Flavin catabolites: identification and quantitation in human urine. *Am. J. Clin. Nutr. 46*:830, 1987.
75. H. E. Sauberlich, J. H. Judd, G. E. Nichoalds, H. P. Broquist, and W. J. Darby, Application of the erythrocyte glutathione reductase assay in evaluating riboflavin nutritional status in a high school student population. *Am. J. Clin. Nutr. 25*:756, 1972.
76. R. S. Rivlin, Riboflavin in *Present Knowledge in Nutrition* (E. E. Ziegler and L. J. Filer, Jr., eds.), ILSI Press, Washington, DC, 1996, p. 167.
77. C. E. Menendez, P. Hacker, M. Sonnenfeld, R. McConnell, and R. S. Rivlin, Thyroid hormone

regulation of glutathione reductase activity in rat erythrocytes and liver. *Am. J. Physiol. 226*: 1480, 1974.

78. T. A. Brun, J. Chen, T. C. Campbell, et al., Urinary riboflavin excretion after a load test in rural China as a measure of possible riboflavin deficiency. *Eur. J. Clin. Nutr. 44*:195, 1990.
79. W. A. Boisvert, I. Mendoza, C. Castenada, et al., Riboflavin requirement of healthy elderly humans and its relationship to macronutrient composition of the diet. *J. Nutr. 123*:915, 1993.
80. L. R. Trebler Winters, J.-S. Yoon, H. J. Kalkwarf, J. C. Davies, M. G. Berkowitz, J. Haas, and D. A. Roe, Riboflavin requirements and exercise adaptation in older women. *Am. J. Clin. Nutr. 56*:526, 1992.
81. A. Z. Belko, E. Obarzanek, R. Roach, M. Rotter, G. Urban, S. Weinberg, and D. A. Roe, Effects of aerobic exercise and weight loss on riboflavin requirements of moderately obese, marginally deficient young women. *Am. J. Clin. Nutr. 40*:553, 1984.
82. M. J. Soares, K. Satyanarayana, M. S. Bamji, C. M. Jacobs, Y. Venkata Ramana, and S. Sudhakar Rao, The effect of exercise on the riboflavin status of adult men. *Br. J. Nutr. 69*: 541, 1993.
83. M. Fogelholm, I. Ruokonen, J. T. Laakso, T. Vuorimaa, and J.-J. Himberg, Lack of association between indices of vitamin B1, B2 and B6 status and exercise-induced blood lactate in young adults. *Int. J. Sport. Nutr. 3*:165, 1993.

8

Thiamine

VICHAI TANPHAICHITR

Mahidol University, Bangkok, Thailand

I. HISTORY

Although *Neiching*, the Chinese medical book, mentioned beriberi in 2697 BC, it was not known for centuries that this illness was due to thiamine deficiency. In 1884, Takaki, a surgeon general of the Japanese navy, concluded from his observations that beriberi was caused by a lack of nitrogenous food components in association with excessive intake of nonnitrogenous food. Though he did not reach the correct conclusion, the dietary changes introduced into the Japanese navy resulted in a drastic reduction in the incidence of beriberi. In 1890, the Dutch physician Eijkman, working in Java, discovered that fowls fed boiled polished rice developed a polyneuritis that resembled beriberi in humans, and this polyneuritis could be prevented or cured by rice bran or polishings. He suggested that the toxic principle was present in polished rice but could be neutralized by some protective factor in rice polishings. His associate, Grijns, in 1901 extracted a water-soluble protective factor for polyneuritis from rice bran. In 1911, Funk, a chemist at the Lister Institute in London, was convinced that he had isolated the antiberiberi principle possessing an amine function from rice bran extracts and named it "vitamine." However, his crystalline substance was shown later to have little antineuritic activity (1–3).

Studies by Frazier and Stanton in 1909–1915 showed that human beriberi was a deficiency disease that responded to extracts from rice polishings. This led Vedder, a U.S. army physician, to recommend to Roger Williams, a chemist working in Manila, that the protective substance be isolated. In 1926, Jansen and Donath, Dutch chemists working in Java, succeeded in isolating and crystallizing antiberiberi factor from rice bran extracts. The trivial name "aneurine" was suggested by Jansen. Later, Roger Williams isolated sufficient quantities of thiamine to elucidate its structure. The chemical synthesis of thiamine was accomplished in 1936 (1–3).

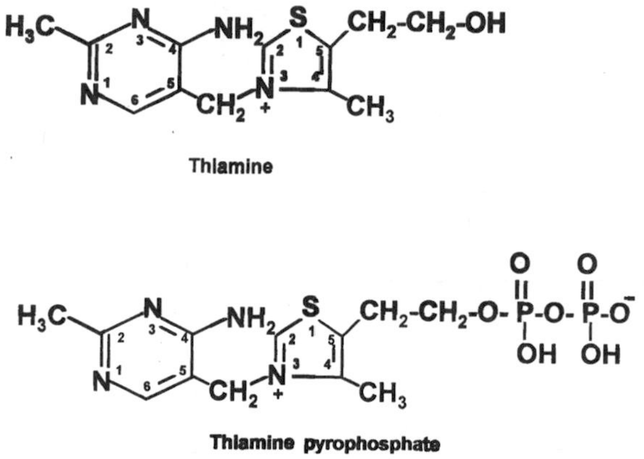

Thiamine

Thiamine pyrophosphate

Fig. 1   Structural formulas of thiamine and thiamine pyrophosphate.

The biochemical functions of thiamine were identified in 1935 by Thompson and Johnson who showed that blood pyruvate levels were elevated in persons with thiamine deficiency. During 1936–1938, Peters and his colleagues at Oxford University demonstrated that thiamine was essential for pyruvate metabolism; the term *biochemical lesion* was coined to describe the failure of thiamine-deficient pigeons to metabolize pyruvate. In 1937, Lohman and Schuster discovered the active coenzyme form of thiamine to be thiamine pyrophosphate (TPP; cocarboxylase) (Fig. 1). In 1953, Horecker and his associates established the coenzymic role of TPP for transkotase (2–5).

II. CHEMISTRY

A.   Structure and Nomenclature

The chemical name of thiamine, formerly known as vitamin B_1, vitamin F, aneurine, or thiamine, is 3-(4-amino-2-methylpyrimidin-5-ylmethyl)-5-(2-hydroxyethyl)-4-methylthiazolium (Fig. 1) (6). The free vitamin is a base. It is isolated or synthesized and handled as a solid thiazolium salt, e.g., thiamine chloride hydrochloride, thiamine mononitrate.

B.   Synthesis

1.   Chemical Synthesis

The original synthesis of thiamine was accomplished almost simultaneously in 1936/37 by Williams et al., Todd and Bergel, and Andersag and Westphal by slightly different procedures that are used today with only slight modifications for commercial production. Either the pyrimidine and thiazole rings can be prepared separately and condensed via the bromide, or the pyrimidine ring can be synthesized and the thiazole ring added to it. Steps of chemical synthesis of thiamine have been reviewed (2).

2.   Biosynthesis

Thiamine cannot be synthesized by animals to any significant extent. Many microbial species can synthesize thiamine if pyrimidine and thiazole are available, and still others have a dependence on thiamine. Higher plants can synthesize thiamine de novo. In these organisms, the pyrimidine precursor of thiamine is synthesized by a different pathway from the pyrimidine precursor of the pyrimidine nucleotides (2). Thiamine biosynthesis requires phosphorylation of pyrimidine and thiazole before synthesis of the thiamine monophosphate (TMP) molecule by thiamine phosphate pyrophosphorylase. TMP is then hydrolyzed by phosphatase to yield free thiamine. Thiamine pyrophosphokinase, which is responsible for the synthesis of TPP from the free thiamine, has been found in yeast and mammalian tissue. *E. coli* can synthesize TPP from TMP. Thiamine triphosphate (TTP) is synthesized in brain by TPP-ATP phosphoryltransferase (Fig. 2). However, studies of ^{14}C-thiamine turnover show very little incorporation of ^{14}C into TTP. TPP must be bound to an endogenous protein to act as a substrate, whereas free TPP does not affect the reaction (3).

C. Chemical Properties

Thiamine hydrochloride is a white crystalline substance. It is readily soluble in water, only partly soluble in alcohol and acetone, and insoluble in other fat solvents. In the dry form, it is stable at 100°C. The thiamine stability in aqueous solutions depends upon pH. Below pH 5 it is quite stable to heat and oxidation; above pH 5 it is rapidly destroyed

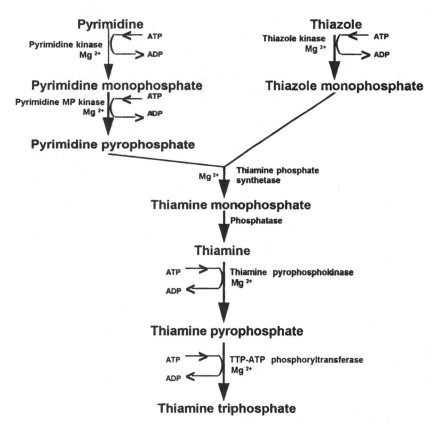

Fig. 2 Biosynthetic pathway of thiamine and its phosphorylated derivatives. Pyrimidine, 2-methyl-4-amino-5-hydroxymethylpyrimidine; thiazole, 4-methyl-5-(2-hydroxymethyl)thiazole; MP, monophosphate; TTP, thiamine triphosphate.

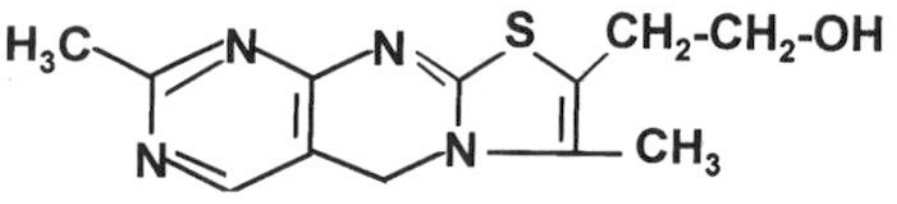

Fig. 3   Structural formula of thiochrome.

by autoclaving, and at pH 7.0 or higher by boiling. Thiamine is readily cleaved at the methylene bridge into 2-methyl-4-amino-5-methylpyrimidylsulfonate and 4-methyl-5-(2-hydroxyethyl)thiazole by sulfite treatment at pH 6.0 or above. At pH 8.0 or above, thiamine turns yellow and is destroyed by a complex series of irreversible reactions. In strong alkaline solution with the presence of oxidizing agents, e.g., potassium ferricyanide, thiamine is converted to thiochrome (Fig. 3), which is fluorescent and is used to determine thiamine in foods, feeds, pharmaceutical preparations, and biological fluids. Thiamine is precipitated by iron and ammonium citrate, tannin, and various alkaloids (2,4,5).

Thiamine forms esters by combining with various acids at the hydroxyethyl side chain. The most important esters are TMP, TPP, and TTP. The chemical synthesis of TPP by phosphorylation of thiamine and the separation from TMP and TTP provide its commercial quantities (2). The two most common commercial preparations of thiamine are thiamine hydrochloride and thiamine mononitrate, which are widely available in various countries. Other commercial forms that are produced by Japanese pharmaceutical industries include thiamine allyl disulfide (TAD), thiamine propyl disulfide (TPD), thiamine tetrahydrofurfuryl disulfide (TTFD), and O-benzoyl thiamine disulfide (Fig. 4). They are more lipid-soluble and less water-soluble, better retained in the body than thiamine hydrochloride, and readily converted to thiamine in vivo (2,7).

D.   Molecular Structure and Biological Activity

Studies in animals using various thiamine analogs have shown that both pyrimidine and thiazole moieties are needed for its vitaminic activity, which is maximal when only one methylene group bridges the two moieties. In the thiazole portion, the quaternary nitrogen and a hydroxyethyl group at C-5 are needed, as is the amino group at C-4 in the pyrimidine portion (Fig. 1) (2,4,5).

E.   Thiamine Analogs with Antagonistic Action

Several synthesized thiamine analogs are shown to be thiamine antagonists, which are employed to produce thiamine deficiency in animals. They are categorized into compounds having substitutions on the pyrimidine or thiazole rings, compounds having modifications of the pyrimidine or thiazole rings, and compounds having anticoccidial action. The representatives of these groups of thiamine antagonists are oxythiamine, pyrithiamine, and amprolium (Fig. 5) (2,4).

1.   Compounds with Substitutions on the Pyrimidine or Thiazole Rings

Extensive studies of thiamine-catalyzed reactions in both nonenzymatic and enzymatic systems by Breslow and Krampitz have shown that the catalytic function of thiamine involves C-2 of the thiazole with its ability to form a carbanion. Thus, substitution at C-2 of thiazole, i.e., 2-methylthiamines completely eliminates both its catalytic and biologic activities (2).

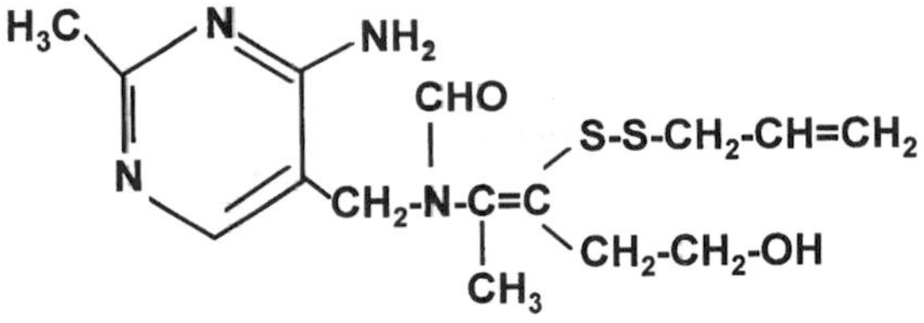

Thiamine allyl disulfide

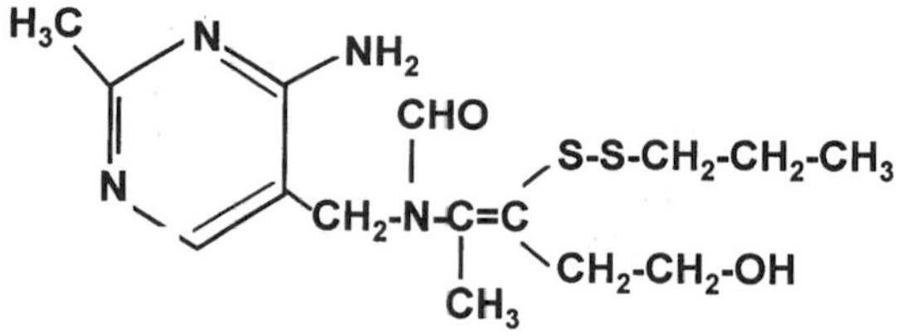

Thiamine propyl disulfide

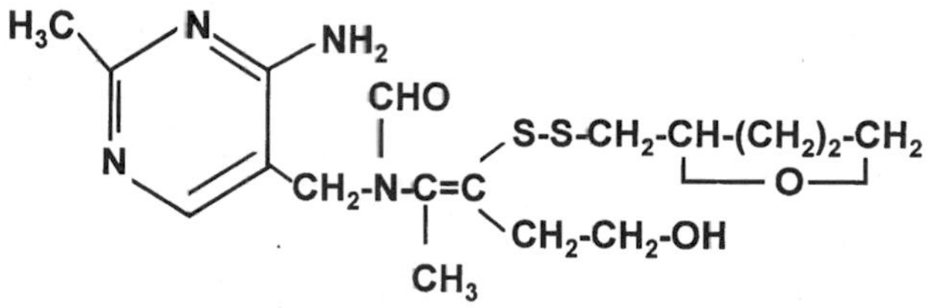

Thiamine terahydrofurfuryl disulfide

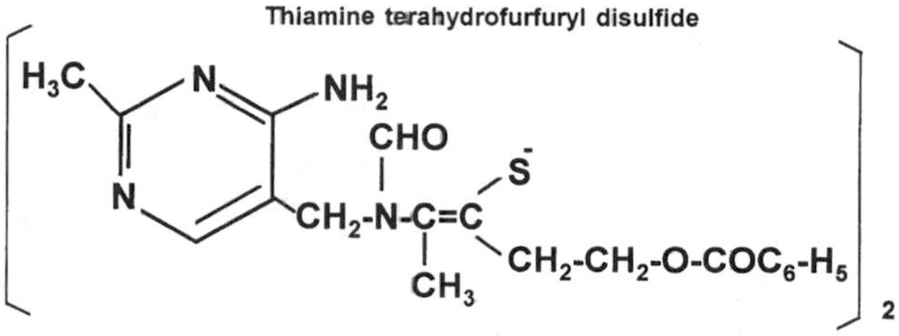

O-Benzoylthiamine disulfide

Fig. 4 Structural formulas of thiamine derivatives.

An amino group at the 4-position on the pyrimidine ring plays an essential role in the release of the aldehyde adduct from the thiazole C-2 after splitting of the —C—C— bond. Oxythiamine is a potent thiamine antagonist because it can form oxythiamine pyrophosphate, which can bind with thiamine-dependent apoenzymes but cannot complete the reaction (2).

2. Compounds with Modifications of the Pyrimidine or Thiazole Rings

Among various compounds having modifications of the pyrimidine or thiazole rings, only pyrithiamine, in which a —CH=CH— has been substituted for S, is a potent thiamine antagonist.

Among the aforementioned two groups of thiamine antagonists, only oxythiamine and pyrithiamine have been studied extensively in animals. Oxythiamine produces most

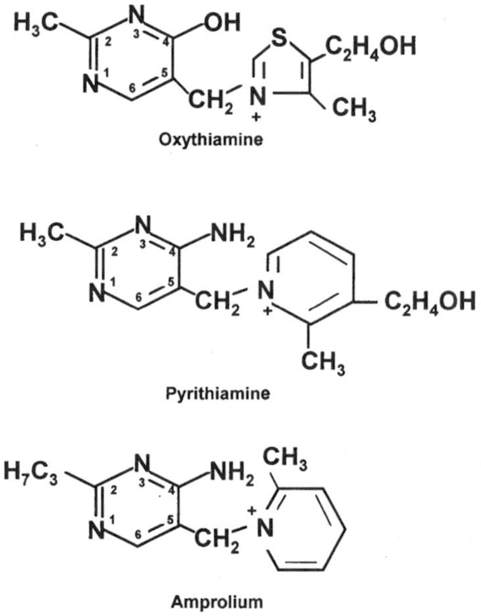

Fig. 5 Structural formulas of thiamine antagonists.

of the usual symptoms and biochemical changes associated with thiamine deficiency. The anorexia and inanition after oxythiamine treatment appear much earlier and more severe than those after pyrithiamine treatment or thiamine deprivation. Since oxythiamine is incapable of crossing the blood–brain barrier, it never produces the neurological manifestations. On the contrary, pyrithiamine produces chiefly neurological manifestations (2).

3. Compounds Having Anticoccidial Action

Chick anticoccidial compounds contain pyrimidine, which is combined through the methylene bridge to a quaternary nitrogen of a pyridine ring. Since they do not contain a hydroxyethyl group they cannot be phosphorylated. Thus, they are not good inhibitors of the TPP-dependent enzymes. A large number of anticoccidial compounds have been produced by various substitutions on the pyrimidine and pyridine rings. In small doses they inhibit chiefly thiamine transport across bacterial cell walls and hence exert anticoccidial action, whereas at larger doses they also inhibit thiamine transport across the chick intestinal wall and hence produce thiamine deficiency in the chicks. Amprolium, 1-(4-amino-2-*n*-propyl-5-pyrimidinylmethyl)-2-picolinium bromide hydrobromide, is most extensively used commercially (2).

III. FOOD-RELATED ISSUES

A. Food Sources

Table 1 shows thiamine content in milligrams per 100 g edible portion of selected foods compiled from the food composition table for use in East Asia published by the U.S. Department of Health, Education and Welfare and the Food and Agricultural Organization

Table 1 Thiamine Content of Some Foods

Food and description	Thiamine (mg/100 g[a])	Food and description	Thiamine (mg/100 g)
Yeast		Mung bean:	
Baker's, dried	2.33	Whole seeds, dried	0.53
Brewer's, debittered	15.61	Strip, dried	0.05
Pork		Peanut:	
Carcass, fresh, lean	0.69	Raw	0.97
Ham, smoked, lean	0.70	Boiled	0.56
Beef		Soybean:	
Carcass, fresh, very lean	0.07	Whole mature seeds, dried	0.66
Chicken, raw		Milk, unsweetened	0.05
Very young birds	0.08	Curd, raw, plain	0.04
Mature birds	0.08	Maize, whole-kernel:	
Liver, raw		Yellow	0.29
Beef	1.68	Popcorn, popped, oil added	0.12
Chicken	1.92	Starch	0
Duck	1.32	Rice:	
Hog	3.57	Bran	1.26
Kidney, raw		Home-pounded	0.20
Beef	1.58	Milled	0.10
Hog	1.31	Cooked, milled	0.02
Abalone, raw	0.24	Noodles, freshly made	0.04
Bass, sea, raw	0.15	Noodles, cooked	Trace
Catfish, freshwater, raw	0.10	Wheat:	
Crab, sea, blue	0.05	Whole grain, hard, red winter	0.37
Lobster, raw	0.01	Germ	2.10
Duck eggs, raw:		Bran	0.54
Whole	0.16	Asparagus: green, raw	0.15
White	Trace	Bamboo shoots, raw	0.09
Yolk	0.54	Cabbage, Chinese, raw	0.07
Hen eggs, raw:		Carrot, raw	0.06
Whole	0.10	Apple, raw	0.02
White	0.01	Banana, ripe	0.03
Yolk	0.24	Guava, raw	0.05
Cheese, blue or cheddar	0.04	Orange, raw	0.04
Ice cream, regular	0.05	Butter, salted	0.01
Milk, cow: fluid, 3.5% fat	0.04	Lard	0
Milk, human, fluid, whole	0.02	Soybean oil	0
		Tallow	0.07

[a] Edible portion.

Source: Compiled from Food Composition Table for Use in East Asia (8).

of the United Nations (8). Though thiamine is found in a large variety of animal and vegetable products, it is abundant in only a few foods. Excellent sources of thiamine are yeast, especially dried brewer's yeast; liver and kidney; lean pork; and legumes, including mung bean, soybean, and peanut. In cereal grains, thiamine is low in the endosperm but high in the germ. Thus, the thiamine contents in rice bran and home-pounded rice is higher than that in milled rice. In eggs, thiamine is also unevenly distributed, being negligible in egg white but present in egg yolk. Thiamine is absent in fats, oils, and refined sugars. Milk and milk products, sea foods, fruits, and vegetables are not good sources of thiamine.

Foods lacking in thiamine are manufactured, i.e., milled rice, cereal flours, refined sugars, separated animal fats, and separated vegetable oils. It should be emphasized that Table 1 shows food sources rich in thiamine without consideration to the amount being regularly consumed. For example, yeast and wheat germ are rich sources of thiamine, but they are normally eaten in small amounts.

In most animal products, 95–98% of thiamine is present as TMP, TPP, and TTP, with about 80–85% as TPP, whereas in plant products thiamine occurs in the nonphosphorylated form (2).

B. Bioavailability

Thiamine status depends not only on the thiamine content of foods but on its bioavailability. Factors influencing bioavailability are thiamine losses after handling and processing, ethanol consumption, presence of antithiamine factors (ATFs), and folate and protein status (2,5,9,10).

1. Thiamine Losses After Handling and Processing of Foods

As noted above, several factors, including pH, temperature, oxidation, solubility, radiation, and the use of sulfites in handling and processing of foods, affect thiamine losses.

a. pH

Thiamine is stable at acid pH and becomes unstable at pH 7.0 or higher. In cooking practice, the addition of sodium bicarbonate to green beans and peas to retain green color or to dried beans to facilitate softening can lead to large losses of thiamine (2,5).

b. Temperature

Thiamine in tissues that is bound to protein is more stable to thermal destruction than is free thiamine. Thiamine losses on cooking and canning of meats, baking of breads, and cooking of vegetables are 25–85%, 5–35%, and 0–60%, respectively. The variation depends on temperature, time, and types of food products. In pasteurization, sterilization, spray-drying, roller-drying, and condensing of milk, thiamine losses are 9–20%, 30–50%, 10%, 15%, and 40%, respectively. Though freezing has no effect on the thiamine content of foods, thawing causes thiamine losses into the drip fluid. Thiamine losses on storage of canned fruits and vegetables at usual room temperatures are 0–20% for a year. Processing foods at higher temperature and pH in the presence of oxygen or other oxidants lead to the formation of thiamine sulfides and disulfides, thiochrome, and other oxidation products. Only thiamine sulfides and disulfides still retain the biological activity of thiamin (2,5).

Thiamine is cleaved by residual chlorine (0.2–2.0 ppm of chlorine) in proportion to the rise in temperature, pH, and concentration of residual chlorine. When rice is boiled in an electric rice cooker, the thiamine in the rice is cleaved by residual chlorine. When rice is cooked in tap water containing 200 ng/mL of chlorine or distilled water without chlorine, thiamine losses from rice are 56.5% and 32.2%, respectively. The degradation of thiamine depends on the amount of ionization of residual chlorine governed by the pH of water. At pH below 5.0, the residual chlorine is present as HClO and the ratio of ClO⁻ increases in proportion to an increase in pH:

$$HClO \rightleftharpoons H^+ + ClO^-$$
$$pH\ 5 \leftarrow pH\ 7.5 \rightarrow pH\ 10$$

Thus, ClO^- plays a key role in the cleavage of thiamine to hydroxymethylpyrimidine and 4-methyl-5-β-hydroxymethythiazide. Though hydroxymethylpyrimidine can induce convulsions in mice, the amount converted from thiamine in rice by chlorine in tap water is so small that its toxic effect should be negligible. However, thiamine loss in polished rice due to residual chlorine during the cooking process needs attention because rice is the major source of dietary thiamine in Asian population (11).

c. Solubility

Since thiamine is highly water-soluble, significant thiamine losses occur during cooking of foods when there is the excessive use of water that is then discarded. About 85% of thiamine is lost by discarding the water after soaking the rice (5,9,10). Though parboiling rice causes the thiamine to move from the outer layers to the inner layers of the rice kernel, thus leading to greater thermal stability and smaller losses due to washing of the thiamine, parboiled rice is not widely available or widely consumed by the Asian population.

d. Oxidation

Processing foods in the presence of oxygen or other oxidants can lead to the formation of thiamine sulfides and disulfides, thiochrome, and other oxidation products. This can lead to irreversible loss of thiamine activity (2).

e. Radiation

Thiamine is also destroyed by x-rays, γ-rays, and UV irradiation. UV irradiation leads to the production of 2-methyl-4-amino-5-aminoethylpyrimidine (2).

f. Sulfites

Sulfites destroy thiamine. Sulfites that form in the treatment of fruits during dehydration with SO_2 destroy most of the thiamine present (5).

2. Ethanol

Excessive alcohol ingestion is associated with thiamine deficiency. Ethanol given orally or intravenously inhibits intestinal thiamine uptake (2,5,12,13). Ethanol reduces the rate of intestinal absorption and the net transmural flux of thiamine. Ethanol inhibits only the active but not the passive component of thiamine transport by impeding the cellular exit of thiamine across the basolateral or serosal membrane. The impairment of thiamine exit from the enterocyte correlates with the Na,K-ATPase activity. This relationship may be at least partly related to the effect of ethanol in increasing the fluidity of enterocyte brush-border and basolateral membranes (13). The maximum absorption of ^{35}S-thiamine in alcoholics may be as low as 1.5 mg, similar to that seen in patients with intestinal resection in whom the reduction is attributable to a decrease in receptor sites (12).

3. ATFs (Antithiamine Factors)

ATFs occurring in foods can alter thiamine structure and reduce the biological activity of thiamine. ATFs are divided into two groups, according to their stability to heat, i.e., thermolabile and thermostable.

a. Thermolabile ATFs

The thermolabile ATFs include thiaminase I (EC 2.5.1.2) and thiaminase II (EC 3.5.99.2). Thiaminase I is found in viscera of freshwater fish (mainly in viscera), shellfish, fern, and a limited number of sea fish and plants, and produced in microorganisms, e.g., *Bacillus thiaminolyticus* and *Clostridium thiaminolyticus*. It catalyzes the cleavage of thiamine by an exchange reaction with an organic base or a sulfhydryl compound involving a nucleophilic displacement on the methylene group of the pyrimidine moiety of thiamine. Thiaminase II is found in several microorganisms, e.g., *Bacillus aneurinolyticus*, *Candida aneurinolytica*, *Trichosporon*, and *Oospora*. It directly catalyzes the hydrolysis of thiamine (Fig. 6) (14,15).

Thiaminases are usually inactive in living cells. They are activated when they are homogenized in a water solution around pH 4–8 or excreted from cells or microorganisms into the medium. The thiamine cleavage in the diets containing thiaminase can occur during storage or preparation prior to ingestion or during passing through the gastrointestinal tract. Continual intake of raw freshwater fish with or without fermentation, raw shellfish, and ferns are risk factors for the development of thiamine deficiency (5,9,10,14–16).

b. Thermostable ATFs

The thermostable ATFs have been demonstrated in fern, tea, betel nut, and a large number of other plants and vegetables and some animal tissues. In animal tissues, it is thought that myoglobin, hemoglobin, and hemin bind thiamine. The ATFs found in plants and vegetables are related to *ortho*- and *para*-polyphenolic compounds, e.g., caffeic acid (3,4-dihydroxycinnamic acid), chlorogenic acid [3-(3,4-dihydroxycinnamoyl) quinic acid], and tannic acid (tannin) (5,9). In vitro studies reveal that the antithiamine activity of polyphenols requires pH 6.5 or higher and oxygen. The high pH is necessary for the ionization of the polyphenols and the opening of the thiazole moiety of thiamine at C-2 to yield the SH form of thiamine. Oxygen helps the oxidation and polymerization of polyphenols to yield active quinones and relatively less active polymerized products. Oxidation of the SH form of thiamine by the active quinones leads to the formation of thiamine disulfide. Further hydrolysis and oxidation yield products that do not exhibit biological activity of thiamine. Ascorbic acid and other reducing agents prevent the formation of quinone, thiamine disulfide, or both. However, chemical modification of thiamine by the oxidation of

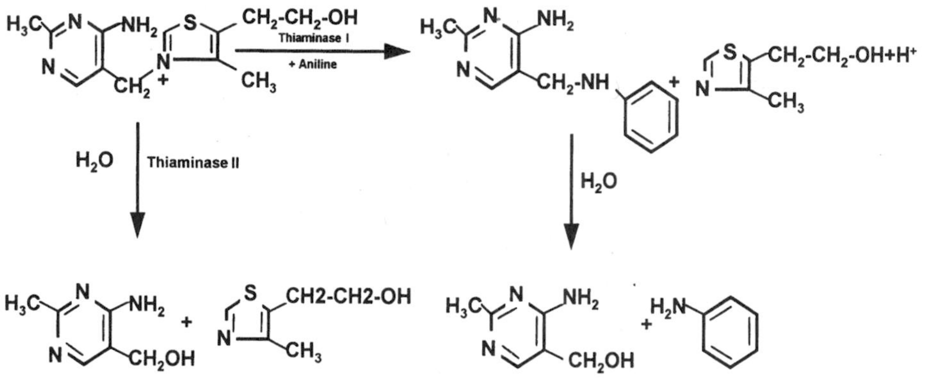

Fig. 6   Clevage of thiamine by thiaminase I and thiaminase II. Thiaminase I activity requires organic bases or thiol compounds.

quinone may not be the only mechanism to reduce thiamine bioavailability. High concentrations of divalent cations, including calcium and magnesium ions present in water, augment the precipitation of thiamine by tannin, which makes thiamine less available for intestinal absorption. Ascorbic, tartaric, and citric acids present in many fruits and vegetables can lower such precipitation presumably by sequestering divalent cations.

Human studies have shown that tea drinking, tea leaf chewing, coffee or decaffeinated coffee drinking, and betel nut chewing lead to biochemical thiamine depletion. Ascorbic acid intake from either pharmaceutical preparations or from foods improves thiamine status of the subjects (16,17).

Studies in rats have demonstrated that prolonged tea consumption, in contrast to consumption of water, lowered (60–80%) blood transketolase, brain total thiamine, brain transketolase activity, as well as α-ketoglutarate and pyruvate dehydrogenase activities. Thiamine deficiency caused by prolonged consumption of tea in rats was reversed by discontinuing tea consumption; brain thiamine and brain and blood transketolase activities were restored to normal levels within a week, whereas the time required for complete restoration of pyruvate dehydrogenase and α-ketoglutarate activities is about 2 weeks. Tea-treated rats receiving intraperitoneal injection of 4 mg of thiamine 3 h prior to decapitation had TPP-dependent enzymic activities exceeding the normal levels, which suggests that the decrease in thiamine-dependent enzymic activities may be due not only to the lack of TPP but also to the decreased synthesis of apoenzymes (18). Impairment of synthesis of whole-brain acetylcholine was also observed in rats consuming tea (19).

4. Folate and Protein Status

Folate-deficient rats absorb thiamine less efficiently than pair-fed controls (20). These data imply that folate may have a role in maintaining the integrity of the active transport process of thiamine (13). Subjects with folate or protein deficiency show a significant reduction in the maximum absorption of ^{35}S-thiamine. The interaction of protein-energy malnutrition and thiamine absorption is demonstrated by an increase in the maximum absorption of ^{35}S-thiamine in malnourished alcoholics, with a decreased maximum absorption of thiamine after correction of protein-energy malnutrition (21).

IV. PHYSIOLOGICAL RELATIONSHIPS

Ingested thiamine is fairly well absorbed, rapidly converted to phosphorylated forms, stored poorly, and excreted in the urine in a variety of hydrolyzed and oxidized products (5).

A. Absorption

The small intestine absorbs thiamine by two mechanisms. At concentrations exceeding 2 μM, thiamine is absorbed by passive diffusion; at concentrations below 2 μM, thiamine is absorbed by an active process. The findings that the lack of Na^+ or inhibition of ATPase with ouabain blocks the uptake of thiamine at low concentrations by the intestinal cells suggest the presence of a specific carrier. This is supported by the isolation of a thiamine-binding protein (TBP) associated with thiamine transport into and out of the cell of *Escherichia coli*. The 4-amino and the imidazole quaternary nitrogen are necessary for thiamine uptake by the rat small intestine, whereas the 2-methyl and 5-hydroxyethyl appear to be necessary for thiamine binding to the carrier protein. Active thiamine absorption is greatest in the jejunum and ileum (2,5,9,13,22,23).

A high percentage of the thiamine in the epithelial cells is phosphorylated, whereas the thiamine arriving on the serosal side of the mucosa is largely free thiamine. Thus, the entry of thiamine into the mucosal cells is linked with a carrier-mediated system that is dependent either on thiamine phosphorylation-dephosphorylation coupling or on some metabolic energetic mechanisms, possibly activated by Na^+. Thiamine exit from the mucosal cell on the serosal side is dependent on Na^+ and on the normal function of ATPase at the serosal pole of the cell (2,5,9,13,23).

Investigations of the absorption of ^{35}S-thiamine in humans indicate that its intestinal transport is rate-limited and behaves according to Michaelis-Menten kinetics, which yielded a V_{max} of 8.3 ± 2.4 mg and a K_m of 12.0 ± 2.4 mg (12). Earlier studies by Japanese investigators have also shown that a single oral dose of thiamine greater than 6 mg does not result in increased urinary thiamine excretion. The results imply that the excessive amount of oral thiamine over 6 mg cannot be absorbed (22) and are consistent with the subsequent studies in six healthy Australian volunteers receiving an oral dose of 10 mg of thiamine. In the Australian study, the mean serum thiamine rose only marginally at 30 min from 5.1 to 5.9 µg/L and peaked at 7.2 µg/L—an increase of 42%. Six hours after the test dose, the serum thiamine concentration had fallen back to its basal level (24). On the contrary, healthy subjects taking each of the following physiological doses of thiamine for a consecutive period of 5 days in the following order: 150, 450, 750, 1050, 1350, 1650, 1950, and 2250 µg had a highly significant correlation between the oral dose and urinary excretion of thiamine ($y = 41.0239 + 0.0465$, df = 68, $r = 0.8618$, $P < 0.001$) (25). Thus, all of the results indicate that passive absorption of thiamine is not significant in humans.

B. Transport

Thiamine is carried by the portal blood to the liver. In normal adults, 20–30% of plasma thiamine is protein-bound, all of which appears to be TPP (24). The transport of thiamine into erythrocytes seems to be a facilitated diffusion process, whereas it enters other cells by an active process (2,26). Erythrocytes contain mainly TPP (24).

Like the intestinal absorption of thiamine, thiamine transport across the blood–brain barrier also involves two different mechanisms. However, the saturable mechanism at the blood–brain barrier may be dependent on membrane-bound phosphatases. In this regard, it differs from the energy-dependent processes described for the gut and cerebral cortex cells. Studies in the rat reveal that a saturable mechanism with a mean K_m of 2.2 nmol/mL and a mean V_{max} of 7.3 nmol/g per hour accounts for 95% of cerebellar and 91% of cerebral cortex uptake at physiological plasma thiamine concentrations. TMP transport rates are 5–10 times lower than those of thiamine. Thiamine uptake rates are 10 times the maximal rate of thiamine loss from the brain (3).

A TBP has been identified in chicken egg (white and yolk) and liver (27) as well as in the serum of pregnant rats (28). The molecular weight of highly purified TBP isolated from egg white was 38,000. The TBP binds to ^{14}C-thiamine with a molar ratio of 1 and an association constant of 0.3 µM (27). Estrogen induces hepatic synthesis of TBP and modulates its plasma level. The vital role of TBP in the transfer of thiamine across the placenta is suggested by the observation that the passive immunization of pregnant rats

(4–16 days) with antibodies to chicken TBP but not ovalbumin resulted in fetal resorption (28).

C. Tissue Distribution and Storage

The total amount of thiamine in a normal adult is approximately 30 mg. High concentrations are found in skeletal muscles, heart, liver, kidneys, and brain. About 50% of the total thiamine is distributed in the muscles. In spinal cord and brain, the thiamine level is about double that of the peripheral nerves. Leukocytes have a 10-fold higher thiamine concentration than erythrocytes. Thiamine has a relatively high turnover rate in the body. The biological half-life of ^{14}C-thiamine is 9–18 days. Besides, thiamine is not stored in large amounts or for any period of time in any tissue (2,5,9,22).

D. Metabolic Modification

Of the total thiamine in the body, about 80% is TPP, 10% is TTP, and the remainder is TMP and thiamine. The three tissue enzymes known to participate in formation of the phosphate esters are thiamine pyrophosphokinase, which catalyzes the formation of TPP from thiamine and ATP; TPP-ATP phosphoryltransferase, which catalyzes the formation of TTP from TPP and ATP (Fig. 2); and thiamine pyrophosphatase, which hydrolyzes TPP to form TMP (2,3,5,29).

At least 25–30 metabolites have been noted to occur in the urine of rats and men given thiamine labeled in either the pyrimidine or thiazole moiety. Those metabolites that have been identified are 2-methyl-4-amino-5-pyrimidine carboxylic acid (pyrimidine carboxylic acid), 2-methyl-4-amino-5-hydroxymethylpyrimidine, 4-methylthiazole-5-acetic acid (thiazole acetic acid), 3-(2'-methyl-4'-amino-5'-pyrimidylmethyl)-4-methylthiazole-5-acetic acid (thiamine acetic acid), 2-methyl-4-amino-5-formylaminomethylpyrimidine, and 5-(2-hydroxyethyl)-4-methylthiazole (thiazole). Of these, pyrimidine carboxylic acid, thiazole acetic acid, and thiamine acetic acid are the major metabolites excreted in the urine of rats and men. Rat liver alcohol dehydrogenase is involved in the in vivo metabolism of thiamine and its thiazole moiety to their corresponding acids. In vitro experiments indicate that although thiamine itself is a poor substrate for the enzyme, its thiazole moiety is oxidized at a faster rate than is ethanol (30).

E. Excretion

Thiamine and its metabolites are mainly excreted in the urine. Very little thiamine is excreted in the bile. Early milk has low thiamine levels. Thiamine administered by oral or parenteral route is rapidly converted to TPP and TTP in the tissues. Thiamine in excess of tissue needs, as well as binding and storage capacity, is rapidly excreted in the urine in the free form (2,4,5,24,25).

V. FUNCTIONS

Functions of thiamine can be categorized into established and plausible functions. The established functions of thiamine are biochemical functions in which TPP serves as the coenzyme of biochemical reactions, whereas the plausible functions are neurophysiological functions.

A. Biochemical Functions

In mammalian systems, TPP functions as the Mg^{2+}-coordinated coenzyme for the active aldehyde transfers, which include the oxidative decarboxylation of α-keto acids and trans-ketolase reaction (2,4,5,31). The key feature of TPP is that the carbon atom between the nitrogen and sulfur atoms in the thiazole ring is much more acidic than most $=CH—$ groups. It ionizes to form a carbanion, which readily adds to the carbonyl group of α-keto acids or ketose. The positively charged ring nitrogen of TPP then acts as an electron sink to stabilize the formation of a negative charge, which is necessary for decarboxylation. Protonation then gives hydroxyethyl TPP (5).

1. Oxidative Decarboxylation of α-Keto Acids

Pyruvic, α-ketoglutaric, and branched-chain α-keto acids undergo oxidative decarboxylation.

a. Oxidative Decarboxylation of Pyruvic Acid

The net reaction of oxidative decarboxylation of pyruvate catalyzed by the pyruvate dehydrogenase complex (PDHC) is

$$Pyruvate + CoA + NAD^+ \rightarrow acetyl\ CoA + CO_2 + NADH + H^+$$

In addition to the stoichiometric coenzymes consisting of CoA and NAD^+, TPP, lipoic acid, and flavin adenine dinucleotide (FAD) also serve as the coenzymes. The PDHC is localized in the mitochondrial inner membrane and is an organized assembly of three enzymes. The conversion of pyruvate to acetyl CoA consists of four steps (2,4,5,9,31). First, pyruvate is decarboxylated after it combines with TPP is catalyzed by pyruvate dehydrogenase. Second, the hydroxyethyl group attached to TPP is oxidized by the disulfide groups of lipoamide to form an acetyl group and concomitantly transferred to lipoamide to yield acetyl lipoamide. This step is catalyzed by the lipoic acid-bound dihydrolipoyltransacetylase. Third, the acetyl group is transferred from acetyl lipoamide to CoA to form acetyl CoA catalyzed by dihydrolipoyl transacetylase. Fourth, the regeneration of the oxidized form of lipoamide is catalyzed by the FAD-dependent dihydrolipoyl dehydrogenase; a hydride ion is transferred to an FAD prosthetic group of the enzyme and then to NAD (Fig. 7) (31,32).

b. Oxidation Decarboxylation of α-Ketoglutaric Acid

The net reaction of oxidative decarboxylation of α-ketoglutarate, taking place in the tricarboxylic acid (TCA) cycle and catalyzed by the α-ketoglutarate dehydrogenase complex (α-KGDHC), is as follows:

$$\alpha\text{-Ketoglutarate} + CoA + NAD^+ \rightarrow succinyl\ CoA + CO_2 + NADH + H^+$$

The coenzyme requirements and the steps in formation of succinyl CoA are analogous to the oxidative decarboxylation of pyruvate (2,4,5,9).

c. Oxidative Decarboxylation of Branched-Chain α-Keto Acids

The oxidative decarboxylation of the three branched-chain α-keto acids, i.e., α-ketoisocaproate, α-keto-β-methylvalerate, and α-ketoisovalerate, to yield isovaleryl CoA, α-methylbutyryl CoA, and isobutyryl CoA, respectively, is catalyzed by the branched-chain α-keto acid dehydrogenase complex (BC α-KADHC), which is analogous to complexes of pyruvate and α-ketoglutarate (2,4,5,9).

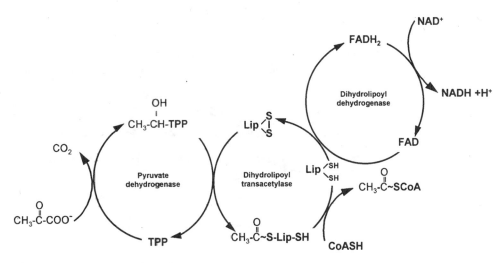

Fig. 7 The pyruvate dehydrogenase complex reactions. TPP, thiamine pyrophosphate; Lip, lipoic acid.

The aforementioned three α-keto acid multienzyme complexes play the significant role in energy generating pathways (Fig. 8). The PDHC irreversibly commits pyruvic acid, a three-carbon intermediate derived from glucose catabolism, to convert to acetyl CoA, which participates in three important metabolic fates: complete oxidation of the acetyl group in the TCA cycle for energy generation; conversion of an excess acetyl CoA into the ketone bodies; and transfer of the acetyl units to the cytosol with subsequent biosynthesis of sterols and long-chain fatty acids. Because PDHC occurs at a significant branch point in the metabolic pathways, its activity is strictly regulated by two separate types of mechanisms. The first mechanism is competitive end-product inhibition of catalytic PDHC by acetyl CoA and NADH. The second mechanism involves covalent modification of the PDHC by phosphorylation/dephosphorylation (inactive PDHC/active PDHC) mechanism, mediated by a specific protein kinase that is tightly bound to the PDHC and by a specific phosphoprotein phosphatase that is much less tightly associated with the PDHC (31,32).

The pyruvate dehydrogenase kinase/phosphatase system is regulated by a number of factors. Inactivation of PDHC is accomplished by an Mg^{2+}-ATP-dependent pyruvate dehydrogenase kinase. Pyruvate dehydrogenase kinase is inhibited by ADP, pyruvate, CoA, NAD^+, TPP, and calcium, whereas it is stimulated by NADH and acetyl CoA; TPP inhibits phosphorylation by binding at the catalytic site of pyruvate dehydrogenase to promote a conformational change, which in turn causes one of the serine hydroxy groups on the subunit of pyruvate dehydrogenase to become less accessible to pyruvate dehydrogenase kinase. The pyruvate dehydrogenase phosphatase reaction requires Mg^{2+}. It is stimulated by Ca^{2+} and inhibited by NADH, and the inhibition is reversed by NAD^+ (31).

The primary metabolic fate of acetyl CoA in most cells is its complete oxidation in the TCA cycle with the generation of 2 CO_2, 2 GTP, and 4 reducing equivalents consisting of 3 NADH and 1 $FADH_2$, which generate 11 ATP by the electron transport system. Thus, the integrity of the TCA cycle is critical for the provision of cellular energy. α-KGDHC is the only enzyme in the TCA cycle that requires TPP as coenzyme. The equilibrium of the α-KGDHC reaction lies strongly toward succinyl CoA formation. In this reaction, the second molecule of CO_2, the second reducing equivalent (NADH + H^+) of the

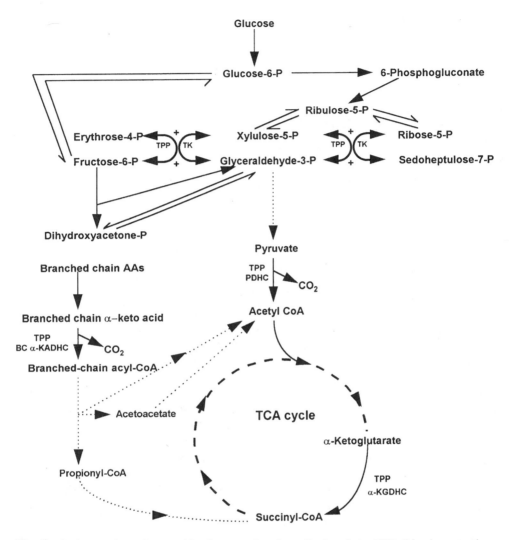

Fig. 8 Pathways dependent on thiamine pyrophosphate. P, phosphate; TPP, thiamine pyrophosphate; PDHC, pyruvate dehydrogenase complex; α-KGDHC, α-ketoglutarate dehydrogenase complex; BC α-KADHC, branched-chain α-keto acid dehydrogenase complex; TK, transketolase.

TCA cycle, and succinyl CoA, an energy-rich thioester compound, are produced. Succinyl CoA is conserved in a substrate level phosphorylation reaction in the next step of the TCA cycle. It is at the level of this reaction where α-ketoglutarate may leave the TCA cycle to be converted to glutamate by glutamate dehydrogenase, a mitochondrial enzyme, in the presence of NADH or NADP and ammonia. Unlike PDHC, the α-KGDHC is not regulated by a protein kinase. ATP, GTP, NADH, and succinyl CoA inhibits the α-KGDHC activity, whereas Ca^{2+} stimulates its activity (32).

Where the BCAAs valine, leucine, and isoleucine are present in excess over what is needed for protein synthesis, they are degraded through several steps to provide energy. The initial step in catabolism is reversible transamination with α-ketoglutarate to form the corresponding α-keto acids, i.e., α-ketoisovalerate, α-ketoisocaproate, and α-keto-βmethylvalerate, respectively. The second step is irreversible oxidative decarboxylation of the three branched-chain α-keto acids by BC α-KADHC to yield CO_2, NADH that feeds electrons to the electron transport system, and isobutyryl CoA, isovaleryl CoA, or α-methylbutyryl CoA. These three analogs of fatty acyl CoA are oxidized by specific dehydrogenases to form the corresponding α,β-unsaturated compounds. The remainder of the isoleucine degradation pathway is identical to that of fatty acid oxidation, with the provision of acetyl CoA and propionyl CoA, whereas that of the valine degradation provides propionyl CoA. The remainder of the leucine degradation is a combination of reactions used in β oxidation and ketone body synthesis with the provision of acetyl CoA and acetoacetate, which may be converted to two acetyl CoA. The acetyl CoA formed from the degradation of the BCAAs enters the acetyl CoA pool, which can be used for any of the aforementioned metabolic functions. Propionyl CoA formed from isoleucine and valine for the most part is not further oxidized as a fatty acid but is utilized for the formation of succinyl CoA, which is further metabolized via the TCA cycle. Thus, the three carbon atoms derived from valine and isoleucine may be completely oxidized to CO_2, but they can be incorporated into carbohydrate by oxidation of succinate to oxaloactate and the reactions of gluconeogenesis. On the contrary, both acetyl CoA and acetoacetate of leucine degradation are characteristic of fatty acid oxidation (33).

2. Transketolase Reaction

A TPP-dependent transketolase found in the cytosol catalyzes the reversible transfer of a glycolaldehyde moiety from the first two carbons of a donor ketose phosphate to the aldehyde carbon of an aldose phosphate in the pentose phosphate pathway. These reactions are as follows:

$$\text{Xylulose-5-phosphate} + \text{ribose-5-phosphate} \rightleftharpoons \text{glyceraldehyde-3-phosphate} + \text{sedoheptulose-7-phosphate}$$

$$\text{Xylulose-5-phosphate} + \text{erythrose-4-phosphate} \rightleftharpoons \text{glyceraldehyde-3-phosphate} + \text{fructose-6-phosphate}$$

The metabolic significance of the pentose phosphate pathway is not to obtain energy from the oxidation of glucose in animal tissues. Its primary purpose is to generate NADPH, which serves as a hydrogen and electron donor in reductive biosynthetic reactions, including the biosynthesis of fatty acids; to convert to pentoses, particularly ribose-5-phosphate, which are components of RNA, DNA, ATP, CoA, NAD, and FAD; and to catalyze the interconversion of C3, C4, C6, and C7 sugars, some of which can enter the glycolytic sequence (2).

B. Neurophysiological Functions

There is increasing evidence of the roles of thiamine in neurotransmitter function and nerve conduction (3,34).

1. Neurotransmitter Function

Though the findings are inconsistent, abnormal metabolism of four types of neurotransmitters (acetylcholine, catecholamine, serotonin, and amino acids) has been reported in thiamine deficiency. Acetylcholine, γ-aminobutyric acid (GABA), glutamate, and aspartate are produced primarily through the oxidative metabolism of glucose (3,34).

a. Acetylcholine

Studies in thiamine-deficient rats have revealed that there is no change in regional brain acetylcholine levels, but there are reductions in acetylcholine turnover and acetylcholine utilization in the cortex, midbrain, diencephalon, and brain stem. These results suggest the depression of central cholinergic mechanisms in thiamine-deficient rats. Since choline acetyltransferase is unaltered in thiamine deficiency, decreased PDHC activity limiting acetyl CoA production may underlie these changes (3).

b. Catecholamines

Decreased synthesis of catecholamines was demonstrated in severely thiamine-deficient rat brain; treatment with thiamine normalized catecholamine synthesis within 2 h, although some neurological signs persist (34). There is also evidence indicating that behavioral deficits in rats persisting after reversal of thiamine deficiency are linked to significant reductions in the norepinephrine content of cortex, hippocampus, and olfactory bulbs (3).

c. Serotonin

Decreased uptake of serotonin by cerebellar synaptosomes and significant increase in 5-hydroxyindoleacetic acid, a serotonin catabolite, occurs in thiamine-deficient rats. These findings are supported by autoradiographic evidence of decreased serotonin uptake in indoleaminergic afferents of cerebellum in thiamine-deficient rats (3,34).

d. Amino Acids

The levels of four amino acids with putative neurotransmitter functions, i.e., glutamate, aspartate, GABA, and glutamine, are decreased in thiamine-deficient rat brain, with marked changes in the cerebellum (3).

2. Nerve Conduction

That thiamine plays a special role in neurophysiology that is independent of its coenzyme function was originally postulated by von Muralt (35). The effect of thiamine antagonists on conduction in peripheral nerves, the localization of thiamine and thiamine pyrophosphatase in peripheral nerve membranes as opposed to axoplasm, the release of thiamine from nerve preparations by electrical stimulation, the association of released thiamine with membrane fragments of brain, spinal cord, and sciatic nerves all support the aforementioned view (2–5,9,35). This thiamine release appears to be due to hydrolysis of TPP and TTP. Tetrodotoxin, local anesthetics, acetylcholine, and serotonin have been shown to cause release of thiamine from nerve and nerve membrane segments (36–38). The finding that thiamine-phosphorylated derivatives are associated with the sodium channel protein has led to the hypothesis that TTP plays a fundamental role in the control of sodium conductance at axonal membranes (39).

VI. FUNCTIONAL CONSEQUENCES OF THIAMINE DEFICIENCY

Cardiac failure, muscle weakness, peripheral and central neuropathy, and gastrointestinal malfunction have been observed in both animals and humans on diets restricted in thiamine (9).

Although the precise biochemical defect responsible for the pathophysiological manifestations of thiamine deficiency is not established, experimental evidence indicates that thiamine has three major roles at the cellular level. The first relates to energy metabolism and concerns the oxidative decarboxylation of α-keto acids, the inhibition of which leads to a failure of ATP synthesis. The second concerns synthetic mechanisms as reflected by the transketolase reaction yielding NADPH and pentose. The third deals with the functions of neurotransmitters and nerve conduction. In Wernicke's disease, failure of energy metabolism predominantly affects neurons and their functions in selected areas of the central nervous system. Glial changes may be caused by biochemical lesions that affect transketolase and nucleic acid metabolism. Membranous structures are altered in a visible manner and secondary demyelination ensues (9).

VII. PATHOGENESIS OF THIAMINE DEFICIENCY AND HIGH-RISK CONDITIONS

Thiamine deficiency in free-living populations and hospitalized patients can be caused by inadequate intake, decreased absorption, and defective transport of thiamine, impaired biosynthesis of TPP, increased requirement, and increased loss of thiamine. These mechanisms may reinforce each other.

A. Inadequate Intake of Thiamine

1. Inadequate Intake of Thiamine with Concomitant High Intake of Carbohydrate

All of the published reports indicate that beriberi, the clinical condition of thiamine deficiency, in Asian populations is always associated with high intake of carbohydrate and insufficient intake of thiamine (5,9,10,30,40–42).

Recent studies in Thailand (10,43–46) and Japan (47–49) have confirmed that inadequate thiamine intake accompanied by high intake of carbohydrate derived from milled rice and refined carbohydrate is still the major cause of thiamine deficiency in Asian populations. For instance, dietary assessment in a village in northeastern Thailand has repeatedly shown that the villagers' protein and energy intakes are mainly supplied from a rice-based diet; 78% of their energy intake is carbohydrate, mainly derived from milled rice. Milled rice is a poor source of thiamine (Table 1), and about 85% of its thiamine content is lost when the water in which it has been soaked is discarded (10,44,50,51). Thus, their thiamine intake is low, e.g., 0.23 mg/1000 kcal (44). These dietary data are consistent with the high prevalence of biochemical thiamine depletion in that population (10,43,44). The adverse effects of the ATFs on thiamine status caused by consumption of raw fermented fish and chewing betel nuts have also been demonstrated in northeastern Thais, whereas inadequate thiamine status is also detected in northern Thais who regularly chew fermented tea leaves (16).

The reappearance of beriberi in Japan in the 1970s is related to high carbohydrate intake derived from milled rice, instant noodles, and sweet carbonated drinks, coupled with low thiamine intake (47–49).

2. Infants Breast-Fed by Thiamine-Deficient Mothers

The mean (± SEM) concentrations of thiamine in breast milk from mothers residing in rural areas of central, northern, and northeastern Thailand were 93 ± 9.9, 102 ± 6.5, and 119 ± 6.3 μg/L. These figures are lower than the concentration of thiamine in breast milk of 142 ± 1.4 μg/L from the North American mothers. The thiamine concentration in mothers with complaints suggestive of thiamine deficiency has been reported to be $87 \pm$

4.8 μg/L (52,53). These findings partly explain the prevalences of infantile beriberi in Thailand (10,54–58).

3. Chronic Alcoholism

The major cause of thiamine deficiency in Western populations is excessive intake of alcohol (ethanol). Chronic alcoholics may eat nothing for days, and when they eat their diet is often high in carbohydrate and low in thiamine (59). Excessive ethanol ingestion creates a metabolic burden in the human body. The liver is primarily responsible for the first two steps of ethanol metabolism:

$$CH_3CH_3OH + NAD^+ \longrightarrow CH_3CHO + NADH + H^+$$
$$\quad\text{Ethanol} \qquad\qquad\qquad \text{Acetaldehyde}$$

$$CH_3CHO + NAD^+ \longrightarrow CH_3COO^- + NADH + H^+$$
$$\text{Acetaldehyde} \qquad\qquad \text{Acetate}$$

The first step, catalyzed by alcohol dehydrogenase, generates acetaldehyde and NADH, and occurs in the cytosol. The second step, catalyzed by aldehyde dehydrogenase, generates acetate and NADH but occurs in the mitochondrial matrix space. The NADH generated by the second step can be used directly by the mitochondrial electron transfer chain, whereas the NADH generated by the first step cannot be used directly and must be oxidized back to NAD^+ by the glycerol phosphate and/or malate-aspartate shuttles. Since the intake of ethanol demands the hepatic disposal of the NADH, intake of even moderate amounts of ethanol results in inhibition of important processes requiring NAD^+, including gluconeogenesis and fatty acid oxidation (60).

Hepatic mitochondria have a limited capacity to oxidize acetate to CO_2 because the activation of acetate to acetyl CoA requires GTP, which is the product of the succinyl CoA synthetase reaction. The TCA cycle, and therefore GTP synthesis, is inhibited by NADH levels during ethanol oxidation. Much of the hepatic acetate generated from ethanol escapes the blood to other cells containing mitochondria, which oxidize it to CO_2 and H_2O through the TCA cycle (Fig. 8). The heart is the most active organ for the disposal of acetate (60). This process cannot be achieved with an inadequate supply of thiamine.

Acetaldehyde can also escape from the liver. Though its concentration in the blood is lower than that of acetate, acetaldehyde is a reactive compound that readily forms covalent bonds with functional groups of biologically important compounds (60).

Not only insufficient thiamine intake but also other factors are involved in the development of thiamine deficiency in chronic alcoholics. These include ethanol inhibition of thiamine absorption, decrease in thiamine absorption secondary to protein or folate deficiency, impaired conversion of thiamine to TPP in the presence of alcoholic liver injury, and reduced hepatic storage of thiamine (2,12,13,21,26,30,61).

4. Patients on Parenteral Nutrition

Inadequate intake of nutrients, including thiamine, is common in hospitalized patients. The cause of their inadequate thiamine intake can be either primary or secondary. Primary inadequate intake of thiamine is caused by insufficient supply of food quantitatively and qualitatively. In secondary inadequate intake of thiamine, illness or treatment affects the thiamine status by altering a person's eating behavior or impairing his or her ingestion. Primary and secondary mechanisms frequently reinforce each other (62). In a dietary assessment in 656 hospitalized patients over a period of 2 years, 57 of the patients had a low daily intake of thiamine (63), and in 19% the intake was less than half the recommended daily allowances (RDAs) for thiamine (64). This result warrants the supplementation of thiamine in hospitalized patients to ensure their adequate thiamine status. Special attention should be paid to patients on parenteral nutrition in which glucose is the major source of energy supply because inadequate or absent thiamine creates beriberi (65–73).

B. Decreased Absorption of Thiamine

The ethanol inhibition of thiamine absorption and the defective absorption of thiamine in folate deficiency states are well documented (2,12,13,21,26,30). Patients with intestinal resection are also prone to develop thiamine deficiency because jejunum and ileum are the sites of thiamine absorption (12,74).

C. Defective Transport of Thiamine

Thus far, there is no acquired condition that is known to cause defective transport of thiamine. However, there is a concern about the adequacy of the transport of thiamine from mothers across the placenta to the fetus. Measurement of erythrocyte or whole-blood transketolase activity (ETKA or WBTKA) and the enhancement from added TPP to basal assays (referred to as the TPP effect, TPPE) is used for thiamine assessment. Using this approach to assessment, thiamine status in cord blood and in mothers within 24 h of delivery (75) or 6–12 h before delivery (76) indicates that newborns have better thiamine status than their mothers. These results are consistent with the findings of previous studies that (a) cord blood has a higher thiamine content than maternal blood (77) and (b) fetal thiamine status may be normal even when the mother is thiamine-deficient (78). These findings imply that there is an efficient transport of thiamine across the placental barrier and that the fetus can sequester thiamine.

D. Impaired Biosynthesis of TPP

The conversion of thiamine to TPP may be impaired in the presence of alcoholic liver injury, possibly because of decreased availability or use of ATP, and may be associated with decreased activities of pyruvate dehydrogenase and transketolase due to apoenzyme or magnesium deficiency (12,13). Besides, the decreases in hepatic and cerebral transketolase activities are accompanied by a parallel fall in thiamine concentrations in liver and brain (79). Since the transketolase activity correlates with TPP concentrations (80), these findings suggest a lowering in the TPP concentrations. However, thiamine deficiency in patients with fulminant hepatic failure can be corrected by intravenous administration of 100 mg of thiamine hydrochloride twice daily. This indicates that the bioconversion of thiamine to TPP is still possible in the presence of acute hepatocellular necrosis when the patient is treated with pharmacological doses of thiamine (81).

E. Increased Requirement of Thiamine

Increased anabolism occurs in various physiological conditions, including pregnancy, lactation, infancy and childhood, adolescence, and increased physical activity, whereas a number of pathological conditions create increased catabolism. Increased anabolism and catabolism raise the requirements of several nutrients, including thiamine.

1. Pregnancy

Studies of urinary excretion of thiamine, blood thiamine levels, and ETKA all indicate that the requirement for thiamine in women increases during pregnancy. This increase appears to occur early in pregnancy and to remain constant throughout (64,75,76,82). Thus women with marginal thiamine status who are pregnant and pregnant women with inadequate intake of thiamine are vulnerable to develop beriberi. This is well illustrated in the report on women with hyperemesis gravidarum presenting with Wernicke's encephalopathy having the symptom of vomiting for a month, and Wernicke's encephalopathy usually occurred following thiamine-free intravenous glucose infusion (83,84).

2. Lactation

Thiamine requirement also increases during lactation due to the loss of thiamine in milk and increased energy consumption during lactation. The healthy lactating woman secretes approximately 0.2 mg of thiamine in her milk each day (85).

3. Infancy and Childhood

Because the metabolic rate of infants and children is greater and the turnover of nutrients more rapid than in adults, the unique nutritional needs for growth and development are superimposed on higher maintenance requirements than those of adults. Whole-blood total thiamine concentrations are higher in infants aged 0–3 months than those aged 3–12 months, i.e., 258 ± 63 (mean $\pm$ SD) and 214 ± 44 nmol/L, respectively. After a large decline in the first year of life, the concentrations become stable and equal to those in adults: 187 ± 39 nmol/L. The overall decrease is due mainly to a decrease in the phosphorylated thiamine. Since these changes occur relatively independently of the changes in hematocrit characteristically seen in infancy, it may be related to changes in metabolic activity, growth rate, or other variables of biological activity (86). The findings that changes in whole-blood thiamine concentrations during infancy and childhood are physiological are supported by nonsignificant differences in WBTKA and TPPE between healthy infants and children aged 1–6 years (55). However, inadequate thiamine status is commonly found in breast-fed infants by thiamine-deficient mothers and children on the rice-based diets (10,54–58).

4. Adolescence

The nutritional requirements of adolescents are influenced primarily by the normal events of puberty and the simultaneous spurt of growth. Adolescence is the only time in extrauterine life when growth velocity increases and exerts a major influence on nutritional requirements. Since thiamine is involved in energy metabolism, adolescents on milled rice as their staple diets are vulnerable to develop thiamine deficiency (10,45,46,48,87,88).

5. Increased Physical Activity

For most people, the second largest component of total energy expenditure is the energy expended in physical activity. Thus, high physical activity can precipitate the development of beriberi in those with marginal thiamine status (10,88).

6. Increased Catabolism

Increased catabolism raises the requirements for energy and thiamine. Thus, catabolic conditions, such as fever, infection, trauma, or surgery, can precipitate the development of beriberi in those with marginal thiamine status (10,89–91).

F. Increased Loss of Thiamine

Renal patients on chronic dialysis (92,93) and patients with congestive heart failure treated with furosemide, a diuretic (94,95), may develop beriberi, which in part is due to an increased renal excretion of thiamine.

VIII. CLINICAL MANIFESTATIONS OF BERIBERI

The clinical manifestations of beriberi vary with age. Infantile beriberi is commonly found in infants aged 2–3 months who are being breast-fed by thiamine-deficient mothers. The clinical manifestations of beriberi in older children and adolescents are similar to those in adults, and as such are classified as adult beriberi (5,9,10,30,41,96–99).

A. Infantile Beriberi

The predominant symptoms of infantile beriberi include edema, dyspnea, oliguria, aphonia, cardiovascular disturbances, and gastrointestinal disorders, including loss of appetite, vomiting, or milk diarrhea. Infantile beriberi may be categorized into four subtypes based on the dominant features, i.e., cardiac, aphonic, pseudomeningitic, or a combination (10,41,96–99).

1. Cardiac or Acute Fulminating Form

Infants with cardiac beriberi usually present with acute onset of manifestations, including a loud piercing cry, cyanosis, dyspnea, vomiting, tachycardia, and cardiomegaly; death may occur within a few hours after onset unless thiamine is administered.

2. Aphonic Form

Aphonic forms of infantile beriberi are generally less acute. The striking feature is the tone of the child's cry, which varies from hoarseness to complete aphonia due to paralysis of the recurrent laryngeal nerve. A decrease in deep tendon reflexes and blepharoptosis are detected in some infants.

3. Pseudomeningitic Form

Infants exhibit vomiting, nystagmus, purposeless movement of the extremities, and convulsion. However, no abnormality of cerebrospinal fluid (CSF) is demonstrated.

4. Mixed Form

Infants with beriberi may present with the combined manifestations of cardiac disturbance, aphonia, and pseudomeningitis.

In a study of 16 infants with beriberi, the prevalence of cardiovascular system disorders was 87% with dyspnea, 81% with tachycardia (heart rate more than 130 beats/min), 73% with cardiomegaly (confirmed by chest x-ray), 63% with vomiting, 62% with hepatomegaly, 56% with edema, 19% with cyanosis, and 6% with shock, whereas those of nervous system disorders included 81% with decreased deep tendon reflexes, 50% with aphonia, 12% with convulsion, and 6% with blepharoptosis (10,56).

B. Adult Beriberi

Older children, adolescents, and adults may present with dry (paralytic or nervous), wet (cardiac), or cerebral beriberi (5,9,10,30,41,96–99).

1. Dry Beriberi

The predominant feature in dry beriberi is peripheral neuropathy, characterized by a symmetrical impairment of sensory, motor, and reflex functions that affects the distal segments of limbs more severely than the proximal ones, calf muscle tenderness, and difficulty in rising from a squatting position (5,10,46,99).

Pathological changes in the peripheral nerve in beriberi neuropathy are unusual axonal degeneration and accumulation of flattened sacs or tubules in the axoplasm of large myelinated fibers in untreated patients. In beriberi patients treated with thiamine, there are numerous Bungner's bands and clusters of thin myelinated fibers, indicating active regeneration, which correspond well with clinical improvement (47). Examination of CSF in patients with dry beriberi reveals normal findings (47,99).

2. Wet Beriberi

In addition to peripheral neuropathy, common signs found in wet beriberi include edema, tachycardia, wide pulse pressure, cardiomegaly, and congestive heart failure (5,10,46,99). In some patients, there is a sudden onset of a cardiac manifestation known as acute fulminant or ''shoshin'' beriberi, with the predominant features of tachycardia, dyspnea, cyanosis, cardiac enlargement, and circulatory collapse (41).

a. Electrocardiographic Findings

Considerable attention has been given to the electrocardiographic changes that occur in wet beriberi (99,100). The serial electrocardiographic study in 11 patients with wet beriberi whose thiamine deficiency was confirmed by ETKA and TPPE revealed the following findings: 8 patients had a prolonged QT interval, which became normal in 5 patients within 2 h to 28 days after receiving the thiamine treatment. Seven patients showed T-wave abnormalities in the precordial leads, consisting of flat or inverted T waves that became normal within 7–40 days following administration of thiamine. Three patients showed relatively low QRS complex voltage, which increased after the appearance of diuresis following the thiamine treatment. The increase of QRS complex voltage and the rapid reduction of heart size were observed concurrently with clinical improvement. Since the pattern of the electrocardiographic voltage in wet beriberi resembles that in pericardial effusion, the transient low QRS voltage in patients with wet beriberi is most likely caused by pericardial effusion (99,101,102).

b. Hemodynamic Findings

Studies in occidental patients with wet beriberi have shown hemodynamic findings of high cardiac output with low peripheral vascular resistance (103,104); an acute response to thiamine administration, i.e., decrease in cardiac output with concomitant increase in total peripheral resistance, was observed within 37 min (104) and 2 h (103).

A hemodynamic study of six Thai nonalcoholic patients with wet beriberi whose diagnosis was confirmed by ETKA and TPPE revealed two types of hemodynamic findings. The first type is consistent with occidental reports (103–105); it consisted of four patients showing high cardiac output and low peripheral and pulmonary vascular resistance. Within 15 min of intravenous administration of thiamine HCl, their cardiac output was reduced and vascular resistance was increased, as was blood pressure (99,106,107). It appears that the major cause of the high cardiac output is vasomotor depression, the precise mechanism of which is not understood, but which leads to a reduced systemic vascular resistance (108).

The second type of hemodynamic finding involved one patient with normal cardiac output and another with low cardiac output. The low cardiac output may be due to myocardial damage sustained by repeated episodes of thiamine deficiency; the increase in cardiac output after thiamine treatment suggests the improvement of myocardial metabolism by the pharmacological dose of thiamine (99,106,107). Thus, cardiac output in wet beriberi may be variable. Normal or low cardiac output does not exclude the diagnosis.

3. Wernicke–Korsakoff Syndrome

The specific factor responsible for most, if not all, of the clinical manifestations of the Wernicke–Korsakoff syndrome is a deficiency of thiamine. Wernicke–Korsakoff syndrome constitutes the most common alcoholic-nutritional affliction of the central nervous system (3,24,108–110). Although alcoholism is the major cause of thiamine deficiency in Wernicke–Korsakoff syndrome, iatrogenic causes, including parenteral glucose administration and chronic dialysis, can aggravate the syndrome in patients with marginal thiamine status (65–68,83,84).

a. Wernicke's Disease or Encephalopathy

Wernicke originally described an illness of acute onset characterized by mental disturbance, paralysis of eye movements, and ataxia of gait. Since Wernicke's time, views regarding Wernicke's disease have undergone considerable modification. The diagnosis is based on the triad of ocular motor signs, ataxia, and derangement of motor functions (3,109).

The ocular motor signs are the most readily recognized abnormalities. They include a paresis or paralysis of abduction accompanied by horizontal diplopia, strabismus, and nystagmus. In advanced cases, there may be complete loss of ocular movement, and the pupils may become miotic and nonreacting. Blepharoptosis is rare. The ocular motor signs are attributable to lesions in the brain stem affecting the abducens nuclei and eye movement centers in the pons and rostral midbrain.

The ataxia affects stance and gait. Persistent ataxia is related to the loss of neurons in the superior vermis of the cerebellum; extension of the lesion into the anterior parts of the anterior lobes accounts for the ataxia of individual movements of the legs.

A derangement of mental function is found in about 90% of patients and takes one of the following three forms. First, a global confusional-apathetic state is most common, characterized by profound listlessness, inattentiveness, indifference to the surroundings, and disorientation. Second, a disproportional disorder of retentive memory, i.e., Korsakoff's amnesic state, occurs in some patients. Third, the symptoms of alcohol withdrawal may be found in a relatively small number of patients. The amnesic defect is related to lesions in the diencephalon, specifically those in the medial dorsal nuclei of the thalamus.

Most patients with Wernicke's disease display peripheral neuropathy and some cardiovascular abnormalities, including tachycardia, exertional dyspnea, and postural hypotension (3,109).

b. Korsakoff's Psychosis

Victor (109) has stated that Wernicke's disease and Korsakoff's psychosis are not separate diseases; Korsakoff's psychosis is the psychic component of Wernicke's disease. Thus, the clinical manifestations should be called Wernicke's disease when the amnesic state is not evident and the Wernicke–Korsakoff syndrome when both the ocular-ataxic and amnesic manifestations are present.

Though the specific role of thiamine in Wernicke–Korsakoff syndrome is established and most patients with Wernicke–Korsakoff syndrome are alcoholics, only a few alcoholics are affected (3,109–111). Genetic abnormalities in ETK may underlie a predisposition to Wernicke–Korsakoff syndrome (112). However, the investigation on the hysteretic properties of human transketolase with emphasis on its dependency on TPP concentration has revealed the substantial lag in formation of active holotransketolase and interindividual differences and cell type variation from the same individual in the lag period; these findings suggest the existence of mechanisms for the loss of transketolase activity during thiamine deficiency and may explain, at least in part, the differential sensitivity to deficiency demonstrated by tissues and individuals (113).

IX. TREATMENT AND CLINICAL RESPONSE

A. Treatment

Thiamine should be promptly administered to beriberi patients. The dosage usually ranges from 50 to 100 mg of thiamine HCl given intravenously or intramuscularly for 7–14 days, after which 10 mg/day can be administered orally until the patient recovers fully. To prevent the recurrence of beriberi, patients should be advised to change their dietary habits and to stop their alcohol drinking (5,10,30,46,56).

The rationale to administer a large dose of thiamine is to replenish the thiamine store, which is consistent with the positive correlation between thiamine concentration in serum and CSF (114), to stimulate the TPP-dependent reactions optimally, which is consistent with the aforementioned thiamine-dependent hysteretic behavior of human transketolase (113), and to improve cardiovascular disorders evidenced by the hemodynamic studies (99,103–107). The parenteral route is used initially to ensure the bioavailability of thiamine.

Several thiamine derivatives, especially thiamine propyl disulfide (TPD) and tetrahydrofurfuryldisulfide (TTFD) (Fig. 4), are useful for oral administration to beriberi patients because even by the oral route they produce a significantly higher thiamine level in the blood, erythrocytes, and CSF than thiamine HCl and TPP at the same dose of 50 mg, and their effectiveness is equivalent to that produced by parenteral administration of thiamine HCl or TPP in healthy subjects (115). The effective intestinal absorption of TTFD is also demonstrated at 100- and 180-mg dose levels in healthy subjects (116). The rapid absorption and transfer to the tissue of TPD is also evident by the increased levels of thiamine in portal vein, hepatic vein, and femoral artery (117). Oral administration of 50 mg of TPD to alcoholics with thiamine deficiency also restored their ETKA to normal and eliminated lateral rectus palsy in patients with Wernicke's disease (117).

The bioavailability of TTFD in multivitamin tablets has also been shown in healthy subjects taking the vitamin preparation containing 10 mg of TTFD daily for 7 days and 10 or 20 mg of TTFD daily for 44 weeks. Their baseline blood thiamine concentrations increased gradually during the long-term administration, with a dose-dependent response (118). The dramatic response of patients with cardiac beriberi to the daily intravenous administration of 50 mg of TTFD has also been reported (119).

B. Response to Thiamine Administration

In wet beriberi, within 6–24 h of thiamine administration improvement can be observed in terms of less restlessness; disappearance of cyanosis; reductions in heart rate, respiratory rate, and cardiac size; and clearing of pulmonary congestion. Dramatic improvement after thiamine treatment is also observed in infants with cardiac beriberi (10,46,55,56,99).

It is difficult to use the response to thiamine administration as a criterion for immediate diagnosis in dry beriberi or infants with aphonia because more time elapses before improvement is observed. In a study of 21 patients with beriberi, impaired sensation disappeared after 7–120 days of thiamine treatment; motor weakness found in 12 patients was recovered within 60 days; but absent or hypoactive ankle and knee jerks remained for several months after treatment (10,46,99).

The response of patients with Wernicke's disease to thiamine treatment shows a characteristic pattern. Occular palsies may begin to improve within hours to several days after the administration of thiamine. Sixth-nerve palsies, ptosis, and vertical gaze palsies recover completely within 1–2 weeks in most cases, but vertical gaze–evoked nystagmus may persist for months. Ataxia improves slowly and approximately half of patients recover incompletely. Apathy, drowsiness, and confusion also recede gradually (109).

The outcome of the Korsakoff amnesic state varies. It ranges from complete or almost complete recovery in less than 20% of patients to slow and incomplete recovery in the remainder. The residual state is characterized by large gaps in memory, usually without confabulation, and an inability of the patient to sort out events in the proper temporal sequence (109).

X. ASSESSMENT OF THIAMINE STATUS

The sequential development of impaired thiamine status can be divided into four stages. It begins with an inadequate thiamine intake. The resulting depletion of thiamine in tissues leads to metabolic derangement of various organs, and clinical manifestations of beriberi appear (Table 2). Thus, the subject's history, laboratory tests, and physical examination are the bases for evaluation of thiamine status in humans (62).

A. Subject's History

In addition to dietary assessment, demographic data, medical history, family history, and psychosocial history in each subject should be recorded because all of these factors can affect the thiamine intake of the subject. Several methods including dietary scan, dietary record, and/or 24-h dietary recall have been employed to assess the thiamine intake in various populations, and comparisons are made between the mean thiamine intakes and RDAs for thiamine. For instance, dietary assessment by dietary record in northeastern Thai villagers revealed a mean daily energy intake of 2441 kcal, with a mean daily thiamine intake of 0.56 mg or 0.23 mg per 1000 kcal (44), which were only 51% or 46% of

Table 2 Sequential Development of Inadequate
Thiamine Status

Stages of inadequate thiamine status	Methods of assessment
I Inadequate thiamine intake	Subject's history
II Tissue desaturation of thiamine	Laboratory assessment
III Metabolic derangement	Laboratory assessment
IV Clinical manifestations	Physical examination

the RDAs for thiamine for the North American population (64). Dietary assessment based on 24-h recall in the North American male population aged 65–75 years showed a mean daily energy intake of 1805 kcal and a mean daily thiamine intake of 1.40 mg or 0.68 mg/1000 kcal (120), which are 116% or 136% of the 1989 RDAs for thiamine (64). It is evident that the northeastern Thai villagers have lower intake of thiamine than the elderly North American population. These dietary data are consistent with the biochemical findings on their thiamine status (44,120).

B. Laboratory Tests

Two crucial properties of laboratory tests must be considered when they are employed for the assessment of thiamine status. The tests must not be overly invasive; they should be sensitive enough to identify inadequate thiamine status prior to the appearance of clinical manifestations (Table 1) and to assess the efficacy of subsequent thiamine treatment.

Various biochemical tests have been developed for detecting thiamine deficiency or assessing the adequacy of thiamine status in humans. These include the measurement of urinary thiamine excretion; blood thiamine level; thiamine concentration in CFS, blood pyruvate, lactate, and α-ketoglutarate levels; and ETKA and TPPE (2,3,5,9,10,26,46,121, 122).

1. Urinary Thiamine Excretion

Prior to the development of the measurement of ETKA and TPPE, the most commonly used approach to thiamine status assessment was the measurement of urinary levels of thiamine by the thiochrome method after a one-step purification by ion exchange chromatography. Microbiological assays were also common and utilized *Lactobacillus viridescens* (121). Currently, in addition to ETKA and TPPE measurement, thiamine levels in urine are estimated by HPLC (122).

There is a reasonably close correlation between the development of a thiamine deficiency and decreased urinary thiamine excretion. Approximately 40–90 and 100 μg or more of thiamine is excreted in the urine daily in adults on intakes of 0.30–0.36 and 0.50 mg of thiamine per 1000 kcal, respectively, whereas daily urinary thiamine excretion falls to 5–25 μg with the daily intake of 0.2 mg of thiamine per 1000 kcal. Thus, measurement of the 24-h urinary thiamine excretion is useful in evaluating human thiamine status. Daily urinary thiamine excretions of ≥100, 40–99, and <40 μg are considered to designate acceptable, low, and deficient thiamine status, respectively. However, it is usually not feasible to collect 24-h urine samples with large groups under survey conditions. Consequently, random urine samples are obtained, preferably during fasting state, with the simultaneous determination of thiamine and creatinine concentrations. A correlation between the urinary thiamine excretion per gram of creatinine and daily thiamine intake in milligrams per 1000 kcal has been observed. Urinary thiamine excretions of ≥66, 27–65, and <27 μg/g of creatinine in adults are considered to designate acceptable, low, or deficient thiamine status. The corresponding figures in pregnant women during the second trimester are ≥55, 23–54, and <23 μg/g creatinine, whereas those during the third trimester are ≥50, 21–49, and <21 μg/g creatinine (121).

Healthy children have markedly higher levels of urinary thiamine excretion per gram creatinine than adults. Thus, children aged 1–3, 4–6, 7–12, and 13–15 years are considered to have acceptable thiamine status when their urinary thiamine excretions are ≥176, ≥121, ≥181, and ≥151 μg/g creatinine, respectively (121).

Though urinary thiamine excretion per gram creatinine may be a useful criterion for surveying thiamine status in a large population, it is not always a good index for diagnosing beriberi, e.g., of 15 patients with beriberi prior to thiamine treatment, 8(53%), 3(20%), and 4(27%) had urinary thiamine excretions of <27, 27–65, and ≥66 µg/g creatinine, respectively (46).

Although a number of thiamine load tests of a known dose of thiamine administered orally or intramuscularly have been of value in evaluating the extent of thiamine depletion in tissue stores, the most commonly used procedure is to administer 5 mg of thiamine parenterally and to measure the urinary thiamine excretion in the following 4-h period. Thiamine-deficient subjects usually excrete less than 20 µg of thiamine following the test dose. However, this procedure is inconvenient in nutrition surveys (121).

2. Blood Thiamine Level

Levels of thiamine in whole blood, plasma or serum, and erythrocytes have been employed to assess human thiamine status. The thiamine concentrations in these biological samples can be determined by chemical methods, including the thiochrome procedure and colorimetric techniques utilizing diazotized *p*-aminoacetophenone or ethyl-*p*-aminobenzoate; microbiological assays using *Ochromonas danica* or a streptomycin-resistant mutant strain of *Lactobacillus fermenti*; and HPLC (2,24,26,86,114,121,122). The specificity, sensitivity, and other limitations of each method must be carefully reviewed whenever employing it for assessing blood thiamine levels because blood contains only about 0.8% of the total body thiamine. Since most of the blood thiamine is in the blood cells, the serum level is still much lower. The thiamine concentration in leukocytes is about 10 times that in erythrocytes. In the interpretation of blood thiamine tests, the hematocrit must be taken into consideration because erythrocyte mass usually contributes 75% or more of the total thiamine (2,86). Although leukocytes contribute 15% or less of the blood thiamine concentration, the interpretation of a low-normal blood thiamine level in a subject with leukocytosis must be done with caution (86).

a. Chemical Methods

The determination of total thiamine concentration in whole blood or serum by the thiochrome method is based on an oxidative conversion of thiamine (by ferricyanide in an alkaline solution) to thiochrome (Fig. 3), which is highly fluorescent and is extracted from the aqueous phase with isobutanol for subsequent fluorometric measurement. Oxidation of phosphorylated thiamine with ferricyanide yields a thiochrome derivative not extractable with isobutanol. Pretreatment of whole blood or serum with phosphatase preparations, e.g., potato acid phosphatase, is necessary for obtaining the total thiamine concentration. Thus, the total and nonphosphorylated thiamine are measured directly, whereas the phosphorylated thiamine is calculated from total thiamine minus nonphosphorylated thiamine (86,123).

The colorimetric method based on the coupling of diazotized *p*-aminoacetophenone with thiamine to produce an insoluble purple-red compound that can be measured photometrically. Uric acid and ascorbic acid interfere with the development of the color. Besides, the method is not sufficiently sensitive to detect the low concentration of blood thiamine (2,123).

b. Microbiological Assays

Baker et al. (124) developed a protozoan method using the phytoflagellate *Ochomonas danica* to measure thiamine concentrations in whole blood, erythrocytes, leukocytes, CSF, and other tissues. *Ochromonas danica* has a sensitive and specific intact thiamine requirement; it can assay 375 pmol/L (100 pg/mL) of thiamine in biological fluids and tissues. It has been shown that reduced blood thiamine levels determined by this protozoan method occurred within 2 days in rats on thiamine-deficient diet. At 15 days there was almost no circulating thiamine, and the thiamine contents in liver and muscle fell to 25% of normal, whereas the ETKA fell after 10 days, and after only 20 days was reduced to nearly 50% of normal (125).

Thiamine concentrations in serum and erythrocytes can also be accurately measured by an automated method using a streptomycin-resistant strain of *Lactobacillus fermenti* as the test organism. This method is sensitive to 1.88 nmol/L (0.5 µg/L) of thiamine. The organism responds to both free and phosphorylated thiamine; its growth response to TPP is approximately 30% greater than that obtained with thiamine HCl (24,26).

c. High-Performance Liquid Chromatography

HPLC determinations of free and phosphorylated thiamine are now frequently carried out to assess thiamine status (114,123,126–128); a sensitivity for the detection of 0.05 pmol of thiamine has been reported (128). Besides, erythrocyte TPP levels in healthy adults determined by HPLC are in general agreement with those determined by the enzymatic method utilizing yeast pyruvic carboxylase apoenzyme (123,126). The validity of assessing thiamine status by erythrocyte TPP levels is supported by the following findings: i.e., erythrocyte TPP levels in rats fed a thiamine deficient diet fell more rapidly than ETKA (123); over an intake range of up to 3.4 mg of thiamine per day in eight hospitalized patients, their erythrocyte TPP levels increased proportionally with the thiamine intake (129); and a significantly positive correlation between ETAK and whole-blood TPP levels in 127 adults ($r = 0.26$, $p < 0.003$) as well as a significantly negative correlation between α-ETKA (ETKA with added TPP/ETKA without added TPP) ($r = -0.44$, $p < 0.0001$) (127).

Unlike urinary thiamine excretion, there is no accepted guideline for the interpretation of blood thiamine levels. Tables 3 and 4 show normal thiamine levels in biological fluids determined by the protozoan (125) and HPLC methods (114), respectively.

3. Thiamine Concentration in CSF

In adults, total thiamine in CSF exists in two forms, i.e., 62% as TMP and 38% as thiamine. The total thiamine, thiamine, and TMP in CSF are four, two, and eight times higher than the corresponding forms in serum. Besides, there is a correlation between serum and CSF concentrations of thiamine. TMP is formed from the dephosphorylation of TPP and is further dephosphorylated into thiamine (Fig. 2). Since TPP is not found in CSF, the high concentration gradient of TMP from serum to CSF suggests the existence of an active

Table 3 Normal Thiamine Levels in Biological Fluids Determined by the Protozoan Method Using *Ochromonas danica* (125)

Specimen	Thiamine (nmol/L)		Thiamine (µg/L)	
	Range	Mean	Range	Mean
Whole blood	94–282	120	25–75	32
Serum	563–957	788	150–255	210
CSF	38–113	60	10–30	16

Table 4 Concentrations of Various Forms of Thiamine in Blood and Cerebrospinal Fluid in Adults Determined by HPLC

Author	Subject[a]			Thiamine forms[d] (nmol/L)			
	Age (y)	N	Specimen	Free	TMP	TTP	Total
Tallaksen et al.	23–58	40	Serum	5 ± 23	2–12	—	NA
(114)	21–83	31[b]	Serum	8.6 ± 3.9	4.0 ± 1.5	—	12.7 ± 4.2
	21–83	45[b,c]	CSF	16.9 ± 8.3	28.1 ± 7.9	—	44.9 ± 14.0
Fidanza et al.	20–82	127	Whole blood			126 ± 35	
(127)						(64–202)	

[a] All of the subjects were healthy except those with superscripts b and c.
[b] Blood and CSF were collected simultaneously from 31 patients with severe back pain but were otherwise healthy during performance of myelogram.
[c] Only CSF samples were collected from another 14 similar patients.
[d] Figures are mean ± SD or range. NA, not available from the reports.

transport of TMP, which is supported by a saturation mechanism (114). Though simultaneous measurements of thiamine in serum and CSF reflect the integrity of blood–brain barrier and thiamine supply to the brain, it is impractical and unjustifiable to assess thiamine status by the measurement of CSF thiamine in a free-living population. Determination of CSF thiamine in hospitalized patients with dry beriberi or Wernicke–Korsakoff syndrome should be based on the risk and benefit of the patients.

4. Blood Pyruvate, α-Ketoglutarate, and Lactate Levels

In beriberi patients, pyruvic and α-ketoglutaric acids accumulate in the tissue (Fig. 8) and their concentration in the blood rises. Part of accumulated pyruvic acid may also be converted to lactic acid. However, these changes lack specificity and consistency. The fasting levels of blood pyruvate have frequently been found to be normal in thiamine deficiency and only rise above normal following a glucose load (99,121). Thus, a glucose load test, termed the carbohydrate metabolic index (CMI), has been developed; blood glucose (G), pyruvate (P), and lactate (L) levels are measured at 1 h after the administration of 1.8 g of glucose per kilogram of body weight and 5 min after completion of a standardized exercise. CMI is then calculated as follows;

$$CMI = \frac{(L - G/10) + (15P - G/10)}{2}$$

The upper normal limit of CMI is 15, based on the studies of experimental thiamine deficiency in humans (130). However, this procedure requires too much cooperation and technical ability to be useful routinely (121).

5. Blood Transketolase Activity

Blood is a feasible specimen to be obtained for the determination of transketolase activity, which can be measured in erythrocytes, leukocytes, or whole blood.

a. ETKA and TPPE

Brin and his co-workers (131) have demonstrated that ETKA and its percent stimulation in vitro by added TPP, called TPPE, are sensitive methods to detect thiamine deficiency both in rats and in humans. Healthy adults fed diets containing 190 μg of thiamine daily for 8 weeks showed a decline in ETKA as early as 8 days, and their ETKA at 17–20 days was markedly lower than that in the control subjects; their TPPE was 15% and 34% at days 8 and 31, respectively, whereas TPPE in the control subjects was less than 10% throughout the study; and following thiamine treatment, their ETKA and TPPE returned to normal values. These findings indicate the specificity of TPP on the depressed ETKA in thiamine deficiency in humans. In this study, the urinary thiamine excretion in these thiamine-deficient subjects was sharply reduced to less than 50 μg/day by the second week and to less than 20 μg/day by the third week, whereas their CMI was below 15 throughout the study. These findings indicate the sensitivity of ETKA and TPPE in detecting human thiamine deficiency. Subsequent works have established the validity of ETKA and TPPE for the diagnosis of various forms of beriberi in infants and adults, and have proved that they are useful indicators of biochemical thiamine depletion prior to the appearance of beriberi (2,3,5,9,10,30,43,46,56,121,130). The measurement of ETKA and TPPE represents a functional test of thiamine adequacy and hence may be a more reliable indicator of thiamine insufficiency than urinary thiamine excretion (121).

ETKA can be measured by a colorimetric procedure involving incubating hemolyzed erythrocyte samples for 30 or 60 min at 37–43°C in a buffered medium with an excess of ribose-5-phosphate with and without added TPP and then determining the amount of ribose-5-phosphate used, sedoheptulose-7-phosphate formed, or the hexoses produced (26,121). TPPE is then calculated as follows:

$$\text{TPPE in \%} = \frac{\text{ETKA with added TPP} - \text{ETKA without added TPP}}{\text{ETKA without added TPP}} \times 100$$

In some reports, the stimulating effect of TPP on ETKA is expressed as the activity coefficient of ETKA (α-ETKA), i.e., ETKA with added TPP/ETKA without added TPP (127).

ETKA obtained without the addition of TPP represents the basal level of enzyme activity and is dependent on the availability of endogenous erythrocyte TPP. The addition of TPP permits estimation of the amount of apoenzyme uncomplexed as well as the maximum potential activity. A thiamine deficiency results in a reduction in TPP, which results in a reduction of the apotransketolase activity (121). This is supported by a correlation between transketolase activity and TPP concentrations (80).

A modification of the Dreyfus method (132) has been employed for the determination of ETKA in our laboratory and ETKA is expressed in international units (IU), which is equivalent to the number of micromoles of sedoheptulose-7-phosphate formed per minute per liter of erythrocytes.

A colorimetric procedure has been developed that distinguishes transketolase activity from overall ribose-5-phosphate utilization by measuring ribose disappearance and sedoheptulose formation in the same color reaction (133). More direct spectrometric methods have evolved based on the change in absorption at 340 nm due to the decrease in NADH. For example, glyceraldehyde-3-phosphate oxidation has been used (134). A semiautomated, continuous-flow procedure for determining unstimulated ETKA and TPP-stimulated ETKA has also been described (135). The sensitivity, reliability, and precision of this approach is improved by eliminating hemoglobin interference.

Table 5 shows Brin's guideline (136) and the guideline of the Interdepartmental Committee on Nutrition for National Defense (ICNND) (121) for interpreting TPPE levels in evaluating thiamine status. Both guidelines are agreeable for the TPPE levels of normal

Table 5 Guidelines for Interpreting TPPE levels in
Evaluating Thiamine Status in All Ages

Brin's guideline (136)		ICNND's guideline (121)	
Category	TPPE (%)	Category	TPPE (%)
Normal	0–14	Acceptable	0–15
Marginally deficient	15–24	Low	16–20
Severely deficient	≥25	Deficient	>20

or acceptable thiamine status, whereas the TPPE levels for grading severity of thiamine deficiency are different. The validity of the TPPE levels of 0–15% for acceptable thiamine status is substantiated by the finding that 94% of beriberi patients prior to thiamine treatment showed TPPE of 16% or higher (46,99). However, both ETKA and TPPE must be considered in assessing the adequacy of thiamine status because the patterns of these two biochemical parameters in acute and chronic thiamine deficiency are different.

Subjects with acute thiamine deficiency have low ETKA with concomitantly high TPPE levels. Their unstimulated ETKA 1 h after the parenteral administration of thiamine is equivalent to or higher than the TPP-stimulated ETKA prior to thiamine treatment and their TPPE is in the acceptable range of 0–15% (10,46,99).

Experimental studies in rats (137) and humans (138) have shown that after a certain period of thiamine deprivation the TPP added in vitro cannot restore ETKA fully. Thus, despite low ETKA, high TPPE is not seen. These data are supported by findings that the TPP added in vitro does not increase ETKA in patients with beriberi prior to thiamine treatment (46,99) and in free-living populations with chronic inadequate thiamine intake (43) to the unstimulated ETKA in normal adults. Also, there is no significant difference in TPPE between subjects with adequate thiamine intake and those with chronically low intake of thiamine, despite significantly lower ETKA in the latter group (43).

The low unstimulated ETKA in cirrotic patients can be due either to the impaired conversion of thiamine to TPP or to apotransketolase deficiency. In those with an impairment of thiamine conversion to TPP, there is no increase in the low ETKA after following thiamine treatment; only the added TPP in vitro raises their ETKA. In those with apotransketolase deficiency, there are no increases in ETKA after thiamine treatment and the addition of TPP in vitro; in some instances, the transketolase level returns to normal when positive nitrogen balance is restored (125).

Apotransketolase deficiency, evidenced by decreases in unstimulated ETKA and TPP-stimulated ETKA with no change in TPPE, has also been shown in patients with insulin-dependent diabetes mellitus, patients with non–insulin-dependent diabetes mellitus (139), and uremic patients (140).

Elevated ETKA with normal TPPE is documented in patients with pernicious anemia (139,141,142), whereas patients with megaloblastic anemia caused by folate deficiency or fish tapeworm do not exhibit elevated ETKA (141). Since TPPE is only rarely increased in pernicious anemia with or without neuropsychiatric manifestations, thiamine deficiency should not be a feature of this disease and does not take part in the genesis of hyperpyruvicacidemia (142). Because transketolase activity is normally greater in young erythrocytes than in older cells, it has been suggested that patients with pernicious anemia have a larger population of young erythrocytes than healthy individuals (139,142). This suggestion is supported by findings of increased ETKA in patients with iron deficiency anemia responding to the iron treatment, in most patients with congenital hemolytic anemia, and in some patients with acute leukemia; and the normalization of unstimulated ETKA after 3–5 weeks of vitamin B_{12} treatment as well as the significantly negative correlation between hemoglobin levels and TPP-stimulated ETKA (142).

b. Leukocyte Transketolase Activity

Erythrocytes have a limited ability to respond to stimuli because they have a long half-life and lack a nucleus, whereas leukocytes would be expected to respond more rapidly to changes in nutritional status and perhaps more nearly reflect the status of other cells in the body. Study in the rats has shown leukocyte transketolase activity to be a reliable indicator of changes in thiamine status. It is not affected by dietary levels of carbohydrate, protein, and fat; it responds well to changes in dietary thiamine levels; it reaches a maximum, which does not increase further when excess thiamine is fed; and there is a parallel change of transketolase activity in leukocytes and liver (143). However, the use of leukocytes in assessing thiamine status in humans imposes greater technical difficulties than the use of erythrocytes.

c. WBTKA and TPPE

Dreyfus (132) has demonstrated that WBTK and TPPE can be employed to assess thiamine status in humans and to confirm the diagnosis of Wernicke's disease. The usefulness of WBTK and TPPE is substantiated by subsequent studies in assessing thiamine status in pregnant women and their cord blood (76), in infants and children (55), in patients with thyrotoxicosis (144), and patients with wet beriberi (145). Thus, in certain circumstances, whenever the separation of erythrocytes is not feasible, a whole-blood specimen can be used to determine transketolase activity and TPPE.

C. Physical Examination

Though physical signs of beriberi appear in the last stage of thiamine deficiency (Table 2), physicians should be able to detect such abnormalities in patients for immediate and long-term nutritional management (62). The delay in diagnosing beriberi affects the morbidity and mortality rates of the patients. Beriberi should be suspected in infants being breast-fed by thiamine-deficient mothers; such infants would have a loud piercing cry, dyspnea, cyanosis, cardiac failure, and aphonia (10,97).

Common suggestive signs in dry beriberi include glove-and-stocking hypoesthesia of pain and touch sensations, loss of ankle and/or knee reflexes, tenderness of calf muscle, difficulty in rising from squatting position, and aphonia. However, other possible known causes of peripheral neuropathy must be carefully ruled out (10,46,97,99).

Patients with wet beriberi exhibit both peripheral neuropathy and edema. Severe cases show tachycardia, wide pulse pressure, cardiac enlargement, and pulmonary congestion (10,46,97,99).

Wernicke's disease must be suspected in chronic alcoholics presenting with the triad of ocular motor signs, ataxia, and derangement of motor functions. The diagnosis of Wernicke–Korsakoff syndrome should be made in those having both the ocular-ataxic and amnestic manifestations (109).

As already mentioned in Sec. IX.B, drastic responses to thiamine treatment are observed in patients with cardiac beriberi or Wernicke's disease, whereas more time elapses before improvement is observed in patients with dry beriberi or Korsakoff's psychosis. Laboratory tests are required in the latter conditions to confirm the diagnosis.

XI. SUBCLINICAL THIAMINE DEFICIENCY

There is concern about the impact of persisting subclinical or borderline thiamine deficiency on the health status of the population. The major hazard for subjects with subclinical
thiamine deficiency is the increased risk of beriberi when faced with extreme physiological
or pathological conditions, e.g., pregnancy, lactation, high physical activity, infection, or
surgery. This is well illustrated in a study in Japanese university students (49). Routine
physical examination in 766 students revealed irritability with mild cardiovascular signs
in 42 students. Of these 42 students, 16 had whole-blood thiamine levels below 50 ng/
mL. A review of 2754 chest x-ray films indicated that 93 students had cardiothoracic
ratios greater than 50%. Of these 93 students, 44 had whole-blood thiamine levels below
50 ng/mL. The analysis of lifestyle in 59 students with low whole-blood thiamine levels
revealed that 50–60% ate their meals in restaurants and 39–47% did not eat breakfast;
86% of students with cardiac enlargement and low blood thiamine levels undertook strenuous daily exercise; only 20% were aware of their own abnormalities. The improvement
of cardiothoracic ratio was obtained in 63% of 83 subjects by eliminating strenuous exercise and administering sound nutritional advice.

   More studies are needed to verify subclinical thiamine deficiency in persons having
psychological, cardiovascular, and neurological symptoms. The application of dietary assessment and laboratory tests should be conducted in populations with chronic low intake
of thiamine and appropriate dietary guidelines be implemented to improve their thiamine
status.

XII. REQUIREMENTS AND RECOMMENDED INTAKES

Thiamine is essential in all phases of metabolism. The recommended thiamine intake is
expressed in terms of total caloric intake. The current U.S. recommended thiamine allowances are 0.5 mg/1000 kcal for children, adolescents, and adults, and 0.4 mg/1000 kcal
for infants. These recommendations are based on the assessment of the effects of varying
levels of dietary thiamine on the occurrence of deficiency signs, on the excretion of thiamine or its metabolites, and on ETKA. A minimum thiamine intake of 1.0 mg/day is
recommended for adults who consume less than 2000 kcal daily. An additional thiamine
intake of 0.4 mg/day is recommended throughout pregnancy to accommodate maternal
and fetal growth and increased maternal caloric intake. To account for both the thiamine
loss in milk and increased energy consumption during lactation, an increment of 0.5 mg/
day is recommended throughout lactation (64).

XIII. THIAMINE-RESPONSIVE DISEASES

Vitamin-responsive or vitamin-dependent diseases are a group of genetically determined
metabolic disorders in which either a vitamin-dependent enzymatic step or a reaction
involving the conversion of a vitamin to its active cofactor form is defective, causing the
abnormal accumulation of metabolites or substrates in the blood. In these diseases, blood
vitamin levels are normal. The basic metabolic defect involves the structure of the apoenzyme, its coenzyme binding sites, or some aspect of coenzyme synthesis (146).

   Thiamine-responsive diseases include maple syrup urine disease, pyruvate decarboxylase deficiency, subacute necrotizing encephalopathy (Leigh's disease), and thiamine-
responsive megaloblastic anemia. Such patients require pharmacological doses of thiamine
to alleviate their clinical manifestations (3,26,146,147).

XIV. TOXICITY

Excess thiamine is easily cleared by the kidneys. Although there is some evidence of toxicity from large doses given parenterally, there is no evidence of thiamine toxicity by oral administration; oral doses of 500 mg taken daily for a month were found to be non-toxic (64).

XV. CONCLUSION

Although much has been accomplished in terms of the nutritional, biochemical, physiological, pharmacological, and molecular aspects of thiamine, humans, especially those on rice-based diets and chronic alcoholics, are still facing the problem of inadequate thiamine status, ranging from subclinical thiamine deficiency to beriberi. To combat the problem of thiamine deficiency, existing nutritional knowledge should be disseminated to motivate the public to consume appropriate foods in their daily life. To achieve this dietary goal, both governmental and private agencies must work closely together and recognize the role of community participation.

REFERENCES

1. R. R. Williams, *Toward the Conquest of Beriberi*, Harvard University Press, Cambridge, MA, 1961.
2. C. J. Gubler, Thiamin, *Handbook of Vitamins: Nutritional, Biochemical, and Clinical Aspects* (L. J. Machlin, ed.), Marcel Dekker, New York, 1984, p. 233.
3. R. H. Haas, Thiamin and the brain, *Annu. Rev. Nutr. 8*:483 (1988).
4. D. B. McCormick, Thiamin, in *Modern Nutrition in Health and Disease*, 7th ed. (M. E. Shils and V. R. Young, eds.), Lea & Febiger, Philadelphia, 1988, p. 355.
5. V. Tanphaichitr, Thiamin, in *Modern Nutrition in Health and Disease*, Vol. 1, 8th ed. (M. E. Shils, J. A. Olson, and M. Shike, eds.), Lea & Febiger, Philadelphia, 1994, p. 359.
6. AIN Committee on Nomenclature, Nomenclature policy: generic descriptors and trivial names for vitamins and related compounds, *J. Nutr. 117*:7 (1987).
7. C. Kawasaki, From natural thiamine to synthetic thiamine: the brief history of thiamine preparations, in *Review of Japanese Literature on Beriberi and Thiamine* (N. Shimazono and E. Katsura, eds.), Igaku Shoin Ltd., Tokyo, 1965, p. 288.
8. U.S. Department of Health, Education, and Welfare, and Food and Agriculture Organization of the United Nations, *Food Composition Table in East Asia*, DHEW Publication No. (NIH) 73-465 (1972), Bethesda, MD, 1972.
9. V. Tanphaichitr and B. Wood, Thiamin, in *Nutrition Review's Present Knowledge in Nutrition*, 5th ed. (R. E. Olson, H. P. Broquist, C. O. Chichester, W. J. Darby, A. C. Kolbye, Jr., and R. M. Stalvey, eds.), Nutrition Foundation, Washington, DC, 1984, p. 273.
10. V. Tanphaichitr, Epidemiology and clinical assessment of vitamin deficiencies in Thai children, in *Child Health in the Tropics* (R. E. Eeckels, O. Ransome-Kuti, and C. C. Kroonenberg, eds.), Martinus Nijhoff, Dordrecht, 1985, p. 157.
11. N. Yagi and Y. Itokawa, Cleavage of thiamine by chlorine in tap water, *J. Nutr. Sci. Vitaminol. 25*:281 (1979).
12. C. M. Leevy, Thiamin deficiency and alcoholism, *Ann. N.Y. Acad. Sci. 378*:316 (1982).
13. A. M. Hoyumpa, Jr., Mechanisms of thiamin deficiency in chronic alcoholism, *Am. J. Clin. Nutr. 33*:2750 (1980).
14. K. Murata, Thiaminase, in *Review of Japanese Literature on Beriberi and Thiamine* (N. Shimazono and E. Katsura, eds.), Igaku Shoin Ltd., Tokyo, 1965, p. 220.

15. K. Murata, Actions of two types of thiaminase on thiamin and its analogues, *Ann. N.Y. Acad. Sci. 378*:146 (1982).

16. S. Vimokesant, S. Kunjara, K. Rungruangsak, S. Nakornchai, and B. Panijpan, Beriberi caused by antithiamin factors in food and its prevention, *Ann. N.Y. Acad. Sci. 378*:123 (1982).

17. D. M. Hilker and J. C. Somogy, Antithiamins of plant origin: their chemical nature and mode of action, *Ann. N.Y. Acad. Sci. 378*:137 (1982).

18. P. Ruenwonsa and S. Pattanavibag, Decrease in the activities of thiamine pyrophosphate dependent enzymes in rat brain after prolonged tea consumption, *Nutr. Rep. Int. 27*:713 (1983).

19. P. Ruenwonsa and S. Pattanavibag, Impairment of acetylcholine synthesis in thiamine deficient rats developed by prolonged tea consumption, *Life Sci. 34*:365 (1984).

20. L. Howard, C. Wagner, and S. Schenker, Malabsorption of thiamine in folate deficient rats, *J. Nutr. 107*:775 (1977).

21. A. D. Thomson, H. Baker, and C. M. Leevy, Patterns of ^{35}S thiamin hydrochloride absorption in the malnourished alcoholic patient, *J. Lab. Clin. Med. 76*:35 (1970).

22. K. Inouye and E. Katsura, Diagnosis, prevention and therapy of beriberi, in *Review of Japanese Literature on Beriberi and Thiamine* (N. Shimazono and E. Katsura, eds.), Igaku Shoin Ltd., Tokyo, 1965, p. 64.

23. G. Rindi and U. Ventura, Thiamine intestinal transport, *Physiol. Rev. 52*:821 (1972).

24. R. E. Davis, G. C. Icke, J. Thom, and W. J. Riley, Intestinal absorption of thiamin in man compared with folate and pyridoxal and its subsequent urinary excretion, *J. Nutr. Sci. Vitaminol. 30*:475 (1984).

25. R. M. Bayliss, R. Brooks, J. McCulloch, J. M. Kuyl, and J. Metz, Urinary thiamin excretion after oral physiological doses of the vitamin, *Int. J. Vitam. Nutr. Res. 54*:161 (1984).

26. R. E. Davis and G. Icke, Clinical chemistry of thiamin, *Adv. Clin. Chem. 23*:93 (1983).

27. K. Muniyappa, U. S. Murphy, and P. R. Adiga, Estrogen induction of thiamin carrier protein in chicken liver, *J. Steroid Biochem. 9*:888 (1978).

28. P. R. Adiga and K. Muniyappa, Estrogen induction and functional importance of carrier proteins for riboflavin and thiamin in the rat during gestation, *J. Steroid Biochem. 9*:829 (1978).

29. N. Shimazono, Metabolism of thiamine in the animal body, *Review of Japanese Literature on Beriberi and Thiamine* (N. Shimazono and E. Katsura, eds.), Igaku Shoin Ltd., Tokyo, 1965, p. 149.

30. V. Tanphaichitr, Thiamin, *Nutrition Review's Present Knowledge in Nutrition*, 4th ed. (D. M. Hegsted, C. O. Chichester, W. J. Darby, K. W. McNutt, R. M. Stalvey, and E. H. Stotz, eds.), Nutrition Foundation, New York, 1976, p. 141.

31. H. Behal, B. B. Buxton, J. G. Robertson, and M. S. Olson, Regulation of the pyruvate dehydrogenase multienzyme complex, *Annu. Rev. Nutr. 13*:497 (1993).

32. M. S. Olson, Bioenergetics and oxidative metabolism, in *Textbook of Biochemistry with Clinical Correlations*, 3rd ed. (T. M. Devlin, ed.), Wiley-Liss, New York, 1992, p. 237.

33. A. H. Mehler, Amino acid metabolism of the individual amino acids, in *Textbook of Biochemistry with Clinical Correlations*, 3rd ed. (T. M. Devlin, ed.), Wiley-Liss, New York, 1992, p. 491.

34. J. P. Blass, Thiamin and the Wernicke-Korsakoff syndrome, in *Vitamins in Human Biology and Medicine* (M. H. Briggs, ed.), CRC Press, Boca Raton, FL, 1981, p. 107.

35. A. von Muralt, The role of thiamine in neurophysiology, *Ann. NY Acad. Sci. 98*:499 (1962).

36. J. R. Cooper and J. H. Pincus, The role of thiamine in nervous tissue, *Neurochem. Res. 4*:223 (1979).

37. Y. Itokawa and J. R. Cooper, Thiamin release from nerve membranes by tetrodoxin, *Science 166*:759 (1969).

38. Y. Itokawa and J. R. Coper, Ion movements and thiamine in nervous tissue. I. Intact nerve preparations, *Biochem. Pharmacol. 19*:985 (1970).

39. E. Schoffeniels, Thiamine phosphorylated derivatives and bioelectrogenesis, *Arch. Int. Physiol. Biochem. 91*:233 (1983).

40. T. Oiso, Thiamine intake and its deficiency signs among the Japanese based on the results of the national nutrition survey, in *Review of Japanese Literature on Beriberi and Thiamine* (N. Shimazono and E. Katsura, eds.), Igaku Shoin Ltd., Tokyo, 1965, p. 275.

41. E. Katsura and T. Oiso, Beriberi, in *Nutrition in Preventive Medicine* (G. H. Beaton and J. M. Bengoa, eds.), World Health Organization, Geneva, 1976, p. 136.

42. M. S. Bamji, Vitamin deficiencies in rice-eating-populations: effects of B-vitamin supplements, *Nutritional Adequacy, Nutrient Availability and Needs* (J. Mauron, ed.), Birkhäuser Verlag, Basel, 1983, p. 245.

43. V. Tanphaichitr, N. Lerdvuthisopon, S. Dhanamitta, and A. Valyasevi, Thiamin status in northeastern Thais, *Intern. Med. 6*:43 (1990).

44. M. Kimura, N. Sato, and Y. Itokawa, Thiamin status of inhabitants on North-East Thailand, *Trace Nutrients Res. 4*:163 (1988).

45. V. Tanphaichitr, Adolescent beriberi, *Thai J. Intern. Med. 4*:42 (1984).

46. V. Tanphaichitr, S. L. Vimokesant, S. Dhanamitta, and A. Valyasevi, Clinical and biochemical studies of adult beriberi, *Am. J. Clin. Nutr. 8*:1017 (1970).

47. K. Takahashi, T. Kittagawa, and M. Shimao, Acute polyneuritis associated with edema: a recent revival of beriberi neuropathy in Japan, *Jpn. J. Med. 15*:214 (1976).

48. C. Kawai, A. Wakabayashi, T. Matsumara, and Y. Yui, Reappearance of beriberi heart disease in Japan, *Am. J. Med. 69*:383 (1980).

49. Y. Hatanaka and K. Ueda, Incidence of subclinical hypovitaminosis of B_1 among university students found by a field study in Ehime, Japan, *J. Osaka Univ. 31*:83 (1981).

50. V. Tanphaichitr, N. Lerdvuthisopon, S. Dhanamitta, and H. P. Broquist, Carnitine status in Thai adults, *Am. J. Clin. Nutr. 33*:876 (1980).

51. A. Valyasevi, S. B. Halstead, S. Pantuwatana, and C. Tankayl IV. Dietary habits, nutritional intake, and infant feeding practices among residents of a hypo-hyperendemic area, *Am. J. Clin. Nutr. 20*:1340 (1967).

52. A. Valyasevi, S. Vimokesant, and S. Dhanamitta, Chemical compositions of breast milk in different locations of Thailand, *J. Med. Assoc. Thailand 51*:348 (1968).

53. I. G. Macy, Composition of human colostrum and milk, *Am. J. Dis. Child. 78*:589 (1949).

54. O. Thanangkul and J. A. Whitaker, Childhood thiamine deficiency in northern Thailand, *Am. J. Clin. Nutr. 18*:275 (1966).

55. B. Pongpanich, N. Srikrikrich, S. Dhanamitta, and A. Valyasevi, Biochemical detection of thiamin deficiency in infants and children in Thailand, *Am. J. Clin. Nutr. 27*:1399 (1974).

56. L. Cheunchit and V. Tanphaichitr, Infantile beriberi, *Supartprasong Hosp. Gaz. 1*:97 (1975).

57. S. Sintarat and S. Pansuravet, Infantile beriberi in Nakornpanom Hospital, *J. Nutr. Assoc. Thailand 14*:184 (1980).

58. V. Nivatvongs and S. Anantkosol, Infantile beriberi: a retrospective study of 52 cases, *Bull. Khon Khaen Univ. Health Sci. 11* (Suppl.):41 (1985).

59. M. Victor, Deficiency diseases of the nervous system secondary to alcoholism, *Postgrad. Med. 50*:75 (1971).

60. R. A. Harris and D. W. Crabb, Metabolic interrelationships, in *Textbook of Biochemistry with Clinical Correlations*, 3rd ed. (T. M. Devlin, ed.), Wiley-Liss, New York, 1992, p. 575.

61. J. P. Bonjour, Vitamins and alcoholism. IV. Thiamin, *Int. J. Vitam. Nutr. Res. 50*:321 (1980).

62. V. Tanphaichitr, Evaluation of nutritional status, *Medical Practice of Preventive Nutrition* (M. L. Wahlqvist and J. S. Vobecky, eds.), Smith-Gordon and Co. Ltd., London, 1994, p. 333.

63. A. Lemonine, C. L. Devahat, J. L. Codacciont, A. Monges, P. Bermound, and R. M. Salkeld, Vitamin B1, B2, B6 and C status in hospital patients, *Am. J. Clin. Nutr. 33*:2595 (1980).

64. Committee on the Tenth Edition of the RDAs, Food and Nutrition Board, *Recommended Dietary Allowances*, 10th ed., National Academy Press, Washington, DC, 1989, p. 125.

65. G. Blennow, Wernicke encephalopathy following prolonged artificial nutrition, *Am. J. Dis. Child. 129*:1456 (1975).

66. A. M. Nadel and P. C. Burger, Wernicke encephalopathy following prolonged intravenous therapy, *JAMA 235*:2403 (1976).

67. F. A. Baughman, Jr., and J. P. Papp, Wernicke's encephalopathy with intravenous hyperalimentation: remarks on similarities between Wernicke's encephalopathy and the phosphate-depletion syndrome, *Mt. Sinai J. Med. 43*:48 (1976).

68. J. Kramer and J. A. Goodwin, Wernicke's encephalopathy: complication of intravenous hyperalimentation, *JAMA 238*:2176 (1976).

69. R. J. Velez, B. Myers, and M. S. Guber, Severe acute metabolic acidosis (acute beriberi): an avoidable complication of total parenteral nutrition, *J. Parent. Ent. Nutr. 9*:216 (1985).

70. P. Laselve, P. Demolin, and L. Hohzapfel, Shoshin beriberi: an unusual complication of prolonged parenteral nutrition, *J. Parent. Ent. Nutr. 10*:102 (1986).

71. D. Oriot, C. Wood, R. Gottesman, and G. Huault, Severe lactic acidosis related to acute thiamine deficiency, *J. Parent. Ent. Nutr. 15*:105 (1991).

72. J. Zak III, D. Burns, T. Lingenfelser, E. Steyn, and I. N. Marks, Dry beriberi: unusual complication of prolonged parenteral nutrition, *J. Parent. Ent. Nutr. 15*:200 (1991).

73. M. Comabella, A. Canton, X. Montalban, and A. Codina, Iatrogenic fulminant beriberi, *Lancet 346*:182 (1995).

74. K. Hiroi, Y. Goto, J. Ischikawa, K. Kida, and H. Matsuda, A case of beriberi accompanying short bowel, *Acta Paediatr. Jpn. 37*:84 (1995).

75. P. Migasena, S. Changbumrung, V. Supawan, and J. Limitrakarn, Erythrocyte transketolase activity in blood of mothers, umbilical cord and newborn babies in Thailand, *J. Nutr. Sci. Vitaminol. 20*:371 (1974).

76. V. Tanphaichitr and N. Lerdvuthisopon, Thiamin status in primigravidas and in cord blood, in *Vitamins and Minerals in Pregnancy and Lactation* (H. Berger, ed.), Raven Press, New York, 1988, p. 71.

77. L. B. Slobody, M. M. Willner, and J. Mestern, Comparison of vitamin B₁ levels in mothers and their newborn infants, *Am. J. Dis. Child. 77*:736 (1949).

78. K. Tripathy, Erythrocyte transketolase activity and thiamin transfer across human placenta, *Am. J. Clin. Nutr. 21*:739 (1968).

79. O. Frank and H. Baker, Vitamin profile in rats fed stock or liquid ethanolic diets, *Am. J. Clin. Nutr. 33*:221 (1980).

80. L. G. Warnock, C. R. Prudhomme, and C. Wagner, The determination of thiamin pyrophosphate in blood and other tissues and its correlation with erythrocyte transketolase activity, *J. Nutr. 108*:421 (1978).

81. D. Labadarios, J. E. Rossouw, J. B. McConnell, M. Davis, and R. Williams, Thiamin deficiency in fulminant hepatic failure and effects of supplementation, *Int. J. Vitam. Nutr. Res. 47*:17 (1977).

82. S. Heller, R. Salkeld, and W. F. Korner, Vitamin B₁ status in pregnancy, *Am. J. Clin. Nutr. 27*:1221 (1974).

83. P. S. Bergin and P. Harvey, Wernicke's encephalopathy and central pontine myelonolysis associated with hyperemesis gravidarum, *Br. Med. J. 35*:517 (1992).

84. P. J. Galloway, Wernicke's encephalopathy and hyperemesis gravidarum, *Br. Med. I. 305*:1096 (1992).

85. P. A. Nail, M. R. Thomas, and B. S. Eakin, The effect of thiamin and riboflavin supplementation on the level of those vitamins in human breast milk and urine, *Am. J. Clin. Nutr. 33*:198 (1980).

86. D. T. Wyatt, D. Nelson, and R. E. Hillman, Age-dependent changes in thiamin concentrations in whole blood and cerebrospinal fluid in infants and children, *Am. J. Clin. Nutr. 53*:530 (1991).

87. D. A. Gans and A. E. Harper, Thiamin status of incarcerated and nonincarcerated adoles-

cent males: dietary intake and thiamin pyrophosphate response, *Am. J. Clin. Nutr. 53*:1471 (1991).

88. V. Pattamakom, P. Rattanasiri, C. Roongpisuthipong, and V. Tanphaichitr, Thiamin malnutrition in institutionalized subjects, *Thai J. Intern. Med. 4*:1 (1984).

89. R. F. Butterworth, C. Gaudreau, J. Vincelette, A. M. Bourgault, F. Lamothe, and A. M. Nutine, Thiamine deficiency and Wernicke's encephalopathy in AIDS, *Metab. Brain Dis. 6*: 207 (1991).

90. R. Boldorini, L. Vago, A. Lechi, F. Tedeschi, and G. R. Tragattoni, Wernicke's encephalopathy: occurrence and pathological aspects in a series of 400 AIDS patients, *Acta Biomed. Ateneo. Parmense. 63*:43 (1992).

91. W. W. Coon and L. S. Bizer, Subclinical thiamine deficiency in postoperative patients, *Surg. Gynecol. Obstet. 121*:315 (1967).

92. E. Descombes, C. A. Dessibury, and G. Fellay, Acute encephalopathy due to thiamine deficiency (Wernicke's encephalopathy) in a chronic hemodialyzed patient: a case report, *Clin. Nephrol. 35*:171 (1991).

93. V. Jayadha, J. H. Deck, W. C. Halliday, and H. S. Smyth, Wernicke's encephalopathy in patients on peritoneal dialysis or hemodialysis, *Ann. Neurol. 2*:78 (1987).

94. H. Seligmann, H. Halkin, S. Rauchfleisch, N. Kaufmann, R. Tal, M. Motro, Z. Vered, and D. Ezra, Thiamine deficiency in patients with congestive heart failure receiving long-term furosemide therapy: a pilot study, *Am. J. Med. 92*:151 (1991).

95. J. A. Brady, C. L. Rock, and M. R. Horneffer, Thiamin status, diuretic medications, and the management of congestive heart failure, *J. Am. Diet. Assoc. 95*:541 (1995).

96. K. Inouye and E. Katsura, Clinical signs and metabolism of beriberi patients, in *Review of Japanese Literature on Beriberi and Thiamine* (N. Shimazono and E. Katsura, eds.), Igaku Shoin Ltd., Tokyo, 1965, p. 29.

97. S. Chaithiraphan, V. Tanphaichitr, and T. O. Cheng, Nutritional heart disease, in *The International Textbook of Cardiology* (T. O. Cheng, ed.), Pergamon Press, New York, 1986, p. 864.

98. J. Salcedo, Jr., Experience in the etiology and prevention of thiamin deficiency in the Philippine Islands, *Ann. N.Y. Acad. Sci. 378*:568 (1982).

99. V. Tanphaichitr, Clinical and biochemical studies in beriberi, A thesis submitted to the Faculty of Graduate Studies in partial fulfillment of the requirements for the degree of master of science, University of Medical Sciences, Bangkok, 1968.

100. R. A. Pallister, Electrocardiogram in Oriental beriberi, *Trans. Soc. Trop. Med. Hyg. 48*:490 (1954).

101. Y. Sukumalchantra, V. Tanphaichitr, V. Tongmitr, and B. Jumbala, Electrocardiographic changes in cardiac beriberi, *J. Med. Assoc. Thailand 57*:80 (1974).

102. Y. Sukumalchantra, V. Tanphaichitr, V. Tongmitr, and B. Jumbala, Electrocardiographic changes in adult beriberi: changes in the QRS voltage, *Mod. Med. Asia 11*:6 (1975).

103. W. J. Lahey, D. B. Arst, M. Silver, C. R. Kleeman, and P. Kunkel, Physiologic observations on a case of beriberi heart disease, *Am. J. Med. 14*:248 (1953).

104. P. I. Wagner, Beriberi heart disease. Physiologic data and difficulties in diagnosis, *Am. Heart J. 69*:200 (1965).

105. M. Akbarian, N. A. Yankapoules, and W. H. Abelmann, Hemodynamic studies in beriberi heart disease, *Am. J. Med. 41*:197 (1966).

106. V. Thongmitr, Y. Sukumalchantra, V. Tanphaichitr, and B. Jumbala, Hemodynamic studies in adult beriberi, *J. Med. Assoc. Thailand 56*:703 (1973).

107. Y. Sukumalchantra, V. Thongmitr, V. Tanphaichitr, and B. Jumbala, Variability of cardiac output in beriberi heart disease, *Mod. Med. Asia 12*:7 (1976).

108. W. S. Colacci and E. Braunwald, Cardiac tumors, cardiac manifestations of systemic diseases, and traumatic cardiac injury, in *Harrison's Principles of Internal Medicine*, Vol. 1, 13th ed. (K. J. Isselbacher, E. Braunwald, J. D. Wilson, J. B. Martin, A. S. Fauci, and D. L. Kasper, eds.), McGraw-Hill, New York, 1994, p. 1101.

109. M. Victor and J. B. Martin, Nutritional and metabolic diseases of the nervous system, in *Harrison's Principles of Internal Medicine*, Vol. 2, 13th ed. (K. J. Isselbacher, E. Braunwald, J. D. Wilson, J. B. Martin, A. S. Fauci, and D. L. Kasper, eds.), McGraw-Hill, New York, 1994, p. 2238.

110. P. M. Dreyfus and M. Seyal, Diet and nutrition in neurologic disorders, in *Modern Nutrition in Health and Disease*, Vol. 2, 8th ed. (M. E. Shils, J. A. Olson, and M. Shike, eds.), Lea & Febiger, Philadelphia, 1994, p. 1349.

111. J. C. M. Brust, Alcoholism, in *Meritt's Textbook of Neurology*, 9th ed. (L. W. Rowland, ed.), Williams & Wilkins, Baltimore, 1995, p. 967.

112. O. E. Pratt, H. K. Rooprii, G. W. Shaw, and A. D. Thomas, The genesis of alcoholic brain tissue injury, *Alcohol Alcoholism 25*:217 (1990).

113. C. K. Singleton, S. R. Pekovich, B. A. McCool, and P. R. Martin. The thiamin-dependent hysteretic behavior of human transketolase: implications for thiamin deficiency, *J. Nutr. 125*: 189 (1995).

114. C. M. E. Tallaksen, T. Bohmer, and H. Bell, Concentrations of the water-soluble vitamins thiamin, ascorbic acid, and folic acid in serum and cerebrospinal fluid of healthy individuals, *Am. J. Clin. Nutr. 56*:559 (1992).

115. H. Baker, A. D. Thomson, O. Frank, and C. M. Leevy, Absorption and passage of fat- and water-soluble thiamin derivatives into erythrocytes and cerebrospinal fluid of man, *Am. J. Clin. Nutr. 27*:676 (1974).

116. N. Kitamori and Y. Itokawa, Pharmacokinetics of thiamin after oral administration of thiamin tetrahydrogurfuryl disulfide to humans, *J. Nutr. Sci. Vitaminol. 39*:465 (1993).

117. H. Baker and O. Farnk, Absorption, utilization and clinical effectiveness of allithiamines compared to water-soluble thiamines, *J. Nutr. Sci. Vitaminol. 22* (Suppl.):63 (1976).

118. Y. Itokawa, M. Kimura, K. Nishino, M. Mino, M. Kitagawa, M. Matsuoka, and H. Otsuka, Blood levels and urinary excretion of thiamin and riboflavin during oral administration of multivitamin tablets to healthy adults, *J. Nutr. Sci. Vitaminol. Special No.*:438 (1992).

119. W. Djoenaidi, H. Notermanus, and G. Dunca, Beriberi cardiomyopathy, *Eu. J. Clin. Nutr. 46*:227 (1992).

120. F. L. Iber, J. P. Blass, M. Brain, and C. M. Leevy, Thiamin in the elderly-relation to alcoholism and neurological degenerative disease, *Am. J. Clin. Nutr. 6*:1067 (1982).

121. H. E. Sauberlich, J. H. Skala, and R. P. Dowdy, *Laboratory Tests for the Assessment of Nutritional Status*, CRC Press, Cleveland, 1974, p. 22.

122. H. E. Sauberlich, Newer laboratory methods for assessing nutriture of selected B-complex vitamins, *Annu. Rev. Nutr. 4*:377 (1984).

123. L. G. Warnock, C. R. Prudhomme, and C. Wagner, Determination of thiamin pyrophosphate in blood and other tissues, and its correlation with erythrocyte transketolase activity, *J. Nutr. 108*:421 (1978).

124. H. Baker, O. Frank, J. J. Fennelly, and C. M. Leevy, A method for assessing thiamine status in man and animals, *Am. J. Clin. Nutr. 14*:197 (1964).

125. H. Baker and O. Frank, *Clinical Vitaminology*, Interscience, New York, 1968, p. 7.

126. L. G. Warnock, The measurement of erythrocyte thiamin pyrophosphate by high-performance liquid chromatography, *Anal. Biochem. 126*:394 (1982).

127. F. Fidanza, M. S. Simonetti, A. Floridi, M. Coddini, and R. Fidanza, Comparison of methods for thiamin and riboflavin nutriture in man, *Int. J. Vitam. Nutr. Res. 59*:40 (1989).

128. M. Kimura, T. Fujita, S. Nishida, and Y. Itokawa, Differential fluorometric determination of picogram levels of thiamine, thiamine monophosphate, diphosphate, and triphosphate using high-performance liquid chromatography, *J. Chromatogr. 188*:417 (1980).

129. J. S. Powers, J. Zimmer, K. Meurer, E. Manske, J. C. Collins, and H. L. Greene, Direct assay of vitamins B_1, B_2 and B_6 in hospitalized patients: relationship to level of intake, *J. Parent. Int. Nutr. 7*:315 (1993).

130. M. K. Horwitt and O. Kreiser, The determination of early thiamine-deficient state by estima-

tion of blood lactic and pyruvic acids after glucose administration and exercise, *J. Nutr. 37*: 411 (1949).

131. M. Brin, Thiamine deficiency and erythrocyte metabolism, *Am. J. Clin. Nutr. 12*:107 (1963).

132. P. M. Dreyfus, Application of blood transketolase determinations, *N. Engl. J. Med. 267*:596 (1968).

133. L. G. Warnock, A new approach to erythrocyte transketolase measurement, *J. Nutr. 100*: 1057 (1978).

134. H. J. Smeets, H. Muller, and J. A. de Weal, NADH dependent transketolase assay in erythrocyte hemolysates, *Clin. Chim. Acta 33*:379 (1971).

135. P. P. Waring, D. Fisher, J. McDonnell, E. L. McGown, and H. E. Sauberlich, A continuous-flow (Auto Analyzer II) procedure for measuring erythrocyte transketolase activity, *Clin. Chem. 28*:2206 (1982).

136. M. Brin, Functional evaluation of nutritional status: thiamin, in *Newer Methods of Nutritional Biochemistry*, Vol. 3 (A. A. Albanese, ed.), Academic Press, New York, 1967, p. 407.

137. M. Brin, Erythrocyte transketolase in early thiamine deficiency, *Ann. N.Y. Acad. Sci. 98*:528 (1962).

138. M. S. Bamji, Transketolase activity and urinary excretion of thiamin in the assessment of thiamin-nutrition status of Indians, *Am. J. Clin. Nutr. 23*:52 (1970).

139. B. Kjosen and S. H. Seim, The transketolase assay of thiamine in some diseases, *Am. J. Clin. Nutr. 30*:1591 (1977).

140. M. Kuriyama, A. Mizuma, R. Yokomine, A. Igata, and Y. Otuji, Erythrocyte transketolase activity in uremia, *Clin. Chim. Acta 108*:169 (1980).

141. T. Markkanen, Transketolase activity of red blood cells in conditions of haematological interest, *Acta Haematol. 39*:321 (1968).

142. D. G. Wells and V. Marks, Anaemia and erythrocyte transketolase activity, *Acta Haematol. 47*:217 (1972).

143. C. H. Cheng, M. Koch, and R. E. Shank, Leukocyte transketolase activity as an indicator of thiamin nutriture in rats, *J. Nutr. 106*:1678 (1976).

144. A. Konttinen and M. Vitherkoski, Blood transketolase and erythrocyte glucose-6-phosphate dehydrogenase activities in thyrotoxicosis, *Clin. Chim. Acta 22*:145 (1968).

145. V. Tanphaichitr, N. Lerdvuthisopon, and P. Boongird, Clinical and nutritional studies in adult beriberi, in *Abstracts of the 4th Asian and Oceanean Congress of Neurology*, Sompong Press, Bangkok, 1975, p. 151.

146. P. M. Dreyfus and M. Seyal, Diet and nutrition in neurologic disorders, in *Modern Nutrition in Health and Disease,* Vol. 2, 8th ed. (M. E. Shils, J. A. Olson, and M. Shike, eds.), Lea & Febiger, Philadelphia, 1994, p. 1349.

147. K. Bartlett, Vitamin-responsive inborn errors of metabolism, *Adv. Clin. Chem. 23*:141 (1983).

9

Pantothenic Acid

NORA S. PLESOFSKY

University of Minnesota, St. Paul, Minnesota

I. INTRODUCTION

In 1933, R. J. Williams gave the name pantothenic acid to a substance that he proved necessary for the growth of the yeast *Saccharomyces cerevisiae* (1). Independently, a factor was isolated from liver that effectively cured dermatitis in chickens caused by a deficiency in B vitamins (2). Pantothenic acid was biochemically separated from pyridoxine (vitamin B_6), a copurifying component of the vitamin B_2 complex, by its lack of adsorption to Fuller's earth (3). Once separated, these two vitamins were shown to reverse different animal B_2 complex deficiency disorders, with pantothenic acid curing chicken dermatitis and vitamin B_6 curing dermatitis in rats.

In 1940, pantothenic acid was successfully synthesized by R. J. Williams and R. T. Major (4). However, its biologically functional form was not discovered until 1947, when F. Lipmann and his colleagues (5) demonstrated that pantothenic acid was contained in coenzyme A (CoA), an essential cofactor for such acetylation reactions as that of sulfonamide in the liver and choline in the brain. The biochemical structure of CoA was published in 1953 (6); and the central roles of CoA in the mitochondrial tricarboxylic acid cycle, fatty acid synthesis and degradation and other metabolic processes were elucidated. Within the past 15 years, there has been increasing evidence for a novel role of CoA in donating acetyl and acyl groups to many cellular proteins. This chapter will discuss the newly discovered role of CoA in the modification of proteins, a role that may ultimately be linked to specific effects of pantothenate deficiency.

II. CHEMISTRY

Microorganisms synthesize pantothenic acid by joining pantoic acid and β-alanine in amide linkage (7). The main route for human biosynthesis of CoA (7) is through phosphorylation of pantothenic acid to pantothenic acid 4'-phosphate, whose condensation with cysteine in an ATP-dependent reaction forms 4'-phosphopantothenoylcysteine. Decarboxylation yields 4'-phosphopantetheine, whose metabolite pantetheine is an essential growth factor for the yogurt-producing bacterium *Lactobacillus bulgaricus*. The anhydride addition of adenosine 5'-monophosphate to 4'-phosphopantetheine, followed by phosphorylation of the ribose 3'-hydroxyl, produces CoA (Fig. 1). The CoA sulfhydryl group, which is derived from cysteine, is the active site of esterification to acetate or acyl groups. As an alternative to being attached to diphosphoadenosine in CoA, 4'-phosphopantetheine may be covalently linked to a protein (8). Proteins that participate in fatty acid metabolism, such as the acyl carrier protein of bacteria and mitochondria and the eukary-

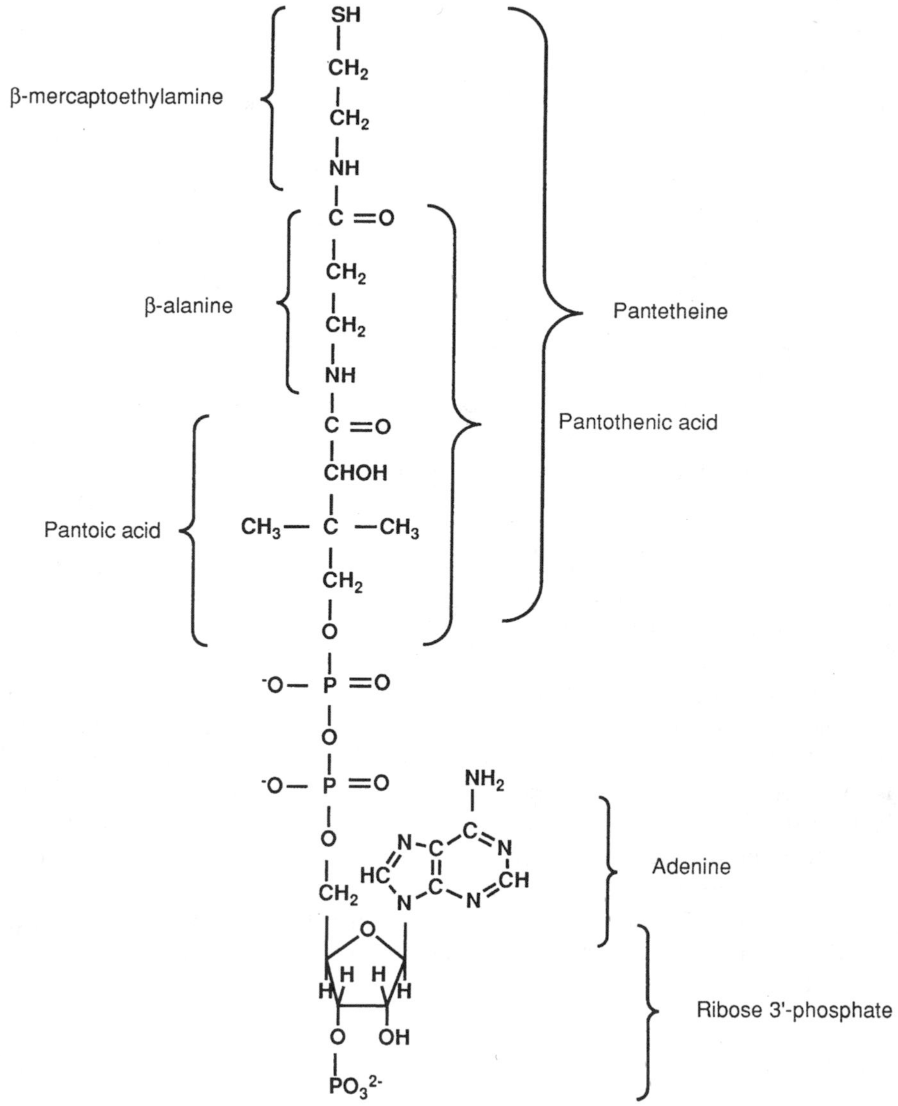

Fig. 1 Coenzyme A and intermediates.

otic fatty acid synthetase, are linked to 4'-phosphopantetheine by a phosphodiester bond. Citrate lyase of anaerobic bacteria and enzymes that are involved in the nonribosomal synthesis of peptide antibiotics, such as tyrocidin and gramicidin S, are also linked to phosphopantetheine.

The two types of chemical reactions in which CoA and other pantothenate-containing molecules participate are acyl group transfer and condensation (9). In acyl group transfer, there is nucleophilic addition to the carbonyl group thioesterified to CoA, followed by new ester bond formation and displacement of CoA. Condensation reactions involve acidification of the α carbon of the acyl group thioesterified to CoA and attachment of the α carbon to an electrophilic center, leading to carbon–carbon bond formation or cleavage. The first step in the tricarboxylic acid cycle of respiratory metabolism is the condensation of acetyl-CoA with oxaloacetic acid to yield citric acid. Fatty acid synthesis involves both types of reactions (10). Malonate is transferred from CoA to the enzyme-linked phosphopantetheine, and acetate is transferred to an enzyme sulfhydryl group. The introduced acetate or the growing fatty acid chain condenses with the pantetheine-linked malonate.

III. FOOD SOURCES AND REQUIREMENT

Since pantothenic acid is essential to all forms of life, it is widely distributed in nature and readily available in food sources. Foods that are particularly rich in pantothenic acid (11) are liver, kidney, yeast, egg yolk, and broccoli, which contain at least 50 µg of the vitamin per gram dry weight. Extremely high levels of pantothenate are found in royal bee jelly (511 µg/g) and ovaries of tuna and cod (2.32 mg/g). The pantothenate content of human milk increases fivefold within the first 4 days after parturition, from 2.2 to 11.2 µmol/L (48 to 245 µg/dL), a level similar to that found in cow's milk. Although pantothenic acid is relatively stable at neutral pH, cooking is reported to destroy 15–50% of the vitamin in meat, and vegetable processing is associated with pantothenate losses of 37–78% (12). Multivitamin preparations commonly contain the alcohol derivative of pantothenic acid, panthenol, because it is more stable than pantothenate to which it is converted by humans (13). Calcium and sodium salts of D-pantothenate are also available as vitamin supplements.

Formal recommended daily allowances (RDAs) have not been established for pantothenic acid, but the recommended daily intake is 18–32 µmol (4–7 mg) for adults. Among younger age groups, 9 µmol (2 mg) pantothenate daily is recommended for infants and 18–23 µmol (4–5 mg) for children 7–10 years of age (14). The pantothenate content of the average American diet, estimated at 5.8 mg/day (15), is consistent with these recommendations (12). Even when the intake of pantothenate was less than 4 mg/day, as indicated by a study of adolescents (16), blood concentrations of the vitamin were in the normal range (0.91–2.74 µmol/L). The excess appeared to be excreted, since pantothenate excretion was highly correlated with dietary intake.

IV. ABSORPTION AND METABOLISM

CoA that is ingested from dietary sources is hydrolyzed in the intestinal lumen to pantothenic acid, which is absorbed into the bloodstream by a sodium-dependent transport mechanism (17). After being circulated in the plasma, pantothenate is taken up into most cells by cotransport with sodium ions; sodium cotransport is also responsible for placental absorption of pantothenate from maternal circulation (18). However, the level of CoA synthesized in tissues does not reflect the amount of pantothenate available (12). Phosphorylation, the first step in the conversion of pantothenic acid to CoA, which is catalyzed by pantothenate kinase, is the primary regulatory site of CoA synthesis in bacteria and rat hearts (19). Feedback inhibition of pantothenate kinase by the products CoA and acyl-CoA was found to be relieved by carnitine, the carrier of fatty acids into mitochondria (20). All of the enzymes required for CoA synthesis are located in the cytoplasm. Nevertheless, mitochondria must also be a final site for CoA synthesis, since 95% of CoA is found in mitochondria and CoA itself does not cross the mitochondrial membranes (19). CoA is ultimately hydrolyzed to pantothenate in multiple steps, with the final, unique step being the hydrolysis of pantetheine to cysteamine and pantothenate, which is excreted in the urine (21).

V. BIOCHEMICAL FUNCTIONS IN METABOLISM AND PROTEIN MODIFICATION

A. Metabolism and Synthesis of Biological Molecules

CoA plays a central role (9) in the energy-yielding oxidation of glycolytic products and other metabolites through the tricarboxylic acid cycle. As acetyl-CoA condenses with oxaloacetate, in the first step of the cycle, to yield citrate, and as succinyl-CoA it provides the energy for substrate level phosphorylation of guanosine diphosphate (GDP). Furthermore, the synthesis of many essential molecules depends on CoA, as does their degradation. Fatty acids and membrane phospholipids, including the regulatory sphingolipids, require CoA for synthesis, and the synthesis of amino acids such as leucine, arginine, and methionine includes a pantothenate-dependent step. CoA participates, as well, in the β-oxidation of fatty acids and the oxidative degradation of amino acids, making the catabolic products available to the respiratory tricarboxylic acid cycle. Pantothenic acid, incorporated in CoA, is also required for synthesis of isoprenoid-derived compounds, such as cholesterol, steroid hormones, dolichol, vitamin A, vitamin D, and heme A. Through succinyl-CoA, pantothenate is essential to the synthesis of δ-aminolevulinic acid, which is a precursor of the corrin ring in vitamin B_{12} and the porphyrin rings in hemoglobin and the cytochromes. CoA donates an essential acetyl group to the neurotransmitter acetylcholine, as well as to the sugars N-acetylglucosamine, N-acetylgalactosamine, and N-acetylneuraminic acid, components of glycoproteins and glycolipids.

B. N-Terminal Acetylation of Proteins

The exposed N-terminal amino acid of 50–90% of soluble eukaryotic proteins is modified with acetate donated by CoA (22). Usually the terminal methionine is cleaved and the second amino acid, typically alanine or serine, is cotranslationally acetylated on its α-amino group. In experiments suggesting that this acetylation may protect proteins from proteolytic degradation, certain acetylated proteins were resistant to ubiquitin-dependent proteolysis in vitro, whereas their unacetylated counterparts were degraded (23). More sophisticated experiments, however, resulted in the proteolysis of acetylated proteins as well. An additional factor found to be required for degradation of N-terminally acetylated proteins was the protein synthesis elongation factor EF-1α (24). Despite its uncertain function, the importance of N-terminal acetylation was demonstrated genetically in *Saccharomyces cerevisiae*, where cell cycle progression and sexual development were disrupted in a strain defective in *N*-terminal acetyltransferase activity (25). The unacetylated proteins in this mutant strain did not evince destabilization due to lack of acetylation. Acetylation alters the structure of certain proteins. For example, it increases the *N*-terminal α-helical content of calpactin I, a calcium-binding protein, which requires acetylation to assemble with its regulatory subunit (26).

A unique type of *N*-terminal acetylation occurs during processing of certain mammalian peptide hormones from their polyprotein precursors, and this acetylation strongly affects hormone activity. Both adrenocorticotropic hormone (ACTH), a steroidogenic hormone, and β-lipotropin, a lipolytic hormone, are processed from the common precursor pro-opiomelanocortin. ACTH is processed, in turn, to α-melanocyte-stimulating hormone (MSH), and β-lipotropin is processed to the opioid β-endorphin (27). Both α-MSH and β-endorphin become acetylated, but they differ in terms of the tissues in which they are acetylated and the effect of acetylation on activity. Both become *N*-terminally acetylated in the intermediate pituitary, but only MSH is acetylated in brain, and neither hormone is acetylated in the anterior pituitary (8). Acetylation stimulates the activity of α-MSH (28) but inactivates β-endorphin by inhibiting its binding to opioid receptors (29). A single *N*-acetyltransferase activity, identified in the intermediate but not the anterior lobe of rat and bovine pituitary, may be responsible for modifying both hormones (30), providing a mechanism for differentially activating two distinct products of a single precursor.

C. Internal Acetylation of Proteins

1. Histones

The histones and α-tubulin are two major groups of proteins that undergo selective, reversible acetylation on the ε-amino group of internal lysine residues. Other DNA-binding proteins, including the high-mobility group proteins HMG1 and HMG2 and protamines, are also subject to acetylation. Within the nucleosome are a tetramer of histones H3 and H4 and two dimers of histones H2A and H2B, around which is wrapped 146 bp of DNA. These core histones of the octamer become acetylated within their basic amino terminal regions, with H3 and H4 each having four possible acetylation sites. A study of preferred acetylation sites in mammalian histones (31) showed that in H3, lysine 14 was acetylated first, followed by lysine 23 and lysine 18. Monoacetylated H4 was modified exclusively at lysine 16, with further additions progressing from lysine 12 to lysine 8 and lysine 5. In histone H2B, lysine 12 and lysine 15 were the preferred acetylation sites over lysine 5 and lysine 20. Histone H2A has only one acetylation site.

Histones that become highly acetylated are associated chiefly with DNA that is either newly replicated or contains genes that are actively transcribed (8,32). Different acetylation sites may be utilized in these two processes. In the ciliated protozoan *Tetrahymena thermophila*, where these activities are easily distinguished, histone H4 becomes diacetylated on lysines 4 and 7 in transcriptionally active chromatin, whereas newly assembled chromatin contains H4 that is diacetylated on lysines 4 and 11 (33). In *Physarum polycephalum*, replicating DNA has acetylation in all four core histones, whereas only histones H3 and H4 have high levels of acetylation in actively transcribed genes (34). The temporal patterns of acetylation also differ. Newly synthesized histones are only transiently acetylated as they assemble with replicating chromatin, whereas histones in transcriptionally active chromatin are continually and dynamically acetylated.

The presence of highly acetylated H3 and H4 noticeably affects nucleosomal structure, in part through charge neutralization of the modified lysines. The more open, unfolded configuration of acetylated chromatin is indicated by its increased sensitivity to nucleases and its increased salt solubility (35). Agarose gel electrophoresis also showed that histone acetylation leads to decompaction of the chromatin fiber (36). Hyperacetylation of H3 and H4 has been demonstrated to lead to a reduction in the linking number change per nucleosome, indicating a decrease in the negative supercoiling within nucleosomes (37). For the H2A/H2B dimer, acetylation makes its nucleosomal association particularly labile, which may facilitate its exchange with newly synthesized histones (38).

Acetylated histones are enriched in genes that are being actively transcribed. The acetylated chromatin fraction from chick embryo erythrocytes, selected by an antibody against the acetylated lysine of histones, was strongly enriched in actively transcribed α-D-globin gene sequences (39). However, histone acetylation does not correspond exactly to the transcribed sequences, either spatially or temporally. Acetylation extends into the interspersed, nontranscribed DNA of the β-globin locus, as well as being present in the transcribed region (40). Furthermore, acetylation does not respond to the switch in globin gene transcription that occurs between 5 and 15 days of embryo development, since acetylated histones were associated with both globin genes at both stages of development (39). Similarly, acetylated histones were associated with the gene for the platelet-derived growth factor B chain before its induction by phorbol esters (41). Histone acetylation, therefore, may be a precondition for transcription, rather than specifying expression of particular genes. In vitro experiments indicate that histone acetylation is required for specific proteins to bind to chromatin. For example, acetylation was required for transcription factor IIIA (TFIIIA) binding to the 5S RNA gene (42). Acetylation of core histones also indirectly facilitated the binding of transcription factors USF and GAL4-AH to nucleosomes, by inhibiting the binding of the internucleosomal histone H1, which represses transcription factor binding (43).

The chromosomal distribution of acetylated histone H4 was observed in the polytene chromosomes of *Drosophila* larvae by immunofluorescence, using antibodies developed against specific, acetylated lysines (44). The single X chromosome in male larvae, which is very actively transcribed, was the chief location of H4 acetylated at lysine 16, an acetylated form that was not observed in male autosomes or in any chromosomes of female larvae. Immunolabeling of mammalian chromosomes showed that acetylated H4 was concentrated in chromosomal regions enriched in coding DNA, but acetylated H4 was largely absent from the inactive X chromosome of female mammalian cells; the few immunolabeled bands corresponded mainly to regions containing known expressed genes (45).

That acetylation of histone H4 affects the transcription of certain genes was demonstrated in *S. cerevisiae*, where mutations were made in H4 that either delete the *N*-terminal tail or substitute its lysine residues. Deletion of the basic tail or an uncharged substitution of lysine 16 led to derepression of the silent mating locus, suggesting that the charge-neutralizing acetylation of lysine 16 may be responsible for activating transcription (46). In contrast, transcription of the *GAL1* and *PHO5* genes was inhibited by similar mutations of H4, in which multiple lysines were substituted (47). However, a neutralizing mutation was still much less inhibitory to transcription than an arginine substitution. Mutations of the *N*-terminal sequences of the other three core histones had no or little effect on transcription of these genes or the mating type locus.

Unlike gene transcription, the replication of DNA does not require histone acetylation for polymerase enzymes to gain access to the nucleosomal DNA. However, newly synthesized histones that assemble with the replicated DNA become transiently acetylated on their *N*-terminal tails, a modification that is detected in the presence of the deacetylase inhibitor butyrate. Histones are deposited onto DNA in a two-step process that begins with the deposition of H3/H4 tetramers, followed by H2A/H2B dimers. Diacetylation of H4 precedes its deposition onto DNA and its assembly with newly synthesized H3, H2A, and H2B (48). Histone H2B is also highly acetylated in newly assembled nucleosomes (49). In the absence of DNA synthesis, newly synthesized H2A and H2B exchange preferentially into acetylated chromatin regions, assembling with acetylated H3 and H4, but newly synthesized H3 and H4 remain cytosolic (48). Deposition-related histone acetylation inhibits the binding of histone H1 to newly assembled chromatin, thereby reducing higher order nucleosome interactions (50).

Other chromatin-associated molecules that undergo acetylation include HMG1 and HMG2, which are acetylated by the same *N*-acetyltransferase as the histones (51). It is apparently the acetylated form of HMG1 that serves as a histone assembly factor and stimulates DNA polymerase activity (52). The same acetyltransferase also acetylates polyamines, which interact with chromatin during DNA synthesis (51).

2. α-Tubulin

A subunit of microtubules is the other class of proteins that becomes acetylated on the ε-amino group of lysine residues. Microtubules are essential structural components of eukaryotic cells that are central to chromosome segregation in nuclei and to the cytoplasmic cytoskeleton, where they affect cell shape, motility, and organelle movement. They also constitute the major component of flagella and cilia. Microtubules are assembled from α-tubulin/β-tubulin dimers that polymerize and depolymerize dynamically. A subset of microtubules has been found to contain α-tubulin that is acetylated on lysine 40 (53). Acetylation occurs on the α-tubulin after it is incorporated into the microtubule, which is a better substrate than the tubulin dimer for the isolated acetylase activity (54), and deacetylation appears to be coupled to microtubule depolymerization. The acetylated microtubules are more stable to depolymerizing agents, such as colchicine, than the unacetylated microtubules; in turn, drugs that stabilize microtubules, such as taxol, induce α-tubulin acetylation (55). Tubulin turnover in neurons was found to be much slower in areas where acetylated α-tubulin was concentrated, such as the neurite shafts, than in the cell body and growth cones (56). These observations suggest that acetylation may help to stabilize microtubules.

The distribution of acetylated α-tubulin in various types of cells has been detected by an antibody specific for the modified tubulin and has been found to be nonrandom. In cerebellar neurons, which have abundant microtubules, acetylated α-tubulin was found to be concentrated in axons, relative to dendrites, during early neuronal development (57), although in mature neurons it was present also in thick dendritic trunks. It was excluded from growth cones at neurite tips in dorsal root ganglion neurons, and acetylated α-tubulin was also absent from leading edge microtubules of migrating 3T3 cells. Acetylated microtubules underlie motor endplates in chick muscle fibers and they contribute to vesicle and organelle transport, whereas growing or motile regions of the cells appear to be depleted of acetylated microtubules (58). Individual axonal microtubules actually contain separate regions of acetylated and unacetylated α-tubulin along their length, with the older regions corresponding to the acetylated domains (59). It was suggested that the acetylation of α-tubulin may be a step in cell commitment. When fibroblast cells were induced to become neuron-like, they formed an acetylated microtubule bundle that subsequently extended into a neurite (60).

Cells undergoing mitosis and meiosis also show nonrandom, varying distribution of acetylated microtubules. During arrest in meiotic metaphase, acetylated α-tubulin was found predominantly at the poles of unfertilized mouse oocytes (61). At meiotic anaphase, the spindle became labeled, but by telophase only the meiotic midbody microtubules were acetylated. Changes in the distribution of acetylated microtubules were also described for the preimplantation development of mouse embryo cells (62). Acetylated microtubules were generally associated with the cell cortex before differentiation. As cellular asymmetry developed, the acetylated microtubule subpopulation relocated to the basal part of the cell cortex and, after asymmetrical cell division, became concentrated in the inside cells rather than the outside cells.

D. Acylation of Proteins

During the past 15 years it has been discovered that a wide variety of cellular proteins are modified with long-chain fatty acids, donated by CoA (8). This type of covalent modification affects the location and activity of many proteins (Table 1), including those that

Table 1 Proteins[a] That Are Modified with Fatty Acids

Palmitate	Myristate	Palmitate + myristate
Ras^v, Ras^c	Transducin	G_o, G_i protein,
G_s protein α subunit	ARF proteins	α subunit
Receptors:	Src^v, Src^c	$p56^{lck}$, $p59^{fyn}$
Rhodopsin	Protein kinase A	Insulin receptor
Dopamine (D_1, D_2)	Calcineurin	ecNOS
α_{2A}-, β_2-Adrenergic	Recoverin	
Thyrotropin-releasing hormone	MARCKS	
Choriogonadotropin	Band 4.2	
Iron-transferrin	IgM heavy chain	
Nicotinic acetylcholine	Interleukin-1α, 1β	
CD4, CD44, CD36, P-selectin	TNF-α	
HLA-B heavy chain	NADH-cytochrome b_5 reductase	
Band 3	NADH dehydrogenase, B18	
Ankyrin	cytochrome c oxidase, su 1	
Spectrin		
Vinculin		
Fibronectin		
Actin		
Gap junction proteins		
Myelin proteolipid subunit		
Cysteine string proteins		
Acetylcholinesterase		
Glutamic acid decarboxylase$_{65}$		
GAP-43, SNAP-25		
Methylmalonate semialdehyde dehydrogenase		
Aldehyde and glutamate dehydrogenases		

[a] Discussed in text.

have a central role in signal transduction. The two fatty acids that are usually added to proteins are myristic acid, a rare 14-carbon saturated fatty acid, and the more common 16-carbon palmitic acid. These two fatty acids are attached to proteins by distinct mechanisms, and they affect proteins differently. The myristoylation that has been best characterized occurs cotranslationally, by amide linkage to the α-amino group of an N-terminal glycine, and it is irreversible. Palmitate, in contrast, is added posttranslationally and forms a reversible ester bond with a cysteine or serine. The greater hydrophobicity of palmitate results in strong membrane association of the acylated protein, whereas myristoylation leads either to weak membrane association or interactions with proteins. The reversibility of palmitoylation allows this modification to be regulatory.

There are stringent sequence requirements for a protein to become N-terminally myristoylated (63). An N-terminal glycine, originally in the second position, is absolutely required, and there is preference for a small neutral subterminal residue and a serine in the sixth position. The myristoyltransferase that has been isolated shows a specific preference for myristoyl-CoA over other acyl-CoA donors (63). The N-terminal myristoylation was shown to be essential for viability in *S. cerevisiae*, where gene disruption of *N*-myristoyltransferase proved to be recessively lethal (64). In contrast, enzymes that transfer palmitate show little substrate specificity, either for the fatty acyl group transferred from CoA or the peptide sequence required for protein acylation.

1. GTP-Binding Proteins

The small GTP-binding proteins comprise an extensive group of influential cellular proteins that become acylated with myristic acid and/or palmitic acid. This group of nucleotide-binding proteins includes the monomeric Ras superfamily of proteins, heterotrimeric G proteins, and ADP-ribosylating factors (ARFs). Several of these proteins are also modified by isoprenoid groups whose synthesis requires CoA, although it is not the lipid donor.

Covalent modification of the Ras proteins, both cellular and viral, has been extensively studied. Most Ras proteins undergo two types of modifications at their carboxyl end, isoprenylation and methylation on the processed terminal cysteine and palmitoylation of a nearby cysteine; in some proteins a basic region functionally replaces palmitoylation (65). The viral Ras proteins, such as those of Harvey murine and Kirsten sarcoma viruses, are deficient in GTP hydrolysis and are oncogenic. Both isoprenylation and palmitoylation are required for viral Ras proteins to bind to the plasma membrane and transform cells, with palmitoylation strengthening the weak activity conferred by isoprenylation (65). A 15-carbon farnesyl group modifies Ras proteins, and farnesyltransferase inhibitors are effective antitumor agents in Ras-transformed cells. However, the Ras-related proteins and most cellular proteins are isoprenylated with the 20-carbon geranylgeranyl (66). The cellular Ras protein is a critical component of the mitogen-activated protein (MAP) signaling kinase cascade, serving to direct the protein kinase Raf to the plasma membrane. When Raf itself was experimentally engineered to contain a farnesylation signal, Ras proved to be unnecessary for signal transduction (67).

The mammalian Rab proteins are Ras-related proteins that are proposed to regulate specific stages of vesicular transport to the cell surface (68). These proteins appear to target vesicles to specific acceptor membranes by binding to the vesicles until their delivery to membranes. The cycling of Rab proteins between membranes and cytosol is related to the membrane budding and fusion that drives vesicular movement. Rab proteins are modified at their C termini with geranylgeranyl, which is required for their association with organellar membranes, as well as for interaction of a Rab protein with its GDP dissociation inhibitor (GDI), which releases the Rab protein from membranes (69). A defect in a protein with sequence homology to GDIs has been implicated in retinal degradation or choroideremia in humans (66).

The α subunits of several heterotrimeric G proteins, which mediate transmembrane signaling, are esterified to palmitate near their N termini on a third-position cysteine. The $G_{s\alpha}$ subunit, which mediates hormonal stimulation of cAMP synthesis, requires palmitoylation both for binding to membranes and for stimulation of adenylyl cyclase (70). Hormonal activation leads to GTP binding by α_s and to its rapid depalmitoylation, which is accompanied by relocalization of α_s to the cytosol. In addition to palmitate, myristic acid is added to the N terminus of those G-protein α subunits that are members of the G_i and G_o subfamilies. Prior myristoylation increases these proteins' membrane affinity, which is required for the addition of palmitate (71). Both types of acylation were found to contribute to the membrane affinity of the α_o subunit and its binding to the $\beta\gamma$ subunits. Mutation of the palmitoylated cysteine of $G_o1\alpha$ disrupted its exclusive localization to membranes, even in the presence of N-myristoylation, whereas mutation of the myristoylated glycine, preventing both fatty acylations, caused α_o to become chiefly cytosolic (72). It has been suggested that dual acylation of these G-protein α subunits and certain Src-related proteins (discussed below) directs their localization to plasma membrane caveolae, possibly via association with glycosylphosphatidylinositol (GPI)-anchored membrane proteins (72). In addition to myristate, heterogeneous fatty acids, such as laurate and unsaturated 14-carbon fatty acids, modify the N terminus of the α subunit of transducin, the photoreceptor G protein that is stimulated by photolyzed rhodopsin. The nature of the modifying fatty acid appears to affect the strength of the interaction of transducin α with $\beta\gamma$ and may influence the speed of visual excitation (73). Since the γ subunits of G proteins are isoprenylated at their C termini (66), increasing their membrane affinity, these heterotrimeric G proteins, which play a central role in cell signaling, undergo all three types of modifications with hydrophobic moieties.

The small GTP-binding proteins that are involved in vesicular transport, known as ARFs, are N-terminally myristoylated, and this modification is essential for their functions in transport. In mammalian cells the ARF protein is the principal coat component of non–clathrin-coated vesicles, and its binding to the Golgi membrane appears to mark the site of budding of a future vesicle. Whereas recombinant myristoylated ARF5 bound to Golgi membranes in a GTP-dependent and temperature-dependent manner, the nonmyristoylated ARF5 did not bind to membranes at all (74). The N-terminal region of ARF1 may be stabilized by myristoylation in an amphipathic α helix, conducive to membrane binding. Myristoylation also apparently stabilizes the interaction of ARF1 with GTP in the presence of membranes and enables ARF1 to release GDP at high physiological concentrations of Mg^{2+} (75).

In addition to the myristoylation of ARFs, reconstitution of in vitro transport through the Golgi stacks requires palmitoyl-CoA (76). The addition of acyl-CoA was necessary for the budding of vesicles from donor Golgi and for the fusion of these vesicles with acceptor Golgi cisternae. A nonhydrolyzable analog of palmitoyl-CoA or an inhibitor of the acyl-CoA synthetase inhibited vesicle budding (76). The source of this acyl group requirement is not known.

2. Protein Kinases and Phosphatases

Like palmitoylation of Ras proteins, myristoylation of the Src protein provided the first model of this type of acylation. Src, first identified in Rous sarcoma virus, is a tyrosine kinase that has both cellular and viral homologues. The viral Src protein is responsible for cell transformation by the oncogenic viruses, and its N-terminal myristoylation is necessary both for the protein's association with the plasma membrane and for its transforming capability. Expression of a mutant, nonmyristoylated pp60^{src} protein resulted in a small group of membrane-associated proteins showing decreased tyrosine phosphorylation (77). pp60^{src} was also prevented from associating with membranes by addition of a less hydrophobic analog of myristic acid, which is a possible candidate for an antitumor agent (78). Mutation of the cellular Src protein at its N-terminal glycine to prevent myristoylation proved inhibitory to its activity in mitosis, presumably because c-Src is activated by a membrane-bound phosphatase (79).

Although pp60^{src} itself is not additionally acylated with palmitate, several Src-related tyrosine kinases, such as p56^{lck} and p59^{fyn}, are dually acylated at their N termini (80). These kinases associate with proteins that are anchored in the outer plasma membrane by glycosylphosphatidylinositol modification, and dual acylation of the kinases is necessary for this association. When the palmitoylated cysteines at positions 3 and 6 in p59^{fyn} and at positions 3 and 5 in p56^{lck} were mutated to serines, these two kinases no longer associated with the GPI-anchored protein DAF (decay-accelerating factor). On the other hand, when the reverse mutations were made in pp60^{src}, from serines at positions 3 and 6 to cysteines, the Src protein uncharacteristically associated with DAF (81). GPI-anchored proteins, present in the membranes of T cells, B cells, monocytes, and granulocytes, are thought to transduce external activation signals through their interaction with tyrosine kinases in the inner membrane layer.

The catalytic subunit of cAMP-dependent protein kinase is also myristoylated at its N terminus. Myristoylation does not appear to affect its association with membranes, since the catalytic subunit is anchored in the membrane by the regulatory subunit until dissociation by cAMP. A comparison of the properties of the myristoylated vs. nonmyristoylated catalytic subunit indicated that acylation increased the thermal structural stability of the kinase, possibly by stabilizing an intramolecular interaction between the N-terminal domain and an internal hydrophobic surface (82). N-terminal myristoylation also occurs in the B subunit of calcineurin (83), a calmodulin-dependent phosphatase that is the target of the immunosuppressant cyclosporin A.

3. Membrane Receptors

Many membrane-anchored receptors are acylated with palmitic acid. Heterotrimeric G-protein–coupled receptors all have a similar domain structure, consisting of an extracellular N terminus, seven transmembrane α helices, and an intracellular C terminus, leading to three intracellular and three extracellular loops. Near the seventh transmembrane region, the C-terminal tail has a conserved cysteine whose palmitoylation has been characterized in rhodopsin, the dopamine (D$_1$ and D$_2$) receptors, the β$_2$-adrenergic receptor, the α$_{2A}$-adrenergic receptor, thyrotropin-releasing hormone receptor, and the choriogonadotropin receptor. This acylation appears to lead to the formation of a fourth intracellular loop. Fluorescence quenching studies with rhodopsin, the retinol-binding photoreceptor of retinal rod cells, showed that two cysteines at positions 322 and 323, modified by palmitic acid, are integrated into the cell membrane (84). Furthermore, the light excitation of rhodopsin is regulated by recoverin, an N-myristoylated protein that responds to changing levels of calcium. Calcium binding alters the conformation of recoverin to expose the N-myristoyl group, which can then bind to membranes (85).

The regulatory functions of receptor palmitoylation have been studied in the dopamine and adrenergic receptors. The human receptor for dopamine (D$_1$) incorporates palmitate shortly after cells are exposed to hormone, concurrent with cellular desensitization to dopamine (86). The receptor also becomes phosphorylated and dissociates from the cell surface. By mutating the palmitoylated cysteine of the β_2-adrenergic receptor, it was determined that acylation is important both for receptor coupling with G_s (the stimulatory G protein) and for agonist-promoted desensitization (87). However, the corresponding mutation in the α_{2A}-adrenergic receptor had a different effect. Mutation of the cysteine abolished the downregulation of receptor number that occurs after prolonged exposure to hormone (88). In contrast, mutational analysis showed that palmitoylation of thyrotropin-releasing hormone and choriogonadotropin receptors slowed their internalization after prolonged agonist exposure (89).

Other transmembrane receptors that are palmitoylated include the iron-transferrin receptor, the insulin receptor, and the nicotinic acetylcholine receptor. In addition to palmitoylation, the insulin receptor is also reported to be linked to myristic acid (90). Rat adipocytes contain a membrane glycoprotein, related to CD36, that becomes palmitoylated on an extracellular domain in response to insulin and energy depletion (91). Several cells of blood and the immune system, such as T lymphocytes, macrophages, and platelets, contain receptors that become palmitoylated. CD4, a surface glycoprotein of T lymphocytes and macrophages that functions in cell adhesion and signaling during antigen recognition, also serves as receptor for the human immunodeficiency virus (HIV). CD4 becomes palmitoylated on two cysteines at the junction of its transmembrane and cytoplasmic domains, but this modification does not apparently affect either membrane localization of the receptor or its interactions with proteins such as p56[lck] (92). However, palmitoylation of a lymphoma glycoprotein, closely related to CD44 of T lymphocytes, was found to be required for its in vitro binding to ankyrin, the protein that mediates attachment of the cell membrane to the cytoskeleton (93). The heavy chain of the human histocompatibility antigen HLA-B is thioesterified to palmitate. An adhesion receptor in human platelets, P-selectin (CD62), is palmitoylated on a cysteine in its cytoplasmic tail, a domain that is responsible for the sorting of P-selectin into secretory granules and its endocytosis into cells (94).

4. Proteins Associated with the Cytoskeleton

An N-myristoylated protein, rich in alanine, that is a major substrate of protein kinase C (MARCKS) has been identified as a protein that cross-links actin filaments. MARCKS is thought to be involved in the cytoskeletal rearrangements that occur during neuronal transmitter release, leukocyte activation, and growth factor–induced mitosis. The initial binding of MARCKS to the plasma membrane requires myristoylation, which provides one of its two membrane binding sites. Phosphorylation within the second membrane binding site occurs upon cell activation and results in dissociation of MARCKS from the membrane (95).

Membrane proteins in erythrocytes are linked to the cytoskeleton by a series of acylated proteins. The anion transport protein Band 3, which is palmitoylated (96), is linked to Band 4.2, which is a major myristoylated protein of the membrane (97). Band 4.2, in turn, is linked to ankyrin, the cytoskeleton attachment protein, which is itself palmitoylated (8). In addition, the cytoskeletal protein spectrin becomes palmitoylated in its β subunit, and this acylated subpopulation is more tightly membrane-associated than the unmodified spectrin (98). Other cytoskeleton-associated proteins whose palmitoylation has been reported are vinculin, fibronectin, and a subpopulation of actin. Gap junction proteins of heart and eye lens also become palmitoylated (99).

5. Neuronal Proteins

One of the first proteins found to be covalently modified by palmitate was the proteolipid subunit of brain myelin, in which palmitate is esterified to a threonine residue (8). Cysteine string proteins (csp) are located in synaptic terminals and are required for neurotransmitter release. Fatty acylation of multiple cysteines, 10 in *Torpedo*, appears to be a mechanism for attachment of Tcsp to vesicles, where they interact with presynaptic calcium channels in the plasma membrane (100). Palmitoylation of acetylcholinesterase, which degrades the neurotransmitter acetylcholine, results in its anchorage in the cell membrane. Palmitate attachment to an esterase subpopulation correlated with its attachment to the outer cell surface, whereas most of the expressed esterase was unmodified and secreted from cultured cells (101). A major isoform of glutamic acid decarboxylase (GAD_{65}) is located in synaptic vesicles of GABA-secreting neurons and in microvesicles of pancreatic β cells. Palmitoylation of GAD_{65} in its N-terminal domain is reported to increase its membrane affinity, possibly regulating its membrane anchorage (102).

Protein acylation appears to play an important role during neuronal development. The reversible palmitoylation of both GAP-43 and SNAP-25, in developing brains, was proposed to influence growth cone motility and process outgrowth (103). GAP-43 is a major component of the growth cone membranes of elongating axons, and SNAP-25 is a synaptic protein involved in later stages of axon growth. Whereas the nonacylated form of GAP-43 stimulates the G protein, G_o, palmitoylation at two cysteines near its N terminus inhibits this activity and keeps GAP-43 membrane-bound (104). Nitric oxide inhibits the acylation of GAP-43 and SNAP-25, along with that of other neuronal proteins (103), and it inhibits the growth of cultured neurites by causing growth cone collapse. It was suggested that these effects of nitric oxide reflects its role in regulating process outgrowth and remodeling in vivo (103). The endothelial form of nitric oxide synthase (ecNOS), which is involved in smooth-muscle relaxation, is itself regulated by dual acylation (105). N-terminal myristoylation is required for the membrane localization of ecNOS and for its reversible modification with palmitate. Depalmitoylation follows addition of the agonist bradykinin and results in ecNOS relocalizing to the cytosol and becoming phosphorylated.

6. Myristoylated Proteins of the Immune System

Several proteins of the immune system are modified with myristate on internal lysine residues. These include the heavy chain of μ immunoglobulins, which becomes acylated during transport to the surface of developing B cells (106). Precursors of the cytokines interleukin (IL)–1α and 1β are myristoylated on lysines (107), as is the tumor necrosis factor (TNF) α precursor (108). After processing, the mature forms of IL-1α and IL-1β are secreted and bind to receptors. The precursor of IL-1α, however, which is the predominant myristoylated species, also exists as a plasma membrane–associated protein (107). The acylated lysines in the propiece of the TNF-α precursor are near a membrane-spanning segment and are thought to facilitate membrane insertion of the precursor, which, like the mature TNF-α, is active in mediating inflammation (108).

7. Mitochondrial Proteins

Mitochondrial proteins involved in oxidation–reduction reactions have been reported to become fatty acylated. The NADH–cytochrome b_5 reductase, located at the outer mitochondrial membranes of animal cells, as well as at microsomes, is N-terminally myristoylated (109). In the inner membrane, there is N-myristoylation of subunit B18 of complex

I of the electron transport chain, the NADH:ubiquinone oxidoreductase, which was detected by cDNA sequencing and electrospray mass spectrometry (110). Complex I also contains the pantothenate-containing mitochondrial acyl carrier protein, identified as an enzyme subunit by its stoichiometry (111). The core catalytic subunit 1 of cytochrome c oxidase, the terminal electron carrier of the respiratory chain, is modified by myristic acid on an internal lysine residue (112). The oxidase subunit 1 is one of only a few proteins that are encoded by mitochondrial genes and synthesized in mitochondria; its modification suggests that there is a mitochondrial N-myristoylating activity.

Dehydrogenases involved in oxidative catabolism also become acylated, a modification that inhibits their enzymatic activity and may be a mechanism for feedback regulation (113). A decrease in mitochondrial energy level was found to correlate with increased acylation of an active-site cysteine in methylmalonate semialdehyde dehydrogenase (MMSDH). It was proposed that long-chain fatty acids accumulate when MMSDH and other dehydrogenases utilize the NAD^+ required to oxidize fatty acids. The accumulation of these fatty acids, in turn, promotes acylation of the dehydrogenases, whose inhibition allows β-oxidation of fatty acids to proceed. Aldehyde dehydrogenase and glutamate dehydrogenase are two other mitochondrial NAD^+-dependent dehydrogenases that become acylated (113).

VI. PHARMACOLOGY

The human effects of pantothenate deficiency have been detected chiefly by the administration, either intentional or unintentional, of pantothenic acid antagonists. Hopantenate is an analog of pantothenic acid, which contains GABA in place of β-alanine. Also an agonist of GABA, hopantenate was used in Japan as a cerebral stimulant for retarded individuals and to alleviate symptoms of tardive dyskinesia induced by tranquilizers. It was found, however, that treatment with hopantenate produced severe side effects in patients, including lactic acidosis, hypoglycemia, and hyperammonemia (114). It ultimately led to acute encephalopathy with hepatic steatosis resembling Reye's syndrome. A study in which dogs were administered hopantenate over several weeks showed that they suffered similar effects (115). This study also characterized the abnormalities that developed in hepatic mitochondria, which became enlarged, had increased numbers of cristae, and contained crystalloid inclusions. These mitochondrial alterations differed from those associated with Reye's syndrome. The side effects of hopantenate treatment are apparently caused by induction of pantothenic acid deficiency. Dogs that were given an equivalent amount of pantothenic acid at the same time as calcium hopantenate did not develop these disorders (115).

When individuals who suffer seizures are treated with the anticonvulsant drug valproic acid, liver damage may follow in association with decreased levels of CoA and acetyl-CoA. These effects appear to be due to sequestration of CoA in esters of valproic acid and its metabolites (20). Coadministration of pantothenic acid, carnitine, and acetylcysteine, along with valproate or its derivative, relieved these metabolic defects in developing mice (20). Skeletal muscles of the murine model of Duchenne's muscular dystrophy show reduced energy metabolism, as indicated by reduced heat production, relative to control muscles. The addition of pantothenate resulted in increased cytoplasmic synthesis of CoA and increased the thermogenic response to glucose in the diseased muscles (116). Pantothenic acid also protected rats against liver damage and peroxidation produced by carbon tetrachloride (117). Furthermore, surgical wound healing was greatly improved by the administration of pantothenic acid, apparently due to its anti-inflammatory properties. In vitro studies show that calcium pantothenate dampens the responses of neutrophils (polymorphonuclear leukocytes) to activation by stimulatory peptides and cytokines (118).

Analogs of pantothenic acid have been used experimentally to induce pantothenate deficiency. The antagonist ω-methylpantothenate, in combination with a pantothenate-deficient diet, produced deficiency symptoms in humans that included headache, fatigue, insomnia, intestinal disturbances, and paresthesia of hands and feet. A decrease in the eosinopenic response to ACTH, loss of antibody production, and increased sensitivity to insulin were also reported (119). Naturally occurring pantothenate deficiency has been detected only under conditions of severe malnutrition. World War II prisoners in the Philippines, Japan, and Burma suffered from a disease in which they experienced numbness in their toes and painful burning sensations in their feet. Pantothenic acid was specifically required to relieve these symptoms of nutritional melalgia (120).

Many studies have documented the physical effects on animals of a pantothenate-restricted diet (11). Rats develop hypertrophy of the adrenal cortex, which is followed by hemorrhage and necrosis; their resistance to certain viral infections is increased. Pantothenate deficiency in dogs produces hypoglycemia, gastrointestinal symptoms, rapid respiration and heart beat, and convulsions. Monkeys show depressed heme synthesis and become anemic. During exercise, deficient mice display lower stamina, and their liver and muscle glycogen levels are reduced (12). Chickens develop dermatitis and poor feathering, as well as axon and myelin degeneration within the spinal cord. Dermatitis and graying in mice, induced by pantothenate deficiency, were reversed by administration of pantothenic acid, but this treatment did not prove successful at restoring hair color in humans. Calcium pantothenate is not toxic to rats, dogs, rabbits, or humans at high doses, but the lethal dose for mice, leading to respiratory failure, was determined to be 42 mmol (10 g)/kg (11).

VII. CONCLUSIONS

Pantothenic acid, through incorporation into phosphopantetheine and CoA, is essential to many metabolic conversions and to all forms of life. It is widely available in dietary sources, and human deficiency in pantothenic acid is rare. Nevertheless, the administration of pantothenate is helpful in counteracting the inhibitory effects certain drugs and diseases have on respiratory metabolism and other activities. It has long been known that pantothenate plays a central role in energy generation and molecular syntheses. It is now clear, in addition, that pantothenic acid participates in regulating numerous proteins by donating acetyl- and fatty acyl-modifying groups, which alter the location and/or activity of the acylated protein. Proteins that are acetylated constitute major structural components of chromatin and microtubules. Myristoylated and palmitoylated proteins are central components of signal transduction systems, vesicular transport, and the cytoskeleton, and they are seemingly common to all eukaryotic cells.

REFERENCES

1. R. J. Williams, C. M. Lyman, G. H. Goodyear, J. H. Truesdail, and D. Holiday, ''Pantothenic acid,'' a growth determinant of universal biological occurrence, *J. Am. Chem. Soc. 55*:2912 (1933).

2. T. H. Jukes, Pantothenic acid and the filtrate (chick antidermatitis) factor, *J. Am. Chem. Soc.* *61*:975 (1939).
3. C. A. Elvehjem and C. J. Koehn, Studies on vitamin B$_2$ (G), *J. Biol. Chem.* *108*:709 (1935).
4. R. J. Williams and R. T. Major, The structure of pantothenic acid, *Science 91*:246 (1940).
5. F. Lipmann, N. O. Kaplan, G. D. Novelli, L. C. Tuttle, and B. M. Guirard, Coenzyme for acetylation, a pantothenic acid derivative, *J. Biol. Chem.* *167*:869 (1947).
6. J. Baddiley, E. M. Thain, G. D. Novelli, and F. Lipmann, Structure of coenzyme A, *Nature 171*:76 (1953).
7. G. Brown, The metabolism of pantothenic acid, *J. Biol. Chem.* *234*:370 (1959).
8. N. Plesofsky-Vig and R. Brambl, Pantothenic acid and coenzyme A in cellular modification of proteins, *Ann. Rev. Nutr.* *8*:461 (1988).
9. D. E. Metzler, *Biochemistry*, Academic Press, New York, 1977, p. 428.
10. S. J. Wakil, J. K. Stoops, and V. C. Joshi, Fatty acid synthesis and its regulation, *Annu. Rev. Biochem.* *52*:537 (1983).
11. F. A. Robinson, *The Vitamin Co-Factors of Enzyme Systems*, Pergamon Press, Oxford, 1966, p. 406.
12. A. G. Tahiliani and C. J. Beinlich, Pantothenic acid in health and disease, *Vitam. Horm. 46*: 165 (1991).
13. O. D. Bird and R. Q. Thompson, Pantothenic acid, in *The Vitamins*, Vol. 7, 2nd ed. (P. Gyorgy and W. N. Pearson (eds.), Academic Press, New York, 1967, p. 209.
14. Food and Nutrition Board, National Research Council, *Recommended Dietary Allowances*, 10th ed., National Academy Press, Washington, D.C., 1989.
15. J. B. Tarr, T. Tamura, and E. L. Stokstad, Availability of vitamin B$_6$ and pantothenate in an average American diet in man, *Am. J. Clin. Nutr. 34*:1328 (1981).
16. B. R. Eissenstat, B. W. Wyse, and R. G. Hansen, Pantothenic acid status of adolescents, *Am. J. Clin. Nutr. 44*:931 (1986).
17. D. K. Fenstermacher and R. C. Rose, Absorption of pantothenic acid in rat and chick intestine, *Am. J. Physiol. 250*:G155 (1986).
18. S. M. Grassl, Human placental brush-border membrane Na$^+$-pantothenate cotransport, *J. Biol. Chem. 267*:22902 (1992).
19. J. D. Robishaw, D. Berkich, and J. R. Neely, Rate-limiting step and control of coenzyme A synthesis in cardiac muscle. *J. Biol. Chem. 257*:10967 (1982).
20. J. H. Thurston and R. E. Hauhart, Reversal of the adverse chronic effects of the unsaturated derivative of valproic acid—2-n-propyl-4-pentenoic acid—on ketogenesis and liver coenzyme A metabolism by a single injection of pantothenate, carnitine, and acetylcysteine in developing mice, *Pediatr. Res. 33*:72 (1993).
21. C. T. Wittwer, D. Burkhard, K. Ririe, R. Rasmussen, J. Brown, B. W. Wyse, and R. G. Hansen, Purification and properties of a pantetheine hydrolysing enzyme from pig kidney, *J. Biol. Chem. 258*:9733 (1983).
22. H. P. C. Driessen, W. W. de Jong, G. I. Tesser, and H. Bloemendal, The mechanism of N-terminal acetylation of proteins, *CRC Crit. Rev. Biochem. 18*:281 (1985).
23. A. Hershko, H. Heller, E. Eytan, G. Kaklij, and I. A. Rose, Role of the α-amino group of protein in ubiquitin-mediated protein breakdown, *Proc. Natl. Acad. Sci. USA 81*:7021 (1984).
24. H. Gonen, C. E. Smith, N. R. Siegel, C. Kahana, W. C. Merrick, K. Chakraburtty, A. L. Schwartz, and A. Ciechanover, Protein synthesis elongation factor EF-1α is essential for ubiquitin-dependent degradation of certain N^α-acetylated proteins and may be substituted for by the bacterial elongation factor EF-Tu, *Proc. Natl. Acad. Sci. USA 91*:7648 (1994).
25. J. R. Mullen, P. S. Kayne, R. P. Moerschell, S. Tsunasawa, M. Gribskov, M. Colavito-Shepanski, M. Grunstein, F. Sherman, and R. Sternglanz, Identification and characterization of genes and mutants for an N-terminal acetyltransferase from yeast, *EMBO J 8*:2067 (1989).

26. N. Johnsson, G. Marriott, and K. Weber, p36, the major cytoplasmic substrate of src tyrosine protein kinase, binds to its p11 regulatory subunit via a short amino-terminal amphiphatic helix, *EMBO J. 7*:2435 (1988).

27. E. Herbert and M. Uhler, Biosynthesis of polyprotein precursors to regulatory peptides, *Cell 30*:1 (1982).

28. T. L. O'Donohue, G. E. Handelmann, R. L. Miller, and D. M. Jacobowitz, *N*-acetylation regulates the behavioral activity of α-melanotropin in a multineurotransmitter neuron, *Science 215*:1125 (1982).

29. D. G. Smyth, D. E. Massey, S. Zakarian, and M. D. A. Finnie, Endorphins are stored in biologically active and inactive forms: isolation of α-*N*-acetyl peptides, *Nature 279*:252.

30. C. C. Glembotski, Characterization of the peptide acetyltransferase activity in bovine and rat intermediate pituitaries responsible for the acetylation of β-endorphin and α-melanotropin, *J. Biol. Chem. 257*:10501 (1982).

31. A. W. Thorne, D. Kmiciek, K. Mitchelson, P. Sautiere, and C. Crane-Robinson, Patterns of histone acetylation. *Eur. J. Biochem. 193*:701 (1990).

32. P. Loidl, Histone acetylation: facts and questions, *Chromosoma 103*:441 (1994).

33. C. D. Allis, L. G. Chicoine, R. Richman, and I. G. Schulman, Deposition-related histone acetylation in micronuclei of conjugating *Tetrahymena, Proc. Natl. Acad. Sci. USA 82*:8048 (1985).

34. J. H. Waterborg and H. R. Matthews, Patterns of histone acetylation in *Physarum polycephalum, Eur. J. Biochem. 142*:329 (1984).

35. J. A. Ridsdale, M. J. Hendzel, G. P. Delcuve, and J. R. Davie, Histone acetylation alters the capacity of the H1 histones to condense transcriptionally active/competent chromatin, *J. Biol. Chem. 265*:5150 (1990).

36. W. A. Krajewski, V. M. Panin, and S. V. Razin, Acetylation of core histones causes the unfolding of 30 nm chromatin fiber: analysis by agarose gel electrophoresis, *Biochem. Biophys. Res. Commun. 196*:455 (1993).

37. V. G. Norton, K. W. Marvin, P. Yau, and E. M. Bradbury, Nucleosome linking number change controlled by acetylation of histones H3 and H4, *J. Biol. Chem. 265*:19848 (1990).

38. W. Li, S. Nagaraja, G. P. Delcuve, M. J. Hendzel, and J. R. Davie, Effects of histone acetylation, ubiquitination and variants on nucleosome stability, *Biochem. J. 296*:737 (1993).

39. T. R. Hebbes, A. W. Thorne, A. L. Clayton, and C. Crane-Robinson, Histone acetylation and globin gene switching, *Nucleic Acids Res. 20*:1017 (1992).

40. T. R. Hebbes, A. L. Clayton, A. W. Thorne, and C. Crane-Robinson, Core histone hyperacetylation co-maps with generalized DNase I sensitivity in the chicken β-globin chromosomal domain, *EMBO J. 13*:1823 (1994).

41. A. L. Clayton, T. R. Hebbes, A. W. Thorne, and C. Crane-Robinson, Histone acetylation and gene induction in human cells, *FEBS Lett. 336*:23 (1993).

42. D. Y. Lee, J. J. Hayes, D. Pruss, and A. P. Wolffe, A positive role for histone acetylation in transcription factor access to nucleosomal DNA, *Cell 72*:73 (1993).

43. L.-J. Juan, R. T. Utley, C. C. Adams, M. Vettese-Dadey, and J. L. Workman, Differential repression of transcription factor binding by histone H1 is regulated by the core histone amino termini, *EMBO J. 13*:6031 (1994).

44. B. M. Turner, A. J. Birley, and J. Lavender, Histone H4 isoforms acetylated at specific lysine residues define individual chromosomes and chromatin domains in *Drosophila* polytene nuclei, *Cell 69*:375 (1992).

45. P. Jeppesen and B. M. Turner, The inactive X chromosome in female mammals is distinguished by a lack of histone H4 acetylation, a cytogenetic marker for gene expression, *Cell 74*:281 (1993).

46. L. M. Johnson, G. Fisher-Adams, and M. Grunstein, Identification of a non-basic domain in the histone H4 *N*-terminus required for repression of the yeast silent mating loci, *EMBO J. 11*:2201 (1992).

47. L. K. Durrin, R. K. Mann, P. S. Kayne, and M. Grunstein, Yeast histone H4 *N*-terminal sequence is required for promoter activation in vivo, *Cell 65*:1023 (1991).

48. C. A. Perry, C. A. Dadd, C. D. Allis, and A. T. Annunziato, Analysis of nucleosome assembly and histone exchange using antibodies specific for acetylated H4, *Biochemistry 32*:13605 (1993).

49. M. J. Hendzel and J. R. Davie, Nucleosomal histones of transcriptionally active/competent chromatin preferentially exchange with newly synthesized histones in quiescent chicken erythrocytes, *Biochem. J. 271*:67 (1990).

50. C. A. Perry and A. T. Annunziato, Histone acetylation reduces H1-mediated nucleosome interactions during chromatin assembly, *Exp. Cell Res. 196*:337 (1991).

51. L. C. Wong, D. J. Sharpe, and S. S. Wong, High-mobility group and other nonhistone substrates for nuclear histone *N*-acetyltransferase, *Biochem. Genet. 29*:461 (1991).

52. S. I. Dimov, E. A. Alexandrova, and B. G. Beltchev, Differences between some properties of acetylated and nonacetylated forms of HMG1 protein, *Biochem. Biophys. Res. Commun. 166*:819 (1990).

53. B. Edde, J. Rossier, J. P. Le Caer, Y. Berwald-Netter, A. Koulakoff, F. Gros, and P. Denoulet, A combination of posttranslational modifications is responsible for the production of neuronal alpha-tubulin heterogeneity, *J. Cell Biochem. 46*:134 (1991).

54. H. Maruta, K. Greer, and J. L. Rosenbaum, The acetylation of alpha-tubulin and its relationship to the assembly and disassembly of microtubules, *J. Cell Biol. 103*:571 (1986).

55. G. Piperno, M. LeDizet, and X.-J. Chang, Microtubules containing acetylated α-tubulin in mammalian cells in culture, *J. Cell Biol. 104*:289 (1987).

56. S. S. Lim, P. J. Sammak, and G. G. Borisy, Progressive and spatially differentiated stability of microtubules in developing neuronal cells, *J. Cell Biol. 109*:253 (1989).

57. S. J. Robson and R. D. Burgoyne, Differential localisation of tyrosinated, detyrosinated, and acetylated alpha-tubulins in neurites and growth cones of dorsal root ganglion neurons, *Cell Motil. Cytoskeleton 12*:273 (1989).

58. B. J. Jasmin, J.-P. Changeux, and J. Cartaud, Compartmentalization of cold-stable and acetylated microtubules in the subsynaptic domain of chick skeletal muscle fibre, *Nature 344*:673 (1990).

59. A. Brown, Y. Li, T. Slaughter, and M. M. Black, Composite microtubules of the axon: quantitative analysis of tyrosinated and acetylated tubulin along individual axonal microtubules, *J. Cell Sci. 104*:339 (1993).

60. M. M. Falconer, U. Vielkind, and D. L. Brown, Establishment of a stable, acetylated microtubule bundle during neuronal commitment, *Cell Motil. Cytoskel. 12*:169 (1989).

61. G. Schatten, C. Simerly, D. J. Asai, E. Szoke, P. Cooke, and H. Schatten, Acetylated α-tubulin in microtubules during mouse fertilization and early development, *Dev. Biol. 130*:74 (1988).

62. E. Houliston and B. Maro, Posttranslational modification of distinct microtubule subpopulations during cell polarization and differentiation in the mouse preimplantation embryo, *J. Cell Biol. 108*:543 (1989).

63. D. A. Towler, S. P. Adams, S. R. Eubanks, D. S. Towery, E. Jackson-Machelski, L. Glaser, and J. I. Gordon, Purification and characterization of yeast myristoyl CoA: protein *N*-myristoyltransferase, *Proc. Natl. Acad. Sci. USA 84*:2708 (1987).

64. R. J. Duronio, D. A. Towler, R. O. Heuckeroth, and J. I. Gordon, Disruption of the yeast *N*-myristoyl transferase gene causes recessive lethality, *Science 243*:796 (1989).

65. J. F. Hancock, A. I. Magee, J. E. Childs, and C. J. Marshall, All ras proteins are polyisoprenylated but only some are palmitoylated, *Cell 57*:1167 (1989).

66. C. J. Marshall, Protein prenylation: a mediator of protein–protein interactions, *Science 259*:1865 (1993).

67. D. Stokoe, S. G. Macdonald, K. Cadwallader, M. Symons, and J. F. Hancock, Activation of Raf as a result of recruitment to the plasma membrane, *Science 264*:1463 (1994).

68. C. C. Farnsworth, M. C. Seabra, and L. H. Ericsson, Rab geranylgeranyl transferase catalyzes the geranylgeranylation of adjacent cysteines in the small GTPases Rab1A, Rab3A, and Rab5A, *Proc. Natl. Acad. Sci. USA 91*:11963 (1994).

69. S. Araki, K. Kaibuchi, T. Sasaki, Y. Hata, and Y. Takai, Role of the C-terminal region of *smg* p25A in its interaction with membranes and the GDP/GTP exchange protein, *Mol. Cell. Biol. 11*:1438 (1991).

70. P. B. Wedegaertner and H. R. Bourne, Activation and depalmitoylation of $G_{s\alpha}$, *Cell 77*:1063 (1994).

71. M. Y. Degtyarev, A. M. Spiegel, and T. L. Z. Jones, Palmitoylation of a G protein α_i subunit requires membrane localization not myristoylation, *J. Biol. Chem. 269*:30898 (1994).

72. S. M. Mumby, C. Kleuss, and A. G. Gilman, Receptor regulation of G-protein palmitoylation, *Proc. Natl. Acad. Sci. USA 91*:2800 (1994).

73. K. Kokame, Y. Fukada, T. Yoshizawa, T. Takao, and Y. Shimonishi, Lipid modification at the N terminus of photoreceptor G-protein α-subunit, *Nature 359*:749 (1992).

74. R. S. Haun, S.-C. Tsai, R. Adamik, J. Moss, and M. Vaughan, Effect of myristoylation on GTP-dependent binding of ADP-ribosylation factor to Golgi, *J. Biol. Chem. 268*:7064 (1993).

75. M. Franco, P. Chardin, M. Chabre, and S. Paris, Myristoylation of ADP-ribosylation factor 1 facilitates nucleotide exchange at physiological Mg^{2+} levels, *J. Biol. Chem. 270*:1337 (1995).

76. N. Pfanner, L. Orci, B. S. Glick, M. Amherdt, S. R. Arden, V. Malhotra, and J. E. Rothman, Fatty acyl-coenzyme A is required for budding of transport vesicles from Golgi cisternae, *Cell 59*:95 (1989).

77. M. E. Linder and J. G. Burr, Nonmyristoylated p60[v-src] fails to phosphorylate proteins of 115–120 kDa in chicken embryo fibroblasts, *Proc. Natl. Acad. Sci. USA 85*:2608 (1988).

78. R. O. Heuckeroth and J. I. Gordon, Altered membrane association of p60[v-src] and a murine 63-kDa *N*-myristoyl protein after incorporation of an oxygen-substituted analog of myristic acid, *Proc. Natl. Acad. Sci. USA 86*:5262 (1989).

79. S. Bagrodia, S. J. Taylor, and D. Shalloway, Myristylation is required for Tyr-527 dephosphorylation and activation of pp60[c-src] in mitosis, *Mol. Cell. Biol. 13*:1464 (1993).

80. M. D. Resh, Myristylation and palmitylation of Src family members: the fats of the matter, *Cell 76*:411 (1994).

81. A. M. Shenoy-Scaria, L. K. T. Gauen, J. Kwong, A. S. Shaw, and D. M. Lublin, Palmitylation of an amino-terminal cysteine motif of protein tyrosine kinases p56[lck] and p59[fyn] mediates interaction with glycosyl-phosphatidylinositol-anchored proteins, *Mol. Cell. Biol. 13*:6385 (1993).

82. W. Yonemoto, M. L. McGlone, and S. S. Taylor, *N*-Myristylation of the catalytic subunit of cAMP-dependent protein kinase conveys structural stability, *J. Biol. Chem. 268*:2348 (1993).

83. A. Aitken, P. Cohen, S. Santikarn, D. H. Williams, A. G. Calder, A. Smith, and C. B. Klee, Identification of the NH_2-terminal blocking group of calcineurin B as myristic acid, *FEBS Lett. 150*:314 (1982).

84. S. J. Moench, J. Moreland, D. H. Stewart, and T. G. Dewey, Fluorescence studies of the location and membrane accessibility of the palmitoylation sites of rhodopsin, *Biochemistry 33*:5791 (1994).

85. J. B. Ames, T. Tanaka, L. Stryer, and M. Ikura, Secondary structure of myristoylated recoverin determined by three-dimensional heteronuclear NMR: implications for the calcium-myristoyl switch, *Biochemistry 33*:10743 (1994).

86. G. Y. K. Ng, B. Mouillac, S. R. George, M. Carori, M. Dennis, M. Bouvier, and B. F. O'Dowd, Desensitization, phosphorylation and palmitoylation of the human dopamine D_1 receptor, *Eur. J. Pharmacol. 267*:7 (1994).

87. S. Moffet, B. Mouillac, H. Bonin, and M. Bouvier, Altered phosphorylation and desensitization patterns of a human β_2-adrenergic receptor lacking the palmitoylated Cys341, *EMBO J. 12*:349 (1993).

88. M. G. Eason, M. T. Jacinto, C. T. Theiss, and S. B. Liggett, The palmitoylated cysteine of

the cytoplasmic tail of α_{2A}-adrenergic receptors confers subtype-specific agonist-promoted downregulation, *Proc. Natl. Acad. Sci. USA 91*:11178 (1994).

89. N. Kawate and K. M. J. Menon, Palmitoylation of luteinizing hormone/human choriogonado-tropin receptors in transfected cells, *J. Biol. Chem. 269*:30651 (1994).

90. J. A. Hedo, E. Collier, and A. Watkinson, Myristyl and palmityl acylation of the insulin receptor, *J. Biol. Chem. 262*:954 (1987).

91. A. Jochen and J. Hays, Purification of the major substrate for palmitoylation in rat adipocytes: N-terminal homology with CD36 and evidence for cell surface acylation, *J. Lipid Res. 34*: 1783 (1993).

92. B. Crise and J. K. Rose, Identification of palmitoylation sites on CD4, the human immunode-ficiency virus receptor, *J. Biol. Chem. 267*:13593 (1992).

93. L. Y. W. Bourguignon, E. L. Kalomiris, and V. B. Lokeshwar, Acylation of the lymphoma transmembrane glycoprotein, GP85, may be required for GP85-ankyrin interaction, *J. Biol. Chem. 266*:11761 (1991).

94. T. Fujimoto, E. Stroud, R. E. Whatley, S. M. Prescott, L. Muszbek, M. Laposata, and R. P. McEver, P-selectin is acylated with palmitic acid and stearic acid at cysteine 766 through a thioester linkage, *J. Biol. Chem. 268*:11394 (1993).

95. H. Taniguchi and S. Manenti, Interaction of myristoylated alanine-rich protein kinase C sub-strate (MARCKS) with membrane phospholipids, *J. Biol. Chem. 268*:9960 (1993).

96. D. Kang, D. Karbach, and H. Passow, Anion transport function of mouse erythroid band 3 protein (AE1) does not require acylation of cysteine residue 861, *Biochim. Biophys. Acta 1194*:341 (1994).

97. M. A. Risinger, E. M. Dotimas, and C. M. Cohen, Human erythrocyte protein 4.2, a high copy number membrane protein, is *N*-myristylated, *J. Biol. Chem. 267*:5680 (1992).

98. M. Mariani, D. Maretzki, and H. U. Lutz, A tightly membrane-associated subpopulation of spectrin is ^{3}H-palmitoylated, *J. Biol. Chem. 268*:12996 (1993).

99. S. Manenti, I. Dunia, and E. L. Benedetti, Fatty acid acylation of lens fiber plasma membrane proteins. MP26 and α-crystallin are palmitoylated. *FEBS Lett. 262*:356 (1990).

100. A. Mastrogiacomo, S. M. Parsons, G. A. Zampighi, D. J. Jenden, J. A. Umbach, and C. B. Gundersen, Cysteine string proteins: a potential link between synaptic vesicles and presynap-tic Ca^{2+} channels, *Science 263*:981 (1994).

101. W. R. Randall, Cellular expression of a cloned, hydrophilic, murine acetylcholinesterase: evidence of palmitoylated membrane-bound forms, *J. Biol. Chem. 269*:12367 (1994).

102. S. Christgau, H.-J. Aanstoot, H. Schierbeck, K. Begley, S. Tullin, K. Hejnaes, and S. Baek-keskov, Membrane anchoring of the autoantigen GAD_{65} to microvesicles in pancreatic β-cells by palmitoylation in the NH_2-terminal domain, *J. Cell Biol. 118*:309 (1992).

103. D. T. Hess, S. I. Patterson, D. S. Smith, and J. H. P. Skene, Neuronal growth cone collapse and inhibition of protein fatty acylation by nitric oxide, *Nature 366*:562 (1993).

104. Y. Sudo, D. Valenzuela, A. G. Beck-Sickinger, M. C. Fishman, and S. M. Strittmatter, Palmi-toylation alters protein activity: blockade of G_o stimulation by GAP-43, *EMBO J. 11*:2095 (1992).

105. L. J. Robinson, L. Busconi, and T. Michel, Agonist-modulated palmitoylation of endothelial nitric oxide synthase, *J. Biol. Chem. 270*:995 (1995).

106. S. Pillai and D. Baltimore, Myristoylation and the posttranslational acquisition of hydropho-bicity by the membrane immunoglobulin heavy-chain polypeptide in B lymphocytes, *Proc. Natl. Acad. Sci. USA 84*:7654 (1987).

107. F. T. Stevenson, S. L. Bursten, C. Fanton, R. M. Locksley, and D. H. Lovett, The 31-kDa precursor of interleukin 1α is myristoylated on specific lysines within the 16-kDa N-terminal propiece *Proc. Natl. Acad. Sci. USA 90*:7245 (1993).

108. F. T. Stevenson, S. L. Bursten, R. M. Locksley, and D. H. Lovett, Myristyl acylation of the tumor necrosis factor α precursor on specific lysine residues, *J. Exp. Med. 176*:1053 (1992).

109. N. Borgese and R. Longhi, Both the outer mitochondrial membrane and the microsomal

forms of cytochrome b_5 reductase contain covalently bound myristic acid, *Biochem. J. 266*: 341 (1990).

110. J. E. Walker, J. M. Arizmendi, A. Dupuis, I. M. Fearnley, S. M. Finel, S. M. Medd, S. J. Pilkington, M. J. Runswick, and J. M. Skehel, Sequences of 20 subunits of NADH:ubiquinone oxidoreductase from bovine heart mitochondria, *J. Mol. Biol. 226*:1051 (1992).

111. U. Sackmann, R. Lensen, D. Rohlen, U. Jahnke, and H. Weiss, The acyl carrier protein in *Neurospora crassa* mitochondria is a subunit of NADH: ubiquinone reductase (complex I), *Eur. J. Biochem. 200*:463 (1991).

112. A. O. Vassilev, N. Plesofsky-Vig, and R. Brambl, Cytochrome *c* oxidase in *Neurospora crassa* contains myristic acid covalently linked to subunit 1, *Proc. Natl. Acad. Sci. USA 92*: 8680–8684 (1995).

113. L. Berthiaume, I. Deichaite, S. Peseckis, and M. D. Resh, Regulation of enzymatic activity by active site fatty acylation, *J. Biol. Chem. 269*:6498 (1994).

114. M. Otsuka, T. Akiba, Y. Okita, K. Tomita, N. Yoshiyama, T. Sasaoka, M. Kanayama, and F. Marumo, Lactic acidosis with hypoglycemia and hyperammonemia observed in two uremic patients during calcium hopantenate treatment, *Jpn. J. Med. 29*:324 (1990).

115. S. Noda, J. Haratake, A. Sasaki, N. Ishii, H. Umezaki, and A. Horie, Acute encephalopathy with hepatic steatosis induced by pantothenic acid antagonist, calcium hopantenate, in dogs, *Liver 11*:134 (1991).

116. P. C. Even, A. Decrouy, and A. Chinet, Defective regulation of energy metabolism in mdx-mouse skeletal muscles, *Biochem. J. 304*:649 (1994).

117. I. Nagiel-Ostaszewski and C. A. Lau-Cam, Protection by pantethine, pantothenic acid and cystamine against carbon tetrachloride–induced hepatotoxicity in the rat, *Res. Commun. Chem. Pathol. Pharmacol. 67*:289 (1990).

118. A. Kapp and G. Zeck-Kapp, Effect of Ca-pantothenate on human granulocyte oxidative metabolism, *Allerg. Immunol. 37*:145 (1991).

119. R. E. Hodges, M. A. Ohlson, and W. B. Bean, Pantothenic acid deficiency in man, *J. Clin. Invest. 37*:1642 (1958).

120. M. Glusman, The syndrome of "burning feet" (nutritional melalgia) as manifestation of nutritional deficiency, *Am. J. Med. 3*:211 (1947).

10

Vitamin B_6

JAMES E. LEKLEM

Oregon State University, Corvallis, Oregon

I. INTRODUCTION AND HISTORY

Vitamin B_6 is unique among the water-soluble vitamins with respect to the numerous functions it serves and its metabolism and chemistry. Within the past few years the attention this vitamin has received has increased dramatically (1–8). Lay publications (9) attest to the interest in vitamin B_6.

This chapter will provide an overview of vitamin B_6 as it relates to human nutrition. Both qualitative and quantitative information will be provided in an attempt to indicate the importance of this vitamin within the context of health and disease in humans. As a nutritionist, my perspective no doubt is biased by these nutritional elements of this vitamin. The exhaustive literature on the intriguing chemistry of the vitamin will not be dealt with in any detail, except as related to the function of vitamin B_6 as a coenzyme. To the extent that literature is available, reference will be made to research in humans, with animal or other experimental work included as necessary.

As we leave the twentieth century behind, there may be a tendency to lose the sense of excitement of discovery that Gyorgy and colleagues experienced when they began to unravel the mystery of vitamin B complex. Some of the major highlights of the early years of vitamin B_6 research are presented in Table 1. Paul Gyorgy was first to use the term vitamin B_6 (10). The term was used to distinguish this factor from other hypothetical growth factors B_3, B_4, B_5 (and Y). Some 4 years later (1938), in what is a fine example of cooperation and friendship, Gyorgy (11) and Lepkovsky (12) reported the isolation of pure crystalline vitamin B_6. Three other groups also reported the isolation of vitamin B_6 that same year (13–15). Shortly after this, Harris and Folkers (16) as well as Kuhn et al. (17) determined that vitamin B_6 was a pyridine derivative and structurally identified it as 3-hydroxy-4,5-hydroxymethyl-2-methylpyridine. The term pyridoxine was first intro-

Table 1 Historical Highlights of Vitamin B_6 Research

1932	A compound with the formula of $C_3H_{11}O_3N$ was isolated from rice polishings.
1934	Gyorgy shows there was a difference between the rat pellagra preventive factor and vitamin B_2. He called this vitamin B_6.
1938	Lepkovsy reports isolation of pure crystalline vitamin B_6. Keresztesky and Stevens, Gyorgy, Kuhn and Wendt, and Ichibad and Michi also report isolation of vitamin B_6.
1939	Chemical structure determined and vitamin B_6 synthesized by Kuhn and associates and by Harris and Folkers.
1942	Snell and co-workers recognize existence of other forms of pyridoxine.
1953	Snyderman and associates observe convulsions in an infant and anemia in an older child fed a vitamin B_6 deficient diet.

duced by Gyorgy in 1939 (18). An important aspect of this early research was the use of animal models in identification of vitamin B_6 (as pyridoxine in various extracts from rice bran and yeast). This early research into vitamin B_6 then provided the ground work for research into the requirement for vitamin B_6 for humans and the functions of this vitamin.

Identification of the other major forms of the vitamin B_6 group, pyridoxamine and pyridoxal, occurred primarily through the use of microorganisms (19,20). In the process of developing an assay for pyridoxine, Snell and co-workers observed that natural materials were more active in supporting the growth of certain microorganisms than predicted by their pyridoxine content as assayed with yeast (20). Subsequently, this group observed enhanced growth-promoting activity in the urine of vitamin B_6 deficient animals fed pyridoxine (20). Treatment of pyridoxine with ammonia also produced a substance with growth activity (21). These findings subsequently led to the synthesis of pyridoxal and pyridoxamine (22,23). The availability of these three forms of vitamin B_6 reduced further research into this intriguing vitamin possible.

II. CHEMISTRY

Since Gyorgy first coined the term vitamin B_6 (10), there has been confusion in the terminology of the multiple forms of the vitamin. "Vitamin B" is the recommended term for the generic descriptor for all 3-hydroxy-2-methylpyridine derivatives (24). Figure 1 depicts the various forms of vitamin B_6, including the phosphorylated forms. Pyridoxine (once referred to as pyridoxal) is the alcohol form and should not be used as a generic

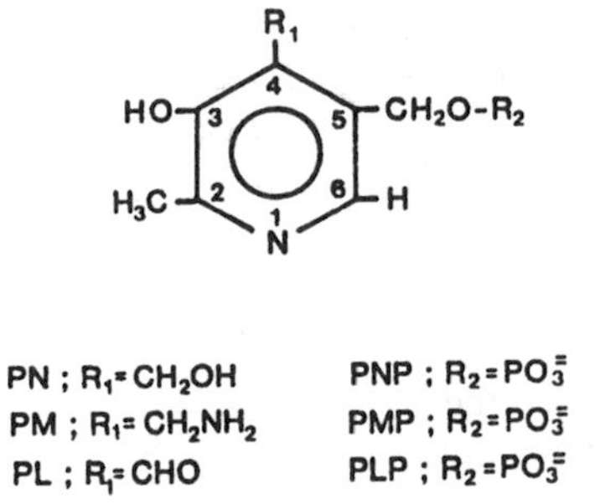

PN ; R_1= CH_2OH PNP ; R_2=$PO_3^=$
PM ; R_1= CH_2NH_2 PMP ; R_2=$PO_3^=$
PL ; R_1= CHO   PLP ; R_2=$PO_3^=$

Fig. 1 Structure of B_6 vitamers.

Table 2 Physical Properties of B$_6$ Vitamers

	Molecular weight	Stability to white light[a] pH 4.5		pk		Fluorescence Maxima[b]		Ultraviolet absorption spectra[e]			
								0.1N HCl		pH 7.0	
		8 hr	15 hr	pk1	pk2	Activation (λ)	Emission (λ)	λ$_{max}$	ε$_{max}$	λ$_{max}$	ε$_{max}$
Pyridoxine	169.1	97%	90%	5.0	8.9	325	400	291	8900	254 324	3760 7100
Pyridoxamine	168.1	81%	57%	3.4	8.1	325	405	293	8500	253 325	4600 7700
Pyridoxal	167.2	97%	68%	4.2	8.7	320	385	288	9100	317	8800
Pyridoxine 5'-phosphate	249.2	—	—	—	—	322	394	290	8700	253 325	3700 7400
Pyridoxal 5'-phosphate	248.2	—	—	2.5	3.5	330	400	293	900	253 325	4700 8300
Pyridoxal 5'-phosphate	247.2	—	—	2.5	4.1	330	375	293 334	7200 1300	388	5500
4-Pyridoxic acid	183.2	—	—	—	—	325[c] 355[d]	425[c] 445[d]	—	—	—	—

[a] Percent stability compared to solution in dark (24). 8 h, 15 h = length of time exposed to light.
[b] From Storvick et al. (25); pH 7.0.
[c] pH 3.4, 0.01 N acetic acid.
[d] pH 10.5, 0.1 N NH$_4$OH, lactone of 4-PA.
[e] Data are for PN-HCL, PL-HCL, PM-2HCl, PLP monohydrate, PMP dihydrate (26).

name for vitamin B_6. The trivial names and abbreviations commonly used for the three principal forms of vitamin B_6, their phosphoric esters, and analogs are as follows: pyridoxine, PN; pyridoxal, PL; pyridoxamine, PM; pyridoxine-5′-phosphate, PNP; pyridoxal-5′-phosphate, PLP; pyridoxamine-5′-phosphate, PMP; 4-pyridoxic acid, 4-PA. As will be discussed later, other forms of vitamin B_6 exist, particularly bound forms.

The various physical and chemical properties of the phosphorylated and nonphosphorylated forms of vitamin B_6 are given in Table 2. Detailed data on fluorescence (28) and ultraviolet (27) absorption characteristics of B_6 vitamers are available. Of importance to researchers as well as to food producers and consumers is the relative stability of the forms of vitamin B_6. Generally, as a group B_6 vitamers are labile, but the degree to which each is degraded varies. In solution the forms are light-sensitive (25,29), but this sensitivity is influenced by pH. Pyridoxine, pyridoxal, and pyridoxamine are relatively heat-stable in an acid medium, but they are heat-labile in an alkaline medium. The hydrochloride and base forms are readily soluble in water, but they are minimally soluble in organic solvents.

The coenzyme form of vitamin B_6, PLP, is found covalently bound to enzymes via a Schiff base with an ε-amino group of lysine in the enzyme. While nonenzymatic reactions with PLP or PL and metal ions can occur (30), in enzymatic reactions the amino group of the substrate for the given enzyme forms a Schiff base via a transimination reaction. Figure 2 depicts the formation of a Schiff base with PLP and an amino acid. Because of the strong electron-attracting character of the pyridine ring, electrons are withdrawn from one of the three substituents (R group, hydrogen, or carboxyl group) attached to the α carbon of the substrate attached to PLP. This results in the formation of a quinonoid structure. There are several structural features of PLP that make it well suited to form a Schiff base and thus act as a catalyst in a variety of enzyme reactions. These features have been detailed by Leussing (31) and include the 2-methyl group, which brings the pK_a of the proton of the ring pyridine closer to the biological range; the phenoxide oxygen (position 3), which aids in expulsion of a nucleophile at the 4-position; the 5-phosphate group, which functions as an anchor for the coenzyme and prevents hemiacetal formation and the drain of electrons from the ring; and the protonated pyridine nitrogen that is para to the aldehyde group aids in delocalizing the negative charge and helps regulate the pK_a of the 3-hydroxyl group. A recent publication has extensively reviewed the chemistry of pyridoxal-5′ phosphate (4).

PLP has been reported to be a coenzyme for over 100 enzymatic reactions (32). Of these, nearly half involve transamination-type reactions. Transamination reactions are but

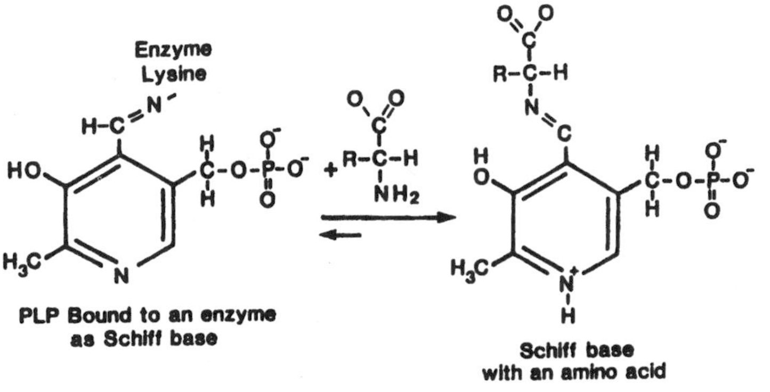

Fig. 2   Schiff base formation between pyridoxal-5′-phosphate and an amino acid.

Table 3 Enzyme Reactions Catalyzed by Pyridoxal-5′-Phosphate

Type of reaction	Typical reaction or enzyme
Reactions involving α carbon	
Transamination	Alamine → pyruvate + PMP
Racemization	D-Amino acid ↔ L-amino acid
Decarboxylation	5-OH tryptophan → T-OH tryptamine + CO_2
Oxidative deamination	Histamine → imidazole-4-acetaldehyde + NH^+
Loss of the side chain	THF + serine → glycine + N5,10-methylene THF
Reactions involving β carbon	
Replacement (exchange)	Cystein synthetase
Elimination	Serine and threonine dehydratase
Reaction involving γ carbon	
Replacement (exchange)	Cystathionine → cysteine + homoserine
Elimination	Homocysteine desulfhydrase
Cleavage	Kynurenine → anthranilic acid

one type of reaction that occur as a result of Schiff base formation. The three types of enzyme reactions catalyzed by PLP are listed in Table 3 and are classified according to reactions occurring at the α, β, or γ carbon.

III. METHODS

The measurement of B$_6$ vitamers and metabolites is important in evaluating vitamin B$_6$ metabolism and status. Methods used in measuring B$_6$ vitamers in foods are complicated not only by the numerous forms but by the various matrices. Reviews of the methods currently used are available (33–35). HPLC techniques are more common today than other methods, such as microbiological (26,36) and enzymatic techniques. B$_6$ vitamers in biological fluids can be determined by a variety of HPLC techniques (35,37). These methods involve nonexchange or paired-ion reversed-phase procedures. Determination of the active coenzyme form, PLP, in plasma and tissue extracts is conveniently done by a radioenzymatic technique (38). The advantage of this type of procedure is that it allows for analyses of a large number of samples in one assay.

Determination of vitamin B$_6$ in foods and biological samples can be done microbiologically (36). Yeast growth assays using *Saccharomyces uvarum* (ATCC 9080) are most commonly used. While it has been reported that the three forms respond differently to yeast (35), in my laboratory we do not observe this if the yeast grows rapidly. In all of the methods mentioned above, adequate extraction of the forms of vitamin B$_6$ is critical. TCA and perchloric acid are effective extractants.

Methods for the determination of the glycosylated form of vitamin B$_6$ PN-glucoside (PNG), in foods are available (35,39). Both microbiological-based (40) and HPLC (35) procedures have been utilized. All procedures for B$_6$ vitamers should be conducted under yellow lights to minimize photodegradation.

IV. OCCURRENCE IN FOODS

To appreciate the role of vitamin B$_6$ in human nutrition, one must first have knowledge of the various forms and quantities found in foods. A microbiological method for determining the vitamin B$_6$ content of foodstuffs was developed by Atkin in 1943 (32). While this

Table 4 Vitamin B_6 Content of Selected Foods and Percentages of the Three Forms

Food	Vitamin B_6[a] (mg/100 g)	Pyridoxine[b] (%)	Pyridoxal[b] (%)	Pyridoxamine[b] (%)
Vegetables				
Beans lima, frozen	0.150	45	30	25
Cabbage, raw	0.160	61	31	8
Carrots, raw	0.150	75	19	6
Peas, green, raw	0.160	47	47	6
Potatoes, raw	0.250	68	18	14
Tomatoes, raw	0.100	38	29	33
Spinach, raw	0.280	36	49	15
Broccoli, raw	0.195	29	65	6
Cauliflower, raw	0.210	16	79	5
Corn, sweet	0.161	6	68	26
Fruits				
Apples, Red Delicious	0.030	61	31	8
Apricots, raw	0.070	58	20	22
Apricots, dried	0.169	81	11	8
Avocados, raw	0.420	56	29	15
Bananas, raw	0.510	61	10	29
Oranges, raw	0.060	59	26	15
Peaches, canned	0.019	61	30	9
Raisins, seedless	0.240	83	11	6
Grapefruit, raw	0.034	—	—	—
Legumes				
Beans, white, raw	0.560	62	20	18
Beans, lima, canned	0.090	75	15	10
Lentils	0.600	69	13	18
Peanut butter	0.330	74	9	17
Peas, green, raw	0.160	69	17	14
Soybeans, dry, raw	0.810	44	44	12
Nuts				
Almonds, without skins, shelled	0.100	52	28	20
Pecans	0.183	71	12	17
Filberts	0.545	29	68	3
Walnuts	0.730	31	65	4
Cereals/grains				
Barley, pearled	0.224	52	42	6
Rice, brown	0.550	78	12	10
Rice, white, regular	0.170	64	19	17
Rye flour, light	0.090	64		14
Wheat, cereal, flakes	0.292	79	11	10
Wheat flour, whole	0.340	71	16	13
Wheat flour, all-purpose white	0.060	55	24	21
Oatmeal, dry	0.140	12	49	39
Cornmeal, white and yellow	0.250	11	51	38
Bread, white	0.040	—	—	—
Bread, whole wheat	0.180	—	—	—

Table 4   Continued

Food	Vitamin B$_6$ (mg/100 g)	Pyridoxine (%)	Pyridoxal (%)	Pyridoxamine (%)
Meat/poultry/fish				
Beef, raw	0.330	16	53	31
Chicken breast	0.683	7	74	19
Pork, ham, canned	0.320	8	8	84
Flounder fillet	0.170	7	71	22
Salmon, canned	0.300	2	9	89
Sardine, Pacific canned, oil	0.280	13	58	29
Tuna, canned	0.425	19	69	12
Halibut	0.430	—	—	—
Milk/eggs/cheese				
Milk, cow, homogenized	0.040	3	76	21
Milk, human	0.010	0	50	50
Cheddar	0.080	4	8	88
Egg, whole	0.110	0	85	15

[a]Values from Ref. 43, Table 1.
[b]Values from Ref. 43, Table 2.

method has been refined (26,42,35), it still stands as the primary method for determining the total vitamin B$_6$ content of foods and has been the basis for most of the data available on the vitamin B$_6$ content of foods. There are various forms of vitamin B$_6$ in foods. In general, these forms are a derivative of the three forms: pyridoxal, pyridoxine, and pyridoxamine. Pyridoxine and pyridoxamine (or their respective phosphorylated forms) are the predominant forms in plant foods. Although there are exceptions, pyridoxal, as the phosphorylated form, is the predominant form in foods. Table 4 contains data for the vitamin B$_6$ content of a representative sample of food commonly consumed in the United States. Data on the amount of each of the three forms are also listed (43). While the phosphorylated forms are usually the predominant forms in most foods, the microbiological methods used to determine the level of each form measure the sum of the phosphorylated and free (nonconjugated) forms.

In addition to the phosphorylated forms, other conjugated forms have been detected in certain foods. A glycosylated form of pyridoxine has been identified in rice bran (44) and subsequently quantitated in several foods (40). The glycosylated form isolated from rice bran has been identified as 5'-O-β-D-pyridoxine (44) (Fig. 3). Suzuki et al. have shown that the 5'-glucoside can be formed in germinating seeds of wheat, barley, and rice cultured on a pyridoxine solution (45). In addition, a small amount of 4'-glucoside was also detected

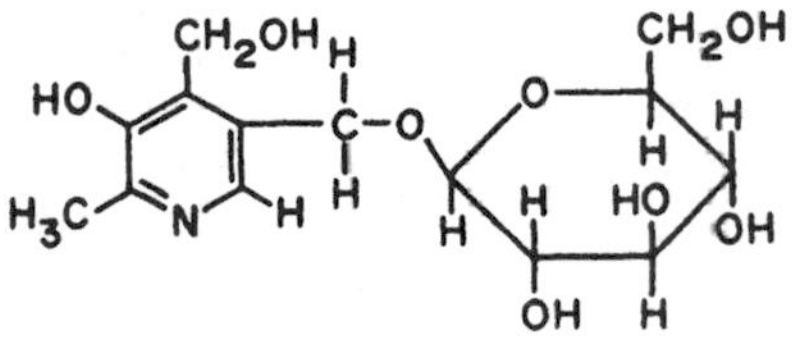

Fig. 3   Structure of 5'-O-(β-D-glucopyranosyl)pyridoxine.

Table 5 Vitamin B$_6$ and Glycosylated Vitamin B$_6$ Content of
Selected Foods

Food	Vitamin B$_6$ (mg/100 g)	Glycosylated vitamin B$_6$ (mg/100g)
Vegetables		
Carrots, canned	0.064	0.055
Carrots, raw	0.170	0.087
Cauliflower, frozen	0.084	0.069
Broccoli, frozen	0.119	0.078
Spinach, frozen	0.208	0.104
Cabbage, raw	0.140	0.065
Sprouts, alfalfa	0.250	0.105
Potatoes, cooked	0.394	0.165
Potatoes, dried	0.884	0.286
Beets, canned	0.018	0.005
Yams, canned	0.067	0.007
Beans/legumes		
Soybeans, cooked	0.627	0.357
Beans, navy, cooked	0.381	0.159
Beans, lima, frozen	0.106	0.039
Peas, frozen	0.122	0.018
Peanut butter	0.302	0.054
Beans, garbanzo	0.653	0.111
Lentils	0.289	0.134
Animal products		
Beef, ground, cooked	0.263	n.d.
Tuna, canned	0.316	n.d.
Chicken breast, raw	0.700	n.d.
Milk skim	0.005	n.d.
Nuts/seeds		
Walnuts	0.535	0.038
Filberts	0.587	0.026
Cashews, raw	0.351	0.046
Sunflower seeds	0.997	0.355
Almonds	0.086	-0-
Fruits		
Orange juice, frozen concentrate	0.165	0.078
Orange juice, fresh	0.043	0.016
Tomato juice, canned	0.097	0.045
Blueberries, frozen	0.046	0.019
Banana	0.313	0.010
Banana, dried chips	0.271	0.024
Pineapple, canned	0.079	0.017
Peaches, canned	0.009	0.002
Apricots, dried	0.206	0.036
Avocado	0.443	0.015
Raisins, seedless	0.230	0.154

Table 5   Continued

Food	Vitamin B$_6$ (mg/100 g)	Glycosylated vitamin B$_6$ (mg/100g)
Cereals/grains		
Wheat bran	0.903	0.326
Shredded wheat cereal	0.313	0.087
Rice, brown	0.237	0.055
Rice, bran	3.515	0.153
Rice, white	0.076	0.015
Rice cereal, puffed	0.098	0.007
Rice cereal, fortified	3.635	0.382

n.d., none detected.
Sources: Data taken from Ref. 40 and Leklem and Hardin, unpublished.

in wheat and rice germinated seeds, but not in soybean seeds. Also of interest is an as-yet-unidentified conjugate of vitamin B$_6$ reported by Tadera and co-workers (46). This conjugate released free vitamin B$_6$ (measured as pyridoxine) only when the food was treated with alkali and then β-glucosidase. Tadera et al. have also identified another derivative of the 5'-glucoside of pyridoxine in seedlings of podded peas (47). This derivative was identified as 5'-*O*(6-*O*-malonyl-β-D-glucopyranosyl)pyridoxine. The role of these conjugates in plants is unknown. Table 5 lists the total vitamin B$_6$ content of pyridoxine 5'-glucoside content of various foods. There is no generalization that can be made at this time as to a given class of foods having high or low amounts of pyridoxine-5'-glucoside. The effect of the 5'-glucoside of vitamin B$_6$ nutrition will be addressed in the section on bioavailability and absorption.

Food processing and storage may influence the vitamin B$_6$ content of food (48–57) and result in production of compounds normally not present. Losses of 10–50% have been reported for a wide variety of foods. Heat sterilization of commercial milk was found to result in conversion of pyridoxal to pyridoxamine (49). Storage of heat-treated milk decreases the vitamin B$_6$ content presumably due to formation of bis-4-pyridoxyldisulfide. The effect of various processes on the vitamin B$_6$ content of milk and milk products has been reviewed (57). Losses range from 0 to 70%. Vanderslice et al. have reported an HPLC method for assessing the various forms of vitamin B$_6$ in milk (58), which aids in understanding the effects of processing on the vitamin B$_6$ content of milk and milk products. DeRitter (59) has reviewed the stability of several vitamins in processed foods, including vitamin B$_6$, and found that the vitamin B$_6$ added to flour and baked into bread is stable. This has been confirmed by Perera et al. (60).

Gregory and Kirk have found that during thermal processing (61) and low-moisture conditions of food storage (54), there is reductive binding of pyridoxal and pyridoxal 5'-phosphate to the ε-amino groups of protein or peptide lysyl residues. These compounds are resistant to hydrolysis and also possess low vitamin B$_6$ activity. Interestingly, Gregory (62) has shown that ε-pyridoxyllysine bound to dietary protein has anti–vitamin B$_6$ activity (50% molar vitamin B$_6$ activity for rats).

V. ABSORPTION AND BIOAVAILABILITY

The questions of how much vitamin B_6 is biologically available (i.e., absorbed and utilizable) and what factors influence this are important in terms of estimating a dietary requirement. Before considering the factors that influence bioavailability, a brief description of absorption of the forms of vitamin B is appropriate.

Absorption of the various forms of vitamin B_6 has been studied most extensively in animals, particularly rats. However, gastrointestinal absorption of pyridoxine has been examined with guinea pig jejunum preparations (63), intestine, cecum, and crop of the chicken (64), and intestine of the hamster (65).

In the rat, Middleton (65–67) and Henderson and co-workers (68) have conducted extensive research on intestinal absorption of B_6 vitamers. The evidence to date indicates that pyridoxine and the other two major forms of vitamin B_6 are absorbed by a nonsaturable, passive process (68). Absorption of the phosphorylated forms can occur (69,70), but to a very limited extent. The phosphorylated forms disappear from the intestine via hydrolysis by alkaline phosphatase (67,70), and a significant part of this takes place intraluminally. Prior intake of vitamin B_6 in rats over a wide range (0.75–100 mg PN-HCl per kg diet) was found to have no affect on in vitro absorption of varying levels of PN-HCl (71). This study provides further support for passive absorption of B_6 vitamers. However, Middelton has questioned the concept of a nonsaturable process (72). Using an in vivo perfused intestinal segment model, he found there was a gradient of decreasing rates of uptake from the proximal to the distal end of the intestine and that there was a saturable component of uptake, especially in the duodenum.

The various forms of vitamin B_6 that are absorbed into the rat intestinal cell (intracellular) can be converted to other forms (i.e., PL to PLP, PN to PLP, and PM to PLP), but that which is ultimately transported to other organs via the circulation system primarily reflects the nonphosphorylated form originally absorbed (69,70). A similar pattern of uptake and metabolism has been observed in mice (73); however, in mice given PN, pyridoxal was the major form detected in the circulation. Portal blood was not examined. The liver was likely the primary organ that further metabolized the PN absorbed and released PL to the circulation.

Bioavailability of a nutrient from a given food is important to an organism in that it is the amount of a nutrient that is both absorbed and available to cells. The word *available* is key here in that the vitamin may not be needed by the cell and simply excreted or metabolized to a nonutilizable form, such as 4-pyridoxic acid in the case of vitamin B_6.

Methods used to evaluate the bioavailability of nutrients such as vitamin B_6 include balance studies in which input and output are determined. Included in these studies is the use of stable isotopic techniques (74). A second approach is to measure an in vivo response, such as growth, after a state of deficiency has been created. The third type of study is the examination of blood levels of the nutrient or a metabolite of the nutrient over a specified period of time after a food is fed. The concentration of a metabolite, such as PLP, is then compared with concentrations after ingestion of graded amounts of the crystalline form of vitamin B_6. Gregory and Ink (74) and Leklem (75) have reviewed vitamin B_6 bioavailability.

One of the early studies that suggested a reduced availability of vitamin B_6 involved feeding canned combat rations that had been stored at elevated temperatures (75). Feeding diets containing 1.9 mg of total vitamin B_6 resulted in a marginal deficiency based on urinary excretion of tryptophan metabolites. Some 18 years later, Nelson et al. observed that the vitamin B_6 in orange juice was incompletely absorbed by humans (77). These authors suggested that a low molecular weight form of vitamin B_6 was present in orange juice and responsible for the reduced availability. Kabir and co-workers (40) subsequently found that approximately 50% of the vitamin B_6 present in orange juice is the pyridoxine-5'-O-glucoside.

Leklem et al. conducted one of the first human studies that directly determined bioavailability of vitamin B_6 (78). In their study, nine men were fed either whole wheat bread, white bread enriched with pyridoxine (0.8 mg), or white bread plus a solution containing 0.8 mg of pyridoxine. After feeding each bread for a week, urinary vitamin B_6 and 4-pyridoxic and fecal vitamin B_6 excretion were measured to assess vitamin B_6 bioavailability. Urinary 4-pyridoxic acid excretion was reduced when whole wheat bread was fed compared to the other two test situations, and the vitamin B_6 from this bread was estimated to be 5–10% less available than the vitamin B_6 from the other two breads. While this relatively small difference in bioavailability may not be nutritionally significant by itself, in combination with other foods of low vitamin B_6 bioavailability, vitamin B_6 status may be compromised.

In other studies in humans, feeding 15 g of cooked wheat bran slightly reduced vitamin B_6 bioavailability (79). Using urinary vitamin B_6 as the sole criterion, Kies and co-workers estimated that 20 g of wheat, rice, or corn bran reduced vitamin B_6 availability 35–40% (80). Since various brans are good sources of vitamin B_6, it is not possible to determine if the vitamin B_6 in the bran itself was unavailable or if the bran may have been binding vitamin B_6 present in the remainder of the diet.

The bioavailability of vitamin B_6 from specific foods or groups of foods has been examined utilizing balance and blood levels (dose response) studies. Tarr et al. estimated a 71–79% bioavailability of vitamin B_6 from foods representing the "average" American diet (81). Using a triple-lumen tube perfusion technique, Nelson et al. found that the vitamin B_6 from orange juice was only 50% as well absorbed as crystalline pyridoxine (82). In our laboratory, Kabir et al. compared the vitamin B_6 bioavailability from tuna, whole wheat bread, and peanut butter (83). Compared to the vitamin B_6 in tuna, the vitamin B_6 in whole wheat bread and peanut butter was 75% and 63% as available, respectively. The level of glycosylated vitamin B_6 in these foods was inversely correlated with vitamin B, bioavailability as based on urinary vitamin B, and 4-pyridoxic acid(84). We have observed an inverse relationship between vitamin B_6 bioavailability as based on urinary 4-pyridoxic acid excretion and the glycosylated vitamin B_6 content of six foods (85). These foods and their respective availabilities were as follows: walnuts (78%), bananas (79%), tomato juice (25%), spinach (22%), orange juice (9%), and carrots (0%). While the glycosylated vitamin B_6 content of foods appears to be a significant contributor to bioavailability, the presence of other forms of vitamin B_6 and/or binding of specific forms of vitamin B_6 to other components in a food may also contribute to availability. The question of the extent to which vitamin B_6 bioavailability affects vitamin B_6 status (and thus requirement) has been studied in women (86). When diets containing 9% of the vitamin B_6 as PNG were compared with diets containing 27% PNG it was observed that vitamin B_6 status was decreased. The decreased bioavailability was consistent with that observed in humans by Gregory et al. (87) who estimated that the bioavailability of PNG may be as low as 58% of the bioavailability of free pyridoxine.

VI. INTERORGAN METABOLISM

Extensive work by Lumeng and Li and co-workers in rats (88) and dogs (89) has shown that the liver is the primary organ responsible for metabolism of vitamin B_6 and supplies the active form of vitamin B_6, PLP, to the circulation and other tissues. The primary interconversion of the B_6 vitamers is depicted in Fig. 4. The three nonphosphorylated forms are converted to their respective phosphorylated forms by a kinase enzyme (pyridoxine kinase EC 2.7.1.35). Both ATP and zinc are involved in this conversion, with ATP serving as a source of the phosphate group. The two phosphorylated forms, pyridoxamine-5'-phosphate and pyridoxine-5'-phosphate, are converted to PLP via a flavin mononucleoticle (FMN)–requiring oxidase (90). A review of the interrelation between riboflavin and vitamin B_6 is available (91).

Dephosphorylation of the 5'-phosphate compounds occurs by action of a phosphatase. This phosphatase is considered to be alkaline phosphatase (92) and is thought to be enzyme-bound in the liver (93). PL arising from dephosphorylation or that taken up from the circulation can be converted to 4-pyridoxic acid by either an NAD-dependent dehydrogenase or an FAD-dependent aldehyde oxidase. As discussed below, in humans only aldehyde oxidase (pyridoxal oxidase) activity has been detected in the liver (94). The conversion of pyridoxal to 4-pyridoxic acid is an irreversible reaction. Thus, 4-pyridoxic acid is an end product of vitamin B_6 metabolism. A majority of ingested vitamin B_6 is converted to 4-pyridoxic acid (95–97).

The interconversion of vitamin B_6 vitamers in human liver has been extensively studied by Merrill et al. (94,98,99). Although only five subjects were examined, this study (94) provides the first detailed work in humans on the activities of enzymes involved in

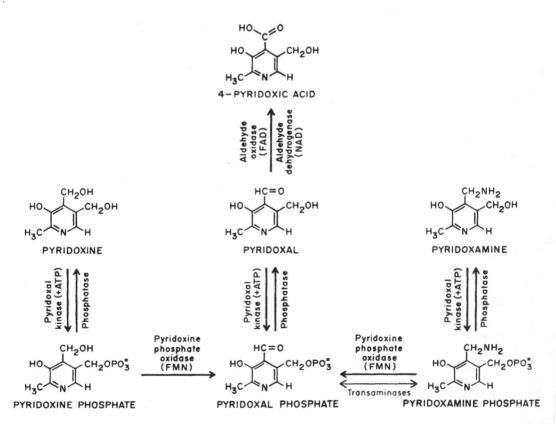

Fig. 4 Metabolic interconversions of the B_6 vitamers.

vitamin B$_6$ metabolism. The activities of pyridoxal kinase, pyridoxine (pyridoxamine)-5'-phosphate oxidase, PLP phosphatase, and pyridoxal oxidase are summarized in Table 6. These activities are optimal ones and, as Merrill et al. have pointed out, at the physiological pH of 7.0 pyridoxal phosphatase activity was less than 1% of the optimal activity at pH 9.0. Considering this, the kinase reaction would be favored and, hence, formation of PLP. The kinase enzyme is a zinc-requiring enzyme. The limiting enzyme in the vitamin B$_6$ pathway appears to be pyridoxine-5'-phosphate oxidase. Since this enzyme requires FMN (108), a reduced riboflavin status may affect the conversion of PN and PM to PLP. Lakshmi and Bamji have reported that whole-blood PLP in persons with oral lesions (presumably riboflavin deficiency) were normal, and supplemental riboflavin had no significant effect on these levels (100). Madigan et al. found that riboflavin supplementation (25 mg/day) of elderly people improved plasma PLP concentration (101). In studies of red cell metabolism of vitamin B$_6$, Anderson and co-workers have shown that riboflavin increases the conversion rate of pyridoxine to PLP (12). In addition, PLP feeds back and inhibits the oxidase. This may be a mechanism by which cells limit the concentration of the highly reactive PLP.

Another riboflavin-dependent (FAD) enzyme, aldehyde oxidase (pyridoxal oxidase), was suggested by Merill et al. (94) to the enzyme that converts pyridoxal to 4-pyridoxic acid. The activity of the aldehyde (pyridoxal) oxidase in humans appears to be sufficient, so that PL which arises from hydrolysis of PLP or that which is taken up into liver would be readily converted to 4-pyridoxic acid. Such a mechanism may prevent large amounts of the highly reactive PLP from accumulating.

The PLP that is formed in liver (and other tissues) can bind via a Schiff base reaction with proteins. The binding of PLP to proteins may be the predominant factor influencing tissue levels of PLP (93). This binding of PLP to proteins is thought to result in metabolic trapping of PLP (vitamin B$_6$ in cells (88,93). PLP synthesized in liver cells is released and found bound to albumin. Whether the PLP is bound to albumin prior to release from the liver or released unbound and subsequently binding to albumin has not been determined.

The binding of PLP to albumin in the circulation serves to protect it from hydrolysis and allows for the delivery of PLP to other tissues (104). This delivery process of PLP to other tissues is thought to involve hydrolysis of PLP and subsequent uptake of PL into the cell (92). Hydrolysis occurs by action of phosphatases bound to cellular membranes. Other forms of vitamin B$_6$ are present in the circulation (plasma). Under fasting conditions, the two aldehyde forms compose 70–90% of the total B$_6$ vitamers in plasma, with PLP making up 50–75% of the total (Table 7). The next most abundant forms are PN, PMP, and PM. Interestingly, pyridoxine-5'-phosphate is essentially absent in plasma.

Table 6 Activity of Human Liver Enzymes Involved in Vitamin B$_6$ Metabolism

Enzyme (activity)	Per gram liver
Pyridoxal kinase (nmol/min)	11.2
Pyridoxine-5'-P-oxidase (nmol/min)	2.4
Pyridoxal-5'-P-phosphatase (nmol/min)	0.1–2.1
Pyridoxal oxidase (nmol/min)	16.5

Source: Data taken from Refs. 94 and 99.

Table 7 B$_6$ Vitamers in Plasma (nmol/L)

	PLP	PL	PNP	PN	PMP	PM
Coburn ($n = 38$)[a]	57 ± 26	23 ± 10	0	19 ± 33	8 ± 8	2 ± 2
Lumeng ($n = 6$)[b]	62 ± 11	13 ± 4	n.d.	32 ± 7	3 ± 3	6 ± 1
Hollins ($n = 10$)[c]	61 ± 34	5 ± 9	n.d.	n.d.	n.d.	n.d.

[a]From Ref. 103.
[b]From Ref. 104.
[c]From Ref. 105.
n.d., none detected.
All data obtained by HPLC methods.

Within the circulating fluid (primarily blood), the erythrocyte also appears to play an important role in the metabolism and transport of vitamin B$_6$. However, the extent of these roles remains controversial (5,106). Both PN and PL are rapidly taken up by a simple diffusion process (107). In the erythrocytes of humans, PL and PN are converted to PLP because both kinase and oxidase activity are present (107). The PLP formed can then be converted to PL by the action of phosphatase; however, this may not be quantitatively important because the phosphatase is considered to be membrane-bound. Any role that the erythrocyte might play in transport of vitamin B$_6$ is complicated by the tight binding of both PLP and PL to hemoglobin (108,109). PL does not bind as tightly as PLP, and each is bound at distinct sites (110). In comparison to the binding to albumin, PL is bound more tightly to hemoglobin (5). As a result, the PL concentration in the erythrocyte is up to four to five times greater than that in plasma (111).

The PLP and PL in plasma, as well as perhaps the PL in erythrocytes, represent the major B$_6$ vitamers available to tissues. To a limited extent, PN would be available following a meal if the uptake was high enough and if the PN escaped metabolism in the liver. Another situation in which PN would be available is following ingestion of vitamin B$_6$ supplements (primarily as PN-HCl). While PN can be converted to PNP in most tissues, conversion to PLP does not take place in many tissues because the oxidase enzyme is absent (112). Human muscle contain PMP oxidase activity (113), but the activity is lower than that of the liver (99). A majority of the vitamin B$_6$ in muscle is present as PLP bound to glycogen phosphorylase (114). Coburn et al. calculated that approximately 66% and 69% of the total vitamin B$_6$ in muscle is present as PLP in males and females, respectively (115). Furthermore, they estimated that the total vitamin B$_6$ pool in muscle was 850 and 900 μmol in males and females, respectively. This pool plus the pool of vitamin in other tissues and circulation would total about 1 nmol (1000 μmol). Previous estimates of total-body pools have been made based on metabolism of a radioactive dose of pyridoxine (116,117). These pools ranged from 100 to 700 μmol.

The precise turnover time of these pools in humans in not known. However, Shane has estimated that there are two pools: one with a rapid turnover of about 0.5 day and a second with a slower turnover of 25–33 days (117). Johansson et al. have also shown that the change in blood levels following administration of tritium-labeled pyridoxine was consistent with a two-compartment model (116). Johansson et al. further suggested that the slow turnover compartment was a storage compartment, but they did not determine the nature of this storage compartment. Figure 5 shows a semilog plot of the decrease in plasma PLP concentration with time in 10 control females and 11 oral contraceptive users fed a diet low in vitamin B$_6$ (0.19 mg, 1.1 μmol) for 4 weeks (95). There was an initial

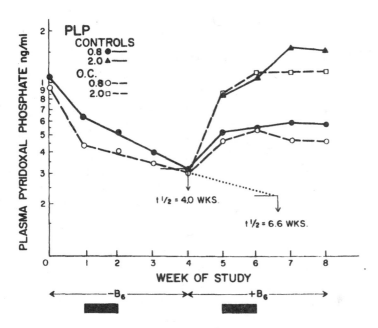

Fig. 5 Semilog plot of plasma pyridoxal-5'-phosphate concentration over 4 weeks of feeding a vitamin B$_6$-deficient diet and 4 weeks of repletion with pyridoxine in control subjects and oral contraceptive users. (From Ref. 86.).

rapid decline in plasma PLP concentration followed by a slower decrease. Extrapolation of the slope for the slowly decreasing portion of the curve for each of the two groups and determination of the plasma $t_{1/2}$ PLP revealed a value of 28 days for the control females and 46 days for the oral contraceptive users. The value for controls is consistent with the data of Shane (117). The longer $t_{1/2}$ for oral contraceptive users may reflect higher levels of enzymes with PLP bound to them (118). Coburn (119) has discussed the turnover and location of vitamin B$_6$ pools, based in part on modeling calculations.

Muscle has been suggested as a possible storage site for vitamin B$_6$. This is based in part on the B$_6$ content of muscle and the total muscle mass of animals. As previously mentioned, in muscle a majority of vitamin B$_6$ is present as PLP bound to glycogen phosphorylase (114,115). In contrast, glycogen phosphorylase accounts for only about 10% of the vitamin B$_6$ content of liver (120). Black and co-workers examined the storage of vitamin B$_6$ in muscle by studying the activity of muscle glycogen phosphorylase (121). In their studies, Black et al. found that feeding rats a diet high in vitamin B$_6$ (70 g of vitamin B$_6$ per kilogram of diet) resulted in a high vitamin B$_6$ content and a high glycogen phosphorylase content in muscle. This increase in content and enzyme level occurred in concert for 6 weeks, whereas the level of alanine and aspartic aminotransferase increased for the first 2 weeks and then plateaued. In subsequent work (122), these same researchers found that muscle phosphorylase content (and thus vitamin B$_6$ content) decreased only when there was a caloric deficit and not necessarily with a deficiency of vitamin B$_6$. This observation of muscle not acting as a mobile reservoir during a vitamin B$_6$ deficiency was also observed in adult swine (123). Coburn et al. (113) observed that human muscle vitamin B$_6$ pools are resistant to depletion. In their study there was a nonsignificant increase in muscle vitamin B$_6$ content when subjects were supplemented with

0.98 µmol pyridoxine HCl per day as compared to the muscle vitamin B_6 content during depletion.

In humans, indirect evidence for muscle serving as a vitamin reservoir has come from my laboratory (124). We have observed an increase in plasma PLP concentration during and immediately after exercise (125,126). Strenuous exercise results in a metabolic state of acute caloric deficit and increased need for gluconeogenesis. Thus, the increased circulating levels of PLP following exercise may reflect PLP released from muscle glycogen phosphorylase. Such a mechanism for this release would mean either that (a) PLP must cross the muscle cell membrane or (b) PLP is hydrolyzed, PL released, and the PL rapidly converted to PLP in liver. Because phosphorylated compounds are thought not to cross membranes easily, the direct release of PLP is considered unlikely by some. Others (127) suggest that this increase in plasma PLP is a release of PLP from liver or interstitial fluid. However, PLP formed in liver is released, and studies of uptake of phosphorylated B_6 vitamers have examined only uptake and not possible transport out of the cell. Further work in my laboratory has shown that in rats starved for 1–3 days there is an increased plasma PLP concentration and an increased PLP concentration in liver, spleen, and heart tissue (unpublished observations). Thus, both direct studies in animals and indirect evidence in humans suggest that vitamin B_6 is stored in muscle and released in times of decreased caloric intake and/or increased need for gluconeogenesis.

VII. ASSESSMENT OF STATUS

The assessment of vitamin B_6 status is central to an understanding of vitamin B_6 nutrition in humans. A variety of methods have been utilized to assess vitamin B_6 status. These methods are given in Table 8 and are divided into direct, indirect, and dietary methods (128–130). Direct indices of vitamin B_6 status are those in which one or more of the B_6 vitamers or the metabolite 4-pyridoxic acid are measured. These are usually measured in plasma, erythrocytes, or urine samples because tissue samples are not normally available. Indirect measures are those in which metabolites of metabolic pathways in which PLP is required for specific enzymes are measured, or in which activities of PLP-dependent enzymes are determined. In this latter case, an activity coefficient is often determined by measuring the enzyme activity in the presence and absence of excess PLP.

Dietary intake of vitamin B_6 itself is not sufficient to assess vitamin B_6 status, especially if only a few days of dietary intake are obtained. In addition to the inherent problems in obtaining accurate dietary intakes, the nutrient databases used in determining vitamin B_6 content of diets are often incomplete with respect to values for vitamin B_6. Thus, reports of vitamin B_6 status based only on nutrient intake must be viewed with caution. Some of the suggested values for the evaluation of status given in Table 8 are based on the relationship of vitamin B_6 and tryptophan metabolism (95). Plasma pyridoxal-5′-phosphate concentration is considered one of the better indicators of vitamin B_6 status (131). Lumeng et al. (104) have shown that plasma PLP concentration is a good indicator of tissue PLP levels in rats. In humans, plasma PLP concentration is significantly correlated with dietary vitamin B_6 intake (97). Table 9 contains mean plasma PLP values reported by several laboratories for males and females. These are selected references drawn from reports in which the sex of the subjects was clearly identified. The means reported range from 27 to 75 mnol/L for males and 26 to 93 µmol/L for females. These ranges should not necessarily be considered as normal since the values given in Table 9 reflect studies in which dietary intake was controlled and other studies in which dietary intake was not assessed.

Table 8 Methods for Assessing Vitamin B$_6$ Status and Suggested Values for Adequate Status

Index	Suggested value for adequate status
Direct	
Blood	
Plasma pyridoxal-5'-phosphate[a]	>30 nmol/L[a]
Plasma pyridoxal	NV
Plasma total vitamin B$_6$	>40 nmol/L
Erythrocyte pyridoxal-5'-phosphate	NV
Urine	
4-Pyridoxic acid	>3.0 μmol/day
Total vitamin B$_6$	>0.5 μmol/day
Indirect	
Blood	
Erythrocyte alanine aminotransferase	>1.25[b]
Erythrocyte aspartate aminotransferase	>1.80[b]
Urine	
2 g Tryptophan load test; xanthurenic acid	<65 μmol/day
3 g Methionine load test; cystathionine	<351 μmol/day
Oxalate excretion	NV
Dietary intake	
Vitamin B$_6$ intake, weekly average	>1.2–1.5 mg/day
Vitamin B$_6$: protein ratio	>0.02
Pyridoxine-β-glucoside	NV
Other	
EEG pattern	NV

[a]Reference values in this table are dependent on sex, age, and protein intake and represent lower limits (130).
[b]For each aminotransferase measure, the activity coefficient represents the ratio of the activity with added PLP to the activity without PLP added.
NV, no value established; limited data available, each laboratory should establish its own reference with an appropriate healthy control population.

As discussed by Shultz and Leklem (97), dietary intake of both vitamin B$_6$ and protein influences the fasting plasma PLP concentration. Miller et al. (136) have shown that plasma PLP and total vitamin B$_6$ concentrations in males were inversely related to protein intake (see Table 9) in males whose protein intake ranged from 0.5 to 2 g/kg per day. Similar results from metabolic studies in women support these findings in men (151).

Other factors that may influence plasma PLP and should be considered when using this index as a measure of vitamin B$_6$ status include the physiological variables of age (133,147,153), exercise (124), and pregnancy (143). Rose et al. determined the plasma PLP concentration in men ranging in age from 18 to 90 years (133). They observed a decrease in plasma PLP with age, especially after 40 years of age. However, one must keep in mind that the PLP concentration was determined 1–2 hours after a meal. The intake of vitamin B$_6$ may have influenced the data. Also, the carbohydrate intake could have resulted in a depressed plasma PLP concentration (124). Hamfelt has reviewed the effect of age on plasma PLP and observed that investigators in several countries (153) have seen decreased vitamin B$_6$ status with increasing age. The mechanism of this decrease

Table 9 Selected Mean Plasma Pyridoxal-5'-Phosphate Concentrations Reported for Healthy Males and Females

Ref.	No. subjects	Age (years)	Diet[a]	PLP (nmol/L)	Method[b]
Males					
Wachstein (132)	27	—[c]	—	35.2 ± 9.3	TDC
Chabner (38)	17	20–34	SS, F	74.8 ± 22.2	TDC
	7	35–49	SS, F	63.9 ± 13.3	
Rose, 1976 (133)	26	18–29	SS, NF	59.1 ± 28.9	TDC
	43	30–39	SS, NF	59.9 ± 29.2	
	82	40–49	SS, NF	53.4 ± 22.0	
	152	50–59	SS, NF	46.9 ± 21.7	
	59	60–69	SS, NF	49.4 ± 24.9	
	65	70–79	SS, NF	47.7 ± 26.1	
	24	80–89	SS, NF	31.1 ± 19.8	
Contractor (134)	5	—	—	54.6 ± 11.7	FL
Wozenski (96)	5	27 ± 3	SS, F	35 ± 14	TDC
Leklem (78)	8	27 ± 4	SS, F	51.5 ± 14.1	TDC
			Met, (1.55), F	33.8 ± 11.2	
Shultz (97)	35	38 ± 14	SS, (2.0 ± 8), F	51.9 ± 19.3	TDC
Shultz (135)	4	22–35	Met, (1.60), F	59.9 ± 41.7	TDC
Leklem (125)	7	16 ± 1	SS, F	47.6 ± 18.7	TDC
Lindberg (79)	5	27 ± 6	Met, (1.60), F	43.3 ± 6.5	TDC
Kabir (83)	9	25 ± 4	SS, F	81.5 ± 36.0	TDC
			Met, F	65.0 ± 23.3	
Miller (136)	8	27 ± 4	Met 1, (1.6)LP, F	43.5 ± 19.4	TDC
			Met, (1.6)MP, F	33.7 ± 9.0	
			Met, (1.6)HP, F	27.9 ± 11.5	
Leklem (137)	8	20–30	Met, (1.6)F	38.8 ± 10.9	TDC
Swift (138)	9	57	SS, (1.9), F	45.5 ± 15.0	TDC
	5	60	SS, (1.5), F	39.2 ± 22.4	
Tarr (81)	6	21–35	Met, (1.1), F	27.5 ± 2.7	TDC
			(2.3), F	55.0 ± 5.7	
			(2.7), F	114.5 ± 7.0	
Ribaya-Mercado (139)	4	63.6 ± 0.8	SS, (1.34), F	25.7 ± 5.1	TDC
			SS, (1.96), F	40.3 ± 7.2	
			SS, (2.88), F	48.0 ± 9.4	
Females					
Wachstein (132)	20	—	—	34.0 ± 10.1	TDC
Chabner (38)	12	20–34	SS, F	67.9 ± 14.6	
	7	35–49	SS, F	46.1 ± 13.8	
Reinken (40)	29	—	SS, —	36.8 ± 8.9	TDC
Miller (141)	11	20–29	SS, F	25.9 ± 15.4	TDC
Lumeng (142)	77	29 ± 8	SS, NF	38.0 ± 17.0	TDC
Brown (95)	6	22 ± 2	Met, (0.8), F	22.9 ± 13.9	TDC
	3	22 ± 2	Met, (1.8), F	60.7 ± 20.2	TDC
Brophy (143)	4	20–34	SS, —	68.4	TDC
Cleary (144)	58	20–34	SS, —	43.4	TDC
Prasad (145)	?	—	SS, (1.19), —	51.8 ± 30.7	TDC
	?	—	SS, (1.02), —	46.5 ± 24.3	TDC
Shultz (135)	4	24–32	SS, F	38.4 ± 15.8	TDC

Table 9   Continued

Ref.	No. subjects	Age (years)	Diet[a]	PLP (nmol/L)	Method[b]
Shultz (97)	41	50 ± 14	SS, (1.6 ± 0.5), F	37.7 ± 14.7	TDC
Guilland (146)	23	27	SS, (1.1), F	92.8 ± 7.3	LC/FL
	29	84	Met, (1.0), F	52.0 ± 4.1	
Lee (147)	5	24 ± 3	SS, F	35.5 ± 14.8	TDC
	5	55 ± 4	SS, F	31.3 ± 13.3	
	5	24 ± 3	Met, (2.3), F	61.7 ± 25.6	
	5	55 ± 4	Met, (2.3), F	40.5 ± 12.2	
	5	24 ± 3	Met, (10.3), F	202 ± 45	
	5	55 ± 4	Met, (10.3), F	168 ± 38	
Driskell (148)	41 (C)[d]	12	SS, (1.23), F	48.1 ± 17.9	TDC
	32 (C)	14	SS, (1.23), F	44.5 ± 15.8	
	23 (C)	16	SS, (1.25), F	43.7 ± 15.3	
	32 (B)	12	SS, (1.30), F	46.1 ± 15.8	
	39 (B)	14	SS, (1.24), F	42.1 ± 15.8	
	19 (B)	16	SS, (1.17), F	46.5 ± 13.9	
Ubbink (149)	9	—	SS, F	31.7 ± 19.4	HPLC
Huang (150)	8	28–34	Met, (1.60), F	58.19 ± 16.28	HPLC
			Met, (0.45), F	32.40 ± 10.50	
			Met, (1.26), F	38.31 ± 9.68	
			Met, (1.66), F	45.43 ± 16.24	
			Met, (2.06), F	53.65 ± 10.94	
Hansen (151)	10	27.5 ± 6.8	Met, (1.03), F	27.9 ± 11.4	TDC
			Met, (1.33), F	32.4 ± 11.6	
			Met, (1.73), F	41.0 ± 14.8	
			Met, (2.39), F	58.9 ± 25.3	
	6	28.2 ± 2.6	Met, (0.84), F	26.5 ± 12.4	
			Met, (1.14), F	29.4 ± 12.5	
			Met, (2.34), F	52.6 ± 22.3	
Ribaya-Mercado (139)	4	63.6 ± 0.8	SS, (0.89), F	21.6 ± 4.4	TDC
			SS, (1.29), F	27.0 ± 5.2	
			SS, (1.90), F	36.6 ± 2.1	
Kretsch (152)	8	21–30	animal + plant protein		TDC
			CF, (0.5), F	8.68 ± 6.13	
			CF, (1.0),(0.5), F	18.66 ± 8.14	
			CF, (1.5)(1.0), F	30.44 ± 14.76	
			CF, (2.0)(1.5), F	42.33 ± 22.11	

[a]The notations for diet indicate if the blood samples were obtained from subjects who self-selected (SS) their diets or were receiving a controlled intake (Met) and the amount of vitamin B$_6$ consumed (value given as mg/day in parentheses), if known. F indicates that the blood sample was collected after a fast of at least 8 h; NF indicates nonfasting. LP, MP, HP refer to grams of protein as 0.5, 1.0, and 2.0 g/kg body weight.

[b]TDC, tyrosine apodecarboxylase; HPLC, high-performance liquid chromatography; FL, fluorimetry.

[c]A dash indicates data was not given in the respective reference.

[d]C, Caucasian; B, black.

Table 10　Urinary 4-Pyridoxic Acid and Vitamin B$_6$ Excretion in Males and Females

Ref.	No. subjects	Age (years)	Diet	4-Pa (nmol/day)	UB6 (μmol/day)
Males					
Kelsay (158)	5	20–25	Met (1.66 mg, P = 150)	5.51 ± 1.75	—
	6	18–35	(1.66 mg, P = 54)	4.80 ± 0.60 (50)[a]	—
Mikai-Devic (159)	10	16–51	—	5.7 ± 1.4	
Leklem (78)	8	27 ± 4	Met (1.55)	4.04 ± 0.85 (44)	0.76 ± 0/17
Wozenski (96)	5	27 ± 3	SS	5.4 ± 0.5	0.8 ± 0.1
Shultz (135)	4	22–35	Met (1.6)	5.71 ± 1.08	0.76 ± 0.10
Shultz (97)	35	35 ± 14	SS (2.0 ± 0.8)	7.46 ± 4.34 (63)	0.92 ± 0.49
Lindberg (79)	10	26 ± 4	SS	4.78 ± 1.40	0.81 ± 0.18
	10	26 ± 4	Met (1.60)	3.62 ± 0.59	0.76 ± 0.10
Kabir (83)	9	25 ± 4	Met (1.55)	4.89 ± 1.10 (53)	1.05 ± 0.20
Dreon (160)	6	28 ± 6	Met (4.2 ± 0.4)	11.15 ± 1.86 (45)	
Miller (136)	8	27 ± 4	Met (1.60)	4.37 ± 0.89 (LP)	0.77 ± 0.14
				3.58 ± 0.54 (MP)	0.71 ± 0.09
				2.74 ± 0.71 (HP)	0.68 ± 0.15
Females					
Mikai-Devic (159)	15	18–47	SS	4.5 ± 0.9	
Contractor (134)	26	—	SS	6.62 ± 4.6	
Reinken (140)	29	25	SS	8.32 ± 1.30	
Brown (95)	6	22 ± 2	Met (0.82)	1.98 ± 0.81 (41)	
	3	22 ± 2	Met (1.81)	6.03 ± 2.04 (56)	
Donald (161)	8	18–23	Met (1.54)	2.4 ± 0.1 (26)	0.33 ± 0.3
			(2.06)	3.78 ± 0.41 (31)	

Reference	n		Condition		
Shultz (97)	41	50 ± 14	SS (1.6 ± 0.5)	5.57 ± 3.09 (59)	0.76 ± 0.24
Lee (147)	5	24 ± 3	Met (2.3)	6.89 ± 0.55 (50)	1.12 ± 0.29
Ubbink (149)	9	—	SS	5.48 ± 2.93	—
Huang (150)	8	28–34	Met, (1.60), F	3.48 ± 0.93 (37)	
			Met, (0.45), F	0.93 ± 0.30 (35)	
			Met, (1.26), F	2.52 ± 0.32 (33 ± 2)	
			Met, (1.66), F	3.01 ± 0.61 (33 ± 2)	
			Met, (2.06), F	4.24 ± 0.74 (33 ± 2)	
Hansen (151)	10	27.5 ± 6.8	Met, (1.03), F	3.23 ± 0.73 (60)	0.54 ± 0.12
			Met, (1.33), F	4.00 ± 0.91 (52)	0.61 ± 0.10
			Met, (1.75), F	5.89 ± 0.77 (54)	0.75 ± 0.09
			Met, (2.39), F	9.51 ± 1.08 (69)	0.95 ± 0.14
	6	28.2 ± 2.6	Met, (0.84), F	2.87 ± 0.47 (58)	0.48 ± 0.08
			Met, (1.14), F	3.35 ± 0.65 (50)	0.54 ± 0.09
			Met, (2.34), F	7.88 ± 0.81 (57)	0.79 ± 0.11
Hansen (86)	5	29 ± 6	High PNG: Met, (1.52), F	3.60 ± 0.84 (40)	0.587 ± 0.082
	4		Low PNG: Met, (1.45), F	4.02 ± 0.58 (47)	0.660 ± 0.116
Kretsch, (152)	4	21–30	animal protein:		
			CF, (0.5), F	0.62 ± 0.28	—
			CF, (1.0)(0.5), F	1.76 ± 0.41 (32)	
			CF, (1.5)(1.0), F	3.55 ± 0.43 (43)	
			CF, (2.0)(1.5), F	5.47 ± 0.71 (50)	
	4		plant protein:		
			CF, (0.5), F	1.35 ± 0.47	—
			CF, (1.0)(0.5), F	2.29 ± 0.27 (39)	
			CF, (1.5)(1.0), F	4.01 ± 0.25 (49)	
			CF, (2.0)(1.5), F	5.98 ± 0.36 (55)	

[a]The number in parentheses refers to the percent of intake excreted as 4-PA.

Abbreviations are as used in Table 9, except for 4-pyridoxic, which is 4-PA and urinary vitamin B₆, which is UB6.

remains to be determined. There is one controlled metabolic study that has evaluated vitamin B_6 status in different age groups. Lee and Leklem (147) studied five women age 20–27 years and eight women age 51–59 years under conditions in which the women received a constant daily vitamin B_6 intake of 2.3 mg for 4 weeks followed 10.3 mg per day for 3 weeks. Compared with the younger women, the older women had a lower mean plasma PLP, plasma and urinary total vitamin B_6, and a slightly higher urinary 4-pyridoxic acid excretion with the 2.3-mg intake. Interestingly, there was no difference in urinary excretion of xanthurenic or kynurenic acid following a 2-g L-tryptophan load. Thus, while there may be age-related differences in vitamin B_6 metabolism, there is no significant age effect on functional activity of vitamin B_6 when intake is adequate. The metabolism of vitamin B_6 has been studied in elderly men and women older than 60 years. While younger individuals were not examined in the same study, the researchers concluded that the elderly had an increased vitamin B_6 requirement, indicative of increased metabolism. Kant et al. (154) observed no age-related impairment in the absorption or phosphorylation of vitamin B_6. However, there was an increase in plasma alkaline phosphatase activity with age that would increase hydrolysis of PLP.

The use of plasma PLP as a status indicator has been questioned (155) and the determination of plasma PL recommended. Others have also suggested that plasma PL may be an important indicator of status. When Barnard et al. (156) studied the vitamin B_6 status in pregnant females and nonpregnant controls, they found that plasma PLP concentration was 50% lower in pregnant females but that the concentration of the total of PLP and PL was only slightly lower. When concentrations of PLP and PL were expressed on a per-gram-albumin basis, there was no difference between groups. In contrast, in pregnant rats both plasma PLP and PL decreased, as did liver PLP, in comparison with nonpregnant control rats (157). These studies are in direct opposition to each other but do provide support for the need to determine several indices of vitamin B_6 status (130,131,155).

Urinary 4-pyridoxic acid excretion is considered a short-term indicator of vitamin B_6 status. In deficiency studies in males (158) and females (159), the decrease in urinary 4-pyridoxic acid paralleled the decrease in plasma PLP concentration. Table 10 lists values for urinary 4-pyridoxic acid and vitamin B_6 in males and females. As reflected in the studies in which dietary intake was assessed or known, 4-pyridoxic acid excretion accounts for about 40–60% of the intake. Because of the design of most studies and the limited number of studies done with females compared with males, it is not possible to determine if there is a significant difference between males and females. The limited data in Table 11 suggest that there is little difference. However, males consistently had higher plasma PLP and total vitamin B_6 concentrations as well as higher excretion of 4-pyridoxic acid and total vitamin B_6. Urinary total vitamin B_6 (all forms, including phosphorylated and

Table 11 Plasma Pyridoxal-5'-Phosphate and Total Vitamin B_6 Concentration, and Urinary 4-Pyridoxic Acid and Vitamin B_6 Excretion in Males and Females Consuming 2.2 mg Vitamin B

Subject	PLP (nmol/L)	TB6 (nmol/L)	4-PA (μmol/day)	UB6 (μmol/day)
Males ($n = 4$)	78.4 ± 27.0[a]	86.2 ± 37.1	7.86 ± 0.74	0.92 ± 0.20
Females ($n = 4$)	58.5 ± 12.6	71.5 ± 15.8	7.02 ± 0.78	0.82 ± 0.19

[a]Mean ± SD.

glycosylated) excretion is not a sensitive indicator of vitamin B$_6$, except in situations where intake is very low (158).

Erythrocyte transaminase activity (alanine and aspartate) has been used to assess vitamin B$_6$ status in a variety of populations (133,142,146,162–168), including oral contraceptive users (95,166). Transaminase activity is considered a long-term indicator of vitamin B$_6$ status. Most often the transaminase activity has been measured in the presence and absence of excess PLP (163). Table 8 indicates suggested norms for activity coefficients for alanine and aspartate aminotransferase. While transaminase activity is used to assess status, there is not unanimous agreement, and some consider this measure to be less reliable than other indices of vitamin B$_6$ status (95,168). The long life of the erythrocyte and tight binding of PLP to hemoglobin may explain the lack of a consistent significant correlation between plasma PLP and transaminase activity or activity coefficient. An additional consideration that complicates the use of aminotransferases is the finding of genetic polymorphism of erythrocyte alanine aminotransferase (169).

Urinary excretion of tryptophan metabolites following a tryptophan load, especially excretion of xanthurenic acid, has been one of the most widely used tests for assessing vitamin B$_6$ status (170,171). Table 8 gives a suggested normal value for xanthurenic acid excretion following a 2-g L-tryptophan load test. The use of the tryptophan load test for assessing vitamin B$_6$ status has been questioned (172,173), especially in disease states or in situations in which hormones may alter tryptophan metabolism independent of a direct effect on vitamin B$_6$ metabolism (174).

Other tests for status include the methionine load (175), oxalate excretion, and electroencephalographic tracings (176). These tests are used less often but under appropriate circumstances provide useful information. The review by Reynolds (155) provides an excellent critique of methods currently in use for assessment of vitamin B$_6$ status.

VIII. FUNCTIONS

A. Immune System

The involvement of PLP in a multiplicity of enzymatic reactions (177) suggests that it would serve many functions in the body. Table 12 lists several of the known functions of PLP and the cellular systems (137) affected. PLP serves as a coenzyme for serine transhydroxymethylase (178), one of the key enzymes involved in one-carbon metabolism. Alteration in one-carbon metabolism can then lead to changes in nucleic acid synthesis. Such changes may be one of the keys to the effect of vitamin B$_6$ on immune function

Table 12 Cellular Processes Affected by Pyridoxal-5'-Phosphate

Cellular process or enzyme	Function/system influenced
One-carbon metabolism, hormone modulation	Immune function
Glycogen phosphorylase, transamination	Gluconeogenesis
Tryptophan metabolism	Niacin formation
Heme synthesis, transamination, O$_2$ affinity	Red cell metabolism and formation
Neurotransmitter synthesis, lipid metabolism	Nervous system
Hormone modulation, binding of PLP to lysine on hormone receptor	Hormone modulation

(179,180). Studies in animals have shown that a vitamin B_6 deficiency adversely affects lymphocyte production (179) and antibody response to antigens (180). Additional studies in animals support an effect of vitamin B_6 on cell-mediated immunity (181). Talbot et al. found in 11 elderly women whose immune response has impaired that treatment with 50 mg pyridoxine per day for 2 months improved their immune system, as judged by lymphocyte response (182). However, in humans a diet-induced marginal vitamin B_6 status for 11 weeks was not found to significantly influence cellular or humoral immunity (183). These two studies differed in their experimental design. The study by van den Berg et al. (183) employed a diet marginally deficient in vitamin B_6 in young adults; that of Talbot et al. (182) utilized a treatment of elderly individuals with an excess of vitamin B_6. This excess intake may be necessary for increased activity of certain cell types of the immune system in the elderly. Meydani et al. examined immune response in healthy elderly adults fed graded levels of vitamin B_6 and found that a deficiency impairs in vitro indices of cell-modulated immunity, especially interleukin-2 production (184). A review of vitamin B_6 and immune competence is available (185).

B. Gluconeogenesis

Gluconeogenesis is key to maintaining an adequate supply of glucose during caloric deficit. Pyridoxal-5'-phosphate is involved in gluconeogenesis via its role as a coenzyme for transamination reactions (177) and for glycogen phosphorylase (114). In animals a deficiency of vitamin B_6 results in decreased activities of liver alanine and aspartate aminotransferase (186). However, in humans (females) a low intake of vitamin B_6 (0.2 mg/ day), as compared with an adequate intake (1.8 mg/day), did not significantly influence fasting plasma glucose concentrations (187). Interestingly, the low vitamin B_6 intake was associated with impaired glucose tolerance in this study.

Glycogen phosphorylase is also involved in maintaining adequate glucose supplies within liver and muscle and, in the case of liver, a source of glucose for adequate blood glucose levels. In rats a deficiency of vitamin B_6 has been shown to result in decreased activities of both liver (188) and muscle glycogen phosphorylase (114,122,188). Muscle appears to serve as a reservoir for vitamin B_6 (114,122,123), but a deficiency of the vitamin does not result in mobilization of these stores. However, Black et al. (122) have shown that a caloric deficit does lead to decreased muscle phosphorylase content. These results suggest that the reservoir of vitamin B_6 (as PLP) is only utilized when there is a need for enhanced gluconeogenesis. In male mice the half-life of muscle glycogen phosphorylase has been shown to be approximately 12 days (189). In contrast to low intake of vitamin B_6, rats given an in injection of a high dose of PN, PL, or PM (300 mg/kg) showed a decrease in liver glycogen and an increase in serum glucose (190). This effect is mediated via increased secretion of adrenal catecholamines. The extent to which lower intake of B_6 vitamers has this effect or if this occurs in humans remains to be determined.

C. Erythrocyte Function

Vitamin B_6 has an additional role in erythrocyte function and metabolism. The function of PLP as a coenzyme for transaminases in erythrocytes has been mentioned. In addition, both PL and PLP bind to hemoglobin (107,108). The binding of PL to the α chain of hemoglobin (191) increased the O_2 binding affinity (192), while the binding of PLP to the β chain of hemoglobin S or A lowers the O_2 binding affinity (193). The effect of PLP and PL on O_2 binding may be important in sickle cell anemia (194).

Pyridoxal-5'-phosphate serves as a cofactor for δ-aminolevulinic acid synthetase (195), the enzyme that catalyzes the condensation between glycine and succinyl-CoA to form δ-aminolevulinic acid. This latter compound is the initial precursor in heme synthesis (196). Therefore, vitamin B$_6$ plays a central role in erythropoiesis. A deficiency of vitamin B$_6$ in animals can lead to hypochromic microcytic anemia. Furthermore, in humans there are several reports of patients with pyridoxine-responsive anemia (197,198). However, not all patients with sideroblastic anemia (in which there is a defect in δ-aminolevulinic acid synthetase) respond to pyridoxine therapy (199).

D. Niacin Formation

One of the more extensive functions of vitamin B$_6$ that has been researched is its involvement in the conversion of tryptophan to niacin (171). This research is in part related to the use of the tryptophan load in evaluating vitamin B$_6$ status. While PLP functions in at least four enzymatic reactions in the complex tryptophan–niacin pathway (Fig. 6), there is only one PLP-requiring reaction in the *direct* conversion of tryptophan to niacin. This step is the conversion of 3-hydroxykynurenine to 3-hydroxyanthranilic acid and is catalyzed by kynureninase. Leklem et al. have examined the effect of vitamin B$_6$ deficiency on the conversion of tryptophan to niacin (200). In this study, the urinary excretion of N'-methylnicotinamide and N'-methyl-2-pyridone-5-carboxamide, two metabolites of niacin, was evaluated in women. After 4 weeks of a low-vitamin B$_6$ diet, the total excretion of these two metabolites following a 2-g L-tryptophan load was approximately half that when subjects received 0.8–1.8 mg vitamin B$_6$ per day. This suggests that low vitamin B$_6$ has a moderate negative effect on niacin formation from tryptophan.

E. Nervous System

In addition to the effect of vitamin B$_6$ on tryptophan-to-niacin conversion, there is another tryptophan pathway that is vitamin B$_6$–dependent. The conversion of 5-hydroxytryptophan

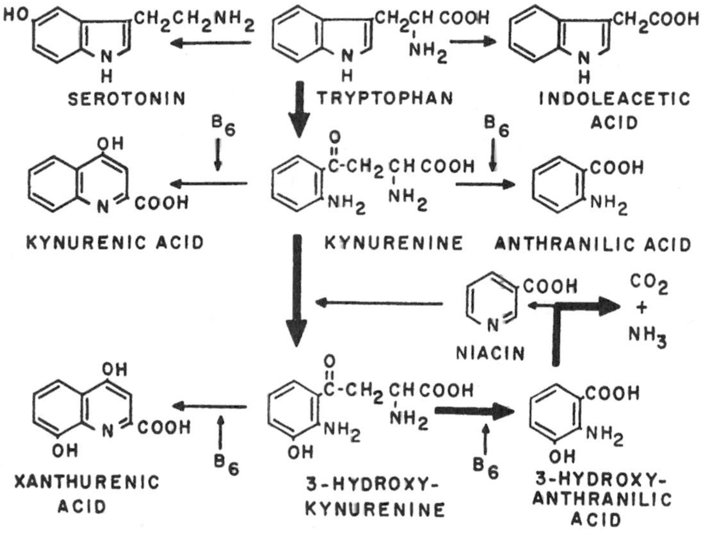

Fig. 6 Tryptophan–niacin pathway, B$_6$ indicates the steps in the pathway in which pyridoxal-5'-phosphate functions as a coenzyme.

to 5-hydroxytryptamine is catalyzed by the PLP-dependent enzyme 5-hydroxytryptophan decarboxylase (201). Other neurotransmitters, such as taurine, dopamine, norepinephrine, histamine, and γ-aminobutyric acid, are also synthesized by PLP-dependent enzymes (201). The involvement of PLP in neurotransmitter formation and the observation that there are neurological abnormalities in human infants (202,203) and animals (204,205) deficient in vitamin B_6 provide support for a role of vitamin B_6 in nervous system function. Reviews on the relationship between nervous system function and vitamin B_6 are available (201,206,207).

In infants fed a formula in which the vitamin B_6 was lost during processing, convulsions and abnormal electroencephalograms (EEGs) were observed (202). Treatment of the infants with 100 mg of pyridoxine produced a rapid involvement in the EEGs. In these studies reported by Coursin, the protein content of the diet appeared to be correlated with the vitamin B_6 deficiency and the extent of symptoms. Other evidence for a role of vitamin B_6 comes from studies of pyridoxine-dependent seizures, disorders, which is an autosomal recessive disorder. Vitamin B_6 dependency, though a rare cause of convulsions, has been reported by several investigators (208–211). The convulsions occur during the neonatal period, and administration of 30–100 mg of pyridoxine is usually sufficient to prevent convulsions and correct an abnormal EEG (211,212). However, there are atypical patients who present a slightly different clinical picture and course but are responsive to pyridoxine (213).

Vitamin B_6 deficiency in adults has also been reported to result in abnormal EEGs (176,214), especially in individuals on a high-protein (100 g/day) intake. In one study (215), subjects were receiving a diet essentially devoid of vitamin B_6 (0.06 mg). In a separate study, Grabow and Linkswiler fed to 11 men a high-protein diet (150 g) and 0.16 mg of vitamin B_6 for 21 days (215). No abnormalities in EEGs were observed, nor were there changes in motor nerve conduction times in five subjects who had this measurement. Kretsch et al. (176) observed abnormal EEG patterns in two of eight women after 12 days of a low (0.05 mg/day) vitamin B_6 diet. Feeding 0.5 mg/day corrected the abnormal pattern. While there were differences in the length of deficiency in these studies, which may explain the differences observed, it appears that long-term very low vitamin B_6 intakes are necessary before abnormal EEGs are observed in humans.

Another aspect of the relationship of vitamin B_6 (as PLP) to the nervous system is the development of the brain under conditions of varying intakes of vitamin B_6. Kirksey and co-workers have conducted numerous well-designed studies in this area. These studies have utilized the rat model to examine the development of the brain, especially during the critical period when cells are undergoing rapid mitosis. Early experiments showed that dietary restriction of vitamin B_6 in the dams was associated with a decrease in alanine aminotransferase and glutamic acid decarboxylase activity and low brain weights of progeny (216). Alterations in fatty acid levels, especially those involved in myelination, were observed in the cerebellum and cerebrum of progeny of dams fed a low (1.2 mg/kg daily) vitamin B_6 diet (217). Kurtz et al. found a 30–50% decrease in cerebral sphingolipids of progeny of dams fed a vitamin B_6 deficient diet (218). In the progeny of dams fed graded levels of vitamin B_6, a decrease in the area of the neocortex and cerebellum as well as reduced molecular and granular layers of the cerebellum were noted (219). Myelination was reduced in the progeny of severely deficient dams (220). Noting in previous studies that Purkinje cells were dispersed from the usual noncellular layer in progeny of severely vitamin B_6-deficient dams, Chang et al. (221) carried out other studies and found that a maternal vitamin B_6 deficiency interferes with normal development of Purkinje cells, seen as reduced total length of Purkinje cell dendrites. At the biochemical level, a vitamin B$_6$ deficiency led to reduced GABA levels, which were associated with impaired function of the extrapyramidal motor system (222). Both Wasynczuk et al. (223) and Kurtz et al. (218) found that with vitamin B$_6$ deficiency amino acid levels were altered in specific regions of the brain. Glycine, leucine, isoleucine, valine, and cystathionine levels were elevated, whereas alanine and serine levels were reduced. Other studies have shown that PLP levels in certain areas of the central nervous system are more dramatically affected by vitamin B$_6$ deficiency than others. Levels in the spinal cord and hypothalamus were less affected than these in the corpus striatum and cerebellum (233). These findings suggested metabolic trapping in a caudal-to-rostral direction.

F. Lipid Metabolism

One of the more intriguing and controversial aspects of vitamin B$_6$ is its role in lipid metabolism (225). Studies conducted more than 60 years ago suggested a link between fat metabolism and vitamin B$_6$ (226). These studies showed a similarity (visual) in symptoms between a lack of essential fatty acids and a deficiency of vitamin B$_6$. Other early studies found that a vitamin B$_6$ deficiency in rats resulted in a decrease in body fats (227). Subsequent research showed that liver lipid levels were significantly lower in vitamin B$_6$-deficient vs. pair-fed rats (228). The changes were due mainly to lower triglyceride levels, whereas cholesterol levels were not different. In contrast, Abe and Kishino showed that rats fed a high-protein (70%), vitamin B$_6$-deficient diet developed fatty livers and suggested that this was due to impaired lysosomal degradation of lipid (229). The synthesis of fat in vitamin B$_6$-deficient rats has been reported to be greater (230), normal (231,232), or depressed (233). The observed differences may be related to the meal pattern of the animals (234).

The effect of vitamin B$_6$ deprivation on fatty acid metabolism has also received attention. A pyridoxine deficiency may impair the conversion of linoleic acid to arachidonic acid (202,235). Cunnane and co-workers (202) found that phospholipid levels of both linoleic and γ-linolenic acid were increased in vitamin B$_6$-deficient rats, but the level of arachidonic acid was decreased as compared with that of control levels in plasma, liver, and skin. They suggested that both linoleic desaturation and γ-linoleic acid elongation may be impaired by a vitamin B$_6$ deficiency. She et al. (236) have observed decreased activity of terminal Δ^6-desaturase in the linoleic acid desaturation system in rats fed a vitamin B$_6$-deficient diet. They also found a positive correlation between phosphatidylcholine (PC) content and Δ^6-desaturase activity in liver microsomes, suggesting that altered PC may affect linoleic acid desaturation and thus decreased arachidonic acid synthesis. Subsequent work by She et al. suggests that altered S-adenosylmethionine (SAM) to S-adenosylhomocysteine is involved in these changes (237). In one of the few studies of vitamin B$_6$ and fatty acid metabolism in humans, desoxypyridoxine was utilized to induce a vitamin B$_6$ deficiency (238). Xanthurenic acid excretion following a 10-g D,L-tryptophan load indicated a moderate vitamin B$_6$-deficient state. Only minor changes in fatty acid levels in plasma and erthyrocytes were observed as a result of the deficiency produced. The pattern of fatty acids observed was interpreted by the authors to support the findings by Witten and Holman (234). Delmore and Lupien also observed a decreased proportion of arachidonic acid in liver phospholipids and an increased level of linoleic acid in vitamin B$_6$-deficient rats (239). They suggested that changes were based on the decrease in PC via methylation of phosphoethanolamine. Support for this comes from the studies of Loo and Smith (240), in which it was found that a deficiency of vitamin B_6 resulted in decreased phospholipid methylation in the liver of rats. The level of SAM in livers of vitamin B_6-deficient rats was nearly five times higher than that in livers of pair-fed animals. This change in SAM is secondary to the inhibition of the catabolism of homocysteine, a PLP-dependent process. Negative feedback of SAM on the conversion of phosphatidyletha-nolamine to PC may thus explain the changes seen in fatty acid metabolism. The work of She et al. (237) supports this. This provides a plausible mechanism because the primary metabolic steps in fatty acid metabolism do not involve nitrogen-containing substrates, a feature common to most PLP-dependent enzymatic reactions.

The change observed in arachidonic acid levels and the role it plays in cholesterol metabolism (241) may have clinical implications (242). The effect, if any, of vitamin B_6 on cholesterol metabolism remains controversial. The increase in plasma cholesterol in monkeys made vitamin B_6-deficient (243) provided much of the impetus for research relating vitamin B_6, cholesterol, and atherosclerosis. Studies by Lupien and co-workers have shown that the rate in incorporation of [^{14}C] acetate into cholesterol was increased in vitamin B_6-deficient rats as compared to controls (244). However, the amount of choles-terol in plasma and liver of rats and other species has been reported to be increased, not changed, or even decreased (244–247). In humans, a deficiency of vitamin B_6 did not result in a significant change in serum cholesterol (248). Significant positive correlation between plasma PLP and high-density lipoprotein (HDL) cholesterol and negative correla-tions with total cholesterol and low-density lipoprotein (LDL) cholesterol have been re-ported in monkeys fed atherogenic Western diets and a ''prudent'' Western diet (249). However, the diets fed to the monkeys contained distinctly different amounts of vitamin B_6. The use of supplemental vitamin B_6 in reduction of blood cholesterol has not been definitively tested. Serfontein and Ubbink reported decreased serum cholesterol (0.8 mmol/L) in 34 subjects given a multivitamin containing 10 mg of pyridoxine (250). The reduction was mainly as LDL cholesterol. In another study, pyridoxine (50 mg/day) ad-ministration prevented the increase in serum cholesterol seen when disulfiram was admin-istered (251). Controlled trials of pyridoxine are needed to resolve the role of vitamin B_6 in modifying serum cholesterol levels.

The role of vitamin B_6 in lipid metabolism remains unclear. Evidence to data sug-gests a role of vitamin B_6 in modifying methionine metabolism and thus an indirect effect on phospholipid and fatty acid metabolism. This effect (240) and an effect of vitamin B_6 on carnitine synthesis (252) appear to be the primary effects of vitamin B_6 on lipid/fatty acid metabolism.

G. Hormone Modulation/Gene Expression

One of the more intriguing functions of PLP is as a modulator of steroid action (253,254). Reviews of this interaction are available (255,256). PLP can be used as an effective tool in extracting steroid receptors from the nuclei of tissues on which the steroid acts (257). Under conditions of physiological concentrations of PLP, reversible reactions occur with receptors for estrogen (258), androgen (259), progesterone (260), and glucocorticoids (261,262). PLP reacts with a lysine residue on the steroid receptor. As a result of the formation of a Schiff base, there is inhibition of the binding of the steroid–receptor com-plex to DNA (253). Holley et al. found that when female rats were made vitamin B_6-deficient and injected with [^{3}H]-estradiol, a greater amount of the isotope accumulated in the uterine tissues of the deficient animal than in the tissues of control rats (263). Bunce and co-workers studied the dual effect of zinc and vitamin B$_6$ deficiency on estrogen uptake by the uterus (264). They found that there was an increased uptake of estrogen in both the vitamin B$_6$- and the zinc-deficient animals. A combined deficiency to the two nutrients resulted in even greater retention of estrogen. The number of estrogen receptors was not altered by the deficiency of vitamin B$_6$. This study suggests that there would be increased sensitivity of the uterus (or other end-target tissues) to steroids when vitamin B$_6$ status was abnormal.

Sturman and Kremzner found enhanced activity of ornithine decarboxylase in testosterone-treated vitamin B$_6$-deficient animals as compared to control animals (265). DiSorbo and Litwack observed increased tyrosine aminotransferase activity in hepatoma cells raised on a pyridoxine-deficient medium and treated with triamcinolone acetonide as compared to pyridoxine-sufficient cells treated with the same steroid (266). Allgood and Cidlowski (267) have used a variety of cell lines and a range of intracellular PLP concentrations to show that vitamin B$_6$ modulates transcriptional activation by several (androgen, progesterone, and estrogen) steroid hormone receptors. This supports the role of vitamin B$_6$ as a physiological modulator of steroid hormone action.

Oka et al. (268) have found that in vitamin B$_6$-deficient rats the level of albumin mRNA was increased sevenfold once that of control rats. They suggest that PLP modulates albumin gene expression by inactivation of tissue-specific transcription factors. Oka and co-workers have also observed a sevenfold increase in the level of mRNA for cystosolic aminotransferase in the level of vitamin B$_6$-deficient rats as compared to that of vitamin B$_6$-sufficient rats (269). Subsequent work by Oka et al. (270) shows an inverse relationship between intracellular PLP concentration and albumin in RNA in rats given amino loads. Thus, PLP may be a modulator of gene expression in animals, especially under conditions of altered amino acid supply. Given the intimate relationship of vitamin B$_6$ and an amino acid metabolism, these studies open a new area of metabolic regulation via altered intracellular nutrient (PLP) concentration.

IX. REQUIREMENTS

Considering the numerous functions in which vitamin B$_6$ is involved, assessment of the requirement for this vitamin becomes important. Reviews of vitamin B$_6$ requirements are available (271–273). Each of the last three recommended dietary allowances (RDA) publications (274–276) has evaluated the requirements for vitamin B$_6$. While vitamin B$_6$ has been known for over 70 years, only in the past 40 years has a requirement been established in the United States. There are numerous factors that may contribute to the requirement for vitamin B$_6$, several of which are listed in Table 13. Many of these have not been experimentally tested, whereas others have received greater attention and been examined in more detail.

Given the central role of PLP in amino acid metabolism, it is not surprising that there is an intimate relationship between vitamin B$_6$ requirements and protein intake. Historically, the establishment of the requirement for vitamin B$_6$ has been based primarily on this protein–vitamin B$_6$ relationship. Early studies by Canham et al. (214), Baker et al. (277), and Miller and Linkswiler (278) in men utilized L-tryptophan load tests and the excretion of tryptophan metabolites as an indicator of vitamin B$_6$ adequacy. In a study by Park and Linkswiler (279), a methionine load was used to assess vitamin B$_6$ adequacy. In all of these studies the intake of vitamin B$_6$ used ranged from 0.1 to 2.2 mg. There are several important observations relative to use of these studies in subsequent estimation

Table 13 Factors Affecting Vitamin B_6 Requirement

1. Dietary
   a. Physical structure of a food
   b. Forms of vitamin B_6 natural; those due to processing
   c. Binding of forms of vitamin B_6
2. Defect in delivery to tissues
   a. Impaired gastrointestinal absorption
   b. Impaired transport—albumin synthesis and binding, impaired phosphatase activity
3. Physiological/biochemical
   a. Physical activity—increased loss, gluconeogenesis
   b. Protein—enzyme induction
   c. Increased catabolism turnover—phosphatase activity, illness
   d. Impaired phosphorylation and/or interconversion, competing pathways, nutrient deficiencies, drugs
   e. Pregnancy—demand of fetus
   f. Growth—increased cell mass, repair
   g. Lactation—adequate levels in milk
   h. Excretion rate—urinary, sweat, menstrual loss
   i. Sex-differences in metabolism
   j. Age differences in metabolism
4. Genetic
   a. Apoenzyme defects—altered binding to apoenzyme
   b. Altered enzyme levels—biochemical individuality
5. Disease prevention/treatment
   a. Which? heart disease, cancer, diabetes, PMS, kidney disease, alcoholism

of a vitamin B_6 requirement: (a) they were done only in men; (b) the men were fed two levels of protein: 30–40 g/day at the low-protein intake levels and 80–100 g/day for the high-protein; intake level; (c) subjects were usually fed a low-vitamin B_6 diet (0.06–0.16 mg/day) for several weeks (4–6 weeks) before vitamin B_6 supplements were given; (d) often the diets were not representative of a typical American diet and could be considered ones in which the vitamin B_6 was highly bioavailable (given that the intake of vitamin B_6 from food was low); (e) the tryptophan load tests were not uniform between studies (2 g of the L form or 10 g of DL form); and (f) most of the studies did not include measurement of plasma PLP concentration.

Based on data from these studies of the vitamin B_6 protein interrelationship and urinary metabolite excretion, the amount of vitamin B_6 considered to be adequate (normalization of amino acid post-load loose metabolite excretion) ranged from 1.0 to 2.0 mg of vitamin B_6. For diets containing 100–150 g of protein per day the requirement was judged to be 1.5–2.0 mg/day. At protein intakes less than 100 g of protein per day the requirement was judged to be 1.0–1.5 mg/day (281).

In a subsequent study by Miller et al. (136), the effect of three levels of protein on vitamin B_6 status indices was evaluated in men. An important feature of this study was that it was not a depletion/repletion study. There was a significant effect of protein (inverse relationship) on plasma PLP and total vitamin B_6 and on urinary 4-pyridoxic acid excretion. The authors pointed out that protein intake should be considered when evaluating requirement.

It should be noted that in some of these early studies one or more of the following indices of B$_6$ adequacy were utilized: urinary B$_6$ excretion, erythrocyte transaminase activity, urinary cystathionine excretion, urinary 4-pyridoxic excretion, and electroencephalograms. Each of these indices varies with respect to its sensitivity to changes in vitamin B$_6$ status and its usefulness in establishing vitamin B$_6$ requirements (130).

Subsequent studies in women focused on evaluating the effect of oral contraceptives on vitamin B$_6$ status (95,200,281). These studies utilized depletion/repletion-type designs. The length of the depletion periods varied (14–43 days) as did the amount of vitamin B$_6$ fed (0.16–0.34 mg/day). In addition, the protein intake ranged from 57–109 g/day. The vitamin B$_6$ repletion levels ranged from 0.6 to 1.85 mg over periods of 3–28 days. Vitamin B$_6$ status (adequacy of vitamin B$_6$ intake) was assessed with a variety of indices, including plasma PLP, plasma vitamin B$_6$, urinary 4-pyridoxic acid excretion, tryptophan metabolite excretion, and erythrocyte transaminase activity and stimulation.

The differences in vitamin B$_6$ intake and protein intake in these studies are critical in evaluating the subsequent vitamin B$_6$ requirements derived from them. Also, these were depletion/repletion experiments. The effect of this experimental design on the length of time for the various vitamin B$_6$ status indices to stabilize during the repletion period could have an effect on the estimation on vitamin B$_6$ requirement. As an example, 4-pyridoxic acid excretion is very sensitive to vitamin B$_6$ intake and establishes a plateau relatively quickly (3–4 days) after B$_6$ intake has been changed. However, plasma PLP takes longer (up to 10 days) before a plateau is achieved. Transaminase activity (and stimulation) would be expected to take longer to achieve a plateau at a given vitamin B$_6$ intake due to the long half-life of the erythrocyte (200).

Several studies have been done since the publication of the 1989 RDAs that are relevant to establishing the next RDA for vitamin B$_6$. These studies have been carried out in both young and elderly adults and in males and females (139,150–152). While some of these studies are similar to previous ones in that they employed depletion/repletion design (139,150,152) and diets with high B$_6$ bioavailability, others have used diets more representative of the usual U.S. diet (151).

What is also different about some of these studies is that they have included additional measurements that may be indicative of intercellular function of PLP. Meydani et al. (184) examined the effect of different levels of vitamin B$_6$ (pyridoxine added to a low-B$_6$ food diet) on immune function. They observed that adequate immune function in elderly women was not achieved until 1.9 mg/day of vitamin B$_6$ was fed. Men required 2.88 mg/day to return function to baseline levels. In addition, several indices of vitamin B$_6$ status were measured. Based on when these values for these indices returned to predepletion levels, the requirement for vitamin B$_6$ was estimated to be 1.96 and 1.90 for men and women, respectively (also see discussion of this study below).

Kretsch et al. (152) fed four graded doses of vitamin B$_6$ to eight young women following a depletion diet (for 11–28 days). A variety of clinical, functional, and biochemical measures were conducted. No abnormal clinical evidence of vitamin B$_6$ deficiency was observed at the 0.5 mg/day intake. Based on this and other studies, less than 0.5 mg/day is needed to observe clinical signs of vitamin B$_6$ deficiency. Functional signs, such as abnormal EEGs, were only seen with an intake lower than 0.5 mg/day. Various biochemical measures, including the functional tests of tryptophan metabolite excretion (xanthurenic acid) and erythrocyte transaminase (EAST) stimulation, were normalized at the 1.5 and 2.0 mg/day level, respectively. The authors stated that if all currently used biochemical measures are to be normalized, then more than 0.020 mg of vitamin B_6 per gram of protein is required.

Hansen et al. (151) used a different approach in evaluating the effect of graded doses of vitamin B_6 on status. First, rather than feeding a diet deficient in vitamin B_6 a diet containing a level that is low but within the realm of what individuals might normally consume was fed. Various levels of pyridoxine (as an oral solution) were then added in addition to the basal diet (range 0.8–2.35 mg B_6/day). Based on both direct and indirect measures (including tryptophan metabolite excretion) it was concluded that a B_6/protein ratio greater than 0.20 was required to normalize all vitamin B_6 status indices. Ribaya-Mercado et al. (139) evaluated the vitamin B_6 requirements of elderly men and women in a depletion/repletion study. Individuals were fed 0.8 or 1.2 g/kg protein and increasing amounts of vitamin B_6. Several indices of vitamin B_6 status were measured and predepletion values used to evaluate when these indices were normalized. Men who ingested about 120 g of protein per day required 1.96 mg (for xanthurenic acid and PLP) to 2.88 mg (for EAST-AC and 4-PA excretion) of vitamin B_6 to normalize these indices. For women ingesting 78 g of protein per day, 1.90 mg/day of vitamin B_6 was needed to normalize the same indices. The authors concluded that the vitamin B_6 requirements of elderly men and women are about 1.96 and 1.90, respectively. The vitamin B_6 (pyridoxine) fed to these subjects was in a highly bioavailable form.

Another metabolic study in young women evaluated the requirement for vitamin B_6 (150). Again, a depletion/repletion design was used and several indices of vitamin B_6 status were measured. These indices included urinary 4-PA excretion, plasma PLP, erythrocyte PLP, and erythrocyte alanine aminotransferase (EALT) and erythrocyte aspartic aminotransferase (EAST) activity coefficients. Using predepletion baseline levels (after 9 days of feeding 1.60 mg/day) of these indices as a basis for comparison in determining adequacy, the amount of vitamin B_6 required to normalize these indices was 1.94 mg/ day (B_6 to protein ratio of 0.019 mg/day).

An important consideration relative to many of these metabolic studies that have been used in establishing the adult vitamin B_6 RDA is the composition of the diets used. Most diets were ones in which the amount of vitamin B_6 from food was low and of relatively high bioavailability. Vitamin B_6 was added back to the diets in the form of pyridoxine hydrochloride and thus is considered 100% bioavailable. Therefore, the total vitamin B_6 in the diets is probably 95–100% bioavailable. Taken together, these four recent metabolic studies support a higher vitamin B_6 requirement for women and men than is currently employed. A value of 1.9 mg/day for women and 2.2 mg/day for men is recommended. Since the vitamin B_6 in these studies was highly available, the inclusion of a factor for bioavailability would further increase the RDA (86). The studies that have been used in determining vitamin B_6 are summarized in Table 14 along with suggested values for requirement.

The above discussion has focused on the vitamin B_6 requirement for adults aged 18–70. There has been little research to support a statement of recommendations for children (aged 1–10) or adolescents (aged 11–18). Extrapolating from adults and taking into account muscle accretion (since this is the major body pool of vitamin B_6, a requirement for these populations could be estimated (119). Of the factors listed in Table 13, pregnancy and lactation are the ones in which additional needs could be estimated from the total vitamin B_6 content of the fetus and vitamin B_6 content of human milk, respectively. While the former is not entirely feasible except in nonhuman primates, the latter is possible because the vitamin B_6 content of human milk has been determined. Table 15 lists reported

Table 14 Studies in Which Vitamin B₆ Requirements Have Been Investigated

Ref.	No. of subjects	Sex	Age (years)	Wt. (kg)	B₆ (mg)	Diet. protein	Energy (kcal)	Food types[c]	Duration[a] Adj.	Duration[a] Defic.	Duration[a] Repletion	Tests[b]	Suggested requirements
Canham (205)	6	M	—	—	0 0	80 40	— —	SS SS	7(2)	28	14(10)	10g DL-Try, XA; SGOT; NME; Urinary B₆	NS
Baker (291)	5 6	M M			0.06 0.06	30 100		Liq. formula	7(4)	49 21	(1.5)	10g DL-Try; XA; Urinary B₆ EEG	1.0 1.5
Harding (76)	9	M	20–28	65–83	4.28 2.76 1.93	149 164 165	3800	SS AR(34) AR(100)	24 24 24			10 g DL-try, XA	1.97–2.76
Canham (277)	—	M	—	—	0.34	100	3000	SS		—	8	5 g L-Try, XA; EEG/EKG GPT (WB)	>0.35; 0.35 submarginal
Swan (292) Yess (293) Brown (294) Baysal (248)	6	M	24–35	61–82	0.16	100	2500–3000	SS	5-SSd	28–41	9–19(0.8 to 1.1)	5 g L-Cysteine; 2 g L-try; TM, NMe, QA; PLP	NS NS
Miller (278)	5	M	21–25	63–73	0.16	54	2800–3600	SS	6(1.66)	16	7(0.76)	2 g L-Try, TM, 4-PA: Urinary B₆	NS(0.8)
Kelsay (158)	6	M	19–31	63–71	0.16	150	2800–3600		18(1.66)	16	16(0.76)	2 g L-Try; TM; EEg; 4-PA; Urinary B₆; QA: NMe	
Cheslock (295)	7 1	F M	18–20 —	— —	0.41 0.50	26 36	2045 2724	Natural Natural	—	52	—	5 g L-Try, XA, Blood B₆	>0.5
Aly (296)	5	F	21–31	—	1.3	81	—	Liquid/solid	6	—	—	2 g L-Try; XA: Urinary B₆; Amino Acids	NS
Donald (297)	8	F	21–31	56	0.34	57	1300+	Natural	—	44	7(0.94) 3(1.54)	Urinary 4-PA, B₆; E-B₆ EGOT; Amino Acids	1.5
Shin (298)	5	F	23	49	0.16	109		SS	7(2.16)	14	14(2.16)	2 g L-Try; TM; Urinary B₆; 3 g L-Meth	NS

Table 14 Continued

Ref.	No. of subjects	Sex	Age (years)	Wt. (kg)	B6 (mg)	Diet. protein	Energy (kcal)	Food types[c]	Duration[a] Adj.	Duration[a] Defic.	Duration[a] Repletion	Tests[b]	Suggested requirements
Leklem (95, 200)	10	F	22	60	0.19	78	1992	SS	4	28	28(0.83)	2 g L-Try, TM; 4-PA; PLP; EGOT, EGPT; 3 g Meth; NMe	>0.83
Hansen (86)	5	F	29 ± 6	55.4 ± 10.1	0.96	72	—	Met	8			Urinary 4-PA; UB6, UPNG, PLP, E-B6 Fecal B6, EALT, EAST	Needs to take into account the PNG content of the diet; PNG ↓15–18% vitamin B6 bioav.
	4				1.52	90			18				
					0.96	72		High PNG	8				
					1.45	84		Low PNG	18				
Ribaya-Mercado (139)	6	M	61–70	94.8 ± 64	—	—	2755 ± 453	Natural	5	20		5 g L-Tryp Urinary 4-PA; PLP; East-AC; XA	1.90–1.96
	6	F	61–71	66.7 ± 4.1	0.003	0.8 g/kgbw	1918 ± 30	Liq. formula			21		
					0.015			SS			21		
					0.0225			SS			21		
					.03375			SS					
	6	M			—	—	3051 ± 232	Natural	5	20			
	6	F			0.003	1.2 g/kgbw	1980 ± 120	Liq. formula			21		
					0.015			SS			21		
					0.0225			SS			21		
					.03375			SS					
Meydani (184)	4	M	63.6 ± 0.8	94.8 ± 6.4	—	1.2 g/kgbw	—	Natural	5	20		5 g L-Try WBC; PLP XA; PBMCs; IL-2	>1.6 M: 2.88 ± 0.17 F: 1.90 ± 0.18
					0.17 ± 0.01		726.4 ± 55.3	Liq. formula			21		
					1.34 ± 0.08			SS			21		
								SS			21		
	4	F		66.7 ± 4.1	1.96 ± 0.11		—	Natural	5	20			
					2.88 ± 0.17		471.4 ± 28.5	Liq. formula			21		
					—						21		
					0.10 ± 0.01			SS			21		
					0.89 ± 0.08			SS			21		
					1.29 ± 0.12			SS					
					1.90 ± 0.18								

Reference	n	Sex	Age	Weight	B6 (mg)	Protein	Energy (kcal)	Food		Duration	Days	Tests	
Huang (150)	8	F	28–34	61.8 ± 5.4	1.60	1.55 g/kgbw	1576 ± 110	Met	9	27		Urinary 4-PA; PLP, PL; E-PLP, PL, PMP; EALT-AC, EAST-AC	1.94
					0.45						21		
					1.26						21		
					1.66						14		
					2.06								
Hansen (151)	10	F	27.5 ± 6.8	73.6 ± 23.2	1.03	85	2000	Met	15	No defic.		2 g L-Try; Urinary 4-PA, UBy; Fecal B6, PLP, TB6-E-PLP EALT, EAST; KA, XA, VA	1.33–1.73 (70.016 mg/g)
					1.33				12			
					1.73				12			
					2.39				12			
	6		28.2 ± 2.6	69.3 ± 12.8	0.84				12			
					1.14				10			
					2.34				10			
Kretsch (152)	4	F	21–30	—	Variable	—	38–42 kcal/kgbw	CF	4			4 g L-Try PLP, PAST, PALT; EAST, EALT; Urinary 4-PA, UTB6 UFB6, XA	0.015–0.020 mg/g protein
					2	animal protein		Formula	3			
					<0.05	1.55 g/kgbw		Formula				
					(2)			CF			14	
					0.5			CF			14	
					1.0 (0.5)			CF			21	
					1.5 (1.0)			CF			14	
					2.0 (1.5)							
	4	F			Variable	plant protein		CF	4	11–28		
					2	1.55 g/kg bw		Formula	3			
					<0.05 (2)			Formula				
					0.5			CF			14	
					1.0 (0.5)			CF			14	
					1.5 (1.0)			CF			21	
					2.0 (1.5)			CF			14	

a Food types: The types of foods used in the metabolic study. SS, semisynthetic; AR, army ration stored at 34°C (100°F).

b Duration refers to length of any adjustment (Adj) period, deficiency (Defic) period, or repletion period given as days. Values in parentheses are milligrams of vitamin B₆ fed.

c Tests: Biochemical tests used to evaluate vitamin B₆ status. Abbreviations: XA, xantherneic acid; SGOT, EGOT serum or erythrocyte glutamic oxalacetate transaminase; NMe, n-methylnicotinamide; EEG, electroencephalogram; EKG, electrocardiogram; GPT (WB), glutamic pyruvate transaminase in whole blood; TM, tryptophan metabolites; QA, quinolinic acid; PLP, pyridoxal-5'-phosphate; 4-PA, 4-pyridoxic acid; Try, tryptophan; Meth, methionine; PNG, pyridoxine glucoside, IL-2, interleukin-2. NS, none suggested by authors.

Table 15   Vitamin B_6 Content of Human Milk

Ref.	No. of subjects[a]	Stage of lactation[b]	Vitamin B_6 intake[c] (mg)	Vitamin B_6 content[d] (μmol/L)
Thomas et al. (282)	6	5–7 d	1.45	0.76
	7	5–7 d	5.69	1.3
	6	43–45 d	0.84	1.21
	7	43–45 d	5.11	1.40
Roepke and Kirksey (283)	9	3 d	2.19	0.05
	42	3 d	7.65	0.10
	9	14 d	2.19	0.24
	38	14 d	7.65	0.33
Sneed et al. (285)	7	5–7 d	1.52	0.72
	9	5–7 d	5.33	1.46
	7	43–45 d	1.41	0.70
	9	43–45 d	5.21	1.42
Styslinger and Kirksey (286)	6	77 d	2.0	0.55
	6	77 d	4.4	1.13
	6	77 d	11.3	1.46
	6	77 d	21.1	2.44
Barnji et al. (287)	27	6–30 d	ND	0.12
	26	7–12 mo	ND	0.42
	18	18 mo	ND	0.37
Borschel et al. (296)	8	1–6 mo	3.6	0.87–1.25
	9	1–6 mo	14.0	2.21–3.16

[a]No. of subjects in which milk samples assayed.
[b]Days after birth.
[c]Intake of mother.
[d]1 μmol/L = 169 μg/L.
ND, not determined.

values for the vitamin B_6 content of human milk and shows the relationship between vitamin B_6 intake and the change in content over time of lactation. While a gradual increase in vitamin B_6 content of milk in the first several months of lactation has been observed (288), a gradual decrease in content from 7 to 25 months of lactation has been reported (289). Intake of vitamin B_6 is reflected in milk vitamin B_6 concentration, especially at intakes above 5 mg/day. When considering the vitamin B_6 requirement for infants and lactating mothers, the recent publication by Borschel (290) provides a comprehensive review. Borschel recommends a vitamin B_6 intake greater than 10 mg/day (as PN-HCl) for lactating mothers. An infant's need in the first 112 days of life when growth is most rapid is estimated to be 0.12–0.16 mg of vitamin B_6 per day. Current requirements during the first year of life may be set too high according to Borschel. The estimate for vitamin B_6 during growth of the infant may be estimated based on the recent recommendation of Coburn, which is 0.02 nmol/g (119).

X.   DISEASE AND TOXICITY

Several books (2,3,7,8) and reviews (6) have examined the relationship between specific diseases and vitamin B_6 nutrition in detail. There are numerous diseases or pathological conditions in which vitamin B_6 metabolism is altered. The primary indicator of an alteration in vitamin B_6 metabolism has been in evaluation of tryptophan metabolism or the plasma PLP concentration. As previously discussed, the first of these is an indirect measure of status and the second is a direct measure. Furthermore, using only PLP as a measure really begs the question of whether vitamin B_6 metabolism is actually altered. Conditions in which tryptophan metabolism has been shown to be altered and in which vitamin B_6 (pyridoxine) administration was used include asthma (299), diabetes (300), certain cancers (173), pellagra (301), and rheumatoid arthritis (302). Diseases and pathological conditions in which plasma PLP levels have been shown to be depressed include asthma (303), diabetes (304), renal disorders (305), alcoholism (306), heart disease (307), pregnancy (132,144,156,308), breast cancer (309), Hodgkin's disease (310), and sickle cell anemia (194). Hypophosphatasia is an example of a condition in which plasma PLP levels are markedly elevated in some individuals (92). Relatively few of these studies have exhaustively evaluated vitamin B_6 metabolism.

Vitamin B_6 in the form of pyridoxine hydrochloride has been used as a therapeutic agent to treat a variety of disorders. Examples of disorders that have been treated with pyridoxine include Down's syndrome (311), autism (312), hyperoxaluria (313), gestational diabetes (314), premenstrual syndrome (315,316), carpal tunnel syndrome (317,318), depression (319), and diabetic neuropathy (329). It should be emphasized that the extent to which pyridoxine was effective in treating these diseases or reducing symptoms has been variable. In addition, in the treatment of these diseases, the amount of pyridoxine given often varied, as did the length of time over which the pyridoxine was given. These two variables, as well as the important consideration of whether a double-blind placebo-controlled design was used, are necessary considerations in evaluating the effectiveness of vitamin B_6 therapy (129). Reynolds (321) has reviewed the use of vitamin supplements, including vitamin B_6, for the treatment of various diseases.

A. Coronary Heart Disease

A relationship between vitamin B_6 and coronary heart disease can be viewed from both an etiological perspective and that of the effect of the disease state on vitamin B_6 metabolism. With respect to an etiological role, an altered sulfur amino acid metabolism has been suggested to result in vascular damage. A poor vitamin B_6 status can result in an increased circulating concentration of homocysteine (322). In the transsulfuration pathway, serine and homocysteine condense to produce cystathionine. This reaction is catalyzed by the PLP-dependent enzyme cystathionine β-synthase. In genetic disorders of this enzyme, homocysteine accumulates in the plasma (323). An increased incidence of arteriosclerosis has been associated with this enzyme defect (324). In addition, elevated levels of homocysteine in the plasma have been observed in people with ischemic heart disease (325–327). There has been an explosion in the number of papers, suggesting that elevated plasma homocysteine is a risk factor for heart disease and stroke. While folic acid is most effective in reducing the plasma concentration of homocysteine (327,328), vitamin B_6 has been shown to be most effective in reducing plasma homocysteine when a methionine load is given. A recent European study of 750 patients with vascular disease and 800 control subjects found that increased fasting homocysteine (more than 12.1 µmol/L) was associated with elevated risk of vascular disease (329). In this study, a plasma concentration of PLP below the 20th percentile (less than 23 nmol/L) for controls was associated with increased risk. This relationship between plasma PLP and atherosclerosis was independent of homocysteine levels. Other studies (326,328) have also found an increase in coronary artery disease risk and low PLP levels in plasma.

While some animal experiments have shown that rhesus monkeys made vitamin B_6-deficient develop atherosclerotic lesions (243,330), other studies did not reveal any pathological lesions (331). In humans at risk for coronary heart disease, a negative correlation between dietary vitamin B_6 and bound homocysteine has been observed (138). For some people with homocystinuria, treatment with high doses of vitamin B_6 reduces the plasma concentration of homocysteine in certain patients but does not totally correct methionine metabolism (332), especially when there is an increased methionine intake. Thus, if vitamin B_6 therapy is to be successful in reducing vascular lesions, diet modification with a lowered methionine intake may be necessary. The extent to which supplemental vitamin B_6 intake (beyond normal dietary intakes) may reduce the risk for coronary heart disease is not known.

A second aspect of coronary heart disease is the relationship between the presence of the disease and vitamin B_6 status. Several recent studies have found that the plasma PLP concentrations in people with coronary heart disease are significantly lower (21–41 mnol/L) than in healthy controls (32–46 mnol/L) (250,307,333,334). However, Vermaak et al. have found that the decrease in plasma PLP concentration is only seen in the acute phase of myocardial infarction (335). Unfortunately, other measures of vitamin B_6 status have not been evaluated in this disease. In one study (334), cardiac patients given vitamin B_6 supplements (amounts not given) resulted in plasma PLP levels well above normal. The effect of long-term vitamin B_6 therapy on recurrence of coronary artery disease has not been evaluated.

Elevated plasma cholesterol concentration has been strongly associated with an increased risk for coronary heart disease. As previously reviewed, vitamin B_6 may influence cholesterol metabolism. Serfontein and Ubbink (250) have found that use of a multivitamin supplement containing about 10 mg of pyridoxine for 22 weeks by hypercholesterolemic adult men resulted in a significant decrease in cholesterol levels, with most of the reduction due to a decreased level of LDL cholesterol. Smoking is an additional risk factor of coronary heart disease. Interestingly, smokers have decreased plasma levels of PLP (250,336). Evidence to date suggests a link between several risk factors for coronary heart disease and altered vitamin B_6 status and a potential beneficial effect of increased vitamin B_6 intake on cholesterol levels. Furthermore, well-controlled studies are needed before the therapeutic effect of vitamin B_6 can be evaluated for this disease.

B. HIV/AIDS

Vitamin B_6 status (337–340) and, to a limited extent, metabolism (341) has been examined in persons with human immunodeficiency virus (HIV). Because of the link between immune function and vitamin B_6 (185), one would expect that maintaining an adequate vitamin B_6 status is critical for HIV patients. Several studies have evaluated vitamin B_6 intake (339,342,343) and the progression of the disease as related to intake of nutrients, including vitamin B_6 (342). These studies generally found low intakes of vitamin B_6, and one study (344) reported an inverse relationship between vitamin B_6 intake and progression.

Biochemical assessment of vitamin B_6 status has been made in several studies (337–339,343) and been found to be low. In most of these studies (337–339), α-EAST activity and stimulation was used as an index of status. In the studies samples were frozen, which may have compromised the data and subsequent evaluation. Although other researchers have measured and reported low levels of "serum vitamin B$_6$," they failed to specify what form was being measured (343,344). Therefore, given the complexities of nutritional well-being in HIV/AIDS patients and methodological problems in these studies, it is difficult to assess the role of vitamin B$_6$ in HIV/AIDS.

In vitro studies suggest that PLP may play a role in HIV/AIDS. Salhany and Schopfer (345) found that PLP binds to the CD4 receptors at a site that is competitive with a known antiviral agent, (4,4'-diisothiocyanato-2; 2'-stilbenedisulfonate). Other investigators have found that PLP is a noncompetitive inhibitor of HIV-1 reverse transcriptase (346,347). Based on these in vitro studies, clinical trials with vitamin B$_6$ appear warranted.

C. Premenstrual Syndrome

Premenstrual syndrome (PMS) is another clinical situation for which vitamin B$_6$ supplementation has been suggested (348). Estimates of 40% of women being affected by this syndrome have been made (349). Using a wide variety of parameters, no difference in vitamin B$_6$ status was observed in women with PMS compared to those not reporting symptoms (172,350,351). Nevertheless, beneficial effects of B$_6$ administration on at least some aspects of PMS have been reported.

Treatment of PMS with vitamin B$_6$ has been based in part on the studies of Adams et al. (352) in which PN was used to manage the depression observed in some women taking oral contraceptives. Of the several studies in which PN was used to treat PMS, there have been open-type studies that were double-blind placebo-controlled. Open studies are prone to a placebo effect error, often as high as 40%. Of the well-controlled type, one study showed no effect of pyridoxine therapy (353), whereas three studies reported significant improvement of at least some of the symptoms associated with PMS. In one study, 21 of 25 patients improved (354). The other study found that about 60% of 48 women showed improvement with pyridoxine (200 mg/day) and 20% showed improvement with placebo (355). The fourth study (356) reported involvement in some symptoms in 55 women treated daily with 150 mg of pyridoxine. Brush (348) reported results of studies he has conducted using vitamin B$_6$ alone and vitamin B$_6$ plus magnesium. His data suggest that doses of 150–200 mg of vitamin B$_6$ are necessary before a significant positive effect is observed. In addition, the combination of vitamin B$_6$ plus magnesium appears to be beneficial. The complexity of PMS and the subjective nature of symptom reporting continue to result in contradictions and controversy in the lay and scientific literature. Kleijnen et al. (357) have reviewed 12 controlled trials in which vitamin B$_6$ was used to treat PMS. They concluded that there is only weak evidence of a positive effect of vitamin B$_6$. There may be a decrease in the availability of vitamin B$_6$ during PMS, possibly due to cell transport competition from fluctuating hormone concentrations. An increase in vitamin B$_6$ concentration could overcome competition and may explain the relief of symptoms seen in some women following high-dose vitamin B$_6$ supplementation.

D. Sickle Cell Anemia

Low levels (18 µmol/L) of plasma PLP have been reported in 16 persons with sickle cell anemia (194). Treatment of 5 of these patients with 100 mg of pyridoxine hydrochloride per day for 2 months resulted in a reduction of severity, frequency, and duration of painful crises in these persons. The mechanism by which vitamin B$_6$ acts is not known, but it may be related to pyridoxal and PLP binding to hemoglobin.

E. Asthma

Depressed levels of plasma and erythrocyte PLP have also been reported in persons with asthma (303). Of significance was the fact that all persons were receiving bronchodilators. Treatment of seven asthmatics with 100 mg of pyridoxine hydrochloride per day resulted in a reduction in the duration, occurrence, and severity of their asthmatic attacks. Subsequent work by one of these authors has not fully supported the earlier findings (358). Treatment of 15 asthmatics with vitamin B_6 did not result in a significant difference in symptom scores, medication usage, or pulmonary function tests as compared to placebo treatment. Ubbink et al. (359) have shown that theophylline lowers plasma and erythrocyte PLP. Pyridoxal kinase is inhibited by theophylline and was responsible for the decreased PLP level in the plasma and presumably intracellularly.

F. Carpal Tunnel Syndrome

At least five placebo-controlled trials from four different laboratories have shown that administration of PN relieved the symptoms of carpal tunnel syndrome (pain and/or numbness in hands) (317,360,361). In one study no significant improvement was observed (362). Since supplementation with vitamin B_6 well in excess of the RDA was required for improvement (generally 50–150 mg), it would seem that individuals with this disorder have a high metabolic demand or that the vitamin is active in some non-coenzyme role. Two recent studies examined the relationship between plasma PLP and carpal tunnel syndrome. One study (363) found no relationship between symptoms of carpal tunnel syndrome and plasma PLP, but a study by Keniston et al. (364) found a significant inverse univariate relationship between plasma PLP concentration and the prevalence of pain, the frequency of tingling, and nocturnal awakening.

G. Drug–Vitamin B_6 Interaction

Treatment of persons with various drugs may also compromise vitamin B_6 status and hence result in an increased need for vitamin B_6. Table 16 lists several drugs and their effect on vitamin B_6 status. Bhagavan has reviewed these interactions in detail (365). A common feature of these drug interactions is their adverse effect on central nervous system function. In addition, many of these drugs react with PLP via a Schiff base formation.

Table 16 Drug–Vitamin B_6 Interactions

Drug or drug	Examples	Mechanism of interaction
Hydrazines	Iproniazid, isoniazid, hydralazine	React with pyridoxal and PLP to form a hydrazone
Antibiotic	Cycloserine	Reacts with PLP to form an oxime
L-DOPA	L-3,4-dihydroxyphenylalanine	Reacts with PLP to form tetrahydroquinoline derivatives
Chelator	Penicillamine	Reacts with PLP to form thiazolidine
Oral contraceptives		Ethinyl estradiol, mestranol, increased enzyme levels in liver and other tissues; retention of PLP
Alcohol	Ethanol	Increased catabolism of PLP, low plasma levels

Table 17 Toxicity Symptoms Reported to Be Associated with Chronic Use of High-Dose Pyridoxine

Ref.	Symptoms
Coleman et al. (311)	Motor and sensory neuropathy; vesicular dermatosis on regions of the skin exposed to sunshine
Schaumburg (376)	Peripheral neuropathy; loss of limb reflexes; impaired touch sensation in limbs; unsteady gait; impaired or absent tendon reflexes; sensation of tingling that proceeds down neck and legs
Brush (348)	Dizziness; nausea; breast discomfort or tenderness
Bernstein (320)	Photosensitivity on exposure to sun

This reaction can result in decreased levels of PLP in tissues, such as the brain, leading to a functional deficiency. In most cases, supplemental vitamin B$_6$ reverses the adverse consequences of the drug. Oral contraceptives do not react directly with PLP but do induce enzyme synthesis. Some of these enzymes are PLP-dependent and as a result PLP is metabolically trapped in tissues. This may then lead to a depressed plasma PLP concentration (366). In addition, the synthetic estrogens specifically affect enzymes of the tryptophan–niacin pathway, resulting in abnormal tryptophan metabolism (200). There may be a need for extra vitamin B$_6$ above the current RDA in a small proportion of women using oral contraceptives and consuming low levels of vitamin B$_6$. Any drug that interacts with the reactive molecule PLP in a Schiff base reaction should be considered an instigator of resultant adverse effects on vitamin B$_6$ status and a subsequent negative influence on central nervous system function.

H. Hazards of High Doses

With the therapeutic use of pyridoxine for various disorders and self-medication has come the potential problem of toxicity. Shaumburg et al. have identified several individuals who developed a peripheral neuropathy associated with chronic high-dose use of pyridoxine (367). Subsequent to this, other reports of toxicity related to pyridoxine ingestion have been made (368,369). The minimal dose at which toxicity develops remains to be determined. Other toxicity symptoms have been identified. These symptoms and those reported by Schaumburg et al. are listed in Table 17. These symptoms are relatively rare, and the use of pyridoxine doses of 2–250 mg/day for extended periods of time appears to be safe (370).

In rats given high doses of pyridoxine hydrochloride for 6 weeks there was a decrease in testis epididymis and prostate gland at the 500 and 1000 mg/kg dose (371). There was also a decrease in mature spermatid counts. This high intake would be equivalent to 1.5–2.0 g of vitamin B$_6$ for a human (372). Thus, the application of these data to human nutrition is not clear.

XI. SUMMARY

In the more than 60 years since vitamin B$_6$ was elucidated, a great deal of information about its functional and metabolic characteristics has been gathered. The involvement of the active form of vitamin B$_6$, PLP, in such a wide spectrum of enzymatic reactions is an indication of the importance of this vitamin. In addition to the involvement of PLP in amino acid metabolism and carbohydrate metabolism, its reactivity with proteins points to the diversity of this vitamin. Further research is needed about the factors controlling the metabolism of vitamin B_6 and determination of vitamin B_6 needs of specific population. With knowledge of the functional properties of vitamin B_6 and quantitation of the metabolism of vitamin B_6 under various physiological and nutritional conditions, the health and well-being of individuals, can be improved.

REFERENCES

1. J. E. Leklem and R. D. Reynolds (eds.), *Methods in Vitamin B_6 Nutrition*, Plenum Press, New York, 1981.
2. G. P. Tryfiates (ed.), *Vitamin B_6 Metabolism and Role in Growth*, Food and Nutrition Press, Westport, CN, 1980.
3. R. D. Reynolds and J. E. Leklem (eds.), *Vitamin B_6: Its Role in Health and Disease*, Alan R. Liss, New York, 1985.
4. D. Dolphin, R. Poulson, and O. Avramovic (eds.), *Coenzymes and Cofactors*, Vol. 1. *Vitamin B_6 pyridoxal phosphate*, John Wiley and Sons, New York, 1986.
5. S. L. Ink and L. M. Henderson, Vitamin B_6 metabolism, *Annu. Rev. Nutr.*, 4:445–470 (1984).
6. A. H. Merrill, Jr. and J. M. Henderson, Diseases associated with defects in vitamin B_6 metabolism or utilization, *Annu. Rev. Nutr.*, 7:137–156 (1987).
7. J. E. Leklem and R. D. Reynolds (eds.), *Clinical and Physiological Applications of Vitamin B_6*. Alan R. Liss, New York, 1988.
8. D. J. Raiten (ed.), *Vitamin B_6 Metabolism in Pregnancy, Lactation and Infancy*, CRC Press, Boca Raton, FL, 1995.
9. E. R. Gruberg and S. A. Raymond, *Beyond Cholesterol*, St. Martin's Press, New York, 1981.
10. P. Gyorgy, Vitamin B_2 and the pellagra-like dermatitis of rats, *Nature*, 133:448–449 (1934).
11. P. Gyorgy, Crystalline vitamin B_6, *J. Am. Chem. Soc.*, 60:983–984 (1938).
12. S. Lepkovsky, Crystalline factor *1, Science*, 87:169–170 (1938).
13. R. Kuhn and Wendt, G. Uber. Das antidermatitische vitamin der hefe. *Ber. Deut. Chem. Ges.* 71B:780–782 (1938).
14. J. C. Keresztesy and J. R. Stevens, Vitamin B_6. *Proc. Soc. Exp. Biol. Med.*, 38:64–65 (1938).
15. A. Ichiba and K. Michi, Isolation of vitamin B_6, *Soc. Papers Inst. Phys. Chem. Res.* (Tokyo), 34:623–626 (1938).
16. S. A. Harris and K. Folkers, Synthesis of vitamin B_6, *J. Am. Chem. Soc.*, 61:1245–1247 (1939).
17. R. Kuhn, K. Westphal, G. Wendt, and O. Westphal, Synthesis of adermin, *Naturwissenschaften*, 27:469–470 (1939).
18. P. Gyorgy and R. E. Eckhardt, Vitamin B_6 and skin lesions in rats, *Nature*, 144:512 (1939).
19. E. E. Snell, Vitamin B_6 analysis: some historical aspects. In: J. E. Leklem and R. D. Reynolds (eds.), *Methods in Vitamin B_6 Nutrition*, Plenum Press, New York, 1981, pp. 1–19.
20. E. E. Snell, B. M. Guirard, and R. J. Williams, Occurrence in natural products of a physiologically active metabolite of pyridoxine, *J. Biol. Chem.*, 143:519–530 (1942).
21. E. E. Snell, The vitamin B_6 group. 1. Formation of additional members from pyridoxine and evidence concerning their structure, *J. Am. Chem. Soc.*, 66:2082–2088 (1944).
22. E. E. Snell, The vitamin activities of pyridoxal and pyridoxamine, *J. Biol. Chem.*, 154:313–314 (1944).
23. S. A. Harris, D. Heyl, and D. Folkers, The structure and synthesis of pyridoxamine and pyridoxal, *J. Biol. Chem.*, 154:315–316; *J. Am. Chem. Soc.*, 66:2089–2092 (1944).
24. IUPAC-IUB Commission on Biochemical Nomenclature. Nomenclature for vitamin B_6 and related compounds. *Eur. J. Biochem.*, 40:325–327 (1973).

25. C. Y. W. Ang, Stability of three forms of vitamin B$_6$ to laboratory light conditions, *J. Assoc. Off. Anal. Chem.*, 62:1170–1173 (1979).

26. C. A. Storvick, E. M. Benson, M. A. Edwards, and M. J. Woodring, Chemical and microbiological determination of vitamin B$_6$, *Methods Biochem. Anal.*, 12:183–276 (1964).

27. S. A. Harris, E. E. Harris, and R. W. Bukrg, Pyridoxine. *Kirk-Othmer Encycl. Chem. Tech.*, 16:806–824 (1968).

28. J. W. Bridges, D. S. Davies, and R. T. Williams. Fluorescence studies on some hydroxypyridines including compounds of the vitamin B$_6$ group, *Biochem. J.*, 98:451–468 (1966).

29. W. E. Schaltenbrand, M. S. Kennedy, and S. P. Coburn, Low-ultraviolet "white" fluorescent lamps fail to protect pyridoxal phosphate from photolysis, *Clin. Chem.*, 33:631 (1987).

30. R. C. Hughes, W. T. Jenkins, and E. H. Fischer, The site of binding of pyridoxal-5′-phosphate to heart glutamic-aspartic transaminase, *Proc. Natl. Acad. Sci. USA*, 48:1615–1618 (1962).

31. Leussing, D. L. Model reactions, in. D. Dolphin, R. Poulson, and O. Avramovic (eds.), *Coenzymes and Cofactors*, Vol. 1. *Vitamins B$_6$ Pyridoxal Phosphate*, John Wiley and Sons, New York, 1986, pp. 69–115.

32. H. E. Sauberlich, Interaction of vitamin B$_6$ with other nutrients. In: R. D. Reynolds and J. E. Leklem (eds.), *Vitamin B$_6$: Its Role in Health and Disease*, Alan R. Liss, New York, 1985, pp. 193–217.

33. J. T. Vanderslice, S. G. Brownlec, M. E. Cortissoz, and C. E. Maire, Vitamin B$_6$ analyses: sample preparation, extraction procedures, and chromatographic separations. In: A. P. Deleenheer, W. E. Lambert, and M. G. M. DeRuyter (eds.), *Modern Chromatographic Analysis of the Vitamins*, Vol. 30, Marcel Dekker, New York, 1985, pp. 436–475.

34. R. D. Reynolds, Vitamin B$_6$. In: A. J. Pesce and L. A. Kaplan (eds.), *Methods in Clinical Chemistry*, Mosby, Washington, D. C., 1987, pp. 558–568.

35. J. F. Gregory, Methods for determination of vitamin B$_6$ in foods and other biological materials: a critical review. *J. Fd. Comp. Analysis* 1:105–123 (1988).

36. C. A. Storvick and J. M. Peters, Methods for the determination of vitamin B$_6$ in biological materials. *Vitamins and Hormones*, 22:833–854 (1964).

37. C. J. Argouldelis, Simple high-performance liquid chromatographic method for the determination of all seven vitamin B$_6$ related compounds. *J. Chromatogr. A.*, 790:83–91 (1997).

38. B. Chabner and D. Livingston, A simple enzymic assay for pyridoxal phosphate. *Anal. Biochem.*, 34:413–423 (1964).

39. K. Tadera and Y. Naka, Isocratic paired-ion high-performance liquid chromatographic method to determine B$_6$ vitamers and pyridoxine glucoside in foods, *Agric. Biol. Chem.*, 55:563–564 (1991).

40. H. Kabir, J. E. Leklem, and L. T. Miller, Measurement of glycosylated vitamin B$_6$ in foods, *J. Food Sci.*, 48:1422–1425 (1983).

41. L. Atkin, A. S. Schultz, W. L. Williams, and C. N. Frey, Yeast microbiological methods for determination of pyridoxine, *Ind. Eng. Chem. Anal. Ed.*, 15:141–144 (1943).

42. M. Polansky, Microbiological assay of vitamin B$_6$ in foods. In: J. E. Leklem and R. D. Reynolds (eds.), *Methods in Vitamin B$_6$ Nutrition*, Plenum Press, New York, 1981, pp. 21–44.

43. M. L. Orr, *Pathothenic Acid, Vitamin B$_6$ and Vitamin B$_{12}$ in Foods*. Home Economics Research Report No. 36, U.S. Dept. of Agriculture, Washington, D.C., 1969.

44. K. Yasumoto, H. Tsuji, K. Iwanii, and H. Metsuda, Isolation from rice bran of a bound form of vitamin B$_6$ and its identification as 5′-*O*-(B.D.-glucopyranosyl) pyridoxine, *Agric. Biol. Chem.*, 41:1061–1067 (1977).

45. Y. Suzuki, Y. Inada, and K. Uchida, β-Glucosylpyridoxines in germinating seeds cultured in the presence of pyridoxine, *Phytochemistry*, 25:2049–2051 (1986).

46. K. Tadera, T. Kaneko, and F. Yagi, Evidence for the occurrence and distribution of a new type of vitamin B$_6$ conjugate in plant foods, *Agric. Biol. Chem.*, 50:2933–2934 (1986).

47. K. Tadera, E. Mori, F. Yagi, A. Kobayashi, K. Imada, and M. Imabeppu, Isolation and structure of a minor metabolite of pyridoxine in seedlings of *Pisum sativwn L, J. Nutr. Sci. Vitaminol*, 31:403–408 (1985).

48. C. H. Lushbough, J. M. Weichman, and B. S. Schweigert, The retention of vitamin B_6 in meat during cooking, *J. Nutr.*, 67:451–459 (1959).

49. F. W. Bemhart, E. D'Amato, and R. M. TomareUi, The vitamin B_6 activity of heat-sterilized milk, *Arch. Biochem. Biophys.*, 88:267–269 (1960).

50. L. R. Richardson, S. Wilkes, and S. J. Ritchey. Comparative vitamin B_6 activity of frozen, irradiated and heat-processed foods, *J. Nutr.*, 73:363–368 (1961).

51. H. N. Daoud, B. S. Luh, and M. W. Miller, Effect of blanching, EDTA and $NaHSO_3$ on color and vitamin B_6 retention in canned garbanzo beans, *J. Food Sci.*, 42:375–378 (1977).

52. J. Augustin, G. I. Nurousek, L. A. Tholen, and B. Bertehi, Vitamin retention in cooked, chilled and reheated potatoes, *J. Food Sci.*, 45:814–816 (1980).

53. J. Augustin, G. I. Marousek, W. E. Artz, and B. G. Swanson, Retention of some water soluble vitamins during home preparation of commercially frozen potato products, *J. Food Sci.*, 46: 1697–1700 (1981).

54. J. F. Gregory and J. R. Kirk, Assessment of roasting effects on vitamin B_6 stability and bioavailability in dehydrated food systems, *J. Food Sci.*, 43:1585–1589 (1978).

55. O. S. Abou-Fadel and L. T. Miller, Vitamin retention, color and texture in thermally processed green beans and Royal Ann cherries packed in pouches and cans, *J. Food Sci.*, 48: 920–923 (1983).

56. C. Y. W. Arig, Comparison of sample storage methods of vitamin B_6 assay in broiler meats, *J. Food Sci.*, 47:336–337 (1981).

57. M. J. Woodring and C. A. Storvick, Vitamin B_6 in milk: review of literature, *J. Assoc. Off. Agric. Chem.*, 43:63–80 (1960).

58. J. T. Vanderslice, S. R. Brownlee, and M. E. Cortissoy, Liquid chromatographic determination of vitamin B_6 in foods, *J. Assoc. Off. Anal. Chem.*, 67:999–1007 (1984).

59. E. DeRitter, Stability characteristics of vitamins in processed foods, *Food Tech.* (Jan) 48: 51–54 (1976).

60. A. D. Perera, J. E. Leklem, and L. T. Miller, Stability of vitamin B_6 during bread making and storage of bread and flour, *Cereal Chem.*, 56:577–580 (1979).

61. J. F. Gregory and J. R. Kirk. Interaction of pyridaxal and pyridoxal phosphate with peptides in a model food system during thermal processing, *J. Food Sci.*, 42:1554–1561 (1977).

62. J. F. Gregory, Effects of ε-pyridoxyllysine bound to dietary protein on the vitamin B_6 status of rats, *J. Nutr.*, 110:995–1005 (1980).

63. S. Yoshida, K. Hayashi, and T. Kawasaki, Pyridoxine transport in brush border membrane vesicles of guinea pig jejunum, *J. Nutr. Sci. Vitaminol.*, 27:311–317 (1981).

64. G. S. Heard and E. F. Annison, Gastrointestinal absorption of vitamin B_6 in the chicken (*Gallus domesticus*), *J. Nutr.*, 116:107–120 (1986).

65. H. A. Serebro, H. M. Solomon, J. H. Johnson, and T. R. Henrix, The intestinal absorption of vitamin B_6 compounds by the rat and hamster, *Bull. Johns Hopkins Hosp.*, 119:166–171 (1966).

66. H. M. Middleton, Uptake of pyridoxine hydrochloride by the rat jejunal mucosa in vitro, *J. Nutr.*, 107:126–131 (1977).

67. H. M. Middleton, Characterization of pyridoxal 5'-phosphate disappearance from in vivo perfused segments of rat jejunum, *J. Nutr.*, 112:269–275 (1982).

68. L. M. Henderson, Intestinal absorption of B_6 vitamers. In: R. D. Reynolds and J. E. Leklem (eds.), *Vitamin B_6: Its Role in Health and Disease*, Alan R. Liss, New York, 1985, pp. 22–23.

69. H. Mehansho, M. W. Hamm, and L. M. Henderson, Transport and metabolism of pyridoxal and pyridoxal phosphate in the small intestine of the rat, *J. Nutr.*, 109:1542–1551 (1979).

70. M. W. Hamm, H. Mehansho, and L. M. Henderson, Transport and metabolism of pyridox-

amine and pyridoxamine phosphates the small intestine of the rat, *J. Nutr.*, 109:1552–1559 (1979).

71. D. A. Roth-Maier, P. M. Zinner, and M. Kirchgessner, Effect of varying dietary vitamin B$_6$ supply on intestinal absorption of vitamin B$_6$, *Int. J. Vitam. Nutr. Res.*, 52:272–279 (1982).

72. H. M. Middleton, Uptake of pyridoxine by in vivo perfused segments of rat small intestine: a possible role for intracellular vitamin metabolism, *J. Nutr.*, 115:1079–1088 (1985).

73. T. Sakurai, T. Asakura, and M. Matsuda, Transport and metabolism of pyridoxine and pyridoxal in mice, *J. Nutr. Sci. Vitaminol.*, 33:11–19 (1987).

74. J. F. Gregory and S. L. Ink, The bioavailability of vitamin B$_6$. In: R. D. Reynolds and J. E. Leklem (eds.), *Vitamin B$_6$: Its Role in Health and Disease*, Alan R. Liss, New York, 1985, pp. 3–23.

75. J. E. Leklem, Bioavailability of vitamins: application of human nutrition. In: A. R. Dobemz, J. A. Milner, and B. S. Schweigert (eds.), *Foods and Agricultural Research Opportunities to Improve Human Nutrition*, University of Delaware, Newark, 1986, pp. A56–A73.

76. R. S. Harding, I. S. Plough, and T. E. Friedemann, The effect of storage on the vitamin B$_6$ content of packaged army ration with a note on the human requirement for the vitamin, *J. Nutr.*, 68:323–331 (1959).

77. E. W. Nelson, C. W. Burgin, and J. J. Cerda, Characterization of food binding of vitamin B$_6$, in orange juice, *J. Nutr.*, 107:2128–2134 (1977).

78. J. E. Leklem, L. T. Miller, A. D. Perera, and D. E. Peffers, Bioavailability of vitamin B$_6$ from wheat bread in humans, *J. Nutr.*, 110:1819–1828 (1980).

79. A. S. Lindberg, J. E. Leklem, and L. T. Miller, The effect of wheat bran on the bioavailability of vitamin B$_6$ in young men, *J. Nutr.*, 113:2578–2586 (1983).

80. C. Kies, S. Kan, and H. M. Fox, Vitamin B$_6$ availability from wheat, rice, corn brans for humans, *Nutr. Rep. Int.*, 30:483–491 (1984).

81. J. B. Tarr, T. Tamura, and E. L. R. Stokstad, Availability of vitamin B$_6$ and pantothenate in an average diet in man, *Am. J. Clin. Nutr.*, 34:1328–1337 (1981).

82. E. W. Nelson, H. J. R. Lane, and J. J. Cerda, Comparative human intestinal bioavailability of vitamin B$_6$ from a synthetic and a natural source, *J. Nutr.*, 106:1433–1437 (1976).

83. H. Kabir, J. E. Leklem, and L. T. Miller, Comparative vitamin B$_6$ bioavailability from tuna, whole wheat bread and peanut butter in humans, *J. Nutr.*, 113:2412–2420 (1983).

84. H. Kabir, J. E. Leklem, and L. T. Miller, Relationship of the glycosylated vitamin B$_6$ content of foods of vitamin B$_6$ bioavailability in humans, *Nutr. Rept. Int.*, 28:709–716 (1983).

85. N. D. Bills, J. E. Leklem, and L. T. Miller, Vitamin B$_6$ bioavailability in plant foods is inversely correlated with percent glycosylated vitamin B$_6$, *Fed. Proc.*, 46:1487 (abst.) (1987).

86. C. M. Hansen, J. E. Leklem, and L. T. Miller, Vitamin B$_6$ status indicators decrease in women consuming a diet high in pyridoxine glucoside, *J. Nutr.*, 126:2512–2418 (1996).

87. J. F. Gregory, P. R. Trumbo, L. B. Bailey, et al., Bioavailability of pyridoxine-5′-β-D-glucoside determined in humans by stable-isotopic methods, *J. Nutr.*, 121:177–186 (1991).

88. L. Lumeng, and T-K. Li, Mammalian vitamin B$_6$ metabolism: regulatory role of protein binding and the hydrolysis of pyridoxal 5′-phosphate in storage and transport. In: G. P. Tryflates (ed.), *Vitamin B$_6$ Metabolism and Role in Growth*, Food and Nutrition Press, Westport, CT, 1980, pp. 27–51.

89. L. Lumeng, R. E. Brashear, and T-K. Li, Pyridoxal 5′-phosphate in plasma: source, protein binding, and cellular transport, *J. Lab. Clin. Med.*, 84:334–343 (1974).

90. H. Wada, and E. E. Snell, The enzymatic oxidation of pyridoxine and pyridoxamine phosphates, *J. Biol. Chem.* 236:2089–2095 (1961).

91. D. B. McCormick, Two interconnected B vitamins: riboflavin and pyridoxine, *Physiol. Rev.*, 69:1170–1198 (1989).

92. S. P. Coburn and M. P. Whyte, Role of phosphatases in the regulation of vitamin B$_6$ metabolism in hypophosphatasia and other disorders. In: J. E. Leklem and R. D. Reynolds (eds.),

Clinical and Physiological Applications of Vitamin B₆, Alan R. Liss, New York, 1988, pp. 65–93.

93. T-K, Li, L. Lumeng, and R. L. Veitch, Regulation of pyridoxal 5′-phosphate metabolism in liver, *Biochem. Biophys. Res. Commun.*, 61:627–634 (1974).

94. A. H. Merrill, J. M. Henderson, E. Wang, B. W. McDonald, and W. J. Millikan, Metabolism of vitamin B₆ by human liver, *J. Nutr.*, 114:1664–1674 (1984).

95. R. R. Brown, D. P. Rose, J. E. Leklem, H. Linkswiler, and R. Arend, Urinary 4-pyridoxic acid, plasma pyridoxal phosphate and erythrocyte aminotransferase levels in oral contraceptive users receiving controlled intakes of vitamin B₆, *Am. J. Clin. Nutr.*, 28:10–19 (1975).

96. J. R. Wozenski, J. E. Leklem, and L. T. Miller, The metabolism of small doses of vitamin B₆ in men, *J. Nutr.*, 110:275–285 (1980).

97. T. D. Shultz and J. E. Leklem, Urinary 4-pyridoxic acid, urinary vitamin B₆ and plasma pyridoxal phosphate as measures of vitamin B₆ status and dietary intake of adults. In: J. E. Leklem and R. D. Reynolds (eds.), *Methods in Vitamin B₆ Nutrition*, Plenum Press, New York, 1981, pp. 297–320.

98. A. H. Merrill, J. M. Henderson, E. Wang, M. A. Codner, B. Hollins, and W. J. Millikan, Activities of the hepatic enzymes of vitamin B₆ metabolism for patients with cirrhosis, *Am. J. Clin. Nutr.*, 44:461–467 (1986).

99. A. H. Merrill and J. M. Henderson, Vitamin B₆ metabolism by human liver, *Ann N. Y. Acad. Sci.*, 585:110–117 (1990).

100. A. V. Lakshmi and M. S. Barnji, Tissue pyridoxal phosphate concentration and pyridoxamine phosphate oxidase activity in riboflavin deficiency in rats and man, *B. J. Nutr.*, 32:249–255 (1974).

101. S. M. Madigan, F. Tracey, H. McNulty, J. Eaton-Evans, J. Coulter, H. McCartney, and J. J. Strain, Riboflavin and vitamin B₆ intake and status and biochemical response to riboflavin supplementation in free-living elderly people, *Am. J. Clin. Nutr.*, 68:389–395 (1998).

102. G. M. Perry, B. B. Anderson, and N. Dodd, The effect of riboflavin on red-cell vitamin B₆ metabolism and globin synthesis, *Biomedicine*, 33:36–38 (1980).

103. S. P. Coburn, and J. D. Mahuren, A versatile cation-exchange procedure for measuring the seven major forms of vitamin B₆ in biological samples, *Anal. Biochem.*, 129:310–317 (1983).

104. L. Lumeng, T-K. Li, and A. Lui, The interorgan transport and metabolism of vitamin B₆. In: R. D. Reynolds and J. E. Leklem (eds.), *Vitamin B₆: Its Role in Health and Disease*, Alan R. Liss, New York, 1985, pp. 35–54.

105. B. Hollins and J. M. Henderson, Analysis of B₆ vitamers in plasma by reversed-phase column liquid chromatography, *J. Chromatogr.*, 380:67–75 (1986).

106. B. B. Anderson, Red-cell metabolism of vitamin B₆. In: G. P. Tryflates (ed.), *Vitamin B₆ Metabolism and Role in Growth*, Food and Nutrition Press, Westport, CT, 1980, pp. 53–83.

107. H. Mehansho and L. M. Henderson, Transport and accumulation of pyridoxine and pyridoxal by erythrocytes, *J. Biol. Chem.*, 255:11901–11907 (1980).

108. M. L. Fonda and C. W. Harker, Metabolism of pyridoxine and protein binding of the metabolites in human erythrocytes, *Am. J. Clin. Nutr.*, 35:1391–1399 (1982).

109. S. L. Ink, H. Mehansho, and L. M. Henderson, The binding of pyridoxal to hemoglobin, *J. Biol. Chem.*, 257:4753–4757 (1982).

110. R. E. Benesch, S. Yung, T. Suzuki, C. Bauer, and R. Benesch, Pyridoxal compounds as specific reagents for the α and β N-termini of hemoglobin, *Proc. Natl. Acad. Sci. USA*, 70:2595–2599 (1973).

111. S. L. Ink and L. M. Henderson, Effect of binding to hemoglobin and albumin on pyridoxal transport and metabolism, *J. Biol. Chem.*, 259:5833–5837 (1984).

112. B. M. Pogell, Enzymatic oxidation of pyridoxamine phosphate to pyridoxal phosphate in rabbit liver, *J. Biol. Chem.*, 232:761–766 (1958).

113. S. P. Coburn, P. J. Ziegler, D. L. Costill, et al., Response of vitamin B₆ content of muscle to changes in vitamin B₆ intake in men, *Am. J. Clin. Nutr.*, 53:1436–1442 (1991).

114. E. G. Krebs and E. H. Fischer, Phosphorylase and related enzymes of glycogen metabolism. In: R. S. Harris, I. G. Wool, and J. A. Lovaine (eds.), *Vitamins and Hormones*, Vol. 22, Academic Press, New York, 1964, pp. 399–410.

115. S. P. Coburn, D. L. Lewis, W. J. Fink, J. D. Mahuren, W. E. Schaltenbrand, and D. L. Costill, Estimation of human vitamin B₆ pools through muscle biopsies, *Am. J. Clin. Nutr.*, 48:291–294 (1988).

116. S. Johansson, D. Lindstedt, U. Register, and L. Wadstrom, Studies on the metabolism of labeled pyridoxine in man, *Am. J. Clin. Nutr.*, 18:185–196 (1966).

117. B. Shane, Vitamin B₆ and blood. In: Human Vitamin B₆ Requirements, *Natl. Acad. Sci.*, Washington, D.C., 1978, pp. 111–128.

118. D. P. Rose and R. R. Brown, The influence of sex and estrogens on liver kynureninase and kynurenine aminotransferase in the rat, *Biochem. Biophys. Acta*, 184:412–419 (1969).

119. S. P. Coburn, Location and turnover of vitamin B₆ pools and vitamin B₆ requirements of humans, *Ann. N. Y. Acad. Sci.*, 585:76–85 (1990).

120. L. Lumeng, M. P. Ryan, and T-K. Li, Validation of the diagnostic value of plasma pyridoxal 5′-phosphate measurements in vitamin B₆ nutrition of the rat, *J. Nutr.*, 108:545–553 (1978).

121. A. L. Black, B. M. Guirard, and E. E. Snell, Increased muscle phosphorylase in rats fed high levels of vitamin B₆, *J. Nutr.*, 107:1962–1968 (1977).

122. A. L. Black, B. M. Guirard, and E. E. Snell, The behavior of muscle phosphorylase as a reservoir for vitamin B₆ in the rat, *J. Nutr.*, 108:670–677 (1978).

123. L. E. Russell, P. J. Bechtel, and R. A. Easter, Effect of deficient and excess dietary vitamin B₆ on amino and glycogen phosphorylase activity and pyridoxal phosphate content in two muscles from postpubertal gilts, *J. Nutr.*, 115:1124–1135 (1985).

124. J. E. Leklem, Physical activity and vitamin B₆ metabolism in men and women: interrelationship with fuel needs. In: R. D. Reynolds and J. E. Leklem (eds.), *Vitamin B₆: Its Role in Health and Disease*, Alan R. Liss, New York, 1985, pp. 221–241.

125. J. E. Leklem and T. D. Shultz, Increased plasma pyridoxal 5′-phosphate and vitamin B₆ in male adolescents after a 4500-meter run, *Am. J. Clin. Nutr.*, 38:541–548 (1983).

126. M. Manore, J. E. Leklem, and M. C. Walter, Vitamin B₆ metabolism as affected by exercise in trained and untrained women fed diets differing in carbohydrate and vitamin B₆ content, *Am. J. Clin. Nutr.* 46:995–1004 (1987).

127. P. Crozier, L. Cordain, and D. Sampson, Exercise induced changes in plasma vitamin B₆ concentration do not vary with intensity, *Am. J. Clin. Nutr.*, 40:552–558 (1994).

128. H. E. Sauberlich, Vitamin B₆ status assessment: past and present. In: J. E. Leklem and R. D. Reynolds (eds.), *Methods in Vitamin B₆ Nutrition*, Plenum Press, New York, 1981, pp. 203–240.

129. J. E. Leklem and R. D. Reynolds, Challenges and direction in the search for clinical applications of vitamin B₆. In: J. E. Leklem and R. D. Reynolds (eds.), *Clinical and Physiological Applications of Vitamin B₆*, Alan R. Liss, New York, 1988, pp. 437–454.

130. J. E. Leklem, Vitamin B₆: a status report, *J. Nutr.*, 120:1503–1507 (1990).

131. J. E. Leklem and R. D. Reynolds, Recommendations for status assessment of vitamin B₆. In: J. E. Leklem and R. D. Reynolds (eds.), *Methods in Vitamin B₆ Nutrition*, Plenum Press, New York, 1981, pp. 389–392.

132. M. Wachstein, I. D. Kellner, and J. M. Orez, Pyridoxal phosphate in plasma and leukocytes of normal and pregnant subjects following B₆ load tests, *Proc. Soc. Exp. Biol. Med.*, 103:350–353 (1960).

133. C. S. Rose, P. Gyorgy, M. Butler, R. Andres, A. H. Norris, N. W. Shock, J. Tobin, M. Brin, and H. Spiegel, Age differences in vitamin B₆ status of 617 men, *Am. J. Clin. Nutr.*, 29:847–853 (1976).

134. S. F. Contractor and B. Shane, Blood and urine levels of vitamin B₆ in the mother and fetus before and after loading of the mother with vitamin B₆, *Am. J. Obstet. Gynecol.*, 107:635–640 (1970).

135. T. D. Shultz and J. E. Leklem, Effect of high dose ascorbic acid on vitamin B_6 metabolism, *Am. J. Clin. Nutr.*, 35:1400–1407 (1982).

136. L. T. Miller, J. E. Leklem, and T. D. Shultz, The effect of dietary protein on the metabolism of vitamin B_6 in humans, *J. Nutr.*, 115:1663–1672 (1985).

137. J. E. Leklem, Vitamin B_6 metabolism and function in humans. In: J. E. Leklem and R. D. Reynolds (eds.), *Clinical and Physiological Applications of Vitamin B_6*, Alan R. Liss, New York, 1988, pp. 1–26.

138. M. E. Swift and T. D. Shultz, Relationship of vitamins B_6 and B_{12} to homocysteine levels: risk for coronary heart disease, *Nutr. Rep. Int.*, 34:1–14 (1986).

139. J. D. Ribaya-Mercado, R. M. Russell, N. Sahyoun, F. D. Morrow, and S. N. Gershoff, Vitamin B_6 requirements of elderly men and women, *J. Nutr.*, 121:1062–1074 (1991).

140. L. Reinken and H. Gant, Vitamin B_6 nutrition in women with hyperemesis gravidarium during the first trimester of pregnancy, *Clin. Chem. Acta*, 55:101–102, (1974).

141. L. T. Miller, A. Johnson, E. M. Benson, and M. J. Woodring, Effect of oral contraceptives and pyridoxine on the metabolism of vitamin B_6 and on plasma tryptophan and α-amino nitrogen, *Am. J. Clin. Nutr.*, 28:846–853 (1975).

142. L. Lumeng, R. E. Cleary, and T.-K. Li, Effect of oral contraceptives on the plasma concentration of pyridoxal phosphate, *Am. J. Clin. Nutr.*, 27:326–333 (1974).

143. M. H. Brophy and P. K. Siiteri, Pyridoxal phosphate and hypertensive disorders of pregnancy, *Am. J. Obstet. Gynecol.*, 121: 1075–1079 (1975).

144. R. E. Cleary, L. Lumeng, and T.-K. Li, Maternal and fetal plasma levels of pyridoxal phosphate at term: adequacy of vitamin B_6 supplementation during pregnancy, *Am. J. Obstet. Gynecol.*, 121:25–28 (1975).

145. A. S. Prasad, K-Y. Lei, D. Oberleas, K. S. Moghissi, and J. C. Stryker, Effect of oral contraceptive agents on nutrients: II. Vitamins, *Am. J. Clin. Nutr.*, 28:385–391 (1975).

146. J. C. Guilland, B. Berekski-Regung, B. Lequeu, D. Moreau, and J. Klepping, Evaluation of pyridoxine intake and pyridoxine status among aged institutionalized people, *Int. J. Vitam Nutr. Res.*, 54:185–193 (1984).

147. C. M. Lee and J. E. Leklem, Differences in vitamin B_6 status indicator responses between young and middle-aged women fed constant diets with two levels of vitamin B_6, *Am. J. Clin. Nutr.*, 42:226–234 (1985).

148. J. A. Driskell and S. W. Moak, Plasma pyridoxal phosphate concentrations and coenzyme stimulation of erythrocyte alanine aminotransferase activities of white and black adolescent girls, *Am. J. Clin. Nutr.*, 43: 599–603 (1986).

149. J. B. Ubbink, W. J. Serfontein, P. J. Becker, and L. S. deVilliers, The effect of different levels of oral pyridoxine supplementation on plasma pyridoxal 5′-phosphate and pyridoxal levels and urinary vitamin B_6 excretion, *Am. J. Clin. Nutr.*, 45:75–85 (1987).

150. Y. C. Huang, W. Chen, M. A. Evans, M. E. Mitchell, and T. D. Shultz, Vitamin B_6 requirement and status assessment of young women fed a high-protein diet with various levels of vitamin B_6, *Am. J. Clin. Nutr.*, 67:208–220 (1998).

151. C. M. Hansen, J. E. Leklem, and L. T. Miller, Changes in vitamin B_6 status indicators of women fed a constant protein diet with varying levels of vitamin B_6, *Am. J. Clin. Nutr.*, 66: 1379–1387 (1997).

152. M. J. Kretsch, H. E. Sauberlich, J. H. Skala, and H. L. Johnson, Vitamin B_6 requirement and status assessment: young women fed a depletion diet followed by plant- or animal-protein diet with graded amounts of vitamin B_6, *Am. J. Clin. Nutr.*, 61:1091–1101 (1995).

153. A. Hamfelt and L. Soderhjelm, Vitamin B_6 and aging. In: J. E. Leklem and R. D. Reynolds (eds.), *Clinical and Physiological Applications of Vitamin B_6*, Alan R. Liss, New York, 1988, pp. 95–107.

154. A. K. Kant, P. B. Moser-Veillon, and R. D. Reynolds, Effect of age on changes in plasma, erythrocyte and urinary B_6 vitamers after an oral vitamin B_6 load, *Am. J. Clin. Nutr.*, 48: 1284–1290 (1988).

155. R. D. Reynolds, Biochemical Methods for status assessment. In: D. J. Raiten (ed.), *Vitamin B₆ Metabolism in Pregnancy, Lactation and Infancy*, CRC Press, Boca Raton, FL, 1995, pp. 41–59.

156. H. C. Barnard, J. J. Dekock, W. J. H. Vermaak, and G. M. Potgieter, A new perspective in the assessment of vitamin B₆ nutritional status during pregnancy in humans, *J. Nutr.*, 117: 1303–1306 (1987).

157. H. van den Berg and J. J. P. Bogaards, Vitamin B₆ metabolism in the pregnant rats: effect of progesterone on the (re)distribution in maternal vitamin B₆ stores, *J. Nutr.*, 117:1866–1874 (1987).

158. J. Kelsay, A. Baysal, and H. Linkswiler, Effect of vitamin B₆ depletion on the pyridoxal, pyridoxamine and pyridoxamine content of the blood and urine of men, *J. Nutr.*, 94:490–494 (1968).

159. D. Mikac-Devic and C. Tomanic, Determination of 4-pyridoxic acid in urine by a fluorimetric method, *Clin. Chem. Acta*, 38:235–238 (1972).

160. D. M. Dreon and G. E. Butterfield, Vitamin B₆ utilization in active and inactive young men, *Am. J. Clin. Nutr.*, 43:816–824 (1986).

161. E. A. Donald and T. R. Bosse, The vitamin in B₆ requirement in oral contraceptive users. II. Assessment by tryptophan metabolites vitamin B₆ and pyridoxic acid levels in urine, *Am. J. Clin. Nutr.*, 32:1024–1032 (1979).

162. W. J. H. Vennaak, H. C. Barnard, E. M. S. P. van Dalen, and G. M. Potgieter, Correlation between pyridoxal 5′-phosphate levels and percentage activation of aspartate aminotransferase enzyme in haemolysate and plasma drug in vitro incubation studies with different B₆ vitamers, *Enzyme*, 35:215–224 (1986).

163. H. E. Sauberlich, J. E. Canhain, E. M. Baker, N. Raica, and Y. F. Herman, Biochemical assessment of the nutritional status of vitamin B₆ in the human, *Am. J. Clin. Nutr.*, 25:629–642 (1972).

164. L. Lumeng, R. E. Cleary, R. Wagner, P-L. Yu, and T-K. Li, Adequacy of vitamin B₆ supplementation during pregnancy: a prospective study, *Am. J. Clin. Nutr.*, 29:1376–1383 (1976).

165. J. A. Driskell, A. J. Clark, and S. W. Moak, Longitudinal assessment of vitamin B₆ status in southern adolescent girls, *J. Am. Diet. Assoc.*, 87:307–310 (1987).

166. B. Shane and S. F. Contractor, Assessment of vitamin B₆ status. Studies on pregnant women and oral contraceptive users, *Am. J. Clin. Nutr.*, 28:739–747 (1975).

167. A. D. Cinnamon and J. R. Beaton, Biochemical assessment of vitamin B₆ status in man, *Am. J. Clin. Nutr.*, 23:696–702 (1970).

168. A. Kirksey, K. Keaton, R. P. Abernathy, and J. L. Greger, Vitamin B₆ nutritional status of a group of female adolescents, *Am. J. Clin. Nutr.*, 31:946–954 (1978).

169. J. B. Ubbink, S. Bisshort, I. van den Berg, L. S. deVilliers, and P. J. Becker, Genetic polymorphism of glutamate-pyruvate transaminase (alanine aminotransferase): influence on erythrocyte activity as a marker of vitamin B₆ nutritional status, *Am. J. Clin. Nutr.*, 50:1420–1428 (1989).

170. J. E. Leklem, Quantitative aspects of tryptophan metabolism in humans and other species: a review, *Am. J. Clin. Nutr.*, 24:659–671 (1971).

171. R. R. Brown, The tryptophan load test as an index of vitamin B₆ nutrition. In: J. E. Leklem and R. D. Reynolds (eds.), *Methods in Vitamin B₆ Nutrition*, Plenum Press, New York, 1985, pp. 321–340.

172. H. van den Berg, E. S. Louwerse, H. W. Bruinse, J. T. N. M., Thissen, and J. Schrijver, Vitamin B₆ status of women suffering from premenstrual syndrome, *Hum. Nutr. Clin. Nutr.*, 40C:441–450 (1986).

173. R. R. Brown, Possible role of vitamin B₆ in cancer prevention and treatment. In: J. E. Leklem and R. D. Reynolds (eds.), *Clinical and Physiological Applications of Vitamin B₆*, Alan R. Liss, New York, 1988, pp. 279–301.

174. D. A. Bender, Oestrogens and vitamin B_6 actions and interactions, RU. *Rev. Nutr. Diet*, 51: 140–188 (1987).

175. H. M. Linkswiler, Methionine metabolite excretion as affected by a vitamin B_6 deficiency. In: J. E. Leklem and R. D. Reynolds (eds.), *Methods in Vitamin B_6 Nutrition*, Plenum Press, New York, 1981, pp. 373–381.

176. M. J. Kretsch, H. E. Sauberlich, and E. Newbrun, Electoencephalographic changes and periodontal status during short-term vitamin B_6 depletion of young nonpregnant women, *Am. J. Clin. Nutr.*, 53:1266–1274 (1991).

177. H. E. Sauberlich, Section IX. Biochemical systems and biochemical detection of deficiency. In: W. H. Sebrell and R. S. Harris (eds.), *The Vitamins: Chemistry, Physiology, Pathology, Assay*, 2nd ed., Vol. 2, Academic Press, New York, 1968, pp. 44–80.

178. L. Schirch and W. T. Jenkins, Serine transhydroxymethylase, *J. Biol. Chem.*, 239:3797–3800 (1964).

179. A. E. Axelrod and A. C. Trakatelles, Relationship of pyridoxine to immunological phenomena, *Vitamin. Horm.*, 22:591–607 (1964).

180. R. K. Chandra and S. Puri, Vitamin B_6 modulation of immune responses and infection. In: R. D. Reynolds and J. E. Leklem (eds.), *Vitamin B_6: Its Role in Health and Disease*, Alan R. Liss, New York, 1985, pp. 163–175.

181. L. C. Robson and M. R. Schwarz, Vitamin B_6 deficiency and the lymphoid system. I. Effect of cellular immunity and in vitro incorporation of ^{3}H-uridine by small lymphocytes, *Cell immunol.*, 16:135–144 (1975).

182. M. C. Talbott, L. T. Miller, and N. I. Kerkvliet, Pyridoxine supplementation: effect on lymphocyte responses in elderly persons, *Am. J. Clin. Nutr.*, 46:659–664 (1987).

183. H. van den Berg, J. Mulder, S. Spanhaak, W. van Dokkum, and T. Ockhuizen, The influence of marginal vitamin B_6 status on immunological indices. In: J. E. Leklem and R. D. Reynolds (eds.), *Clinical and Physiological Applications of Vitamin B_6* Alan R. Liss, New York, 1988, pp. 147–155.

184. S. N. Meydani, J. D. Ribaya-Mercado, R. M. Russell, et al. Vitamin B_6 deficiency impairs interleukin-2 production and lymphocyte proliferation in elderly adults, *Am. J. Clin. Nutr.*, 53:1275–1280 (1991).

185. L. C. Rall and S. N. Meydani, Vitamin B_6 and immune competence, *Nutr. Rev.*, 51:217–225 (1993).

186. J. F. Angel, Gluconeogenesis in meal-fed, vitamin B_6 deficient rats, *J. Nutr.*, 110:262–269 (1980).

187. D. P. Rose, J. E. Leklem, R. R. Brown, and H. M. Linkswiler, Effect of oral contraceptives and vitamin B_6 deficiency on carbohydrate metabolism, *Am. J. Clin. Nutr.*, 28:872–878 (1975).

188. J. F. Angel and R. M. Mellor, Glycogenesis and gluconeogenesis in meal-fed pyridoxine-deprived rats, *Nutr. Rep. Int.*, 9:97–107 (1974).

189. P. E. Butler, E. J. Cookson, and R. J. Beyon, The turnover and skeletal muscle glycogen phosphorylase studied using the cofactor, pyridoxal phosphate, as a specific label, *Biochem. Biophys.*, A847:316–323 (1985).

190. C. A. Lau-Cam, K. P. Thadikonda, and B. F. Kendall, Stimulation of rat liver glucogenolysis by vitamin B_6: A role for adrenal catecholamines, *Res. Commun. Chem. Path. Pharm.*, 73: 197–207 (1991).

191. J. A. Kark, R. Bongiovanni, C. U. Hicks, G. Tarassof, J. S. Hannah, and G. Y. Yoshida, Modification of intracellular hemoglobin with pyridoxal and pyridoxal 5'-phosphate, *Blood Cells*, 8:299–314 (1982).

192. R. Benesch, R. E. Benesch, R. Edalji, and T. Suzuki, 5'-Deoxypyridoxal as a potential antisickling agent, *Proc. Natl. Acad. Sci. USA*, 74:1721–1723 (1977).

193. N. Maeda, K. Takahashi, K. Aono, and T. Shiga, Effect of pyridoxal 5'-phosphate on the oxygen affinity of human erythrocytes, *Br. J. Haematol*, 34:501–509 (1976).

194. R. D. Reynolds and C. L. Natta, Vitamin B$_6$ and sickle cell anemia. In: R. D. Reynolds and J. E. Leklem (eds.), *Vitamin B$_6$: Its Role and in Health and Disease*, Alan R. Liss, New York, 1985, pp. 301–306.

195. G. Kikuchi, A. Kumar, and P. Talmage, The enzymatic synthesis of γ-aminolevulinic acid, *J. Biol. Chem.*, 233:1214–1219 (1958).

196. S. S. Bottomley, Iron and vitamin B$_6$ metabolism in the sideroblastic anemias. In: J. L. Lindenbaum (ed.), *Nutrition in Hematology*, Churchill Livingstone, New York, 1983, pp. 203–223.

197. J. W. Harris, R. M. Wittington, R. Weisman, Jr., and D. L. Horrigan, Pyridoxine responsive anemia in the human adult, *Proc. Soc. Exp. Biol. Med.*, 91:427–432 (1956).

198. D. L. Horrigan and J. W. Harris, Pyridoxine responsive anemia in man, *Vitam. Horm.*, 26: 549–568 (1968).

199. A. V. O. Pasanen, M. Salmi, R. Tenhunen, and P. Vuopio, Haem synthesis during pyridoxine therapy in two families with different types of hereditary sideroblastic anemia, *Ann. Clin. Res.*, 14:61–65 (1982).

200. J. E. Leklem, R. R. Brown, D. P. Rose, H. Linkswiler, and R. A. Arend, Metabolism of tryptophan and niacin in oral contraceptive users receiving controlled intakes of vitamin B$_6$, *Am. J. Clin. Nutr.* 28:146–156 (1975).

201. K. Dakshinamurti, Neurobiology of pyridoxine. In: H. H. Draper (ed.), *Advances in Nutritional Research*, Vol. 4, Plenum Press, New York, 1982, pp. 143–179.

202. D. B. Coursin, Convulsive seizures in infants with pyridoxine-deficient diet, *J. Am. Med. Assoc.*, 154:406–408 (1954).

203. C. J. Maloney and A. H. Parmalee, Convulsions in young infants as a result of pyridoxine deficiency, *J. Am. Med. Assoc.*, 154:405–406 (1954).

204. M. G. Alton-Mackey and B. L. Walker, Graded levels of pyridoxine in the rat during gestation and the physical and neuromotor development of offspring, *Am. J. Clin. Nutr.*, 26:420–428 (1973).

205. M. C. Stephens, V. Havlicek, and K. Dakshinamurti, Pyridoxine deficiency and development of the central nervous system in the rat, *J. Neurochem.*, 18:2407–2416 (1971).

206. D. A. Bender, B vitamins in the nervous system, *Neurochem. Int.* 6:297–321 (1984).

207. T. R. Guilarte, The role of vitamin B$_6$ in central nervous system development: neurochemistry and behavior. In: D. J. Raiten (ed.) *Vitamin B$_6$ metabolism in Pregnancy, Lactation, and Infancy*, CRC Press, Boca Raton, FL, 1995, pp. 77–92.

208. A. D. Hunt, J. Stokes, W. W. McCrory, and H. H. Stroud, Pyridoxine dependency: report of a case of intractable convulsions in an infant controlled by pyridoxine, *Pediatrics*, 13: 140–145 (1954).

209. C. R. Scriver, Vitamin B$_6$ dependency and infant convulsions, *Pediatrics*, 26:62–74 (1960).

210. R. Garry, Z. Yonis, J. Brahain, and K. Steinitz, Pyridoxine-dependent convulsions in an infant, *Arch. Dis. Child*, 37:21–24 (1962).

211. K. Iinuma, K. Narisawa, N. Yamauchi, T. Yoshida, and T. Mizuno, Pyridoxine dependent convulsion: effect of pyridoxine therapy on electroencephalograms, *Tokohu J. Exp. Med.*, 105:19–26 (1971).

212. A. Baniker, M. Turner, and I. J. Hopkins, Pyridoxine dependent seizures—a wider clinical spectrum, *Arch. Dis. Child.*, 58:415–418 (1983).

213. F. Goutieres and J. Aicardi, A typical presentations of pyridoxine-dependent seizures: a treatable cause of intractable epilepsy in infants, *Ann. Neurol.*, 17:117–120 (1985).

214. J. E. Canhwn, E. M. Baker, R. S. Harding, H. E. Sauberlich, and I. C. Plough, Dietary protein: its relationship to vitamin B$_6$ requirements and function, *Ann. N. Y. Acad. Sci.*, 166:16–29 (1969).

215. J. D. Grabow and H. Linkswiler, Electroencephalographic and nerve-conduction studies in experimental vitamin B$_6$ deficiency in adults, *Am. J. Clin. Nutr.*, 22:1429–1434 (1969).

216. J. E. Aycock and A. Kirksey, Influence of different levels of dietary pyridoxine on certain parameters of developing and mature brains in rats, *J. Nutr.*, 106:680–688 (1976).

217. M. R. Thomas and A. Kirksey, A postnatal patterns of fatty acids in brain of progeny for vitamin B_6 deficient rats before and after pyridoxine supplementation, *J. Nutr.*, 106:1415–1420 (1976).

218. D. J. Kurtz, H. Levy, and J. N. Kanfer, Cerebral lipids and amino acids in the vitamin B_6 deficient suckling rat, *J. Nutr.*, 102:291–298 (1972).

219. D. M. Morre, A. Kirksey, and G. D. Das, Effects of vitamin B_6 deficiency on the developing central nervous system of the rat. Gross measurements and cytoarchitectural alterations, *J. Nutr.*, 108:1250–1259 (1978).

220. D. M. Morre, A. Kirksey, and G. D. Das, Effects of vitamin B_6 on the developing central nervous systems of the rat. Myelination, *J. Nutr.*, 108:1260–1265 (1978).

221. S-J., Chang, A. Kirksey, and D. M. Morre, Effects of vitamin B_6 deficiency on morphological changes in dendritic trees of Purkinje cells in developing cerebellum of rats, *J. Nutr.*, 111: 848–857 (1981).

222. A. Wasynczuk, A. Kirksey, and D. M. Morre, Effects of vitamin B_6 deficiency on specific regions of developing rat brain: the extrapyramidal motor system, *J. Nutr.*, 113:746–754 (1983).

223. A. Wasynczuk, A. Kirksey, and D. M. Morre, Effect of maternal vitamin B_6 deficiency on specific regions of developing rat brain: amino acid metabolism, *J. Nutr.*, 113:735–745 (1983).

224. S. Groziak, A Kirksey, and B. Hamaker, Effect of maternal vitamin B_6 restriction on pyridoxal phosphate concentrations in developing regions of the central nervous system in rats, *J. Nutr.*, 114:727–732 (1984).

225. J. F. Mueller, Vitamin B_6 in fat metabolism, *Vitam. Horm.*, 22:787–796 (1964).

226. T. W. Birch, The relations between vitamin B_6 and the unsaturated fatty acid factor, *J. Biol. Chem.*, 124:775–793 (1938).

227. E. W. McHenry and G. Gauvin, The B vitamins and fat metabolism, 1. Effects of thiamine, riboflavin and rice polish concentrate upon body fat, *J. Biol. Chem.*, 125:653–660 (1938).

228. A. Audet and P. J. Lupien, Triglyceride metabolism in pyridoxine-deficient rats, *J. Nutr.*, 104:91–100 (1974).

229. M. Abe and Y. Kishino, Pathogenesis of fatty liver in rats fed a high protein diet without pyridoxine, *J. Nutr.*, 112:205–210 (1982).

230. D. J. Sabo, R. P. Francesconi, and S. N. Gershoff, Effect of vitamin B_6 deficiency on tissue dehydrogenases and fat synthesis in rats, *J. Nutr.*, 101:29–34 (1971).

231. H. S. R. Desikachar and E. W. McHenry, Some effects of vitamin B_6 deficiency on fat metabolism in the rat, *Biochem. J.*, 56:544–547 (1954).

232. J. F. Angel, Lipogenesis by hepatic and adipose tissues from meal-fed pyridoxine-deprived rats, *Nutr. Rept. Int.*, 11:369–378 (1975).

233. J. F. Angel and G-W. Song, Lipogenesis in pyridoxine-deficient nibbling and meal-fed rats, *Nutr. Rept. Int.*, 8:393–403 (1973).

234. P. W. Witten and R. T. Holman, Polyethenoid fatty acid metabolism, VI. Effect of pyridoxine on essential fatty acid conversions, *Arch. Biochem. Biophys.*, 41:266–273 (1952).

235. S. C. Cunnane, M. S. Manku, and D. F. Horrobin, Accumulation of linoleic and γ-linolenic acids in tissue lipids of pyridoxine-deficient rats, *J. Nutr.*, 114:1754–1761 (1984).

236. Q. B. She, T. Hayakawa, and H. Tsuge, Effect of vitamin B_6 deficiency on linoleic acid desaturation in arachidonic acid biosynthesis of rat liver microsomes, *Biosci. Biochem.*, 58: 459–463 (1994).

249. J. E. Fincnam, M. Faber, M. J. Weight, D. Labadarious, J. J. F. Taljaard, J. G. Steytler, P. Jacobs, and D. Kritchevsky, Diets realistic for westernized people significantly effect lipoproteins, calcium, zinc, vitamin-C, vitamin-E, vitamin B_6 and hematology in vervet monkeys, *Atherosclerosis*, 66:191–203 (1987).

250. W. J. Serfontein and J. B. Ubbink, Vitamin B₆ and myocardial infarction. In: J. E. Leklem and R. D. Reynolds (eds.), *Clinical and Physiological Applications of Vitamin B₆*, Alan R. Liss, New York, 1988, pp. 201–217.

251. L. F. Major and P. F. Goyer, Effects of disulfiram and pyridoxine on serum cholesterol, *Ann. Intern. Med.*, 88:53–56 (1978).

252. Y-O. Cho and J. E. Leklem, In vivo evidence of vitamin B₆ requirement in carnitine synthesis, *J. Nutr.*, 120:258–265 (1990).

253. G. Litwack, A. Miller-Diener, D. M. DiSorbo, and T. J. Schmidt, Vitamin B₆ and the glucocorticoid receptor. In: R. D. Reynolds and J. E. Leklem (eds.), *Vitamin B₆: Its Role in Health and Diseases*, Alan R. Liss, New York, 1985, pp. 177–191.

254. J. A. Cidlowski and J. W. Thanassi, Pyridoxal phosphate: a possible cofactor in steroid hormone action, *J. Steroid Biochem.*, 15:11–16 (1981).

255. D. B. Tully, V. E. Allgood, and J. A. Cidlowski, Modulation of steroid receptor–mediated gene expression by vitamin B₆, *FASEB J.*, 8:343–349 (1994).

256. R. Brandsch, Regulation of gene expression by cofactors derived from B vitamins, *J. Nutr. Sci. Vitaminol.*, 40:371–399 (1994).

257. M. M. Compton and J. A. Cidlowski, Vitamin B₆ and glucocorticoid action, *Endocr. Rev.*, 7:140–148 (1986).

258. T. G. Muldoon and J. A. Cidlowski, Specific modification of rat uterine estrogen receptor by pyridoxal 5′-phosphate, *J. Biol. Chem.*, 255:3100–3107 (1980).

259. R. A. Hiipakka and S. Liao, Effect of pyridoxal phosphate on the androgen receptor from rat prostate: inhibition of receptor aggregation and receptor binding to nuclei and to DNA cellulose, *J. Steroid Biochem.*, 13:841–846 (1980).

260. H. Nishigori, V. K. Moudgil, and D. Taft, Inactivation of avian progesterone receptor binding to ATP-sepharose by pyridoxal 5′-phosphate, *Biochem. Biophys. Res. Commun.*, 80:112–118 (1978).

261. D. M. DiSorbo, D. S. Phelps, V. S. Ohl, and G. Litwack, Pyridoxine deficiency influences the behavior of the glucocorticoid receptor complex, *J. Biol. Chem.*, 255:3866–3870 (1980).

262. V. E. Allgood, F. E. Powell-Oliver, and J. A. Cidlowski, Vitamin B₆ influences glucocorticoid receptor–dependent gene expression, *J. Biol. Chem.*, 265:12424–12433 (1990).

263. J. Holley, D. A. Bender, W. F. Coulson, and E. K. Symes, Effects of vitamin B₆ nutritional status on the uptake of (³H) oestradiol into the uterus, liver and hypothalamus of the rat, *J. Steroid. Biochem.*, 18:161–165 (1983).

264. G. E. Bunce and M. Vessal, Effect of zinc and/or pyridoxine deficiency upon oestrogen retention and oestrogen receptor distribution in the rat uterus, *J. Steroid. Biochem.*, 26:303–308 (1987).

265. J. A. Sturman and L. T. Kremzner, Regulation of ornithine decarboxylase synthesis: effect of a nutritional deficiency of vitamin B₆, *Ufe Sci.*, 14:977–983 (1974).

266. D. M. DiSorbo and G. Litwack, Changes in the intracellular levels of pyridoxal 5′-phosphate affect the induction of tyrosine aminotransferase by glucocorticoids, *Biochem. Biophys. Res. Commun.*, 99:1203–1208 (1981).

267. V. E. Allgood, and J. A. Cidlowski, Vitamin B₆ modulates transcriptional activation by multiple members of the steroid hormone receptor superfamily, *J. Biol. Chem.*, 267:3819–3824 (1992).

268. T. Oka, N. Komori, M. Kuwahata, M. Okada, and Y. Natori, Vitamin B₆ modulates expression of albumin gene by inactivating tissue-specific DNA-binding protein in rat liver, *Biochem. J.*, 309:243–248 (1995).

269. T. Oka, N. Komori, M. Kuwahata, et al. Pyridoxal 5′-phosphate modulates expression of cytosolic aspartate aminotransferase gene by inactivation of glucocorticoid receptor, *J. Nutr. Sci. Vitaminol*, 41:363–375 (1995).

270. T. Oka, M. Kuwahata, H. Sugitatsu, H. Tsuge, K. Asagi, H. Kohri, S. Horiuchi, and Y. Natori, Modulation of albumin gene expression by amino acid supply in rat liver is mediated

through intracellular concentration of pyridoxal 5'-phosphate, *J. Nutr. Biochem.*, 8:211–216 (1997).

271. C. M. Hansen and J. E. Leklem, Vitamin B$_6$ status and requirements of women of childbearing age. In: D. J. Paiten (ed.), *Vitamin B$_6$ Metabolism in Pregnancy, Lactation, and Infancy*, CRC Press, Boca Raton, FL, 1995, pp. 41–59.

272. J. A. Driskell, Vitamin B$_6$ requirements of humans, *Nutr. Res.*, 14:293–324 (1994).

273. *Human Vitamin B$_6$ Requirements*, Natl. Acad. Sci. Natl. Res. Council, Washington, D.C., 1978.

274. *Recommended Dietary Allowances*, 8th ed., Natl. Acad. Sci.-Natl. Res. Council, Washington, D.C., 1974.

275. *Recommended Dietary Allowances*, 9th ed., Natl. Acad. Sci.-Natl. Res. Council, Washington, D.C., 1980.

276. *Recommended Dietary Allowances*, 10th ed., Natl. Acad. Sci.-Natl. Res. Council, Washington, D.C., 1989.

277. J. E. Canharn, E. M. Baker, N. Raica, Jr., and H. E. Sauberlich, Vitamin B$_6$ requirement of adult men. In: *Proc. 7th Int. Congress Nutr.*, Hamburg 1966, Vol. 5: *Physiology and Biochemistry of food Components*, Pergamon Press, New York, 1967, pp. 555–562.

278. L. T. Miller and H. Linkswiler, Effect of protein intake on the development of abnormal tryptophan metabolism by men during vitamin B$_6$ depletion, *J. Nutr.*, 93:53–67 (1967).

279. Y. K. Park and H. Linkswiler, Effect of vitamin B$_6$ depletion in adult man on the excretion of cystathionine and other methionine metabolites, *J. Nutr.* 100:110–116 (1970).

280. H. Linkswiler, Vitamin B$_6$ requirements of men. In: *Human Vitamin B$_6$ Requirements*. Natl. Acad. Sci., Washington, D.C., 1978, pp. 279–290.

281. E. A. Donald, L. D. McBean, M. H. W. Simpson, M. F. Sun, and H. E. Aly, Vitamin B$_6$ requirements of young adult women, *Am. J. Clin. Nutr.* 24:1028–1041 (1971).

282. M. R. Thomas, J. Kawamoto, S. M. Sneed, and R. Eakin, The effects of vitamin C, vitamin B$_6$ and vitamin B$_{12}$ supplementation on the breast milk and maternal status of well-nourished women, *Am J. Clin. Nutr.*, 32:1679–1685 (1979).

283. J. L. Roepke and A. Kirksey, Vitamin B$_6$ nutriture during pregnancy and lactation. l. Vitamin B$_6$ intake levels of the vitamin in biological fluids, and condition of the infant at birth, *Am. J. Clin. Nutr.*, 32:2249–2256 (1979).

284. S. M. Sneed, C. Zane, and M. R. Thomas, The effect of ascorbic acid, vitamin B$_6$, vitamin B$_{12}$, and folic acid supplementation on the breast milk and maternal nutritional status of low socioeconomic lactating women, *Am. J. Clin. Nutr.*, 34:1338–1346 (1981).

285. L. Styslinger and A. Kirksey, Effects of different levels of vitamin B$_6$ supplementation on vitamin B$_6$ concentrations in human milk and vitamin B$_6$ intakes of breast fed infants, *Am. J. Clin. Nutr.*, 41:21–31 (1985).

286. M. L. S. Bainji, K. Prema, C. M. Jacob, B. A. Ramalakshmi, and R. Madhavapeddi, Relationship between maternal vitamins B$_2$ and B$_6$ status and the levels of these vitamins, in milk at different stages of lactation, *Hum. Nutr. Clin. Nutr.*, 40C:119–124 (1986).

287. M. W. Borschel, A. Kirksey, and R. E. Hannemann, Effects of vitamin B$_6$ intake on nutriture and growth of young infants, *Am. J. Clin. Nutr.*, 43:7–15 (1986).

288. M. B. Andon, M. P. Howard, P. B. Moser, and R. D. Reynolds, Nutritionally relevant supplementation of vitamin B$_6$ in lactating women: effect on plasma prolactin, *Pediatrics*, 76:769–773 (1986).

289. M. V. Karra, S. A. Udipi, A. Kirksey, and J. L. B. Roepke, Changes in specific nutrients in breast milk during extended lactation, *Am. J. Clin. Nutr.*, 43:495–503 (1986).

290. M. W. Borschel, Vitamin B$_6$ in infancy: requirements and current feeding practices, in: D. J. Raiten (ed.), *Vitamin B$_6$ Metabolism in Pregnancy, Lactation, and Infancy*, CRC Press, Boca Raton, FL, 1995, pp. 109–124.

291. E. M. Baker, J. E. Canham, W. T. Nunes, H. E. Sauberlich, and M. E. McDowell, Vitamin B$_6$ requirement for adult men, *Am. J. Clin. Nutr.*, 15:59–66 (1964).

292. P. Swan, J. Wentworth, and H. Linkswiler, Vitamin B$_6$ depletion in man: urinary taurine and sulfate excretion and nitrogen balance, *J. Nutr.*, 84:220–228 (1964).

293. N. Yess, J. M. Price, R. R. Brown, P. Swan, and H. Linkswiler, Vitamin B$_6$ depletion in man: urinary excretion of tryptophan metabolites, *J. Nutr.*, 84:229–236 (1964).

294. R. R. Brown, N. Yess, J. M. Price, H. Linkswiler, P. Swan, and L. V. Hankes, Vitamin B$_6$ depletion of man: urinary excretion of quinolinic acid and niacin metabolites, *J. Nutr.*, 87:419–423 (1965).

295. K. E. Cheslock and M. T. McCully, Response of human beings to a low-vitamin B$_6$ diet, *J. Nutr.*, 70:507–513 (1960).

296. H. E. Aly, E. A. Donald, and M. H. W. Simpson, Oral contraceptives and vitamin B$_6$ metabolism, *Am. J. Clin. Nutr.*, 24:297–303 (1971).

297. E. A. Donald, L. D. McBean, M. H. W. Simpson, M. F. Sun, and H. E. Aly, Vitamin B$_6$ requirement of young adult women, *Am. J. Clin. Nutr.*, 24:1028–1041 (1971).

298. H. K. Shin and H. Linkswiler, Tryptophan and methionine metabolism of adult females as affected by vitamin B$_6$ deficiency, *J. Nutr.*, 104:1348–1355 (1974).

299. P. J. Collip, S. Goldzier, N. Weiss, Y. Soleyman, and R. Snyder, Pyridoxine treatment of childhood bronchial asthma, *Ann. Allergy*, 35:93–97 (1975).

300. L. Musajo and C. A. Benassi, Aspects of disorders of the kynurenine pathway of tryptophan metabolism in man. In: H. Sobotka and C. P. Stewart (eds.), *Advances in Clinical Chemistry*, Vol. 7, Academic Press, New York, 1964, pp. 63–135.

301. L. V. Hankes, J. E. Leklem, R. R. Brown, and R. C. P. M. Mekel, Tryptophan metabolism of patients with pellagra: problem of vitamin B$_6$ enzyme activity and feedback control of tryptophan pyrrolase enzyme, *Am. J. Clin. Nutr.*, 24:730–739 (1971).

302. J. H. Flinn, J. M. Price, N. Yess, and R. R. Brown, Excretion of tryptophan metabolites by patients with rheumatoid arthritis, *Arth. Rheum.*, 7:201–210 (1964).

303. R. D. Reynolds and C. L. Natta, Depressed plasma pyridoxal phosphate concentrations in adult asthmatics, *Am. J. Clin. Nutr.*, 41:684–688 (1985).

304. C. B. Hollenbeck, J. E. Leklem, M. C. Riddle, and W. E. Connor, The composition and nutritional adequacy of subject-selected high carbohydrate, low fat diets in insulin-dependent diabetes mellitus, *Am. J. Clin. Nutr.*, 38:41–51 (1983).

305. W. J. Stone, L. G. Warnock, and C. Wagner, Vitamin B$_6$ deficiency in uremia, *Am. J. Clin. Nutr.*, 28:950–957 (1975).

306. L. Lumeng and T.-K. Li, Vitamin B$_6$ metabolism chronic alcohol abuse. Pyridoxal phosphate levels in plasma and the effects of acetaldehyde on pyridoxal phosphate synthesis and degradation in human erythrocytes, *J. Clin. Invest.*, 53:693–704 (1974).

307. W. J. Serfontein, J. B. Ubbink, C. S. DeVilliers, C. H. Rapley, and P. J. Becker, Plasma pyridoxal-5'-phosphate level as risk index for coronary artery disease, *Atherosclerosis*, 55:357–361 (1985).

308. A. Hamfelt and T. Tuvemo, Pyridoxal phosphate and folic acid concentration in blood and erythrocyte aspartate aminotransferase activity during pregnancy, *Clin. Chem. Acta.*, 41:287–298 (1972).

309. C. Potera, D. P. Rose, and R. R. Brown, Vitamin B$_6$ deficiency in cancer patients, *Am. J. Clin. Nutr.*, 30:1677–1679 (1977).

310. V. T. Devita, B. A. Chabner, D. M. Livingston, and V. T. Oliverio, Anergy and tryptophan metabolism in Hodgkin's disease, *Am. J. Clin. Nutr.*, 24:835–840 (1971).

311. M. Coleman, Studies of the administration of pyridoxine to children with Down's syndrome. In: J. E. Leklem and R. D. Reynolds (eds.), *Clinical and Physiological Applications of Vitamin B$_6$*, Alan R. Liss, New York, 1988, pp. 317–328.

312. C. Barthelemy, J. Martineau, N. Bruneau, J. P. Muh, G. Lelord, and E. Callaway, Clinical and biological effects of pyridoxine plus magnesium in autistic subjects. In: J. E. Leklem and R. D. Reynolds (eds.), *Clinical and Physiological Applications of Vitamin B$_6$*, Alan R. Liss, New York, 1988, pp. 329–356.

313. R. W. E. Watts, Hyperoxaluria. In: J. E. Leklem and R. D. Reynolds (eds.), *Clinical and Physiological Aspects of Vitamin B_6*, Alan R. Liss, New York, 1988, pp. 245–261.

314. H. J. T. Coelingh-Bennink and W. H. P. Schreurs, Improvement of oral glucose tolerance in gestational diabetes by pyridoxine, *Br. Med. J.*, 3:13–15 (1975).

315. J. B. Day, Clinical trials in the premenstrual syndrome, *Curr. Med. Res. Opin.*, 6: (Suppl. 5):40–45 (1979).

316. R. S. London, L. Bradley, and N. Y. Chiamori, Effect of a nutritional supplement on premenstrual symptomatology in women with premenstrual syndrome: a double-blind longitudinal study, *J. Am. Coll. Nutr.*, 10:494–499 (1991).

317. J. Ellis, K. Folkers, T. Watanabe, M. Kaji, S. Saji, J. W. Caldwell, C. A. Temple, and F. S. Wood, Clinical results of a cross-over treatment with pyridoxine and placebo of the carpal tunnel syndrome, *Am. J. Clin. Nutr.*, 32:2040–2046 (1979).

318. A. L. Berstein and J. S. Dinesen, Brief communication: effect of pharmacologic doses of vitamin B_6 on carpal tunnel syndrome, electroencephalographic results, and pain, *J. Am. Coll. Nutr.*, 12:73–76 (1993).

319. P. W. Adams, V. Wynn, M. Seed, and J. Folkard, Vitamin B_6, depression and oral conception, *Lancet*, 2:516–517 (1974).

320. A. L. Bernstein and C. S. Lobitz, A clinical and electrophysiologic study of the treatment of painful diabetic neuropathies with pyridoxine. In: J. E. Leklem and R. D. Reynolds (eds.), *Clinical and Physiological Applications and Vitamin B_6*, Alan R. Liss, New York, 1988, pp. 415–423.

321. R. D. Reynolds, Vitamin supplements: current controversies, *J. Am. Coll. Nutr.*, 13:118–126 (1994).

322. K. S. McCully, Homocysteine theory of arteriosclerosis development and current status, *Atherosclerosis Rev.*, 11:157–246 (1983).

323. S. H. Mudd and H. L. Levy, Disorders of transsulfuration. In: J. B. Stanbury, et al. (eds.), *The Metabolic Basis of Inherited Disease*, McGraw-Hill, New York, 1983, pp. 522–559.

324. R. T. Wall, J. M. Harlan, L. A. Harker, et al. Homocysteine-induced endothelial cell injury in vitro: a model for the study of vascular injury, *Thromb. Res.*, 18:113–121 (1980).

325. A. J. Olszewski and W. B. Szostak, Homocysteine content of plasma proteins in ischemic heart disease, *Atherosclerosis*, 69:109–113 (1988).

326. K. Robinson, E. L. Mayer, D. Miller, et al., Hyperhomocysteinemia and low pyridoxal phosphate: common and independent reversible risk factors for coronary artery disease, *Circulation*, 92:2825–2830 (1995).

327. H. I. Morrison, D. Schaubel, M. Desmeules, and D. T. Wigle, Serum folate and risk of fatal coronary heart disease. *J. Am. Med. Assoc.*, 275:1893–1896 (1996).

328. J. V. Woodside, J. Yarnell, D. McMaster, I. S. Young, D. L. Harmon, E. E. McCrum, C. C. Patterson, K. F. Gey, A. S. Whitehead, and A. Evans, Effect of B-group vitamins and antioxidant vitamins on hyperhomocysteinemia: a double-blind, randomized, factorial-design, controlled trial, *Am. J. Clin. Nutr.*, 67:858–866 (1998).

329. K. Robinson, K. Arheart, H. Refsum, L. Brattstrom, G. Boers, P. Ueland, P. Rubba, R. Palma-Reis, R. Meleady, L. Daly, J. Witteman, and I. Graham, Low circulating folate and vitamin B_6 concentrations: risk factors for stroke, peripheral vascular disease, and coronary artery disease, *Circulation*, 97:437–443 (1998).

330. J. F. Rinehart and L. D. Greenberg, Pathogenesis of experimental arteriosclerosis in pyridoxine deficiency with notes on similarities to human arteriosclerosis, *Arch. Pathol.*, 51:12–18 (1951).

331. K. Krishnaswarmy and S. B. Rao, Failure to produce atherosclerosis in Macca radiata on a high-methionine, high fat, pyridoxine-deficient diet, *Atherosclerosis*, 27:253–258 (1977).

332. G. H. Boers, A. G. H. Smals, et al. Pyridoxine treatment does not prevent homocystinemia after methionine loading in adult homocystinuria patients, *Metabolism*, 32:390–397 (1983).

333. A. M. Gressner and D. Sittel, Plasma pyridoxal 5′-phosphate concentrations in relation to

apoaminotransferase levels in normal, uraemic, and post-myocardial infarct sera, *J. Clin. Chem. Clin. Biochem.*, 23:631–636 (1985).

334. W. J. H. Vermaak, H. C. Barnard, G. M. Potgieter, and J. D. Marx, Plasma pyridoxal-5'-phosphate levels in myocardial infarction, *South Afr. Med. J.*, 70:195–196 (1986).

335. W. J. H. Vermaak, H. C. Barnard, G. M. Potgieter, and H. du T. Theron, Vitamin B$_6$ and coronary artery disease. Epidemiological observations and case studies, *Atherosclerosis*, 63: 235–238 (1987).

336. W. J. Serfontein, J. B. Ubbink, L. S. DeViUiers, and P. J. Becker, Depressed plasma pyridoxal-5'-phosphate levels in tobacco-smoking men, *Atherosclerosis*, 59:341–346 (1986).

337. J. D. Bogden, H. Baker, O. Frank, et al. Micronutrient status and human immunodeficiency virus (HIV) infection, *Ann. N. Y. Acad. Sci.*, 587:189–195 (1990).

338. M. K. Baum, G. Shor-Posner, P. Bonvchi, et al. Influence of HIV infection on vitamin status requirements, *Ann. N. Y. Acad. Sci.*, 669:165–173 (1992).

339. M. K. Baum, E. Mantero-Atienza, G. Shor-Posner, et al. Association of vitamin B$_6$ status with parameters of immune function in early HIV-1 infection, *J. AIDS*, 4:1122–1132 (1991).

340. R. S. Beach, E. Mantero-Atienza, G. Shor-Posner, et al., Specific nutrient abnormalities in asymptomatic HIV-1 infection, *J. AIDS*, 6:701–708 (1992).

341. J. Pease, M. Niewinski, D. Pietrak, J. E. Leklem, and R. D. Reynolds, Vitamin B$_6$ metabolism and status in HIV positive and HIV negative high-risk patients, *FASEB J.*, 12:A510 (1998).

342. A. M. Tang, N. M. H. Graham, A. J. Kirby, et al., Dietary micronutrient intake and risk of progression to acquired immunodeficiency syndrome (AIDS) in human immunodeficiency virus type 1 (HIV-1)–infected homosexual men. *Am. J. Epidemiol.*, 138:937–951 (1993).

343. G. O. Coodley, M. K. Coodley, H. D. Nilson, and M. D. Lovdess, Micronutrient concentrations in the HIV wasting syndrome, *J. AIDS*, 7:1595–1600 (1993).

344. A. M. Tang, N. M. H. Graham, R. K. Chandra, and A. J. Saah, Low serum vitamin B$_{12}$ concentrations are associated with faster human immunodeficiency virus type 1 (HIV-1) disease progression, *J. Nutr.*, 127:345–351 (1997).

345. J. M. Salhany and L. M. Schopfer, Pyridoxal 5'-phosphate binds specifically to soluble CD4 protein, the HIV-1 receptor, *J. Biol. Chem.*, 268:7643–7645 (1993).

346. L. L. W. Mitchell and B. S. Cooperman, Active site studies of human immunodeficiency virus reverse transcriptase, *Biochemistry*, 31:7707–7713 (1992).

347. L. K. Moen, I. C. Bathurst, and P. J. Barr, Pyridoxal 5'-phosphate inhibits the polymerase activity of a recombinant RNase H-deficient mutant HIV-1 reverse transcriptase, *AIDS Res. Human Retroviruses*, 8:597–604 (1992).

348. M. G. Brush, Vitamin B$_6$ treatment of premenstrual syndrome. In: J. E. Leklem and R. D. Reynolds (eds.), *Clinical and Physiological Applications of Vitamin B$_6$*, Alan R. Liss, New York, 1988, pp. 363–379.

349. P. M. S. O'Brien, The premenstrual syndrome: a review of the present status of therapy, *Drugs*, 24:140–151 (1982).

350. C. D. Ritchie and R. Singkamani, Plasma pyridoxal 5'-phosphate in women with the premenstrual syndrome, *Hum. Nutr. Clin. Nutr.* 40C:75–80 (1986).

351. M. Mira, P. M. Stewart, and S. F. Abraham, Vitamin and trace element status in premenstrual syndrome, *Am. J. Clin. Nutr.*, 47:636–641 (1988).

352. P. W. Adams and D. P. Rose, et al. Effects of pyridoxine hydrochloride (vitamin B$_6$) upon depression associated with oral contraception, *Lancet*, 1:897 (1973).

353. J. Stokes and J. Mendels, Pyridoxine and premenstrual tension, *Lancet*, 1:1177–1178 (1972).

354. G. E. Abraham and J. T. Hargrove, Effect of vitamin B$_6$ on premenstrual symptomatology in women with premenstrual tension syndromes: a double blind crossover study, *Infertility*, 3:155–165 (1980).

355. W. Barr, Pyridoxine supplements in the premenstrual syndrome, *Practitioner*, 228:425–428 (1984).

356. K. E. Kendall and P. P. Schurr, The effects of vitamin B_6 supplementation on premenstrual syndromes, *Obstet. Gynecol.*, 70:145–149 (1987).

357. J. Kleijen, G. T. Riet, and P. Knipschild, Vitamin B_6 in the treatment of the premenstrual syndrome—a review, *Br. J. Obst. Gynecol.*, 97:847–852 (1990).

358. R. A. Simon and R. D. Reynolds, Vitamin B_6 and asthma. In: J. E. Leklem and R. D. Reynolds (eds.), *Clinical and Physiological Applications of Vitamin B_6*, Alan R. Liss, New York, 1988, pp. 307–315.

359. J. B. Ubbink, R. Delport, S. Bissbort, W. J. Vermaak, and P. J. Becker, Relationship between vitamin B_6 status and elevated pyridoxal kinase levels induced by theophylline therapy in humans, *J. Nutr.*, 120:1352–1359 (1990).

360. J. A. Driskell, R. L. Wesley, and I. E. Hess, Effectiveness of pyridoxine hydrochloride treatment on carpal tunnel syndrome patients, *Nutr. Rep. Int.*, 34:1031–1040 (1986).

361. M. L. Kasdan and C. James, Carpal tunnel syndrome and vitamin B_6, *Plast. Reconstr. Surg.*, 79:456–459 (1987).

362. G. P. Smith, P. J. Rudge, and J. J. Peters, Biochemical studies of pyridoxal and pyridoxal phosphate status and therapeutic trial of pyridoxine in patients with carpal tunnel syndrome, *Ann. Neurol.*, 15:104–107 (1984).

363. A. Franzblau, C. L. Rock, R. A. Werner, et al., The relationship of vitamin B_6 status to median nerve function and carpal tunnel syndrome among active industrial workers, *J. Occup. Environ. Med.*, 38:485–491 (1996).

364. R. C. Keniston, P. A. Nathan, J. E. Leklem, and R. S. Lockwood, Vitamin B_6, vitamin C, and carpal tunnel syndrome, *J. Occup. Environ. Med.*, 39:949–959 (1997).

365. H. N. Bhagavan, Interaction between vitamin B_6 and drugs. In: R. D. Reynolds and J. E. Leklem (eds.), *Vitamin B_6: Its Role in Health and Disease*, Alan R. Liss, New York, 1985, pp. 401–415.

366. J. E. Leklem, Vitamin B_6 requirement and oral contraceptive use-a concern? *J. Nutr.*, 116:475–477 (1986).

367. H. Schaumburg, J. Kaplan, A. Windebank, N. Vick, S. Rasmus, D. Pleasure, and M. J. Brown, Sensory neuropathy from pyridoxine abuse: a new megavitamin syndrome, *N. Engl. J. Med.*, 309:445–448 (1983).

368. G. J. Parry and D. E. Bredesen, Sensory neuropathy with low-dose pyridoxine, *Neurology*, 35:1466–1468 (1985).

369. K. Dalton and M. J. T. Dalton, Characteristics of pyridoxine overdose neuropathy syndrome, *Acta. Neurol. Scand.*, 76:81–11 (1987).

370. M. Cohen and A. Bendich, Safety of pyridoxine—a review of human and animal studies, *Toxicol. Lett.* 34:129–139 (1986).

371. K. Mori, M. Kaido, K. Fujishiro, N. Inoue, and O. Koide, Effects of megadoses of pyridoxine on spermatogenesis and male reproductive organs in rats, *Arch. Toxicol.*, 66:198–203 (1992).

372. P. A. Cohen, K. Schneidman, F. Ginsberg-Fellner, et al. High pyridoxine diet in the rat: possible implications for megavitamin therapy, *J. Nutr.*, 103:143–151 (1973).

11

Biotin

DONALD M. MOCK

University of Arkansas for Medical Sciences, Little Rock, Arkansas

I. HISTORY OF DISCOVERY

Although a growth requirement for the "bios" fraction had been demonstrated in yeast, Boas was the first to demonstrate the requirement for biotin in a mammal (1). In rats fed protein derived from egg white, Boas observed a syndrome of severe dermatitis, hair loss, and neuromuscular dysfunction known as "egg-white injury." A factor present in liver cured the egg-white injury and was named protective factor X. It is now recognized that the critical event in this egg-white injury of both the human and the rat is the highly specific and very tight binding ($K_d = 10^{-15}$ M) of biotin by avidin, a glycoprotein found in egg white. Native avidin is resistant to intestinal proteolysis in both the free and biotin-combined form. Thus, dietary avidin (e.g., in diets containing uncooked egg white) is thought to bind and prevent the absorption of both dietary biotin and any biotin synthesized by intestinal bacteria.

II. CHEMISTRY OF BIOTIN

A. Structure

As reviewed by Bhatia et al. (2) and Bonjour (1), the structure of biotin (Fig. 1) was established by Köyl and his group in Europe and by du Vignraud and his collaborators in the United States between 1940 and 1943. Because biotin has three asymmetrical carbons in its structure, eight stereoisomers exist; of these, only one, designated d-(+)-biotin, is found in nature and is enzymatically active. This compound is generally referred to simply as biotin or D-biotin. Biocytin (ε-N-biotinyl-L-lysine) is about as active as biotin on a mole basis in mammalian growth studies.

     Biotin is a bicyclic compound. One of the rings contains a ureido group (—N—CO—N—), and the other contains sulfur and is termed a tetrahydrothiophene ring. The

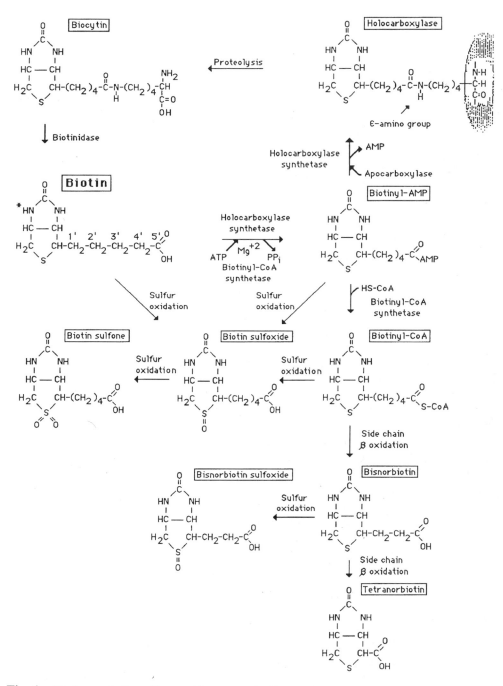

Fig. 1 Biotin metabolism. The specific systems leading to the sulfoxides have not been defined. ATP, adenosine triphosphate; AMP, adenosine monophosphate; HS-CoA, coenzyme A; PPᵢ pyrophosphate; *, site of attachment of carboxyl moiety. (Courtesy of International Life Sciences Institute, Washington, D.C., M. Brown, ed.)

two rings have a boat configuration with respect to each other. The tetrahydrothiophene ring has a valeric acid side chain. Based on binding of biotin analogs by avidin and on x-ray crystallography of the biotin–avidin complex (3), the ureido ring of the molecule is the most important region in terms of the extraordinarily tight binding of biotin to avidin and to streptavidin, a protein similar to avidin that is excreted by *Streptomyces avidinii*. Other studies (3) suggest that the length of the side chain or the apolar nature of the —CH$_2$— moieties in the side chain also play a role in the binding of biotin to the hydrophobic site on avidin.

B. Chemical Synthesis of Biotin

The structure of biotin was confirmed by de novo chemical synthesis by Harrison and co-workers in the 1940s (1). As reviewed recently (1,4), all of the early synthetic methods suffered from the disadvantage that either the yield of the proper stereospecific isomer was low or that special intermediates were required to obtain stereospecificity. These shortcomings are avoided by the stereospecific synthesis developed by Goldberg and Sternbach in 1949 in the laboratories of Hoffman-LaRoche (5). The Goldberg–Sternbach synthesis, or modifications thereof, is the method by which biotin is synthesized commercially (1). Additional stereospecific methods of synthesis have been published recently (6,7).

III. PHYSIOLOGY OF BIOTIN

A. Digestion of Protein-Bound Biotin

Neither the mechanism of intestinal hydrolysis of protein-bound biotin nor the relationship of the digestion of protein-bound biotin to its bioavailability has been clearly defined. The content of free biotin and protein-bound biotin in foods is variable, but the majority of biotin in meats and cereals appears to be protein-bound. Biotinidase is the enzyme that catalyzes the cleavage of biotin from its covalent attachment to protein (e.g., the biotin-dependent enzymes) during cellular turnover of these proteins. Wolf et al. have postulated that biotinidase plays a critical role in the release of biotin from dietary protein (8). Biotinidase in pancreatic juice might be responsible for release of biotin during the luminal phase of proteolysis. Mucosal biotinidase might release biotin from biotinyl oligopeptides, the presumed products of intestinal proteolysis. The role of mucosal biotinidase is not certain because the activity is not enriched in intestinal brush-border membranes (8).

In view of the observations suggesting that a free carboxyl group is necessary for binding to the intestinal biotin transporter (see discussion below), significant uptake of biotinyl oligopeptides by the biotin transporter seems unlikely. This conclusion is supported by recent studies on the uptake of biocytin (9) but is not universally accepted (10). Alternatively, biotinyl oligopeptides might be absorbed directly by a nonspecific pathway for peptide absorption.

In patients with biotinidase deficiency, doses of free biotin that do not greatly exceed the estimated dietary intake (e.g., 50–150 µg/day) appear adequate to prevent the symptoms of biotinidase deficiency, presumably by preventing biotin deficiency (11). These observations are consistent with a mechanism in which biotinidase deficiency contributes to biotin deficiency through impaired intestinal digestion of protein-bound biotin or impaired renal salvage of ultrafiltered biotin or both.

B. Intestinal Absorption of Biotin

Early studies using low-specific-activity radiolabeled biotin and intact tissues, such as everted gut sacs, concluded that intestinal and renal transport of biotin occurred by simple diffusion (3). However, given the small amounts of biotin in foodstuffs and animal tissues, and the efficient conservation of biotin by the body under most circumstances, simple diffusion is teleologically unattractive (12). Within the last decade, studies using higher specific activity radiolabeled biotin, brush-border membrane vesicles, and cell culture systems have greatly expanded current knowledge of intestinal and renal transport of biotin (3–17). A biotin transporter is present in the intestinal brush-border membrane. According to most studies, the carrier is structurally specific, requiring both a free carboxyl group on the valeric acid side chain and an intact ureido ring (12,17); however, not all structure–activity studies (18) have confirmed the need for a free carboxyl group. Based on studies with normal human tissue, human intestinal cell lines, and intestinal tissue from rabbit and rat, the transport of biotin is temperature-dependent and occurs against a concentration gradient. Biotin transport is electroneutral because of the 1:1 coupling of biotin ($R-COO^-$) with Na^+. Biotin transport also occurs by simple diffusion; diffusion predominates at higher (pharmacologic) concentrations. These recent studies of the biotin transporter cast into doubt the conclusion drawn from earlier studies that biotinidase is the principal biotin-binding protein in the intestinal brush-border membrane (3).

In rats, biotin transport is upregulated with maturation from the suckling to adult rat, and the site of maximal transport by the biotin transporter increases aborally with age, shifting from the ileum to the jejunum. Although carrier-mediated transport of biotin was most active in the proximal small bowel of the rat, Bowman and Rosenberg (19) concluded that the absorption of biotin from the proximal colon was still significant, supporting the potential nutritional significance of biotin synthesized by enteric flora. Clinical studies have also provided some evidence that biotin is absorbed from the human colon (20,21).

In intact rats and Caco-2 cells, upregulation of biotin transport occurs in response to biotin deficiency; the mechanism for most of the change appears to be an increased V_{max} (presumably mediated by an increased number of carriers) rather than a change in carrier affinity. Based on studies of rabbit intestinal brush-border membrane transport, histidine residues and sulfhydryl groups are important in the normal function of the transporter. The histidine residues are probably located at or near the biotin binding site (22).

In contrast to most other investigators, Leon-Del-Rio and co-workers observed only passive diffusion of biotin in the rat, but saturable transport for biocytin. Moreover, they reported saturable transport of biotin but passive diffusion of biocytin in the hamster (10). They have proposed that protein-bound biotin is absorbed mainly in its free form in the hamster but at least partially as biocytin in the rat.

The exit of biotin from the enterocyte (i.e., transport across the basolateral membrane) is also carrier-mediated, but basolateral transport is independent of Na^+, is electrogenic, and does not accumulate biotin against a concentration gradient (23). To investigate the mechanism leading to reduced plasma biotin concentration in a substantial portion of alcoholics, Said and co-workers (24) studied the effect of chronic ethanol feeding and acute ethanol exposure on intestinal transport of biotin using everted intestinal sacs from the rat. Acute ethanol exposure inhibited intestinal transport in biotin; chronic ethanol feeding reduced intestinal transport of biotin and decreased plasma concentrations of biotin. These authors speculated that the effects of ethanol on intestinal transport of biotin may contribute to the impaired biotin status associated with chronic alcoholism.

C. Transport of Biotin from the Intestine to Peripheral Tissues

Little has been definitively established concerning the transport of biotin to the liver and peripheral tissues from the site of absorption in the intestine. Investigation of the binding of biotin to proteins in plasma and serum has proceeded along two distinct lines: (a) investigation of the biotin binding properties of biotinidase and (b) empirical assessment of the covalent and reversible binding of biotin to whole plasma and fractionated plasma proteins.

Wolf et al. (25) originally hypothesized that biotinidase might serve as a biotin-binding protein in plasma or perhaps even as a carrier protein for the transport of biotin into the cell. Chuahan and Dakshinamurti (26) provided evidence to support that hypothesis. Biotin binding to purified biotinidase, albumin, α- and β-globulins, and fractionated human serum was assessed using [^{3}H]biotin, ammonium sulfate precipitation, and equilibrium dialysis. These investigators concluded that biotinidase is the only protein in human serum that specifically binds biotin. Others reached a different conclusion. Using [^{3}H]biotin, centrifugal ultrafiltration, and dialysis to assess reversible binding in plasma from the rabbit, pig, and human (27,28), Mock and co-workers found that less than 10% of the total pool of free plus reversibly bound biotin is reversibly bound to plasma macromolecules (presumably proteins). A similar biotin-binding system was detected by Mock et al. (28) in experiments with physiological concentrations of human serum albumin. Additional studies determined the proportion of biotin covalently bound to plasma protein (27). Using acid hydrolysis and [^{3}H]biotinylalbumin to assess the completeness of biotin release, these investigators found that an additional 12% of biotin was released by hydrolysis after extensive dialysis. Those investigators concluded that the percentages of free, reversibly bound, and covalently bound biotin in human serum are approximately 81% to 7% to 12% (27).

Results of the two approaches discussed above apparently conflict; the conflict may arise from differences in the experimental approach, the definition of binding, or both. The importance of either type of biotin binding to the transport of biotin from the intestine to the peripheral tissues is not yet clear. The binding detected by Mock and co-workers may represent a structurally nonspecific interaction between the hydrophobic portions of the biotin molecule and one or more of the hydrophobic binding sites on a serum protein, such as albumin. This binding may be analogous to the binding of apolar optical and fluorescent dyes to the hydrophobic binding sites on albumin and avidin (29–34).

D. Transport of Biotin to the Liver

The uptake of biotin by liver and peripheral tissues from mammals has been the subject of several investigations. Studies of 3T3-L1 fibroblasts (35), rat hepatocytes isolated by collagenase profusion (17,36), basolateral membrane vesicles from human liver (37,38), and Hep G$_2$ human hepatoma cells (39) indicate that uptake of free biotin is mediated both by diffusion and by a specialized carrier system that is dependent on an Na$^+$ gradient, temperature, and energy. Transport is electroneutral (Na$^+$/biotin = 1:1) and specific for a free carboxyl group, but transport is not strongly specific for the structure in the region of the thiophene ring (39). Different results were found in studies of isolated cultured hepatocytes (40); these cells did not exhibit a carrier-mediated transport system.

Additional studies in cultured rat hepatocytes demonstrated trapping of biotin, presumably covalently bound in holocarboxylase enzymes (41). These studies confirm earlier studies by McCormick and recapitulate the importance of metabolic trapping of water-soluble vitamins as a mechanism for an intracellular accumulation (17). After entering the hepatocyte, biotin diffuses into the mitochondria via a pH-dependent process (42). Said and co-workers have postulated that biotin enters the mitochondria in the neutral protonated form and dissociates into the anionic form in the alkaline mitochondrial environment, thus becoming trapped by the charge (42).

E.   Transport of Biotin into the Central Nervous System

Using an in situ rat brain perfusion technique and tracer [^{3}H]biotin, Spector and Mock (43) demonstrated that biotin is transported across the blood–brain barrier by a saturable system; the apparent K_m was about 100 µmol/L (a value several orders of magnitude greater than the concentration of free biotin in plasma). Inhibition of transport by structural analogs suggested that the free carboxylate group in biotin was important to transport, presumably due to structural specificity of a biotin transport protein. Transfer of biotin directly into the cerebral spinal fluid (CSF) via the choroid plexus did not appear to be an important mechanism of biotin entry into the central nervous system. In additional studies using either intravenous or intraventricular injection of [^{3}H]biotin into rabbits (44), Spector and Mock found that [^{3}H]biotin was cleared from the CSF more rapidly than mannitol, suggesting specific transport systems for biotin uptake into the neurons after biotin crosses the blood–brain barrier. Whether infused intravenously or injected intraventricularly, unlabeled biotin competed with [^{3}H]biotin for transport. Two hours after intraventricular injection, little metabolism of biotin or covalent binding to brain proteins was observed; however, after 18 h, approximately one-third of the biotin had been incorporated into brain protein. These findings suggest that biotin enters the brain by a saturable transport system that does not depend on the subsequent metabolism of biotin or immediate ''trapping'' by incorporation into brain proteins.

Mock and co-workers have measured the concentrations of free ''biotin'' (i.e., total avidin-binding substances) in human CSF and ultrafiltrates of plasma; the ratio was 0.85 ± 0.5 for 11 subjects (3). This result is similar to the CSF plasma ratios determined for biotin by Spector and Mock in the rabbit, an animal that has a specific system for biotin transport across the blood–brain barrier (44).

F.   Renal Handling of Biotin

Specific systems for the reabsorption of water-soluble vitamins from the glomerular filtrate may make an important contribution to conservation of the water-soluble vitamins (12). A biotin transport system has been identified by Podevin and Barbarat in both brush-border and basolateral membrane vesicles from rabbit kidney cortex (45). Uptake by brush-border membrane vesicles was saturable, occurred against a biotin concentration gradient, and was dependent on an inwardly directed Na$^+$ gradient. The K_m was 28 µmol/L, and transport exhibited structural specificity. In contrast, the uptake of biotin by basolateral membrane vesicles was not sensitive to an Na$^+$ gradient. In rat kidney, Spencer and Roth demonstrated a similar system with a K_m of 0.2 µmol/L that was inhibited by equimolar concentrations of biocytin (46).

In vitro studies of biotin transport by renal tissue preparations in humans have not been published. Baumgartner and co-workers have measured the renal clearance of biotin in vivo and have calculated biotin/creatinine clearance ratios (47–49). In normal adults and children who are not receiving biotin supplementation, the clearance ratio is approximately 0.4. In patients with biotinidase deficiency, renal wasting of biotin and biocytin occurs; biotin/creatinine clearance ratios typically exceed 1, and half-lives for biotin clearance are about half of the normal value. The mechanism for the increased renal excretion of biotin in biotinidase deficiency has not been defined, but this observation suggests that there may be a role for biotinidase in the renal handling of biotin. For example, abnormal plasma biotinidase might (a) bind biotin less tightly, increasing the glomerular sieving coefficient; (b) serve less effectively as a reclamation transporter in the renal tubule; or (c) alter cellular salvage of biotin during turnover of renal holocarboxylases.

G. Placental Transport of Biotin

Specific systems for transport of biotin from the mother to the fetus have recently been reported (50–52). Studies using microvillus membrane vesicles and cultured trophoblasts (50,51) detected a saturable transport system for biotin that was dependent on Na^+ and actively accumulated biotin within the placenta with slower release into the fetal compartment. However, in the isolated perfused single cotyledon (50,51), transport of biotin across the placenta was slow relative to placental accumulation. Little evidence of accumulation on the fetal side was detected, suggesting that the overall placental transfer of biotin is most consistent with a passive process. Membrane vesicle transport was sensitive to short-term exposure to ethanol, but overall transfer was not (51). Further studies using fetal facing (basolateral) membrane vesicles detected a saturable, Na^+-dependent, electroneutral, carrier-mediated uptake process that was not as active as the biotin uptake system in the maternal facing (apical) membrane vesicles (52).

H. Transport of Biotin to Human Milk

Using an avidin-binding assay, Mock and co-workers have concluded that greater than 95% of the biotin (i.e., total avidin-binding substance) is present in the skim fraction of human milk rather than in the cell pellet or fat fraction (53). Less than 3% of the biotin was reversibly bound to macromolecules, and less than 5% was covalently bound to macromolecules. Thus, almost all of the biotin in human milk is free in the aqueous compartment of the skim fraction. The concentration of biotin in human milk remains fairly constant between the fore, middle, and hind milk in the same feeding but varies substantially over 24 h in some women (54). Though no single postpartum pattern is followed by all women, a steady increase in the biotin concentration was observed during the first 18 days post partum in half of the women studied. However, rather than reaching a stable plateau in mature milk, biotin concentrations vary substantially in most women after 18 days post partum. The cause of this variation is not known.

When separated from other biotin catabolites by high-performance liquid chromatography (HPLC) and measured by an avidin-binding assay, the concentration of biotin in the aqueous phase of human milk exceeds the concentration in serum by one to two orders of magnitude (55). It seems likely that a transport system exists that conveys biotin from the plasma to human milk against a concentration gradient.

Measurements of the biotin and metabolite in human milk using an HPLC separation and avidin-binding assay indicate that bisnorbiotin accounts for approximately 50% and biotin sulfoxide about 10% of the total biotin plus metabolites in early and transitional human milk. With maturation post partum the biotin concentration increases, but the bisnorbiotin and biotin sulfoxide concentrations still account for 25% and 8% at 5 weeks post partum (55). Studies to date provide no evidence for trapping by a biotin-binding protein.

IV. SPECIFIC FUNCTIONS

In mammals, biotin serves as an essential cofactor for four carboxylases, each of which catalyzes a critical step in intermediary metabolism (see Chapter 5). All four of the mammalian carboxylases catalyze the incorporation of bicarbonate into a substrate as a carboxyl group. Four similar carboxylases, two other carboxylases, two decarboxylases, and a transcarboxylase are found in nonmammalian organisms. All of these biotin-dependent enzymes appear to work by a similar mechanism.

A. Incorporation into Carboxylases

Attachment of the biotin to the apocarboxylase (Fig. 1) is a condensation reaction catalyzed by holocarboxylase synthetase. The holocarboxylase synthetase reaction is driven thermodynamically by hydrolysis of ATP to inorganic phosphate. An amide bond is formed between the carboxyl group of the valeric acid side chain of biotin and the ε-amino group of a specific lysyl residue in the apocarboxylase; these regions of the protein backbone contain sequences of amino acids that tend to be highly conserved within and between species for the individual carboxylases.

Holocarboxylase synthetase (EC 6.3.4.10) is present in both the cytosol and the mitochondria. Studies of human mutant holocarboxylase synthetase indicate that both the mitochondrial and cytoplasmic forms are encoded by one gene (57–60). However, other investigators have concluded that two different holocarboxylase synthetases catalyze the biotinylation of the mitochondrial and cytosolic carboxylases in rat and chicken liver (61–63). This conclusion is based on differences in pH optima, nucleotide specificity, and tetrapeptide sequence in the areas containing the lysine residues, as well as on site-specific mutagenesis studies.

B. Regulation of Intracellular Carboxylase Activity by Biotin

Currently, the role of biotin in regulating activity of the four mammalian carboxylases at the gene level remains to be elucidated. However, the interaction of biotin synthesis and production of holoacetyl-CoA carboxylase in *Escherichia coli* has been extensively studied and reviewed (64,65). The biotin-protein ligase (specifically a holoacetyl-CoA carboxylase synthetase in *E. coli*) catalyzes formation of the covalent bond between biotin and a specific lysine residue in the biotin carboxylase carrier protein (BCCP) of acetyl-CoA carboxylase. The biotin binding domain of BCCP of *E. coli* acetyl-CoA carboxylase has been sequenced (66). As with the four mammalian carboxylases (Fig. 1), the biotinylation of the apocarboxylase proceeds in two steps. First, holocarboxylase synthetase reacts with biotin and ATP to form a complex between the synthetase and biotinyl-AMP, releasing pyrophosphate. The stoichiometric amounts of biotinyl-AMP are synthesized quickly and diffuse off the enzyme slowly in the absence of apocarboxylase to complete the two-step reaction (67). If a suitable amount of the BCCP portion of acetyl-CoA carboxylase is present, the holocarboxylase is formed and AMP is released. If insufficient apocarboxylase is present, the holocarboxylase synthetase:biotinyl-AMP complex acts to repress further synthesis of biotin by binding to the promoter regions of the biotin operon (''bio''). These promoters control a cluster of genes that encode enzymes that catalyze biotin synthesis; these enzymes include biotin synthetase, the enzyme complex that converts dethiobiotin to biotin. In its role as a repressor of the bio operon, the holocarboxylase synthetase has been named BirA; this name arose from initial observations on *b*iotin *i*ntracellular *r*etention properties and was found to be allelic to *r*epression of *bio*tin synthesis by biotin (''bioR''). Biotinyl-AMP acts as a corepressor through its role in the BirA:biotinyl-AMP complex. Thus, the rate of biotin synthesis is responsive to both the supply of apo-BCCP and the supply of biotin as reflected in the biotinyl-AMP concentration.

Additional research has focused on conversion of dethiobiotin (or an earlier precursor pimelic acid) to biotin (68). This is an unusual enzymatic reaction that closes the tetrahydrothiophene ring by inserting a sulfur. These studies used *E. coli* or *Bacillus sphaericus* bioB transformants that overproduce biotin. Ifuku and co-workers have sequenced several point mutations leading to bioB transformants in *E. coli* (69). Such studies have provided evidence that biotin synthetase is a two-iron/two-sulfur enzyme that requires NADPH, *S*-adenosylmethionine, and Fe^{3+} or Fe^{2+} (70–72). Flavodoxin is also required, probably as an electron donor (73). The source of the sulfur in the thiophene ring is probably cysteine or a derivative (74–76). Additional studies used cell-free systems to establish the cofactor and metal ion requirements (74,75,77). Alanine and acetate serve as carbon sources; carbon dioxide liberated via the tricarboxylic acid cycle may serve as well (78). The first eukaryotic biotin synthetase has now been cloned and sequenced from the yeast *Saccharomyces cerevisiae* (79).

C. The Four Mammalian Carboxylases

In the carboxylase reaction, the carboxyl moiety is first attached to biotin at the ureido nitrogen opposite the side chain; next, the carboxyl group is transferred to the substrate. The reaction is driven by the hydrolysis of ATP to ADP and inorganic phosphate. Subsequent reactions in the pathways of the four mammalian carboxylases release CO_2 from the products of the various carboxylase reactions. Thus, these reaction sequences rearrange the substrates into more useful intermediates but do not violate the classic observation that mammalian metabolism does not result in the *net* fixation of carbon dioxide. The mechanism for the carboxylase reaction is discussed in detail in Chapter 5, ''Bioorganic Mechanisms Important to Coenzyme Functions,'' by Donald McCormick.

Three of the four biotin-dependent carboxylases are mitochondrial; the fourth (acetyl-CoA carboxylase, ACC) is found in both the mitochondria and the cytosol. Based on a series of observations by Allred and co-workers (63,80–83), ACC (EC 6.4.1.2) exists in an active cytosolic form and two largely inactive mitochondrial forms with molecular weights of approximately 264 and 234 kDa, respectively. The mitochondrial forms are postulated to serve as storage forms that leave the mitochondria and are transformed into the active cytosolic ACC during periods of restricted biotin availability. The timing of decrease of the mitochondrial form during maintenance of the activity of cytosolic ACC is consistent with the hypothesis of Allred and co-workers (63). However, the reported maintenance of normal amounts of the three mitochondrial enzymes after 4 weeks of egg-white feeding is not consistent with the reports of abnormal organic aciduria after as little as 2 weeks of egg-white feeding in rats (84) and humans (85).

ACC catalyzes the incorporation of bicarbonate into acetyl-CoA to form malonyl-CoA (Fig. 2). This three-carbon compound then serves as a substrate for the fatty acid synthetase complex. The net result is the elongation of the fatty acid substrate by two carbons and the loss of the third carbon as CO_2.

Pyruvate carboxylase (PC; EC 6.4.1.1) catalyzes the incorporation of bicarbonate into pyruvate to form oxaloacetate, an intermediate in the Kreb's tricarboxylic acid cycle (Fig. 2). Thus, PC catalyzes an anapleurotic reaction. In gluconeogenic tissues (i.e., liver

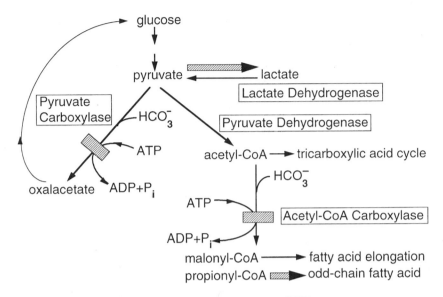

Fig. 2   Lactate and pyruvate metabolism. Deficiencies ▨ of PC and ACC can redirect intermediates into unusual pathways ▧▸. Abbreviations as in Fig. 1. ADP, adenosine diphosphate; P_i, inorganic phosphate.

and kidney), the oxaloacetate can be converted to glucose. Deficiency of PC has been proposed as the cause of the lactic acidemia, central nervous system lactic acidosis, and abnormalities in glucose regulation observed in biotin deficiency and biotinidase deficiency as discussed below.

Methylcrotonyl-CoA carboxylase (MCC; EC 6.4.1.4) catalyzes an essential step in the degradation of the branch-chained amino acid leucine (Fig. 3). Deficient activity of this enzyme (whether due to the isolated genetic deficiency, multiple carboxylase deficiency, or biotin deficiency per se) leads to metabolism of its substrate 3-methylcrotonyl-CoA by an alternate pathway to 3-hydroxyisolvaleric acid, 3-methylcrotonylglycine, or both (3,86). Thus, increased urinary excretion of these abnormal metabolites in the urine reflects deficient activity of MCC and can reflect biotin depletion at the tissue level in genetically normal individuals.

Propionyl-CoA carboxylase (PCC; EC 6.4.1.3) catalyzes the incorporation of bicarbonate into propionyl-CoA to form methylmalonyl-CoA, which undergoes isomerization to succinyl-CoA and enters the tricarboxylic acid cycle (Fig. 4). The three-carbon propionic acid moiety originates from several sources:

1.   Catabolism of the branch-chained amino acids isoleucine, valine, methionine, and threonine
2.   The side chain of cholesterol
3.   Oxidation of odd chain length saturated fatty acids, and
4.   Metabolism of dietary carbohydrate by intestinal flora

In a fashion analogous to MCC deficiency, deficiency of PCC leads to increased urinary excretion of 3-hydroxypropionic acid and 2-methylcitric acid (3,86).

PCC consists of 12 subunits: 6 α subunits and 6 β subunits (87,88). Each subunit is transported into the mitochondria, cleaved to a smaller "mature" subunit, and assem-

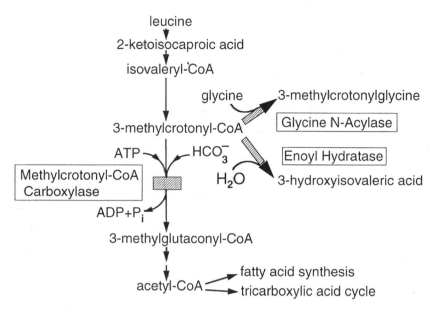

Fig. 3 Leucine degradation. A deficiency ▨ of MCC causes increased urinary excretion of 3-methylcrotonylglycine and 3-hydroxyisovaleric acid. Abbreviations as in Fig. 1 and 2.

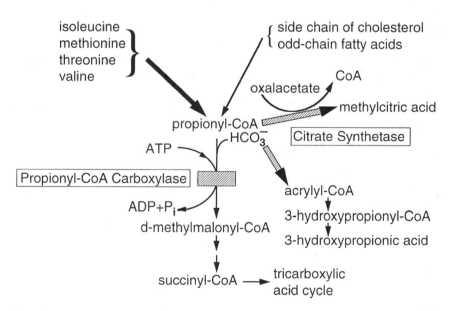

Fig. 4 Propionate metabolism. Propionate is derived primarily from amino acid degradation and to a lesser extent from cholesterol and odd-chain fatty acids, a deficiency ▨ of PCC causes increased urinary excretion of methylcitric and 3-hydroxypropionic acids. Abbreviations as in Fig. 1–3.

bled into the active enzyme. Inborn errors affecting the two subunits lead to the two major complementation groups (pcc A and pcc BC) of propionic acidemia (89). Differential rates of degradation of the α and β subunits explain observations concerning the presence or absence of the corresponding mRNA, synthesis rates of the subunits, and steady-state levels of the subunits (90). The α subunit can be transported to the mitochondria before or after biotinylation in the cytosol, providing evidence that biotinylation is not required for transport of carboxylase subunits into mitochondria and that holocarboxylase synthetase is active in both cytosol and mitochondria (91). Glycine and valine residues near the biotinylated lysine residue are highly conserved among all known biotinylated peptides and among many lipoylated proteins as well. One interpretation concerning conservation of this amino acid sequence is that these residues allow the biotinylated (or lipoylated) peptide to swing the carboxyl (or acetyl) group from the site of activation to the receiving substrate (91). Indeed, there is a close functional analogy between these two enzyme cofactors. Each has a valeric acid side chain joined to a lysine residue on one end; the other end joins to a complex sulfur containing ring(s) that accept and transfer the activated moiety (i.e., carboxyl for the carboxylases and acetyl for the pyruvate dehydrogenase complex). Avidin binds to the lipoyl domains in the pyruvate dehydrogenase complex (92). However, binding of lipoic acid analogs to avidin appears to be weak compared to biotin binding to avidin because the interaction is not detectable in our sequential, solid-phase avidin-binding assay (Zempleni, Mock, and McCormick, unpublished observations).

Genetic deficiencies of holocarboxylase synthetase and biotinidase cause the two distinct types of multiple carboxylase deficiency that were previously designated as the neonatal and juvenile forms. Biotinidase deficiency is discussed in several sections of this chapter because studies of these patients have provided important insights into biotin nutrition and the pathogenesis of the clinical findings of biotin deficiency. In the normal turnover of cellular proteins, holocarboxylases are degraded to biotin linked to lysine (biocytin) or biotin linked to an oligopeptide containing at most a few amino acid residues (Fig. 1). Because the amide bond between biotin and lysine is not hydrolyzed by cellular proteases, a specific hydrolase is required to release biotin for recycling. This enzyme is biotinidase (biotin–amide hydrolase, EC 3.5.1.12). Current opinion is that serum biotinidase and lipoamidase (the enzyme that cleaves lipoic acid from the dihydrolipoyl dehydrogenase component of the various multienzyme α-keto acid dehydrogenase complexes) are the same enzyme. This conclusion is based on the following observations: (a) greater than 95% inactivation of lipoamidase by a monospecific polyclonal antibody against biotinidase (93) and (b) less than 2% of normal lipoamidase activity in the serum of patients with biotinidase deficiency (93,94).

The gene for human biotinidase has been cloned, sequenced, and characterized (95). The biotinidase gene is a single-copy gene of 1629 bases encoding a 543-amino-acid protein; the mRNA is present in multiple tissues, including heart, brain, placenta, liver, lung, skeletal muscle, kidney, and pancreas. Highest biotinidase activities are found in serum, liver, kidney, and adrenal gland. On the basis of decreased serum concentrations of biotinidase in patients with impaired liver function, Grier et al. have concluded that the liver is the source of serum biotinidase (96).

D. Biotin Catabolism

Instead of being incorporated into carboxylases after entering the pools of biotin and its intermediary metabolites, dietary biotin or biotin released by carboxylase turnover may be catabolized. For example, biotinyl-AMP can be converted to biotinyl-CoA by biotinyl-CoA synthetase (97,98). Biotinyl-CoA synthetase also catalyzes the formation of biotinyl-AMP from biotin and ATP; thus, biotinyl-CoA synthetase catalyzes a two-step process converting biotin to biotinyl-CoA. The relation between biotinyl-CoA synthetase and holo-carboxylase synthetase, as well as the existence and location of intracellular pools of biotinyl-AMP and biotinyl-CoA (if any), remains unclear. Biotinyl-CoA is oxidized to bisnorbiotin and tetranorbiotin (metabolites with two and four fewer carbons in the valeric acid side chain, respectively; Fig. 1).

Contrary to the tacit assumption of many early biotin balance studies (3), it now appears that about half of biotin undergoes metabolism before excretion, and thus a significant proportion of the total avidin-binding substances in human urine and plasma and rat urine is attributable to biotin metabolites rather than to biotin per se. The findings of a pioneering study of biotin metabolites in human urine (99) using paper chromatography and bioassays have been confirmed by recent studies that take advantage of the greater sensitivity and reproducibility of HPLC separation and avidin-binding assays (3,100,101). Such studies have indicated that biotin, bisnorbiotin, and biotin sulfoxide are present in mole ratios of approximately 3:2:1 in human urine and plasma. Recent observations provide evidence that biotin catabolism is induced in some individuals during pregnancy and by anticonvulsants, thereby increasing the ratio of biotin catabolites to biotin (102,103). As discussed below, these observations emphasize the importance of distinguishing biotin from its catabolites when assaying physiological fluids.

E. Other Roles for Biotin

Effects of biotin on cell growth, glucose homeostasis (104,105), DNA synthesis, and expression of the asialoglycoprotein receptor (106) have been reported. For these effects, a direct relationship to biotin's role as a cofactor for the four carboxylases has not been defined. Whether one or more of the effects will ultimately prove to be the indirect result of carboxylase deficiency remains unclear.

V. REQUIREMENT AND ASSESSMENT

A. Circumstances Leading to Deficiency

That the normal human has a requirement for biotin has been clearly documented in two situations: (a) prolonged consumption of raw egg white and (b) parenteral nutrition without biotin supplementation in patients with short-gut syndrome and other causes of malabsorption (3). Biotin deficiency also has been clearly demonstrated in biotinidase deficiency (107). The mechanism by which biotinidase deficiency leads to biotin deficiency probably involves several processes:

1. Gastrointestinal absorption of biotin may be decreased because deficiency of biotinidase in pancreatic secretions leads to inadequate release of protein-bound biotin.
2. Salvage of biotin at the cellular level may be impaired during normal turnover of proteins to which biotin is linked covalently.
3. Renal loss of biocytin and biotin is abnormally increased.

The clinical findings and biochemical abnormalities caused by biotinidase deficiency are similar to those of biotin deficiency; the common findings include periorificial dermatitis, conjunctivitis, alopecia, ataxia, and developmental delay (107,108). These clinical similarities support the hypothesis that the pathogenesis of biotinidase deficiency involves a secondary biotin deficiency. However, the reported signs and symptoms of biotin deficiency and biotinidase deficiency are not identical. Seizures, irreversible neurosensory hearing loss, and optic atrophy have been observed in biotinidase deficiency (109–111) but have not been reported in biotin deficiency. However, cerebral atrophy and apparent stretching of the optic nerve have been reported in one patient with biotin deficiency (112). Moreover, Heard et al. have reported that biotin deficiency causes impaired auditory brain stem function in young rats (113).

On the basis of lymphocyte carboxylase activity and plasma biotin levels, Velazquez and co-workers have reported that biotin deficiency occurs in children with severe protein energy malnutrition (114,115). These investigators have speculated that the effects of biotin deficiency may be responsible for part of the clinical syndrome of protein energy malnutrition.

Accumulating data provide evidence that long-term anticonvulsant therapy in adults can lead to biotin depletion and that the depletion can be severe enough to interfere with amino acid metabolism. In the initial reports, Krause and co-workers (116,117) reported decreased plasma concentrations of biotin (as determined using the *Lactobacillus plantarum* bioassay). The later demonstration of increased urinary excretion of 3-hydroxyisovaleric acid in some adults receiving long-term anticonvulsant therapy (118) has been confirmed by Mock and Dyken (103) and provides evidence that biotin is depleted at the tissue level.

The mechanism of biotin depletion during anticonvulsant therapy is not known. The anticonvulsants implicated include phenobarbital, phenytoin, carbamazepine, and primidone. These drugs each have a carbamide (—NH—CO—) moiety in their structures, as does biotin; in some cases, they incorporate a full ureido group (—NH—CO—NH—). Said and co-workers (16,119) have demonstrated that therapeutic concentrations of primidone and carbamazepine specifically inhibit biotin uptake by brush-border membrane vesicles from human intestines; these investigators have suggested that these anticonvulsants compete with biotin for binding to the intestinal transporter leading to biotin malabsorption and biotin deficiency. Mock and co-workers have recently reported substantial increases in the urinary excretion of biotin catabolites, especially bisnorbiotin, in these patients (103,120). These urinary losses are sufficient to waste a substantial proportion of dietary intake and thus eventually deplete total-body pools of biotin; hence, these investigators have suggested that accelerated catabolism of biotin may also contribute to reduced biotin status in these individuals (103). Chuahan and Dakshinamurti (26) have also demonstrated that phenobarbital, phenytoin, and carbamazepine displace biotin from biotinidase. In this manner, these anticonvulsants could conceivably affect plasma transport of biotin, renal handling of biotin, or cellular uptake of biotin.

Biotin deficiency has also been reported or inferred in several other circumstances:

1. *Leiner's disease*: A severe form of seborrheic dermatitis that occurs in infancy. Although a number of studies have reported prompt resolution of the rash with biotin therapy (3), biotin was ineffective in the only double-blind therapeutic trial (121).

2. *Sudden infant death syndrome*: Biotin deficiency in the chick produces a fatal hypoglycemic disease dubbed "fatty liver–kidney syndrome"; impaired gluconeogenesis due to deficient activity of PC is the cause of the hypoglycemia. Hood et al. have proposed that biotin deficiency may cause sudden infant death syndrome (SIDS) by an analogous pathogenic mechanism (122,123). They have supported their hypothesis by demonstrating

that hepatic biotin is significantly lower at autopsy in SIDS infants than in infants dying from other causes. Additional studies (e.g., levels of hepatic PC, urinary organic acids, and blood glucose) are needed to confirm or refute this hypothesis.

3. *Pregnancy*: Concerns about the teratogenic effects of biotin deficiency (see below) have led to studies of biotin status during human gestation. Some of these studies have detected low plasma concentrations of biotin; others have not (3). Recent studies by Mock et al. have detected increased urinary excretion of 3-hydroxyisovaleric acid in more than half of normal women by the third trimester of pregnancy; urinary excretion of biotin was abnormally decreased in about half of the pregnant women studied (102).

4. *Dialysis*: Patients undergoing chronic hemodialysis have been reported to have low plasma concentrations of biotin (124). Yatzidis et al. (125) have reported nine patients on chronic hemodialysis who developed encephalopathy (four patients) or peripheral neuropathy (five patients); all responded to biotin therapy. Blood concentrations of biotin and lactic acid and urinary excretion rates of the characteristic organic acids were not reported. Braguer and co-workers have reported that microtubule formation, at least in vitro, is inhibited by toxins partially purified from the plasma of uremic patients (126). This inhibition is completely reversed by the addition of biotin. These investigators speculated that impaired tubulin formation might be involved in uremic neuropathy and that reversal of impaired tubulin formation could be a partial explanation for the clinical improvement seen with biotin in uremic neuropathy. However, other investigators have reported that plasma and red cell concentrations of biotin are significantly increased in patients receiving chronic hemodialysis (127). The etiological role of biotin in uremic neurological disorders and the general applicability of these results remain to be determined.

5. *Gastrointestinal diseases or alcoholism*: Reduced blood or liver concentrations of biotin or urinary excretion of biotin have been reported in alcoholism (1,128), gastric disease (1), and inflammatory bowel disease (129).

6. *Brittle nails*: Because pathological hoof changes in horses and swine have responded to biotin supplementation, Colombo and co-workers treated women with brittle fingernails with 2.5 mg biotin per day orally (130). They reported a 25% increase in nail thickness and improved morphology by electron microscopy. The biotin status of the subjects was not assessed.

B. Clinical Findings of Frank Deficiency

Whether caused by egg-white feeding or omission of biotin from total parenteral nutrition, the clinical findings of frank biotin deficiency in adults and older children have been similar to those reported by Sydenstricker in his pioneering study of egg-white feeding (3,131). Typically, the findings began to appear gradually after weeks to several years for egg-white feeding Six months to 3 years typically elapsed between the initiation of total intravenous feeding without biotin and the onset of the findings of biotin deficiency (3,132). Thinning of hair, often with loss of hair color, was reported in most patients. A skin rash described as scaly (seborrheic) and red (eczematous) was present in the majority; in several, the rash was distributed around the eyes, nose, and mouth. Depression, lethargy, hallucinations, and paresthesias of the extremities were prominent neurological symptoms in the majority of adults.

In infants who developed biotin deficiency, the signs and symptoms of biotin deficiency began to appear 3–6 months after initiation of total parenteral nutrition. This earlier onset may reflect an increased biotin requirement because of growth. The rash typically appeared first around the eyes, nose, and mouth; ultimately, the ears and perineal orifices were involved (''periorificial''). The appearance of the rash was similar to that of cutaneous candidiasis (i.e., an erythematous base and crusting exudates); typically, *Candida* could be cultured from the lesions. The rash of biotin deficiency is similar in appearance and distribution to the rash of zinc deficiency. In infants, hair loss was noted after 6–9 months of parenteral nutrition; within 3–6 months of the onset of hair loss, two infants had lost all hair, including eyebrows and lashes. These cutaneous manifestations, in conjunction with an unusual distribution of facial fat, have been dubbed ''biotin deficiency facies.'' The most striking neurological findings in biotin-deficient infants were hypotonia, lethargy, and developmental delay. A peculiar withdrawn behavior was noted and may reflect the same central nervous system dysfunction diagnosed as depression in adult patients.

C. Laboratory Findings of Biotin Deficiency

1. Methodology for Measuring Biotin

Methods for measuring biotin at pharmacological and physiological concentrations have been reviewed (3). In this chapter, assays for biotin and metabolites will be discussed briefly with two objectives in mind: (a) to provide the reader with sufficient knowledge of methods to understand the implications of recent studies of biotin nutritional status and (b) to review recent analytical advances.

For measuring biotin at physiological concentrations (i.e., 100 pmol/L to 100 nmol/L), a variety of assays have been proposed, and a limited number have been used to study biotin nutriture. All the published studies of biotin nutriture have used one of three basic types of biotin assays: bioassays (most studies), avidin-binding assays (several recent studies), or fluorescent derivative assays (two published studies).

Bioassays generally have adequate sensitivity to measure biotin in blood and urine. Precision is limited for the turbidity methods, but a recent modification of the *L. plantarum* assay uses agar plates previously injected with *L. plantarum* to obtain better precision (133). Radiometric bioassays offer both sensitivity and precision. However, the bacterial bioassays (and perhaps the eukaryotic bioassays as well) suffer interference from unrelated substances and variable growth response to biotin analogs; these bioassays can give conflicting results if biotin is bound to protein (3). For some bioassay organisms, prior acid or enzymatic hydrolysis (or both) is required to release the biotin from protein and thus make the biotin available to the assay organism. For other organisms (e.g., *Klockera brevis*), the detectable biotin decreases with enzymatic hydrolysis (134), suggesting that some destruction of biotin may occur during acid hydrolysis (53).

Avidin-binding assays generally measure the ability of biotin to do one of the following: (a) to compete with [³H]biotin, [¹²⁵I]biotin, or [¹⁴C]biotin for binding to avidin (isotope dilution assays), (b) to bind to [¹²⁵I]avidin and thus prevent [¹²⁵I]avidin or enzyme-coupled avidin from binding to a biotinylated protein adsorbed to plastic (sequential, solid phase assay), or (c) to prevent the binding of a biotinylated enzyme to avidin and thereby prevent the consequent inhibition of an enzyme activity. Other methods detect the postcolumn enhancement of fluorescence activity caused either by the mixing of the column eluate with fluorescent-labeled avidin or derivatization of biotin and metabolites by a fluorescent agent before separation by HPLC (135,136). Avidin-binding assays using novel detection systems, such as electrochemical detection (137), bioluminescence linked through glucose-6-phosphate dehydrogenase (138), or a double-antibody technique (139), have been reported recently and may offer some advantages in terms of sensitivity. Avidin-binding assays have been criticized for being cumbersome, requiring highly specialized equipment or reagents, or performing poorly when applied to biological fluids. Avidin-binding assays detect all avidin-binding substances, although the relative detectability of biotin and analogs varies between analogs and between assays, depending on how the assay is conducted (e.g., competitive vs. sequential). In a manner analogous to the pioneering work of Wright et al. (99), assays that couple chromatographic separation of biotin analogs with subsequent avidin-binding assays of the chromatographic fractions are both sensitive and chemically specific. These assays have been used in several recent studies that provide new insights into biotin nutrition.

A problem in the area of biotin analytical technology that remains unaddressed is the disagreement among the various bioassays and avidin-binding assays concerning the true concentration of biotin in human plasma. Reported mean values range from approximately 500 pmol/L to more than 10,000 pmol/L.

Although commonly used to assess biotin status in a variety of clinical populations, the putative indices of biotin status had not been previously studied during progressive biotin deficiency. To address this issue, Mock and co-workers (85) induced progressive biotin deficiency by feeding egg white; then the urinary excretion of biotin, bisnorbiotin, and 3-hydroxyisovaleric acid was measured. The urinary excretion of biotin declined dramatically with time on the egg-white diet, reaching frankly abnormal values in eight of nine subjects by day 20 of egg-white feeding. Bisnorbiotin excretion declined in parallel, providing evidence for regulation of catabolism of the biotin metabolic pools. By day 14 of egg-white feeding, 3-hydroxyisovaleric acid excretion was abnormally increased in all nine subjects, providing evidence that biotin depletion decreases the activity of the biotin-dependent enzyme MCC and alters intermediary metabolism of leucine earlier in the course of experimental biotin deficiency than previously appreciated. The time course for development of metabolic abnormalities was similar to that observed in the egg-white–fed rat (84,140). Serum concentrations of free biotin as measured by HPLC separation and avidin-binding assay decreased to abnormal values in less than half of the subjects. Thus, these studies provide objective confirmation of the impression of many investigators in this field (141) that blood biotin concentration is not an early or sensitive indicator of impaired biotin status.

Plasma concentrations of biotin (i.e., total avidin-binding substances) are higher in term infants than older children and, for reasons that are not simply related to dietary intake, decline after 3 weeks of breast-feeding or feeding a formula containing 11 µg/L of biotin. Infant formulas supplemented with 300 µg/L produce plasma concentrations approximately 20-fold greater than normal (142); consequences of these higher levels, if any, are unknown.

Odd-chain fatty acid accumulation is a marker of biotin deficiency, as shown by the work of Kramer et al. (143), Suchy et al. (144), and Mock et al. (145). These groups independently demonstrated increases in the percentage composition of odd-chain fatty acids (e.g., 15:0, 17:0, etc.) in hepatic, cardiac, or serum phospholipids in the biotin-deficient rat; similar accumulation has been reported in the liver of the biotin-deficient chick (146). Further, Mock and co-workers reported the accumulation of these odd-chain fatty acids in the plasma of patients who developed biotin deficiency during parenteral nutrition (147). The accumulation of odd-chain fatty acid is thought to result from propionyl-CoA carboxylase (PCC) deficiency, based on the observation that the isolated genetic deficiency of PCC and related disorders cause an accumulation of odd-chain fatty acids in plasma, red blood cells, and liver (148,149). Apparently, the accumulation of propionyl-CoA leads to the substitution of propionyl-CoA moiety for acetyl-CoA in the ACC reaction and hence to the ultimate incorporation of a three-carbon, rather than a two-carbon, moiety during fatty acid elongation (150).

2. Biochemical Pathogenesis

The mechanisms by which biotin deficiency produces specific signs and symptoms remain to be completely delineated. However, several studies have given new insights into the biochemical pathogenesis of biotin deficiency. The tacit assumption of most of these studies is that the clinical findings of biotin deficiency result directly or indirectly from deficient activities of the four biotin-dependent carboxylases.

Sander and co-workers initially suggested that the central nervous system (CNS) effects of biotinidase deficiency (and, by implication, biotin deficiency) might be mediated through deficiency of PC and the attendant CNS lactic acidosis (151). Because brain PC activity declined more slowly than hepatic PC activity during progressive biotin deficiency in the rat, these investigators discounted this mechanism. However, subsequent studies suggest their original hypothesis is correct. Diamantopoulos et al. (11) expanded the hypothesis by proposing that deficiency of brain biotinidase (which is already low in the normal brain) combined with biotin deficiency leads to a deficiency of brain pyruvate carboxylase and, in turn, to CNS accumulation of lactic acid. This CNS lactic acidosis is postulated to be the primary mediator of the hypotonia, seizures, ataxia, and delayed development seen in biotinidase deficiency. Additional support for the CNS lactic acidosis hypothesis has come from direct measurements of CSF lactic acid in children with either biotinidase deficiency or isolated pyruvate carboxylase deficiency and from the rapid resolution of lactic acidemia and CNS abnormalities in patients who have developed biotin deficiency during parental nutrition (3). The work of Suchy, Wolf, and Rizzo has provided evidence against an etiologic role for disturbances in brain fatty acid composition in the CNS dysfunction (144,152).

Several studies have demonstrated abnormalities in the metabolism of fatty acids in biotin deficiency and have suggested that these abnormalities are important in the pathogenesis of the skin rash and hair loss. Initially, the similarity between cutaneous manifestations of biotin deficiency and essential fatty acid deficiency as well as the established role of biotin in lipid synthesis suggested a relationship that has led to dietary intervention with polyunsaturated fatty acid. For example, Munnich et al. (153) have described a 12-year-old boy with multiple carboxylase deficiency; in retrospect, the enzymatic defect was almost certainly biotinidase deficiency (154). The child presented with alopecia and periorificial scaly dermatitis. Oral administration of "unsaturated fatty acids composed of 11% C18:1, 71% C18:2, 8% C18:3, and 0.3% C20:4" at a rate of 200–400 mg/day plus twice-daily topical administration of the same mixture of fatty acids "resulted in dramatic improvement of the dermatologic condition" and hair growth. Lactic acidosis and organic aciduria remained the same. These investigators speculated that the ACC deficiency led to impaired synthesis or metabolism of long-chain polyunsaturated fatty acids (PUFAs), which was treated by the topical and oral administration of PUFAs.

Three studies in the rat support the possibility of abnormal PUFA metabolism as a result of biotin deficiency and as a cause of the cutaneous manifestations. Kramer et al. (143) and Mock et al. (145) reported significant abnormalities in the n-6 phospholipids of blood, liver, and heart. Watkins and Kratzer also found abnormalities of n-6 phospholipids in liver and heart of biotin-deficient chicks (146). It has been speculated (145–156) that these abnormalities in PUFA composition might result in abnormal composition or metabolism of the prostaglandins and related substances derived from these PUFAs. However, these studies do not directly address the question of an etiological role. To address that question, Mock (157) examined the effect of supplementation of the n-6 PUFAs (as Intralipid) on the cutaneous manifestations of biotin deficiency in a nutrient interaction experiment. Supplementation of n-6 PUFAs prevented the development of the cutaneous manifestations of biotin deficiency in a group of rats that were as biotin-deficient (based on biochemical measurements) as the biotin-deficient control group. The rats not receiving the supplemental n-6 fatty acids did develop the classic rash and hair loss. These investigators concluded that an abnormality in n-6 PUFA metabolism plays a pathogenic role in the cutaneous manifestations of biotin deficiency and that the effect of the n-6 PUFAs cannot be attributed to biotin sparing.

D. Other Effects of Deficiency

Subclinical biotin deficiency has been shown to be teratogenic in several species, including chicken, turkey, mouse, rat, and hamster (3). Fetuses of mouse dams with biotin deficiency too mild to produce the characteristic cutaneous or CNS findings developed micrognathia, cleft palate, and micromelia (158–160). The incidence of malformation increased with the degree of biotin deficiency to a maximum incidence of approximately 90%. Differences in teratogenic susceptibility among rodent species have been reported; a corresponding difference in biotin transport from the mother to the fetus has been proposed as the cause (161). Bain et al. have hypothesized that biotin deficiency affects bone growth by affecting synthesis of prostaglandins from n-6 fatty acid (162). This effect on bone growth might be the mechanism for the skeletal malformations caused by biotin deficiency.

On the basis of studies of cultured lymphocytes in vitro and of rats and mice in vivo, biotin is required for normal function of a variety of immunological cells. These functions include production of antibodies, immunological reactivity, protection against sepsis, macrophage function, differentiation of T and B lymphocytes, afferent immune response, and cytotoxic T-cell response (3). In humans, Okabe et al. (129) have reported that patients with Crohn's disease have depressed natural killer all activity caused by biotin deficiency and are responsive to biotin supplementation. In patients with biotinidase deficiency, Cowan et al. (163) have demonstrated defects in both T-cell and B-cell immunity.

E. Diagnosis of Biotin Deficiency

The diagnosis of biotin deficiency has been established by demonstrating reduced urinary excretion of biotin, increased urinary excretion of the characteristic organic acids discussed earlier, and resolution of the signs and symptoms of deficiency in response to biotin supplementation. Plasma and serum levels of biotin, whether measured by bioassay or by avidin-binding assay, have not uniformly reflected biotin deficiency (141). The clinical response to administration of biotin has been dramatic in all well-documented cases of biotin deficiency. Within a few weeks, healing of the rash was striking, and by 1–2 months, growth of healthy hair was generally present. Within 1–2 weeks in infants, hypotonia, lethargy, and depression generally resolved; accelerated mental and motor development followed.

F. Requirements and Allowances

Data providing an accurate estimate of the biotin requirement for infants, children, and adults are lacking (164); as a result, recommendations often conflict (3). Data providing an accurate estimate of the requirement for biotin administered parenterally are also lacking. For parenteral administration, uncertainty about the true metabolic requirement for biotin is compounded by a lack of information concerning the effects of infusing biotin systemically and continuously (rather than the usual postprandial absorption into intestinal portal blood). Despite these limitations, recommendations for biotin supplementation have been formulated for oral and parenteral intake from preterm infants through adults (164–166). These recommendations are given in Table 1. One published study of parenterally supplemented infants (167) found normal plasma levels of biotin in term infants supplemented at 20 µg/day and increased plasma levels of biotin in preterm infants supplemented at 13 µg/day. [Note that the unit for plasma biotin should be picograms per milliliter in this publication (167).]

An important factor in the current uncertainty concerning the biotin requirement is the possibility that biotin synthesized by intestinal bacteria (referred to hereafter as ''bacterial biotin'') may contribute significantly to absorbed biotin. If so, the required intake would be reduced and might be dependent on factors that influence the density and species distribution of intestinal flora. For example, it is conceivable that interruption of absorption of bacterial biotin is a critical event in both biotin deficiency caused by egg-white ingestion and biotin deficiency that develops during biotin-free parenteral feeding. Binding of bacterial biotin by avidin may be occurring in the former, and reduced intestinal surface for absorption, rapid transit time, and antibiotic suppression of gut bacteria may lead to reduced absorption of bacterial biotin in the latter. Unfortunately, few data are available for assessing the actual magnitude of the absorbed bacterial biotin.

G. Dietary Sources of Biotin

There is no published evidence that biotin can be synthesized by mammals; thus, the higher animals must derive biotin from other sources. The ultimate source of biotin appears

Table 1 Recommended Daily Intake of Biotin (µg/kg)

Age	Safe and adequate oral intakes[a]	Parenteral intakes[b]
Preterm infants[c]	5	5–8 µg · kg^{-1}
Infants up to 6 mo	5	20
Infants 0.5–1 y	6	20
Children 1–3 y	8	20
Children 4–8 y	12	20
Older children, 9–13 y	20	20
Older children 14–18 y	25	60
Adults	30	—
Pregnancy	30	—
Lactation	35	—

[a] Ref. 164
[b] Ref. 165.
[c] Ref. 166.

to be de novo synthesis by bacteria; primitive eukaryotic organisms, such as yeast, molds, and algae; and some plant species.

Most measurements of the biotin content of various foods have used bioassays. Despite the limitations due to interfering substances, protein binding, and lack of chemical specificity as discussed above, there is reasonably good agreement among the published reports (168–172), and some worthwhile generalizations can be made. Biotin is widely distributed in natural foodstuffs, but the absolute content of even the richest sources is low when compared with the content of most other water-soluble vitamins. Foods relatively rich in biotin include egg yolk, liver, and some vegetables. Based on the data of Hardinge (168), the average dietary biotin intake has been estimated to be approximately 70 μg/day for the Swiss population. This result is in reasonable agreement with the estimated dietary intake of biotin in a composite Canadian diet (62 μg/day) and the actual analysis of the diet (60 μg/day) (173). Calculated intake of biotin for the British population was 35 μg/day (174,175).

VI. PHARMACOLOGY AND TOXICITY

A. Treatment of Biotin Deficiency

Pharmacological doses of biotin (e.g., 1–10 mg) have been used to treat most patients. For two patients, parenteral administration of 100 μg of biotin per day was adequate to cause resolution of the signs and symptoms of biotin deficiency and to prevent their recurrence (3). However, abnormal organic aciduria persisted for at least 10 weeks in one patient receiving 100 mg/day, suggesting that this dose may not have been adequate to restore tissue biotin levels to normal over that time. Could organic aciduria be an indication that biotin status at the tissue level has not been restored to normal? Could this degree of deficiency be less severe but sufficient to cause significant, subtle morbidity? If so, should a loading dose of 1 or 10 mg for 1 or 2 weeks be given as initial therapy for acquired biotin deficiency? There are currently no data on which to base answers to these questions.

B. Toxicity

Daily doses up to 200 mg orally and up to 20 mg intravenously have been given to treat biotin-responsive inborn errors of metabolism and acquired biotin deficiency; toxicity has not been reported.

C. Pharmacology

Mounting reports of biotin deficiency in commercial animals and humans have led to several studies of plasma levels, pharmacokinetics, and bioavailability after acute or chronic oral, intramuscular, or intravenous administration of biotin in cattle (176), swine (177,178), and human subjects (179–181).

Studies using bioassays to measure biotin in blood and urine suggest that doses of biotin less than 150 mg are suitable as loading doses to assess biotin status (180). Doses greater than 300 mg result in high biotin concentrations in blood and urinary excretion of a large proportion of the administered biotin as the unchanged vitamin (179–181). When metabolites are measured separately from biotin per se, increased blood concentrations (181) and urine excretion rates (unpublished observations) of bisnorbiotin and biotin sulfoxide as well as biotin are observed. These observations are consistent with the metabolites originating from human tissues rather than from enteric bacteria.

ACKNOWLEDGMENTS

Many thanks to Nell Mock and Celia Bernheimer for preparing the art work and to Gwyn Hobby for typing the manuscript.

REFERENCES

1. J.-P. Bonjour, Biotin, in *Handbook of Vitamins* (L. J. Machlin, ed.), Marcel Dekker, New York, 1991, pp. 393–427.
2. D. Bhatia, B. Borenstein, S. Gaby, H. Gordon, A. Iannarone, L. Johnson, L. Machlin, W. Mergens, J. Scheiner, J. Scott, E. Waysek, in *Encyclopedia of Food Science and Technology*, Y. U. Hui, ed. John Wiley and Sons, New York, 1992, pp. 2764–2770.
3. D. M. Mock, Biotin, in *Present Knowledge in Nutrition* (M. Brown, ed.), International Life Sciences Institute–Nutrition Foundation, Washington, D.C., 1989, pp. 189–207.
4. A. Marquet, New aspects of the chemistry of biotin and some analogs, *Pure Appl. Chem. 49*:183–196 (1977).
5. L. H. Sternbach, Biotin, in *Comprehensive Biochemistry* (M. Florkin and E. G. Stotz, eds.), Elsevier, New York, 1963, pp. 66–81.
6. D. Miljkovic, S. Velimirovic, J. Csanadi, and V. Popsavin, Studies directed towards stereospecific synthesis of oxybiotin, biotin, and their analogs. Preparation of some new 2,5 anhydro-xylitol derivatives, *J. Carbohydr. Chem. 8*:457–467 (1989).
7. F. D. Deroose and P. J. DeClercq, Novel enantioselective syntheses of (+)−biotin, *J. Org. Chem. 60*:321–330 (1995).
8. B. Wolf, G. Heard, J. R. S. McVoy, and H. M. Raetz, Biotinidase deficiency: the possible role of biotinidase in the processing of dietary protein-bound biotin, *J. Inher. Metab. Dis. 7*:121–122 (1984).
9. H. M. Said, L. P. Thuy, L. Sweetman, and B. Schatzman, Transport of the biotin dietary derivative biocytin (N-biotinyl-L-lysine) in rat small intestine, *Gastroenterology 104*:75–80 (1993).
10. A. Leon-Del-Rio, G. Vizcaino, G. Robles-Diaz, and A. Gonzalex-Noriega, Association of pancreatic biotinidase activity and intestinal uptake of biotin and biocytin in hamster and rat, *Ann. Nutr. Metab. 34*:266–272 (1990).
11. N. Diamantopoulos, M. J. Painter, B. Wolf, G. S. Heard, and C. Roe, Biotinidase deficiency: accumulation of lactate in the brain and response to physiologic doses of biotin, *Neurology 36*:1107–1109 (1986).
12. B. B. Bowman, D. B. McCormick, and I. H. Rosenberg, Epithelial transport of water-soluble vitamins, *Annu. Rev. Nutr. 9*:187–199 (1989).
13. T. Y. Ma, D. L. Dyer, and H. M. Said, Human intestinal cell line Caco-2: a useful model for studying cellular and molecular regulation of biotin uptake, *Biochim. Biophys. Acta 1189*: 81–88 (1994).
14. H. M. Said and I. Derweesh, Carrier-mediated mechanism for biotin transport in rabbit intestine—studies with brush-border membrane vesicles, *Am. J. Physiol. 261*:R94–R97 (1991).
15. H. M. Said, D. W. Horne, and D. M. Mock, Effect of aging on intestinal biotin transport in the rat, *Exp. Gerontol.* 67–73 (1989).
16. H. M. Said, D. M. Mock, and J. C. Collins, Regulation of biotin intestinal transport in the rat: effect of biotin deficiency and supplementation, *Am. J. Physiol. 256(2)P1*:G306–G311 (1989).

17. D. McCormick, and Z. Zhang, Cellular assimilation of water-soluble vitamins in the mammal: riboflavin, B$_6$, biotin, and C, *Proc. Soc. Exp. Biol. Med. 202*:265–270 (1993).

18. K.-Y. Ng and R. T. Borchardt, Biotin transport in a human intestinal epithelial cell line (Caco-2), *Life Sci. 53*:1121–1127 (1993).

19. B. B. Bowman, and I. Rosenberg, Biotin absorption by distal rat intestine, *J. Nutr. 117*:2121–2126 (1987).

20. M. F. Sorrell, O. Frank, A. D. Thomson, H. Aquino, and H. Baker, Absorption of vitamins from the large intestine in vivo, *Nutr. Rep. Int. 3*:143–148 (1971).

21. T. W. Oppel, Studies of biotin metabolism in man: IV. Studies of the mechanism of absorption of biotin and the effect of biotin administration on a few cases of seborrhea and other conditions, *Am. J. Med. Sci.* 76–83 (1948).

22. H. M. Said and R. Mohammadkhani, Involvement of histidine residues and sulfhydryl groups in the function of the biotin transport carrier of rabbit intestinal brush-border membrane, *Biochim. Biophys. Acta 1107*:238–244 (1992).

23. H. M. Said, R. Redha, and W. Nylander, Biotin transport in basolateral membrane vesicles of human intestine, *Gastroenterology 94*:1157–1163 (1988).

24. H. M. Said, A. Sharifian, A. Bagherzadeh, and D. Mock, Effect of chronic ethanol feeding and acute ethanol exposure in vitro on intestinal transport of biotin, *Am. J. Clin. Nutr. 52*:1083–1086 (1990).

25. B. Wolf, R. E. Grier, J. R. S. McVoy, and G. S. Heard, Biotinidase deficiency: a novel vitamin recycling defect, *J. Inher. Metab. Dis. 8*:53–58 (1985).

26. J. Chuahan, and K. Dakshinamurti, Role of human serum biotinidase as biotin-binding protein, *Biochem. J. 256*:265–270 (1988).

27. D. M. Mock and M. I. Malik, Distribution of biotin in human plasma: most of the biotin is not bound to protein, *Am. J. Clin. Nutr. 56*:427–432 (1992).

28. D. M. Mock and G. Lankford, Studies of the reversible binding of biotin to human plasma, *J. Nutr. 120*:375–381 (1990).

29. N. M. Green, Avidin, *Adv. Protein Chem. 29*:85–133 (1975).

30. N. M. Green, Spectrophotometric determination of avidin and biotin, in *Methods in Enzymology* (D. B. McCormick and L. D. Wright, eds.), Academic Press, New York, 1970, pp. 418–424.

31. N. M. Green, A spectrophotometric assay for avidin and biotin based on binding of dyes by avidin, *Biochem. J. 94*:23c–24c (1965).

32. D. M. Mock and P. Horowitz, A fluorometric assay for avidin-biotin interaction, in *Methods in Enzymology* (M. Wilchek and E. A. Bayer, eds.), Academic Press, San Diego, 1990, pp. 234–240.

33. D. M. Mock, G. L. Lankford, and P. Horowitz, A study of the interaction of avidin with 2-anilinonaphthalene-6-sulfonic acid as a probe of the biotin binding site, *Biochim. Biophys. Acta 956*:23–29 (1988).

34. D. M. Mock, G. Lankford, D. DuBois, and P. Horowitz, A fluorometric assay for the biotin-avidin interaction based on displacement of the fluorescent probe 2-anilinonaphthalene-6-sulfonic acid., *Ann. Biochem. 151*:178–181 (1985).

35. N. D. Cohen and M. Thomas, Biotin transport into fully differentiated 3T3-L1 cells, *Biochem. Biophys. Res. Commun. 108*:1508–1516 (1982).

36. D. M. Bowers-Komro, and D. B. McCormick, Biotin uptake by isolated rat liver hepatocytes, in *Biotin* (K. Dakshinamurti and H. N. Bhagavan, eds.), Annals of the New York Academy of Sciences, New York, 1985, pp. 350–358.

37. H. M. Said, J. Hoefs, R. Mohammadkhani, and D. Horne, Biotin transport in human liver basolateral membrane vesicles: a carrier-mediated, Na$^+$ gradient–dependent process, *Gastroenterology 102*:2120–2125 (1992).

38. H. J. Said, S. Korchid, and D. W. Horne, Transport of biotin in basolateral membrane vesicles of rat liver, *Am. J. Physiol. 259*:G865–G8720 (1990).

39. H. M. Said, T. Y. Ma, and V. S. Kamanna, Uptake of biotin by human hepatoma cell line, Hep G(2): a carrier-mediated process similar to that of normal liver, *J. Cell. Physiol. 161*: 483–489 (1994).

40. D. Weiner and B. Wolf, Biotin uptake in cultured hepatocytes from normal and biotin-deficient rats, *Biochem. Med. Metab. Biol. 44*:271–281 (1990).

41. D. Weiner and B. Wolf, Biotin uptake, utilization, and efflux in normal and biotin-deficient rat hepatocytes, *Biochem. Med. Metab. Biol. 46*:344–363 (1991).

42. H. M. Said, L. McAlister-Henn, R. Mohammadkhani, and D. W. Horne, Uptake of biotin by isolated rat liver mitochondria, *Am. J. Physiol. 263*:G81–G86 (1992).

43. R. Spector and D. M. Mock, Biotin transport through the blood–brain barrier, *J. Neurochem. 48*:400–404 (1987).

44. R. Spector and D. M. Mock, Biotin transport and metabolism in the central nervous system, *Neurochem. Res. 13*:213–219 (1988).

45. R.-A. Podevin and B. Barbarat, Biotin uptake mechanisms in brush-border and basolateral membrane vesicles isolated from rabbit kidney cortex, *Biochim. Biophys. Acta 856*:471–481 (1986).

46. P. D. Spencer and K. S. Roth, On the uptake of biotin by the rat renal tubule, *Biochem. Med. Metab. Biol. 40*:95–100 (1988).

47. E. R. Baumgartner, T. Sourmala, and H. Wick, Biotin-responsive multiple carboxylase deficiency (MCD): deficient biotinidase activity associated with renal loss of biotin, *J. Inher. Metab. Dis. 8*:59–64 (1985).

48. E. R. Baumgartner, T. Sourmala, and H. Wick, Biotinidase deficiency: factors responsible for the increased biotin requirement, *J. Inher. Metab. Dis. 8*:59–64 (1985).

49. E. R. Baumgartner, T. Sourmala, and H. Wick, Biotinidase deficiency associated with renal loss of biocytin and biotin, *J. Inher. Metab. Dis. 7*:123–125 (1985).

50. P. Karl and S. E. Fisher, Biotin transport in microvillous membrane vesicles, cultured trophoblasts and the isolated perfused cotyledon of the human placenta, *Am. J. Physiol. 262*:C302–308 (1992).

51. S. Schenker, Z. Hu, R. F. Johnson, Y. Yang, T. Frosto, B. D. Elliott, G. I. Henderson, and D. M. Mock, Human placental biotin transport: normal characteristics and effect of ethanol, *Alcoholism: Clin. Exp. Res. 17*:566–575 (1993).

52. Z.-Q. Hu, G. I. Henderson, S. Schenker, and D. M. Mock, Biotin uptake by basolateral membrane of human placenta: normal characteristics and role of ethanol, *Proc. Soc. Biol. Exp. Med. 206*:404–408 (1994).

53. D. M. Mock, N. I. Mock, and S. E. Langbehn, Biotin in human milk: Methods, location, and chemical form, *J. Nutr. 122*:535–545 (1992).

54. D. M. Mock, N. I. Mock, and J. A. Dankle, Secretory patterns of biotin in human milk, *J. Nutr. 122*:546–552 (1992).

55. S. Stratton, N. Mock, and D. Mock, Biotin and biotin metabolites in human milk: The metabolites are not negligible, *J. Invest. Med.* (1996).

56. Reference deleted.

57. L. Sweetman and W. L. Nyhan, Inheritable biotin-treatable disorders and associated phenomena, *Annu. Rev. Nutr. 6*:317–343 (1986).

58. G. L. Feldman and B. Wolf, Deficient acetyl CoA carboxylase activity in multiple carboxylase deficiency, *Clin. Chim. Acta 111*:147–151 (1981).

59. K. Narisawa, N. Arai, Y. Igarashi, T. Satoh, K. Tada, and Y. Hirooka, Clinical and biochemical findings on a child with multiple biotin-responsive carboxylase deficiencies, *J. Inher. Metab. Dis. 2*:67–68 (1982).

60. S. Packman, N. Caswell, M. C. Gonzalez-Rios, T. Kadlecek, H. Cann, D. Rassin, and C. McKay, Acetyl CoA carboxylase in cultured fibroblasts: differential biotin dependence in the two types of biotin-responsive multiple carboxylase deficiency, *Am. J. Hum. Genet.* 80–92 (1984).

61. P. N. Murthy and S. P. Mistry, Some aspects of biotin binding to protein catalized by biotin-deficient chicken liver preparations, *Proc. Soc. Exp. Biol. Med. 145*:565–570 (1974).

62. D. H. Bai, T. W. Moon, F. Lopez-Casillas, P. C. Andrews, and K. H. Kim, Analysis of the biotin-binding site on acetyl-CoA carboxylase from rat, *Eur. J. Biochem. 182*:239–245 (1989).

63. B. J. Shriver, C. Roman-Shriver, and J. B. Allred, Depletion and repletion of biotinyl enzymes in liver of biotin-deficient rats: evidence of a biotin storage system, *J. Nutr. 123*:1140–1149 (1993).

64. R. Brandsch, Regulation of gene expression by cofactors derived from B vitamins, *J. Nutr. Sci. Vitaminol. 40*:371–399 (1994).

65. J. E. Cronan Jr., The *E. coli* bio operon: transcriptional repression by an essential protein modification enzyme, *Cell 58*:427–429 (1989).

66. A. Chapman-Smith, D. L. Turner, J. E. Cronan, T. W. Morris, and J. C. Wallace, Expression, biotinylation and purification of a biotin-domain peptide from the biotin carboxy carrier protein of *Escherichia coli* acetyl-CoA carboxylase, *Biochem. J. 302*:881–887 (1994).

67. Y. Xu, and D. Beckett, Kinetics of biotinyl-5′-adenylate synthesis catalyzed by the *Escherichia coli* repressor of biotin biosynthesis and the stability of the enzyme–product complex, *Biochemistry 33*:7374–7360 (1994).

68. I. Ohsawa, T. Kisou, K. Kodama, I. Yoneda, D. Speck, R. Gloeckler, Y. Lemoine, and K. Kamogawa, Bioconversion of pimelic acid into biotin by *Bacillus sphaericus* bioB transformants, *J. Ferment. Bioeng. 73*:121–124 (1992).

69. O. Ifuku, S. Haze, J. Kishimoto, N. Koga, M. Yanagi, and S. Fukushima, Sequencing analysis of mutation points in the biotin operon of biotin-overproducing *Escherichia coli* mutants, *Biosci. Biotech. Biochem. 57*:760–765 (1993).

70. I. Sanyal, D. Cohen, and G. Flint, Biotin synthase: purification, characterization as a (2FE-2S) cluster protein, and in vitro activity of the *Escherichia coli* bioB gene product, *Biochemistry 33*:3625–2631 (1993).

71. S. Bower, J. Perkins, R. R. Yocum, P. Serror, A. Sorokin, P. Rahaim, C. L. Howitt, N. Prasad, S. D. Ehrlich, and J. Pero, Cloning and characterization of the *Bacillus subtilis* birA gene encoding a repressor of the biotin operon, *J. Bacteriol. 177*:2572–2575 (1995).

72. T. Ohshiro, M. Yamamoto, B. T. S. Bui, D. Florentin, A. Marquet, and Y. Izumi, Stimulatory factors for enzymatic biotin synthesis from dethiobiotin in cell-free extracts of *Escherichia coli*, *Biosci. Biotech. Biochem. 59*:943–944 (1995).

73. O. Ifuku, N. Koga, S. Haze, J. Kishimoto, and Y. Wachi, Flavodoxin is required for conversion of dethiobiotin to biotin in *Escherichia coli*, *Eur. J. Biochem. 224*:173–178 (1994).

74. A. Fujisawa, T. Abe, I. Ohsawa, S. Shiozaki, K. Kamogawa, and Y. Izumi, Bioconversion of dethiobiotin into biotin by resting cells and protoplasts of *Bacillus sphaericus* bioB transformant, *Biosci. Biotech. Biochem. 57*:740–744 (1993).

75. T. Ohshiro, M. Yamamoto, Y. Izumi, B. T. S. Bui, D. Florentin, and A. Marquet, Enzymatic conversion of dethiobiotin to biotin in cell-free extracts of a *Bacillus sphaericus* bioB transformant, *Biosci. Biotech. Biochem. 58*:1738–1741 (1994).

76. D. Florentin, B. T. S. Bui, A. Marquet, T. Ohshiro, and Y. Izumi, On the mechanism of biotin synthase of *Bacillus sphaericus*, *Biochimie 317*:485–488 (1994).

77. A. Fujisawa, T. Abe, I. Ohsawa, K. Kamogawa, and Y. Izumi, Bioconversion of dethiobiotin to biotin by a cell-free system of a bioYB transformant of *Bacillus sphaericus*, *FEMS Microbiol. Lett. 110*:1–4 (1993).

78. O. Ifuku, H. Miyaoka, N. Koga, J. Kishimoto, S. Haze, Y. Wachi, and M. Kajiwara, Origin of carbon atoms of biotin ^{13}C-NMR studies on biotin biosynthesis in *Escherichia coli*, *Eur. J. Biochem. 220*:585–591 (1994).

79. S. Zhang, I. Sanyal, G. H. Bulboaca, A. Rich, and D. H. Flint, The gene for biotin synthase from *Saccharomyces cerevisiae*: cloning, sequencing, and complementation of *Escherichia coli* strains lacking biotin synthase, *Arch. Biochem. Biophys. 309*:29–35 (1994).

80. J. Allred and C. R. Roman-Lopez, Enzymatically inactive forms of acetyl-CoA carboxylase in rat liver mitochondria, *Biochem. J. 251*:881–885 (1988).

81. J. Allred, C. Roman-Lopez, R. Jurin, and S. McCune, Mitochondrial storage forms of acetyl-CoA carboxylase: mobilization/activation accounts for increased activity of the enzyme in liver of genetically obese Zucker rats, *J. Nutr. 119*:478–483 (1989).

82. J. B. Allred, C. R. Roman-Lopez, T. S. Pope, and J. Goodson, Dietary dependent distribution of acetyl-CoA carboxylase between cytoplasm and mitochondria of rat liver, *Biochem. Biophys. Res. Commun. 129*:453–460 (1985).

83. C. Roman-Lopez, B. Shriver, C. Joseph, and J. Allred, Mitochondrial acetyl-CoA carboxylase: time course of mobilization/activation in liver of refed rats, *Biochem. J. 260*:927–930 (1989).

84. N. I. Mock and D. M. Mock, Biotin deficiency in rats: disturbances of leucine metabolism is detectable early, *J. Nutr. 122*:1493–1499 (1992).

85. N. I. Mock, D. M. Mock, M. Malik, and M. W. Bishop, Urinary excretion of biotin and 3-hydroxyisovaleric acid (3-HIA) are early indicators of biotin deficiency, *FASEB J. 9*:A985 (1995).

86. Y. Liu, Y. Shigematsu, A. Nakai, Y. Kikawa, M. Saito, T. Fukui, I. Hayakawa, J. Oizumi, and M. Sudo, The effects of biotin deficiency on organic acid metabolism: increase in propionyl coenzyme A–related organic acids in biotin-deficient rats, *Metab. Clin. Exp. 42*:1392–1397 (1993).

87. F. C. Haase, H. Beegen, and S. H. K. Allen, Propionyl coenzyme A carboxylase of *Mycobacterium smegmatis*, *Eur. J. Biochem. 140*:147–151 (1984).

88. G. J. Goodall, W. Johannssen, J. C. Wallace, and D. B. Keech, Sheep liver propionyl-CoA carboxylase: purification and some molecular properties, *Ann. N.Y. Acad. Sci. 47*: (1985).

89. R. A. Gravel, K.-F. Lam, K. J. Scully, and Y. E. Hsia, Genetic complementation of propionyl-CoA carboxylase deficiency in cultured human fibroblasts, *Am. J. Hum. Genet. 29*:378–388 (1977).

90. T. Ohura, J. P. Kraus, and L. E. Rosenberg, Unequal synthesis and differential degradation of propionyl-CoA carboxylase subunits in cells from normal and propionic acidemia patients, *Am. J. Hum. Genet. 45*:33–40 (1989).

91. M. F. Browner, F. Taroni, E. Sztul, and L. B. Rosenberg, Sequence analysis, biogenesis, and mitochondrial import of the α-subunit of rat liver propionyl-CoA carboxylase, *J. Biol. Chem. 264*:12680–12685 (1989).

92. G. Hale, N. G. Wallis, and R. N. Perham, Interaction of avidin with the lipoyl domains in the pyruvate dehydrogenase multienzyme complex: three-dimensional location and similarity to biotinyl domains in carboxylases, *Proc. R. Soc. Lond. B. Biol. Sci. 249*:247–253 (1992).

93. C. L. Garganta and B. Wolf, Lipoamidase activity in human serum is due to biotinidase, *Clin. Chim. Acta 189*:313–326 (1990).

94. L. Nilsson and E. Ronge, Lipoamidase and biotinidase deficiency: evidence that lipoamidase and biotinidase are the same enzyme in human serum, *Eur. J. Clin. Chem. Clin. Biochem. 30*:119–126 (1992).

95. H. Cole, T. R. Reynolds, J. M. Lockyer, G. A. Buck, T. Denson, J. E. Spence, J. Hymes, and B. Wolf, Human serum biotinidase cDNA cloning, sequence, and characterization, *J. Biol. Chem. 269*:6566–6570 (1994).

96. R. E. Grier, G. S. Heard, P. Watkins, and B. Wolf, Low biotinidase activities in the sera of patients with impaired liver function: evidence that the liver is the source of serum biotinidase, *Clin. Chim. Acta 186*:397–400 (1989).

97. R. M. Baxter and J. H. Quastel, The enzymatic breakdown of *d*-biotin in vitro, *J. Biol. Chem. 201*:751–764 (1953).

98. J. E. Christner, M. J. Schlesinger, and M. J. Coon, Enzymatic activation of biotin, *J. of Biol. Chem. 239*:3997–4005 (1964).

99. L. D. Wright, E. L. Cresson, and C. A. Driscoll, Biotin derivatives in human urine, *Proc. Soc. Exp. Biol. Med. 91*:248–252 (1956).

100. D. M. Mock, G. L. Lankford, and J. Cazin Jr., Biotin and biotin analogs in human urine: biotin accounts for only half of the total, *J. Nutr. 123*:1844–1851 (1993).

101. D. M. Mock, G. L. Lankford, and N. I. Mock, Biotin accounts for only half of the total avidin-binding substances in human serum, *J. Nutr. 125*:941–946 (1995).

102. D. M. Mock, N. I. Mock, S. L. Stratton, and D. D. Stadler, Urinary excretion of biotin decreases during pregnancy, providing evidence of decreased biotin status, *FASEB J. 9*:A155 (1995).

103. D. M. Mock and M. E. Dyken, Biotin deficiency results form long-term therapy with anticonvulsants, *Gastroenterology 108*:A740 (1995).

104. M. Maebashi, Y. Makino, Y. Furukawa, K. Ohinata, S. Kimura, and T. Sato, Therapeutic evaluation of the effect of biotin on hyperglycemia in patients with non-insulin dependent diabetes mellitus, *J. Clin. Biochem. Nutr. 14*:211–218 (1993).

105. A. Reddi, B. DeAngelis, O. Frank, N. Lasker, and H. Baker, Biotin supplementation improves glucose and insulin tolerances in genetically diabetic KK mice, *Life Sci. 42*:1323–1330 (1988).

106. J. C. Collins, E. Paietta, R. Green, A. G. Morell, and R. J. Stockert, Biotin-dependent expression of the asialoglycoprotein receptor in HepG2, *J. Biol. Chem. 263*:11280–11283 (1988).

107. B. Wolf, G. S. Heard, J. R. S. McVoy, and R. E. Grier, Biotinidase Deficiency, *Ann. NY Acad. Sci. 447*:252–262 (1985).

108. D. M. Mock, A. A. Delorimer, W. M. Liebman, L. Sweetman, and H. Baker, Biotin deficiency complicating parenteral alimentation, *Ann. N.Y. Acad. Sci. 447*:314–334 (1985).

109. B. A. Salbert, J. Astruc, and B. Wolf, Ophthalmologic abnormalities associated with biotinidase deficiency, *Ophthalmologica 206*:177–181 (1993).

110. B. A. Salbert, J. M. Pellock, and B. Wolf, Characterization of seizures associated with biotinidase deficiency, *Neurology 45*:1351–1354 (1993).

111. D. M. Bousounis, P. R. Camfield, and B. Wolf, Reversal of brain atrophy with biotin therapy in biotinidase deficiency, *Neuropediatrics 24*:214–217 (1993).

112. D. M. Mock, D. L. Baswell, H. Baker, R. T. Holman, and L. Sweetman, Biotin deficiency complicating parenteral alimentation: Diagnosis, metabolic repercussions, and treatment, *J. Pediatr. 106*:762–769 (1985).

113. G. S. Heard, M. L. Lenhardt, R. M. Bowie, A. M. Clarke, S. W. Harkins, and B. Wolf, Increased central conduction time (CTT) but no hearing loss (HL) in young biotin deficient rats, *FASEB J. 3*:A1242, #5901 (1989).

114. A. Velazquez, C. Martin-del-Campo, A. Baez, S. Zamudio, M. Quiterio, J. L. Aguilar, B. Perez-Ortiz, M. Sanchez-Ardines, Guzman-Hernandez, and E. Casanueva, Biotin deficiency in protein-energy malnutrition, *Eur. J. Clin. Nutr. 43*:169–173 (1988).

115. A. Velazquez, M. Teran, A. Baez, J. Gutierrez, and R. Rodriguez, Biotin supplementation affects lymphocyte carboxylases and plasma biotin in severe protein-energy malnutrition, *Am. J. Clin. Nutr. 61*:385–391 (1995).

116. K.-H. Krause, P. Berlit, and J.-P. Bonjour, Impaired biotin status in anticonvulsant therapy, *Ann. Neurol. 12*:485–486 (1982).

117. K.-H. Krause, P. Berlit, and J.-P. Bonjour, Vitamin status in patients on chronic anticonvulsant therapy, *Int. J. Vitam. Nutr. Res. 52*:375–385 (1982).

118. K.-H. Krause, W. Kochen, P. T. Berli, and J.-P. Bonjour, Excretion of organic acids associated with biotin deficiency in chronic anticonvulsant therapy, *Int. J. Vitam. Nutr. Res. 54*:217–222 (1984).

119. H. M. Said, R. Reyadh, and W. Nylander, Biotin transport and anticonvulsant drugs, *Am. J. Clin. Nutr. 49*:127–131 (1989).

120. K. A. Lombard, D. M. Mock, and R. P. Nelson, Does long-term anticonvulsant therapy predictably cause biotin deficiency in children? *Am. J. Clin. Nutr. 51*:518 (abstr.) (1990).

121. M. Erlichman, R. Goldstein, E. Levi, A. Greenberg, and S. Freier, Infantile flexural seborr-hoeic dermatitis. Neither biotin nor essential fatty acid deficiency, *Arch. Dis. Child. 567*: 560–562 (1981).

122. A. R. Johnson, R. L. Hood, and J. L. Emery, Biotin and the sudden infant death syndrome, *Nature 285*:159–160 (1980).

123. G. S. Heard, R. L. Hood, and A. R. Johnson, Hepatic biotin and the sudden infant death syndrome, *Med. J. Austr. 2*:305–306 (1983).

124. E. Livaniou, G. P. Evangelatos, D. S. Ithakissios, H. Yatzidis, and D. C. Koutsicos, Serum biotin levels in patients undergoing chronic hemodialysis, *Nephron 46*:331–332 (1987).

125. H. Yatzidis, D. Koutisicos, B. Agroyannis, C. Papastephanidis, M. Frangos-Plemenos, and Z. Delatola, Biotin in the management of uremic neurologic disorders, *Nephron 36*:183–186 (1984).

126. D. Braguer, P. Gallice, H. Yatzidis, Y. Berland, and A. Crevat, Restoration by biotin in the in vitro microtubule formation inhibited by uremic toxins, *Nephron 57*: 192–196 (1991).

127. V. DeBari, O. Frank, H. Baker, and M. Needle, Water soluble vitamins in granulocytes, erythrocytes, and plasma obtained from chronic hemodialysis patients, *Am. J. Clin. Nutr. 39*: 410–415 (1984).

128. J.-P. Bonjour, Biotin-dependent enzymes in inborn errors of metabolism in human, *World Rev. Nutr. Diet. 38*:1–88 (1981).

129. N. Okabe, K. Urabe, K. Fujita, T. Yamamoto, and T. Yao, Biotin effects in Crohn's disease, *Dig. Dis. Sci. 33*:1495–1496 (1988).

130. V. E. Colombo, F. Gerber, M. Bronhofer, and G. L. Floersheim, Treatment of brittle finger-nails and onychoschizia with biotin: scanning electron microscopy, *J. Am. Acad. Dermatol. 23*:1127–1132 (1990).

131. V. P. Sydenstricker, S. A. Singal, A. P. Briggs, N. M. DeVaughn, and H. Isbell, Observations on the ''egg white injury'' in man, *J. Am. Med. Assoc. 118*:1199–1200 (1942).

132. D. M. Mock, Water-soluble vitamin supplementation and the importance of biotin, in *Textbook on Total Parenteral Nutrition in Children: Indications, Complications, and Pathophysiological Considerations* (E. Lebenthal, ed.), Raven Press, New York, 1986, pp. 89–108.

133. T. Fukui, K. Iinuma, J. Oizumi, and Y. Izumi, Agar plate method using Lactobacillus plantarum for biotin determination in serum and urine, *J. Nutr. Sci. Vitam. 40*:491–498 (1994).

134. T. R. Guilarte, Measurement of biotin levels in human plasma using a radiometric-microbiological assay, *Nutr. Rep. Int. 31*:1155–1163 (1985).

135. A. Przyjazny, N. G. Hentz, and L. G. Bachass, Sensitive and selective liquid chromatographic postcolumn reaction detection system for biotin and biocytin using a homogeneous fluorophore-linked assay, *J. Chromatogr. 654*:79–86 (1993).

136. J. Stein, A. Hahn, B. Lembcke, and G. Rehner, High-performance liquid chromatographic determination of biotin in biological materials after crown ether–catalyzed fluorescence derivatization with panacyl bromide, *Ann. Biochem. 200*:89–94 (1992).

137. K. Sugawara, S. Tanaka, and H. Nakamura, Electrochemical determination of avidin-biotin binding using an electroactive biotin derivative as a marker, *Bioelectrochem. Bioenerg. 33*: 205–207 (1994).

138. B. Terouanne, M. Bencheich, P. Balaguer, A. M. Boussioux, and J. C. Nicolas, Bioluminescent assays using glucose-6-phosphate dehydrogenase: application to biotin and streptavidin detection, *Ann. Biochem. 180*:43–49 (1989).

139. L. P. Thuy, L. Sweetman, and W. L. Nyhan, A new immunochemical assay for biotin, *Clin. Chim. Acta 202*:191–198 (1991).

140. D. M. Mock, H. Jackson, G. L. Lankford, N. I. Mock, and S. T. Weintraub, Quantitation of urinary 3-hydroxyisovaleric acid using deuterated 3-hydroxyisovaleric acid as internal standard, *Biomed. Environ. Mass Spectrosc. 18*:652–656 (1989).

141. J.-P. Bonjour, Biotin in human nutrition, in *Biotin* (K. Dakshinamurti and H. Bhagavan, eds.), New York Academy of Sciences, New York, 1985, pp. 97–104.

142. E. Livaniou, S. Mantagos, S. Kakabakos, V. Pavlou, G. Evangelatos, and D. S. Ithakissios, Plasma biotin levels in neonates, *Biol. Neonate 59*:209–212 (1991).

143. T. R. Kramer, M. Briske-Anderson, S. B. Johnson, and R. T. Holman, Effects of biotin deficiency on polyunsaturated fatty acid metabolism in rats, *J. Nutr. 114*:2047–2052 (1984).

144. S. F. Suchy, W. B. Rizzo, and B. Wolf, Effect of biotin deficiency and supplementation on lipid metabolism in rats: saturated fatty acids, *Am. J. Clin. Nutr. 44*:475–480 (1986).

145. D. M. Mock, N. I. Mock, S. B. Johnson, and R. T. Holman, Effects of biotin deficiency on plasma and tissue fatty acid composition: evidence for abnormalities in rats, *Pediatr. Res. 24*:396–403 (1988).

146. B. A. Watkins and F. H. Kratzer, Tissue lipid fatty acid composition of biotin-adequate and biotin-deficient chicks, *Poult. Sci. 66*:306–313 (1987).

147. D. M. Mock, S. B. Johnson, and R. T. Holman, Effects of biotin deficiency on serum fatty acid composition: evidence for abnormalities in humans, *J. Nutr. 118*:342–348 (1988).

148. U. Wendel, R. Baumgartner, S. B. v. d. Meer, and L. J. M. Spaapen, Accumulation of odd-numbered long-chain fatty acids in fetuses and neonates with inherited disorders of propionate metabolism, *Pediatr. Res. 29*:403–405 (1991).

149. Y. Kishimoto, M. Williams, H. W. Moser, C. Hignite, and K. Biemann, Branched-chain and odd-numbered fatty acids and aldehydes in the nervous system of a patient with deranged vitamin B_{12} metabolism, *J. Lipid Res. 14*:69–77 (1973).

150. W. Fenton and L. E. Rosenberg, Disorders of propionate and methylmalonate metabolism, in *The Metabolic and Molecular Bases of Inherited Disease* (C. R. Scriver, A. L. Beaudet, W. S. Sly, and D. Valle, eds.), McGraw-Hill, New York, 1995, pp. 1423–1449.

151. J. Sander, S. Packman, and J. Townsend, Brain pyruvate carboxylase and the pathophysiology of biotin-dependent diseases. *Neurology. 32*:878–880 (1982).

152. S. F. Suchy and B. Wolf, Effect of biotin deficiency and supplementation on lipid metabolism in rats: cholesterol and lipoproteins, *Am. J. Clin. Nutr. 43*:831–838 (1986).

153. A. Munnich, J. M. Saudubray, F. K. Coude, C. Charpentier, J. H. Saurat, and J. Frezal, Fatty-acid-responsive alopecia in multiple carboxylase deficiency, *Lancet 1*:1080–1081 (1980).

154. B. Wolf, R. E. Grier, R. J. Allen, S. I. Goodman, and C. L. Kien, Biotinidase deficiency: the enzymatic defect in late-onset multiple carboxylase deficiency, *Clin. Chim. Acta 131*: 273–281 (1983).

155. M. W. Marshall, The nutritional importance of biotin—an update, *Nutr. Today 22*:26–30 (1987).

156. B. A. Watkins and F. H. Kratzer, Dietary biotin effects on polyunsaturated fatty acids in chick tissue lipids and prostaglandin E_2 levels in freeze-clamped hearts, *Poult. Sci. 66*:1818–1828 (1987).

157. D. M. Mock, Evidence for a pathogenetic role of fatty acid (FA) abnormalities in the cutaneous manifestations of biotin deficiency, *FASEB J. 2*:A1204 (1988).

158. T. Watanabe and A. Endo, Teratogenic effects of maternal biotin deficiency in mouse embryos examined at midgestation, *Teratology 42*:295–300 (1990).

159. T. Watanabe, Dietary biotin deficiency effects reproductive function and prenatal development in hamsters, *J. Nutr. 123*:2101–2108 (1993).

160. T. Watanabe, K. Dakshinamurti, and T. V. N. Persaud, Biotin influences palatal development of mouse embryos in organ culture, *J. Nutr. 125*:2114–2121 (1995).

161. T. Watanabe and A. Endo, Species and strain differences in teratogenic effects of biotin deficiency in rodents, *Am. Inst. Nutr. 119*:255–261 (1989).

162. S. D. Bain, J. W. Newbrey, and B. A. Watkins, Biotin deficiency may alter tibiotarsal bone growth and modeling in broiler chicks, *Poult. Sci. 67*:590–595 (1988).

163. M. J. Cowan, D. W. Wara, S. Packman, M. Yoshino, L. Sweetman, and W. Nyhan, Multiple biotin-dependent carboxylase deficiencies associated with defects in T-cell and B-cell immunity, *Lancet July 21*:115–118 (1979).

164. National Research Council, *Biotin.* In: Dietary Reference Intakes for thiamin, riboflavin, nia-

cin, vitamin B_6, folate, vitamin B_{12}, pantothenic acid, biotin, and choline, National Academy Press, Washington, DC, 1998.

165.   H. L. Greene, K. M. Hambridge, R. Schanler, and R. C. Tsang, Guidelines for the use of vitamins, trace elements, calcium, magnesium, and phosphorus in infants and children receiving total parenteral nutrition: report of the Subcommittee on Pediatric Parenteral Nutrient Requirements for the Committee on Clinical Practice Issues of the American Society for Clinical Nutrition, *Am. J. Clin. Nutr. 48*:1324–1342 (1988).

166.   H. Greene and L. Smidt, Nutritional needs of the preterm infant: water soluble vitamins: C, B_1, B_2, B_6, niacin, pantothenic acid, and biotin, in *Nutritional Needs of the Preterm Infant* (R. C. Tsang, A. Lucas, R. Uauy, and S. Zlotkin, eds.), Williams & Wilkins, Baltimore, 1993, pp. 121–133.

167.   M. C. Moore, H. L. Greene, and B. Phillips, Evaluation of a pediatric multiple vitamin preparation for total parenteral nutrition in infants and children: I. Blood levels of water-soluble vitamins, *Pediatrics 77*:530–538 (1986).

168.   M. G. Hardinge and H. Crooks, Lesser known vitamins in food, *J. Am. Diet. Assoc. 38*:240–245 (1961).

169.   J. Wilson and K. Lorenz, Biotin and choline in foods—nutritional importance and methods of analysis: a review, *Food Chem. 4*:115–129 (1979).

170.   K. Hoppner and B. Lampi, The biotin content of breakfast cereals, *Nutr. Rep. Int. 284*:793–798 (1983).

171.   J. A. T. Pennington and H. N. Church, Biotin, in *Bowes and Church's Food Values of Portions Commonly Used*. Harper and Row, New York, 1985.

172.   T. R. Guilarte, Analysis of biotin levels in selected foods using a radiometric-microbiological method, *Nutr. Rep. Int. 324*:837–845 (1985).

173.   K. Hoppner, B. Lampi, and D. C. Smith, An appraisal of the daily intakes of vitamin B12, pantothenic acid and biotin from a composite Canadian diet, *Can. Inst. Food Sci. Technol. J. 11*:71–74 (1978).

174.   N. L. Bull and D. H. Buss, Biotin, panthothenic acid and vitamin E in the British household food supply, *Hum. Nutr.: Appl. Nutr. 36A*:125–129 (1982).

175.   J. Lewis and D. H. Buss, Trace nutrients: minerals and vitamins in the British household food supply, *Br. J. Nutr. 60*:413–424 (1988).

176.   M. Frigg, C. Straub, and D. Hartmann, The bioavailability of supplemental biotin in cattle, *Int. J. Vitam. Nutr. Res. 63*:122–128 (1993).

177.   K. L. Bryant, E. T. Kornegay, J. W. Knight, and D. R. Notter, Uptake and clearance rates of biotin in pig plasma following biotin injections, *Int. J. Vitam. Nutr. Res. 60*:52–57 (1989).

178.   R. Misir and R. Blair, Biotin bioavailability from protein supplements and cereal grains for weanling pigs, *Can. J. Anim. Sci. 68*:523–532 (1988).

179.   R. Bitsch, I. Sal, and D. Hotzel, Studies on bioavailability of oral biotin doses for humans, *Int. J. Vitam. Nutr. Res. 59*:65–71 (1988).

180.   B. Clevidence and M. Marshall, Biotin levels in plasma and urine of healthy adults consuming physiological levels of biotin, *Nutr. Res. 8*:1109–1118 (1988).

181.   D. Mock and N. Mock, Serum concentrations of biotin and biotin analogs increase during acute and chronic biotin supplementation, *FASEB J. 8(5)*:A921 (1994).

12

Folic Acid

TOM BRODY and BARRY SHANE

University of California, Berkeley, California

I. HISTORY

A dietary factor that prevented megaloblastic anemia of pregnancy and animal growth factors and lactic acid bacteria growth factors was studied during the 1930s. Liver, yeast, and spinach proved to be good sources of these factors. Because of the relative ease in performing bacterial growth assays, the bacterial growth factors were extensively purified from liver and yeast and crystallized. Eventually, it was determined that the various anti-anemia and growth factors all had a common structure, and they were named folate. Folic acid (Fig. 1) was identified and synthesized in 1946 using techniques previously used to study other pteridines, such as butterfly pigments. The isolation of folate allowed its use for the treatment of anemia. Because folate and vitamin B_{12} deficiency lead to an identical and indistinguishable megaloblastic anemia in which blood cells are enlarged due to a derangement of DNA synthesis, folate was originally thought to be the only factor required for the treatment of megaloblastic anemia, and was used to "cure" pernicious anemia. With the isolation of vitamin B_{12} several years later, it became clear that vitamin B_{12} was a second anti-anemia factor and it also became clear that while folate could ameliorate the anemia of vitamin B_{12} deficiency, it could not prevent the neurological manifestations of vitamin B_{12} deficiency.

The biochemical functions of folate were first determined with bacteria. Purines or thymine could partly replace the nutritional requirements for folate or for p-aminobenzoic acid (PABA). Similarly, the inhibition of bacterial growth by folate analogues could be overcome by adding purines or thymine to the growth medium. Studies with [14C]formate and [14C]formaldehyde disclosed the role of folate in transferring 1-carbon units. For example, [14C]formate was incorporated into specific positions of the purine ring in studies of

427

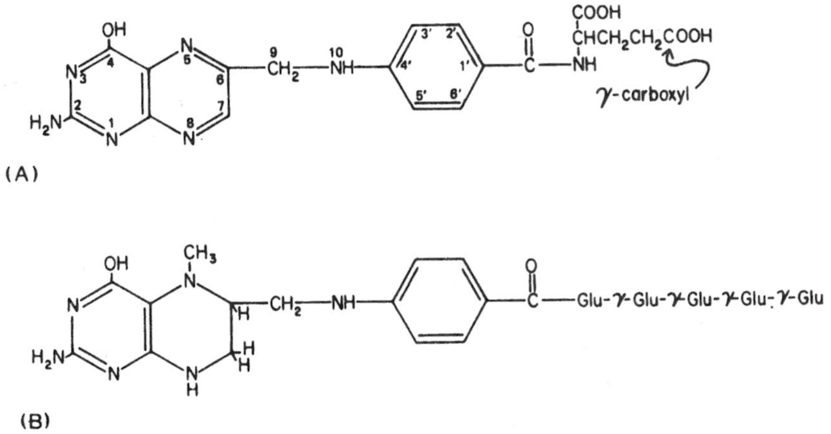

Fig. 1 Structure of (A) folic acid (PteGlu) and (B) 5-methyl-tetrahydrofolate pentaglutamate (5-Methyl-H$_4$PteGlu$_5$). Although folic acid is not naturally occurring, it is readily transported and reduced to the natural forms of the vitamin. One-carbon substituents can be at the N-5 and/or N-10 positions of the reduced folate molecule.

animals. In studies with liver extracts, it was shown that tetrahydrofolate stimulated the incorporation of [^{14}C]formaldehyde into the amino acid serine.

Folates were originally isolated as polyglutamates (Fig. 1), and these forms were found to be the major intracellular form of the vitamin. The polyglutamate had the "property" of supporting animal growth but not bacterial growth. The monoglutamate supported growth of both. The value of the poly-γ-L-glutamyl chain was not recognized until the 1960s and 1970s, when it was determined that the folate polyglutamates were the preferred substrates for folate-dependent enzymes and appeared to be more active metabolically in cells and also were better retained in cells than folate monoglutamates.

Soon after the isolation of folate, various chemically forms were shown to be folate antagonists, and one of these, methotrexate, proved to be very effective in the treatment and cure of childhood leukemia. Because of the role of folate coenzymes in the synthesis of DNA precursors, a multitude of folate antagonists have found clinical use as anticancer and antimicrobial agents. More recently, the demonstration that periconceptional folic acid supplementation reduces the incidence of birth defects in humans has led to fortification of the U.S. food supply with folic acid.

An excellent comprehensive multivolume series on folate and pterins was published by Blakley et al. in the 1980s (1,2), and is a valuable source of background information and references. A more recent book by Brody (3) details the metabolic pathways that require folate.

II. CHEMISTRY

A. Isolation

Folates in natural sources should be extracted and isolated under conditions that preserve the oligo-γ-glutamyl chain and the reductive state of the folate. For maximal recovery of intact folates, the biological sample may be minced, then heated for 5–10 min in a boiling water bath in the presence of antioxidants to denature folate-metabolizing enzymes, such as γ-glutamyl hydrolase, as well as folate-binding proteins. The boiled sample should then be cooled and homogenized (4). Suitable antioxidants for preserving reduced folates are 0.2 M 2-mercaptoethanol or 1.0% sodium ascorbate. One might be cautioned that heating folates in ascorbate alone can alter the folates themselves, though this should not be a problem where sulfhydryl agents are also present (5).

Although folates covalently bound to macromolecules have not yet been discovered, heat treatment may be needed to release folates bound to such components as folate-binding proteins (6,7), membranes (8), and viruses (9). The selective absorption to charcoal and solvent extractions were originally used to purify folates. These have given way to ionic exchange column chromatography, molecular sieve column chromatography, and high-pressure liquid chromatography.

B. Structure and Nomenclature

Folic acid (pteroylmonoglutamate, PteGlu) consists of a 2-amino-4-hydroxy-pteridine (pterin) moiety linked via a methylene group at the C-6 position to a p-aminobenzoyl-glutamate moiety (Fig. 1). Natural folates occur in the reduced, 7,8-dihydro- and 5,6,7,8-tetrahydro-forms. Folates bearing one 1-carbon units are 5-methyl-, 10-formyl-, 5-formyl-, 5,10-methenyl-, 5,10-methylene-, and 5-formiminotetrahydrofolate. Folates in nature occur as pteroyloligo-γ-L-glutamates (PteGlu$_n$) of from one to nine or more glutamates long. The subscript "n" indicates the total number of glutamate residues. "Folate" is a generic term for the above compounds.

The C-6 position of the pterin ring of tetrahydrofolates is an isometric center. This carbon in the naturally occurring forms of H$_4$PteGlu, 5-formyl-H$_4$PteGlu, and 5-methyl-H$_4$PteGlu has the S configuration, while 10-substituted folates such as 10-formyl-H$_4$Pte-Glu$_n$, 5,10-methylene-H$_4$PteGlu and 5,10-methenyl-H$_4$PteGlu have the R configuration (10,11).

C. Synthesis

Folic acid was first synthesized at Lederle Laboratories by the condensation of triamino-6-hydroxypyrimidine, dibromopropionaldehyde, and p-aminobenzoylglutamic acid (12). Folic acid and pterins in general have been synthesized by the method of Taylor et al. (13). This method uses 2-amino-3-cyano-5-chloromethylpyrazine as a key intermediate and avoids a contaminating isomer formed in earlier procedures. The L-glutamyl group of folic acid may racemize in the above method, and a strategy for avoiding this problem has been suggested (14). Pteroyloligo-L-γ-glutamates (folate polyglutamates) were first isolated by the Lederle group (15) and were synthesized by Meienhofer et al. (16). Baugh et al. (17), introduced a solid phase synthesis method.

Folic acid can be reduced to the dihydro- form (18) with dithionite or to the tetrahydro form with hydrogen using a platinum catalyst. Folic acid can be formylated (19) followed by reduction to 10-formyl-H$_4$PteGlu (20) and reduced further to 5-methyl-H$_4$PteGlu (21). The natural isomers of reduced folates can be made enzymatically. These derivatives, including the polyglutamate forms, are now commercially available.

D. Chemical Properties

Folic acid is yellow, has a molecular weight of 441.4, and is slightly soluble in water in the acid form but quite soluble in the salt form. Tetrahydrofolates in solutions are sensitive to oxygen, light, and pH extremes, and can break down to p-aminobenzoylglutamic acid and dihydroxanthopterin, pterin-6-carboxaldehyde, or pterin-6-carboxylic acid (22). One oxidation product of 5-methyl-H_4PteGlu is 5-methyl-H_2PteGlu (23). When acidified, 10-formyl- and 5-isomerize to the relatively oxygen stable folate, 5,10-methenyl-H_4PteGlu (24). When neutralized, the 5,10-methenyl- H_4PteGlu isomerizes to 10-formyl-H_4PteGlu, one of the less stable folate compounds. The formyl group can be removed under anaerobic conditions (19). 5-Formimino- H_4PteGlu, although stable to oxygen, is readily hydrolyzed with the production of ammonia (25). Formaldehyde condenses reversibly with H_4PteGlu to form 5,10-methylene-H_4PteGlu (26). The concentration of this folate can be maintained only when excess formaldehyde is in the solution. 5,10-Methylene-H_4PteGlu is stable at pH 9.5 but is in somewhat rapid equilibrium with formaldehyde at neutral and lower pH (27). This equilibrium can be disturbed by the reaction of thiols with formaldehyde (26).

E.   Chemical Degradation of Folates

Tetrahydrofolates and N10-substituted tetrahydrofolates are unstable to oxygen. In contrast, folic acid and tetrahydrofolates substituted at N5 (or N5 and N10) are relatively stable to oxygen. All folates are degraded by light. H_4PteGlu in oxygenated solutions breaks down to form 6-formyl-pterin (pterin-6-carboxyaldehyde), H_2pterin, pterin, and small amounts of xanthopterin. Cleavage at the C9-N10 bond is rapid, forming p-amino-benzoylglutamic acid (pABAGlu) as a by-product. In mild acid, H_2pterin is the initial major product while pterin is the eventual major product. At neutral and alkaline pH, 6-formyl-pterin is the eventual major product. H_2PteGlu accumulates momentarily as a major product only at alkaline pH (28), while 6-formyl-H_2pterin may be the major breakdown product of H_2PteGlu (29). PteGlu is only a minor breakdown product of H_4PteGlu (1), but under some conditions it may be a major breakdown product of H_2PteGlu (5). 10-Formyl-H_4PteGlu$_n$ breaks down under oxygen to produce 10-formyl-H_2PteGlu and 10-formyl-PteGlu. Folates are usually extracted from biological sources by boiling (5–10 min) in 0.2 M 2-mercaptoethanol, with or without ascorbate. Most forms of folate are fairly stable under these conditions, where recoveries are 70–95 percent (5).

III.   BIOSYNTHESIS

Folates are synthesized from GTP by microorganisms and plants as the 7,8-dihydrofolate form, and all naturally occurring folates are reduced derivatives. The biosynthesis pathway was elucidated by G. M. Brown, T. Shiota, and others (reviewed in Refs. 30,31). GTP cyclohydrolase I catalyzes the conversion of GTP to 7,8-dihydro-neopterin triphosphate and formic acid. The reaction is complex, apparently occurring in four steps, one of which is an Amadori rearrangement. The E. coli cyclohydrolase is a 210 kDal protein and contains four apparently identical subunits. GTP cyclohydrolase I also occurs in mammals where it is used for the first step in biopterin synthesis. However, mammals do not contain the subsequent enzymes of the folate biosynthetic pathway. The second step of folate synthesis involves an unidentified phosphatase, which converts dihydroneopterin-PPP to H_2neopterin. Dihydroneopterin aldolase then converts dihydroneopterin to 6-hydroxymethyl-dihydropterin and glycoaldehyde, a 2-carbon by-product. Hydroxymethyl-dihydropterin pyrophosphokinase (32) catalyzes the fourth step in the pathway, the ATP-dependent conversion of 6-hydroxymethyl-7,8-dihydropterin to 6-hydroxymethyl-7,8-dihydropterin pyrophosphate, with AMP as the by-product. Dihydropteroate synthase

(34 kDal) catalyzes the condensation of 6-hydroxymethyl-7,8-dihydropterin-pyrophosphate and pABA to form 7,8-dihydropteroic acid, with the release of PP_i. Dihydrofolate synthetase (46 kDal) catalyzes the ATP-dependent condensation of 7,8-dihydropteroic acid and glutamate to form $H_2PteGlu$. In *E. coli*, the latter enzyme is bifunctional and occurs as dihydrofolate synthase/folylpoly-γ-glutamyl synthase (33), where the latter activity catalyzes the addition of additional glutamate residues to folate derivatives. Some microorganisms contain separate dihydrofolate synthetase and folylpolyglutamates synthetase proteins. Mammals contain a folylpolyglutamate synthetase activity but lack the dihydrofolate synthetase enzyme. Some bacteria also contain a second folylpolyglutamate synthetase enzyme that adds glutamates residues in α-peptide linkages to H_4folyltriglutamate to produce longer chain length folates (34), although the role of these α-linked forms is not known. In bacteria, hydroxymethyl-dihydroneopterin pyrophosphokinase, dihydropteroate synthase, and dihydrofolate synthase occur as separate proteins. However, in *Pneumocystis carinii*, the enzymes responsible for carrying out these sequential reactions occur as a single trifunctional polypeptide of 84 kDal (35).

Archaea may utilize unusual folate-like coenzymes for mediating one-carbon metabolism. H_4Methanopterin, rather than H_4PteGlu, mediates one-carbon metabolism in methanogens. Methanopterin is a cofactor used in the reduction of CO_2 through formyl, methenyl, methylene, and methyl stages to methane (36). Certain thermophilic *archaea* contain unique versions of folate bearing an oligosaccharide chain, rather than an oligoglutamyl chain. Sarcinapterin is a monoglutamated version of methanopterin and occurs in unique species of methanogens.

IV. ANALYTICAL PROCEDURES

The method used to identify folates depends on whether one needs to know the identities of the one-carbon unit and reductive state of the cofactor, the length of the oligo-glutamyl chain of the folate, or both. For identification of polyglutamate distributions, folates can be cleaved at the C9-N10 bond to yield p-aminobenzoylpolyglutamate derivatives, which can be separated and identified by HPLC analysis (37–39). If the one-carbon distribution is required, folates can be converted to monoglutamates by treatment with a neutral γ-glutamyl hydrolase and the monoglutamates separated by reverse phase HPLC (5). Care must be taken to ensure that the experimental conditions used do not cause interconversion of folate one carbon forms (40). Identification of one-carbon distribution and polyglutamate chain length of all folates is complicated by the large number of potential derivatives. Selhub (41) devised a method for the affinity purification of extracted folates using immobilized milk folate binding protein prior to chromatographic analysis by reverse phase HPLC. Although complete resolution of all folates is not obtained, the high purity of the folate sample applied to the HPLC column allows quantitation of individual folates in overlapping peaks by their spectral properties. This method works well for tissues, such as liver, that contain relatively high levels of folate. Alternately, eluted folates can be detected by microbiological assay or by fluorescence.

The folate content of natural sources, as well as that in fractions recovered after chromatographic procedures, can be measured by microbiological assay (42). The hydrolysis of the oligo-γ-glutamyl chain by treatment with γ-glutamyl hydrolase is required to support a maximal growth response of the test microorganism (43), usually *Lactobacillus casei*, as this organism responds to all folate one carbon forms. The method involving treatment with γ-glutamyl hydrolase followed by assay with *L. casei* has been called an assay for ''total'' folate. The amount of 5-methyl-tetrahydrofolates can be assessed by subtracting the value obtained with *Streptococcus fecium* from that obtained with *L. casei*, where the sample was treated with γ-glutamyl hydrolase. 5-Methyl-H_4PteGlu does not support growth of *S. fecium*. Another lactic acid bacterium, *Pediococcus cerevisiae*, does not respond to nonreduced folate, i.e., PteGlu, or to 5-methyl-H_4PteGlu. The recent development of a microtiter plate assay procedure has greatly simplified the microbiological assay of folate (42).

V. CONTENT IN FOOD AND BIOAVAILABILITY

The major source of folate in the U.S. diet is fortified cereals, followed by vegetables, bread and bread products, citrus fruits and juices, and meat, poultry and fish. Fully oxidized folic acid is only found in the diet when foodstuffs are fortified with folic acid or when dietary folates are oxidized. Reduced folates are less stable than folic acid and large losses in food folate can occur during food preparation such as heating, particularly under oxidative conditions. Additional losses can also occur by leaching out folate during food preparation. When using food tables, attention should be given to the manner in which the test food was prepared, i.e., cooked or raw.

Absorption of orally administered folate mono- and polyglutamates and of dietary folate has been followed by measuring the appearance of vitamin appearing in the mesenteric vein (44) or in the urine (45). The availability of food folate can range from 30–80% that of PteGlu. This was found to be the case for folates extracted from yeast (46) and folates in the presence of certain foods, such as yeast or cabbage (47). Synthetic PteGlu in cooked food may be absorbed at a lesser rate than PteGlu administered in water (48). These rates were determined from increases in serum folates levels in subjects presaturated with folic acid prior to test doses (48). Stronger data on the biological availability of folic acid are those (49,50) showing that synthetic folic acid in cooked food was 55% as available as folic acid in tablet form, as determined by changes in red blood cell folate content. The test foods used by Colman et al. (48–50) were maize, rice, and bread. Certain foods such as cabbage and legumes contain glutamyl hydrolase inhibitors, which can decrease the availability of folylpolyglutamates. The bioavailability of folic acid when given as a supplement or in fortified food is high (51). However, the bioavailability of food folate is less than 50% and may be significantly lower than this because recent studies have suggested that methods commonly used for the analysis of folate in foodstuffs may have underestimated the folate content (52). The general consensus of these studies is that food folate bioavailability averages about 50%, while that of folic acid, either given as a supplement or added to fortified foods, is in excess of 90%. Pharmacological doses of folate are well absorbed but most of the vitamin is not retained in the body due to a limited capacity of tissues to retain large amounts of folate.

Many of the values shown in food tables for folate content of foods were obtained after treating food extracts with glutamyl hydrolase. Some recent studies suggest many of these values may be underestimates as treatment of food extracts with three enzymes, γ-glutamyl hydrolase, α-amylase and protease, resulted in significant increases in folate values, up to twofold, increases that were dependent on the food stuff being analyzed (52). There is some disagreement whether differences in extraction procedures may have confounded these results and validation of the tri-enzyme procedure is currently underway. If these new data hold up, this would suggest that folate bioavailability from unfortified food may be significantly less than 50%.

VI. METABOLISM

A. Absorption

Dietary folates are predominantly pteroylpolyglutamate derivatives and are hydrolyzed to monoglutamates by a γ-glutamylhydrolase activity (sometimes called conjugase) prior to their absorption across the intestinal mucosa. In some species, such as in humans and pigs, the hydrolase activity is a membrane-bound exopeptidase (53) while in other species, such as the rat, the enzyme activity is secreted in the bile. This latter activity is similar to a hydrolase activity found in the lysosomes of all tissues. The human jujenal membrane-bound enzyme has recently been cloned (54). In common with many proteases, γ-glutamyl-hydrolase can be shown to catalyze the reverse of the hydrolytic reaction (ligase), as well as a transpeptidase reaction (55).

The mechanism by which folate crosses the mucosal cell and is released across the basolateral membrane into the portal circulation is not well understood. Intestinal folate absorption occurs mainly in the jejunum. Absorption of folate monoglutamate is via a saturable carrier-mediated process although at high folate concentrations a diffusion-like process occurs. Transport is maximal at pH 5–6 with a rather sharp pH optimum. The intestinal transporter, which is encoded by the reduced folate carrier gene (*RFC1*), is a transmembrane protein that is expressed in most, if not all, tissues (56,57) but the specificity of this transporter for various folates differs between tissues and between the apical and basolateral membranes of tissues. Affinities for reduced folates are in the low micromolar range while affinities for folic acid are similar for some tissues such as the intestine but can be 100-fold lower in other tissues. These differences may reflect tissue specific differences in posttranslational modification of the reduced folate carrier protein. A folate binding protein (sometimes called the folate receptor) has also been detected in the intestine. This protein, which has been shown to mediate endocytosis of folate in the kidney and other tissues, has high affinity for folates, in the nanomolar range (57). It has been shown that the binding protein and the membrane transporter are located on opposing membranes in some polarized cells (58). It is attractive to speculate that the high affinity folate binding protein is involved in the movement of folate into the intestinal cell across the apical membrane while transport across the basolateral membrane to the portal circulation is mediated by the lower affinity RFC transmembrane protein.

Most dietary folate is metabolized to 5-methyl-H_4PteGlu during its passage through the intestinal mucosa although this metabolism is not required for transport (59–61). The degree of metabolism in the intestinal mucosa is dependent on the folate dose given. When pharmacological doses of folic acid, or other folates, are given, most of the transported vitamin appears unchanged in the portal circulation. During passage through the liver, immediate conversion of the PteGlu to 5-methyl-H_4PteGlu occurs, with release of much of the converted vitamin back to the bloodstream (60,61). With large oral doses of PteGlu, substantial amounts of the vitamin are recovered, unchanged, in the urine.

B. Transport and Tissue Accumulation

Pteroylmonoglutamates, primarily 5-methyl-H_4PteGlu, are the circulating forms of folate in plasma, and mammalian tissues cannot transport polyglutamates of chain length three and above (62). After folate absorption into the portal circulation, much of the folate can be taken up by the liver via the reduced folate carrier. In the liver, it is metabolized to polyglutamate derivatives and retained, or it may be released into blood. Plasma folate levels in humans are usually in the 10–30 nM range, while hepatic levels, practically all polyglutamates, are about 20 μM. Some folate is secreted in bile, but this can be reabsorbed in the intestine via an enterohepatic circulation (63). The predominance of 5-methyl-H_4PteGlu in plasma probably reflects that this is the major cytosolic folate in mammalian tissues. The extent of release of short chain folylpolyglutamates from tissues is unknown. Plasma contains a soluble γ-glutamylhydrolase activity, and any polyglutamate released into plasma would be hydrolyzed to the monoglutamate.

Folic acid, naturally occurring reduced folates and folate analogs such as methotrexate may share a common transport system in mammalian cells (64), which is thought to be the low affinity reduced folate carrier. However, the specificity and kinetics of transporters in different tissues varies considerably, although thus far only one RFC gene has been identified (65–69). The hepatic transporter has similar affinity for folic acid and reduced folate monoglutamates, while many other tissues have a greatly decreased affinity for folic acid.

Some plasma folate is bound to low-affinity protein binders, primarily albumin. Plasma also contains low levels of a high-affinity folate binder, the levels of which are increased in pregnancy and in some leukemia patients. The high-affinity binder is a soluble form of a second membrane-associated folate transporter known as the folate-binding protein or the folate receptor (70–72). The folate receptor is encoded by at least three distinct genes in humans and two genes in mice with most tissues expressing the α form. The encoded protein is usually attached to the plasma membrane of cells via a glycosyl-phosphatidylinositol anchor. Levels are highest in the choroid plexus, kidney proximal tubes, placenta, and in a number of human tumors, while lower levels have been found in many other tissues. As described above, in polarized cells the binding protein is often found on the opposing membrane to the reduced folate carrier (58). Although the physiological function of this binding protein has not been established for all tissues, its role in the receptor-mediated reabsorption of folate in the kidney has been well documented (73). The function of the soluble form of folate binding protein, which is expressed at high levels in milk, is not understood, but may also play a role in folate transport.

Red blood cells contain higher levels of folate than plasma (normally 0.5–1 μM). Mature red cells do not transport folate and their folate stores are formed during erythropoiesis and are retained, probably due to binding to hemoglobin, through the life span of the human red cell. Red cell folate levels are often used as a measure of long-term folate status.

Most of the folate in tissues is found in the mitochondrion and cytosol. Mitochondria contain a transporter that is specific for reduced folates and which differs from the plasma membrane transporter in that it will not transport folic acid or methotrexate (74,75). This transporter has only been characterized kinetically. The major hepatic mitochondrial folates are 10-formyl-H_4PteGlu$_n$and H_4PteGlu$_n$, and much of these are found bound to two folate enzymes, dimethylglycine dehydrogenase and sarcosine dehydrogenase. 5-Methyl-H_4PteGlu$_n$, is the major cytosolic folate, and much of this is bound to glycine N-methyltransferase in liver, while much of the cytosolic H_4PteGlu$_n$ is bound to 10-formyltetrahydofolate dehydrogenase.

C. Intracellular Metabolism and Storage

Folate metabolism involves the reduction of the pyrazine ring of the pterin moiety to the coenzymatically active tetrahydro form, the elongation of the glutamate chain by the addition of L-glutamate residues in an unusual γ-peptide linkage, and the acquisition and oxidation or reduction of one-carbon units at the N-5 and/or N-10 positions. Folates are substrates for multiple enzymes and the interconversion of folate one-carbon forms is intertwined with the metabolic roles of folate as described below. Folates in cells and tissues occur almost exclusively as reduced folate oligo-γ-glutamates. The polyglutamates are more effective substrates than pteroylmonoglutamates of most folate-dependent enzymes and usually exhibit greatly increased affinities for these enzymes. For most folate-dependent enzymes the major kinetic advantages are achieved by elongation of the glutamate chain to the triglutamate (reviewed in Ref. 76). Longer polyglutamate forms are required for the enzymes involved in the methionine re-synthesis cycle. Folates are metabolized to polyglutamates of chain lengths considerably longer than the triglutamate form required for folate retention. The glutamate chain length varies with the species with the pentaglutamate predominating in the rat, hexaglutamate in the mouse, and hexa- and heptaglutamates as well as longer chain length derivatives, up to the decaglutamate, in human cells (77).

The accumulation of folate in the mitochondrion and cytosol of tissues requires their conversion to polyglutamates, which is catalyzed by the enzyme folylpolyglutamate synthetase (76, reaction 17 Fig. 2).

$$\text{MgATP} + \text{folate}(\text{glu}_n) + \text{glutamate} \rightarrow \text{MgADP} + \text{folate}(\text{glu}_{n+1}) + \text{phosphate}$$

Glutamate residues are added one at a time and the products of the reaction have to be released from the enzyme following each catalytic cycle before rebinding and further chain extension. Tetrahydrofolate and its polyglutamate forms are the preferred substrates for human folylpolyglutamate synthetase, while 5-substituted folates such as 5-methyl-H_4PteGlu are poor substrates. The affinity of folates for the human enzyme drops off as the chain length is extended beyond the diglutamate. Because 5-methyl-H_4PteGlu, the major folate transported into most tissues, is a very poor substrate and its diglutamate derivative is almost inactive, the extent of folate accumulation is dependent on the tissue's ability to metabolize 5-methyl-H_4PteGlu to H_4PteGlu via the methionine synthase reaction (Fig. 4). The cell can rapidly release any folate that is not converted to the triglutamate (78). Folylpolyglutamate synthetase is a low abundance protein and its activity is rate-limiting for folate retention and accumulation by some tissues, and also limits the ability of tissues to accumulate large stores of folate. With pharmacological doses of folate, competition by the entering monoglutamate limits the glutamate chain extension of cellular folates and much of the folate is converted to the diglutamate and released from the tissue, while the predominant chain length of folate that is retained by the tissue is shortened.

Folylpolyglutamate synthetase is encoded by a single human gene, and cytosolic and mitochondrial isozymes are generated by alternate transcription start sites for the gene (79,80) and by alternate translational start sites for its mRNA (81). Cells that lack folylpolyglutamate synthetase activity are unable to accumulate folate and are auxotrophic for products of one carbon metabolism (82,83). Cells that lack mitochondrial folylpolyglutamate synthetase activity are defective in mitochondrial one carbon metabolism and are unable to accumulate folate in the mitochondria despite possessing normal cytosolic folate pools.

Polyglutamylation of antifolates such as methotrexate is required for their accumulation and cytotoxic efficacy and clinical resistance to methotrexate can result from decreased folylpolyglutamate synthetase activity (84,85). The level of folylpolyglutamate synthetase activity in human leukemia blasts varies over a very wide range and this may explain some of the differences in sensitivity of tumors to antifolate agents. Tumor cells

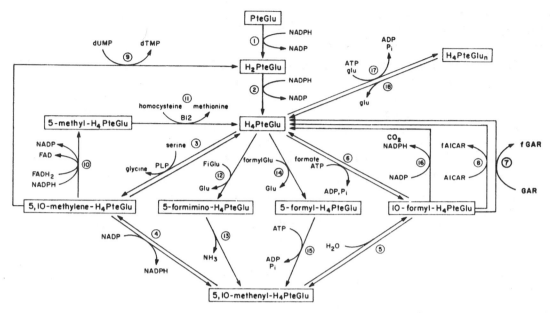

Fig. 2 Metabolic cycles of one-carbon metabolism in the cytosol of mammalian tissues. The numbers refer to reactions catalyzed by the following enzymes: (1) dihydrofolate reductase, (2) dihydrofolate reductase, (3) serine hydroxymethyltransferase, (4) 5,10-methylene-tetrahydrofolate dehydrogenase, (5) 5,10-methenyltetrahydrofolate cyclohydrolase, (6) 10-formyl-tetrahydrofolate synthetase, (7) glycinamide ribonucleotide transformylase, (8) aminoimidazole carboxamide ribonucleotide transformylase, (9) thymidylate synthase, (10) 5,10-methylene-tetrahydrofolate reductase, (11) methionine synthetase, (12) tetrahydrofolate formiminotransferase, (13) formiminotetrahydrofolate cyclodeaminase, (14) glutamate transformylase, (15) 5-formyl-tetrahydrofolate isomerase, (16) 10-formyl-tetrahydrofolate dehydrogenase, (17) folyl-polyglutamate synthetase, and (18) γ-glutamyl hydrolase. In mammals, reactions 12 and 13 are catalyzed by a single bifunctional enzyme, and reactions 4, 5, and 6 are catalyzed by a trifunctional enzyme. dUMP, deoxyuridine monophosphate; dTMP, deoxythymidine monophosphate (thymidine monophosphate); PLP, pyridoxal phosphate; FIGLU, formiminoglutamic acid; fAICAR, formyl-aminoimidazole carboxyamide ribonucleotide; AICAR, aminomidazole carboxamide ribonucleotide; GAR, glycinamide ribonucleotide; FGAR, formyl-glycinamide ribonucleotide; and B_{12}, vitamin B_{12} (cobalamin).

that are especially sensitive to methotrexate appear to have an enhanced ability to form methotrexate polyglutamates (86).

    Tissues contain a soluble lysosomal γ-glutamylhydrolase activity that converts folylpolyglutamates to the mono- or diglutamate form (reaction 18, Fig. 2). The enzyme also hydrolyzes p-aminobenzoylpolyglutamates, and can hydrolyze poly(γ-glutamate) although this is much poorer substrate. Lysosomes contain a transporter that can transport folate and methotrexate polyglutamates (87). This transporter may be involved in the subsequent hydrolysis of folylpolyglutamates with their subsequent release from the tissue. However, the physiological role of this lysosomal system may be related to the catabolism of folate (see Section D, below).

D. Catabolism and Excretion

The daily urinary excretion of intact folates is only between 1 and 12 μg, which accounts for only a small fraction of the total folate intake. Fecal folate levels are quite high, sometimes much higher than the estimated intake, which presumably reflects folate production by the gut microflora. The major urinary excretory product of folate is N-acetyl-p-amino-benzoylglutamate (88) with smaller amounts of p-aminobenzoylglutamate. These compounds can arise from cleavage of the labile C9-C10 bond of reduced folates, which would yield p-aminobenzoylpolyglutamates and pterin derivatives, followed by their subsequent hydrolysis by γ-glutamylhydrolase, and N-acetylation of the resulting p-aminobenzoylglutamate (89).

Although this catabolic pathway was initially thought to be initiated by a nonenzymatic cleavage of labile folate derivatives, recent studies have suggested that several enzyme-mediated systems may be involved in this process and that formyl derivatives of folate may be the immediate substrates for the cleavage reactions (P. Stover, personal communication). Metabolic conditions or manipulations that cause an accumulation of 5- or 10-formyl-$H_4PteGlu_n$ result in increase folate catabolism and heavy chain ferritin has been isolated as a activity that cleaves folates to pterin derivatives (P. Stover, personal communication).

VII. BIOCHEMICAL FUNCTIONS

Folates are used as cofactors and serve as acceptors and donors of one-carbon units in a variety of reactions involved in amino acid and nucleotide metabolism. The one-carbon units can be at the oxidation levels of methanol, formaldehyde, and formate but not carbon dioxide. These reactions, known as one-carbon metabolism, are shown in Fig. 2, which emphasizes the cyclical nature of folate metabolism. Although the major pathways of methionine, thymidylate, and purine synthesis occur in the cytosol, folate metabolism also occurs in the mitochondria and mitochondrial folate metabolism plays an important role in glycine metabolism and in providing one-carbon units for cytosolic one-carbon metabolism (90–93, Fig. 3). Folate coenzymes are cosubstrates in these reactions. Consequently, folate metabolism and its regulation is interwoven with the regulation of the synthesis of products of one-carbon metabolism and factors that regulate any one cycle of one-carbon metabolism would be expected to influence folate availability for the other cycles of one-carbon metabolism. The C-3 of serine is the major source of one-carbon units for folate metabolism. Other sources include formate, much of which is derived from serine metabolism in the mitochondria, and the C-2 of histidine. Many of the enzymes involved in folate metabolism are multifunctional or are part of multiprotein complexes which allows channeling of polyglutamate intermediates between active sites without release of intermediate products from the complex (94).

Some folate-requiring enzymes are used for biosynthetic purposes, whereas others are used only for interconversion of the various forms of the vitamin. The folate molecule does not remain enzyme bound, but acts rather as a cosubstrate. The available studies have shown that, within the cytosol, the folylpentaglutamates and folyhexaglutamates have an equal tendency to participate in one-cabon metabolism (209,210).

A. Amino Acid Metabolism

1. Serine-Glycine Interconversion and Metabolism

Serine hydroxymethyltransferase (E.C. 2.1.2.1; reaction 3, Fig. 2) catalyzes the transfer of formaldehyde from serine to $H_4PteGlu_n$ as follows:

$$serine + H_4PteGlu_n \leftrightarrow glycine + 5,10\text{-methylene-}H_4PteGlu_n$$

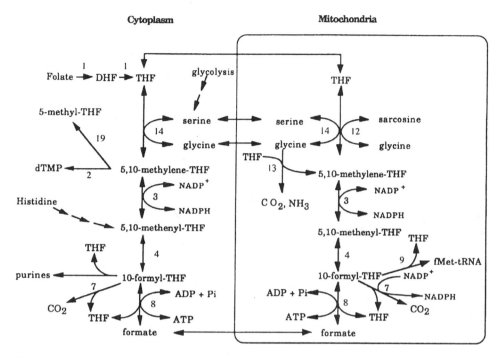

Fig. 3 Compartmentalization of folate-dependent one-carbon metabolism between mitochondria and cytoplasm. (From Ref. 93, adapted from Ref. 145.)

The enzyme contains bound pyridoxal 5′-phosphate and is present as two distinct isozymes in the cytosol and mitochondria of all tissues (95,96, reaction 14, Fig. 3). The cytosolic enzyme is the predominant species in liver although the mitochondrial form predominates in cultured cells. The highest levels are in the liver and kidney. In mammalian tissues, the β-carbon of serine is the major source of one-carbons for folate metabolism. The 5,10-methylene-$H_4PteGlu_n$ formed in this reaction plays a central role in one-carbon metabolism because it can be directed into the three cytoplasmic one-carbon cycles of methionine, de novo purine, and thymidylate synthesis (Fig. 2). The hydroxymethyltransferase reaction, which is freely reversible, is a major pathway for serine catabolism. In kidney and in liver under gluconeogenic conditions net synthesis of serine from glycine may occur via serine hydroxymethyltransferase catalyzed reactions.

The role of the two serine hydroxymethyltransferase isozymes is not completely understood. Serine, a nonessential amino acid, is derived from glucose. Some tissues are net producers of glycine while others, such as kidney, are net producers of serine from glycine. Glycine is a gluconeogenic amino acid and, in liver and kidney at least, the net flux through one of the hydroxymethyltransferase isozymes should be in the direction of serine synthesis under normal conditions of net gluconeogenesis. Mammalian cell mutants that lack mitochondrial serine hydroxymethyltransferase enzyme activity but have normal levels of cytosolic activity require exogenous glycine for growth (97–100) and overexpression of the cytosolic isozyme increases serine synthesis (unpublished data). Most of the one carbons used in cytosolic one-carbon metabolism in cultured cells are derived from serine metabolism in the mitochondria (unpublished data) indicating that the mitochondrial isozyme is required for net glycine synthesis, while the cytosolic enzyme catalyzes a net flux from glycine to serine, in mammalian cells at least. Whether this is true of all normal tissues remains to be ascertained.

Different genes encode the two serine hydroxymethyltransferase isozymes. The cDNA for human cytosolic serine hydroxymethyltransferase codes for a 483 amino acid protein of 53 kDal. The cytosolic and mitochondrial enzymes share about 63% amino acid identity. The genes coding for the cytosolic and mitochondrial enzymes lie on human chromosomal regions 17p11.2 and 12q13, respectively (101–103).

Serine hydroxymethyltransferase also catalyzes the irreversible hydrolysis of 5,10-methenyl-H_4PteGlu$_n$ to 5-formyl-H_4PteGlu$_n$ (104). 5-Formyl-H_4PteGlu$_n$ was originally thought to be an artifact of folate extraction procedures. However, it is not found in cells that lack serine hydroxymethyltransferase activity, and overexpression of the hydroxymethyltransferase increases the level of this folate derivative, and this reaction probably accounts for the low level of 5-formyl-H_4PteGlu$_n$ found in many biological sources. 5-Formyl-H_4PteGlu$_n$ is a potent inhibitor of some folate-dependent enzymes, including serine hydroxymethyltransferase (95), and is the substrate for the folate catabolism enzyme (P. Stover, personal communication). However, it is not used directly as a substrate in one-carbon transfer reactions.

5-Formyl-H_4PteGlu$_n$ can be reconverted to 5,10-methenyl-H_4PteGlu$_n$ by methenyl-tetrahydrofolate synthetase (EC 6.3.3.2; reaction 15, Fig. 2):

$$\text{5-formyl-}H_4\text{PteGlu}_n + \text{MgATP} \rightarrow [\text{5,10-methenyl-}H_4\text{PteGlu}_n]^+ + \text{MgADP} + P_i$$

A human cDNA for by methenyl-tetrahydrofolate synthetase has been isolated and encodes a 27 kDal cytosolic enzyme (104). 5-Formyl-H_4PteGlu is used clinically and experimentally as a source of reduced folate because it is more stable than other reduced folates.

The major pathway of glycine catabolism is via the glycine cleavage system (reaction 13, Fig. 3) which is located in the mitochondria and catalyzes the following reaction:

$$\text{glycine} + H_4\text{PteGlu}_n + \text{NAD}^+ \rightarrow \text{5,10-methylene-}H_4\text{PteGlu}_n$$
$$+ \text{NADH} + CO_2 + NH_4^+$$

This complex, which is present in high concentrations in liver and kidney, can provide an additional one carbon derived from C-2 of glycine to the folate pool (105,106). The glycine cleavage system is composed of four associated proteins, P,T,L, and H. P protein, which contains pyridoxal phosphate, catalyzes glycine decarboxylation and transfer of methylamine to lipoic acid on H protein. The lipoic acid is reduced and the carbon moiety from glycine is oxidized to the level of formaldehyde. T protein catalyzes the transfer of formaldehyde to H_4PteGlu$_n$, and the reduced lipoate on H protein is reoxidized by NAD^+ in a reaction catalyzed by L protein. Although potentially reversible, the glycine cleavage system does not appear to play a role in the synthesis of glycine. Coupling of the serine hydroxymethyltransferase and glycine cleavage systems provides a mechanism for the net synthesis of serine from two molecules of glycine, with the C-1 and C-2 of serine derived directly from one molecule of glycine and the C-3 derived from 5,10-methylene-H_4PteGlu$_n$ arising glycine cleavage.

The cDNA encoding the proteins of the complex have been isolated and contain mitochondrial leader sequences (107,108). A human genetic disease involving a mutation in T-protein results in a disease called "nonketotic hyperglycinemia" and results in the accumulation of glycine in body fluids and early death.

2. Homocysteine Methylation and Methionine Synthesis

A major cytosolic cycle of one-carbon utilization involves the reduction of 5,10-methylene-$H_4PteGlu_n$ to 5-methyl-$H_4PteGlu_n$ and the transfer of the methyl group to homocysteine to form methionine and to regenerate $H_4PteGlu_n$ (Fig. 4). This cycle is catalyzed by two enzymes, methylenetetrahydrofolate reductase and methionine synthase, although serine hydroxymethyltransferase can also be considered part of the cycle.

5,10-Methylenetetrahydrofolate reductase (E.C. 1.1.1.68; reaction 10, Figure 2; reaction 9, Figure 4) catalyzes the conversion of one-carbon units at the oxidation level of formaldehyde to that of methanol as follows:

$$5,10\text{-methylene-}H_4PteGlu_n + NADPH + H^+ \rightarrow 5\text{-methyl-}H_4PteGlu_n + NADP^+$$

The enzyme is a flavoprotein and catalyzes the committed step in methionine synthesis and folate-dependent homocysteine remethylation in mammalian tissues (109,110). The reaction is physiologically irreversible under in vivo conditions due to the redox state of the NADPH/NADP couple (111).

The human gene for methylenetetrahydrofolate reductase has been cloned (112). A number of common polymorphisms have been described, one of which (C677T) is associated with lower tissue levels of the enzyme (113). This polymorphism results in an A to V change in the coding region of the protein. Although the kinetic properties of the protein are unchanged, the affinity for the flavin cofactor is reduced and the apoenzyme is unstable and degraded (114). The mammalian protein is a dimer and consists of two domains, an N terminal catalytic domain that shares extensive homology with the analogous bacterial proteins, and a C terminal regulatory domain that binds adenosylmethionine, an allosteric inhibitor (109).

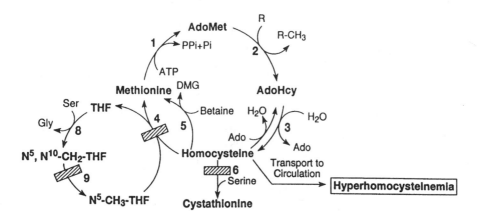

Fig. 4 The folate-dependent methionine resynthesis cycle and its relationship to transmethylation and transsulfuration cycles in tissues. The numbers refer to reactions catalyzed by the homocysteine remethylation cycle (8, serine hydroxymethyltransferase; 9, methylenetetrahydrofolate reductase;4, B_{12}-dependent methionine synthase; 5, betaine methyltransferase), the transmethylation cycle (1, adenosylmethionine synthetase; 2, R-methyltransferase; 3, adenosylhomocysteine hydrolase), and the transsulfuration pathway (6, cystathionine β-synthase). Genetic defects in key enzymes of homocysteine metabolism that lead to hyperhomocysteinemia and homocystinuria are indicated by the blocks. (From Ref. 125.)

The next enzyme in this cycle, methionine synthase (E.C.2.1.1.13; reaction 11, Fig. 2; reaction 4, Fig. 4), is one of only two B_{12}-dependent mammalian enzymes and catalyzes the transfer of the methyl group from 5-methyl-H_4PteGlu$_n$ to homocysteine (115):

$$5\text{-methyl-}H_4\text{PteGlu}_n + \text{homocysteine} \rightarrow H_4\text{PteGlu}_n + \text{methionine}$$

The methionine synthase reaction is the only reaction in which the methyl group of 5-methyl-H_4PteGlu$_n$ can be metabolized in mammalian tissues. Although methionine is an essential amino acid and is required in the diet, the methionine synthase reaction plays a major role in methyl group metabolism as it allows the reutilization of the homocysteine backbone as a carrier of methyl groups derived primarily from the C-3 of serine. The enzyme contains tightly bound cob(I)alamin, and the reaction proceeds via methylation of the cofactor to the methylcob(III)alamin intermediate by 5-methyl-H_4PteGlu$_n$, and then by transfer of the methyl group from methylcob(III)alamin to homocysteine in a heterolytic cleavage to generate methionine and H_4PteGlu$_n$ and regenerate the enzyme cob(I)alamin form. The enzyme is a Zn metalloprotein.

Human cDNAs and the gene for human methionine synthase have been cloned (116–118). The cDNA encodes a 140 kDal protein. The enzyme is a monomer and has a three domain structure. The N terminal domain encodes the catalytic site with homocysteine, folate, and Zn binding sites, the central domain binds the B_{12} cofactor, and the C terminal domain interacts with adenosylmethionine and accessory proteins (see below).

Methionine synthase in mammalian tissues is normally present as the holoenzyme form, containing a tightly bound B_{12}-cofactor. The cob(I)alamin cofactor is highly reactive and the cofactor is occasionally oxidized to the nonfunctional cob(II)alamin form during catalysis. The enzyme is reactivated by one or several poorly characterized accessory proteins that catalyze the AdoMet and NADPH-dependent reductive methylation of enzyme bound cob(II)alamin to methylcob(III)alamin. Bacteria possess two methionine synthase accessory proteins, flavodoxin and flavodoxin reductase, that use NADH, FAD, and FMN as cofactors. Flavodoxins are not present in mammalian tissues. Mammalian methionine synthase, when isolated, contains tightly bound soluble cytochrome b5 and, in an in vitro anaerobic assay system, soluble cytochrome b5 and cytochrome P450 reductase catalyzed the AdoMet and NADPH dependent reactivation of the enzyme (119). The gene for a single putative human methionine synthase reductase protein that contains binding sites for NADPH, FAD, and FMN has recently been cloned (120) by searching the databases for a novel gene that contained binding sites for all three cofactors. The gene encodes a novel member of the P450 reductase and NO synthase family. While it has not yet been demonstrated that the gene product reactivates methionine synthase in a cell free assay system, genetic evidence strongly points to this protein as the physiological reactivator of methionine synthase. DNA from patients with severe genetic disease due to a failure to reactivate methionine synthase contains mutations in conserved regions of the putative methionine synthase reductase (121). It is difficult to reconcile two systems for the reactivation of methionine synthase with a single genetic defect that causes loss of activity. It remains to be established which of these systems is the one that functions under physiological conditions.

Although cytosolic methionine synthase and mitochondrial methylmalonyl CoA mutase are the only enzymes that use B_{12} as a cofactor in mammalian tissues, a large number of other proteins are involved in the transport and metabolism of vitamin B_{12}. Genetic defects in many of these as well as in the methionine synthase structural gene can result in defective methionine synthase activity.

Methionine synthase can also catalyze the reduction of the anesthetic gas nitrous oxide to nitrogen. During this process, a hydroxyl radical is formed, which can lead to destruction of the polypeptide backbone of the protein and inactivation of the enzyme. Nitrous oxide is sometimes used to inactivate methionine synthase in experimental animals to generate a model for the metabolic effects of vitamin B_{12} deficiency (209).

Homocysteine arises from hydrolysis of adenosylhomocysteine (AdoHcy;reaction 3, Figure 4), the product of adenosylmethionine (AdoMet)-dependent methylation reactions (reaction 2, Fig. 4), and is not normally found in the diet. Homocysteine can be metabolized to cysteine via reactions catalyzed by PLP-dependent cystathionine β-synthase (reaction 6, Fig. 4) and cystathioninase in a transsulfuration pathway. Alternatively, homocysteine can be remethylated back to methionine via the methionine synthase reaction or by betaine-homocysteine methyltransferase, which catalyzes the transfer of one of the methyl groups of betaine to homocysteine to generate methionine and dimethylglycine (122; reaction 5, Fig. 4). Betaine is a product of choline oxidation in liver mitochondria. Although cytosolic betaine methyltransferase is a high abundance protein in liver, it has a very limited tissue distribution in other tissues and is present in the kidney of humans but not in the rat, while methionine synthase is present in all tissues. The high abundance probably reflects the low catalytic activity of this enzyme, which does not use a cobalamin cofactor. Many microorganisms, including yeast, express a sluggish B_{12}-independent methionine synthase that is absolutely specific for 5-H_4PteGlu polyglutamates and normally found at very high concentrations in the cell.

The dimethylglycine product of the betaine methyltransferase reaction is converted to glycine in the mitochondria via folate-dependent reactions. The flavoproteins dimethylglycine dehydrogenase and sarcosine dehydrogenase catalyze the oxidative demethylation of dimethylglycine to sarcosine and sarcosine to glycine, respectively, with the generation of 5,10-methylene-H_4PteGlu$_n$ (reaction 12, Fig. 3). Most of the folate in mitochondria is associated with these proteins (123). Thus, although the betaine methyltransferase reaction is folate-independent, the product of the reaction can generate two one-carbon moieties that can potentially be used for folate-dependent homocysteine remethylation. Human betaine methyltransferase has been cloned and the deduced protein sequence shares some limited homology with the N terminal domain of methionine synthase (124), reflecting that both enzymes bind homocysteine and Zn.

Tissue levels of homocysteine are normally in the low micromolar range and increased homocysteine causes elevated AdoHcy, an inhibitor of many methylation reactions. The extent of homocysteine remethylation or transsulfuration is tissue dependent, and many tissues export homocysteine and cystathionine into the circulation. The major sites of homocysteine remethylation and for transsulfuration are believed to be the liver and kidneys, but this also appears to be species specific, and has not been completely clarified (125). Remethylation, in liver at least, is also dependent on the methyl status of the animal. The major regulator of the folate-dependent methionine cycle is AdoMet, which is a potent allosteric inhibitor of methylenetetrahydrofolate reductase (Fig. 3). Liver contains a high K_m adenosylmethionine synthetase (reaction 1, Fig. 4), the product of the *MATI* gene, and hepatic levels of AdoMet vary with hepatic methionine status. Elevated levels of AdoMet inhibit the reductase, reducing 5-methyl-H_4PteGlu$_n$ formation and remethylation of homocysteine, and activate cystathionine β-synthase, stimulating transsulfuration of homocysteine to cysteine. When methionine levels are low, AdoMet levels are reduced relieving the inhibition of methylenetetrahydrofolate reductase, and remethylation of homocysteine is favored and transsulfuration is inhibited. The adenosylmethionine synthetase in nonhepatic tissues is the product of the *MATII* gene and has a lower K_m for methionine. Consequently, AdoMet levels and the methionine synthesis cycle in nonhepatic tissues are less sensitive to changes in methionine levels.

Liver, kidneys, and pancreas also contain a very high abundance cytosolic protein, glycine N-methyltransferase, that acts as a sink for excess methyl groups. This enzyme catalyzes the AdoMet-dependent methylation of glycine to sarcosine. Although folate is not a substrate, 5-methyl-H_4PteGlu$_n$ is a potent inhibitor and the protein is a major cytosolic folate binding protein (126,127). When AdoMet levels are high, methylenetetrahydrofolate reductase is inhibited, which reduces 5-methyl-H_4PteGlu$_n$ formation and relieves inhibition of glycine methyltransferase, allowing removal of excess methyl groups. At low methionine and AdoMet concentrations, methylenetetrahydrofolate reductase is more active, 5-methyl-H_4PteGlu$_n$ accumulates, and glycine methyltransferase is inhibited. Glycine N-methyltransferase binds polycyclic aromatic hydrocarbons and may also function as a receptor protein for these compounds as small amounts of the protein are found in the nucleus (128).

Methionine synthesis requires higher levels of folate than other metabolic cycles of one-carbon metabolism and longer polyglutamate forms than the other metabolic cycles (129). Although methylenetetrahydrofolate reductase is considered the major regulatory enzyme in the folate-dependent methionine cycle, 5-methyl-H_4PteGlu$_n$ is the major cytosolic folate in tissues. This suggests that the methionine synthase reaction is also partially rate limiting this cycle. Both methylenetetrahydrofolate reductase and methionine synthase are present at considerably lower concentrations than most of the other enzymes involved in the metabolism of folate coenzymes. 5-Methyl-H_4PteGlu is the major form of folate taken up by tissues. Removal of the methyl group, via the methionine synthase reaction, is required before the entering folate can be utilized in other reactions of one-carbon metabolism, or can be metabolized to the polyglutamates that are retained by cells. As the entering monoglutamate has to compete for methionine synthase with the preferred 5-methyl-H_4PteGlu polyglutamate substrate present in tissues, accumulation of exogenous folate by tissues is repressed by high intracellular folate or by expansion of the 5-methyl-H_4PteGlu$_n$ pool.

The ubiquitous nature of the folate-dependent methionine cycle in all tissues probably reflects its importance in maintaining methyl group status despite normal intakes of methionine, and possibly because most tissues lack any other mechanism for disposing of homocysteine, although the homocysteine can be exported to the circulation. As discussed below, the elimination of this cycle in the methylenetetrahydrofolate knockout mouse results in viable but sick offspring with many defects including birth defects (R. Rozen, personal communication). The methionine synthase knockout is embryonically lethal, with embryos dying very early in gestation (unpublished data). This would suggest that although this cycle is clearly of physiological importance in homocysteine remethylation and/or methionine synthesis, the role of methionine synthetase in the conversion of 5-methyl-H_4PteGlu$_n$ to H_4PteGlu$_n$, is of more importance to the tissue.

3. Histidine Catabolism

C-2 of the imidazole ring of histidine provides one-carbon units at the oxidation level of formate for one-carbon metabolism. Cytosolic formiminotransferase (E.C.2.1.2.5; reaction 12, Fig. 2) catalyzes one of the final steps in histidine catabolism. The formimino group of formiminoglutamate is transferred to H_4PteGlu$_n$ as follows:

$$\text{formiminoglutamate} + H_4\text{PteGlu}_n \rightarrow 5\text{-formimino-}H_4\text{PteGlu}_n + \text{glutamate}$$

The formimino moiety is converted to 5,10-methenyl-$H_4PteGlu_n$ in a formimino-tetrahydrofolate cyclodeaminase (E.C.4.3.1.4; reaction 13, Fig. 2) catalyzed reaction:

$$5\text{-formimino-}H_4PteGlu_n + H^+ \rightarrow [5,10\text{-methenyl-}H_4PteGlu_n]^+ + NH_3$$

The transferase and cyclodeaminase activities are found on a single bifunctional protein in mammalian tissues (94). The protein contains a single binding site for the polyglutamate tail of folate and channeling between active sites has been demonstrated for polyglutamate substrates, with the most effective channeling occurring with the longer polyglutamates normally found in mammalian tissues (94,130). Channeling is abolished with the pteroylmonoglutamate substrate. Under folate or vitamin B_{12} deficiency conditions, formiminoglutamate is excreted in urine.

B. Nucleotide Metabolism

1. Thymidylate Synthesis

Folate in not involved in the de novo synthesis of pyrimidines but is required for the synthesis of thymidylate. Thymidylate synthase (E.C.2.1.1.45) catalyzes the conversion of dUMP to dTMP, where the donated one-carbon unit is at the oxidation level of formaldehyde (reaction 9, Fig. 2):

$$5,10\text{-methylene-}H_4PteGlu_n + dUMP \rightarrow H_2PteGlu_n + dTMP$$

The reaction is unique among folate-dependent reactions in that it not only involves the transfer of the hydroxymethyl group from folate to the 5 position of dUMP, it also involves the transfer of the hydride from 6 position of $H_4PteGlu_n$ to the 5 position of the nucleotide. During catalysis, a binary covalent complex is formed between enzyme and nucleotide through an active site cysteine, and then a ternary complex is formed with folate (131,132). The use of the tetrahydrofolate molecule as a reductant results in its conversion to $H_2PteGlu_n$. This is the only reaction in which the oxidation state of folate changes from the tetrahydro to the dihydro form. $H_2PteGlu_n$ is inactive as a coenzyme and has to be reduced back to $H_4PteGlu_n$ in a reaction catalyzed by dihydrofolate reductase (E.C.1.5.1.3; reaction 2, Fig. 2) before it can play a further role in one-carbon metabolism:

$$H_2PteGlu_n + NADPH + H^+ \rightarrow H_4PteGlu_n + NADP^+$$

The sole physiological role of dihydrofolate reductase in mammalian tissues is the reduction of $H_2PteGlu_n$ formed as a result of thymidylate synthesis. In microorganisms, the enzyme is also required for the reduction of $H_2PteGlu$ formed as the end product of folate biosynthesis. Fortuitously, the enzyme will also reduce unnatural folic acid to $H_2PteGlu$ (reaction 1, Fig. 2), thus making it available for folate metabolism, although folic acid is a poorer substrate than $H_2PteGlu$.

Thymidylate synthase activity is only expressed in replicating tissues and expression of the synthase and dihydrofolate reductase mRNA is highest during the S phase of the cell cycle. Many folate antagonists that inhibit these enzymes have been developed and used widely as anticancer agents. The drug 5-fluoro-deoxyuridylate acts initially as a substrate for the thymidylate synthase reaction and forms a covalent complex with enzyme and folate. Because the fluorine atom cannot be abstracted to complete the reaction, this drug acts as a covalent inhibitor of the enzyme. Folate binding enhances the stability of the complex and drug efficacy is improved if pharmacological doses of folate, usually as 5-formyl-$H_4PteGlu$, are also given. Drug resistance often develops due to increased thymidylate synthase enzyme. The mechanism for this is somewhat unusual as it involves translational activation (133). Thymidylate synthase protein binds to its cognitive mRNA and regulates its own translation. 5-Fluoro-deoxyuridylate binding to thymidylate synthase inhibits mRNA binding, which leads to derepression of translation. Methotrexate, a 4-aminofolic acid analog, is a potent inhibitor of dihydrofolate reductase. Treatment of rapidly growing cells with this drug causes trapping of folate in the nonfunctional dihydrofolate form. Slowly growing tissues, which have negligible or low levels of thymidylate synthase activity, do not convert reduced folate to the dihydrofolate form as rapidly, so are less affected by a dihydrofolate reductase inhibitor. Clinical resistance to methotrexate often develop by a number of mechanisms. These include mutations in the reduced folate carrier that result in decreased methotrexate uptake, amplification of the dihydrofolate reductase gene, which results in a corresponding amplification of reductase mRNA and protein levels, and decreased folylpolyglutamate synthetase activity, which reduces accumulation of the drug by the tissue.

2. Purine Synthesis

The C-8 and C-2 positions of the purine ring are derived from 10-formyl-$H_4PteGlu_n$ in reactions catalyzed by glycinamide ribonucleotide transformylase (GART, E.C.2.1.2.2; reaction 7, Fig. 2) and aminoimidazolecarboxamide ribonucleotide transformylase (AICART, E.C.2.1.2.3; reaction 8, Fig. 2), as follows:

$$10\text{-formyl-}H_4PteGlu_n + GAR \rightarrow H_4PteGlu_n + \text{formyl-GAR}$$

$$10\text{-formyl-}H_4PteGlu_n + AICAR \rightarrow H_4PteGlu_n + \text{formyl-AICAR}$$

GART catalyzes the third step in purine biosynthesis. GART is part of a 110 kDal mammalian trifunctional protein that also catalyzes two other steps in the purine biosynthetic pathway (GAR synthase and AIR synthase, 134), while AICART is part of a bifunctional protein that also possess inosinicase, which catalyzes the final ring closure step of purine synthesis. The gene for human GAR transformylase is on the q22 band of chromosome 21 (135). A second novel GAR transformylase occurs in *E. coli* that utilizes formate and ATP (and not folate) for the synthesis of formyl-GAR (134).

Aminoimidazolecarboxamide (AIC) is excreted in elevated amounts by animals and humans with a deficiency in folate or vitamin B_{12} (136). In fact, AIC was discovered because it accumulated when bacteria were treated with folate antagonists (137).

The 10-formyl-$H_4PteGlu_n$ required for purine synthesis can be formed by the oxidation of 5,10-methylene-$H_4PteGlu_n$, which is reversibly, catalyzed by methylene-tetrahydrofolate dehydrogenase (E.C. 1.5.1.5; reaction 4, Fig. 2; reaction 3, Fig. 3) and methenyl-tetrahydrofolate cyclohydrolase (reaction 5, Fig. 2; reaction 4, Fig. 3):

$$5,10\text{-methylene-}H_4PteGlu_n + NADP^+ \leftrightarrow [5,10\text{-methenyl-}H_4PteGlu_n]^+ + NADPH$$

$$[5,10\text{-methenyl-}H_4PteGlu_n]^+ + H_2O \leftrightarrow 10\text{-formyl-}H_4PteGlu_n + H^+$$

Alternatively, 10-formyl-$H_4PteGlu_n$ can be obtained by the direct formylation of $H_4PteGlu_n$ (reaction 6, Fig. 2; reaction 8, Fig. 3), catalyzed by formyltetrahydrofolate synthetase:

$$\text{formate} + MgATP + H_4PteGlu_n \rightarrow 10\text{-formyl-}H_4PteGlu_n + MgADP + P_i$$

The dehydrogenase, cyclohydrolase and synthetase activities are associated on a single trifunctional protein in mammalian tissues that is called C_1 synthase (138,139). The synthase consists of two separate domains: One contains the dehydrogenase and cyclo-hydrolase activities and the other the synthetase activity.

This activity plays a central role in one-carbon metabolism as it controls the oxidation levels of one-carbon units that can be channeled into purine biosynthesis or into the synthesis of thymidylate and methionine. It is widely distributed, with very high levels in liver and kidney. In common with many folate enzymes, it is absolutely specific for NADP(H). The cDNA for the human protein encodes a 935 amino acid protein (101.5 kDal, 140). The gene for the human enzyme lies on chromosome 14q24. The K_m for formic acid for the synthetase is more favorable when the folylpolyglutamate substrate is used, demonstrating that the length of the polyglutamyl tail can affect the kinetic properties of substrates other than the folate itself (141). Yeast also possess a monofunctional methylenetetrahydrofolate dehydrogenase activity that uses NAD(H) as a substrate, but this activity has not been observed in mammalian tissues (142).

Mitochondria can also interconvert 5,10-methylene-$H_4PteGlu_n$ and 10-formyl-$H_4PteGlu_n$, and this may be catalyzed by a mitochondrial C_1 synthase (Fig. 3). This activity has been well established in yeast, and very low levels of this activity have been reported in rat liver mitochondria (91), although some investigators have not been able to observe this activity. No mitochondrial C_1 synthase gene has thus far been identified. A separate bifunctional 5,10-methylenetetrahydrofolate dehydrogenase-cyclohydrolase that uses NAD^+ rather than $NADP^+$ as the acceptor has also been described and its gene cloned (143). This mitochondrial enzyme activity is found in embryonic, undifferentiated, or transformed tissues and cells. One suggested role for this enzyme is to increase the one-carbon flux into purine biosynthesis and away from other one-carbon cycles such as methionine synthesis.

Isolated mammalian liver mitochondria can oxidize the β-carbon of serine to formate and CO_2, the proportion of which is dependent on the respiratory state of the organelle (144). Formate formed can leave the mitochondria and be reincorporated into the folate one carbon pool by 10-formyltetrahydrofolate synthetase. NMR studies in yeast suggest that mitochondrial formate generated from mitochondrial serine oxidation is the major provider of one-carbon units for purine synthesis in the cytosol and that the cytosolic C_1 synthase operates in the reductive direction (145,146). Mammalian cells with defects in mitochondrial folate metabolism are glycine auxotrophs, have depleted cytosolic one-carbon pools, require higher levels of folate to support purine synthesis, and are defective in homocysteine remethylation (unpublished data). Recent mass spectrometry studies with mammalian cells have suggested that over 95% of the one-carbon moieties that are incorporated into methionine via the cytoplasmic folate-dependent methionine cycle are derived from mitochondrial serine metabolism to formate. This mimics the directionality for one-carbon pathways suggested for yeast. Whether this is also true for normal mammalian tissues remains to be explored.

C. Disposal of One-Carbon Units

One-carbon moieties are oxidized to CO_2 in a reaction catalyzed by 10-formyl-tetrahydrofolate dehydrogenase (E.C.1.5.1.6; reaction 16, Fig. 2; reaction 7, Fig. 3):

$$\text{10-formyl-}H_4PteGlu_n + NADP^+ + H_2O \rightarrow H_4PteGlu_n + CO_2 + NADPH + H^+$$

The purified enzyme also catalyzes the hydrolysis of 10-formyl-$H_4PteGlu_n$ to $H_4PteGlu_n$ and formate (147,148). $H_4PteGlu_n$ is a potent product inhibitor of the enzyme. The metabolic flux through this reaction in liver may be regulated by the 10-formyl-$H_4PteGlu_n$/ $H_4PteGlu_n$ ratio, rather than by the tissue concentration of 10-formyl-$H_4PteGlu_n$, as the concentrations of these folates are in excess of their K_m and K_i values for the enzyme under physiological conditions. The physiological role of this protein would appear to be to regulate the proportion of folate present in the $H_4PteGlu_n$ form, presumably to make it available for other reactions of one-carbon metabolism, and to dispose of excess one-carbon moieties. The enzyme is a major $H_4PteGlu_n$ cytosolic folate binding protein in liver (149,150). It appears to bind 10-formyl-$H_4PteGlu_n$, which turns over to $H_4PteGlu_n$ when the protein is isolated.

The rat cDNA encodes a 902 amino acid protein of 100 kDal (151). The N-terminal 200 amino acids share some homology with GAR transformylase while the final C-terminal 400 amino acids share homology with aldehyde dehydrogenases and the protein possesses a low NADP-dependent aldehyde dehydrogenase activity.

Mice lacking this activity have been generated by random neutron bombardment (152). These animals have elevated 10-formyl-$H_4PteGlu_n$ levels and reduced tissue folate, the latter being consistent with recent studies suggesting that formyl folate derivatives are substrates for the enzymes that catabolize folates to inactive derivatives. These animals should serve as a useful model for studying the metabolic role of this protein, although the possibility of other lesions in these animals can not be eliminated.

D.   Mitochondrial Protein Synthesis

Protein biosynthesis in mitochondria, chloroplasts, and bacteria is initiated by a special type of transfer RNA, N-formyl-methionyl—$tRNA_fmet$. The formylation of the methionine residue bound to the tRNA is catalyzed by methionyl-tRNA transformylase (reaction 9, Fig. 3):

$$10\text{-formyl-}H_4PteGlu_n + \text{methionyl—}tRNA_fmet$$
$$\rightarrow H_4PteGlu_n + \text{N-formyl-met-}tRNA_fmet$$

This enzyme is essential in bacteria and interruption of the gene in yeast causes defects in mitochondrial protein synthesis. Although it is assumed that an analogous protein is required for normal mitochondrial function in mammalian cells, it is not clear whether this is the case. Mammalian cells will grow at normal rates in the complete absence of folate if glycine, purines, thymidine, and methionine, the major products of one-carbon metabolism, are provided, which would suggest that formylation of methionyl-$tRNA_fmet$ is not required in mammalian cells.

VIII.   DEFICIENCY EFFECTS

The classical symptom of folate deficiency in humans is a megaloblastic anemia that is indistinguishable from that caused by vitamin B_{12} deficiency. Folate deficiency symptoms are usually due to a dietary insufficiency, although they can arise from other causes such as malabsorption syndromes or drug treatment. Cases of increased requirement due to genetic variation have been identified, including some individuals who require increased folate in early pregnancy to reduce the risk of birth defects. Many of the clinical effects of folate deficiency can be explained by the metabolic role of folate coenzymes in pathways leading to DNA precursor synthesis and methyl group homeostasis. Because of these roles, symptoms of deficiency are often expressed first in rapidly growing tissues. Depressed folate status has also been associated with increased cancer and vascular disease risk (153,154). Neurological manifestations of deficiency have been suggested, although the evidence for this is not convincing. However, in rare cases of genetic disease resulting in severe defects in enzymes of one-carbon metabolism, neurological symptoms have been clearly documented with many cases of mental retardation.

Serum folate falls below normal after three weeks of experimental folate deprivation in humans. After seven weeks there is an increase in the average number of lobes of the nuclei of the neutrophils (hypersegmentation). Red blood cell folates gradually fall and reach subnormal levels after four months of folate deprivation, reflecting the life span of the red cell. At about 4.5 months the bone marrow becomes megaloblastic and anemia occurs (155). Lesions throughout the intestinal tract may occur in subjects receiving anti-folate chemotherapy (156).

Folate deficiency can be induced in experimental animals by feeding diets lacking in or deficient in the vitamin. Deficiency symptoms appear earlier in the young growing animal than in the adult. In some animals, such as the rat, folate deficiency is difficult to achieve unless the diet contains an antibacterial drug to prevent synthesis of folate by the gut microflora (157). Although coprophagy may explain the need for antibacterials, it appears that rats are able to absorb folate biosynthesized toward the distal end of the intestine. Addition of labeled p-aminobenzoic acid to the diet of the rat resulted in labeled folates in the liver (158). The extent to which bacterially synthesized folates can contribute to the animal's vitamin stores has not been evaluated. The signs of folate deficiency in animals includes anorexia, diarrhea, cessation of growth, weakness, decreased leukocyte count, and low red blood cell counts (anemia), and eventually death. Although leukopenia develops universally in rats fed folate-free sulfa drug-containing diets, anemia occurs less frequently.

A. Megaloblastic Anemia

Megaloblastic anemia is a reflection of deranged DNA synthesis in blood cells. It is characterized by enlarged red cells and hypersegmentation of the nuclei of circulating polymorphonuclear leukocytes with reduced cell number. Megaloblastic changes also occur in other tissues such as the small intestine but the condition is usually detected clinically by the anemia. Cellular DNA content is increased, but the DNA contains strand breaks suggesting a defect in DNA synthesis or repair. Cells growth is arrested in the G2 phase of the cell cycle just prior to mitosis, preventing cell division. If cell division occurs, many of the cells undergo apoptosis (159,160).

These defects are thought to be a result of uracil misincorporation into DNA in place of thymidylate. dUTP is not normally incorporated into DNA but can arise by deamination of cytosine. This is a potentially mutagenic event as uracil is recognized as thymine and can form a base pair with adenine. Replication of the DNA will change the C-G base pair to a U(T)-A base pair. This change is repaired by uracil-DNA glycosylase, which removes the uracil base (211). A few additional bases are removed on either side of the damage and the DNA is repaired by complementary base pairing with a C being reinserted opposite the G. Direct dUTP incorporation into DNA is also minimized by a dUTPase activity, which hydrolyzes dUTP to dUMP and keeps cellular dUTP pools very low.

In severe experimental folate deficiency induced by antifolate drugs, thymidylate pools are depressed and dUTP pools are increased leading to increased dUTP incorporation instead of dTTP (160). Increased repair by the glycosylase would lead to more transient single-stranded breaks. In addition, repair of the damage by reinsertion of thymidine is defective due to the lower thymidylate pools and the probability of uracil being reinserted by mistake is higher, which would lead to prolongation of the single-stand breaks. A double-strand break can occur when uracil is misincorporated on both DNA strands in close proximity. Although this mechanism has been established in cell culture with antifolate drugs and is believed to be responsible for inducing the megaloblastosis of human folate deficiency, technical problems in the measurement of uracil in DNA have prevented, until recently, confirmation that this occurs in folate-depleted humans. Damaged blood cells, which might be expected to demonstrate an increased uracil content, are normally removed by the spleen. In a recent study using a more sensitive assay for uracil in DNA and involving splenectomized subjects, individuals with low red cell folate levels had an increased uracil content and double-strand breaks in their DNA, and folate supplementation reversed these abnormal findings (161).

B. Vitamin B_{12} Interactions

The pernicious anemia that results from vitamin B_{12} deficiency is identical to that observed in folate deficiency, and vitamin B_{12} deficiency induces many metabolic changes in one-carbon metabolism that are identical to those observed in folate deficiency. The interrelationship between these two vitamins is best explained by the methyl trap hypothesis (162). The two vitamins are substrates or cofactors for the methionine synthase reaction. A block in this enzyme would cause accumulation of folate in the 5-methyl-H_4PteGlu$_n$ form. As 5-methyl-H_4PteGlu$_n$ cannot be metabolized by any other mechanism, this would result in the trapping of folate in a nonfunctional form with a concomitant reduction in the level of other folate coenzymes required for other reactions of one-carbon metabolism. Because 5-methyl-H_4PteGlu is a poor substrate for folypolyglutamate synthetase, the ability of tissues to polyglutamate and retain entering folate would be greatly diminished (163,164), and a true folate deficiency would be superimposed on the functional deficiency caused by the methyl trap.

Methionine synthase activity in bone marrow of pernicious anemia patients is reduced over 85%, and most of the protein is present in the apoenzyme form (165). Patients with severe genetic defects in methionine synthase or methionine synthase reductase that cause gross impairment of methionine synthase activity develop early onset megaloblastic anemia (166,167). A defect in methionine synthesis or methyl group status cannot explain this anemia as patients with severe genetic disease involving methylenetetrahydrofolate reductase, a defect that would block the methionine cycle but not result in a methyl trap, do not develop megaloblastosis (168). As discussed previously, deletion of both copies of the mouse methionine synthase gene causes embryonic lethality while the methylenetetrahydrofolate reductase knockout is viable. This illustrates that the methyl trap can cause a very severe derangement of one-carbon metabolism and that no mechanism exists that can compensate for impaired methionine synthase activity.

Vitamin B_{12} deficiency can be induced in the rat by feeding a vitamin-free diet and the metabolic effects of a severe deficiency can be mimicked by nitrous oxide exposure, which causes essentially total loss of methionine synthase activity. Nitrous oxide treatment causes a gross impairment in tissue folate levels due to an inability to retain folate, an increase in plasma folate levels, and an increase in the proportion of hepatic folate in the 5-methyl-H_4PteGlu$_n$ form, all of which are consistent with the methyl trap mechanism (210). Although not all hepatic folate is trapped as 5-methyl-H_4PteGlu$_n$, essentially all cytosolic folate is converted to this form, and the nonmethyl folate remaining in the tissue is located in the mitochondria (unpublished data). Addition of high levels of methionine or ethionine to the diet, or their injection, ameliorates the effects of nitrous oxide (169) on hepatic folate levels and one-carbon distributions, consistent with adenosyl-methionine or adenosyl-ethionine inhibition of methylenetetrahydrofolate reductase. Megaloblastic changes also occur in the blood profiles of subjects treated with the anesthetic nitrous oxide as a result of destruction of methionine synthase. After treatment is terminated, enzyme levels gradually return to normal due to synthesis of new protein.

C. Cancer

Epidemiological studies have suggested that folate deficiency is associated with increased risk for certain types of cancer, including colon cancer (154). While the mechanism for this has not been established, uracil misincorporation arising from defective thymidylate synthesis has been hypothesized as one possibility. Transcription of many genes is turned off during development by methylation of their promoter regions' and changes in gene methylation, both hyper and hypomethylation, have been observed in tumors. As folate deficiency impairs the remethylation of homocysteine to methionine and alters AdoMet/ AdoHcy ratios, it has also been proposed that the increased cancer risk in folate deficiency may be due to hypomethylation of DNA. It has been shown that methionine deficiency causes hypomethylation of DNA (170). A clear demonstration that folate deficiency results in DNA hypomethylation remains to be carried out.

The increased cancer risk in subjects with poorer folate status is reduced in subjects homozygous for a common polymorphism in methylenetetrahydrofolate reductase (154) and, somewhat surprisingly, this polymorphism had an even greater beneficial effect on risk in subjects with good folate status. This polymorphism causes decreased enzyme activity and presumably impaired conversion of 5,10-methylene-H_4PteGlu$_n$ to 5-methyl-H_4PteGlu$_n$. It is speculated that this may allow a redirection of more of the folate one-carbons into the cycles of nucleotide biosynthesis in these subjects, but this remains to be established.

D. Folate Antagonists—Cancer Treatment and Other Diseases

Folate antagonists that are inhibitors of thymidylate synthase, dihydrofolate reductase, and de novo purine biosynthetic enzymes have been used extensively for the treatment of a variety of cancers. These metabolic inhibitors cause a functional folate deficiency and generally show a selective toxicity for rapidly growing tumors because of the increased rates of DNA synthesis in these tumors. The most widely used drug is methotrexate, a 4-amino-folate, which is a potent inhibitor of dihydrofolate reductase. This drug was not a 'rationally designed drug' as it was first synthesized and used for chemotherapy almost 60 years ago, long before dihydrofolate reductase was identified. ''Rescue'' therapy is sometimes employed to increase the efficacy of methotrexate. In this treatment, a large toxic dose of the drug is given and this is followed by giving a large dose of folinic acid (5-formyl-H_4PteGlu) as an antidote to rescue normal cells. Another commonly used drug is 5-flurouracil, an inhibitor of thymidylate synthase described earlier.

In experimental animals, toxicity and drug effectiveness is reduced by the provision of purines and thymidine. Uracil misincorporation and apoptosis have been demonstrated as the mechanism for cell death. Uracil misincorporation is also potentially mutagenic, and successful treatment with antifolates, or with many other drugs used for cancer chemotherapy, carries the risk of further cancers after ten to twenty years.

Antagonists of folate biosynthesis, such as sulfa drugs, have been used extensively as antimicrobiol agents with limited side effects, as most of these drugs do not interfere with mammalian folate metabolism.

Recently there has been a renewed interest in the treatment of rheumatoid arthritis using low dose methotrexate treatment (171). Here, as with the treatment of cancer and other diseases such as psoriasis, the patient should be monitored for anemia and other forms of methotrexate toxicity.

E. Hyperhomocysteinemia and Vascular Disease

Patients with severe genetic disease involving enzymes in the homocysteine remethylation and transsulfuration pathways are homocystinuric, display very marked hyperhomocysteinemia, and suffer from a variety of clinical symptoms including early onset occlusive cardiovascular and cerebrovascular disease (172). These genetic diseases include methylenetetrahydrofolate reductase deficiency and cystathionine β- synthase deficiency (Fig. 4). The beneficial effects of folate and betaine on disease progression in patients with severe B_6-nonresponsive cystathionine-β-synthase deficiency is the strongest evidence that elevated homocysteine or a homocysteine metabolite is the primary cause of the vascular complications in these patients (173). These treatments can not rescue the metabolic defect but they can divert homocysteine into methionine. Elevated homocysteine increases proliferation of smooth muscle cells and inhibits proliferation of endothelial cells by a mechanism that is not understood (174). Many potential reasons for the adverse effects of homocysteine have been described. These include effects on transcription factors involved in regulation of cell growth, covalent modification via disulfide bond formation of proteins such as apolipoprotein B100, modulation of NO synthase activity, and increased cellular adenosyl homocysteine levels. It has not been established which, if any, of these potential adverse changes is responsible for the vascular disease.

Recently, it has been recognized that chronic mild hyperhomocysteinemia may also be a major risk factor for occlusive vascular disease (175). Most of the evidence for this has come from case control studies and prospective studies have been less convincing. Plasma homocysteine concentrations in patients with vascular disease were about 30% higher than in controls, and carotid artery stenosis was positively correlated with plasma homocysteine concentrations over the entire range of normal and abnormal homocysteine values (176). Prospective assessment of vascular disease risk in men with higher homocysteine concentrations indicated that plasma homocysteine levels only 12% above the upper limit of normal levels were associated with a threefold increase in acute myocardial infarction (177).

Fasting homocysteine levels have been inversely correlated with both plasma folate levels and food folate intake (178). Increased folate intake lowers the mean homocysteine of groups, and the lowering effect is greatest in subjects with the highest plasma homocysteine levels. Folate is less effective in reducing elevated homocysteine in renal disease patients, suggesting that the kidney is a major site of homocysteine metabolism. A common polymorphism (Ala to Val) in methylenetetrahydrofolate reductase that results in a heat-labile enzyme, and decreased enzyme activity in tissues has been implicated as one reason for the folate-responsiveness of a subset of hyperhomocysteinemic subjects (113). The incidence of the val/val homozygote (around 10 to 15% in most populations) is significantly increased in subjects with the highest deciles of homocysteine levels. Although the contribution of this polymorphism to elevated homocysteine levels has varied from study to study, and the incidence of the polymorphism varies between different population groups, the val/val polymorphism may at most contribute to or be associated with 30% of hyperhomocysteinemia. While case control studies indicate an association between this polymorphism and vascular disease risk, this relationship was not observed in a large prospective study (179). It is also interesting to note that this potential genetic risk factor for vascular disease is the same polymorphism that epidemiological studies have suggested lowers the risk of cancer (described above). About 50% of the general population are heterozygous or homozygous for this polymorphism, and it is particularly interesting that a simple nutritional intervention may ameliorate at least some of the adverse effects of a potentially deleterious genetic trait.

Impaired vitamin B_{12} status has also been associated with the highest decile of fasting plasma homocysteine in the general population, but B_{12} is quantitatively a less important risk factor than folate for hyperhomocysteinemia (178). Again, this is consistent with the role of cobalamin as a cofactor for methionine synthase. Dietary vitamin B_{12} content had little effect on homocysteine levels, reflecting that defects in B_{12} absorption rather than dietary vitamin B_{12} content play a greater role in the development of impaired B_{12} status. PLP is a cofactor for cystathionine β-synthase. However, vitamin B_6 status has little effect on fasting homocysteine levels (178), but improved vitamin B_6 status reduces the increase in plasma homocysteine following a methionine load or a meal, and also nonfasting homocysteine levels.

Very high levels of homocysteine are clearly associated with severe vascular disease, but it remains to be established whether the increased risk associated with mildly elevated homocysteine is due directly to homocysteine. The increased risk may be due to some other metabolic disturbance or change in vitamin status for which homocysteine acts as an indicator. Consequently, although homocysteine is associated with risk for vascular disease, and increased intake of folate reduces circulating homocysteine levels, it remains to be determined whether increased folate intake reduces vascular disease risk. A number of clinical intervention trials are currently in progress to ascertain whether vitamin supplementation influences the incidence of stroke and of cardiovascular disease, which should help in answering some of these questions.

F. Pregnancy

Megaloblastic anemia due to folate insufficiency has long been recognized as a complication of pregnancy (180) with an incidence of 3–5% in developed countries and much higher incidence in Africa, Southeast Asia, and South America. The development of this syndrome is due to the increased nutritional need due to the growth of the fetus. The anemia presents as hematocrit values below the accepted norm for pregnancy, and folate deficiency as indicated by low red blood cell folates.

G. Birth Defects

Neural tube defects (NTDs), including most forms of spina bifida, are the most common birth defects in humans, affecting about 0.1% of births, although in some regions the numbers are much higher. The reoccurrence rate for this condition is about 2%. Neural tube defects are a family of birth defects involving the brain or spinal cord. They arise from incomplete closure of the neural tube during the fourth week of pregnancy. Neurulation is the first organogenetic process to be initiated and completed in humans, and occurs during the fourth week of pregnancy. In this process, a flat structure called the neural plate forms two parallel ridges. These ridges fold over and move towards each other and seal and form the neural tube. Closure of the tube begins separately at three sites, the cervical-hind brain boundary, the forebrain-midbrain boundary, and the rostral extremity of the forebrain. Closure spreads to the intervening areas and a failure of closure at any of the regions can lead to a NTD. Any drug, nutrient imbalance, or genetic defect that interferes with this stage of development may prevent normal closure of the tube, resulting in defects in the newborn. At its least severe, an NTD may be undetectable except by examination of the spine by the fingers of a physician. At its most severe, an NTD involves the absence of the brain. The most common NTDs are spina bifida (spinal cord at the lumbar vertebra not covered with bone) and anencephaly (no brain).

The observation that peri-conceptual folate supplementation with folic acid reduced the incidence of NTDs by about two-thirds (181), which has now been confirmed in a number of studies (182,183), has led to fortification of the U.S. food supply with folic acid (184). Folate supplementation is only useful if given very early in pregnancy at a time when many women do not realize they are pregnant. Although folate status affects the risk for neural tube defects, this condition is not thought to be a result of folate deficiency per se. It appears to be a genetic disease, probably multigenetic, with a phenotype that can be modified by increased folate in the subset of individuals that are folate responsive. As the mechanism behind the disease in not known, there is currently no screening technique to identify individuals at risk.

A defect in homocysteine metabolism has been proposed as a mechanism, although the evidence supporting this proposal is far from conclusive. Plasma homocysteine levels are slightly higher in affected mothers (185). Homocysteine can cause teratogenic effects in embryo culture, but very high levels were used in these studies. An increased incidence of the homozygous Ala to Val polymorphism in methylenetetrahydrofolate reductase has been reported in a number of studies but this could account for, at most, 15% of neural tube defect risk (186). One epidemiological study suggested that vitamin B_{12} status is an independent risk factor for neural tube defects (187), which would implicate methionine synthase, or a gene encoding a product that influences methionine synthase activity, as another possible locus for the defect. A common polymorphism in the methionine synthase gene has a modest influence on homocysteine levels (188). However, no polymorphisms or mutations in the methionine synthase gene or in the methionine synthase reductase gene have been identified thus far that track with neural tube defects.

A number of folate responsive mouse models of neural tube defects have been developed. Disruption of the FBP1 gene, which encodes one of the two mouse folate binding proteins, causes NTDs, the incidence and severity of which are reduced by folate supplementation (189). However, no polymorphisms in the equivalent human FBPα gene were observed in human NTD cases (190). The methylenetetrahydrofolate reductase knockout also exhibits neural tube defects that are alleviated to some extent by high folate. The methionine synthase knockout embryo does not survive to day 8.5 of gestation when the neural tube closes in the mouse, even when the mother is supplemented with very high levels of folate and products of one-carbon metabolism. Because of the role of folate in pathways involved in cell growth and DNA metabolism, it is likely that disruption of many of the genes encoding folate enzymes in mice would cause birth defects. Which, if any, of the genes is responsible for human birth defects remains to be discovered. None of the polymorphisms that are known for human cystathionine β-synthase track with NTD cases and total disruption of this gene in mice does not have any apparent effect on fetal development (191). The animals become quite sick soon after birth when homocysteine levels become greatly elevated. However, they die of hepatic failure rather than vascular disease. The animals can be maintained for longer periods if the diet is fortified with choline. Although this might suggest that homocysteine is not involved in the etiology of NTDs, the synthase enzyme is not expressed during early fetal development and the fetus appears to be dependent on remethylation for metabolism of homocysteine.

The folate intervention trials that established the protective effect of folic acid, coupled with other dietary surveys, have indicated that 400 μg of supplemental folic acid in addition to customary dietary folate intake is sufficient to provide the maximum benefit in reducing the incidence of folate-responsive birth defects (reviewed in Ref. 192). It is not known whether lower levels would be as effective, or whether disease risk could be reduced by dietary folate alone, as food folate is less bioavailable than folic acid. The fortification of the U.S. food supply with folic acid is designed to provide an average intake of 100 μg of supplemental folic acid in addition to normal dietary folate intake. This represents a compromise between the needs of the relatively small population at risk for birth defects and the relatively larger population at risk for masking of symptoms of vitamin B_{12} deficiency by high folate (see below). Preliminary studies have suggested that folate supplementation may also have a beneficial effect on other pregnancy outcomes. Although the fortification of the food supply with folic acid has just started, it is already clear that this had a major impact on indicators of folate status in the U.S. population. The average additional intake of folic acid is considerably higher than 100μg because of overage by food producers; in part to allow for losses in storage. As folic acid is almost twice as bioavailable as food folate, the average increased intake is equivalent to about 250 to 300 μg food folate, or about a doubling of the folate intake. This has led to large increases in plasma folate levels and a significant drop in plasma homocysteine levels, particularly in individuals who were at the higher end of the homocysteine distribution previously (193). Whether this fortification has a positive effect of birth defect incidence or on vascular disease incidence remains to be seen.

While the beneficial effect of increased folate is probably due to its correction of a metabolic defect(s), other possibilities cannot be excluded. One of the intervention trials that demonstrated a protective effect of folate for neural tube defects also noted an increased spontaneous abortion rate in supplemented subjects. A large proportion of embryos abort very early in pregnancy, and it has been suggested the folate effect may be due to an increased abortion rate of impaired embryos in early pregnancy rather than a correction of a metabolic imbalance.

H. Anticonvulsants

A small percentage of epileptics treated with diphenylhydantoin develop a folate deficiency and anemia (194,195). Studies with mice (196) and humans (197) suggest that increased folate excretion may be a contributing factor. Large doses of folic acid (5–30 mg/day) have been used to reverse the hematological signs of folate deficiency (198), though the folic acid may exacerbate the seizures in certain patients.

I. Ethanol

Low serum folates are commonly encountered in alcoholics, largely because of poor nutritional habits. Chronic alcoholism is probably the major cause of folate deficiency in the

United States. An antifolate effect of ethanol was shown in studies of humans (199). Megaloblastic anemia could be induced with a low-folate diet after 6–10 weeks, whereas the inclusion of alcohol with this diet provoked anemia at an earlier time, 2–3 weeks (199).

J. Malabsorption Syndromes

Diseases of the intestinal tract such as tropical sprue and nontropical sprue (gluten enteropathy) lead to general malabsorption syndromes including folate deficiency. Folate is normally absorbed in the jejunum. In some sprue cases, the malabsorption in the jejunum is partially alleviated by increased absorption lower down the intestine. Folate deficiency can also occur in Crohn's disease and ulcerative colitis because of the malabsorption (200). Congenital disorders associated with folate malabsorption have been reported, implying a specific transport system for the vitamin (201). The resulting deficiency can be relieved by large oral doses of the vitamin (40–100 mg).

IX. METHODS OF NUTRITIONAL ASSESSMENT

A. Assay of Folate Derivatives in Plasma and Serum

Folate status is most commonly assessed by plasma or serum folate levels. This can be measured by microbiological assay using *L. casei* as the test organism, but this test can be confounded if the subject is on antibiotic treatment. Other bacteria such as *S. fecium* or *P. cerevisiae* are not suitable for the assay of serum folates as they do not respond to 5-methyl-H_4PteGlu. Serum folate levels of less than 3 ng/ml indicate a folate deficiency, levels of 3–6 ng/ml a marginal deficiency, and levels above 6 ng/ml adequate folate status. One problem in interpreting serum folate values is that they reflect recent dietary intake, and a vitamin deficiency can be ascribed only where serum folate remains low over a period of time.

Radioassay procedures for the measurement of serum folate have been developed and radioassay kits are used extensively in clinical laboratories. These competitive protein-binding assays are easier to perform than microbiological assays and are not affected by antibiotics, which give false low values in microbiological assays. However, the affinities of different folate monoglutamates for the binding proteins vary considerably, making this assay method useful only for those tissues in which one form of folate predominates, i.e.,serum or plasma (202). In addition, serum folate values obtained by using the different commercially available kits fluctuate considerably. Because of this, no absolute values can be given to indicate folate deficiency. Each laboratory has to define its own lower limits based on a large number of sample assays from a representative population of normal subjects. In the measurement of serum folate levels by both microbiological and radioassay methods, great care must be taken to prevent hemolysis of the samples, as red cell levels of folate are considerably higher than serum levels.

B. Assay of Red Cell Folates

Red cell folate levels reflect the body folate stores at the time of red cell formation and hence reflect a more accurate and less variable index of folate status than plasma folate levels. Folates in the red cell are polyglutamate derivatives and must be hydrolyzed to monoglutamates prior to their assay by microbiological or radioassay techniques. As practically all the whole blood folate is located in the red cell, this is usually accomplished by lysing whole blood and allowing plasma glutamyl hydrolase to hydrolyze the polyglutamate derivatives. As in the assay of serum folates, it is essential to protect the vitamin with reducing agents, such as ascorbate. Red blood cell folate levels of less than 140 ng/ml packed cells, measured by microbiological assay, indicate folate deficiency, levels of 140–160 ng/ml suggest marginal status, and levels above 160 ng/ml indicate normal folate status.

Subjects homozygous for the C677T polymorphism in methylenetetrahydrofolate reductase exhibit a changed folate one-carbon distribution in their red cells (203). As the radioassay procedure responds differently to different folate forms (202), it may be confounded under conditions such as this that alter folate one-carbon distributions.

C. Histidine and Methionine Load Tests

The levels of metabolites such as urinary formiminoglutamate and plasma homocysteine following a loading dose of histidine or methionine can be used a measure of folate status (193). However, the levels of these metabolites are also abnormal under conditions of vitamin B_{12}, and in some cases, vitamin B_6 deficiency.

D. Deoxyuridine Suppression Test

Labeled thymidine incorporation into the DNA of bone marrow cells is reduced in the presence of deoxyuridine due to the conversion of dUMP to dTMP, and the consequent competition between labeled and unlabeled dTTP (204). The extent of the competition is a measure of thymidylate synthase activity, which depends on the level of functional folate in the cell. Under conditions of low cellular folate, the conversion of dUMP to dTMP is reduced and the suppression of labeled thymidine incorporation into DNA is reduced. This test, known as the dU suppression test, can distinguish between folate and vitamin B_{12} deficiency as, in the former case, addition of any folate to the medium increases the suppression rate. In the case of a vitamin B_{12} deficiency, addition of methylcobalamin or any folate (with the exception of 5-methyl-H_4PteGlu) increases the suppression. Although this test has been used experimentally, it has not found widespread usage, partly because of the difficulty in obtaining marrow cells from patients on a routine basis. Attempts to extend this test to other cells, such as lymphocytes, have met with mixed success.

X. NUTRITIONAL REQUIREMENTS

The current RDA is 150–300 µg for children, 400 µg for the adolescent, adult, and the elderly, 600 µg for pregnant women, and 500 µg for lactating women (192). Higher levels may be required to minimize the risk of birth defects. The recommendation for women of child bearing age, who are capable of becoming pregnant, is to take 400 µg folic acid per day, derived from supplements plus fortified food, in additional to their normal food folate intake. Requirements for the infant are an Adequate Intake (AI) of 65–80 µg, based on the folate content of milk of well-nourished mothers. Increases in dietary intake of folate do not affect maternal milk levels.

XI. HAZARDS OF HIGH DOSES

No toxicity of high doses of folate has been reported. In adults no adverse effects were noted after 400 mg/day for 5 months or after 10 mg/day for 5 years (205). The acute toxicity (LD_{50}) is about 500 mg/kg body weight for rats and rabbits (206). Chronic doses of 10–75 mg PteGlu/kg body weight (IP) can injure the kidneys, probably because of precipitation of the PteGlu at acidic pH (207,208).

Large doses of folic acid can produce a hematological response in subjects with megaloblastic anemia due to vitamin B_{12} deficiency, although it does not correct the severe neurological symptoms of vitamin B_{12} deficiency. As large doses of folate may prevent or delay the development and diagnosis of anemia in vitamin B_{12}-deficient subjects, there is an increased risk that these subjects are only recognized when they develop frank neurological symptoms. Megadose levels of folate should be avoided and an upper limit of 1 mg has been suggested (192). The intakes of many individuals now exceed this level due to fortification of the food supply and vitamin pill intake. Whether this leads to an increased incidence of neurological disease in the elderly will be closely watched, particularly as there continues to be on ongoing debate about whether the current levels of fortification is sufficient to eliminate all folate-responsive birth defects.

LIST OF ABBREVIATIONS

PteGlu, pteroylglutamic acid, folic acid
H_4PteGlu, tetrahydrofolate
H_4PteGlu$_n$, tetrahydropteroylpolyglutamate, n indicating the number of glutamate residues
AdoMet, S-adenosylmethionine
AdoHcy, S-adenosylhomocysteine

REFERENCES

1. R. L. Blakley, S. J. Benkovic. *Folates and Pterins, Vol. 1: Chemistry and Biochemistry of Folate*. New York: Wiley, 1984.
2. R. L. Blakley, V. M. Whitehead. *Folates and Pterins, Vol. 3: Nutritional, Pharmacological, and Physiological Aspects*. New York: Wiley, 1986.
3. T. Brody. *Nutritional Biochemistry*. New York: Academic Press, 1999, pp 493–524, 550–554.
4. O. D. Bird, V. M. McGlohon, J. W. Vaitkus. *Can. J. Microbiol. 15*:465, 1969.
5. S. D. Wilson, D. W. Horne. *Proc. Natl. Acad. Sci. U.S.A., 80*: 6500, 1983.
6. T. Markkanen, S. Virtanen, P. Himanen, R. Pajula. *Acta. Haematol. 48*:213, 1972.
7. T. Brody. *Pteridines, 1*: 159, 1989.
8. G. B. Henderson, E. M. Zevely, F. M. Huennekens. *J. Biol. Chem. 252*: 3760, 1977.
9. L. M. Kozloff, M. Lute, L. K. Crosby. *J. Virol. 16*:1391, 1975.
10. M. Poe, S. J. Benkovic. *Biochemistry 19*: 4576, 1980.
11. L. J. Slieker, S. J. Benkovic. *Mol. Pharmacol. 25*: 294, 1984.
12. C. W. Waller, B. L. Hutchings, J. H. Mowat, E. L. R. Stokstad, J. H. Boothe, R. B. Angier, J. Semb, Y. Subbarow, D. B. Cosulich, M. J. Fahrenbach, M. E. Hultquist, E. Kuh, E. H. Northey, D. R. Seeger, J. P. Sickels, J. M. Smith. *J. Am. Chem. Soc. 70*:19, 1948.
13. E. C. Taylor, R. Portnoy, D. Hochsetter, T. Kobayashi. *J. Org. Chem. 40*:2347, 1975.
14. H. Mautner, Y. Kim. *J. Org. Chem. 40*:3447, 1975.
15. J. H. Boothe, J. H. Mowat, B. L. Hutchings, R. B. Angier, C. W. Waller, E. L. R. Stokstad, J. Samb, A. L. Gazzola, Y. Subbarow. *J. Am. Chem. Soc. 70*:1099, 1948.
16. J. Meienhofer, P. M. Jacobs, H. A. Godwin, I. H. Rosenberg. *J. Org. Chem., 35*:4137, 1970.
17. C. M. Baugh, J. Stevens, and C. Krumdieck. *Biochem. Biophys. Acta. 212*:116, 1970.

18. R. L. Blakley. *Nature 188*:231, 1960.
19. B. Roth, M. E. Hulquist, M. J. Fahrenbach, D. B. Cosulich, H. P. Broquist, J. A. Brokman, J. M. Smith, R. P. Parker, E. L. R. Stokstad, T. H. Jukes. *J. Am. Chem. Soc. 74*:3247, 1952.
20. R. L. Slakley. *Biochem. J. 65*: 331, 1957.
21. I. Chanarin, J. Perry. *Biochem. J. 105*:633, 1967.
22. D. Chippel, K. G. Scrimgeour. *Can. J. Biochem. 48*:999, 1970.
23. G. Gapski, J. Whiteley, F. M. Huennekens. *Biochem. 10*:2930, 1971.
24. P. B. Rowe, G. P. Lewis. *Biochem. 12*:1962, 1973.
25. J. C. Rabinowitz, W. E. Pricer. *J. Am. Chem. Soc. 78*:5702, 1956.
26. R. G. Kallen, W. P. Jencks. *J. Biol. Chem. 241*: 5851, 1966.
27. M. J. Osborn, P. T. Talbert, F. M. Huennekens. *J. Am. Chem. Soc. 82*:4921, 1960.
28. M. C. Archer, L. S. Reed. *Methods Enzymol. 66*:452, 1980.
29. J. M. Whiteley, J. Drais, J. Kirchner, F. M. Huennekens. *Arch. Biochem. Biophys. 126*:55, 1968.
30. R. L. Blakley. In: A Neuberger and E. L. Tatum, eds. *The Biochemistry of Folic Acid and Related Pteridines, Frontiers of Biology, Vol. 13*. Amsterdam: North-Holland, 1969.
31. R. L. Blakley, S. J. Benkovio. *Folates and Pterins, Vol. 2: Chemistry and Biochemistry of Pterins*. New York: Wiley, 1985.
32. P. Lopez, B. Greenberg, S. A. Lacks. *J. Bacteriol. 172*:4766, 1990.
33. A. L. Bognar, C. Osborne, B. Shane. *J. Biol. Chem. 262*:12337, 1987.
34. R. Ferone, M. H. Hanlon, S. C. Singer, D. F. Hunt. *J. Biol. Chem. 261*:16356, 1986.
35. P. C. Raemakers-Franken, R. Bongaerts, R. Fokkens, C. van der Drift, G. D. Vogels. *Eur. J. Biochem. 200*:783, 1991.
36. J. T. Keltjens, G. D. Vogels. *Biofactors 1*:95, 1988.
37. B. Shane. *Am. J. Clin. Nutr. 35*: 599, 1982.
38. I. Eto, C. L. Krumdieck. *Analyt. Biochem. 120*:323, 1982.
39. T. Brody, B. Shane, E. L. R. Stokstad. *Analyt. Biochem. 92*:501, 1979.
40. S. D. Wilson, D. W. Horne. *Anal. Biochem. 142*:529, 1984.
41. G. Varela-Moreiras, E. Seyoum, J. Selhub. *J. Nutr. Biochem. 2*:44, 1991.
42. D. W. Horne, D. Patterson. *Clin. Chem. 34*:2357, 1988.
43. O. D. Bird, M. V. McGlohon, J. W. Vaitkus. *Analyt. Biochem. 12*:18, 1965.
44. C. M. Baugh, C. L. Krumdieck, H. J. Baker, C. E. Butterworth. *J. Clin. Invest. 50*:2009, 1971.
45. C. J. Chandler, T. T. Y. Wang, C. H. Halsted. *J. Biol. Chem. 261*:928, 1986.
46. N. Grossowicz, M. Rachmilewitz, G. Itzak. *Proc. Soc. Exp. Biol. Med. 150*:77, 1975.
47. T. Tamura, E. L. R. Stokstad. *Br. J. Haematol. 25*:513, 1973.
48. N. Colman, R. Green, J. Metz. *Am. J. Clin. Nutr. 28*:459, 1975.
49. N. Colman, M. Barker, R. Green, J. Metz. *Am. J. Clin. Nutr. 27*:339, 1974.
50. N. Colman, J. Larsen, M. Barker, E. Barker, R. Green, J. Metz. *Am. J. Clin. Nutr. 28*:465, 1975.
51. J. F. Gregory, 3rd. *Eur. J. Clin. Nutr. 51 Suppl. 1*: S54, 1997.
52. K. Aiso, T. Tamura. *J. Nutr. Sci. Vitaminol (Tokyo) 44*: 361, 1998.
53. T. T. Y. Wang, A. M. Reisenaur, C. H. Halsted. *J. Nutr. 115*:814, 1985.
54. C. H. Halsted, E. Ling, R. Luthi-Carter, J. A. Villanueva, J. M. Gardneri, J. T. Coyle. *J. Biol. Chem. 273*:20417, 1998.
55. T. Brody, E. L. R. Stokstad. *J. Biol. Chem. 257*:14271, 1982.
56. J. A. Moscow, M. Gong, R. He, M. K. Sgagias, K. H. Dixon, S. L. Anzick, P. S. Meltzer, K. H. Cowan. *Cancer. Res. 55*:3790, 1995.
57. H. M. Said, T. T. Nguyen, D. L. Dyer, K. H. Cowan, S. A. Rubin. *Biochim. Biophys. Acta. 1281*:164, 1996.
58. C. D. Chancy, R. Kekuda, W. Huang, P. D. Prasad, J. Kuhnel, F. M. Sirotnak, P. Roon, V. Ganapathy, S. B. Smith. *J. Biol. Chem. 275*:20676, 2000.

59. R. F. Pratt, B. A. Cooper. *J. Clin. Invest. 50*: 455, 1971.

60. H. Dencker, M. Jägerstaad, A.-K. Westesson. *Acta. Hepato-Gastroenterol. 23*:140, 1970.

61. J. Kiil, M. Jägerstaad, L. Elsborg. *Internat. J. Vit. Res. 49*:296, 1979.

62. L. L. Samuels, D. M. Moccio, F. M. Sirotnak, F. M. *Cancer Res. 45*:1488, 1985.

63. S. E. Steinberg, C. L. Campbell, R. S. Hillman. *J. Clin. Invest. 64*:83, 1979.

64. G. B. Henderson, M. R. Suresh, K. S. Vitols, F. M. Huennekens. *Cancer Res. 46*:1639, 1986.

65. C.-H. Yang, M. Dembo, F. M. Sirotnal. *J. Membrane Biol. 79*:285, 1984.

66. J. Galivan. *Arch. Biochem. Biophys. 206*:113, 1981.

67. C.-H. Yang, F. M. Sirotnak, M. Dembo. *J. Membrane Biol. 79*:285, 1984.

68. D. W. Horne, W. Y. Briggs, C. Wanger. *J. Biol. Chem. 253*:3529, 1978.

69. G. B. Henderson, J. M. Tsuji, H. P. Kumar. *Cancer Res. 46*:1633, 1986.

70. H. Wang, J. F. Ross, M. Ratnam. *Nucleic Acids Research 26*:2132, 1998.

71. E. Sadasivan, M. Cedeno, S. P. Rothenberg. *Biochimica et Biophysica Acta. 1131*:91, 1992.

72. A. C. Antony. *Annu. Rev. of Nutr. 16*:501, 1996.

73. J. T. Hjelle, E. I. Christensen, F. A. Carone, J. Selhub. *Am. J. Physiol. 260*:C338, 1991.

74. D. W. Horne, R. S. Holloway, and H. M. Said. *J. Nutr. 122*:2204, 1992.

75. J. S. Kim, B. Shane. *J. Biol. Chem. 269*:9714, 1994.

76. B. Shane. *Vitam. Horm. 45*:263, 1989.

77. S. K. Foo, R. M. McSloy, C. Rousseau, B. Shane. *J. Nutr. 112*:1600, 1982.

78. C. B. Osborne, K. E. Lowe, B. Shane. *J. Biol. Chem. 268*:21657, 1993.

79. T. A. Garrow, A. Admon, B. Shane. *Proc. Natl. Acad. Sci. U.S.A. 89*:9151, 1992.

80. S. J. Freemantle, R. G. Moran. *J. Biol. Chem. 272*:25373, 1997.

81. H. Qi, I. Atkinson, S. Xiao, Y. Choi, T. Tobimatsu, B. Shane. *Adv. Enz. Reg. 39*:263, 1999.

82. M. W. McBurney, G. F. Whitmore. *Cell 2*:173, 1974.

83. D. Sussman, G. Milman, C. Osborne, B. Shane. *Anal. Biochem. 158*:371, 1986.

84. J. S. Kim, K. E. Lowe, B. Shane. *J. Biol. Chem. 268*:21680, 1993.

85. J. J. McGuire, W. H. Haile, C. A. Russell, J. M. Galvin, B. Shane. *Oncol. Res. 7*:535, 1995.

86. L. H. Matherly, C. K. Barlowe, I. D. Goldman. *Cancer Res. 46*:588, 1986.

87. J. R. Barrueco, F. M. Sirotnak. *J. Biol. Chem. 266*:11732, 1991.

88. M. Murphy, M. Keating, P. Boyle, D. G. Weir, J. M. Scott. *Biochem. Biophys. Res. Commun. 71*:1017, 1976.

89. B. Stea, P. S. Bachlund, P. B. Berkey, A. K. Cho, B. C. Halpern, R. M. Halpern, R. A. Smith. *Cancer Res. 38*:2378, 1978.

90. W. Pfendner, L. I. Pizer. *Arch. Biochem. Biophys.* 1980.

91. C. K. Barlowe, D. R. Appling. *Biofactors 1*:171, 1988.

92. B. F. Lin, J. S. Kim, J. C. Hsu, C. Osborne, K. Lowe, T. Garrow, B. Shane. *Advances in Food and Nutrition Research 40*:95, 1996.

93. C. Wagner. In: L. B. Bailey, ed. *Folate in Health and Disease*. New York: Marcel Dekker, Inc., 1995, pp 23–42.

94. J. Paquin, C. M. Baugh, R. E. MacKenzie. *J. Biol. Chem. 260*:14925, 1985.

95. L. Schirch, M. Ropp. *Biochem. 6*:253, 1967.

96. R. G. Matthews, J. Ross, C. M. Baugh, J. D. Cook, L. Davis. *Biochem. 21*:1230, 1982.

97. W. Pfender, L. I. Pizer. *Arch. Biochem. Biophys. 200*:503, 1980.

98. L. A. Chasin, A. Feldman, M. Konstam, G. Urlaub. *Proc. Natl. Acad. Sci. U.S.A. 71*:718, 1974.

99. R. T. Taylor, M. L. Hanna. *Arch. Biochem. Biophys. 217*:609, 1982.

100. B. F. Lin, B. Shane. *J. Biol. Chem. 269*:9705, 1994.

101. T. A. Garrow, A. A. Brenner, V. M. Whitehead, X. N. Chen, R. G. Duncan, J. R. Korenberg, B. Shane. *J. Biol. Chem. 268*:11910, 1993.

102. P. J. Stover, L. H. Chen, J. R. Suh, D. M. Stover, K. Keyomarsi, B. Shane. *J. Biol. Chem. 272*:1842, 1997.

103. S. Girgis, J. R. Suh, J. Jolivet, P. J. Stover. *J. Biol. Chem. 272*:4729, 1997.

104. P. Stover, V. Schirch. *Biochem. 31*:2155, 1992.

105. K. Okamura-Ikeda, K. Fujiwara, Y. Motokawa. *J. Biol. Chem. 257*:135, 1982.

106. K. Hiraga, G. Kikuchi. *J. Biol. Chem. 255*:11664, 1980.

107. J. L. Van Hove, F. Lazeyras, S. H. Zeisel, T. Bottiglieri, K. Hyland, H. C. Charles, L. Gray, J. Jaeken, S. G. Kahler. *J. Inherited Metabolic Dis. 21*:799, 1998.

108. K. Fujiwara, K. Okamura-Ikeda, K. Hayasaka, Y Motokawa. *Biochem. Biophys. Res. Commun. 176*:711, 1991.

109. C. Kutzbach, E. L. R. Stokstad. *Biochim. Biophys. Acta 250*:459, 1971.

110. R. G. Matthews, C. M. Baugh. *Biochemistry 19*:2040, 1980.

111. J. M. Green, D. P. Ballou, R. G. Matthews. *Faseb J. 2*:42, 1988.

112. P. Goyette, J. S. Sumner, R. Milos, A. M. V. Duncan, D. S. Rosenblatt, R. G. Matthews, R. Rozen. *Nature Genetics 7*:195, 1994.

113. P. Frosst, H. J. Blom, R. Milos, P. Goyette, C. A. Sheppard, R. G. Matthews, G. J. Boers, M. den Heijer, L. A. Kluijtmans, L. P. van den Heuvel, et al. *Nature Genet 10*:111, 1995.

114. B. D. Guenther, C. A. Sheppard, P. Tran, R. Rozen, R. G. Matthews, M. L. Ludwig. *Nature Struct. Biol. 6*:359, 1999.

115. R. G. Matthews. In: R. L. Blakley, S. J. Benkovic, eds. *Folates and Pterins Vol. 1*: Chemistry and Biochemistry of Folates. New York: John Wiley & Sons, 1984, pp 497–554.

116. L. H. Chen, M. L. Liu, H. Y. Hwang, L. S. Chen, J. Korenberg, B. Shane. *J. Biol. Chem. 272*:3628, 1997.

117. D. Leclerc, E. Campeau, P. Goyette, C. E. Adjalla, B. Christensen, M. Ross, P. Eydoux, D. S. Rosenblatt, R. Rozen, R. A. Gravel. *Hum. Mol. Genet. 5*:1867, 1996.

118. Y. N. Li, S. Gulati, P. J. Baker, L. C. Brody, R. Banerjee, W. D. Kruger. *Hum. Mol. Genet. 5*:1851, 1996.

119. Z. Chen, R. Banerjee. *J. Biol. Chem. 273*:26248, 1998.

120. D. Leclerc, A. Wilson, R. Dumas, C. Gafuik, D. Song, D. Watkins, H. H. Heng, J. M. Rommens, S. W. Scherer, D. S. Rosenblatt, R. A. Gravel. *Proc. Natl. Acad. Sci. U.S.A. 95*:3059, 1998.

121. D. Leclerc, M. H. Odievre, Q. Wu, A. Wilson, J. J. Huizenga, R. Rozen, S. W. Scherer, R. A. Gravel. *Gene*, 1999.

122. J. L. Emmert, T. A. Garrow, D. H. Baker. *J. Nutr. 126*:2050, 1996.

123. R. J. Cook, K. S. Misono, C. Wagner. *J. Biol. Chem. 260*:12998, 1985.

124. T. A. Garrow. *J. Biol. Chem. 271*:22831, 1996.

125. R. Green, D. W. Jacobsen. In: L. B. Bailey, ed. *Folate in Health and Disease*, New York: Marcel Dekker, 1995, p. 75–122.

126. R. J. Cook, C. Wagner. *Proc. Natl. Acad. Sci. U.S.A. 81*:3631, 1984.

127. C. Wagner, W. T. Briggs, R. J. Cook. *Biochem. Biophys. Res. Commun. 127*:746, 1985.

128. R. Bhat, C. Wagner, E. Bresnick. *Biochem. 36*:9906, 1997.

129. K. E. Lowe, C. B. Osborne, B. F. Lin, J. S. Kim, J. C. Hsu, B. Shane. *J. Biol. Chem. 268*: 21665, 1993.

130. R. E. Mackenzie, C. M. Baugh. *Biochim. Biophys. Acta 611*:187, 1980.

131. D. V. Santi, C. S. McHenry. *Proc. Natl. Acad. Sci. USA 69*:1855, 1972.

132. P. V. Danenberg, R. J. Langenbach, C. Heidelberger. *Biochemistry 13*:926, 1974.

133. E. Chu, T. Cogliati, S. M. Copur, A. Borre, D. M. Voeller, C. J. Allegra, S. Segal. *Nucleic Acids Res. 24*:3222, 1996.

134. S. C. Daubner, J. L. Schrimsher, F. J. Schendel, M. Young, S. Henikoff, D. Patterson, J. Stubbe, S. J. Benkovio. *Biochem. 24*:7059, 1985.

135. V. T. Chan, J. E. Baggott. *Biochim. Biophys. Acta 702*:99, 1982.

136. K. Tarczy-Hornoch, E. L. R. Stokstad. *J. Nutr. 95*:445, 1968.

137. W. Shive, W. W. Ackermann, M. Gordon, M. E. Getzendaner, R. E. Eakin. *J. Am. Chem. Soc. 69*:725, 1947.

138. J. L. Paukert, L. D. 'Ari Straus, J. C. Rabinowitz. *J. Biol. Chem. 251*:5104, 1976.

139. L. U. L. Tan, E. J. Drury, R. E. MacKenzie. *J. Biol. Chem. 252*:1117, 1977.

140. F. A. Hol, N. M. van der Put, M. P. Geurds, S. G. Heil, F. J. Trijbels, B. C. Hamel, E. C. Mariman, H. J. Blom. *Clin. Genet. 53*:119, 1998.

141. W. B. Strong, S. J. Tendler, R. L. Seither, I. D. Goldman, V. Schirch. *J. Biol. Chem. 265*: 12149, 1990.

142. M. G. West, D. W. Horne, D. R. Appling. *Biochem. 35*:3122, 1996.

143. D. R. Appling. *FASEB J. 5*:2645, 1991.

144. N. R. Mejia, R. E. MacKenzie. *J. Biol. Chem. 260*:14616, 1985.

145. L. F. García-Martínez, D. R. Appling. *Biochem. 32*:4671, 1993.

146. L. B. Pasternack, L. E. Littlepage, D. A. Laude, Jr., D. R. Appling. *Archives of Biochemistry and Biophysics 326*:158, 1996.

147. C. Kutzbach, E. L. R. Stokstad. *Methods Enzymol, Part B, 18*:793, 1971.

148. N. R. Mejia, R. E. MacKenzie. *J. Biol. Chem. 260*:14616, 1985.

149. H. Min, B. Shane, E. L. Stokstad. *Biochim. Biophys. Acta 967*:348, 1988.

150. C. Wagner, W. T. Briggs, D. W. Horne, R. J. Cook. *Archives of Biochemistry and Biophysics 316*:141, 1995.

151. S. A. Krupenko, C. Wagner, R. J. Cook. *Journal of Biological Chemistry 272*:10273, 1997.

152. K. M. Champion, R. J. Cook, S. L. Tollaksen, C. S. Giometti. *Proc. Natl. Acad. Sci. U.S.A. 91*:11338, 1994.

153. E. B. Rimm, W. C. Willett, F. B. Hu, L. Sampson, G. A. Colditz, J. E. Manson, C. Hennekens, M. J. Stampfer. *JAMA 279*:359, 1998.

154. J. Ma, M. J. Stampfer, E. Giovannucci, C. Artigas, D. J. Hunter, C. Fuchs, W. C. Willett, J. Selhub, C. H. Hennekens, R. Rozen. *Cancer Res. 57*:1098, 1997.

155. V. Herbert. *Trans. Assoc. Am. Physicians 75*:307, 1962.

156. A. Nirenberg. *Am. J. Nursing 76*:1776, 1976.

157. A. J. Clifford, D. S. Wilson, N. D. Bills. *J. Nutr. 119*:1956, 1989.

158. N. Rong, J. Selhub, B. R. Goldin, I. H. Rosenberg. *J. Nutr. 121*:1955, 1991.

159. M. J. Koury, D. W. Horne. *Proc. Natl. Acad. Sci. U.S.A. 91*:4067, 1994.

160. M. Goulian, B. Bleile, B. Y. Tseng. *Proc. Natl. Acad. Sci. U.S.A. 77*:1956, 1980.

161. B. C. Blount, M. M. Mack, C. M. Wehr, J. T. MacGregor, R. A. Hiatt, G. Wang, S. N. Wickramasinghe, R. B. Everson, B. N. Ames. *Proc. Natl. Acad. Sci. U.S.A. 94*:3290, 1997.

162. B. Shane, E. L. Stokstad. *Annu. Rev. Nutr. 5*:115, 1985.

163. D. J. Cichowicz, B. Shane. *Biochem. 26*:513, 1987.

164. L. Chen, H. Qi, J. Korenberg, T. A. Garrow, Y. J. Choi, B. Shane. *J. Biol. Chem. 271*:13077, 1996.

165. R. T. Taylor, M. L. Hanna, J. J. Hutton. *Arch. Biochem. Biophys. 165*:787, 1974.

166. A. Wilson, D. Leclerc, F. Saberi, E. Campeau, H. Y. Hwang, B. Shane, J. A. Phillips, 3rd, D. S. Rosenblatt, R. A. Gravel. *Am. J. Hum. Genet. 63*:409, 1998.

167. A. Wilson, D. Leclerc, D. S. Rosenblatt, R. A. Gravel. *Hum. Mol. Genet. 8*:2009, 1999.

168. P. Goyette, P. Frosst, D. S. Rosenblatt, R. Rozen. *Am. J. Hum. Gen. 56*:1052, 1995.

169. T. Brody, J. E. Watson, E. L. R. Stokstad. *Biochemistry 21*:276, 1982.

170. S. J. James, D. R. Cross, B. J. Miller. *Carcinogenesis 13*:2471, 1992.

171. M. E. Weinblatt, J. S. Coblyn, D. A. Fox, P. A. Fraser, D. E. Holdsworth, D. N. Glass, D. E. Trentham. *New Engl. J. Med. 312*:818, 1985.

172. B. Fowler. *J. Inher. Met. Dis. 20*:270, 1997.

173. N. P. Dudman, D. E. Wilcken, J. Wang, J. F. Lynch, D. Macey P. Lundberg. *Arteriosclerosis Thrombosis 13*:1253, 1993.

174. J. C. Tsai, M. A. Perrella, M. Yoshizumi, C. M. Hsieh, E. Haber, R. Schlegel, M. E. Lee. *Proc. Natl. Acad. Sci. U.S.A. 91*:6369, 1994.

175. H. Refsum, P. M. Ueland, O. Nygard, S. E. Vollset. *Ann. Rev. Med. 49*:31, 1998.

176. J. Selhub, P. F. Jacques, A. G. Bostom, R. B. D'Agostino, P. W. Wilson, A. J. Belanger, D. H. O'Leary, P. A. Wolf, E. J. Schaefer, I. H. Rosenberg. *New Engl. J. Med. 332*:286, 1995.

177. C. Boushey, S. Beresford, G. Omenn, A. G. Motulsky. *J. Am. Med. Assoc. 274*:1049, 1995.

178. J. Selhub, P. F. Jacques, P. W. F. Wilson, D. Rush, I. H. Rosenberg. *JAMA 270*:2693, 1993.

179. J. Ma, M. J. Stampfer, C. H. Hennekens, P. Frosst, J. Selhub, J. Horsford, M. R. Malinow, W. C. Willett, R. Rozen. *Circulation 94*:2410, 1996.

180. D. Rothman. *Am. J. Obstetr. Gynecol. 108*:149, 1970.

181. N. Wald, J. Sneddon, J. Densem, C. Frost, R. Stone. *Lancet 338*:131, 1991.

182. A. E. Czeizel, I. Dudas. *New Engl. J. Med. 327*:1832, 1992, p. 183.

183. R. J. Berry, Z. Li, J. D. Erickson, S. Li, C. A. Moore, H. Wang, J. Mulinare, P. Zhao, L. Y. Wong, J. Gindler, S. X. Hong, A. Correa, *N. Engl. J. Med. 341*:1485, 1999.

184. J. M. Scott, P. N. Kirke, D. G. Weir. *Annu. Rev. Nutr. 10*:277, 1990.

185. J. L. Mills, J. M. McPartlin, P. N. Kirke, Y. J. Lee, M. R. Conley, D. G. Weir, J. M. Scott. *Lancet 345*:149, 1995.

186. A. S. Whitehead, P. Gallagher, J. L. Mills, P. N. Kirke, H. Burke, A. M. Molloy, D. G. Weir, D. C. Shields, J. M. Scott. *QJM 88*:763, 1995.

187. P. N. Kirke, A. M. Molloy, L. E. Daly, H. Burke, D. G. Weir, J. M. Scott. *Quar. J. Med. 86*:703, 1993.

188. D. L. Harmon, D. C. Shields, J. V. Woodside, D. McMaster, J. W. G. Yarnell, I. S. Young, K. Peng, B. Shane, E. A. Evans, A. S. Whitehead. *Genet. Epidemiol. 17*:298, 1999.

189. J. A. Piedrahita, B. Oetama, G. D. Bennett, J. van Waes, B. A. Kamen, J. Richardson, S. W. Lacey, R. G. Anderson, R. H. Finnell. *Nat. Genet. 23*:228, 1999.

190. R. Barber, S. Shalat, K. Hendricks, B. Joggerst, R. Larsen, L. Suarez, and R. Finnell. *Molec. Genet. Metab. 70*:45, 2000.

191. M. Watanabe, J. Osada, Y. Aratani, K. Kluckman, R. Reddick, M. R. Malinow, N. Maeda. *Proc. Natl. Acad. Sci. U.S.A. 92*:1585, 1995.

192. Food and Nutrition Board. In: *Dietary reference intakes: thiamin, riboflavin, niacin, vitamin B_6, folate, vitamin B_{12}, pantothenic acid, biotin, and choline,* Institute of Medicine, National Academy, Washington, D.C., 2000.

193. P. F. Jacques, J. Selhub, A. G. Bostom, P. W. Wilson, I. H. Rosenberg. *N. Engl. J. Med. 340*:1449, 1999.

194. M. P. Rivey, D. D. Schottelius, M. J. Berg. *Drug Intelligence and Clinical Pharmacol. 18*: 292, 1984.

195. E. H. Reynolds. *Lancet 1*:1376, 1973.

196. D. Kelly, D. Weir, B. Reid, J. Scott. *J. Clin. Invest. 64*:1089, 1979.

197. C. L. Krumdieck, K. Fukushima, T. Fukushima, T. Shiota, C. E. Butterworth. *Am. J. Clin. Nutr. 31*:88, 1978.

198. D. B. Smith, E. Obbens. In: M. I. Botez, E. H. Reynolds eds. *Folic Acid in Neurology, Psychiatry, and Internal Medicine.* New York: Raven Press, 1979, p 267.

199. C. H. Halsted. *Am. J. Clin. Nutr. 33*:2736, 1980.

200. C. H. Halsted. *Annu. Rev. Med. 31*:79, 1980.

201. J. Selhub, G. J. Dhar, I. H. Rosenberg. *Pharmacol. Ther. 20*:397, 1983.

202. B. Shane, T. Tamura, E. L. R. Stokstad. *Clin. Chem. Acta. 100*:13, 1980.

203. P. J. Bagley, J. Selhub. *Proc. Natl. Acad. Sci. U.S.A. 95*:13217, 1998.

204. K. C. Das, V. Herbert. *Br. J. Haematol. 38*:219, 1978.

205. T. Spies, R. Hillman, S. Cohlan, B. Kramer, A. Kanof. In: G. Duncan, ed. *Diseases of Metabolism,* Saunders, Philadelphia, 1959.

206. S. O. Schwartz, S. R. Kaplan, B. E. Armstrong. *J. Lab. Clin. Med. 35*:894, 1950.

207. B. Horned, R. Cunningham, H. Smith, M. Clark. *Ann. N. Y. Acad. Sci. 48*:289, 1946.

208. J. Dawson, C. Woodruff, W. Darby. *Proc. Soc. Exp. Biol. Med. 73*:646, 1950.

209. T. Brody, E. L. R. Stokstad. *J. Nutr. 120*:71, 1990.

210. T. Brody, E. L. R. Stokstad. *J. Nutr. Biochem. 2*:492, 1991.

211. T. Brody. *Nutritional Biochemistry.* New York: Acadeemic Press, 1999, pp 897–899.

13

Cobalamin (Vitamin B$_{12}$)

WILLIAM S. BECK

Massachusetts General Hospital and Harvard University, Cambridge, Massachusetts

I. HISTORY

Cobalamin, the preferred name for the family of derivatives familiarly known as vitamin B$_{12}$, is a distinctive molecular species that shares many structural features with other compounds. Nonetheless, it possesses numerous biochemically unique attributes. The chronicle of discoveries leading to current knowledge of cobalamin and its deficiency states embraces a distinguished chapter in the history of biomedical science, one that still needs an ending. It is worth noting that the influence of these discoveries has reached far into many and diverse fields, ranging through evolutionary biology, taxonomy, microbial metabolism, organic chemistry, nutrition, animal husbandry, and clinical medicine.

The familiar story of the discovery of cobalamin (1,2) began with Addison's description in 1835 of the disorder known in the English-speaking world as pernicious anemia (3,4). Until Minot's demonstration in 1926 of the successful treatment of pernicious anemia by liver feeding (5,6), this disease was often fatal. In 1929, Castle discovered intrinsic factor in a study still viewed as a model clinical investigation (7).

There ensued the unsuccessful efforts of two decades to isolate and identify the liver principle by crude methods—mainly in the laboratory of E. J. Cohn at Harvard University [though a stimulating new account vividly recounts the contribution of Y. A. Subbarow at Lederle Laboratories (8)]. As a result of these efforts, parenterally administered liver extracts early replaced ingested liver in the treatment of pernicious anemia. But great difficulties plagued investigators attempting to isolate and identify the antipernicious anemia principle of liver. For example, their failure to discover a naturally occurring animal disease resembling pernicious anemia meant that purification studies could be guided only by tiresome assays performed on pernicious anemia patients in relapse.

Soon bacterial and animal nutritionists independently found three factors that led to the discovery of cobalamin (2): (a) LLD factor, a factor in yeast and liver extracts essential in the nutrition of *Lactobacillus lactis* Dorner and other microorganisms; (b) an animal protein factor obtained from tissue extracts and animal feces that promotes growth of pigs and poultry receiving only vegetable rations; and (c) a ruminant factor, whose lack causes a wasting disease of ruminants grazing in cobalt-poor pastures and whose replacement is effected by oral feeding of cobalt salts (or by dusting cobalt on pastures), parenteral cobalt being ineffective.

The discovery and crystallization of cobalamin in 1948 by Rickes and associates of Merck Laboratories (9) followed the astute observation by Mary Shorb of proportionality between the nutrient activity of liver extracts in cultures of *L. lactis* Dorner and their therapeutic activity in pernicious anemia (10,11). A swiftly devised, simple microbiology assay hastened the final purification and identification of the vitamin. Shortly thereafter, cobalamin was isolated from liver by H. G. Wijmenga and associates, following the unwitting use of cyanide-activated papain as a proteolytic ferment, with cyanide converting various forms of the vitamin to the stable and readily crystallizable cyano derivative (12). Isolation was also achieved in 1948 by E. L. Smith in England, without benefit of microbiological assay and cyanide (13). Animal protein factor and cobalt-dependent ruminant factor were promptly identified with cobalamin. West (14) was the first to show that injections of cobalamin, supplied to him by Rickes et al., induced a dramatic beneficial response in patients with pernicious anemia. Folkers has presented a stimulating review of these events (15).

The elucidation of the structure of cobalamin (16) by Dorothy Hodgkin at Oxford University was a pioneering contribution of the x-ray crystallographer. The discovery by Barker, Weissbach, and Smyth of the first cobalamin coenzyme (17) was also a tour de force of biochemical ingenuity. However compelling these works, we can here consider them only briefly. Historical reviews will be found elsewhere (1,2,6,15).

Early workers isolated two crystalline ''vitamin B_{12}'' preparations from cultures of *Streptomyces aureofaciens* (18). One had an absorption spectrum similar to that earlier reported for vitamin B_{12} (19,20); the other had a different spectrum and was termed vitamin B_{12}. Because vitamin B_{12}, in contrast to vitamin B_{12} (21), lacked cyanide, it was named hydroxocobalamin (22). There were no differences in the antipernicious anemia effects and microbiological nutritional activity of hydroxo- and cyanocobalamin (23). As noted below, adventitious cyanide arising in the preparatory process probably converted hydroxocobalamin to cyanocobalamin.

II. CHEMICAL ASPECTS

A number of reviews dealing with aspects of cobalamin chemistry (24–29) may be consulted for additional references on topics in the following sections.

A. Structure

Despite intensive investigation, elucidation of cobalamin's unusual chemical structure required 7 years (16). The structure has several unique features (**I**, Fig. 1). The cyanocobalamin molecule ($C_{83}H_{88}O_{14}N_{14}PCo$; mol wt 1355) has two major portions; a planar group, which bears a close but imperfect resemblance to the porphyrin macro ring, and a nucleotide, which lies nearly perpendicular to the planar group (**II**). The porphyrin-like moiety contains four reduced pyrrole rings (designated A–D) that link to a central cobalt atom, the two remaining coordination positions of which are occupied by a cyano group (above)

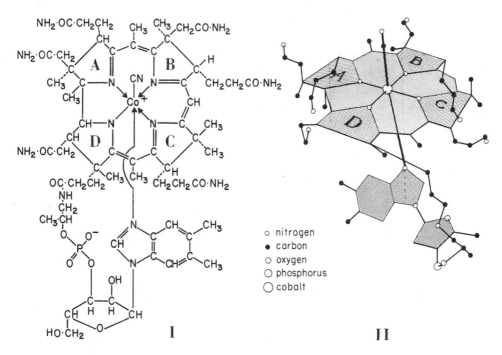

o nitrogen
• carbon
○ oxygen
○ phosphorus
○ cobalt

I   **II**

Fig. 1   Cobalamin. **I**, chemical structures of cyanocobalamin. **II**, semidiagrammatic representation of three-dimensional structure showing relations of planar and nucleotide moieties. Hydrogen atoms and a number of oxygen atoms are omitted.

and a 5,6-dimethylbenzimidazolyl moiety (below the planar group). With one exception, the pyrrole rings are connected to one another by methylene carbon bridges similar to those found in porphyrin precursors. The exception is the direct linkage between the carbon of rings A and D. Another dissimilarity from the porphin structure is the relatively saturated character of the pyrrole rings, which are extensively substituted with methyl groups or longer acetamide and propionamide residues.

The macro ring of cobalamin and related compounds is termed *corrin*; the major corrin derivatives are known as *corrinoid* compounds. The term corrin was proposed originally to refer to the "core" of the vitamin B_{12} structure. Its first two letters do not denote the presence of cobalt. That fact is implied by the "cob" of cobalamin.

A biological relation between corrin and porphin macro rings is seen in the fact that both are synthesized from δ-aminolevulinic acid (30,31). Porphobilinogen, a precursor of uroporphyrin, coprophyrin, and protoporphyrin, is incorporated into the corrin system by certain microorganisms.

The several brightly colored forms taken by cyclic tetrapyrroles in nature (Fig. 2), including heme, chlorophyll, bacteriochlorophyll, siroheme, corrinoids, and F_{430}, the recently discovered coenzyme of bacterial methanogenesis (32), give striking notice of the many diverse assignments given to this molecular form in the course of evolution. Indeed, the cobalt in corrinoids is but one of five metals now known to occur in cyclic tetrapyrroles with biological functions. The others are iron in hemes and siroheme; magnesium in chlorophyll; copper in turacin, a pigment of turaco bird feathers; and nickel in F_{430}. Functionally, as described below, the corrinoids most closely resemble F_{430} in that both can accept

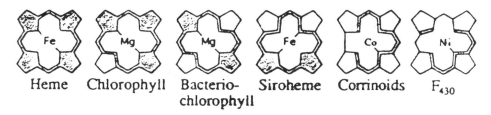

Fig. 2 Cyclic tetrapyrroles with biological functions. Stippled areas show double bonds and resonating conjugated bond systems. (From Ref. 32.)

a methyl group when the metal is in the +1 oxidation state. However, the stability of these organometallic bonds differs greatly.

Among the biochemically unusual features of cobalamin is a nucleotide, whose base—5,6-dimethylbenzimidazole—is not found elsewhere in nature. Others include its base–ribose linkage, which sterically is an α-glycoside unlike the β linkages typical of nucleic acid and coenzyme nucleotides. The ribose is phosphorylated at C-3, one of the rare natural occurrences of a ribose 3-phosphate structure. There are two connections between the planar and nucleotide moieties; (a) an ester linkage between the nucleotide phosphate and a 1-amino-2-propanol moiety that is joined in turn in amide linkage with a propionic side chain in ring D; and (b) the coordinate linkage between cobalt and the glyoxalinium nitrogen atom of benzimidazole.

B. Nomenclature

Of many known corrinoid compounds, a certain number occur naturally, whereas others are prepared by chemical transformation or by manipulation of microbial biosynthetic systems. A few received trivial names before their structures were known. To abate confusion, systematic nomenclature was introduced in 1959 (33). Additional rules for naming corrinoids were proposed in 1966 (34) and 1973 (35). Numbering and ring designations of the corrin system are summarized in Fig. 3. Nucleotide derivatives of the corrinoids are named by adding ''-yl'' to the name of the nucleotide base; e.g., α-(5,6-dimethylbenzimidazolyl)cyanocobamide. Since in ordinary parlance complex systematic names cannot displace trivial or semisystematic names, the semisystematic term *cobalamin*, introduced before its chemical structure was known, came to be used to refer to the combining form of cobalamin that lacks a ligand in the cobalt-β position (i.e., above the plane). Once lacking in precise meaning, cobalamin is now defined as a cobamide containing a 5,6-dimethylbenzimidazolyl moiety. In recent years, editors have preferred cobalamin to the inexact term vitamin B_{12}.

In chemical relatives of cobalamin (e.g., cyanocobalamin), various ligands are covalently bound to cobalt above the plane. Other such compounds are hydroxocobalamin (or the basic product of its combination with H^+, aquacobalamin) and nitritocobalamin. The ligand below the plane can also be replaced. In strong acid, a second H_2O^- displaces the 5,6-dimethylbenzimidazolyl moiety to form diaquocobalamin. Pseudovitamin B_{12}, a cyanocobalamin analog containing adenine in place of 5,6-dimethylbenzimidazole, is active in some microorganisms but inert in animals.

C. Cobalamins in Animal Cells

Four cobalamins are of major importance in animal cell metabolism (36). Two are equivalents of the natural vitamin—cyanocobalamin and its analog hydroxocobalamin—abbrevi-

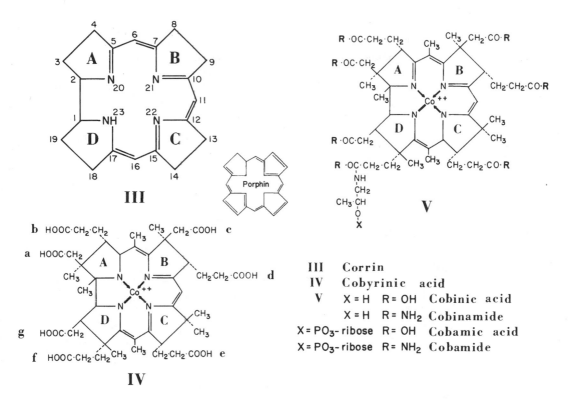

Fig. 3 Systematic nomenclature of cobalamin and related compounds. Inset shows porphyrin structure for comparison. Note that pyrrole rings of the corrin macro rings are designated A–D. Substituent acetamide and propionamide groups are designated a–g.

ated CNCbl and OHCbl (35). The other two are alkyl derivatives that are synthesized from the vitamin and serve as coenzymes. In one, a 5′-deoxy-5′-adenosyl (in short form, adenosyl) moiety replaces CN as the ligand of cobalt above the plane (Fig. 4). This compound is usually called adenosylcobalamin (AdoCbl). The other coenzymatic derivative is methylcobalamin (MeCbl).

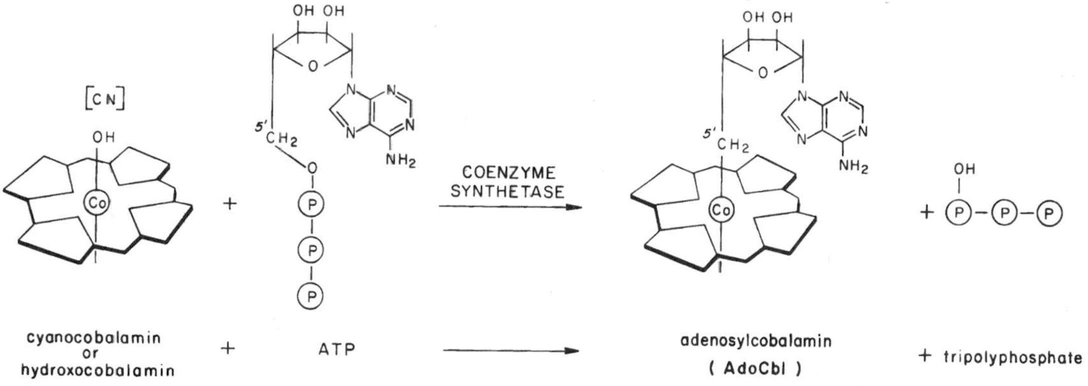

Fig. 4 The coenzyme synthetase reaction. ATP adenosylates cobalamin to form adenosylcobalamin.

Knowledge of AdoCbl derived from classic studies of Barker and associates on the conversion of glutamate to β-methylasparate by extracts of *Clostridium tetanomorphum* (17,24). Purified preparations of the glutamate mutase contained a coenzyme shown to be a novel derivative of pseudovitamin B_{12}, α-(7-adenyl)adenosylcobamide. When *C. tetanomorphum* was cultivated in the presence of free exogenous 5,6-dimethylbenzimidazole, the biosynthetic product was AdoCbl. *Proplonibacterium shermanii* was found to produce AdoCbl naturally and the same compound was soon found in mammalian liver. In AdoCbl the 5'-methylene carbon atom of the 5'-deoxy-5'-adenosyl moiety is linked directly to the cobalt atom (37).

In the second cobalamin with coenzyme activity, the ligand of cobalt is a methyl group (38). MeCbl is the major form of cobalamin in human blood plasma (39,40). As noted below, the C−Co bond in both coenzymes is labile to light, cyanide, and acid. It is likely that the cyanocorrinoids encountered in nature, including CNCbl, arise mainly from the cyanolysis of corrinoid coenzymes.

CNCbl and OHCbl are converted to AdoCbl in tissues by a "coenzyme synthetase" system (Fig. 4) (41–43). The reaction requires a thiol or dithiol, a reduced flavin, and ATP—the biological alkylating agent. The 5'-deoxy-5'-adenosyl moiety of ATP is transferred to the vitamin, and the three phosphates of ATP are released as inorganic triphosphate. Reducing agents are required in a preliminary step that converts the trivalent cobalt of cobalamin [cyanocob(III)alamin] through the bivalent state [vitamin B_{12r} or cob(II) alamin] to the univalent state, vitamin B_{12} (or cob(I)alamin, which has nucleophilic properties).

D. Biosynthesis

A detailed summary of the biosynthetic pathway of cobalamin in microorganisms is beyond the scope of this chapter. We note here only that the basic precursor, uroporphyrinogen III (also the precursor of other natural pigments), is channeled toward cobalamin by successive methylations yielding precorrins 1, 2, 3A, 4, 5, and 6A. A reduction step produces precorrin 6B. The last two of the required eight methylations plus a decarboxylation step yields precorrin 8x, an isomer of hydrogenobyrinic acid. A 1,5-methyl shift from C-11 to C-12 then creates the conjugated system typical of the corrin macrocycle. Several amidation steps followed by cobalt insertion, cobalt adenosylation, and creation of the nucleotide loop generate the coenzyme AdoCbl. A review, vividly describing both the step-by-step pathway and the remarkable developments that led to its elucidation, has appeared (29).

E. Chemical Properties

The extensively studied chemical properties of cobalamin have been well reviewed (14,33,44,45).

The absorption spectrum of cyanocobalamin shows three characteristic maxima that are relatively independent of pH. Extinction coefficients are $E_{278} = 16.3 \times 10^2$; $E_{361} = 28.1 \times 10^3$; and $E_{550} = 8.7 \times 10^3$.

From the biological and clinical viewpoint it is significant that the cyano moiety of cyanocobalamin is readily replaced by other groups to form hydroxocobalamin, chlorocobalamin, nitrocobalamin, sulfitocobalamin, adenosylcobalamin, methylcobalamin, and others. Each of these derivatives is freely reconverted to cyanocobalamin on exposure to CN^-. A purple compound formed on addition of excess cyanide to alkaline solutions of cyanocobalamin is called dicyanocobalamin. In this labile compound, two cyano groups are coordinated to the cobalt atom.

The two coenzymes AdoCbl and MeCbl are strikingly unstable to light, which rapidly causes spectral changes and loss of coenzyme activity, owing to the homolytic cleavage of the C—Co bond. In contrast to the photolysis of AdoCbl, the photolysis of MeCbl and other alkylcorrinoids is accelerated by ambient oxygen. The reaction proceeds slowly under anaerobic conditions. CNCbl is slowly decomposed by strong visible light. The cyano group is split off, yielding OHCbl. Prolonged exposure to light causes irreversible decomposition and inactivation.

Mild acid hydrolysis of CNCbl induces removal of the nucleotide, whereas vigorous acid hydrolysis liberates ammonia, 5,6-dimethylbenzimidazole, D-1-amino-2-propanol, and cobyrinic acid. Dilute acids split amides from the side chains to yield mono- and polycarboxylic acids.

CNCbl, a relatively durable cobalamin derivative, is stable in air and, in dry form, is relatively stable at 100°C for a few hours. Aqueous solutions at pH 4–7 can be autoclaved at 120°C. Crystalline CNCbl can be safely mixed with a wide variety of therapeutic and nutritional agents. In solution, thiamine and nicotinamide, or nicotinic acid, slowly destroy CNCbl; the addition of small amounts of iron or thiocyanate appears to protect it.

III. ASSAY METHODS

Assays of serum cobalamin came into use in the 1950s, but for a number of years they were performed only in specialized laboratories. For a time, cobalamin levels could be measured only by microbiological methods, using *Lactobacillus leichmannii* ATCC 7830 (46–49), *Eugelena gracilis* strain Z (50–52), or, less frequently, *Ochromonas malhamensis* (53), or the cobalamin/methionine auxotrophic mutant *Escherichia coli* 113-3 (54–56). Of the several microbiological procedures, the one employing *O. malhamensis* is most specific, since this organism is supported almost exclusively by cobalamins in the growth medium, whereas cobalamins and certain other corrins (or nucleosides) support the growth of the other three organisms—*L. leichmannii*, *E. gracilis*, and the *E. coli* mutant. Nonetheless, comparative studies show that the several assay organisms yield nearly identical results in serum assays in a clinical setting (57–60). In the writer's view, the most reliable of the microbiological procedures employs *L. leichmannii*. Details of this procedure are presented elsewhere (61).

The radioisotope dilution (RID) principle was first applied to the assay for serum cobalamin in 1961 (62). Though performed in many variations, all of these methods rest on the following principles: (a) extraction of serum cobalamins from binding proteins; (b) addition to the extract of a standard amount of ^{57}Co-CNCbl; (c) exposure of the mixture of native and labeled cobalamin to a cobalamin-binding agent; and (d) removal by centrifugation of the saturated binder and radioassay of the pellet. The level of serum cobalamin is then read from an appropriate calibration curve (63).

In the early years, these assays regularly yielded higher values than microbiological assays. It was shown (64,65) that serum contains corrinoid compounds, termed cobalamin analogs, that are not utilized by assay organisms but are bound by R protein, then the commonly used binder in RID assays. When intrinsic factor (a more specific cobalamin binder) replaced R proteins in assay protocols, the two procedures—microbiological and RID methods—gave comparable results.

Interpretation of serum cobalamin levels is discussed below.

IV. NUTRITIONAL ASPECTS

A. Sources

Cobalamin is synthesized only by certain microorganisms; hence, it is a unique vitamin. Wherever it occurs in nature, it can be traced to bacteria or other microorganisms growing in soil, sewage, water, intestine, or rumen. Animals depend ultimately on microbial synthesis for their cobalamin supply; foods in the human diet that contain cobalamin are essentially those of animal origin—liver, seafood, meat, eggs, and milk. Although the nitrogen-fixing bacteria associated with leguminous plants are cobalamin-dependent, cobalamin is not found in plant tissues.

The most intensive natural synthesis occurs in rumen bacteria (66). Of the microorganisms that synthesize cobalamin, many do so in quantities just sufficient for their own needs. However, species such as the rumen organism *P. shermanii* and the antibiotic-producing molds *Streptomyces griseus* and *Streptomyces aureofaciens* synthesize amounts sufficient to make them feasible commercial sources. Some microorganisms that cannot synthesize cobalamin (e.g., *L. lactis*, *L. leichmannii*) require an exogenous supply and hence are useful for the microbiological assay of cobalamin. Others cannot synthesize cobalamin and appear not to require it (e.g., *E. coli*).

B. Daily Requirements

The daily diet in Western countries contains 5–30 µg of cobalamin—with an average of 7–8 µg/day ingested by adult men, 4–5 µg/day by adult women, and 3–4 µg/day by children 1–5 years of age. An additional nondietary source of small amounts of absorbable cobalamin may be synthesis by intestinal microorganisms. Only 1–5 µg is absorbed (67). Less than 250 ng of cobalamin and cobalamin analogs appears in the urine; the unabsorbed remainder appears in the faces. Long-term studies, employing radioactive cobalamin, showed a total daily loss of 0.66–2.1 µg, with a mean of 1.3 µg (68). Of an administered oral dose of 0.5–2 µg of pure cobalamin, 60–80% is absorbed. As the oral dose increases, the percent absorbed decreases; at a dose of 5 µg, 30% or less is absorbed.

Total-body content is 2–5 mg in an adult man (69,70). Of this approximately 1 mg is in the liver. Thus, the concentration in adult liver is about 0.7 µg per gram wet weight. Kidney is also rich in cobalamin (approximately 0.4 µg per gram wet weight) (71).

It has been suggested that cobalamin has a daily rate of obligatory loss approximating 0.1% (range 0.05–0.2) of the total body pool, irrespective of its size (67). This conclusion implies (a) that the daily dietary requirement is 2–5 µg and (b) that a deficiency state will not develop for several years after cessation of cobalamin intake. The officially recommended daily allowance for adults is 2 µg (72); for infants during the first year, the recommended allowance is 0.3 µg/day. It had earlier been held that the minimal daily requirement was 0.6–1.2 µg. However, this amount is adequate to maintain nutritional balance only in subjects with low body stores (in whom daily obligatory losses are proportionately low).

Growth, hypermetabolic states, and pregnancy increase daily requirements. Because of the buffering effects of body stores, it has been difficult to obtain precise nutritional data in these conditions. A diet containing 15 µg/day will gradually replenish depleted body stores (72). Average U.S. diets appear to meet this requirement.

C. Absorption

Intrinsic factor (IF) is the name given long ago by Castle (7) to a normal constituent of gastric juice needed to facilitate the hematopoietic effects of the essential dietary ingredient (i.e., "extrinsic factor") now recognized as cobalamin. Later work showed that IF facilitates absorption in the ileum of cobalamins ingested at physiological dose levels. Knowledge of the molecular and functional properties of IF has expanded dramatically in recent years (73–75).

For decades after its discovery, efforts to purify IF encountered difficulty. Substantial purifications, based on affinity chromatography, were not achieved until 1973 (76–78). This elegant work opened a significant new era in the study of IF and other cobalamin-binding proteins. Human IF was shown to be an alkali-stable glycoprotein [or class of proteins (79)] that avidly binds cobalamin (cyano, hydroxo, or adenosyl derivative). Its properties are summarized in Table 1. Bound vitamin alters the conformation of IF, producing a more compact form that resists proteolysis.

In fact, gastric juice contains several cobalamin-binding proteins. Only one possesses IF activity (such activity being defined as the capacity to promote intestinal absorption of cobalamin). Others lack it. In zone electrophoresis of gastric juice at least two immunologically nonidentical binders or classes of binders are observed, one with slow and one with rapid mobility (74). These were designated, respectively, S-type and R-type protein(s). IF activity resides in the S-type class. R-type proteins are a large class found in serum, leukocytes, saliva, and virtually all body cells. The name *cobalophilin* has been proposed for all R-type proteins (80). Currently under active study, they are discussed below in the context of plasma cobalamin transport proteins.

IF is secreted by the parietal cells of the fundic mucosa in man, guinea pig, cat, rabbit, and monkey; by the chief cells in the rat; and by glandular cells of the pylorus and duodenum in the hog. Secretion of IF usually parallels that of HCl. It is enhanced by histamine, methacholine, and gastrin.

Food cobalamins are liberated by peptic digestion in the stomach and bound there to IF. Proteins or peptides bound to naturally occurring cobalamins are competitively displaced by IF at the low pH of gastric juice. The stable IF-Cbl complex subsequently encounters specific mucosal receptors of the microvilli of the ileum. A specific site on

Table 1   Properties of Human Intrinsic Factor

Property	Value
M_r (approximate)	44,000
$E_{1\,cm}^{1\%}$ at 279 nm	9.5
$S_{20,w}^0$	5.75
Cyanocobalamin-binding capacity, µg/mg	30.1
	18.6
Association constant for cyanocobalamin, M^{-1}	1.5×10^{10}
Composition:	
Carbohydrate content, %	15.0
Hexoses, including fucose, %	6.9
Hexosamine, residues/mol	4.1
Sialic acid, residues/mol	1.7

the IF molecule (other than the cobalamin binding site) attaches to a receptor; binding requires neutral pH, Ca^{2+}, or other divalent cations, but no energy (81). Mucosal receptors accept IF-Cbl in preference to free IF (82). The specific mucosal receptor has been isolated in soluble form (83). It appears to be a polypeptide containing critical disulfide bonds. Pancreatic secretions may contain a factor that promotes cobalamin absorption (84,85). The factor appears to be a proteolytic enzyme that acts to protect IF-Cbl (86).

Following binding of IF-Cbl to receptor, the vitamin either passes into the cell leaving IF behind, in analogy with the passage of transferrin-bound iron into a normoblast, or the complex enters a mucosal cell by pinocytosis and then dissociates. Unlike the rapid attachment of IF-Cbl to surface receptors, passage of vitamin into the mucosal cell is a slow, energy-requiring process (87). Cobalamin entering the ileal cell is bound there to a macromolecule that is immunologically similar to IF (88). It is of interest that in the guinea pig cobalamin in the ileal cell accumulates in the mitochondrial fraction (89).

Cobalamin (without IF) is eventually transferred to portal blood. After a small oral dose (e.g., 10–20 µg), cobalamin first appears in the blood in 3–4 h, reaching a peak level at 8–12 h. Larger oral doses are absorbed by simple diffusion that is not mediated by IF. In these instances, vitamin appears in blood within minutes.

At least two types of anti-IF antibodies are recognized: (a) blocking antibodies, which prevent binding of cobalamin by IF, and (b) binding antibodies (AB II), which combine with IF-Cbl complex or with free IF without impairing its ability to bind cobalamin.

D. Transport of Cobalamin in Plasma

Normal plasma contains 150–450 pg of cobalamin per mL (the range of normal varying with the assay method and laboratory, as noted below). All plasma cobalamin is bound to transport proteins of unusual physiological complexity.

1. Properties of Plasma Cobalamin–Binding Proteins

Plasma and extracellular fluids contain two major cobalamin-binding proteins that have been known by different names over the years. The original terms *transcobalamin I* (TC I) and *transcobalamin II* (TC II) (90–93) were criticized by workers who urged that "transcobalamin" be reserved for the protein previously termed TC II and that TC I be renamed "haptocorrin" (94). Other synonyms for this protein in the literature include cobalophilin (80), R binder, granulocyte vitamin B_{12}-binding protein, and salivary binder. None of the newer terms has found general acceptance. We here use the terms haptocorrin and transcobalamin (or TC).

Plasma also contains a quantitatively minor protein (or class of proteins) that has also been variously named. A component of what was once termed TC III (95–97) was given the noncommittal name "void volume binder" (91). Properties of these proteins, summarized in Table 2, have been extensively reviewed (73,74,94,98–102).

Cobalamin-binding proteins in plasma are either complexed with cobalamin (holo binder) or free of cobalamin (apo binder). Holo binder is assayed by the mass of endogenous cobalamin bound to it (101). Apo binder is assayed by the amount of additional cobalamin which it binds in vitro.

The R-type binders compose a class of immunologically similar glycoproteins that are found in various tissues and body fluids (including plasma, tears, gastric juice, saliva,

Table 2 Properties of the Cobalamin-Binding Proteins of Plasma

Property	Haptocorrin	Transcobalamin
Electrophoretic mobility (pH 8.6)	α_1	$\alpha_2\beta$
M_r (approximate)	120,000[a]	38,000[a]
Cyanocobalamin binding capacity, μg/mg	12.2	28.6
Protein type (R or S)	R	S
Composition:		
Carbohydrate content, %	33–40	0
Sialic acid, residues/mol	18	0
Fucose, residues/mol	9	0
Portion of plasma vitamin B_{12} bound, % (approximate)	75	25
Portion of binder unsaturated, % (approximate)	50	98
$T_{1/2}$ of TC-B_{12} complex	9–12 days	60–90 min
Reacts with:		
Anti-TC II	No	Yes
Anti-TC I	Yes	No
Anti-saliva R protein	Yes	No

[a] M_r is 150,000 dalton by gel filtration and 95,000–100,000 by sodium dodecyl sulfate electrophoresis.

and milk). All have the same amino acid sequence in their polypeptide portions, with differences within the class being attributable to qualitative and quantitative differences in their carbohydrate content (103). Notably, IF and haptocorrin contain the same carbohydrates (94).

Haptocorrin was long thought to derive from granulocytes, investigators having observed increased levels of plasma R-type binders in association with myeloproliferative and other granulocytic diseases (95,99,103–107). Some held that R-binder accumulating in these diseases is different from haptocorrin and transcobalamin; hence, the names "granulocyte binder" and TC III (109). So-called "granulocytic binder" is now considered indistinguishable from haptocorrin (97,102). Some plasma R binder appears to arise in vitro from granulocytes after blood has been collected, i.e., during clotting (96,104). Thus the pattern of R binders may differ in plasma and in serum. It now seems likely that haptocorrin has many sources, including erythrocyte precursors, hepatoma cells (105,106), salivary gland cells, and granulocytes (107). Unsaturated haptocorrin is found both intracellularly and in all extracellular fluids.

The source of human TC is not certain. Some data suggest hepatic synthesis (110,111); others suggest that synthesis occurs in small intestinal mucosa cells (112,113). There is also evidence of secretion by endothelial cells (114).

Although the nature of the cobalamin binding sites in cobalamin-binding proteins has attracted interest (115,116), most past studies on the types of corrinoids bound were performed with inadequately purified binding proteins. A recent study of the binding of corrinoids to human intrinsic factor, transcobalamin, and haptocorrin escapes this criticism (117). p-Cresolyl cobamide and 2-aminovitamin B_{12} (complete corrinoids, whose nucleotide at the lower face of the corrin ring is not coordinated to Co) are $\geq 10^3$ times less efficiently recognized by intrinsic factor or transcobalamin than vitamin B_{12} itself, which contains a Co-coordinated nucleotide.

In summary, evidence suggests that cobamide binding to IF and TC is affected by the Co—N coordination bonds of their lower cobalt nucleotide ligands, probably because this bond positions the nucleotide at a critical distance to the corrin ring, which is recognized by the binding proteins. However, haptocorrin reveals greater selectivity among differing corrinoid structures. This protein binds all corrinoids with comparable avidity, independent of the strength of their Co—N coordinations or the structures of their lower Co—α ligands. Hence, the corrin ring, rather than a structural feature induced by Co—N coordination, appears responsible for corrinoid binding to haptocorrin.

Workers have recently cloned and characterized the human gene encoding transcobalamin (118). The gene spans a minimum of 18 kbp and contains nine exons and eight introns, with a polyadenylation signal sequence located 509 bp downstream from the termination codon and a transcription initiation site beginning 158 bp upstream from the ATG translation start site. The 5′ flanking DNA does not have a TATA or CCAAT regulatory element, but a 34-nucleotide stretch beginning just upstream of the CAP site contains four tandemly organized 5′-CCCC-3′ tetramers. This sequence is a motif for a trans-active transcription factor that regulates expression of the epidermal growth factor receptor gene, which also lacks TATA and CCAAT regulatory elements. A number of the exon/intron splice junctions of human transcobalamin, haptocorrin, and IF genes are located in homologous regions of these proteins, suggesting that these genes have evolved by duplication of an ancestral gene. Characterization of the TC gene should facilitate identification of the mutation(s) responsible for the genetic abnormalities of TC expression (see below).

2. Functions of Plasma Cobalamin-Binding Proteins

Haptocorrin and TC are present in plasma in trace quantities. The concentration of haptocorrin is approximately 60 μg/L (400×10^{-12} mol/L). TC concentration is about 20 μg/L (600×10^{-12} mol/L). In fasting plasma, three-quarters of the circulating cobalamin (mainly MeCbl) is bound to haptocorrin. Nonetheless, haptocorrin has substantial unsaturated binding capacity, ranging range from 79 to 939 pg/mL of plasma (mean = 330) (119). Thus, in normal plasma more than half of the binding capacity of haptocorrin is saturated with cobalamin; the remaining 8.5–47.1% is free. Apo-R binder in plasma has a more basic isoelectric point than holo-binder owing to differing sialic acid levels.

TC, the only true transport protein of the several plasma cobalamin-binding proteins, binds only 10–25% of the total plasma cobalamin (97,119,120). The unsaturated binding capacity of normal plasma is 611–1505 pg/mL (mean = 986) (91). A small fraction of plasma TC is saturated at a given moment; TC accounts for at least two-thirds of the unsaturated cobalamin-binding power of plasma. Hence, it is the major apo-binder of plasma.

Known functions of the cobalamin binders include (a) transport of cobalamins through cell membranes; (b) protection of bound MeCbl from photolysis (121); and (c) prevention of loss of cobalamin in urine, sweat, and other body secretions. When a small dose of cobalamin (injected or orally administered) enters the blood, it is initially bound by transcobalamin. Indeed, more than 90% of recently absorbed cobalamin is carried by TC (120). TC-Cbl complex is then cleared from plasma in minutes (98,122), with ligand and protein moiety disappearing at comparable rates (123). In contrast, haptocorrin, which carries most of the cobalamin in plasma, clears slowly from plasma (98,124), with a half-life of 9–10 days (125). Despite the rapid disappearance of most TC-bound freshly absorbed cobalamin, a small portion of the circulating cobalamin continues to be carried by

TC long after its intestinal absorption (119). Cobalamin evidently recirculates to an extent, and recycling vitamin is bound to TC.

Interest in the mechanism of membrane transport facilitation has stimulated studies of the TC receptor (126). In hog kidney membranes binding of TC-Cbl complex to solubilized TC receptor requires bivalent cations and is influenced by salt concentration. The highly specific binding of TC-Cbl to TC receptor has an association constant of 4.6×10^9 L/mol. Studies on subcellular fractions supported the view that the receptor was located on the brush-border membrane. Similar cation requirements and kinetic constants were found in earlier studies of the binding of transcobalamins to subcellular fractions of rat liver (127).

A persisting conundrum is the failure of haptocorrin to behave as an effective transport protein. Indeed, congenital absence of haptocorrin is seemingly innocuous (128); in contrast, a severe, even lethal, megaloblastic anemia occurs in infants lacking TC (129,130). Only TC promotes cellular uptake of cobalamin (131,132). Cobalamin is assimilated in many cells; liver cells have a notably high affinity for TC-bound cobalamin.

In summary, current evidence suggests that (a) TC turns over rapidly and is the major transport protein of recently absorbed cobalamin; (b) haptocorrin turns over slowly and transports cobalamin only after it has been circulating for some time; (c) there are no other functional transport proteins; (d) TC is a membrane transferase that binds to a specific membrane receptor and functions uniquely in promoting entry of cobalamin into cells; and (e) R-type proteins include many immunologically indistinguishable proteins that arise in many tissues and differ widely in their carbohydrate content and hence in their molecular weight and isoelectric focusing properties.

The following abnormalities of plasma cobalamin-binding proteins have been described in various disease states: (a) plasma TC and haptocorrin levels rise in myeloproliferative disorders, e.g., chronic granulocytic leukemia (73,95,119,133,134) and polycythemia vera (108,135), and occasionally in hepatocellular carcinoma (105,136) and other solid tumors (137); (b) haptocorrin is undersaturated in pernicious anemia and other cobalamin deficiency states, occurring early in patients with impaired cobalamin absorption (138); (c) the sum of unsaturated haptocorrin and transcobalamin—sometimes termed serum unsaturated B_{12} binding capacity, or UBBC—may be decreased in cirrhosis and infectious hepatitis (139); (d) TC is an acute-phase reactant that rises nonspecifically in infectious or inflammatory states (140,141); and (e) UBBC tends to rise in transient neutropenia (142); however, the situation is unclear in neutropenia. No strict correlation has been found between serum cobalamin level, unsaturated or total cobalamin binding capacity, or any serum binder, on the one hand, and neutrophil count, bone marrow findings, total blood granulocyte pool, or granulocyte turnover rate, on the other (142).

The fact that TC contains less sialic acid than haptocorrin (Table 2) may explain its low level in plasma. Asialoglycoproteins have been shown to be cleared more rapidly from plasma by the liver than sialoglycoproteins (143), which may explain why it is cleared more rapidly than haptocorrin.

Cobalamin undergoes enterohepatic circulation. In humans, 0.5–9 µg of cobalamin daily enters the gut via the bile stream (144,145). Of this, 65–75% is reabsorbed. Reabsorption is IF-dependent (146).

In rabbits, cobalamin bound to R-type proteins is processed by hepatocytes before reentering the plasma or being excreted in the bile (143). If human hepatocytes also preferentially process cobalamin bound to R proteins, such as haptocorrin (147,148), the liver may play a major role in clearing the circulation of inactive cobalamin analogs produced by bacteria in infectious foci and then transported on R proteins derived from granulocytes in those foci. Conceivably, the liver disposes of such analogs by secreting them preferentially into the bile. Since IF binds a narrow range of cobalamin analogs (115,116), most are probably not reabsorbed from the intestine.

V. METABOLIC FUNCTIONS

The two coenzymatic cobalamin derivatives described above—adenosylcobalamin (AdoCbl) and methylcobalamin (MeCbl)—participate in the functioning of a dozen or more separate enzyme systems. It is a conspicuous fact that only two—AdoCbl-dependent methylmalonyl CoA mutase and the MeCbl-dependent methyltetrahydrofolate-homocysteine methyltransferase (now termed methionine synthase)—occur in animal cells. All other known systems are found in monerans and protista. Advances in understanding of the metabolic functions of cobalamin coenzymes have been surveyed extensively (149–158).

A. Adenosylcobalamin-Dependent Reactions

As noted above, the first enzyme shown unequivocally to require cobalamin was the glutamate mutase of the unusual bacterium *Clostridium tetanomorphum* (Fig. 5)—a notable achievement in 1958 of H. A. Barker and co-workers at Berkeley (17). At that time, parallel studies of propionic acid metabolism in animal tissues (159–162) revealed the existence of methylmalonyl CoA isomerase (later termed methylmalonyl CoA mutase). Earlier work had demonstrated an ATP-dependent conversion of propionate to succinate by mitochondrial enzymes of rat liver (163). The observation that extracts of pig heart converted propionate to methylmalonate (a compound then virtually unknown

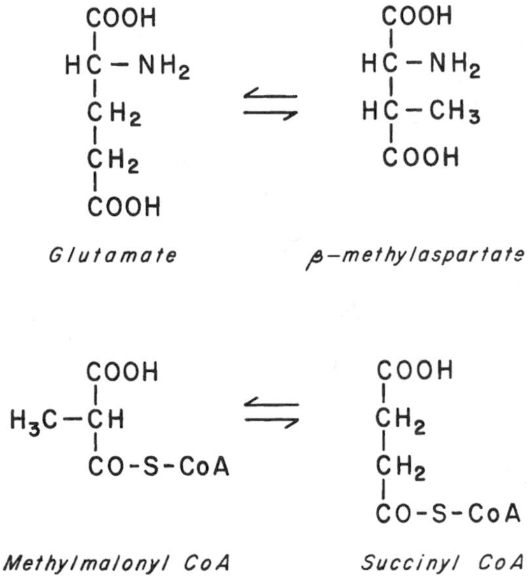

Fig. 5 Glutamate mutase, the first enzyme shown to require AdoCbl as coenzyme. It is here compared with methylmalonyl CoA mutase.

to biochemists), whereas rat liver or sheep kidney converted it to methylmalonate and succinate, led to the resolution of a remarkable two-step pathway: a carboxylation of propionate to methylmalonate followed by a novel isomerization of methylmalonate to succinate. Then Barker and co-workers showed not only that glutamate isomerase is cobalamin-dependent; they discovered three forms of the first known cobalamin coenzyme—5′-deoxyadenosyl cobalamin (or adenosylcobalamin)—then abbreviated DBCC and later AdoCbl (164).

The resemblance of the reactions catalyzed by that enzyme and by methylmalonyl CoA mutase was striking and obvious (Fig. 5). Workers soon demonstrated the AdoCbl dependence of methylmalonyl CoA mutases from vitamin B_{12}-deficient rats (165,166) and propionibacteria (166,168). It was now possible to write a pathway of propionic acid catabolism (Fig. 6) that in a few steps embodies important novelties. Among them is the fact revealed by later work (169,170) that carboxylation of propionyl CoA (a biotin-dependent reaction) yields an epimer of methylmalonyl CoA that must then be enzymatically racemized to another epimer, which is the true substrate of the mutase. In propionibacteria, traffic along this pathway is heavily in the opposite direction. Indeed, these organisms excrete propionic acid—hence, their genus name. Cox and White soon observed striking methylmalonic aciduria in cobalamin-deficient patients (171). Were it not for difficulties noted below in assaying methylmalonate, this would have been a more widely used clinical test for cobalamin deficiency.

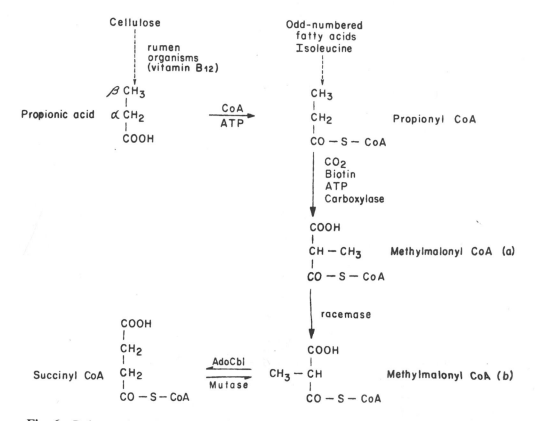

Fig. 6   Pathway of propionate metabolism.

Table 3 Adenosylcobalamin-Dependent Enzymes

$$R^1-\underset{\underset{H}{|}}{C}-\overset{\overset{R^2\diagdown H}{}}{C}-R^3$$

Enzyme	Mode of hydrogen transfer	R^1	R^2	R^3	Biological occurrence
Glutamate mutase	Intramolecular	H	$CH(NH_2)COOH$	COOH	*Clostridium*
Methylmalonyl CoA mutase	Intramolecular	H	COCoA	COOH	Animal cells, propionibacteria
Leucine 2,3-amino mutase	Intramolecular	$(CH_3)_3CH$	NH_2	COOH	Animal cells, bacteria
Dioldehydrase	Intramolecular	CH_3	OH	OH	*Aerobacter*
		H	NH_2	OH	
Glycerol dehydrase	Intramolecular	CH_2OH	OH	OH	*Aerobacter*
Ethanolamine deaminase	Intramolecular		NH_2	OH	*Clostridium*
β-Lysine mutase	Intramolecular	—	NH_2	$CH_2CH(NH_2)CH_2COOH$	*Clostridium*
Ribonucleoside triphosphate reductase	Intermolecular	Ribonucleotide + $R(SH)_2 \rightarrow$ deoxyribonucleotide + $R(S)_2$			Lactobacilli, protists

The discovery of adenosylcobalamin, the first known cobalamin coenzyme, inspired important observations, summarized above, on the mechanism of cobalamin coenzyme synthesis by adenosylating enzyme (Fig. 4). In lactobacilli, this enzyme is on the ribosomal surface (43), a deft method of gathering the small number of cobalamin molecules present in these cells (172,173) to a restricted activation site. In liver cells, cobalamin is more plentiful, but the adenosylating enzyme is still localized. Here it is mainly in mitochondria (174).

A dozen or so AdoCbl-dependent enzymes have been identified (Table 3). In most, hydrogen transfer occurs intramolecularly; in one, hydrogen transfer occurs intermolecularly (Fig. 7). Reactions in which there is a cobalamin-mediated intramolecular transfer of hydrogen forms a new carbon–hydrogen bond. The apparent diversity of these reactions was rationalized by recognition that in all the coenzyme is an acceptor–donor of hydrogen, with the locus of the transfer being the 5′-deoxyadenosyl carbon atom next to the cobalt atom (175,176).

The one known AdoCbl-dependent reaction in which hydrogen is transferred *inter*-molecularly—hence it is a reduction—is ribonucleoside triphosphate reductase, which converts ribonucleotides to deoxyribonucleotides at the triphosphate level (177–180). In this reaction, so far demonstrated only in lactobacilli, *Euglena gracilis*, and other cobalamin-requiring protists, hydrogen for the reduction of the ribonucleotide substrate comes from a dithiol (181,182). Later work showed that this reductase catalyzes a reversible cleavage of the unique C-Co bond (183,184).

The strategic significance of the reductase reaction for the synthesis of DNA precursors in a cobalamin-requiring lactobacillus is evident in the scheme in Fig. 8. In lactobacilli, the cobalamin-dependent reaction is the link between ribosyl and deoxyribosyl nucleotides. This explains the occurrence of unbalanced growth, i.e., impaired DNA synthesis

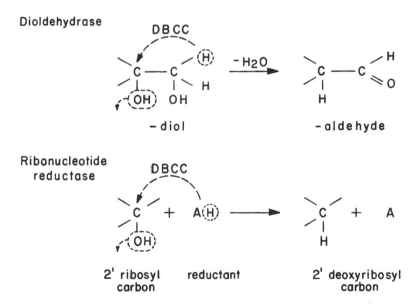

Fig. 7 Parallel mechanisms in reactions catalyzed by dioldehydrase and ribonucleotide reductase. In both, an H displaces an OH. The H derives from an outside reductant only in the reductase reaction. Note: DBCC is AdoCbl.

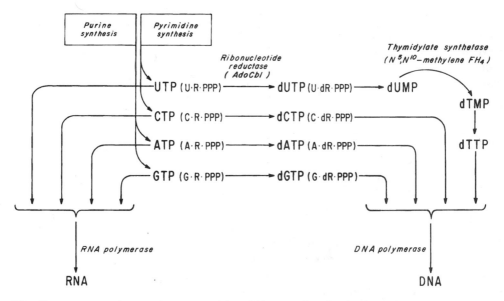

Fig. 8 Pathways of nucleotide and nucleic acid in *Lactobacillus leichmannii*.

and inability to divide, and unimpaired RNA and protein synthesis (185,186). The resulting forms are elongated filaments with elevated RNA/DNA ratios that are analogous to those of the megaloblasts of human cobalamin delivery.

Great biological interest attaches to the fact that evolution has produced at least three major classes of ribonucleotide reductases, of which only one is cobalamin-dependent (187). Animal cells and aerobically grown *E. coli* have a cobalamin-independent ribonucleoside *di*phosphate reductase. This enzyme, unlike the *tri*phosphate reductase of *L. leichmannii* (a single polypeptide chain), consists of two separate homodimers named R1 and R2 (188). R1 contains the catalytic sites and two types of allosteric sites that bind nucleoside triphosphate effectors. R2 contains two dinuclear iron centers with associated stable tyrosyl free radicals, generation of which is oxygen-dependent. Surprisingly, anaerobically grown *E. coli* was recently shown to contain a ribonucleotide *tri*phosphate reductase requiring strict anaerobiosis. A homodimer, the active form of this enzyme contains an oxygen-sensitive glycyl free radical within its protein structure (Gly-681) and an iron–sulfur center (189–191). Radical generation requires the presence of *S*-adenosylmethionine, NADPH, and a reducing enzyme system (flavodoxin and its reductase) (192).

Aside from their important evolutionary implications, these enzymes have in common their dependence on free-radical generators. It is now clear that the cobalamin coenzymes are effective and facile generators of free radicals. The reason, it appears, is the unique and notably fragile C−Co bond. Its bond energy is only 26 ± 2 kcal/mol (193), and many agents, including light, can rupture it. In an enzyme-catalyzed homolytic cleavage reaction, one electron of the C−Co bond remains with C and the other with Co. As a result, an unusual free radical ($-CH_2 \cdot$) forms around a temporarily free carbon atom and efficiently abstracts H from a substrate molecule, thus triggering a chain of events that ends in the homolytic cleavage of the 3′ C−H bond on ribonucleotides followed by cleavage of the 2′ C−OH bond, i.e., a stereospecific exchange of H for OH. Similar mechanisms explain the two other types of cobalamin-dependent reactions: intramolecular rearrangements and methylations. Significantly, a similar mechanism appears to underlie to role of S-adenosylmethionine in the anaerobic reductase reaction (192). The structural homology of this compound and adenosylcobalamin has not escaped notice. For that reason, one reviewer termed S-adenosylmethionine "a poor man's adenosylcobalamin" (194).

B. Methylcobalamin-Dependent Reactions

Participation of methylcobalamin in the reactions listed in Table 4 have been well reviewed (28,155,195). Unlike the carbon skeleton rearrangements catalyzed by AdoCbl-dependent enzymes, the salient feature of the MeCbl-dependent enzymes is removal of a methyl group from a tertiary amine. Nonetheless, the mechanism rests on the ability of a cobalamin prosthetic group to form a C-Co bond. In cases at hand, this bond links cobalt to a methyl group in MeCbl; in those discussed above, the bond links to a cyano group in CNCbl or to a 5′-deoxyadenosyl group in AdoCbl. AdoCbl-linked enzymes catalyze group migrations that are ordinarily initiated by homolytic cleavage of the C-Co bond to form an adenosyl radical and cob(II)alamin. In contrast, MeCbl-linked enzymes catalyze methyl group transfers involving heterolytic cleavage of the C-Co bond to form cob(I)alamin, with transfer of the methyl to an acceptor substrate.

Methionine synthase, the prototypic and best studied MeCbl-dependent enzyme (Fig. 9), in essence catalyzes two methyl transfers in which enzyme-bound MeCbl is successively demethylated and remethylated:

$$CH_3\text{-cob(III)alamin} + \text{homocysteine} \rightarrow \text{cob(I)alamin} + \text{methionine} \tag{1}$$

$$Cob(I)\text{alamin} + N^5\text{-methyltetrahydrofolate} \rightarrow CH_3\text{-cob(III)alamin} \tag{2}$$
$$+ \text{tetrahydrofolate}$$

A remaining issue concerns the means by which the reactivity of cobalamin is made to favor heterolytic cleavage of the C-Co in one class of enzymes and homolytic cleavage in the other. Some light may have been shed on this question by a remarkable recent achievement (196)—the first resolution by x-ray diffraction of how MeCbl binds to one domain (27 kDa) of this 236-kDa protein, earlier crystallized from *E. coli* (197). As noted by Stubbe (198), the ability of methionine synthase to use cob(II)alamin might provide a link with the class of AdoCbl enzymes, which use cob(II)alamin as the catalytically active form of the cofactor. Studies of methionine synthase and two AdoCbl-dependent mutases suggest a common binding motif for the cobalamin cofactor.

Having shown that cobalamin-dependent methionine synthase is a large enzyme composed of structurally and functionally distinct regions, workers have recently begun to define the roles of several regions (199). X-ray crystallographic determination of the

Table 4 Reactions Requiring Methylcobalamin

Enzyme	Reaction catalyzed
1. N^5-Methyltetrahydrofolate homocysteine methyltransferase (methionine synthetase)	$CH_3\text{-THFA} + HSCH_2CH_2CH(NH_2)$ $COOH \rightarrow CH_3SCH_2CH_2CH(NH_2)COOH + THFA$
2. Methane synthetase	$2CH_3OH + H_2 \rightarrow CH_4 + 2H_2O$
3. Acetate synthetase	$2CO_2 + 4H_2 \rightarrow CH_3COOH + 2H_2O$

THFA, tetrahydrofolic acid.

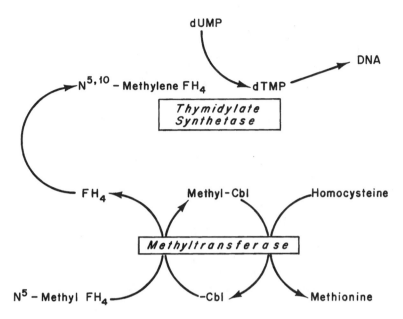

Fig. 9 Methylcobalamin-dependent pathway of methionine synthesis, showing essential role of cobalamin-dependent methyltransferase in conversion of N^5-methyl FH$_4$ to FH$_4$. If folate is "trapped" as N^5-methyl FH$_4$, it cannot be converted to $N^{5,10}$-methylene FH$_4$, the cofactor of thymidylate synthetase.

structure of a 27-kDa cobalamin-binding fragment of the enzyme from *E. coli* has revealed the motifs and interactions responsible for recognition of the cofactor. The amino acid sequences of several AdoCbl-dependent enzymes, the methylmalonyl coenzyme A mutases and glutamate mutases, show homology with the cobalamin binding region of methionine synthase and retain conserved residues that are determinants for the binding of the prosthetic group, suggesting that these mutases and methionine synthase share common three-dimensional structures.

MeCbl-dependent methionine synthesis occurs in bacteria (38,200) and animal cells (201,202). In humans, this pathway, one of several means by which the body acquires methionine, serves also as a mechanism for converting N^5-methyltetrahydrofolate to tetrahydrofolate. The level of this pathway in a cell appears related to its proliferative capacity (203,204).

Methylcobalamin functions in many exotic bacterial species in the curious reactions on which their survival depends. For example, it is an intermediary in the biosynthesis of methane by the several methane bacteria. In *Clostridium aceticum* and other anaerobic clostridia, it participates in acetate synthesis.

C. Pending Claims

Another proposed role for cobalamin in animal cells merits brief mention. An interesting, if fragmented, literature has suggested its possible participation in the metabolism of cyanide (205) for at least three reasons: (a) cyanide readily converts the various cobalamins with coenzymatic functions to metabolically inert cyanocobalamin; (b) chronic cyanide exposure (like cobalamin deficiency) has certain neuropathological effects (to be discussed below); and (c) there is evidence of the participation of cobalamin in the conversion of cyanide to thiocyanate (206). These considerations gain importance when it is realized that the body normally must deal with a continuing burden of cyanide from sources such as tobacco smoke and foods containing cyanogenic glycosides (e.g., fruit, beans, and nuts).

VI. ASPECTS OF COBALAMIN DEFICIENCY

Cobalamin deficiency, whatever its cause, induces two clinical complexes: the megaloblastic state (with anemia and related phenomena) and a characteristic neuropathy (156,207,208). Each complex evidently reflects distinctive and incompletely understood pathogenetic mechanisms, the study of which has engendered much controversy. One source of contention has concerned laboratory tests used in the diagnosis of cobalamin deficiency.

A. Causes of Cobalamin Deficiency

The many disorders leading to megaloblastic anemia occur in three broad etiological categories: (a) those due to cobalamin deficiency that respond to cobalamin therapy, (b) those due to folate deficiency that respond to folic acid therapy, and (c) refractory marrow disorders not due to cobalamin or folate deficiency and not reversed by their administration. It is rarely possible to infer the underlying cause from clinical features of the anemia alone.

Cobalamin deficiency, the subject at hand, has many underlying causes. As in all vitamin deficiencies, it may result from (a) inadequate dietary intake; (b) increased metabolic requirements; or (c) impaired vitamin activation or utilization in tissues. Poor diet, a rare cause of cobalamin deficiency, occurs mainly in vegetarians who abstain from dairy products and eggs as well as from meat.

Most cobalamin deficiencies result from diminished intestinal absorption of various etiologies. In pernicious anemia, a gastric mucosal defect diminishes intrinsic factor synthesis. Other causes include total (occasionally subtotal) gastrectomy; pancreatic disease; overgrowth of intestinal bacteria in the "blind-loop" syndrome, anastomoses, diverticula, and other conditions producing intestinal stasis; HIV infection; advanced age (209–211); nitrous oxide anesthesia (212,213); infestation with the cobalamin-utilizing fish tapeworm *Diphyllobothrium latum*; and organic disease of the ileum (or other digestive organs) that interferes with cobalamin absorption despite the adequate presence of intrinsic factor.

Cobalamin deficiency resulting from increased requirements occurs mainly in pregnancy (214), especially when fetal demands supervene in a setting of poor nutrition. Impaired utilization of cobalamin occurs in various genetic defects, involving deletions or defects of methylmalonyl CoA mutase, transcobalamin, and enzymes in the pathway of cobalamin adenosylation (215–217).

All of these disorders cause tissue deficiencies of cobalamin coenzymes that are usually correctable by cobalamin repletion. Hematopoiesis then reverts from megaloblastic from normoblastic. Diagnostic study of megaloblastic anemia is imperative because it guides the choice of therapy and often discloses a significant underlying disorder.

B. Diagnostic Approach

In a clinical setting, after recognition of the presence of megaloblastic changes (e.g., bone marrow failure with characteristic morphological changes), diagnosis requires; first, serum vitamin assays and other tests to be described to elucidate the broad etiological category; second, appropriate studies aimed at elucidating the underlying cause; and third, specific treatment and observation of response (218). Too often, patients with megaloblastic anemia are given cobalamin and folic acid and dismissed without further investigation.

1. Is There a Deficiency of Cobalamin?

Serum Cobalamin Levels. Leaving aside the well-known difficulties of establishing meaningful ranges of normal, which are discussed elsewhere (61), and the problem of precision and comparability of various assay kits (219,220), it is important to recognize that in differing settings the term ''vitamin deficiency'' may imply (a) total-body vitamin deficiency; (b) low serum vitamin level; or (c) decreased levels of the intracellular coenzymes derived from the vitamin in question.

There are no practical methods for measuring total-body cobalamin. Clearly, serum cobalamin levels are not measures of total-body levels. A corollary of defining ''deficiency'' by serum levels is the practical need of physicians to designate a given level as a cutoff, below which they will initiate further diagnostic studies. A problem with this definition is that low serum levels (assuming the validity of the ''range of normal'') may occur without clinical morbidity (221,222)—and vice versa.

When deficiency is defined as a decrease in coenzyme levels within body cells sufficient to inhibit relevant coenzyme-dependent enzymes, one is dealing with the deficiency that leads to clinical signs and symptoms. Newer diagnostic tests are now available to assess cobalamin and folate coenzyme levels in a clinical setting.

These considerations are relevant in considering available diagnostic tests, both old and new. Confronted with clinical signs suggestive of cobalamin deficiency (e.g., megaloblastosis, myelopathy, etc.), one asks first whether or not cobalamin deficiency exists. If it does, then one seeks its cause. The first question initially requires assay of the serum cobalamin level. This may be considered a screening test. Basic principles of the assay procedures were summarized above. However, aspects of this familiar clinical test merit additional comment.

Assay of serum cobalamin has long been a diagnostic mainstay (223), despite the discovery in 1978 that normal human serum contains cobalamin analogs, which were assayed as cobalamin by commercial radioisotope dilution assay (RIDA) kits containing R protein as binder (64,65). These analogs do not support the growth of *L. leichmannii* or other cobalamin-dependent microorganisms. Hence, laboratories, which performed routine assays by microbiological methods, were obtaining lower but more accurate results. After 1978, kit makers replaced R-protein binders with IF. Thereafter, the two methods yielded comparable results (61). The nature and possible pathophysiological significance of serum cobalamin analogs remain unclear.

As noted above, cited ranges of normal serum cobalamin levels vary with the assay method and the laboratory. The following points merit emphasis: (a) serum cobalamin levels generally fall before the appearance of megaloblastosis or neuropathy; (b) these clinical manifestations are commonly (but not invariably) present when the serum level has fallen below 100 pg/mL*; (c) since a level of less than 150 pg/mL probably indicates a deficiency state, many physicians regard 200 pg/mL as the level below which further

---

* Cobalamin levels are expressed here in picograms per milliliter, though some prefer picomoles per liter in accordance with the SI system (224). In this system, a level of 200 pg/mL would be equivalent to 147 pmol/L.

studies are initiated; (d) several conditions (e.g., pregnancy, transcobalamin deficiency, etc.) can lower serum cobalamin without inducing intracellular cobalamin deficiency (225); and (e) several conditions (e.g., coexisting myelocytic leukemia, other myeloproliferative, disease, etc.) can elevate serum cobalamin levels enough to invalidate them as indicators of intracellular cobalamin levels.

Studies of the relation between serum and tissue cobalamin levels in normal subjects (226), postgastrectomy patients (227), and patients with pernicious anemia (228–230) indicate that although the correlation is fairly good in normal subjects, serum levels are often poor indicators of tissue levels. One reason is the poorly understood influence of serum cobalamin–binding proteins. Also, kinetic studies confirm that as deficiency develops serum levels are maintained at the expense of the tissue level. Because cobalamin reserves are normally large, a low serum cobalamin level implies a long-term abnormality. It is, therefore, a compelling reason for further investigation.

Although serum cobalamin levels may be normal in the presence of intracellular cobalamin deficiency (and the clinical signs it produces), the opposite pattern—low serum cobalamin levels in the absence of megaloblastic anemia—occurs. Some of these patients display neurological disturbances (231).

To better assess the state of cobalamin nutrition, workers have studied the relation of plasma holotranscobalamin (holoTC) (232) with levels of red blood cell holoTC and cobalamin (RBC-Cb1), accepted measures of tissue cobalamin content, and studied the relationship between RBC-B12 and plasma holoTC II levels. Plasma holoTC and RBC-Cbl concentrations were concomitantly assayed in 20 hematologically normal controls and cancer patients. In normal controls, the mean value of RBC-Cbl was 241 ± 51 pg/mL of packed erythrocytes (range 180–355). In cancer patients, values of holoTC and RBC-Cbl were subnormal. RBC-Cbl was greater than 180 pg/mL packed erythrocytes only when the holoTC level was above 70 pg/mL.

Other Tests for Cobalamin Deficiency. Methylmalonic aciduria indicates cobalamin deficiency (171,233) except in rare cases in which it is due to an inborn metabolic error (215–217). Urinary methylmalonate has been assayed colorimetrically (234), by paper (235), thin-layer (236), or gas (237) chromatography, or by mass spectrometry (238,239). Unfortunately, none of these methods is ideally suited for many diagnostic laboratories.

Normal subjects excrete only trace amounts of methylmalonate: 0–3.5 mg (0–38 μmol) per 24 h. Excretion increases in cobalamin deficiency, sometimes to 300 mg (3260 μmol) or more per 24 h. In many studies, urinary methylmalonic acid (measured with and without an oral loading dose of DL-valine (240)) has proved to be normal in folate deficiency and elevated in cobalamin deficiency, often rising before cobalamin levels has fallen below 200 pg/mL. Hence, elevated urinary methylmalonic acid may be considered an earlier indication of cobalamin deficiency than depressed serum cobalamin (241) and thus a better indicator of intracellular cobalamin deficiency.

For years, the closest approach to a test for cobalamin coenzyme levels was the urinary methylmalonic acid. Lacking that datum, which was hard to obtain, it was necessary to guess whether intracellular cobalamin deficiency exists on the basis of serum cobalamin levels and clinical signs. A pitfall of this reasoning is that the clinical signs of intracellular cobalamin deficiency (megaloblastosis, neuropathy) have other possible causes. For example, megaloblastic anemia can be caused by folate deficiency, various drugs, and so forth, and the neuropathy is mimicked by other disorders.

Two new tests—serum methylmalonic acid and serum homocysteine—appear to go far toward permitting a direct assessment of intracellular cobalamin levels, though these tests are still not readily available.

Proceeding on the premise that intracellular cobalamin deficiency should be reflected most sensitively by elevated concentrations of the unmetabolized substrates of the two cobalamin-dependent enzymes, and on the basis of earlier work showing elevations of urinary methylmalonic acid and intracellular homocysteine (242) in cobalamin deficiency, stabler and co-workers in the laboratories of Allen and Lindenbaum demonstrated the diagnostic value of serum methylmalonic acid and total homocysteine (free plus bound) assays performed by innovative methods employing capillary gas chromatography and mass spectrometry (243–247). Although these techniques are still beyond the scope of routine laboratories, the assays are increasingly available in commercial and reference laboratories. Significantly, their precision and careful exploration has redefined the inquiry and in consequence opened a new chapter in the study of an old problem.

These studies revealed the following normal ranges (248):

	Mean ± 2 SD	Mean ± 3 SD
Serum methylmalonic acid (nmol/L)	73–271	53–376
Serum total homocysteine (μmol/L)	5.4–16.2	4.1–21.3

Data suggest that these tests are more sensitive than the serum cobalamin level in detecting intracellular cobalamin deficiency. Early studies (246,247) of normal and deficient subjects showed that about 95% of cobalamin-deficient patients had elevated serum methylmalonic acid and total homocysteine levels. Of 86 consecutive cobalamin-deficient patients with serum cobalamin levels below 200 pg/mL, 77% had marked elevations (more than 3 SD above the normal mean) of both metabolites, whereas 9% had elevated methylmalonic acid alone and 8% had a marked elevation of homocysteine alone. Only 6% had normal levels of both metabolites.

In a study of 40 nonanemic cobalamin-deficient patients (231), only 22 had serum cobalamin levels under 100 pg/mL, yet all but one had elevated serum methylmalonic acid levels and all but two had elevated homocysteine levels. Later work (247) indicated that 95% of patients who relapse because of suboptimal therapy display early elevations of serum methylmalonic acid, total homocysteine, or both metabolites, compared to 69% with low serum cobalamin levels. Of 419 consecutive patients with recognized significant cobalamin deficiency, 12 had serum cobalamin levels greater than 200 pg/mL, mild or absent anemia, and (in 5) prominent neurological signs that responded to cobalamin. In all 12 cases, both serum methylmalonic acid and total homocysteine were increased. Observations by others (248,249) confirm the value of total homocysteine assays in cobalamin deficiency. Such data, it would appear, establish that serum cobalamin is normal in a significant minority of cobalamin-deficient patients and that assay of these serum metabolites is diagnostically essential.

The so-called deoxyuridine suppression test is an isotopic procedure that assays the ability of nonradioactive deoxyuridine to suppress the incorporation of labeled thymidine into DNA via a pathway that is impaired in folate and cobalamin deficiency. The test has proved useful in investigative settings, particularly in detecting the effect of various added metabolites or temporal changes. Though it is said to become abnormal prior to the emergence of clinical signs (250), the complexity of its metabolic basis has been cause for concern (251,252).

The diagnostic approach to suspected cobalamin deficiency has been usefully summarized by reviewers (218,225,253).

2. Determining the Cause of Cobalamin Deficiency

Malabsorption, by far the most common cause of cobalamin deficiency, was opened to investigation by classic studies of Schilling (254). His demonstration that a large parenteral dose (1 mg) of nonradioactive cobalamin promotes renal excretion of an ingested tracer dose of radioactive cobalamin, presumably by blocking cobalamin binding sites in plasma and liver, led to a standard procedure for assessing cobalamin absorption. It is emphasized that the Schilling test estimates cobalamin absorption, not cobalamin repletion. Thus, it may explicate the basis of cobalamin deficiency by revealing the typical absorptive patterns of such disorders as pernicious anemia and ileal disease. Though a bedrock of diagnosis, the Schilling test was long more popular in the United States than in Europe, where assays of IF in gastric juice were generally preferred. Today, it is universally used despite several potential sources of error.

The Schilling test procedure is as follows: After voiding, a fasting patient ingests 0.5 µCi (0.5–2.0 µg) ^{57}Co-cyanocobalamin in water at time zero when a 24-h urine collection is begun. At 2 h, 1 mg of nonradioactive cyanocobalamin—the so-called flushing dose—is administered intramuscularly. The subject may then take food. An adequate sample of pooled urine is assayed for radioactivity and the percentage of administered radioactivity excreted in the first 24 h is calculated. With a 1-µg dose of labeled cobalamin, normal subjects excrete 7% or more of the administered radioactivity in the first 24 h. A greater percentage is excreted when a smaller dose is given. It is important that laboratories performing this test establish a normal range in suitable control subjects.

If excretion of radioactivity is slow, the second part of the Schilling test is performed in no less than 5 days. The procedure is the same except that 60 mg of demonstrably active hog IF (equivalent to 1 NF unit) is given orally with the radioactive cobalamin. If poor excretion in the first part was due to IF deficiency, the result in the second part should be normal. If excretion in the second part is still abnormal, other explanations must be found for cobalamin malabsorption.

The major source of error is incomplete urine collection. Completeness of urine collection may be assessed by a determination of total creatinine content in a 24-h urine specimen. The lower limit of normal is 15 mg/kg of body weight per day.

The kidneys excrete cyanocobalamin and inulin in a similar manner. Indeed, radioactive cobalamin is useful in measurements of the glomerular filtration rate (255). Renal disease associated with impaired glomerular filtration may delay excretion of radioactivity in the Schilling test (256). To circumvent that difficulty, various modifications have been proposed, e.g., a 72-h urine collection with flushing doses of cobalamin every 24 h. Whole-body counting may be the only satisfactory technique in severe renal insufficiency (257,258). In this procedure, the flushing dose is omitted. Normal subjects retain 45–80% of administered radioactivity following an oral dose of radioactive cobalamin. In one such procedure (259), the plasma of normal subjects 8 h after an oral dose of ^{57}Co-cobalamin contained 1.4–4.1% of administered radioactivity per liter of plasma, with a mean of 2.3%. The pernicious anemia group had a range of 0.0–0.6%, with a mean of 0.2%.

Over the years, laboratories performing both Schilling tests and serum cobalamin assays accumulated data on two seemingly anomalous patterns unlike the majority pattern that displayed a generally linear relation in the lower range of values between Schilling test results and serum cobalamin levels (61). One anomalous pattern—a low Schilling test with a normal serum cobalamin—is seen in subjects with early pernicious anemia. The other—a normal Schilling test and a low serum cobalamin—proved more interesting. Some patients in this group had parasitic competitors (fetus or tumor); others were revealed by later work to malabsorb food cobalamin, while absorbing normally the pure cobalamin used in the standard Schilling test (260–267).

This pattern, first described by Doscherholmen (266), is common in atrophic gastritis, post vagotomy (268), and in subtotal gastrectomy (268). It is also seen in subjects taking cimetidine or other inhibitor of gastric secretion (270) and in the elderly. In many, there is no overt evidence of gastric dysfunction.

In later years, modified Schilling tests were developed in which the labeled cobalamin is bound to egg yolk (264). Variously known as the food Schilling test or egg-yolk cobalamin absorption test, the procedure at first involved ingestion of labeled cobalamin derives from eggs produced by hens injected with labeled cobalamin (266). Later workers substituted labeled cobalamin bound in vitro to dried egg proteins. Now generally available in most major medical centers, this procedure has accounted for otherwise unexplained low serum cobalamin levels and frank cobalamin deficiency in many cases. Indeed, this patient group is at least as large as the group with pernicious anemia or other classic causes of cobalamin deficiency. One study (264) reported 47 patients with low serum cobalamin levels and normal Schilling test results. Egg test results were significantly lower than normal, whereas routine Schilling test results were normal. Twenty subjects had egg test excretions below 1.5%. No other clinical features distinguished them from the 27 who excreted more than 1.5% other than the presence of lower ratios of pepsinogen I:II. Interestingly, 60% of tested patients had neurologic, cerebral, or psychiatric abnormalities. If food cobalamin malabsorption is often associated with otherwise unexplained low cobalamin levels, low cobalamin levels in the presence of normal Schilling test results should not be dismissed without testing for food cobalamin malabsorption, whether or not there is known gastric dysfunction.

Assays of IF in gastric juice can also be based on IF-mediated (a) enhancement of cobalamin uptake by intestinal cell preparations in vitro (271,272), (b) inhibition of an AdoCb1-dependent enzyme (273), and (c) various immunological attributes (274,275). The use of such techniques led to discovery of a physiologically inert IF molecule in a child with cobalamin malabsorption (276). Studies of the abnormal IF molecule isolated by affinity chromatography (277) demonstrated nonidentity of the cobalamin- and ileum receptor–binding sites on the IF molecule. The genetic abnormality was confined to the latter site.

C. Clinical and Metabolic Features

1. Megaloblastic Anemia and Related Phenomena

The metabolic derangement that leads to megaloblastic anemia is a defect of DNA synthesis—without impairment of RNA synthesis (150). The ribonucleotide reductase of bone marrow and other animal cells was shown to be cobalamin-independent (278–280); hence, it became necessary to seek other explanations for the observable fact that cobalamin deficiency leads to impaired DNA synthesis.

"Methylfolate Trap" Hypothesis.   Of the hypotheses offered to account for the role of cobalamin in DNA synthesis, one (with the vernacular name "methylfolate trap hypothe-

sis'') was proposed in 1961 to account for abnormalities of folate metabolism in cobalamin-deficient rats (281). Because methionine supplementation decreases the proportion of N^5-methyltetrahydrofolate in rat liver from 80–90% of total folate to about 50%, it was suggested that a similar shift might follow suppression of MeCb1-dependent methionine synthesis in cobalamin deficiency (Fig. 9). It was then suggested that in human cobalamin deficiency (282) accumulation (''trapping'') of folate as N^5-methyltetrahydrofolate sequesters folate in this form, thus blocking its conversion to tetrahydrofolate, a precursor of N^5,N^{10}-methyltetrahydrofolate, the cofactor of thymidylate synthetase. Hence, conversion of dUMP to dTMP—and thus DNA synthesis—is impaired.

The following evidence seemed to favor this view: (a) changes in folate metabolism often, but not always, occurring in cobalamin deficiency, e.g., the altered partitioning of tissue folate compounds referred to above (281); elevation of the N^5-methyltetrahydrofolate in serum (282); increased urinary excretion of formininoglutamic acid after an oral histidine loading dose (283); and elevated urinary excretion of aminoimidazole carboxamide (284); (b) megaloblastosis in a patient with congenital methyltransferase deficiency (285); (c) diminished methyltransferase activity in cobalamin-deficient rats (286); (d) correction of defective DNA synthesis in cobalamin deficiency with tetrahydrofolate (287); and (e) apparent impairment of the conversion of dUMP to dTMP in bone marrow cells in cobalamin-deficient subjects, the so-called dU suppression test (250–252).

Despite this evidence, opposition to this hypothesis stemmed from certain contrary findings, the following among them: (a) lack of urinary excretion of formiminoglutamic acid and aminoimidazole carboxamide in many cobalamin-deficient subjects (288); (b) clearance studies by some investigators (289,290)—but not others (291)—showing normal rates of N^5-methyltetrahydrofolate utilization in cobalamin deficiency, suggesting that elevation of serum N^5-methyltetrahydrofolate is due to translocation rather than ''trapping''; (c) normal $^{14}CO_2$ production from (^{14}C-methyl)-N^5-methyltetrahydrofolate in cobalamin-deficient rats (292); (d) evidence that a cobalamin-independent pathway exist for the release of methyl groups from N^5-methyltetrahydrofolate in neural tissue (293,294) and blood cells (295); (e) depression in cobalamin deficiency of total folate in red cells (296) and liver cells (294), contrary to predictions of the ''methylfolate trap hypothesis''; (f) questions about the validity of the so-called dU suppression test (250–253), which has been widely employed (297–302) without critical scrutiny; and (g) evidence in children with severe homocystinuria, cystathioninuria, hypomethionimemia, and methylmalonic aciduria (303–305) of defective formation of both AdoCbl and MeCbl, possibly resulting from a defect in membrane transport of cobalamin-binding protein. Significantly, these children did not have megaloblastic anemia. Cultured skin fibroblasts showed depressed methyltransferase activity and correspondingly impaired conversion of homocysteine to methionine—precisely the defect envisioned in the ''methylfolate trap hypothesis.''

Evidence that intracellular folates are in the form of polyglutamates (a device preventing leakage from cells) and that N^5-methyltetratetrahydrofolate is, among folylmonoglutamates, a uniquely poor substrate for folylpolyglutamate synthetase added another facet to this story. If cobalamin deficiency traps folate as N^5-methyltetrahydrofolate, this may account for unpredicted decreases of tissue folates in this setting. In sum, the trap hypothesis is now widely accepted, though reservations persist.

Clinical Features. Although anemia is a prominent feature of the megaloblastic state, there is often pancytopenia associated with a familiar morphological pattern in blood and bone marrow cells—and indeed in all proliferating cells—that includes gigantism of these cells and various signs of impaired cell division (207,208). The anemia is typically macrocytic, though not all macrocytic anemias are megaloblastic and not all megaloblastic anemias are macrocytic. Megaloblastic blood cell precursors (megaloblasts) contain a normal or increased amount of DNA and an increased amount of RNA per cell, which accounts for cytoplasmic basophilia in Wright's-stained smears.

Misincorporation of Uracil into DNA. Defective DNA replication generally reflects impaired conversion of dUMP to dTMP, as a result of which there is decreased intracellular dTMP and dTTP and increased dUMP and dUTP. Thus, the dUTP/dTTP ratio rises (306–308) and dUTP is misincorporated into DNA (309,310). DNA uracil is removed by uracil-DNA-glycosylase (311), but dTTP is unavailable for repair and DNA becomes increasingly fragmented. This irreversibly impairs cell division and causes eventual cell death.

2. Neurological Abnormalities

Aspects of the neuropathy of cobalamin deficiency have been reviewed (312–317).

Classic Syndrome. In its classic form, the neurologic syndrome of cobalamin deficiency (which does not occur in folate deficiency) consists of symmetrical paresthesias in feet and fingers, with associated disturbances of vibratory sense and proprioception, progressing to spastic ataxia with "subacute combined system" disease of the spinal cord, i.e., degenerative changes of the dorsal and lateral columns (318). In fact, the picture is more often chronic than subacute—and, as noted below, the clinical perimeter of this syndrome may be wider than previously thought.

The earliest recognizable pathological changes in the cord consist of small vacuolated areas in the myelin with focal swellings of individual myelinated fibers. Later, the lesions coalesce into larger foci involving many fiber systems, but not in a systematic manner (319).

Usually, so we once believed, the spinal cord is initially affected and predominantly involved throughout the clinical course. Typically, the process begins at the cervicothoracic junction of the spinal cord, attacking the posterior columns first. From there, it spreads up and down the cord as it takes in the more anterior portions.

In a common onset, patients experience distressing and persistent paresthesias of the hands and feet. Often, when something is touched with hands or feet, a burst of paresthetic sensations make their way up the limb. Characteristically, paresthesias begin first in the hands and then the feet. Without treatment, all limbs and body parts are eventually affected until the full-blown classic picture emerges: posterior column, corticospinal tract, spinothalamic tract, and ultimately peripheral nerve disease, which occurs late and in a relatively few cobalamin-deficient patients.

A patient is often brought to a physician by persistence of the paresthesias throughout the day and night. This is a manifestation of posterior column disease. The patient soon notices ataxia, as he brusquely places widely spread feet upon the floor and totters from side to side, sometimes falling. This sensory ataxia, resulting from lost proprioception, is to be distinguished from cerebellar ataxia, which is vertiginous. By the time the gait is affected, impairment of vibratory disturbance is advanced.

Corticospinal tract disease is characterized in advanced cases by weakness or paralysis of voluntary movement, with impaired motor function and spasticity, increased tendon reflexes, even clonus and a positive Babinski's sign.

Occasionally, the disease begins by affecting the optic nerve with impaired vision, scotomata, and sometimes blindness. That is a rare but well-established syndrome, typically occurring in heavy smokers; hence, the name tobacco amblyopia. It was traditionally taught that in advanced disease there can be deranged mental function and in some cases structural changes of the white matter of the brain resembling those in the white matter of the spinal cord. More will be said of this syndrome later.

Differential diagnosis of the clinical neurological syndrome includes cerebellar disease, which typically displays dysmetria on heel-to-shin testing, nystagmus, and impaired tandem walking; multiple sclerosis and other demyelinating diseases; subacute myelo-optic neuropathy; tropical ataxic neuropathy; Leber's hereditary optic atrophy; and a variety of other disorders. An essential diagnostic datum, obviously, is proof of cobalamin deficiency.

Typically, patients with untreated pernicious anemia display megaloblastic anemia first and then a complex of neurological signs that, if untreated, progress to spastic ataxia and beyond. If treated early, these changes are generally reversible, but every experienced hematologist will recall cases in which therapy came late and neuropathies became tragically irreversible. Significant neuropathy was once observed in virtually every new patient with pernicious anemia. Today, perhaps because of earlier diagnosis and treatment, the classic neurological syndrome is seen less frequently.

The appearance of neurological signs before the appearance of anemia (318) has been attributed to administration of folate (without cobalamin). This stimulates marrow function enough to maintain adequate erythropoiesis while inciting neuropathy or failing to prevent its progression. Those who develop neurological signs as a first manifestation will invariably show hematologic signs within weeks or months of the onset if not treated with cobalamin.

Newer Perspectives. Recently, workers have claimed that newer, more sensitive diagnostic tests justify the attribution of such common complaints as memory loss, fatigue and weakness, and personality and mood changes to cobalamin deficiency, even in the absence of depressed serum cobalamin levels and megaloblastic anemia.

It is an old idea that cobalamin deficiency can cause a variety of neuropsychiatric symptoms beyond the classic ones described above. A huge literature of several decades, consisting mainly of clinical reports (320–328), described cerebral dysfunction; EEG changes; irritability; somnolence; perversion of taste and smell; and all manner of psychological derangements, even "megaloblastic madness" (329), in cobalamin-deficient patients. However, much of this work lacked adequate controls, i.e., double-blinded therapeutic studies on groups of deficient and nondeficient subjects with similar symptoms, and adequate diagnostic testing.

In 1988, Lindenbaum, Allen, and co-workers published an interesting study (231) of 141 consecutive patients in whom various neuropsychiatric abnormalities had been attributed to cobalamin deficiency. Of these, 40 patients (28%) had no anemia or macrocytosis. Although these patients had serum cobalamin levels that were often equivocal, serum methylmalonic acid and homocysteine levels were markedly elevated in many— and patients benefited from cobalamin therapy, displaying improvement of neuropsychiatric abnormalities, of abnormal blood counts, and of elevated levels of serum methylmalonic acid and/or total homocysteine.

The new element in this work was the sensitivity of diagnostic testing and the evidence of therapeutic response. However, double-blinded controls were lacking, a possible significant omission because such symptoms wax and wane, and are thus notoriously difficult to deal with quantitatively. Despite this reservation, it seems clear that determination of serum methylmalonic acid and total homocysteine provides a new diagnostic standard against which other procedures are henceforth to be compared and evaluated. The use of these tests has provided compelling evidence that intracellular cobalamin deficiency can and often does exist in the absence of megaloblastic anemia or low serum cobalamin levels.

Pathogenesis.   The neuropathy has proved a difficult target for investigators. The main impediment to its study has been the unavailability of affected tissues or suitable animal models, although recent efforts appear to have produced neurological lesions in cobalamin-deficient fruit bats (330) and primates (331,332). Unfortunately, such materials are as hard to come by as incisive hypotheses.

It is of interest that cobalamin deficiency in most laboratory animals produces a picture quite different from that in deficient humans. It does not include megaloblastic anemia, even when body cobalamin stores have been totally consumed. Perhaps hemato-poietic precursors in these species are richer in cobalamin-binding proteins than human marrow cells. Neurological abnormalities in deficient laboratory animals are also peculiar in character or difficult to recognize.

Although it is generally accepted that neurologic involvement is associated with a defect in myelin synthesis, its mechanism remains unknown. Nor is it clear whether this is the primary neuropathic event or whether that putative event is a consequence of impaired myelin synthesis, destruction of existing myelin, or another process. Three major theories have been advanced to explain the neuropathy of cobalamin deficiency: (a) depressed myelin synthesis and incorporation into myelin of abnormal fatty acids; (b) impaired DNA synthesis; and (c) chronic cyanide intoxication.

Abnormal Fatty Acid Synthesis.   In view of evidence that cobalamin deficiency impairs one of the body's two cobalamin-dependent enzymes—methyltransferase—in causing the defect in DNA synthesis underlying megaloblastosis, it seemed logical that impairment of the other cobalamin-dependent enzyme—methylmalonyl CoA mutase—might account for abnormalities of myelin synthesis associated with neural lesions, especially since affected neurons are not dividing cells engaged in DNA synthesis.

Some experimental work supports that view. The lipids of myelin turn over rapidly (333,334), and myelin replacement is heavily dependent on fatty acid synthesis. Normally, the de novo synthesis of fatty acids consists of the repetitive sequential addition of two-carbon units (deriving from malonyl CoA, a three-carbon compound) to a growing chain that begins with acetyl CoA, a two-carbon compound. The substantial accumulation of methylmalonyl CoA in cobalamin-deficient tissues suggested that methylmalonyl CoA (a branched four-carbon molecule) might inhibit fatty acid synthesis or compete with malonyl CoA or acetyl CoA in the initiation of fatty acid synthesis and thus generate abnormal branched fatty acids or fatty acids with an odd number of carbon atoms. This hypothesis gained credibility from the observed toxicity of valine, a precursor of methylmalonyl CoA, in cobalamin-deficient pigs. Doses of valine that have no effect on normal pigs can be fatal to cobalamin-deficient pigs (335).

Further support is found in the observation that rat glial cells cultured in cobalamin-deficient medium produce increasing amounts of two unusual fatty acids (336)—unbranched acids with 15 and 17 carbon atoms—as cells progressively lose the ability to metabolize propionic acid. Levels of 15- and 17-carbon acids return to normal when cultures are supplemented with cobalamin. Studies of the in vitro effect of methylmalonyl CoA on fatty acid synthesis in rat liver fractions showed that methylmalonyl CoA inhibits fatty acid synthesis and can be incorporated into branched fatty acids (335). Similar results were obtained in unpublished studies of lipid extracts of plasma and red cells from patients with untreated pernicious anemia (337).

Further confirmation came from experiments (338,339) demonstrating synthesis of abnormal 15- and 17-carbon fatty acids from [^{14}C]propionate in the nerves excised from patients with pernicious anemia. Finally, branched and odd-numbered fatty acids were found in the neural tissues of a patient with methylmalonic aciduria and deranged cobalamin metabolism (340). Additional evidence of a role for accumulated methylmalonyl CoA comes from recent studies, discussed below, of methylmalonic acid levels in cerebrospinal fluid (341).

These results, which curiously have stimulated little further work, are provocative, yet they fail to prove that such fatty acids in myelin directly account for the neuropathy of cobalamin deficiency. Moreover, several considerations may argue against the pathogenetic importance if this pathway: (a) the relatively low level of propionate metabolism in humans, especially since valine and other amino acids are not major precursors of methylmalonyl CoA (342); (b) failure to demonstrate such changes in animal models in which neuropathy is induced by exposure to nitrous oxide (315); and (c) the unimpressive correlation between methylmalonic aciduria and the presence or severity of neurological disease. In the various genetically induced methylmalonic acidurias in infants and children, the only neurological signs are lethargy and mental retardation (216).

Impaired DNA Synthesis. It is generally believed that blocked DNA synthesis, which causes megaloblastosis in cobalamin deficiency, does not account for damaged neurons, which are nondividing. However, the cells that synthesize myelin divide, and recent preliminary evidence has implicated impairment of cobalamin-dependent methyltransferase in a neuropathy resulting from N$_2$O exposure (213,343,344).

A suggestive case study (345) reported presumed genetic depression of methyltransferase that was associated with homocystinuria, mild megaloblastic anemia, and neurological deficits appearing at age 21. Methylmalonyl CoA mutase activity was normal. In this case, decreased methyltransferase activity may have induced the neuropathy, but proof of causality and mechanistic details remain to be demonstrated.

Impaired Methylation. According to Scott, Weir, and co-workers (346,347), the neuropathy of cobalamin deficiency may be determined by abnormalities in the ratio of the methyl donor, S-adenosylmethionine, to the coproduct, S-adenosylhomocysteine. This so-called methylation ratio is thought to control the activity of tissues methyltransferases. They found that inactivation of cobalamin-dependent methionine synthase reduces the methylation ratio in rats and pigs in vivo. Methylation ratios found in the neural tissues of cobalamin-inactivated pigs (i.e., after prolonged exposure to nitrous oxide) significantly inhibits protein methyltransferases of pigs and humans, whereas the altered methylation ratio in deficient rats only marginally inhibits the equivalent rat methyltransferases. This may be consistent with the induction of a myelopathy by such treatment in pigs and humans, but not in rats. Dietary supplements of methionine are given to cobalamin-inactivated pigs to prevent the myelopathy in vivo by elevating neural S-adenosylmethionine levels and normalizing the methylation ratio. It is proposed that reduction of the methylation ratio in the brain of pigs as a consequence of methionine synthase inhibition leads to brain hypomethylation, which could affect critical neural components and induce the vacuolar myelopathic changes and demyelination seen in the spinal cord of these animals, which mimic those of human subacute combined degeneration. The validity of this thesis remains to be established (315).

Possible Role of Cyanide.   The hypothesis that neuropathy can occur in cobalamin deficiency as a consequence of chronic occult cyanide intoxication rests on fragmentary evidence that includes (a) the relatively high incidence of tobacco amblyopia, retrobulbar neuritis, and optic atrophy in cobalamin-deficient tobacco smokers (348–350) and carriers of the fish tapeworm *Diphyllobothrium latum* (351); (b) the relatively elevated cyanocobalamin fraction reported in the plasma cobalamin of smokers (352,353); and (c) the reported association of neurological disorders with chronic cyanide exposure or abnormalities of cyanide metabolism (350,354,355).

Tobacco smoke contains 6 ppm of cyanide. Far higher cyanide levels arise from the heating or cooking of many plant materials, notably including cassava, a root tuber (*Manihot esculenta*) widely used as food in the tropics, which is rich in the cyanogenic glycoside linamarin (356,357). Some circumstantial evidence links these cyanogens to various tropical neuropathies (350,358). Conceivably, an exogenous cyanide load would convert cobalamin coenzymes to cyanocobalamin, which is metabolically inactive, less firmly held by binding proteins, and hence more readily lost to renal excretion that other cobalamins. Any such theory might require the assumption that neurons are more vulnerable to such effects than bone marrow cells. It has been suggested, but not proved (359,360), that certain neural elements, especially those of the eye, are uniquely sensitive to cyanide exposure.

Much of the literature proposing a neuropathic role for cyanide comes from British and African writers of the 1960s and 1970s. Though speculative, these ideas are evocative enough to justify further investigation with current methods.

Role of Folate.   A word is in order on the curious role of folic acid in the neuropathy of cobalamin deficiency. Despite long contention (361,362), a convincing case has yet to be made for a neuropathy due to folate deficiency alone. However, it is well established that administration of folate to a cobalamin-deficient patient can induce or worsen neurological abnormalities, even as a partial (and temporary) hematological remission is taking place. Folate has no neuropathic effect in cobalamin-repleted individuals (363).

An explanation for this phenomenon still eludes us. In the author's view, it is plausible to suppose that pharmacological doses of folic acid stimulate DNA biosynthesis previously limited by reduced methyltransferase activity. That would account for a partial hematological remission. Increased cell division in marrow and other cells would likely increase the cobalamin uptake of these cells from already depleted plasma carriers and storage depots. This would probably promote transfer of cobalamin from nondividing cells (such as neurons) to dividing cells in bone marrow. The result would be an abrupt appearance or worsening of neuropathy. Though a defensible hypothesis, it would be difficult to test without a suitable animal model.

D. Response to Treatment

The clinical response to cobalamin therapy is another useful diagnostic datum. Following parenteral administration of cobalamin to deficient subjects, elevated plasma bilirubin, iron, and lactic dehydrogenase levels fall promptly (Fig. 10). Decreasing plasma iron turnover and fecal urobilinogen excretion reflects cessation of ineffective erythropoiesis. Within 8–12 h, the appearance of a bone marrow aspirate begins to convert from megalo-

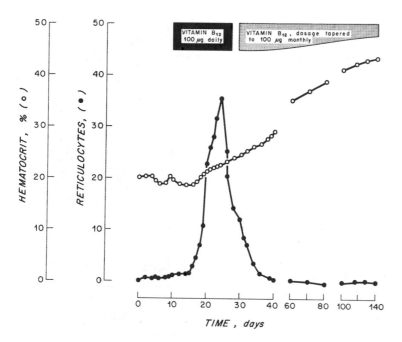

Fig. 10 Effect of cyanocobalamin on reticulocyte count, serum iron, serum bilirubin, stool urobilinogen, and plasma iron turnover.

blastic to normoblastic. Transformation is complete in 48–72 h. The population of accumulated megaloblasts is probably converted to normoblasts ineffectively, i.e., a majority of the megaloblasts die within the marrow or soon after delivery into the circulation. Abrupt reticulocytosis begins on the third to fifth day, reaching a climax on the fourth to tenth day. Cells released at this time apparently come from new normoblasts, not from converted megaloblasts. The intensity of the reticulocyte crisis is roughly proportional to the severity of the anemia.

Hemoglobin levels begin to rise, though the increase during the reticulocyte crisis is smaller than the increase in circulating red cells as judged by the reticulocyte count. The discrepancy is attributable to the fact that the unusually young reticulocytes delivered after sudden remission of a megaloblastic process undergo prolonged maturation in the blood. Hence, the count at a given moment reflects the accumulated reticulocyte output of several days.

When maturation delay in the marrow is corrected by cobalamin, a new condition is established in which a still severe anemia elicits an intensive erythropoietin-mediated stimulation of erythropoiesis. In later stages of hemoglobin restoration, hypochromia and other signs of iron deficiency may appear. In such instances, the plasma iron level decreases as the cobalamin level becomes normal. A second reticulocyte response may then be produced by iron administration.

As documented elsewhere (208), other changes produced by repletion of cobalamin deficiency include the following: (a) striking and prompt improvement in sense of well-being; (b) rise in serum alkaline phosphatase (which is often depressed in cobalamin deficiency; (c) positive nitrogen balance; (d) sharp rise in serum and urine uric acid (variably depressed in cobalamin deficiency) within 24 h of the start of therapy, a peak occurring

24 h before the peak of the reticulocyte crisis; (e) decrease in serum folate; (f) decrease in urine phosphorous after cobalamin administration, increase during reticulocytosis, and then normal; (g) rise in serum cobalamin; and (h) sharp drop in serum potassium, in some cases severe enough to warrant replacement therapy. Failure to provide such replacement has occasionally led to sudden death during treatment for cobalamin deficiency.

A comment should be made on the therapy of the neuropathy of cobalamin deficiency. Basic principles of therapy are well known: early administration of sufficient cobalamin at a slow rate to replete reserves, followed by administration to meet ongoing needs. Several writers argued in the past that a monthly cobalamin dose of 100 µg should be more than adequate. However, this author has seen dramatic evidence that that is bad advice, e.g., a middle-aged person presenting with classic hematological signs of pernicious anemia without neuropathy then developed neuropathy for the first time during the first year of therapy at a monthly dose of 100 µg, a dose that produced hematological remission. Reinducing therapy and increasing the monthly dose to 1000 µg reversed the neurological syndrome.

Physicians have been urged to look for cobalamin deficiency in all patients with ''unexplained'' neuropsychiatric disorders or, if that is not possible, to administer cobalamin therapy anyway. This, of course, would add new costs to medical care, but if real benefits await these patients the costs are justified. At least one can hope that physicians will try to gather meaningful data before giving cobalamin uncritically—and that they will observe the results of cobalamin therapy. Although recent findings are provocative and encouraging, they need extensive confirmation.

It is of interest that methylcobalamin (Methylcobal, Eisai) is widely prescribed in Japan for various peripheral nerve disorders. A large Japanese literature amply demonstrates its value in many cases but fails to show its superiority over cyano- or hydroxocobalamin.

Some have claimed that hydroxocobalamin is superior to cyanocobalamin in the therapy of cobalamin deficiency neuropathy. To the extent that this is true it has been speculatively attributed to the ability of hydroxocobalamin to bind cyanide, which is a putative neurotoxin. This has periodically led those who are persuaded of the neuropathic role of cyanide in cobalamin deficiency to demand the withdrawal of cyanocobalamin from the market. They may be right.

ADDENDUM

My manuscript, as it appears in print, was originally submitted in 1995. For circumstances that appear not to have been under anyone's control, I was not informed until November 2000 of the opportunity to update this chapter. Regrettably, such notice was insufficient. Consequently, this chapter is outdated in some respects wih regard to recent developments in this field. Although my preference was to withdraw the chapter, with reluctance, I append this addendum to provide the diligent reader with a road map to major develops in this field.

Given that space is limited, I will cite only two bodies of work. First, there are the published proceedings of the fourth international Innsbruck symposium *Vitamin B_{12} and B_{12}-proteins.* B. Kräutler, D. Arigoni, and B. T. Golding, editors. Weinheim: Wiley-VCH 1998. This useful volume authoritatively updates readers on the current state of B_{12} research—with emphasis on what we of medicine call basic science. There is a fine overview of the field by Kräutler, and excellent presentations on: (1) biosynthesis; (2) enzymatic methyl transfer; (3) B_{12}-dependent enzymatic rearrangements (wherein I was relieved to learn from JoAnne Stubbe that adenosylcobalamin-dependent ribonucleotide reductase is still amazing, but no longer confusing); (4) B_{12} structure and reactivity; (5) reflections on the role of B_{12} and like molecules in the ancient world; and finally (5) a small concession to medical aspects.

This leads to my second suggestion regarding important recent biological developments, not anticipated in my chapter, i.e., the work of Moestrup and colleagues on cubilin. See, for example, (1) Renata Kozyraki, Mette Kristiansen, Asli Silahtaroglu, Claus Hansen, Christian Jacobsen, Niels Tommerup, Pierre J. Verroust, and Søren K. Moestrup, The human intrinsic factor-vitamin B_{12} receptor, *cubilin:* molecular characterization and chromosomal mapping of the gene to 10p within the autosomal recessive megaloblastic anemia (*MGA1*) region, *Blood 91*: 3593 (1998); and (2) Mette Kristiansen, Maria Aminoff, Christian Jacobsen, Albert de la Chapelle, Ralf Krahe, Pierre J. Verroust, and Søren K. Moestrup, *Cubilin* P1297L mutation associated with hereditary megaloblastic anemia 1 causes impaired recognition of intrinsic factor-vitamin B_{12} by *cubilin, Blood 96*: 405 (2000). Ebba Nexø's excellent work on cobalamin-binding proteins also deserves mention.

To readers, I apologize for circumstances that appear not to have been under anyone's control.

REFERENCES

1. W. B. Castle, Treatment of pernicious anemia; historical aspects, *Clin. Pharmacol. Ther. 7*: 147 (1966).
2. W. B. Castle, The history of corrinoids, in *Cobalamin Biochemistry and Pathophysiology* (B. M. Babior, ed.), Wiley-Interscience, New York, 1975, p. 1.
3. T. Addison, The constitutional and local effects of disease of the surparenal capsules, in *A Collection of the Published Writings of the Late Thomas Addison, M.D.*, New Sydenham Society, London, 1868, p. 211.
4. S. Wilks and G. T. Bettany, *Biographical History of Guy's Hospital*, Ward, Lock, Bowden, London, 1892.
5. G. R. Minot and W. P. Murphy, Treatment of pernicious anemia by a special diet, *JAMA 87*:470 (1926).
6. W. B. Castle, The contributions of George Richards Minot to experimental medicine, *N. Engl. J. Med. 247*:585 (1952).
7. W. B. Castle, Observations on the etiologic relationship of achylia gastrica to pernicious anemia. 1. The effect of administration to patients with pernicious anemia of the contents of the normal human stomach recovered after the ingestion of beef muscle, *Am. J. Med. Sci. 178*:748 (1929); (reprint) *267*:2 (1974).
8. S. P. K. Gupta and E. L. Milford, *In Quest of Panacea: Successes and Failures of Yellapragada Subbarow*, Evelyn Publishers, Nanuet, NY, 1987.
9. E. L. Rickes, N. G. Brink, F. R. Koniuszy, T. R. Wood, and K. Polkers, Crystalline vitamin B_{12}, *Science 107*:396 (1948).
10. M. S. Shorb, Unidentified growth factors for *Lactobacillus lactis* in refined liver extracts, *J. Biol. Chem. 169*:455 (1947).
11. M. S. Shorb, Activity of vitamin B_{12} for growth of *Lactobacillus lactis, Science 107*:397 (1948).
12. H. G. Wijmenga, J. Lens, and A. Middlebeck, Some properaties of vitamin B_{12}, *Chem. Week 45*:342 (1949).

13. E. L. Smith, Purification of anti-pernicious anemia factors from liver, *Nature (Lond.) 161*: 638 (1948).

14. R. West, Activity of vitamin B_{12} in Addisonian pernicious anemia, *Science 107*: 393 (1948).

15. K. Folkers, Historical perspectives on the isolation of crystalline vitamin B_{12}, in *Vitamin B_{12}. Proceedings of the Third European Symposium on Vitamin B_{12} and Intrinsic Factor. University of Zurich, March 5–8, 1979, Zurich, Switzerland (B. Zagalak and W. Friedrich, eds.),* Walter de Gruyter, Berlin, 1979, p. 7.

16. D. C. Hodgkin, J. Kamper, M. MacKay, J. Pickworth, K. N. Trueblood, and J. G. White, Structure of Vitamin B_{12}, *Nature (Lond.) 178*:64 (1956).

17. H. A. Barker, H. Weissbach, and R. D. Smyth, A coenzyme containing pseudovitamin B_{12}, *Proc. Natl. Acad. Sci. USA 44*:1093 (1958).

18. W. L. C. Veer, J. H. Edelhaus, H. G. Wijmenga, and J. Lens, Vitamin B_{12}. I. The relation between vitamin B_{12} and vitamin B_{12b}, *Biochim. Biophys. Acta 6*:225 (1950).

19. B. Ellis, V. Petrow, and G. R. Snook, Chemistry of anti-pernicious anaemia factors; liberation of phosphorus as phosphate by acid hydrolysis, *J. Pharm. Pharmacol. 1*:287 (1949).

20. E. A. Kaczka, D. E. Wolf, and K. Folkers, Vitamin B_{12}, V. Identification of crystalline vitamin B_{12}, *J. Am. Chem. Soc. 71*:1514 (1949).

21. N. G. Brink, F. A. Kuehl, Jr., and K. Folkers, Vitamin B_{12}: identification of vitamin B_{12} as cyano-cobalt coordination complex, *Science 112*:354 (1950).

22. E. A. Kaczka, D. E. Wolf, F. A. Kuehl, Jr., and K. Folkers, Vitamin B_{12}. XVI. Modifications of cyano-cobalamin, *J. Am. Chem. Soc. 73*:3569 (1951).

23. H. Lichtman, J. Watson, V. Ginsberg, J. V. Pierce, E. L. R. Stokstad, and T. H. Jukes, Vitamin B_{12}; some properties and its therapeutic use, *Proc. Soc. Exp. Biol. Med. 72*:643 (1949).

24. H. A. Barker, Biochemical functions of corrinoid compounds, *Biochem. J. 105*:1 (1967).

25. B. Zagalak and W. Friedrich, eds., *Vitamin B_{12}. Proceedings of the Third European Symposium on Vitamin B_{12} and Intrinsic Factor. University of Zurich, March 5–8, 1979, Zurich, Switzerland,* Walter de Gruyter, Berlin, 1979.

26. D. Dolphin, ed., B_{12} Vol. 1: Chemistry; Vol. 2: Biochemistry and Medicine, Wiley-Interscience, New York, 1982.

27. B. M. Babior, ed., *Cobalamin: Biochemistry and Pathophysiology.* Wiley-Interscience, New York, 1975.

28. T. C. Stadtman, Vitamin B_{12}, *Science 171*:859 (1971).

29. A. R. Battersby, How nature builds the pigments of life: the conquest of vitamin B_{12}, *Science 264*:1551 (1994).

30. J. W. Corcoran and D. Shemin, Biosynthesis of the porphyrin-like moiety of vitamin B_{12}; mode of utilization of V-aminolevulinic acid, *Biochim. Biophys. Acta 25*:661 (1957).

31. H. C. Friedmann, Biosynthesis of corrinoids, in *Cobalamin: Biochemistry and Pathophysiology* (B. M. Babior, ed.), Wiley-Interscience, New York, 1975, p. 75.

32. D. S. Weiss and R. K. Thauer, Methanogenesis and the unity of biochemistry, *CelL 72*:819 (1993).

33. E. L. Smith, *Vitamin B_{12}.* 3d ed., Wiley, New York, 1965.

34. IUPAC-IUB, Tentative rules, *J. Biol. Chem. 241*:2991 (1966).

35. IUPAC-IUB, The nomenclature of corrinoids, *Biochemistry 13*:1555 (1974).

36. P. Gimsing and E. Nexø, The forms of cobalamin in biological materials, in *The Cobalamins* (C. A. Hall, ed.), Churchill, Livingstone, Edinburgh, 1983, p. 7.

37. P. G. Lenhert and D. C. Hodgkin, Structure of the 5,6-dimethylbenzimidazyl cobamide coenzyme, *Nature (Lond.) 192*:937 (1961).

38. J. R. Guest, S. Friedman, D. D. Woods, and E. L. Smith, A methyl analogue of cobamide coenzyme in relation to methionine synthesis by bacteria, *Nature (Lond.) 195*:340 (1962).

39. K. G. Stahlberg, Studies on methyl-B_{12} in man. *Scand. J. Haematol. (Suppl. 1)*:1 (1967).

40. J. C. Linnell, The fate of cobalamins in vivo, in *Cobalamin: Biochemistry and Pathophysiology* (B. M. Babior, ed.), Wiley-Interscience, New York, 1975, p. 287.

41. A Peterkofsky and H. Weissbach, Release of inorganic tripolyphosphate from adenosine triphosphate during vitamin B_{12} coenzyme biosynthesis, *J. Biol. Chem. 238*:1491 (1963).

42. E. Vitols, G. A. Walker, and F. M. Huennekens, Enzymatic synthesis of a vitamin B_{12} coenzyme, *Biochem. Biophys. Res. Commun. 15*:372 (1964).

43. H. Ohta and W. S. Beck, Studies of the ribosome-associated vitamin B_{12} adenosylating enzyme of *Lactobacillus leichmannii, Arch. Biochem. Biophys. 174*:713 (1976).

44. J. Halpern, Chemistry and significance of vitamin B_{12} model systems, in *B_{12}*. Vol. 1: Chemistry (D. Dolphin, ed.), Wiley-Interscience, New York, 1982, p. 501.

45. H. P. C. Hogenkamp, The chemistry of cobalamins and related compounds, in *Cobalamin: Biochemistry and Pathophysiology* (B. M. Babior, ed.), Wiley-Interscience, New York, 1975, p. 21.

46. H. R. Skeggs, J. W. Huff, L. D. Wright, and D. K. Bosshardt, The use of *Lactobacillus leichmannii* in the microbiological assay of the "animal protein factor," *J. Biol. Chem. 176*: 1450 (1948).

47. H. T. Thompson, L. S. Dietrich, and C. A. Elvehjem, The use of *Lactobacillus leichmanuii* in the estimation of vitamin B_{12} activity, *J. Biol. Chem. 184*:175 (1950).

48. H. L. Rosenthal and H. P. Sarrett, The determination of vitamin B_{12} activity in serum, *J. Biol. Chem. 199*:433–442 (1952).

49. G. H. Spray, An improved method for the rapid estimation of vitamin B_{12} in serum, *Clin. Sci. 14*:661 (1955).

50. G. I. M. Ross, Vitamin B_{12} assay in body fluids, *Nature (Lond.), 166*:270 (1950).

51. D. L. Mollin and G. I. M. Ross, The vitamin B_{12} concentrations of serum and urine of normals and of patients with megaloblastic anaemias and other diseases, *J. Clin. Pathol. 5*:129 (1952).

52. A. Killander, The serum vitamin B_{12} levels at various ages, *Acta Paediat. (Uppsala) 46*:585 (1957).

53. J. E. Ford, The microbiological assay of "vitamin B_{12}". The specificity of the requirement of *Ochromonas malhamensis* for cyanocobalamin, *Br. J. Nutr. 7*:299 (1953).

54. B. D. Davis and E. S. Mingioli, Mutants of *Escherichia coli* requiring methionine or vitamin B_{12}, *J. Bacteriol. 60*:17 (1950).

55. P. R. Burkholder, Determination of vitamin B_{12} with a mutant strain of *Escberichia coli, Science 114*:459 (1951).

56. N. Grossowicz, J. Aronovitch, and M. Rachmilewitz, Determination of vitamin B_{12} in human serum by a mutant of *Escherichia coli, Proc. Soc. Exp. Biol. Med. 87*:513 (1954).

57. E. Nexø and H. Olesen, Quantitation of cobalamins in human serum, in *B_{12} Biochemistry and Medicine* (D. Dolphin, ed.), Wiley-Interscience, New York, 1982, p. 88.

58. J. L. Raven, M. B. Robson, J. O. Morgan, and A. V. Hoffbrand, Comparison of three methods for measuring vitamin B_{12} in serum: radioisotopic, *Euglena gracilis* and *Lactobacillus leichmannli, Br. J. Haematol. 22*:21 (1972).

59. N. A. Sourial, Use of an improved *E. coli* method for the measurement of cobalamin in serum: comparison with the *E. gracilis* assay results, *J. Clin. Pathol. 34*:351 (1981).

60. H. Baker, O. Frank, I. Pasher, and H. Sobotka, Vitamin B_{12} in human blood and serum. I. Comparison of microbiological assays using normal subjects, *Clin. Chem. 6*:578 (1960).

61. W. S. Beck, The assay of serum cobalamin by *Lactobacillus leichmannii* and the interpretation of serum cobalamin levels, in *The Cobalamins* (C. A. Hall, ed.), Churchill Livingstone, Edinburgh, 1983, p. 31.

62. R. M. Barakat and R. P. Ekins, Assay of vitamin B_{12} in blood. A simple method, *Lancet 2*: 25 (1961).

63. M. Muir and I. Chanarin, The assay of serum cobalamin by solid phase saturation analysis, in *The Cobalamins* (C. A. Hall, ed.), Churchill Livingstone, Edinburgh, 1983, p. 65.

64. J. F. Kolhouse, H. Kondo, N. C. Allen, E. Podell, and R. H. Allen, Cobalamin analogues are present in human plasma and can mask cobalamin deficiency because current radioisotope dilution assays are not specific for true cobalamin, *N. Engl. J. Med. 299*:785 (1978).

65. B. A. Cooper and M. V. Whitehead, Evidence that some patients with pernicious anemia are not recognized by radiodilution assay for cobalamin in serum, *N. Engl. J. Med. 299*:816 (1978).

66. D. Perlman, Microbial synthesis of cobamides, in *Advances in Applied Microbiology,* Vol. 1 (W. W. Umbreit, ed.), Academic Press, New York, 1959, p. 87.

67. R. M. Heyssel, R. C. Bozian, W. J. Darby, and M. C. Bell, Vitamin B_{12} turnover in man: the assimilation of vitamin B_{12} from natural foodstuff by man and estimates of minimal daily dietary requirements, *Am. J. Clin. Nutr. 18*:176 (1966).

68. C. A. Hall, Long-term excretion of ^{57}Co-B_{12} and turnover within the plasma, *Am. J. Clin. Nutr. 14*:156 (1964).

69. J. F. Adams, Consideration governing the maintenance treatment of patients with pernicious anemia, in *Vitamin B_{12} und Intrinsic Faktor, 2nd Europäisches Symposion* (H. C. F. Heinrich, ed.), Enke, Stuttgart, 1962, p. 628.

70. R. Gräsbeck, Calculations on vitamin B_{12} turnover in man, *Scand. J. Clin. Lab. Invest. 11*: 250 (1959).

71. J. M. Hsu, B. Kawin, P. Minor, and J. A. Mitchell, Vitamin B_{12} concentration in human tissues, *Nature (Lond.) 210*:1264 (1966).

72. Food and Nutrition Board, National Research Council, *Recommended Dietary Allowances*, 10th ed., National Academy of Sciences, Washington, DC, 1989, p. 162.

73. W. B. Castle, Development of knowledge concerning the gastric intrinsic factor and its relation to pernicious anemia, *N. Engl. J. Med., 249*:614 (1953).

74. R. Gräsbeck, Intrinsic factor and the other vitamin B_{12} transport proteins, *Prog. Hematol. 6*: 223 (1969).

75. L. Ellenbogen, Absorption and transport of cobalamin: intrinsic factor and the transcobalamins, in *Cobalamin: Biochemistry and Pathophysiology* (B. M. Babior, ed.), Wiley-Interscience, New York, 1975, p. 215.

76. R. H. Allen and C. S. Mehlman, Isolation of gastric vitamin B_{12}-binding proteins using affinity chromatography. 1. Purification and properties of human intrinsic factor, *J. Biol. Chem. 248*: 3660 (1973).

77. J. M. Christensen, E. Hippe, H. Olesen, M. Rye, E. Haber, L. Lee, and J. Thomsen, Purification of human intrinsic factor by affinity chromatography, *Biochim. Biophys. Acta 303*:319 (1973).

78. K. Visuri and R. Gräsbeck, Human intrinsic factor: isolation by improved conventional methods and properties of the preparation, *Biochim. Biophys. Acta 310*:508 (1973).

79. G. Marcoullis and R. Gräsbeck, Vitamin B_{12}-binding-proteins in human gastric mucosa: general pattern and demonstration of intrinsic isoproteins typical of mucosa, *Scand. J. Clin. Lab. Invest. 35*:5 (1975).

80. U.-H. Stenman, Vitamin B_{12}-binding proteins of R-type cobalophilin: characterization and comparison of cobalophalin from different sources, *Scand. J. Haematol. 14*:91 (1975).

81. V. Herbert and W. B. Castle, Divalent cation and pH dependence of rat intrinsic factor action in everted sacs and mucosal homogenates of rat small intestine, *J. Clin. Invest. 40*:1978 (1961).

82. V. I. Mathan, B. M. Babior, and R. M. Donaldson, Kinetics of the attachment of intrinsic factor-bound cobamides to ileal receptors, *J. Clin. Invest. 54*:498 (1974).

83. M. Katz and B. A. Cooper, Solubilized receptor for vitamin B_{12}-intrinsic factor complex from human intestine, *Br. J. Haematol. 26*:569 (1974).

84. P. A. McIntyre, M. V. Sachs, J. R. Krevans, and C. L. Conley, Pathogenesis and treatment of macrocytic anemia, *Arch. Intern. Med. 98*:541 (1956).

85. G. Perman, R. Gullberg, P. G. Reizensten, P. G. Snellman, and L.-G. Allgen, A study of absorption patterns in malabsorption syndromes, *Acta Med. Scand. 168*:117 (1960).

86. P. P. Toskes, J. Hansell, J. Cerda, and J. J. Deren, Vitamin B_{12} malabsorption in chronic pancreatic insufficiency, *N. Engl. J. Med. 264*:627 (1973).

87. J. D. Hines, A. Rosenberg, and J. W. Harris, Intrinsic factor-mediated radio-B_{12} uptake in sequential incubatior studies using everred sacs of guinea pig small intestine: evidence that IF is not absorbed into the intestinal cell, *Proc. Soc. Exp. Biol. Med. 129*:653 (1968).

88. S. P. Rothenberg, H. Weisberg, and A. Ficarra, Evidence for the absorption of immunoreactive intrinsic factor into the intestinal epithelial cell during vitamin B_{12} absorption, *J. Lab. Clin. Med. 79*:587 (1972).

89. T. J. Peters and A. V. Hoffbrand, Absorption of vitamin B_{12} by the guinea pig. Subcellular localization of vitamin B_{12} in the ileal enterocyte during absorption, *Br. J. Haematol. 19*:369 (1970).

90. C. A. Hall and A. E. Finkler, A second vitamin B_{12}-binding substance in human plasma, *Biochim. Biophys. Acta 78*:233 (1963).

91. B. L. Hom and B. K. Ahluwalia, The vitamin B_{12}-binding capacity of transcobalamin I and in normal human serum, *Scand J. Haematol. 5*:64 (1965).

92. R. H. Allen and P. W. Majerus, Isolation of vitamin B_{12}-binding proteins using affinity chromatography. III. Purification and properties of human plasma transcobalamin II, *J. Biol. Chem. 247*:7709 (1972).

93. C. A. Hall and A. E. Finkler, Function of transcobalamin II: a vitamin B_{12}-binding protein in human plasma, *Proc. Soc. Exp. Biol. Med. 123*:55 (1966).

94. E. Nexø and H. Olesen, Intrinsic factor, transcobalamin, and haptocorrin, in B_{12} *Biochemistry and Medicine* (D. Dolphin, ed.), Wiley-Interscience, New York, ch 3, p 57.

95. C. A. Hall and A. E. Finkler, Abnormal transport of vitamin B_{12} in plasma in chronic myelogenous leukemia, *Nature 204*:1207 (1964).

96. M. M. England, H. G. M. Clarke, M. C. Down, and I. Chanarin, Studies on the transcobalamins, *Br. J. Haematol. 25*:737 (1973).

97. R. I. Burger, C. S. Mehlman, and R. H. Allen, Human plasma R-type vitamin B_{12}-binding proteins. I. Isolation and characterization of transcobalamin I, transcobalamin III, and the normal granulocyte vitamin B_{12}-binding protein, *J. Biol. Chem. 250*:7700 (1975).

98. H. Olesen, Serum transcobalamins, *Scand J. Gastroenterol. 9 (Suppl. 29)*:13 (1974).

99. R. H. Allen, Human vitamin B_{12} transport proteins, *Prog. Hematol. 9*:57 (1975).

100. E. Jacob, S. J. Baker, and V. Herbert, Vitamin B_{12}-binding proteins. *Physiol. Rev. 60*:918 (1980).

101. J. A. Begley, The materials and processes of plasma transport in *The Cobalamins* (C. A. Hall, ed.), Churchill Livingstone, Edinburgh, 1983, p. 109.

102. U.-H Stenman, K. Simons, and R. Gräsbeck, Vitamin B_{12} binding proteins in normal and leukemic leukocytes and sera, *Scand. J. Clin. Lab. Invest. 21 (Suppl. 101)*:13 (1974).

103. R. I. Burger and R. H. Allen, Characterization of vitamin B_{12}-binding proteins isolated from human milk and saliva by affinity chromatography, *J. Biol. Chem. 249*:7220 (1974).

104. J. M. Scott, F. J. Bloomfield, R. Stebbins, and V. Herbert, Studies on derivation of transcobalamin III from granulocytes: enhancement by lithium and elimination by fluoride of in vitro increments in vitamin B_{12}-binding capacity, *J. Clin. Invest. 53*:228 (1974).

105. S. Waxman and H. W. Gilbert, A tumor-related vitamin B_{12} binding protein in adolescent hepatoma, *N. Engl. J. Med. 289*:1053 (1973).

106. E. Nexø, H. Olesen, K. Nørredam, and M. Schwartz, A rare case of megaloblastic anemia caused by disturbances in the plasma cobalamin binding proteins in a patient with hepatocellular carcinoma, *Scand. J. Haematol. 14*:320 (1975).

107. B. Rachmilewitz, M. Rachmilewitz and J. Gross, A vitamin B_{12} binder with transcobalamin I characteristics synthesized and released by human granulocytes in vitro, *Br. J. Haematol. 26*:557 (1974).

108. H. S. Gilbert, Demonstration of the "PV" B_{12}-binding protein in leukocytes of normals and patients with myeloproliferative disorders (MPD), *Blood 38*:605 (1971).

109. R. Carmel, Vitamin B_{12}-binding protein abnormality in subject without myeloproliferative disease. II. The presence of a third vitamin B_{12}-binding protect in serum, *Br. J. Haematol.* *22*:53 (1972).

110. C. E. Tan and H. J. Hansen, Studies on the site of synthesis of transcobalamin-II, *Proc. Soc. Exp. Biol. Med. 127*:740 (1968).

111. D. W. Sonneborn, G. Abouna, and G. Mendez-Picon, Transcobalamin II synthesis in totally hepatectomized dogs, *Fed. Proc. 20*:240 (1971).

112. S. P. Rothenberg, J. P. Weiss, and R. Cotter, Formation of transcobalamin II—vitamin B_{12} complex by guinea-pig ileal mucosa in organ culture after in vivo incubation with intrinsic factor—vitamin B_{12}, *Br. J. Haematol. 40*:401 (1978).

113. I. Chanarin, M. Muir, A. Hughes, and A. V. Hoffbrand, Evidence for intestinal origin of transcobalamin during vitamin B_{12} absorption, *Br. Med. J. 1*:1453 (1978).

114. R. Carmel, S. M. Neely, and R. B. Francis, Jr. Human umbilical vein endothelial cells secrete transcobalamin II, *Blood 75*:251 (1990).

115. E. Hippe, E. Haber, and H. Olesen, Nature of vitamin B_{12} binding. II. Steric orientation of vitamin B_{12} on binding and number of combining sites of human intrinsic factor and the transcobalamins, *Biochim. Biophys. Acta 243*:75 (1971).

116. C. W. Gottlieb, F. P. Retief, and V. Herbert, Blockade of vitamin B_{12}-binding sites in gastric juice, serum and saliva by analogues and derivatives of vitamin B_{12} and by antibody to intrinsic factor, *Biochim. Biophys. Acta 141*:560 (1967).

117. E. Stupperich and E. Nexø, Effect of the cobalt-N coordination on the cobamide recognition by the human vitamin B_{12} binding proteins intrinsic factor, transcobalamin and haptocorrin, *Eur. J. Biochem. 199*:299 (1991).

118. A. Regec, E. V. Quadros, O. Platica, and S. P. Rothenberg, The cloning and characterization of the human transcobalamin II gene. *Blood 85*:2711 (1995).

119. R. E. Benson, M. E. Rappazzo, and C. A. Hall, Late transport of vitamin B_{12} by transcobalamin II, *J. Lab. Clin. Med. 90*:455 (1972).

120. C. A. Hall, Transcobalamins I and II as natural transport proteins of vitamin B_{12}, *J. Clin. Invest. 56*:1125 (1973).

121. S. M. Frisbie, and M. R. Chance, Human cobalophilin: the structure of bound methylcobalamin and a functional role in protecting methylcobalamin from photolysis. *Biochemistry 32*: 13886 (1993).

122. B. L. Hom and H. A. Olesen, Plasma clearance of [57]cobalt-labeled vitamin B_{12} bound in vitro and in vivo transcobalamin, *Scand. J. Clin. Lab. Invest. 23*:201 (1969).

123. R. J. Schneider, R. L. Burger, C. S. Mehlman, and R. H. Allen, The role and fate of rabbit and human transcobalamin II in the plasma transport of vitamin B_{12} in the rabbit, *J. Clin. Invest. 56*:27 (1976).

124. E. J. Gizis, S. N. Arkun, J. F. Miller, G. Choi, M. F. Dietrich, and L. M. Meyer, Plasma clearance of transcobalamin I- and transcobalamin II-bound Co[57] and vitamin B_{12}, *J. Lab. Clin. Med. 74*:574 (1969).

125. E. Nexø and P. Gimsing, Turnover in humans of iodine- and cobalamin-labeled transcobalamin I and of iodine-labeled albumen, *Scand. J. Clin. Lab. Invest. 35*:391 (1975).

126. S. Yamada, L. Riitinen, R. Majuri, M. Fukuda, and R. Gräsbeck, Studies on the transcobalamin receptor in hog kidney, *Kidney Int. 39*:289 (1991).

127. C. M. Becker and W. S. Beck, Calcium dependencies in the binding of transcobalamins to subcellular particles of liver cells, in *Vitamin B_{12} Proceedings of the Third European Symposion on Vitamin B_{12} and Intrinsic Factor, University of Zurich, March 5–8, 1979, Zurich, Switzerland* (B. Zagalak and W. Friedrich, eds.), Walter de Gruyter, New York, 1979, p. 833.

128. R. Carmel and V. Herbert, Deficiciency of vitamin B_{12}-binding alpha globulin in two brothers, *Blood 33*:1 (1969).

129. N. Hakami, P. E. Neiman, G. P. Canellos, and J. Lazerson, Neonatal megaloblastic anemia

due to inherited transcobalamin II deficiency in two siblings, *N. Engl. J. Med. 285*:1163 (1971).

130. W. H. Hitzig, U. Dohmann, H. J. Pluss, and D. Vischer, Hereditary transcobalamin II deficiency: clinical findings in a new family, *J. Pediatr. 85*:622 (1074).

131. A. E. Finkler and C. A. Hall, Nature of the relationship between vitamin B_{12} binding and cell uptake, *Arch. Biochem. 120*:79 (1967).

132. F. P. Retief, C. W. Gottlieb, and V. Herbert, Delivery of Co $^{57}B_{12}$ to erythrocytes from α and β globulin of normal, B_{12}-deficient, and chronic myeloid leukemia serum, *Blood 29*:837 (1967).

133. C. A. Hall and A. E. Finkler, Isolation and evaluation of the various B_{12}-binding proteins in human plasma, *Meth. Enzymol. 18C*:108 (1971).

134. J. Zittoun, R. Zittoun, J. Marquet, and C. Sultan, The three transcobalamins in myeloproliferative disorders and acute leukemia, *Br. J. Haematol. 31*:187 (1975).

135. C. A. Hall and A. E. Finkler, Vitamin B_{12}-binding protein in polycythemia vera plasma, *J. Lab. Clin. Med. 73*:60 (1969).

136. R. L. Burger, S. Waxman, H. S. Gilbert, C. S. Mehlman, and R. H. Allen, Isolation and characterization of a novel vitamin B_{12}-binding protein associated with hepatocellular carcinoma, *J. Clin. Invest. 56*:1262 (1975).

137. R. Carmel, Extreme elevation of serum transcobalamin I in patients with metastatic cancer, *N. Engl. J. Med. 292*:282 (1975).

138. V. Herbert, W. Fong, V. Gulle and T. Stopler, Low holotranscobalamin II is the earliest serum marker for subnormal vitamin B_{12} (cobalamin) absorption in patients with AIDS, *Am. J. Hematol., 34*:132 (1990).

139. F. S. Jorgensen, Vitamin B_{12} and its binding proteins in cirrhosis and infectious hepatitis, *Scand. J. Haematol. 7*:322 (1970).

140. H. S. Gilbert and N. Weinreb, Increased circulating levels of transcobalamin II in Gaucher's disease, *N. Engl. J. Med. 295*:1096 (1976).

141. R. Carmel and D. Hollander, Extreme elevations of transcobalamin II levels in multiple myeloma and other disorders, *Blood 51*:1057 (1978).

142. R. Carmel, C. A. Coltman, Jr., and L. H. Brubaker, Serum vitamin B_{12}-binding proteinis in neutropenia, *Proc. Soc. Exp. Biol. Med. 148*:1217 (1975).

143. R. L. Burger, R. J. Schneider, C. S. Mehlman, and R. H. Allen, Human plasma R-type vitamin B_{12}-binding protein in the plasma transport of vitamin B_{12}, *J. Biol. Chem. 250*:7707 (1975).

144. R. Gräsbeck, W. Nyberg, and P. Reizenstein, Biliary and fecal vitamin B_{12} excretion in man: an isotope study, *Proc. Soc. Exp. Biol. Med. 97*:780 (1958).

145. P. G. Reizenstein, Excretion of non-labeled vitamin B_{12} in man, *Acta Med. Scand. 165*:313 (1959).

146. M. A. Booth and G. H. Spray, Vitamin B_{12} activity in the serum and liver of rats after total gastrectomy, *Br. J. Haematol. 6*:288 (1960).

147. K. Simons, Vitamin B_{12} binders in human body fluids and blood cells, *Soc. Sci. Fenn. Commun. Biol. 27 (Suppl. 1)*:1 (1964).

148. K. Simons and T. Weber, The vitamin B_{12}-binding protein in human leukocytes, *Biochim. Biophys. Acta 117*:201 (1966).

149. W. S. Beck, The metabolic functions of vitamin B_{12}, *N. Engl. J. Med. 266*:708, 765, 814 (1962).

150. W. S. Beck, Deoxyribonucleotide synthesis and the role of vitamin B_{12} in erythropoiesis, *Vitam. Horm. 26*:413 (1968).

151. H. P. C. Hogenkamp, Enzymatic reactions involving corrinoids, *Annu. Rev. Biochem. 37*:225 (1968).

152. F. M. Huennekens, Biochemical functions and interrelationships of folic acid and vitamin B_{12}, *Prog. Hematol. 5*:83 (1966).

153. H. Weissbach and R. T. Taylor, Metabolic role of vitamin B_{12}, *Vitam. Horm. 26*:395 (1968).

154. H. A. Barker, Corrinoid-dependent enzymatic reactions, *Annu. Rev. Biochem. 41*:55 (1972).

155. J. M. Poston, and T. C. Stadtman, Cobamides as cofactors: methylcobamides and the synthesis of methionine, methane, and acetate, in *Cobalamin: Biochemistry and Pathophysiology* (B. M. Babior, ed.) Wiley-Interscience, New York, 1975, p. 111.

156. B. M. Babior, Cobamides as cofactors: adenosylcobamide-dependent reactions, in *Cobalamin: Biochemistry and Pathophysiology* (B. M. Babior, ed.), Wiley-Interscience, New York, 1975, p. 141.

157. W. S. Beck, Metabolic aspects of vitamin B_{12} and folic acid, in *Hematology*, 3rd ed. (W. J. Williams, E. Beutler, A. J. Erslev, and M. A. Lichtman, eds.), McGraw-Hill, New York, 1983, p. 311.

158. W. S. Beck, Cobalamin as coenzyme: a twisting trail of research, *Am. J. Hematol. 34*:83 (1990).

159. M. Flavin, P. J. Ortiz, and S. Ochoa, Metabolism of propionic acid in animal tissues, *Nature (Lond.) 176*:823 (1955).

160. M. Flavin and S. Ochoa, Metabolism of propionic acid in animal tissues. I. Enzymatic conversion of propionate to succinate, *J. Biol. Chem. 229*:965 (1957).

161. W. S. Beck, M. Flavin, and S. Ochoa, Metabolism of propionic acid in animal tissues, III. Formation of succinate, *J. Biol. Chem. 229*:997 (1957).

162. W. S. Beck and S. Ochoa, Metabolism of propionic acid in animal tissues, IV. Further studies on the enzymatic isomerization of methylmalonyl coenzyme A, *J. Biol. Chem. 232*:931 (1958).

163. H. A. Lardy and J. Adler, Synthesis of succinate from propionate and bicarbonate by soluble enzymes from liver mitochondria, *J. Biol. Chem. 219*:933 (1956).

164. H. A. Barker, R. D. Smyth, H. Weissbach, J. I. Toohey, J. N. Ladd, and B. E. Volcani, Isolation and properties of crystalline cobamide coenzymes containing benzimidazole or 5, 6-dimethylbenzimidazole, *J. Biol. Chem. 235*:480 (1960).

165. R. M. Smith and K. J. Monty, Vitamin B_{12} and propionate metabolism, *Biochem. Biophys. Res. Commun. 1*:105 (1959).

166. S. Gurnani, S. P. Mistry, and B. C. Johnson, Function of vitamin B_{12} in methylmalonate metabolism. I. Effect of a cofactor form of B_{12} on the activity of methylmalonyl-CoA isomerase, *Biochim. Biophys. Acta 38*:187 (1959).

167. E. R. Stadtman, P. Overath, H. Eggerer and F. Lynen, The role of biotin and vitamin B_{12} coenzyme in propionate metabolism, *Biochem. Biophys. Res. Commun. 2*:1 (1960).

168. R. W. Kellermeyer, S. H. G. Allen, R. Stjernholm, and H. G. Wood, Methylmalonyl isomerase. IV. Purification and properties of the enzyme from propionibacteria, *J. Biol. Chem. 239*:2562 (1964).

169. P. Lengyel, R. Mazumder, and S. Ochoa, Mammalian methylmalonyl isomerase and vitamin B_{12} coenzymes, *Proc. Natl. Acad. Sci. USA 46*:1312 (1960).

170. R. Mazumder, T. Sasakawa, Y. Kaziro, and S. Ochoa, A new enzyme in the conversion of propionyl coenzyme A to succinyl coenzyme A, *J. Biol. Chem. 236*:PC53 (1961).

171. E. V. Cox and A. M. White, Methylmalonic acid excretion: an index of vitamin B_{12} deficiency, *Lancet 2*:853 (1962).

172. S. Kashket, J. T. Kaufman, and W. S. Beck, The metabolic functions of vitamin B_{12}. III, Vitamin B_{12} binding in *Lactobacillus leichmannii* and other lactobacilli, *Biochim. Biophys. Acta 64*:447 (1962).

173. S. Kashket, J. T. Kaufman, and W. S. Beck, The metabolic functions of vitamin B_{12}. IV. Binding of vitamin B_{12} by ribosomes in *Lactobacillus leichmannii*, *Biochim. Biophys. Acta 64*:458 (1962).

174. A. Peterkofsky and H. Weissbach, Release of inorganic tripolyphosphate from adenosine triphosphate during vitamin B_{12}-coenzyme biosynthesis, *J. Biol. Chem. 238*:1491 (1963).

175. J. Retey and D. Arigoni, Coenzyme B_{12} als gemeinsamer Wasserstoffubertrager der Dioldehydrase and der Methylmalonyl-CoA-Mutase-Reaktion, *Experimentia 22*:783 (1966).

176. R. H. Abeles and H. A. Lee, Tr., An intramolecular oxidation–reduction requiring a vitamin B_{12} coenzyme, *J. Biol. Chem. 236*:2347 (1961).

177. R. L. Blakley and H. A. Barker, Cobamide stimulation of the reduction of ribotides to deoxyribotides in *Lactobacillus leichmannii, Biochem. Biophys. Commun. 16*:391 (1964).

178. W. S. Beck and J. Hardy, Requirement of ribonucleotide reductase for cobamide coenzyme: a product of ribosomal activity, *Proc. Natl. Acad. Sci. USA 54*:286 (1965).

179. R. Abrams and S. Duraiswami, Deoxycytidylate formation from cytidylate without glycosidic cleavage in *Lactobacillus leichmannii* extracts containing vitamin B_{12} coenzyme, *Biochem. Biophys. Res. Commun. 18*:409 (1965).

180. M. Goulian and W. S. Beck, Purification and properties of cobamide-dependent ribonucleotide reductase from *Lactobacillis leichmannii J. Biol. Chem. 241*:4233 (1966).

181. M. M. Gottesman and W. S. Beck, Transfer of hydrogen in the cobamide-dependent ribonucleotide reductase reaction, *Biochem. Biophys. Res. Commun. 24*:353 (1966).

182. R. H. Abeles and W. S. Beck, The mechanism of action of cobamide coenzyme in the ribonucleotide reductase reaction, *J. Biol. Chem. 242*:3589 (1967).

183. J. A. Stubbe, Mechanism of B_{12}-dependent ribonucleotide reductase, *Mol. Cell. Biochem. 50*: 25 (1983).

184. G. W. Ashley, G. Harris, and J. Stubbe, The mechanism of *Lactobacillus leichmannii* ribonucleotide reductase. Evidence for the 3' C—H bond cleavage and a unique role for coenzyme B_{12}, *J. Biol. Chem. 261*:3958 (1986).

185. W. S. Beck, S. Hook, and B. H. Barnett, The metabolic functions of vitamin B_{12}. I. Distinctive modes of unbalanced growth behavior in *Lactobacillus leichmannii Biochim. Biophys. Acta 55*:455 (1962).

186. W. S. Beck, The metabolic basis of megaloblastic erythropoiesis, *Medicine (Baltimore) 43*: 715 (1964).

187. P. Reichard, From RNA to DNA, why so many reductases? *Science 260*:1773 (1993).

188. L. Thelander and P. Reichard, Reduction of ribonucleotides, *Annu. Rev. Biochem. 48*:133 (1979).

189. E. Mulliez, M. Fontecave, J. Gaillard, and P. Reichard, An iron-sulfur center and a free radical in the active anaerobic ribonucleotide reductase of *Escherichia coli, J. Biol. Chem. 268*:2296 (1993).

190. X. Sun, J. Harder, M. Krook, H. Jörnvall, B.-M. Sjöberg, and P. Reichard, A possible glycine radical in anaerobic ribonucleotide reductase from *Escherichia coli*: nucleotide sequence of the cloned *nrdD* gene, *Proc. Natl. Acad. Sci. USA 90*:577 (1993).

191. R. Eliasson, E. Pontis, X. Sun, and P. Reichard, Allosteric control of the substrate specificity of the anaerobic ribonucleotide reductase from *Escherichia coli, J. Biol. Chem. 269*:26052 (1994).

192. J. Harder, R. Eliasson, E. Pontis, M. D. Ballinger, and P. Reichard, Activation of the anaerobic ribonucleotide reductase from Escherichia coli by *S*-adenosylmethionine, *J. Biol. Chem. 267*:25548 (1992).

193. J. Halpern, S.-H. Kim, and T. W. Leung, Cobalt-carbon bond dissociation energy of coenzyme B_{12}, *J. Am. Chem. Soc. 106*:8317 (1984).

194. P. A. Frey, Lysine 2,3-aminomutase: is adenosylmethionine a poor man's adenosylcobalamin? *FASEB J. 7*:662 (1993).

195. R. T. Taylor, B_{12}-dependent methionine biosynthesis, in B_{12}. *Biochemistry and Medicine* (D. Dolphin, ed.), New York, John Wiley and Sons, 1982, p. 307.

196. C. L. Drennan, S. Huang, J. T. Drummond, R. G. Matthews, and M. L. Ludwig, How a protein binds B_{12}: a 3.0 Å x-ray structure of B_{12}-binding domains of methionine synthase, *Science 266*:1669 (1994).

197. R. V. Banarjee, N. L. Johnston, J. K. Sobeske, P. Datta, and R. G. Mathews, Cloning and sequence analysis of the *Escherichia coli metH* gene encoding cobalamin-dependent methionine synthase and isolation of a tryptic fragment containing the cobalamin-binding domain, *J. Biol. Chem. 264*:13888 (1989).

198. J. Stubbe, Binding site revealed of nature's most beautiful cofactor, *Science 266*:1663 (1994).

199. C. L. Drennan, R. G. Matthews, and M. L. Ludwig, Cobalamin-dependent methionine syn-

thase: the structure of a methylcobalamin-binding fragment and implications for other B_{12}-dependent enzymes, *Curr. Opin. Struct. Biol. 4*:919 (1994).

200. R. T. Taylor and H. Weissbach, N^5-Methyltetrahydrofolate-homocysteine transmethylase, *J. Biol. Chem. 242*:1502 (1967).

201. R. E. Loughlin, H. I. Elford, and J. M. Buchanan, Enzymatic synthesis of the methyl group of methionine. VIII, Isolation of a cobalamin-containing transmethylase (5-methyltetrahydrofolate homocysteine) from mammalian liver, *J. Biol. Chem. 239*:2888 (1964).

202. H. Dickerman, B. G. Redfield, J. Bieri, and H. Weissbach, The role of vitamin B_{12} in methionine synthesis in avian liver, *J. Biol. Chem. 239*:2545 (1964).

203. H.-J. Sauer, K. Wilms, W. Wilmanns, and L. Jaenicke, Die Aktivitat der, methionin-synthetase (5-methyl-5,6,7,8-tetrahydrofolsäure: homocystein methyltransferase) als proliferationsparameter in wachsenden Zellen, *Acta Haematol. (Basel) 49*:200 (1973).

204. R. Peytremann, J. Thorndike, and W. S. Beck, Studies on N^5-methyltetrahydrofolate-homocysteine methyltransferase in normal and leukemic leukocytes, *J. Clin. Invest. 56*:1293 (1975).

205. R. G. S. Banks, R. J. Henderson, and J. M. Pratt, Reactions of gases in solution. III. Some reactions of nitrous oxide with transition-metal complexes, *J. Chem. Soc. (A)*: 2886 (1968).

206. D. M. Matthews and J. Wilson, Cobalamins and cyanide metabolism in neurological diseases, in *The Cobalamins: A Glaxo Symposium* (H. R. V. Arnstein and R. J. Wrighton, eds.), Williams & Wilkins, Baltimore, 1971, p. 115.

207. W. S. Beck, Megaloblastic anemias, in *Cecil's Textbook of Medicine*, 18th ed. (J. B. Wyngaarden and L. H. Smith, Jr., eds.), W. B. Saunders, Philadelphia, 1985, p. 893.

208. W. S. Beck, The megaloblastic anemias, in *Hematology*, 3rd ed. (W. J. Williams, E. Beutler, A. J. Erslev, and M. A. Lichtman, eds.), McGraw-Hill, New York, 1983, p. 434.

209. E. Joosten, W. Pelemans, P. Devos, E. Lesaffre, W. Goossens, A. Criel,and R. Verhaeghe, Cobalamin absorption and serum homocysteine and methylmalonic acid in elderly subjects with low serum cobalamin, *Eur. J. Haematol. 51*:25 (1993).

210. E. Joosten, A. Van den Berg, R. Riezler, H. J. Naurath, J. Lindenbaum, S. P. Stabler, and R. H. Allen, Metabolic evidence that deficiencies of vitamin B-12 (cobalamin), folate, and vitamin B-6 occur commonly in elderly people, *Am. J. Clin. Nutr. 58*:468–476, 1993.

211. E. J. Norman and J. A. Morrison, Screening elderly populations for cobalamin (vitamin B_{12}) deficiency using the urinary methylmalonic acid assay by gas chromatography mass spectrometry, *Am. J. Med. 94*:589–594, 1993.

212. I. Chanarin, Cobalamins and nitrous oxide: a review, *J. Clin. Pathol. 33*:909 (1980).

213. T. S. Flippo, and W. D. Holder, Jr. Neurologic degeneration associated with nitrous oxide anesthesia in patients with vitamin B_{12} deficiency, *Arch. Surg. 128*:1391 (1993).

214. J. Metz, K. McGrath, M. Bennett, K. Hyland, and T. Bottiglieri, Biochemical indices of vitamin B_{12} nutrition in pregnant patients with subnormal serum vitamin B_{12} levels, *Am. J. Hematol. 48*:251 (1995).

215. Y. Kano, S. Sakamoto, Y. Miura, and F. Takaku, Disorders of cobalamin metabolism, *CRC Crit. Rev. Oncol. Hematol. 3*:1 (1985).

216. L. E. Rosenberg and W. A. Fenton, Disorders of propionate and methylmalonate metabolism, in *The Metabolic Basis of Inherited Disease*, 5th ed. (C. R. Scriver, A. L. Beaudet, W. S. Sly, and D. Valle, eds.), McGraw-Hill, New York, 1989, p. 821.

217. D. S. Rosenblatt and B. A. Cooper, Interited disorders of vitamin B_{12} metabolism, *Bioessays 12*:331 (1990).

218. W. S. Beck, Diagnosis of megaloblastic anemia, *Annu. Rev. Med. 42*:311 (1991).

219. W. G. Thompson, L. Babitz, C. Cassino, M. Freedman, and M. Lipkin, Jr., Evaluation of current criteria used to measure vitamin B_{12} levels, *Am. J. Med. 82*:291 (1987).

220. P. Gimsing and W. S. Beck, Cobalamin analogues in plasma: an in vitro phenomenon, *Scand. J. Clin. Lab. Invest. 194 (Suppl. 194)*:37 (1989).

221. R. Carmel, Pernicious anemia. The expected findings of very low serum cobalamin levels, anemia, and macrocytosis are often lacking, *Arch. Intern. Med. 148*:1712 (1988).

222. V. Herbert, Don't ignore low serum cobalamin (vitamin B$_{12}$) levels, *Arch. Intern. Med. 148*: 1705 (1988).
223. D. L. Mollin, B. B. Anderson, and J. F. Burman, The serum vitamin B$_{12}$ level: its assay and significance, *Clin. Haematol. 5*:521 (1976).
224. D. S. Young, Normal laboratory values in SI units, *N. Engl. J. Med. 292*:795–797 (1975).
225. J. Lindenbaum, Status of laboratory testing in the diagnosis of megaloblastic anemia, *Blood 61*:624 (1983).
226. J. F. Adams, H. I. Tankel, and F. MacEwen, Estimation of the total body vitamin B$_{12}$ in the live subject, *Clin. Sci. 39*:107 (1970).
227. J. F. Adams, The clinical and metabolic consequences of total gastrectomy. 2. Anaemia, metabolism of iron, vitamin B$_{12}$ and folic acid, *Scand. J. Gastroenterol. 2*:137 (1967).
228. J. F. Adams and K. Boddy, Metabolic equilibrium of tracer and natural vitamin B$_{12}$, *J. Lab. Clin. Med. 72*:392 (1968).
229. K. Boddy and J. F. Adams, The long-term relationship between serum vitamin B$_{12}$ and total body vitamin B$_{12}$, *Am. J. Clin. Nutr. 25*, 395 (1972).
230. P. K. Pruthi and A. Tefferi, Pernicious anemia revisited, *Mayo Clin. Proc. 69*:144 (1994).
231. J. Lindenbaum, E. B. Healton, D. G. Savage, J. C. M. Brust, T. J. Garrett, E. R. Podell, P. D. Marcell, S. P. Stabler, and R. H. Allen, Neuropsy-chiatric disorders caused by cobalamin deficiency in the absence of anemia or macrocytosis, *N. Engl. J. Med. 318*, 1720 (1988).
232. G. Tisman, T. Vu, J. Amin, et al., Measurement of red blood cell-vitamin B$_{12}$: a study of the correlation between intracellular B$_{12}$ content and concentrations of plasma holotranscobalamin II, *Am. J. Hematol. 43*:226 (1993).
233. S. B. Kahn, W. J. Williams, L. A. Barness, D. Young, B. Shafer, R. J. Vivacqua, and E. M. Beaupre, Methylmalonic acid excretion, a sensitive indicator of vitamin B$_{12}$ deficiency in man, *J. Lab. Clin. Med. 66*:75 (1965).
234. P. M. Dreyfus and V. E. Dubé, The rapid detection of methylmalonic acid in urine—a sensitive index of vitamin B$_{12}$ deficiency, *Clin. Chim. Acta 15*:525 (1967).
235. A. J. Giorgio and G. W. E. Plaut, A method for the colorimetric determination of urinary methylmalonic acid in pernicious anemia, *J. Lab. Clin. Med. 66*:667 (1965).
236. H. V. Bashir, H. Hinterberger, and B. P. Jones, Methylmalonic acid excretion in vitamin B$_{12}$ deficiency, *Br. J. Haematol. 12*:704 (1966).
237. N. E. Hoffman and J. J. Barboriak, Gas chromatographic determination of urinary methylmalonic acid, *Anal. Biochem. 18*:10 (1967).
238. E. J. Norman, O. J. Martelo, and M. D. Denton, Cobalamin (vitamin B$_{12}$) deficiency detection by urinary methylmalonic acid quantitation, *Blood 59*:1128 (1982).
239. E. J. Norman, Detection of cobalamin deficiency using the urinary methylmalonic acid test by gas chromatography mass spectrometry, *J. Clin. Pathol. 46*:382 (1993).
240. D. Gompertz, J. H. Jones, and J. P. Knowles, Metabolic precursors of methylmalonic acid in vitamin B$_{12}$ deficiency, *Clin. Chim. Acta 18*, 197 (1967).
241. M. Brozovic, A. V. Hoffbrand, A. Dimitriadou, and D. L. Mollin, The excretion of methylmalonic acid and succinic acid in vitamin B$_{12}$ and folate deficiency, *Br. J. Haematol. 13*:1021 (1967).
242. L. Kass, Cytochemical detection of homocysteine in pernicious anemia and chronic erythremic myelosis, *Am. J. Clin. Pathol. 67*:53 (1977).
243. P. D. Marcell, S. P. Stabler, E. R. Podell, and R. H. Allen, Quantitation of methylmalonic acid and other dicarboxylic acids in normal serum and urine using capillary gas chromatography–mass spectrometry, *Anal. Biochem. 150*:58 (1985).
244. S. P. Stabler, P. D. Marcell, E. R. Podell, and R. H. Allen, Quantitation of total homocysteine, total cysteine, and methionine in normal serum and urine using capillary gas chromatography–mass spectrometry, *Anal. Biochem. 162*:185 (1987).
245. S. P. Stabler, P. D. Marcell, E. R. Podell, R. H. Allen, and J. Lindenbaum, Assay of methylmalonic acid in the serum of patients with cobalamin deficiency using capillary gas chromatography–mass spectrometry, *J. Clin. Invest. 77*:1606 (1986).

246. S. P. Stabler, P. D. Marcell, E. R. Podell, R. H. Allen, D. G. Savage, and J. Lindenbaum, Elevation of total homocysteine in the serum of patients with cobalamin or folate deficiency detected by capillary gas chromatography–mass spectrometry, *J. Clin. Invest. 81*:466 (1988).

247. J. Lindenbaum, D. G. Savage, S. P. Stabler, and R. H. Allen, Diagnosis of cobalamin deficiency II. Relative sensitivities of serum cobalamin, methylmalonic acid, and total homocysteine concentrations, *Am. J. Hematol. 34*:99 (1990).

248. R. C. Chu and C. A. Hall, The total serum homocysteine as an indicator of vitamin B_{12} and folate status, *Am. J. Clin. Pathol. 90*, 446 (1988).

249. C. A. Hall, Function of vitamin B_{12} in the central nervous system as revealed by congenital defects, *Am. J. Hematol. 34*:121 (1990).

250. S.-A. Killmann, Effect of deoxyuridine on incorporation of triated thymidine: difference between normoblasts and megaloblasts, *Acta Med. Scand. 175*:483 (1964).

251. T.-T. Pelliniemi and W. S. Beck, Biochemical mechanisms in the Killmann experiment: critique of the deoxyuridine suppression test, *J. Clin. Invest. 65*:449 (1980).

252. M. B. Van Der Weyden, Deoxyuridine metabolism in human megaloblastic marrow cells, *Scand. J. Haematol. 23*:37 (1979).

253. E. Nexø, M. Hansen, K. Rasmussen, A. Lindgren, and R. Gräsbeck, How to diagnose cobalamin deficiency, *Scand. J. Clin. Lab. Invest. 54* Suppl. 219:61 (1994).

254. R. Schilling, Intrinsic factor studies. II. The effect of gastric juice on the urinary excretion of radioactivity after the oral administration of radioactive vitamin B_{12}, *J. Lab. Clin. Med. 42*, 860–865 (1953).

255. W. B. Nelp, H. N. Wagner, Jr., and R. C. Reba, Renal excretion of vitamin B_{12} and its use in measurements of glomerular filtration rate in man, *J. Lab. Clin. Med. 63*:480 (1964).

256. C. E. Rath, P. R. McCurdy, and B. J. Duffy, Effect of renal disease on the Schilling test, *N. Engl. J. Med. 256*:111 (1956).

257. S. T. Callender, L. J. Witts, G. T. Warner, and R. Oliver, The use of simple whole-body counter for haematological investigations, *Br. J. Haematol. 12*:276 (1966).

258. N. D. C. Finlayson, D. J. C. Shearman, J. D. Simpson, and R. H. Girdwood, Determination of vitamin B_{12} absorption by a simple whole body counter, *J. Clin. Pathol. 21*:595 (1968).

259. S. N. Akun, J. F. Miller, and L. M. Meyer, Vitamin B_{12} absorbtion test, *Acta Haematol. 41*: 341 (969).

260. G. Marcoullis, Y. Parmentier, J.-P. Nicolas, M. Jimenez, and P. Gerard, Cobalamin malabsorption due to nondegradation of R proteins in the human intestine, *J. Clin. Invest. 66*:430 (1980).

261. D. W. Dawson, A. H. Sawers, and R. K. Sharma, Malabsorption of protein bound vitamin B_{12}, *Br. Med. J. 288*:675 (1984).

262. A. Doscherholmen, S. Silvis, and J. McMahon, Dual isotope Schilling test for measuring absorption of food-bound and free vitamin B_{12} simultaneously, *Am. J. Clin. Pathol. 80*:490 (1983).

263. R. Carmel, R. M. Sinow, and D. S. Karnaze, Atypical cobalamin deficiency. Subtle biochemical evidence of deficiency is commonly demonstrable in patients without megaloblastic anemia and is often associated with protein-bound cobalamin malabsorption, *J. Lab. Clin. Med. 109*:454 (1987).

264. R. Carmel, R. M. Sinow, M. E. Siegel, and I. M. Samloff, Food cobalamin malabsorption occurs frequently in patients with unexplained low serum cobalamin levels, *Arch. Intern. Med. 148*:1715 (1988).

265. A. Miller, D. Furlong, B. A. Burrows, and D. W. Slingerland, Bound vitamin B_{12} absorption in patients with low serum B_{12} levels, *Am. J. Hematol. 40*:163 (1992).

266. A. Doscherholmen and W. R. Swain, Impaired assimilation of egg Co^{57} vitamin B_{12} in patients with achlorhydria and after gastric resection, *Gastroenterology 64*:913 (1973).

267. A. Doscherholmen, J. McMahon, and D. Ripley, Inhibitory effect of eggs on vitamin B_{12} absorption: description of a simple ovalbumin ^{57}Co-vitamin B_{12} absorption test, *Br. J. Haematol. 33*:261 (1976).

268. A. M. Streeter, B. Duraiappah, R. Boyle, B. J. O'Neill, and M. T. Pheils, Malabsorption of vitamin B_{12} after vagotomy, *Am. J. Surg. 128*:340–343 (1974).

269. K. Mahmud, M. E. Kaplan, D. Ripley, W. R. Swaim, and A. Doscherholmen, The importance or red cell B_{12} and folate levels after partial gastrectomy, *Am. J. Clin. Nutr. 27*:51–54 (1974).

270. A. M. Streeter, K. J. Goulston, F. A. Bathur, R. S. Hilmer, G. G. Crane, and M. T. Pheils, Cimetidine and malabsorption of cobalamin, *Dig. Dis. Sci. 27*:13–16 (1982).

271. E. W. Strauss and T. H. Wilson, Factors controlling B_{12} uptake by intestinal sacs in vitro, *Am. J. Physiol. 195*:103 (1960).

272. H. Schonsby and T. J. Peters, The estimation of intrinsic factor using guinea pig intestinal brush borders, *Scand. J. Gastroenterol., 65*:441 (1971).

273. L. Ellenbogen, D. R. Highley, H. A. Barker, and R. D. Smyth, Inhibition of cobamide coenzyme activity by intrinsic factor, *Biochem. Biophys. Res. Commun. 3*:178 (1960).

274. C. Gottlieb, K. S. Lau, L. R. Wasserman, and V. Herbert, Rapid charcoal assay for intrinsic factor (IF), gastric juice unsaturated B_{12} binding capacity, antibody to IF, and serum unsaturated B_{12} binding capacity, *Blood 25*:875 (1965).

275. P. Rødbro, P. M. Christiansen, and M. Schwartz, Intrinsic factor secretion in stomach diseases, *Lancet 2*:1200 (1965).

276. M. Katz, S. K. Lee, and B. A. Cooper, Vitamin B_{12} malabsorption due to a biologically inert intrinsic factor, *N. Engl. J. Med. 287*:425 (1972).

277. M. Katz, C. S. Mehlman, and R. H. Allen, Isolation and characterization of an abnormal human intrinsic factor, *J. Clin. Invest. 53*:1274 (1974).

278. E. C. Moore and P. Reichard, Enzymatic synthesis of deoxyribanucleotides. VI. The cytidine diphosphate reductase system from Novikoff hepatoma, *J. Biol. Chem. 239*:3453 (1964).

279. S. Fujioka and R. Silber, Ribonucleotide reductase in human bone marrow: lack of stimulation by 5′-deoxyadenosyl B_{12}, *Biochem. Biophys. Res. Commun. 35*:759 (1969).

280. S. Hopper, Ribonucleotide reductase of rabbit bone marrow. I. Purification properties and separation into two protein fractions, *J. Biol. Chem. 247*:3336 (1972).

281. J. M. Norohna and M. Silverman, On folic acid, vitamin B_{12} methionine and formiminoglutamic acid metabolism, in *Vitamin B_{12} und Intrinsic Faktor* (H. C. F. Heinrich, ed.), Enka, Stuttgart, 1962, p. 728.

282. V. Herbert and R. Zalusky, Interrelations of vitamin B_{12} and folic acid metabolism: folic acid clearance studies, *J. Clin. Invest. 41*:1263 (1962).

283. M. Silverman and A. S. Pitney, Dietary methionine and the excretion of formiminoglutamic acid by the rat, *J. Biol. Chem. 233*:1179 (1958).

284. A. L. Luhby and J. M. Cooperman, Aminoimidazolecarboxamide excretion in vitamin B_{12} and folic acid deficiencies, *Lancet 2*:1381 (1962).

285. I. Arakawa, Congenital defects in folate utilization, *Am. J. Med. 48*:594 (1970).

286. C. Kutzbach, E. Galloway, and E. L. R. Stokstad, Influence of vitamin B_{12} and methionine on levels of folic acid compounds and folate enzymes in rat liver, *Proc. Soc. Exp. Biol. Med. 124*:801 (1967).

287. A. V. Hoffbrand and B. F. A. Jackson, Correction of the DNA synthesis defect in vitamin B_{12} deficiency by tetrahydrofolate: evidence in favour of the methyl-folate trap hypothesis as the cause of megaloblastic anaemia in vitamin B_{12} deficiency, *Br. J. Haematol. 83*:643–647 (1993).

288. W. S. Beck, unpublished results.

289. I. Chanarin and J. Perry, Metabolism of 5-methyltetralhydrofolate in pernicious anaemia, *Br. J. Haematol. 14*:297 (1968).

290. P. F. Nixon and J. R. Bertino, Impaired utilization of serum folate in pernicious anemia: a study with radiolabeled 5-methyltetrahydrofolate, *J. Clin. Invest. 51*:1431 (1972).

291. V. Herbert, Recent developments in cobalamin metabolism, in *The Cobalamins: A Glaxo Symposium* (H. R. V. Arnstein and R. J. Wrighton, eds.), Williams & Wilkins, Baltimore, 1971, p. 20.

292. S. W. Thenen, J. M. Hawthorne, and E. L. R. Stokstad, The oxidation of 5-methyl ^{14}C-

tetrahydrofolate and histidine-2-^{14}C to $^{14}CO_2$ in vitamin B_{12}-deficient rats, *Proc. Soc. Exp. Biol. Med. 134*:199 (1970).

293. P. Laduron, N-Methylation of a dopamine to epinine in brain tissue using N^5-methyltetrahydrofolic acid as the methyl donor, *Nature (New Biol.) 238*:212 (1972).

294. R. T. Taylor and M. L. Hanna, 5-methyltetrahydrofolate aromatic alkylamine N-methyltransferase: an artefact of 5,10-methylenetetrahydrofolate reductase activity, *Life Sci. 17*:111 (1975).

295. J. Thorndike and W. S. Beck, Enzymatic production of formaldehyde from N^5-methyltetrahydrofolate in normal and leukemic leukocytes, *Cancer Res. 37*:1125 (1977).

296. B. A. Cooper and L. Lowenstein, Relative folate deficiency of erythrocytes in pernicious anaemia and its correction with cyanocobalamin, *Blood 24*:502 (1964).

297. R. M. Smith and W. S. Osborn-White, Folic acid metabolism in vitamin B_{12} deficient sheep, *Biochem. J. 136*:279 (1973).

298. J. Metz, A. Kelly, V. C. Swath, S. Waxman, and V. Herbert, Deranged DNA synthesis by bone marrow from vitamin B_{12}-deficient humans, *Br. J. Haematol. 14*:575 (1968).

299. S. Waxman, J. Mertz, and V. Herbert, Defective DNA synthesis in human megaloblastic bone marrow: effects of hamocysteine and methionine, *J. Clin. Invest. 48*:284 (1969).

300. R. Stebbins, J. Scott, and V. Herbert, Therapeutic trial in the test tube: The ''dU suppression test'' using ''physiologic'' doses of B_{12} and folic acid to replace therapeutic trial in vivo for diagnosis of B_{12} and folate deficiency, *Blood 40*:927 (1972).

301. M. B. Van der Weyden, M. Cooper, and B. G. Firkin, Defective DNA synthesis in human megaloblastic bone marrow: effects of hydroxy-B_{12} 5'-deoxyadenosyl-B_{12} and methyl-B_{12}, *Blood 41*:299 (1973).

302. V. Herbert, Laboratory aids in the diagnosis of folic acid and vitamin B_{12} deficiencies, *Ann. Clin. Lab. Sci. 1*:193 (1971).

303. H. L. Levy, S. H. Mudd, J. D. Schulman, P. M. Dryfus, and R. H. Abeles, A derangement in B_{12} metabolism associated with homocytinemia, cystathioninemia, hypomethioninemia and methylmalonic aciduria, *Am. J. Med. 48*:390 (1970).

304. S. H. Mudd, B. W. Uhlendorf, K. R. Hinds, and H. L. Levy, Deranged B_{12} metabolism: studies of fibroblasts grown in tissue culture, *Biochem. Med. 4*:215 (1970).

305. S. I. Goodman, P. C. Moe, K. B. Hammond, S. H. Mudd, and B. W. Uhlendorf, Homocystinuria with methylmalonic aciduria: two cases in a sibship, *Biochem. Med. 4*:500 (1970).

306. R. H. Grafstrom, B. Y. Tseng, and M. Goulian, The incorporation of uracil into animal cell DNA in vitro, *Cell 15*:131 (1978).

307. M. Goulian, B. Bleile, and B. Y. Tseng, Methotrexate-induced misincorporation of uracil into DNA, *Proc. Natl. Acad. Sci. USA 77*, 1956 (1980).

308. M. Goulian, B. Bleile, and B. Y. Tseng, The effect of methotrexate on levels of dUTP in cells, *J. Biol. Chem. 255*:10630 (1980).

309. W. S. Beck, G. E. Wright, N. J. Nusbaum, J. D. Chang, and E. M. Isselbacher, Enhancement of methotrexate cytotoxicity by uracil analogues that inhibit deoxyuridine triphosphate nucleotidylhydrolase (dUTPase) activity, in *Purine and Pyrimidine Metabolism in Man,* Vol. 5, Part B (W. L. Nyhan, L. F. Thompson, and R. W. E. Watts, eds.). Plenum Press, New York, 1986, p. 97.

310. S. N. Wickramasinghe and S. Fida, Misincorporation of uracil into the DNA of folate- and B_{12}-deficient HL60 cells, *Eur. J. Haematol. 50*:127 (1993).

311. T. Lindahl, New class of enzymes acting on damaged DNA, *Nature (Lond.) 259*:64 (1976).

312. E. J. Fine, E. Soria, M. W. Parowski, and L. Thomasula, The neurophysiological profile of vitamin B_{12} deficiency, *Muscle Nerve 13*:158 (1990).

313. W. S. Beck, Neuropsychiatric consequences of cobalamin deficiency, *Adv. Int. Med. 36*:33 (1991).

314. R. D. Adams and M. Victor, *Principles of Neurology*, 4th ed., McGraw-Hill, New York, 1989, p. 833.

315. J. Metz, Cobalamin deficiency and the pathogenesis of nervous system disease, *Annu. Rev. Nutr. 12*:59 (1992).

316. M. I. Shevell and D. S. Rosenblatt, The neurology of cobalamin, *Can. J. Neurol. Sci. 19*: 472 (1992).

317. R. Surtees, Biochemical pathogenesis of subacute combined degeneration of the spinal cord and brain, *J. Inher. Metab. Dis. 16*:762 (1993).

318. M. Victor and A. A. Lear, Subacute combined degeneration of the spinal cord. Current concepts of the disease process. Value of serum vitamin B_{12} determinations in clarifying some of the common clinical problems. *Am. J. Med. 20*:896 (1956).

319. S. A. Pant, A. K. Asbury, and E. P. Richardson, Jr., The myelopathy of pernicious anemia: a neuropathological reappraisal, *Acta Neurol. Scand. 44 (Suppl. 35)*:1, (1968).

320. J. S. Wiener and J. M. Hope, Cerebral manifestations of vitamin B_{12} deficiency, *JAMA 170*: 1038 (1959).

321. T. N. Fraser, Cerebral manifestations of addisonian pernicious anaemia, *Lancet 2*:458 (1960).

322. R. W. Strachan and J. G. Henderson, Psychiatric syndromes due to avitaminosis B_{12} with normal blood and marrow, *Q. J. Med. 135*:303 (1965).

323. E. Birket-Smith, Neurological disorders in vitamin B_{12} deficiency with normal B_{12} absorption, *Dan. Med. Bull. 12*:158 (1965).

324. R. Shulman, Psychiatric aspects of pernicious anemia, *Br. Med. J. 3*:266 (1967).

325. O. Abramsky, Common and uncommon neurological manifestations as presenting symptoms of vitamin-B_{12} deficiency, *J. Am. Geriatr. Soc. 20*:93, 1972.

326. S. D. Shorvon, M. W. Carney, I. Chanarin, and E. H. Reynolds, The neuropsychiatry of megaloblastic anaemia, *Br. Med. J. 281*:1036, 1980.

327. R. L. Ruff and J. Scally, Reversible neurological deficits in patients with normal serum vitamin B_{12} levels [corresp], *Arch. Neurol. 37*:255, 1980.

328. M. Rose, Why assess vitamin-B_{12} status in patients with known neuropsychiatric disorder? [corresp], *Lancet 2*:1191, 1976.

329. A. D. M. Smith, Megaloblastic madness, *Br. Med. J. 2*:1840, 1960.

330. R. Green, S. V., Van Tonder, G. J. Oettle, G. Cole, and J. Metz, Neurological changes in fruit bats deficient in vitamin B_{12}, *Nature 254*:148 (1975).

331. D. P. Agamanolis, M. Victor, J. W. Harris, J. D. Hines, E. M. Chester, J. A. Kark, An ultrastructural study of subacute combined degeneration of the spinal cord in vitamin B_{12}-deficient rhesus monkeys, *J. Neuropathol. Exp. Neurol. 37*:273 (1978).

332. A. M. Goodman, J. M. Harris, and D. Kiraly, Studies in B_{12}-deficient monkeys with combined system disease. I. B_{12}-deficient patterns in bone marrow deoxyuridine suppression tests without morphologic or functional abnormalities, *J. Lab. Clin. Med. 96*:722 (1980).

333. M. E. Smith and L. F. Eng, The turnover of the lipid components of myelin, *J. Am. Oil Chem. Soc. 42*:1013 (1965).

334. M. E. Smith, The turnover of myelin in the adult rat, *Biochim. Biophys. Acta 164*:285 (1968).

335. G. J. Cardinale, T. J. Carty, and R. H. Abeles, Effect of methylmalonyl coenzyme A, a metabolite which accumulates in vitamin B_{12} deficiency, on fatty acid synthesis, *J. Biol. Chem. 245*:3771 (1970).

336. F. W. Barley, G. H. Sato, and R. H. Abeles, An effect of vitamin B_{12}deficiency in tissue culture, *J. Biol. Chem. 247*:4270 (1972).

337. S. Kashket and W. S. Beck, unpublished results.

338. E. P. Frenkel, Abnormal fatty acid metabolism in peripheral nerves of patients with pernicious anemia, *J. Clin. Invest. 52*:1237 (1973).

339. E. P. Frenkel, R. L. Kitchens, L. B. Hersh, and R. Frenkel, Effect of vitamin B_{12} deprivation on the in vivo levels of coenzyme A intermediates associated with propionate metabolism, *J. Biol. Chem. 249*:6984 (1974).

340. Y. Kishimoto, M. Williams, H. W. Moser, C. Hignite, and K. Biemann, Branched-chain and odd-numbered fatty acids and aldehydes in the nervous system of a patient with deranged vitamin B_{12} metabolism, *J. Lipid Res. 14*:69 (1973).

341. S. P. Stabler, R. H. Allen, D. G. Savage, and J. Lindenbaum, Marked elevation of methylma-

lonic acid (MMA) in cerebral spinal fluid (CSF) of patients with cobalamin (Cbl) deficiency [abstract], *Clin. Res. 37*:550A (1989).

342. J. A. DeGrazia, M. B. Fish, M. Pollycove, R. O. Wallerstein, and L. Hollander, The role of propionic acid as a precursor of methylmalonic acid in normal and vitamin B_{12}-deficient man, *J. Lab. Clin. Med. 73*:917 (1969).

343. R. Green, D. W. Jacobsen, and C. Sommer, Effect of nitrous oxide on methionine synthetase activity in brain and kidney of vitamin B_{12} deficienct fruit bats, *Proc. 18th Congress Int. Soc. Hematol. 148* (1980).

344. A. B. Guttormsen, H. Refsum, and P. M. Ueland, The interaction between nitrous oxide and cobalamin. Biochemical effects and clinical consequences, *Acta Anaesthesiol. Scand. 38*:753 (1994).

345. R. Carmel, D. Watkins, S. I. Goodman, and D. S. Rosenblatt, Hereditary defect of cobalamin metabolism (*cblG* mutation) presenting as a neurologic disorder in adulthood, *N. Engl. J. Med. 318*:1738 (1988).

346. M. McKeever, A. Molloy, D. G. Weir, P. B. Young, D. G. Kennedy, S. Kennedy, and J. M. Scott, An abnormal methylation ratio induces hypomethylation in vitro in the brain of pig and man, but not in rat, *Clin. Sci. 88*:73 (1995).

347. M. McKeever, A. Molloy, P. Young, S. Kennedy, D. G. Kennedy, J. M. Scott, and D. G. Weir, Demonstration of hypomethylation of proteins in the brain of pigs (but not in rats) associated with chronic vitamin B_{12} inactivation, *Clin. Sci. 88*:471 (1995).

348. F. Wokes, Tobacco amblyopia, *Lancet 2*:526, 1958.

349. A. D. M. Smith and S. Duckett, Cyanide, vitamin B-12, experimental demyelination and tobacco amblyopia, *Br. J. Exp. Pathol. 46*:615 (1965).

350. A. G. Freeman, Optic neuropathy and chronic cyanide intoxication: a review, *J. R. Soc. Med. 81*:103 (1988).

351. B. Björkenheim, Optic neuropathy caused by vitamin B_{12} deficiency in carriers of the fish tapeworm, *Diphyllobothrium latum*, *Lancet 1*:688 (1966).

352. J. C. Linnell, A. D. M. Smith, C. L. Smith, J. Wilson, and D. M. Matthews, Effects of smoking on metabolism and excretion of vitamin B_{12}, *Br. Med. J. 2*:215 (1968).

353. J. C. Linnell, The fate of cobalamin in vivo, in *Cobalamin: Biochemistry and Pathophysiology* (B. M. Babior, ed.), New York, John Wiley and Sons, 1975, p. 287.

354. D. M. Matthews and J. Wilson, Cobalamins and cyanide metabolism in neurological disease, in *The Cobalamins: A Glaxo Symposium* (H. R. V. Arnstein and R. J. Wrighton, eds), Williams & Wilkins, Baltimore, 1971, p. 115.

355. A. Dorozynksi, Cassava may lead to mental retardation, *Nature 272*:121 (1978).

356. R. D. Cooke and D. G. Coursey, Cassavs: a major cyanide-containing food crop, in *Cyanide in Biology* (B. Vennesland, E. E. Conn, C. J. Knowles, J. Westley, and F. Wissing, eds.), Academic Press, New York, 1981, p. 93.

357. F. Nartley, Cyanogenesis in tropical feeds and foodstuffs, in *Cyanide in Biology* (B. Vennesland, E. E. Conn, C. J. Knowles, J. Westley, and F. Wissing, eds.), Academic Press, New York, 1981, p. 115.

358. B. O. Osuntokun, Cassava diet, chronic cyanide intoxication and neuropathy in the Nigerian Africans, *World Rev. Nutr. Diet 36*:141 (1981).

359. C. I. Phillips, R. G. Ainley, P. Van Peborgh, E. J. Watson-Williams, and A. C. Bottomley, Vitamin B_{12} content of aqueous humour, *Nature 217*:67 (1968).

360. R. G. Ainley, C. I. Phillips, A. Gibbs, R. R. Acheson, E. J. Watson-Williams, and A. C. Bottomley, Aqueous humour vitamin B_{12} and intramuscular cobalamins, *Br. J. Ophthalmol. 53*:854 (1969).

361. E. H. Reynolds, Neurological aspects of folate and vitamin B_{12} metabolism, *Clin. Haematol. 5*:661 (1976).

362. Editorial, Folic acid and the nervous system. *Lancet 2*:836 (1976).

363. E. A. Harvey, I. Howard, and W. P. Murphy, Absence of a toxic effect of folic acid on the central nervous system of persons without pernicious anemia, *N. Engl. J. Med. 242*:446 (1950).

14

Choline

STEVEN H. ZEISEL and MINNIE HOLMES-McNARY

University of North Carolina at Chapel Hill, Chapel Hill, North Carolina

I. HISTORY

Choline is a dietary component that is important for normal functioning of all cells (1). We believe that humans require dietary choline for sustaining normal life (2). It is ubiquitous in foods, is required for synthesis of essential components of membranes, is a precursor for biosynthesis of the neurotransmitter acetylcholine, and is an important source of labile methyl groups (3). Choline was discovered in 1862 and was chemically synthesized in 1866 (4). It was known to be a component of phospholipids, but the pathway for its biosynthesis was first described in 1941 by duVigneaud (5). The route for its incorporation into phosphatidylcholine (lecithin) was not elucidated until 1956 (6).

The importance of choline as a nutrient was first appreciated during early research on the functions of insulin (7,8) when it was found to be the nutrient needed to prevent fatty liver and death in dogs lacking a pancreas. The term "lipotropic" was coined to describe choline and other substances that prevented deposition of fat in the liver. In 1946, Copeland (9) first observed that rats fed diets deficient in choline develop liver cancer. In the subsequent 50 years, this model has been used to develop an understanding of the mechanisms which initiate and promote carcinogenesis (10–12). In 1975, it was discovered that administration of choline accelerated the synthesis and release of acetylcholine by neurons (13–18). This led to an increased interest in dietary choline and brain function.

II. ASSAY

Choline can be measured using gas chromatography–mass spectrometry (19). If choline, acetylcholine, phosphocholine, glycerophosphocholine, cytidine diphosphocholine, lyso-phosphatidylcholine, and phosphatidylcholine are to be determined, extraction and isolation of these metabolites can be accomplished using high-pressure liquid chromatography (20). This assay can be used with a nitrogen-phosphorus or a flame ionization detector instead of with the mass spectrometer (21), but internal standards are not as easy to use. Other methods for measurement of choline include high-performance liquid chromatography combined with continuous-flow fast-atom bombardment mass spectrometry (22) or with a postcolumn reaction converting the choline to betaine and then forming hydrogen peroxide detected by an electrochemical detector (23–26), a biological assay using the thermophilic enteric yeast *Torulopsis pintolopessi* (27), a chemiluminescence method (28), and a radioenzymatic method (29,30).

III. CONTENT IN FOOD

Many foods eaten by humans contain significant amounts of choline and esters of choline. No information is available about the phosphocholine or glycerophosphocholine content of foods, but it is likely that they are present in significant amounts. Some of this choline is added during food processing [especially during preparation of infant formula (31)]. Average choline dietary intake (as choline and choline esters) in the adult human (as free choline and the choline in phosphatidylcholine and other choline esters) is more than 7–10 mmol/day (32,33). A kilogram of beef liver contains 50 mmol of choline moiety; thus, it is easy to consume a diet of normal foods (eggs, liver, etc.) that delivers much more choline per day than the calculated average intake or 700 mg/day (34). The human infant consumes a choline-rich diet. Breast milk contains approximately 1.5 mmol/L choline and choline esters; only recently did we discover that the major forms of choline in mature human and rat milk are phosphocholine and glycerophosphocholine (35,36). Thus, an infant consuming 500 mL breast milk ingests 750 µmol choline. Choline is routinely added to commercially available infant formulas (approximately 1 mmol free choline per L).

IV. METABOLISM

There are several comprehensive reviews of the metabolism and functions of choline (33,37,38) (Figs. 1–3; Table 1). A small fraction of dietary choline is acetylated (39,40) in cholinergic neurons (41) and in such nonnervous tissues as the placenta (42). In brain it is unlikely that choline acetyltransferase is saturated with either of its substrates, so that choline (and possibly acetyl-CoA) availability determines the rate of acetylcholine synthesis (43). Some investigators report that administration of choline or phosphatidylcholine results in the accumulation of acetylcholine within brain neurons (13–16), whereas others observe that such acceleration of acetylcholine synthesis by choline administration can only be detected after pretreatments with agents that cause cholinergic neurons to fire rapidly (17,44–48). Increased brain acetylcholine synthesis is associated with an augmented release of this neurotransmitter into the synapse.

The metabolisms of choline, methionine, and methyl folate are closely interrelated, and the use of choline molecules as methyl donors is probably the major factor that determines how rapidly a diet deficient in choline will induce pathology (10). The pathways of choline and one-carbon metabolism intersect at the formation of methionine from homocysteine. Betaine-homocysteine methyltransferase catalyzes the methylation of homocysteine using the choline metabolite betaine as methyl donor (49–51). In an alternative pathway, 5-methyltetrahydrofolate-homocysteine methyltransferase regenerates methionine using a methyl group derived de novo from the one-carbon pool (50,52). Perturbing metabolism of one of the methyl donors results in compensatory changes in the other

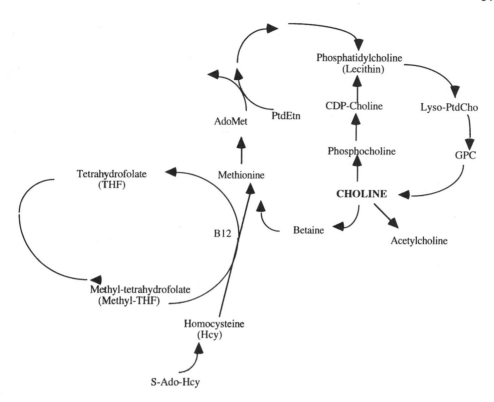

Fig. 1 Choline metabolic pathway is interrelated with folate, methionine, and B_{12} metabolism. S-AdoMet, S-adenosylmethionine; CDP-Choline, choline–cytidine diphosphocholine; PtdEtn, phosphatidylethanolamine; GPC, glycerophospocholine; LysoCho, lysophosphatidylcholine; S-AdoHcy, S-adenosylhomocysteine.

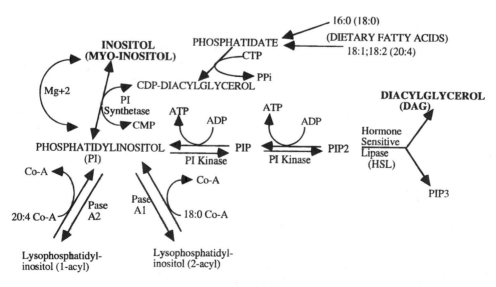

Fig. 2 Myoinositol metabolic pathway. Mg+2, magnesium; PI kinase, phosphoinositol kinase; Pase A1, phospholipase A1; Pase A2, phospholipase A2; Co-A, coenzyme A; PIP, phosphoinositol phosphate; PIP2, phosphoinositol biphosphate; PIP3, phosphoinositol triphosphate.

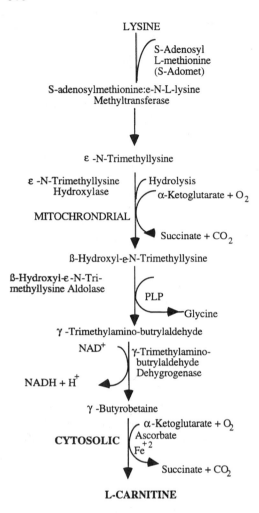

Fig. 3 Carnitine metabolic pathway. CO_2, carbon dioxide; O_2, oxygen; NAD^+ ($NADH$), nicotin-amide dehydrogenase; Fe^{+2}, ferrous iron; PLP, pyridoxyl phosphate.

Table 1 Pathophysiology Associated with Required Carnitine Supplementation

Altered biosynthesis	Increased excretion	Altered transport
Malnutrition	Immaturity	Malnutrition
Immaturity	Organic aciduria	Immaturity
Genetic carnitine deficiency	Genetic carnitine deficiency	Genetic carnitine deficiency
Liver disease	Liver disease	Liver disease
Renal disease	Renal disease	
	Drug treatment (e.g., valproate)	

Complied from P. R. Borum (1991) in *The Handbook of Vitamins.*

methyl donors due to the intermingling of these metabolic pathways (53–61). Redundant parallel pathways protect the capacity to donate methyl groups. Clearly, if we had established that choline was essential first, we would believe that methionine is not an essential amino acid because it can be replaced by homocysteine and choline. The amounts of choline, methyl folate, or methionine required in the diet are dependent on the availability of the other two. Does this mean that other methyl donors can completely substitute for choline? Many investigators assume that this is true, but this question has never been adequately investigated.

Synthesis of phosphatidylcholine occurs by two pathways. In the first, choline is phosphorylated, then converted to cytidine diphosphocholine (CDP-choline) in the regulated step in phosphatidylcholine biosynthesis (62). This high-energy intermediate, in combination with diacylglycerol, forms phosphatidylcholine and cytidine monophosphate (6). In the alternative pathway, phosphatidylethanolamine is sequentially methylated to form phosphatidylcholine, using S-adenosylmethionine as methyl donor (63). This is the only source of choline other than diet.

V. FUNCTIONS OF CHOLINE AND CHOLINE PHOSPHOLIPIDS

The numerous functions of choline as a methyl donor and of choline phospholipids as structural elements of cells have been reviewed extensively elsewhere (33). Choline and its relationships to acetylcholine synthesis also have been thoroughly reviewed elsewhere (64). We focus on new observations about physiologic roles for these compounds.

A. Hepatic Secretion of Very Low Density Lipoprotein

The triacylglycerol produced by liver is delivered to other tissues mainly in the form of very low density lipoprotein (VLDL). Phosphatidylcholine is a required component of the VLDL particle (65,66), and in choline deficiency the diminished capacity of liver cells to synthesize new phosphatidylcholine molecules results in the intracellular accumulation of triglycerides. Methionine can substitute for choline in VLDL secretion, but only as long as phosphatidylethanolamine-N-methyltransferase activity (see earlier discussion) is active (65). Secretion of high-density lipoprotein (HDL) from hepatocytes does not require the synthesis of new phosphatidylcholine molecules (67). Choline-deficient humans have diminished plasma low-density lipoprotein cholesterol (LDL; derived from VLDL) (2). This observation is consistent with the hypothesis that in humans, as in other species, choline is required for VLDL secretion.

B. Platelet-Activating Factor: A Choline-Containing Phospholipid with Hormonal Functions

Platelet-activating factor (PAF; 1-O-alkyl-2-acetyl-sn-glycero-3-phosphocholine) is a choline phospholipid characterized by a fatty alcohol (usually hexadecanol) at the sn-1 position of the glycerol backbone and an acetate residue at the sn-2 position. It has diverse biological functions that are exerted through specific receptors located on cells of many different types. Receptor-mediated synthesis of PAF requires calcium and is mediated by a G-protein. The effects of PAF include the activation of phosphoinositide-specific phospholipase C, and the release of fatty acids from phospholipids, possibly by activation of phospholipase A_2 (68). Besides activating platelet aggregation, the actions of PAF include lowering of blood pressure; increasing of vascular permeability; activation of monocytes, macrophages, and polymorphonuclear neutrophils; and stimulation of hepatic glyco-genolysis. PAF may be one of the mediators of parturition, and it mediates many processes of inflammation and allergy. Many excellent reviews describe the biosynthesis, turnover, and actions of PAF (68–71).

C. Signal Transduction

Our understanding of choline phospholipid–mediated signal transduction has been vastly improved during the past decade. Stimulation of membrane-associated receptors activates neighboring phospholipases, resulting in the formation of breakdown products that are signaling molecules either by themselves (i.e., they stimulate or inhibit the activity of target macromolecules), or after conversion to signaling molecules by specific enzymes. Much of signaling research focused on minor membrane phospholipid components, partic-ularly phosphatidylinositol derivatives [signal transduction via receptor-mediated hydroly-sis of phosphatidylinositol biphosphate has been extensively reviewed elsewhere (72,73)]. However, metabolism of choline phospholipids, especially phosphatidylcholine and sphin-gomyelin, results in biologically active molecules that can amplify external signals or that can terminate the signaling process by generating inhibitory second messengers (74).

For example, activation of heptahelical receptors leads to altered conformation of the receptor so that it can activate a GTP-binding protein (G-protein). The activation of the G-protein results in the subsequent activation of phospholipase C activity within the plasma membrane. Phospholipase C hydrolyzes intact phospholipids to generate 1,2-sn-diacylglycerol and an aqueous soluble head group. The action of phospholipase C triggers the next event in the signal cascade, which is activation of protein kinase C (PKC; serine/threonine kinase). Products generated by phosphatidylinositol-bis-phosphate-phospho-lipase C include inositol-1,4,5-trisphosphate (Ins-1,4,5-P_3) and diacylglycerol. Ins-1,4,5-P_3 is a water soluble product, which acts to release calcium from stores in the endoplasmic reticulum. Some PKC isotypes are Ca^{2+}-dependent (PKC α, β_2, and γ). PKC δ, ϵ, ζ, θ, and η lack a calcium binding domain (75). Calcium, by increasing membrane occupancy of PKC, places the enzyme in close proximity to phosphatidylserine, a cofactor for PKC activation. Diacylglycerol, the other product of phospholipase C, remains in the plasma membrane, and is both a messenger molecule and an intermediate in the metabolism of lipids. Normally, PKC is folded so that an endogenous ''pseudosubstrate'' region on the protein is bound to the catalytic site, thereby inhibiting activity. The combination of diacyl-glycerol and Ca^{2+} causes a conformational change in PKC, causing flexing at a hinge region that heads to withdrawal of the pseudosubstrate and unblocking of the PKC catalytic site. The appearance of diacylglycerol in membranes is usually transient, and therefore PKC is activated for a short time after a receptor has been stimulated. The generation of diacylglycerol from membrane phosphatidylcholine amplifies the signal. Phospholipase C and, indirectly, phospholipase D generate this diacylglycerol (76–79). Other products of phosphatidylcholine hydrolysis, such as phosphatidic acid, lysophosphatidylcholine, and free fatty acids, also are second messengers (76,80). Phosphatidic acid can act as a mitogen (81). Lysophosphatidylcholine stimulates PKC activity (75), but it is a membrane-lytic detergent with potential toxic effects. Lysophosphatidylcholine generation is impor-tant in chemotaxis, relaxation of smooth muscle, and activation of T lymphocytes (75). Recent data indicate that phosphatidylcholine may synergize with diacylglycerol and cal-cium to activate PKC in lymphocytes (82) and that modulation of PKC isozymes by phos-phatidylcholine may be isoform-specific (83).

The characterization of events that occur downstream from PKC is just beginning. Serine-threonine kinases and tyrosine kinases catalyze phosphorylation of target proteins distal to PKC. Phosphorylation alters the biochemical properties of these substrates, resulting in a range of cellular responses. These phosphorylation cascades serve to enhance amplification of the original signal. PKC signals impinge on several known intracellular control circuits (84). The targets for phosphorylation by PKC include receptors for insulin, epidermal growth factor, and many proteins involved in control of gene expression (85,86).

Although choline sphingolipids are ubiquitous components of mammalian cells, it has only recently been proven that they are necessary for cellular survival and growth (87). While phosphatidylcholine hydrolysis generates a series of messengers that sustain the PKC phosphorylation cascade, hydrolysis of sphingomyelin generates messengers that terminate the cascade. Sphingomyelin metabolism has been reviewed extensively (88). The activation of sphingomyelinase may constitute one of the signal transduction pathways for triggering differentiation (89,90). Activation of sphingomyelinase causes elevations in cellular ceramide levels, a compound with potent biological activities including the triggering of programmed cell death (apoptosis) (91). The formation of sphingosine from ceramide may be critical for regulating PKC signal transduction.

VI.   DEFICIENCIES

A.   Choline Deficiency in Animals

The liver dysfunction and fatty liver associated with choline deficiency has been discussed earlier. Choline deficiency compromises renal function, with abnormal concentrating ability, free water reabsorption, sodium excretion, glomerular filtration rate, renal plasma flow, and gross renal hemorrhage (92–95). Diets low in choline content also cause infertility, growth impairment, bone abnormalities, decreased hematopoiesis, and hypertension (96–99).

B.   Choline Deficiency in Humans

Humans require choline for sustaining normal life. Is a dietary source of choline required? As discussed above, normal diets deliver sufficient choline. When healthy humans were fed a choline-deficient diet for 3 weeks they developed biochemical changes consistent with choline deficiency (2). These included diminished plasma choline and phosphatidylcholine concentrations, as well as diminished erythrocyte membrane phosphatidylcholine concentrations. Serum alanine transaminase (ALT) activity, a measure of hepatocyte damage, increased significantly during choline deficiency. This experiment established a requirement for choline in the diet of normal humans.

This requirement may be more apparent in special human populations. The demand for choline in normal adults is likely to be smaller than the demand for choline in the infant, as large amounts of choline must be used to make phospholipids in growing organs (37). The observed changes that occurred in choline-deficient adult humans might have been greater had we studied growing children. Malnourished humans, in whom stores of choline, methionine, and folate have been depleted (100,101), are also likely to need more dietary choline than did our healthy adult subjects. The liver is the primary site for endogenous synthesis of choline. Alcoholics with liver cirrhosis have diminished plasma choline concentration, alone with fatty liver which resolves when patients are supplemented with choline (101). In baboons, treatment with phosphatidylcholine can prevent alcohol-induced liver fibrosis (102).

C.   Carcinogenesis

Choline is the only nutrient for which dietary deficiency causes development of hepatocarcinomas in the absence of a known carcinogen (10). It is interesting that choline-deficient rats not only have a higher incidence of spontaneous hepatocarcinoma but are markedly sensitized to the effects of administered carcinogens (10). Choline deficiency is therefore considered to have both cancer-initiating and promoting activities.

There are several mechanisms suggested for the cancer-promoting effect of a choline-devoid diet. In the choline-deficient liver there is a progressive increase in cell proliferation, related to regeneration after parenchymal cell death (103). Cell proliferation, with associated increased rate of DNA synthesis, could be the cause of greater sensitivity to chemical carcinogens (104). Stimuli for increased DNA synthesis increase carcinogenesis; hepatectomy and necrogenic chemicals are examples. However, the overall rate of liver cell proliferation could be dissociated from the rate at which preneoplastic lesions formed during choline deficiency (105), suggesting that cell proliferation is not the sole condition acting as a promoter of liver cancer. Methylation of DNA is important for the regulation of expression of genetic information. Undermethylation of DNA, observed during choline deficiency (despite adequate dietary methionine), may be responsible for carcinogenesis (106,107). Another proposed mechanism derives from the observation that, when rats eat a choline-deficient diet, increased lipid peroxidation occurs within liver (108). Lipid peroxides in the nucleus could be a source of free radicals that could modify DNA and cause carcinogenesis. We have proposed that choline deficiency perturbs PKC signal transduction, thereby promoting carcinogenesis. Recently, we reported that a defect in cell suicide (apoptosis) mechanisms may contribute to the carcinogenesis of choline deficiency.

As discussed earlier, choline deficiency causes massive fatty liver (see Sec. V.A). We have observed that 1, 2-sn-diacylglycerol accumulates in this fatty liver (11,109). In plasma membrane from livers of choline-deficient rats, diacylglycerol reaches values higher than those occurring after stimulation of a receptor linked to phospholipase C activation (e.g., vasopressin receptor). This results in a stable activation of PKC and/or an increase in the total PKC pool in the cell (11) with changes in several PKC isotypes (at 6 weeks of choline deficiency, amounts of PKC α and δ increased 2-fold and 10-fold, respectively). The accumulation of diacylglycerol and subsequent activation of PKC within liver during choline deficiency may be the critical abnormality that eventually contributes to the development of hepatic cancer in these animals (11). Abnormalities in PKC-mediated signal transduction may trigger carcinogenesis (86).

The process of carcinogenesis involves an initiating event that induces genetic damage, followed by survival and progression of selected clones of the mutant cells to form tumors. In order to study the underlying mechanisms involved in progression of carcinogenesis after initiation, we developed a cell culture model using immortalized CWSV-1 rat hepatocytes, in which p53 protein is inactivated by SV40 large T antigen. Many cancers are p53-defective (110), suggesting that their precursor (initiated) cells also share this defect. Previously, we showed that choline deficiency (CD) medium induced apoptosis in CWSV-1 cells via a p53-independent pathway (111). In normal tissues, apoptosis provides a physiological way to eliminate terminally differentiated, damaged, or genetically altered cells, thus facilitating tissue remodeling following cell injury (112). Apoptosis is an important defensive barrier that inhibits carcinogenesis by eliminating initiated cells, usually via p53-dependent mechanisms (113). However, in p53-defective cells, alternative, p53-independent apoptosis pathways may serve as a mechanism(s) for eliminating initiated cells. When both p53-dependent and p53-independent apoptosis are inactivated, an environment is created in which initiated cells may have a high survival rate, significantly enhancing carcinogenesis. We observed that CWSV-1 cells that are gradually deprived of choline can adapt and become resistant to CD apoptosis. These adapted cells express a tumorigenic phenotype, and treatment with an antioxidant during the CD adaptation process can substantially abrogate the acquisition of a transformed phenotype (114).

D. Brain Development and Function

Given the central role of choline as a precursor of acetylcholine and phosphatidylcholine, nature has developed a number of mechanisms to ensure that a developing animal gets adequate amounts of choline. In mammals, the placenta transports choline to the fetus (115). At birth, all mammals studied, including the human, have plasma choline concentrations that are much higher than those in adults (116). The capacity of brain to extract choline from blood is greatest during the neonatal period (117). There is a novel phosphatidylethanolamine-*N*-methyltransferase (synthesizes choline de novo) in neonatal rat brain that is extremely active (118). Human and rat milk contain especially large amounts of choline (35,36).

There are two sensitive periods in development of rat brain during which treatment with choline results in long-lasting facilitation of spatial memory. The first occurs during embryonic days 12–17 and another during postnatal days 16–30 (119–121). Choline supplementation during these critical periods elicits a constant percentage improvement in choice performance at all stages of training for a visuospatial task (12-arm radial maze), suggesting a memory rather than a learning rate effect. This effect is apparent months after the brief exposure to extra choline. The two sensitive periods correlate with neurogenesis of cholinergic cells (prenatal) and with synaptogenesis (postnatal) (122–124).

The observation that perinatal supplementation with choline results in improved memory performance of adult rats suggests that choline availability is critical for brain development. The choline content of rat chow has been established based on choline requirements of nonpregnant adult rats. We do not know what the minimal requirement is for the pregnant or lactating female, or for the fetus. Perhaps changes in eating behaviors during pregnancy allow rats in the wild to adjust their choline intake. These considerations are also important when considering the choline requirements for humans.

VII. NUTRITIONAL STATUS

There are no established approaches for determining nutritional status for choline. Plasma choline and phosphatidylcholine concentrations fall when humans are fed a choline-deficient diet (2), but this also occurs in marathon runners (125). Even in severe deficiency, plasma choline concentrations do not fall below 50% of normal. We have observed that liver phosphocholine concentrations are the most accurate indicator of acute changes in the choline content of diet (60).

VIII. FACTORS THAT MAY INFLUENCE NUTRITIONAL STATUS

Amino acid–glucose solutions used in total parenteral nutrition of humans contain no choline (100,126). The lipid emulsions used to deliver extra calories and essential fatty acids during parenteral nutrition contain choline in the form of phosphatidylcholine (20% emulsion contains 13.2 mmol/L). Humans treated with parenteral nutrition required 1–1.7 mmol of choline-containing phospholipid all day during the first week of parenteral nutrition therapy to maintain plasma choline levels (100). Others (127) reported that plasma choline concentrations decreased in parenteral nutrition patients at the same time that they detected liver dysfunction. Conditions that enhance hepatic triglyceride synthesis (such as carbohydrate loading) increase the requirement for choline needed for export of triglyceride from liver (128). Thus, treatment of malnourished patients with high-calorie parenteral nutrition solutions at a time of depleted choline stores might enhance the likelihood of hepatic dysfunction. A recent clinical trial supports the requirement for supplemental choline during total parenteral nutrition (129,130).

ACKNOWLEDGMENTS

Some of the work described herein was supported by grants from the National Institutes of Health (AG09525, DK56350, DK55865) and the American Institute for Cancer Research.

REFERENCES

1. Zeisel, S. H. and Blusztajn, J. K. (1994). Choline and human nutrition. *Annu. Rev. Nutr. 14*, 269–296.
2. Zeisel, S. H., daCosta, K.-A., Franklin, P. D., Alexander, E. A., Lamont, J. T., Sheard, N. F. and Beiser, A. (1991). Choline, an essential nutrient for humans. *FASEB J. 5*, 2093–2098.
3. Zeisel, S. H. (1988). ''Vitamin-like'' molecules: choline. In *Modern Nutrition in Health and Disease* (M. Shils and V. Young, eds.), Lea & Febiger, Philadelphia, pp. 440–452.
4. Strecker, A. (1862). *Ann. Chem. Pharmacie. 123*, 353–360.
5. duVigneaud, V., Cohn, M., Chandler, J. P., Schenck, J. R. and Simmonds, S. (1941). The utilization of the methyl group of methionine in the biological synthesis of choline and creatine. *J. Biol. Chem. 140*, 625–641.
6. Kennedy, E. P. and Weiss, S. B. (1956). The function of cytidine coenzymes in the biosynthesis of phospholipids. *J. Biol. Chem. 222*, 193–214.
7. Best, C. H. and Huntsman, M. E. (1932). The effects of the components of lecithin upon the deposition of fat in the liver. *J. Physiol. 75*, 405–412.
8. Best, C. H. and Huntsman, M. E. (1935). Effect of choline on liver fat of rats in various states of nutrition. *J. Physiol. 83*, 255–274.
9. Copeland, D. H. and Salmon, W. D. (1946). The occurrence of neoplasms in the liver, lungs, and other tissues of rats as a result of prolonged choline deficiency. *Am. J. Pathol. 22*, 1059–1081.
10. Newberne, P. M. and Rogers, A. E. (1986). Labile methyl groups and the promotion of cancer. *Annu. Rev. Nutr. 6*, 407–432.
11. da Costa, K., Cochary, E. F., Blusztajn, J. K., Garner, S. C. and Zeisel, S. H. (1993). Accumulation of 1,2-*sn*-diradylglycerol with increased membrane-associated protein kinase C may be the mechanism for spontaneous hepatocarcinogenesis in choline deficient rats. *J. Biol. Chem. 268*, 2100–2105.
12. da Costa, K.-A., Garner, S. C., Chang, J. and Zeisel, S. H. (1995). Effects of prolonged (1

year) choline deficiency and subsequent refeeding of choline on 1,2,-sn-diradylglycerol, fatty acids and protein kinase C in rat liver. *Carcinogenesis 16*, 327–334.

13. Cohen, E. L. and Wurtman, R. J. (1975). Brain acetylcholine: increase after systemic choline administration. *Life Sci. 16*, 1095–1102.

14. Cohen, E. L. and Wurtman, R. J. (1976). Brain acetylcholine: control by dietary choline. *Science 191*, 561–562.

15. Haubrich, D. R., Wang, P. F., Clody, D. E. and Wedeking, P. W. (1975). Increase in rat brain acetylcholine induced by choline or deanol. *Life Sci. 17*, 975–980.

16. Haubrich, D. R., Wedeking, P. W. and Wang, P. F. (1974). Increase in tissue concentration of acetylcholine in guinea pigs in vivo induced by administration of choline. *Life Sci. 14*, 921–927.

17. Wecker, L. (1986). Neurochemical effects of choline supplementation. *Can. J. Physiol. Pharmacol. 64*, 329–333.

18. Wood, J. L. and Allison, R. G. (1982). Effects of consumption of choline and lecithin on neurological and cardiovascular systems. *Fed. Proc. 41*, 3015–3021.

19. Zeisel, S. H. and daCosta, K. A. (1990). Choline determination using gas chromatography/ mass spectrometry. *J. Nutr. Biochem. 1*, 55–59.

20. Pomfret, E. A., daCosta, K., Schurman, L. L. and Zeisel, S. H. (1989). Measurement of choline and choline metabolite concentrations using high-pressure liquid chromatography and gas chromatography-mass spectrometry. *Anal. Biochem. 180*, 85–90.

21. Maruyama, Y., Kusaka, M., Mori, J., Horikawa, A. and Hasegawa, Y. (1979). Simple method for the determination of choline and acetylcholine by pyrolysis gas chromatography. *J. Chromatog. 164*, 121–127.

22. Ishimaru, H., Ikarashi, Y. and Maruyama, Y. (1993). Use of high-performance liquid chromatography continuous-flow fast atom bombardment mass spectrometry for simultaneous determination of choline and acetylcholine in rodent brain regions. *Biol. Mass Spectrom. 22*, 681–686.

23. Barnes, N. M., Costall, B., Fell, A. F. and Naylor, R. J. (1987). An HPLC assay procedure of sensitivity and stability for measurement of acetylcholine and choline in neuronal tissue. *J. Pharm. Pharmacol. 39*, 727–731.

24. Damsma, G. and Flentge, F. (1988). Liquid chromatography with electrochemical detection for the determination of choline and acetylcholine in plasma and red blood cells. Failure to detect acetylcholine in blood of humans and mice. *J. Chromatog. 428*, 1–8.

25. Kaneda, N., Asano, M. and Nagatsu, T. (1986). Simple method for the simultaneous determination of acetylcholine, choline, noradrenaline, dopamine and serotonin in brain tissue by high-performance liquid chromatography with electrochemical detection. *J. Chromatogr. 360*, 211–218.

26. Ikarashi, Y., Sasahara, T. and Maruyama, Y. (1984). A simple method for determination of choline (Ch) and acetylcholine (ACh) in rat brain regions using high-performance liquid chromatography with electrochemical detection (HPLC-ED) [Jpn]. *Nippon Yakurigaku Zasshi Folia Pharmacologica Japonica 84*, 529–536.

27. Baker, H., Frank, O., Tuma, D. J., Barak, A. J., Sorrell, M. F. and Hutner, S. H. (1978). Assay for free and total choline activity in biological fluids and tissues of rats and man with Torulopsis pintolopessi. *Am. J. Clin. Nutr. 31*, 532–540.

28. Das, I., de Belleroche, J., Moore, C. J. and Rose, F. C. (1986). Determination of free choline in plasma and erythrocyte samples and choline derived from membrane phosphatidylcholine by a chemiluminescence method. *Analyt. Biochem. 152*, 178–82.

29. Gilberstadt, M. L. and Russell, J. A. (1984). Determination of picomole quantities of acetylcholine and choline in physiologic salt solutions. *Analyt. Biochem. 138*, 78–85.

30. Muma, N. A. and Rowell, P. P. (1985). A sensitive and specific radioenzymatic assay for the simultaneous determination of choline and phosphatidylcholine. *J. Neurosci. Meth. 12*, 249–257.

31. FASEB Life Sciences Research Office (1975). *Evaluation of the health aspects of choline chloride and choline bitartrate as food ingredients*. Report # PB-223 845/9. Bureau of Foods, Food and Drug Administration, Department of Health, Education, and Welfare, Washington, DC.

32. FASEB Life Sciences Research Office (1981). *Effects of consumption of choline and lecithin on neurological and cardiovascular systems*. Report # PB-82-133257. Bureau of Foods, Food and Drug Administration, Department of Health, Education, and Welfare, Washington, DC.

33. Zeisel, S. H. (1981). Dietary choline: biochemistry, physiology, and pharmacology. *Annu. Rev. Nutr. 1*, 95–121.

34. Zeisel, S. H., Growdon, J. H., Wurtman, R. J., Magil, S. G. and Logue, M. (1980). Normal plasma choline responses to ingested lecithin. *Neurology 30*, 1226–1229.

35. Rohlfs, E. M., Garner, S. C., Mar, M.-H. and Zeisel, S. H. (1993). Glycerophosphocholine and phosphocholine are the major choline metabolites in rat milk. *J. Nutr. 123*, 1762–1768.

36. Zeisel, S. H., Char, D. and Sheard, N. F. (1986). Choline, phosphatidylcholine and sphingomyelin in human and bovine milk and infant formulas. *J. Nutr. 116*, 50–58.

37. Zeisel, S. H. (1990). Choline deficiency. *J. Nutr. Biochem. 1*, 332–349.

38. Zeisel, S. H. (1993). Choline. In *Modern Nutrition in Health and Disease* (M. E. Shils, J. A. Olson, and M. Shike, eds.), Lea & Febiger, Philadelphia, pp. 449–458.

39. White, H. L. and Cavallito, C. J. (1970). Choline acetyltransferase. Enzyme mechanism and mode of inhibition by a styrylpyridine analogue. *Biochim. Biophys. Acta 206*, 343–58.

40. Haubrich, D. R., Wang, P. F. and Wedeking, P. W. (1975). Distribution and metabolism of intravenously administered choline [methyl-3-H] and synthesis in vivo of acetylcholine in various tissues of guinea pigs. *J. Pharmacol. Exp. Ther. 193*, 246–55.

41. Malthe, S. D. and Fonnum, F. (1972). Multiple forms of choline acetyltransferase in several species demonstrated by isoelectric focusing. *Biochem. J. 127*, 229–236.

42. Rama Sastry, B. V. and Henderson, G. I. (1972). Kinetic mechanisms of human placental choline acetyltransferase. *Biochem. Pharmacol. 21*, 787–802.

43. White, H. L. and Wu, J. C. (1973). Kinetics of choline acetyltransferases (EC 2.3.1.6) from human and other mammalian central and peripheral nervous tissues. *J. Neurochem. 20*, 297–307.

44. Wecker, L. and Dettbarn, W. D. (1979). Relationship between choline availability and acetylcholine synthesis in discrete regions of rat brain. *J. Neurochem. 32*, 961–967.

45. Trommer, B. A., Schmidt, D. E. and Wecker, L. (1982). Exogenous choline enhances the synthesis of acetylcholine only under conditions of increased cholinergic neuronal activity. *J. Neurochem. 39*, 1704–1709.

46. Wecker, L. (1988). Influence of dietary choline availability and neuronal demand on acetylcholine synthesis by rat brain. *J. Neurochem. 51*, 497–504.

47. Miller, L. G., Greenblatt, D. J., Roy, R. B., Lopez, F. and Wecker, L. (1989). Dietary choline intake modulates benzodiazepine receptor binding and gamma-aminobutyric acid A receptor function in mouse brain. *J. Pharmacol. Exper. Ther. 248*, 1–6.

48. Wecker, L., Cawley, G. and Rothermel, S. (1989). Acute choline supplementation in vivo enhances acetylcholine synthesis in vitro when neurotransmitter release is increased by potassium. *J. Neurochem. 52*, 568–575.

49. Mudd, S. H. and Poole, J. R. (1975). Labile methyl balances for normal humans on various dietary regimens. *Metab. Clin. Exp. 24*, 721–735.

50. Finkelstein, J. D., Martin, J. J., Harris, B. J. and Kyle, W. E. (1982). Regulation of the betaine content of rat liver. *Arch. Biochem. Biophys. 218*, 169–173.

51. Wong, E. R. and Thompson, W. (1972). Choline oxidation and labile methyl groups in normal and choline-deficient rat liver. *Biochim. Biophys. Acta. 260*, 259–271.

52. Finkelstein, J. D., Martin, J. J. and Harris, B. J. (1988). Methionine metabolism in mammals. The methionine-sparing effect of cystine. *J. Biol. Chem. 263*, 11750–4.

53. Zeisel, S. H., Zola, T., daCosta, K. and Pomfret, E. A. (1989). Effect of choline deficiency

on S-adenosylmethionine and methionine concentrations in rat liver. *Biochem. J. 259*, 725–729.

54. Horne, D. W., Cook, R. J. and Wagner, C. (1989). Effect of dietary methyl group deficiency on folate metabolism in rats. *J. Nutr. 119*, 618–621.

55. Selhub, J., Seyoum, E., Pomfret, E. A. and Zeisel, S. H. (1991). Effects of choline deficiency and methotrexate treatment upon liver folate content and distribution. *Cancer Res. 51*, 16–21.

56. Barak, A. J. and Kemmy, R. J. (1982). Methotrexate effects on hepatic betaine levels in choline-supplemented and choline-deficient rats. *Drug Nutr. Interact. 1*, 275–278.

57. Barak, A. J., Tuma, D. J. and Beckenhauer, H. C. (1984). Methotrexate hepatotoxicity. *J. Am. Coll. Nutr. 3*, 93–96.

58. Svardal, A. M., Ueland, P. M., Berge, R. K., Aarsland, A., Aarsaether, N., Lonning, P. E. and Refsum, H. (1988). Effect of methotrexate on homocysteine and other compounds in tissues of rats fed a normal or a defined, choline-deficient diet. *Cancer Chemother. Pharmacol. 21*, 313–318.

59. Freeman-Narrod, M., Narrod, S. A. and Custer, R. P. (1977). Chronic toxicity of methotrexate in rats: partial to complete projection of the liver by choline: brief communication. *J. Natl. Cancer Inst. 59*, 1013–1017.

60. Pomfret, E. A., da Costa, K. and Zeisel, S. H. (1990). Effects of choline deficiency and methotrexate treatment upon rat liver. *J. Nutr. Biochem. 1*, 533–541.

61. Kim, Y.-I., Miller, J. W., da Costa, K.-A., Nadeau, M., Smith, D., Selhub, J., Zeisel, S. H. and Mason, J. B. (1995). Folate deficiency causes secondary depletion of choline and phosphocholine in liver. *J. Nutr. 124*, 2197–2203.

62. Vance, D. E. (1990). Boehringer Mannheim Award lecture. Phosphatidylcholine metabolism: masochistic enzymology, metabolic regulation, and lipoprotein assembly. *Biochem. Cell. Biol. 68*, 1151–1165.

63. Vance, D. E. and Ridgway, N. D. (1988). The methylation of phosphatidylethanolamine. *Prog. Lipid Res. 27*, 61–79.

64. Blusztajn, J. K. and Wurtman, R. J. (1983). Choline and cholinergic neurons. *Science. 221*, 614–620.

65. Yao, Z. M. and Vance, D. E. (1988). The active synthesis of phosphatidylcholine is required for very low density lipoprotein secretion from rat hepatocytes. *J. Biol. Chem. 263*, 2998–3004.

66. Yao, Z. M. and Vance, D. E. (1989). Head group specificity in the requirement of phosphatidylcholine biosynthesis for very low density lipoprotein secretion from cultured hepatocytes. *J. Biol. Chem. 264*, 11373–11380.

67. Yao, Z. M. and Vance, D. E. (1990). Reduction in VLDL, but not HDL, in plasma of rats deficient in choline. *Biochem. Cell. Biol. 68*, 552–558.

68. Prescott, S. M., Zimmerman, G. A. and McIntyre, T. M. (1990). Platelet-activating factor. *J Biol Chem. 265*, 17381–17384.

69. Snyder, F. (1987). Enzymatic pathways for platelet-activating factor, related alkyl glycerolipids and their precursors. In *Platelet-Activating Factor and Related Lipid Mediators* (F. Syder, ed.), Plenum Press, New York, pp. 89–113.

70. Kumar, R. and Hanahan, D. J. (1987). Diversity of the biochemical and biological behavior of platelet-activating factor. In *Platelet-Activating Factor and Related Lipid Mediators* (F. Syder, ed.), Plenum Press, New York, pp. 239–254.

71. Chao, W. and Olson, M. S. (1993). Platelet-activating factor: receptors and signal transduction. *Biochem J. 292*, 617–629.

72. Berridge, M. J. (1989). Inositol trisphosphate, calcium, lithium, and cell signaling. *JAMA 262*, 1834–1841.

73. Taylor, C. W. and Marshall, I. (1992). Calcium and inositol 1,4,5-trisphosphate receptors: a complex relationship. *Trends Biochem Sci. 17*, 403–407.

74. Zeisel, S. H. (1993). Choline phospholipids: signal transduction and carcinogenesis. *FASEB J. 7*, 551–557.

75. Nishizuka, Y. (1992). Intracellular signaling by hydrolysis of phospholipids and activation of protein kinase C. *Science 258*, 607–614.

76. Exton, J. H. (1990). Signaling through phosphatidylcholine breakdown. *J. Biol. Chem. 265*, 1–4.

77. Qian, Z. and Drewes, L. R. (1990). A novel mechanism for acetylcholine to generate diacylglycerol in brain. *J. Biol. Chem. 265*, 3607–3610.

78. Conricode, K. M., Brewer, K. A. and Exton, J. H. (1992). Activation of phospholipase D by protein kinase C. Evidence for a phosphorylation-independent mechanism. *J. Biol. Chem. 267*, 7199–7202.

79. Liscovitch, M. (1992). Crosstalk among multiple signal-activated phospholipases. *Trends Biochem. Sci. 17*, 393–399.

80. Besterman, J. M., Duronio, V. and Cuatrecasas, P. (1986). Rapid formation of diacylglycerol from phosphatidylcholine: a pathway for generation of a second messenger. *Proc. Natl. Acad. Sci. USA. 83*, 6785–6789.

81. Wakelam, M. J. O., Cook, S. J., Currie, S., Plamer, S. and Plevin, R. (1991). Regulation of the hydrolysis of phosphatidylcholine in Swiss 3T3 cells. *Biochem. Soc. Trans. 19*, 321–324.

82. Asaoka, Y., Oka, M., Yoshida, K., Sasaki, Y. and Nishizuka, Y. (1992). Role of lysophosphatidylcholine in T-lymphocyte activation: involvement of phospholipase A2 in signal transduction through protein kinase C. *Proc Natl Acad Sci USA 89*, 6447–6451.

83. Sasaki, Y., Asaoka, Y. and Nishizuka, Y. (1993). Potentiation of diacylglycerol-induced activation of protein kinase C by lysophospholipids: Subspecies difference. *FEBS Lett. 320*, 47–51.

84. Stable, S. and Parker, P. J. (1991). Protein kinase C. *Pharmacol. Ther. 51*, 71–95.

85. Nishizuka, Y. (1986). Studies and perspectives of protein kinase C. *Science 233*, 305–312.

86. Weinstein, I. B. (1990). The role of protein kinase C in growth control and the concept of carcinogenesis as a progressive disorder in signal transduction. *Adv. Second Messenger Phosphoprotein Res. 24*, 307–316.

87. Hanada, K., Nishijima, M., Kiso, M., Hasegawa, A., Fujita, S., Ogawa, T. and Akamatsu, Y. (1992). Sphingolipids are essential for the growth of Chinese hamster ovary cells. Restoration of the growth of a mutant defective in sphingoid base biosynthesis by exogenous sphingolipids. *J. Biol. Chem. 267*, 23527–23533.

88. Merrill, A. H. and Jones, D. D. (1990). An update of the enzymology and regulation of sphingomyelin metabolism. *Biochim. Biophys. Acta. 1044*, 1–12.

89. Kim, M. Y., Linardic, C., Obeid, L. and Hannun, Y. (1991). Identification of sphingomyelin turnover as an effector mechanism for the action of tumor necrosis factor α and gamma-interferon. Specific role in cell differentiation. *J. Biol Chem. 266*, 484–489.

90. Dressler, K. A., Mathias, S. and Kolesnick, R. N. (1992). Tumor necrosis factor-α activates the sphingomyelin signal transduction pathway in a cell-free system. *Science 255*, 1715–1718.

91. Obeid, L. M., Linardic, C. M., Karolak, L. A. and Hannun, Y. A. (1993). Programmed cell death induced by ceramide. *Science 259*, 1769–1771.

92. Michael, U. F., Cookson, S. L., Chavez, R. and Pardo, V. (1975). Renal function in the choline deficient rat. *Proc. Soc. Exp. Biol. Med. 150*, 672–676.

93. Baxter, J. H. (1947). A study of hemorrhagic-kidney syndrome of choline deficiency. *J. Nutr. 34*, 333.

94. Best, C. H. and Hartroft, W. S. (1949). Symposium on nutrition in preventative medicine: Nutrition, renal lesions and hypertension. *Fed. Proc. 8*, 610.

95. Griffith, W. H. and Wade, N. J. (1939). The occurance and prevention of hemorrhagic degeneration in young rats on a low choline diet. *J. Biol. Chem. 131*, 567–573.

96. Chang, C. H. and Jensen, L. S. (1975). Inefficacy of carnitine as a substitute for choline for normal reproduction in Japanese quail. *Poultry Sci. 54*, 1718–1720.

97. Jukes, T. H. (1940). The prevention of perosis by choline. *J. Biol. Chem. 134*, 789–792.

98. Kratzing, C. C. and Perry, J. J. (1971). Hypertension in young rats following choline deficiency in maternal diets. *J. Nutr. 101*, 1657–1661.

99. Caniggia, A. (1950). Effect of choline on hemopoiesis. *Haematologica 34*, 625–627.

100. Sheard, N. F., Tayek, J. A., Bistrian, B. R., Blackburn, G. L. and Zeisel, S. H. (1986). Plasma choline concentration in humans fed parenterally. *Am. J. Clin. Nutr. 43*, 219–224.

101. Chawla, R. K., Wolf, D. C., Kutner, M. H. and Bonkovsky, H. L. (1989). Choline may be an essential nutrient in malnourished patients with cirrhosis. *Gastroenterology 97*, 1514–1520.

102. Lieber, C. S., Robins, S. J., Li, J., DeCarli, L. M., Mak, K. M., Fasulo, J. M. and Leo, M. A. (1994). Phosphatidylcholine protects against fibrosis and cirrhosis in the baboon. *Gastroenterology 106*, 152–159.

103. Chandar, N. and Lombardi, B. (1988). Liver cell proliferation and incidence of hepatocellular carcinomas in rats fed consecutively a choline-devoid and a choline-supplemented diet. *Carcinogenesis 9*, 259–263.

104. Ghoshal, A. K., Ahluwalia, M. and Farber, E. (1983). The rapid induction of liver cell death in rats fed a choline-deficient methionine-low diet. *Am. J. Pathol. 113*, 309–314.

105. Shinozuka, H. and Lombardi, B. (1980). Synergistic effect of a choline-devoid diet and phenobarbital in promoting the emergence of foci of γ-glutamyltranspeptidase-positive hepatocytes in the liver of carcinogen-treated rats. *Cancer Res. 40*, 3846–3849.

106. Locker, J., Reddy, T. V. and Lombardi, B. (1986). DNA methylation and hepatocarcinogenesis in rats fed a choline devoid diet. *Carcinogenesis 7*, 1309–1312.

107. Dizik, M., Christman, J. K. and Wainfan, E. (1991). Alterations in expression and methylation of specific genes in livers of rats fed a cancer promoting methyl-deficient diet. *Carcinogenesis 12*, 1307–1312.

108. Rushmore, T., Lim, Y., Farber, E. and Ghoshal, A. (1984). Rapid lipid peroxidation in the nuclear fraction of rat liver induced by a diet deficient in choline and methionine. *Cancer Lett. 24*, 251–255.

109. Blusztajn, J. K. and Zeisel, S. H. (1989). 1,2-*sn*-diacylglycerol accumulates in choline-deficient liver. A possible mechanism of hepatic carcinogenesis via alteration in protein kinase C activity? *FEBS Lett. 243*, 267–270.

110. Hollstein, M., Sidransky, D., Vogelstein, B. and Harris, C. C. (1991). p53 mutations in human cancers. *Science. 253*, 49–53.

111. Albright, C. D., Lui, R., Bethea, T. C., da Costa, K.-A., Salganik, R. I. and Zeisel, S. H. (1996). Choline deficiency induces apoptosis in SV40-immortalized CWSV-1 rat hepatocytes in culture. *FASEB J. 10*, 510–516.

112. Arends, M. J. and Wyllie, A. H. (1991). Apoptosis mechanisms and role in pathology. *Int. Rev. Exp. Pathol. 32*, 223–254.

113. Lane, D. P., Lu, X., Hupp, T. and Hall, P. A. (1994). The role of the p53 protein in the apoptotic response. *Phil. Trans. 8*, 277–280.

114. Zeisel, S. H., Albright, C. D., Shin, O.-K., Mar, M.-H., Salganik, R. I. and da Costa, K.-A. (1997). Choline deficiency selects for resistance to p53-independent apoptosis and causes tumorigenic transformation of rat hepatocytes. *Carcinogenesis.* (in press).

115. Welsch, F. (1976). Studies on accumulation and metabolic fate of (N-Me^{3}H)choline in human term placenta fragments. *Biochem. Pharmacol. 25*, 1021–1030.

116. Zeisel, S. H., Epstein, M. F. and Wurtman, R. J. (1980). Elevated choline concentration in neonatal plasma. *Life Sci. 26*, 1827–1831.

117. Braun, L. D., Cornford, E. M. and Oldendorf, W. H. (1980). Newborn rabbit blood-brain barrier is selectively permeable and differs substantially from the adult. *J. Neurochem. 34*, 147–152.

118. Blusztajn, J. K., Zeisel, S. H. and Wurtman, R. J. (1985). Developmental changes in the activity of phosphatidylethanolamine N-methyltransferases in rat brain. *Biochem. J. 232*, 505–511.

119. Meck, W. H., Smith, R. A. and Williams, C. L. (1988). Pre- and postnatal choline supplementation produces long-term facilitation of spatial memory. *Dev. Psychobiol. 21*, 339–353.

120. Loy, R., Heyer, D., Williams, C. L. and Meck, W. H. (1991). Choline-induced spatial memory facilitation correlates with altered distribution and morphology of septal neurons. *Adv. Exp. Med. Biol. 295*, 373–382.

121. Meck, W. H., Smith, R. A. and Williams, C. L. (1989). Organizational changes in cholinergic activity and enhanced visuospatial memory as a function of choline administered prenatally or postnatally or both. *Behav. Neurosci. 103*, 1234–1241.

122. Durkin, T. (1989). Central cholinergic pathways and learning and memory processes: presynaptic aspects. *Comp. Biochem. Physiol. A: Comp. Physiol. 93*, 273–280.

123. Mandel, R. J., Gage, F. H. and Thal, L. J. (1989). Spatial learning in rats: correlation with cortical choline acetyltransferase and improvement with NGF following NBM damage. *Exp. Neurol. 104*, 208–217.

124. Fibiger, H. C. (1991). Cholinergic mechanisms in learning, memory and dementia: a review of recent evidence. *Trends NeuroSci. 14*, 220–223.

125. Conlay, L. A., Wurtman, R. J., Blusztajn, K., Coviella, I. L., Maher, T. J. and Evoniuk, G. E. (1986). Decreased plasma choline concentrations in marathon runners [letter]. *N. Engl. J. Med. 315*, 892.

126. Chawla, R. K., Berry, C. J., Kutner, M. H. and Rudman, D. (1985). Plasma concentrations of transsulfuration pathway products during nasoenteral and intravenous hyperalimentation of malnourished patients. *Am. J. Clin. Nutr. 42*, 577–584.

127. Burt, M. E., Hanin, I. and Brennan, M. F. (1980). Choline deficiency associated with total parenteral nutrition. *Lancet 2*, 638–639.

128. Carroll, C. and Williams, L. (1982). Choline deficiency in rats as influenced by dietary energy. *Nutr. Rep. Int. 25*, 773–782.

129. Buchman, A. L., Dubin, M., Jenden, D., Moukarzel, A., Roch, M. H., Rice, K., Gornbein, J., Ament, M. E. and Eckhert, C. D. (1992). Lecithin increases plasma free choline and decreases hepatic steatosis in long-term total parenteral nutrition patients. *Gastroenterology 102*, 1363–1370.

130. Buchman, A. L., Moukarzel, A., Jenden, D. J., Roch, M., Rice, K. and Ament, M. E. (1993). Low plasma free choline is prevalent in patients receiving long term parenteral nutrition and is associated with hepatic aminotransferase abnormalities. *Clin. Nutr. 12*, 33–37.

15

Ascorbic Acid

CAROL S. JOHNSTON

Arizona State University East, Mesa, Arizona

FRANCENE M. STEINBERG and ROBERT B. RUCKER

University of California, Davis, California

I. INTRODUCTION AND HISTORY

Ascorbic acid (vitamin C) functions as a redox cofactor and catalyst in a broad array of biochemical reactions and processes. In humans, vitamin C cures and prevents scurvy, hence the designation ascorbic acid (1). Scurvy or "scurfy" (Old English) was probably derived from the Scandinavian terms, skjoerberg and skorbjugg, meaning rough skin. Aside from famine, scurvy has caused the most suffering of nutritional origin in human history.

Those without access to fresh fruits and vegetables are susceptible to scurvy. One-half of the original 60 colonists of Plymouth died of scurvy in 1628. The military of the seventeenth and eighteenth centuries often adhered to dietary protocols that promoted scurvy. During the Civil War, poor nutrition resulting in scurvy, as well as pellagra, also took its toll (2,3). Excellent accounts of the history of scurvy have been presented by Carpenter (4) and Clemetson (5), and the *Treatise on the Scurvy* by James Lind published in 1753 (6) is considered the first recorded account of a controlled clinical trial.

By the end of the 1800s, the connection between scurvy and diet was established. The observation in 1907 that guinea pigs were susceptible to scurvy was an important breakthrough in the understanding of scurvy. It was also one of the earliest examples of the use of an animal model to study a nutritional disease. By 1915, Zilva and his associates of the Lister Institute in London had isolated antisorbutic activity from a crude fraction of lemon. Work using animal assays demonstrated that the activity was destroyed by oxidation and protected by reducing agents (8). Important to the evolving nomenclature for vitamins, it was suggested that the new antiscorbic factor be designated "factor or vitamin C" since "A" and "B" had been previously designated as potential health and growth factors or vitamins (9).

Throughout the 1930s, work progressed rapidly with validation and identification of vitamin C in a number of foods. Early papers by Szent-Gyorgyi, Haworth, King, and their colleagues document this effort and provide chemical identification and elucidation of ascorbic acid structure (10–15).

II. CHEMICAL FEATURES

A. Nomenclature and Structure

Ascorbic acid or L-ascorbic acid is the IUPAC-IUB Commission designation for vitamin C (2-oxo-L-theo-hexono-4-lactone-2, 3-enediol). The chemical structures of ascorbic acid and selected derivatives are given in Fig. 1. Ascorbic acid has a near planar five-member ring; the two chiral centers at positions 4 and 5 determine the four stereoisomers. Dehydroascorbic acid, the oxidized form of ascorbic acid, retains some vitamin C activity and can exist as a hydrated hemiketal, or as a dimer (16).

B. Physical and Chemical Properties

Table 1 summarizes physical and chemical features of ascorbic acid. Data are also available from x-ray crystallographic analysis, [^{1}H] and [^{13}C] NMR spectroscopy, IR- and UV-spectroscopy, and mass spectroscopy (see 16–19 and references cited). The reversible oxidation to semidehydro-L-ascorbic acid (and further to dehydro-L-ascorbic acid) is the most important chemical property of ascorbic acid. This property is the basis for its known physiological activities. The proton on oxygen-3 is acidic ($pK_1 = 4.17$) and contributes to the acidic nature of the vitamin.

Degradation reactions of L-ascorbic acid in aqueous solutions depend on a number of factors such as pH, temperature, and the presence of oxygen or metals. Alkali-catalyzed degradation results in over 50 compounds, mainly mono-, di-, and tricarboxylic acids (15,18,19). The vitamin can be stabilized in biological samples with trichloracetic acid or metaphosphoric acid (20). In general, ascorbic acid is not stable in aqueous media, e.g., culture media or parenteral/enteral solutions, at typical room temperatures. Ascorbic acid is reasonably stable in blood, when stored at or below −20 C (20–22).

The ascorbate radical is an important intermediate in redox reactions (15,18,23,24). The physiologically dominate ascorbic acid mono anion ($pK_1 = 4.1$) and dianions ($pK_2 = 11.79$) are shown in Fig. 1.

The ascorbate/ascorbate monoanion reduction potential is low in comparison to other redox systems (25). Ascorbic acid readily scavenges reactive oxygen species; rate constants 10^4–$10^8 \cdot mol^{-1}/L \cdot s^{-1}$ (23,24). Moreover, since the ascorbate radical has a high disproportion rate under physiological conditions and converts rapidly to ascorbic acid and dehydroascorbic acid without reacting with adjacent molecules, ascorbic acid is well suited as a physiological antioxidant (25). In biological systems, accessory enzymatic systems are present to reduce dehydroascorbic acid. This recycling of ascorbic acid helps to maintain ascorbic acid in tissues. Further, the process diminishes the possibility of an abnormal excess of ascorbic acid radical and nonspecific oxidations.

In plants, NADH: monodehydroascorbate reductase (EC 1.6.5.4) evolved to maintain ascorbic acid in its reduced form, and plays a major role in the stress response of

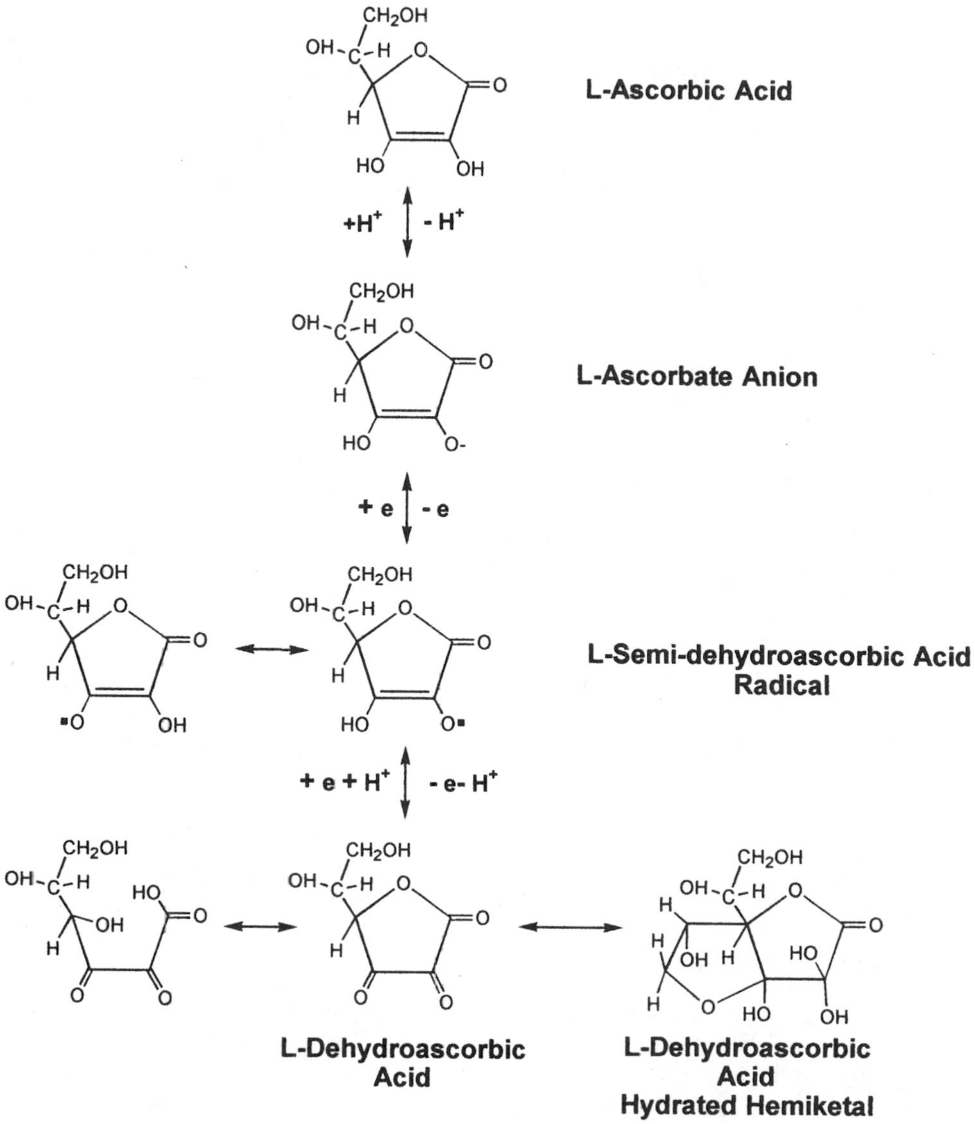

Fig. 1 Ascorbic acid and various oxidation products. The two predominant forms of ascorbic acid and their associated oxidation products are shown. In solution, ascorbic acid probably exists as the hydrated hemiketal.

plants. In animal tissues, glutathione dehydroascorbate reductase (EC 1.8.5.1) is the primary enzyme that serves this purpose (26). Dehydroascorbic acid (DHA) is a more stable oxidation product of the ascorbyl radical. Dehydroascorbic acid is thought to exist mainly in vivo as the hydrated hemiketal and has two possible metabolic fates (Fig. 1). It may either be reduced to ascorbate or irreversibly hydrolyzed to a variety of metabolites including oxalate. The reduction of dehydroascorbic acid is therefore very important in maintaining cellular ascorbate levels.

Table 1 Selected Physical Properties of Ascorbic Acid

Empirical formula	$C_6H_8O_6$
Molar mass	176.13
Crystalline form	Monoclinic, mix of platelets and needles
Melting point	190–192°C
Optical rotation	[α] 25/D +20.5° to 21.51°
	(cm = 1 in water)
pH, at 5 mg/ml	~3
at 50 mg/ml	~2
pK_1	4.17
pK_2	11.57
Redox potential (dehydroascorbic acid/ascorbate)	−174 mV
(ascorbate •—, H+/ascorbate−)	+282 mV
Solubility, g/ml	
water	0.33
ethanol, abs.	0.02
ether, chloroform, benzene	Insoluble
Absorption spectra	
at pH 2	E_{max} (1%, 10 mm) 695 at 245 nm
at pH 6.4	E_{max} (1%, 10 mm) 940 at 265 nm

C. Isolation

Ascorbic acid is stable in many organic and inorganic acids. Metaphosphoric acid-containing ethylenediamine tetra-acetic acid (0.5–2%), oxalic acid, dilute trichloracetic acid or dilute perchloric acid containing reducing agents, such as 2,3-dimercaptopropanol, are often used as solvent for tissue extraction (15,20). Extraction of ascorbic acid should be carried out under subdued light and an inert atmosphere to avoid the potential for degradation (20).

D. Chemical Synthesis

The chemical pathway for the industrial synthesis of ascorbic acid from glucose was first developed in the 1930s and continues to be used (Fig. 2). Initially, glucose is converted to L-sorbitol by hydrogenation; subsequent fermentation with Acetobacter suboxidans yields L-sorbose. Next, a carboxyl group is introduced at the C_1 position with derivatization to diacetone-2-keto-L-gulonic acid. Removal of acetone and heating under acid conditions yields L-ascorbic acid, which may be crystallized in high purity.

The need for and utilization of commercial sources of ascorbic acid is high. In the United States, the annual per capita use is 25–30 mg of ascorbic acid, 9–10 mg of sodium ascorbate, 6 mg of erythorbic acid, the D-isomer of ascorbic acid, and 6–8 mg sodium erythorbic acid (27). Erythorbic acid is a common food additive. Both forms of ascorbic acid are inhibitors of enzymatic browning reactions and are used as preservative antioxidants. Note that erythorbic acid does not possess antiscorbutic activity, nor does it act as an ascorbic acid antagonist in vivo (28).

Selective derivatization of ascorbic acid can be difficult, because of delocation of the negative charge of ascorbate in its anionic form. If the C-2 and C-3 hydroxyl groups are protected, however, base-promoted alkylation or acylation can take place at the more sterically accessible primary hydroxyl group on C-6, but not at C-5. Reactions at the C-

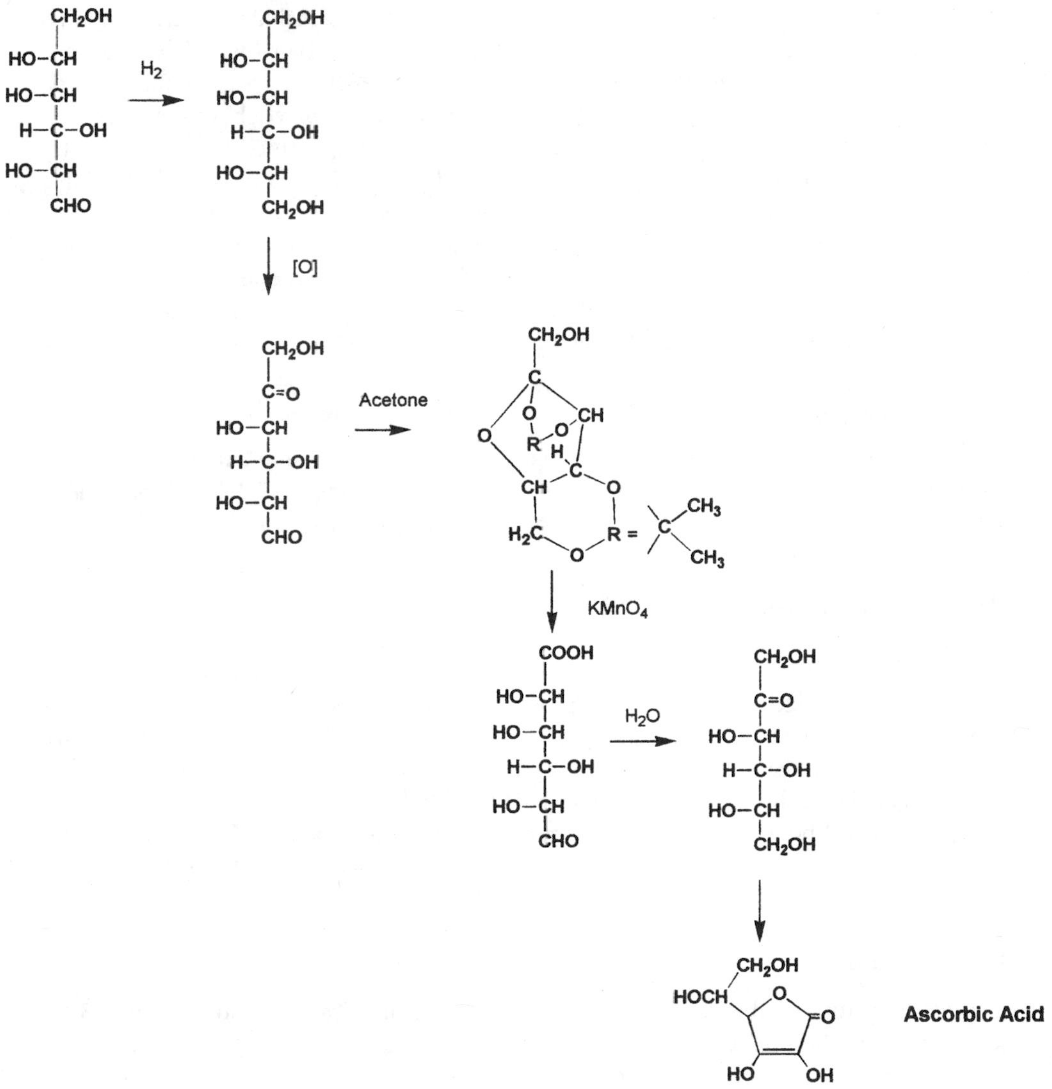

Fig. 2 Chemical synthesis of ascorbic acid from D-glucose.

5 position often require derivatizations at the C-2, C-3, and C-6 positions (25,26). The formation of acetates or ketals of ascorbic acid is useful for protection of the molecule while reactions at the other carbons are carried out.

E. Analysis

Numerous high-performance liquid chromatographic methods have been developed for determination of ascorbic acid and related isomers or derivatives. Electrochemical detection is generally used for measuring ascorbic acid and derivatives. For example, chromatographic approaches include both the ion exchange, gas, and reversed-phase chromatography (29–36).

Colorimetric assays of ascorbic acid in crude mixtures include the 2,2′-dipytidyl calorimetric method, which is based on the reduction of Fe (III) to Fe (II) by ascorbic acid (33), and the 2,4-dinitrophenylhydrazine method (37). Methods based on fluorometric and chemiluminescence detection also provide highly sensitive approaches for the determination ascorbic acid (33,34). Further, conventional and isotope ratio mass spectrometry techniques have been used to analyze ascorbic acid when ^{13}C ascorbic acid is available for use as a reference or standard in the analysis of complex matrices (35).

As a final point, since ascorbic acid is a reductant and can cause nonspecific color formation, the presence of ascorbic acid may interfere with many chemical tests, including the analysis of glucose, uric acid, creatinine, bilirubin, glycohemoglobin, hemoglobin A, cholesterol, triglycerides, leukocytes, and inorganic phosphate (36,38–41).

F. Sources of Ascorbic Acid

Ascorbic acid occurs in significant amounts in vegetables, fruits, and animal organs such as liver, kidney, and brain. In many European cultures, potatoes and cabbage have been important sources of vitamin C. Typical values are given in Table 2.

III. BIOCHEMICAL FUNCTIONS

A. Plants

Ascorbic acid is detected in yeast and prokaryotes, with the exception of cyanobacteria (15). In higher plants, ascorbic acid is synthesized from D-glucose and functions chiefly as a reductant, protecting and participating in many metabolic processes. For example, ascorbic acid is crucial for the scavenging of H_2O_2 in plants (42) and functions to maintain the α-tocopherol pool, which scavenges radicals in plant membranes (43). Ascorbic acid is the cofactor for violaxanthin depoxidase, which forms zeaxanthin, an important factor in the stress response of plants (44).

B. Animals

The pathway for ascorbic acid synthesis from glucose in animals is shown in Fig. 3. A key enzyme in the synthesis, L-gulonolactone oxidase (GLO, EC 1.1.3.8), resides in the kidney of reptiles, but during the course of evolution, the activity was transferred to the liver of mammals. For reasons that are not clear, the ability to express L-glulonolactone oxidase is absent in the guinea pig, some fruit-eating bats, and most primates, including man. In these animals, signs of ascorbic acid deprivation include a range of symptoms that can be related to the failure of specific enzymatic steps and processes that require ascorbic acid as cofactor (Table 3). As given in Fig. 3, ascorbic acid can be converted to ascorbic acid 2-sulfate and methylated ascorbic acid derivatives. Although the exact role of such derivatives is unknown, they may be important to intracellular partitioning, transport, or storage or protection from cellular excesses of ascorbic acid. Decarboxylation at the C-1 position or cleavage to oxalic acid and four carbon fragments occur when ascorbic acid is in excess (5,15).

1. Ascorbic Acid and Glutathione Interrelationships

Glutathione L-glutamyl-L-cysteine-glycine (GSH) is widely distributed in plant and animal cells and functions predominately as an antioxidant scavenging reactive oxygen radicals.

Table 2 Vitamin C in Selected Foods

Animal products	mg/100 g of edible portion
Cow's milk	0.5–2
Human milk	3–6
Beef	1–2
Pork	1–2
Veal	1–1.5
Ham	20–25
Liver, chicken	15–20
Beef	10
Kidney, chicken	6–8
Heart, chicken	5
Gizzard, chicken	5–7
Crab muscle	1–4
Lobster	3
Shrimp muscle	2–4
Fruits	**mg/100 g of edible portion**
Apple	3–30
Banana	8–16
Blackberry	8–10
Cherry	15–30
Currant, red	20–50
Currant, black	150–200
Grape	2–5
Grapefruit	30–70
Kiwi fruit	80–90
Lemon	40–50
Melons	9–60
Mango	10–15
Orange	30–50
Pear	2–5
Pineapple	15–25
Plums	2–3
Rose hips	250–800
Strawberry	40–70
Tomato	10–20
Vegetable	**mg/100 g of edible portion**
Asparagus	15–30
Avocado	10
Broccoli	80–90
Beet	6–8
Beans, various	10–15
Brussels sprout	100–120
Cabbage	30–70
Carrot	5–10
Cucumber	6–8
Cauliflower	50–70
Eggplant	15–20
Chive	40–50
Kale	70–100
Lettuce, various	10–30

Table 2   Continued

Onion	10–15
Pea	8–12
Potato	4–30
Pumpkin	15
Radish	25
Spinach	35–40
Spices and condiments	**mg/100 g of edible portion**
Chicory	33
Coriander (spice)	90
Garlic	16
Horseradish	45
Leek	15
Parsley	200–300
Papaya	39
Pepper, various	150–200

GSH is synthesized by a two-step reaction involving L-glutamyl cysteine synthetase and GSH synthetase. In addition, GSH-dependent reduction of intracellular dehydroascorbic acid to ascorbate is believed to play a crucial role in ascorbic acid recycling in vivo, thereby enhancing the antioxidant reserve as it relates to ascorbic acid (45–48).

When GSH synthesis is blocked, e.g., by use of inhibitors, such as L-buthionine-(SR)-sulfoximine, newborn animals die within a few days due to oxidative stress (45). The cellular damage involves mostly abnormal alterations in mitochondria that lead to proximal renal tubular damage, liver damage, and disruption of lamellar bodies in lung (45). Furthermore, the buthionine-(SR)-sulfoximine-induced loss in GSH is associated with a marked increase in dehydroascorbic acid. Interestingly, ascorbic acid administration ameliorates most signs of chemically induced GSH deficiency in rats (45). In guinea pigs, treatment with GSH ester significantly delays the onset of scurvy (49). The sparing effect is probably due to the need for both ascorbic acid and GSH in counteracting the deleterious effects of reactive oxidant species (Fig. 4).

2.   Norepinephrine Synthesis

The dopamine-β-hydroxylase mediated conversion of dopamine to norepinephrine (Fig. 5) is dependent on ascorbic acid (50,51) and this may explain in part the high concentration of ascorbic acid in adrenal tissue. Dopamine β-hydroxylase (EC 1.14.17.1) is present in catecholamine storage granules in nervous tissues and in chromaffin cells of adrenal medulla, the site of the final and rate-limiting step in the synthesis of norepinephrine. Dopamine-β-hydroxylase is a tetramer containing two Cu (I) ions per monomer that consumes ascorbate stoichiometrically with O_2 during its cycle. At a steady state the predominant enzyme form is an enzyme-product complex, and the primary function of ascorbate is to maintain copper in a reduced state. Only the reduced enzyme seems to be catalytically competent, with bound cuprous ions as the only reservoir of reducing equivalents.

3.   Hormone Activation

Many hormones and hormone-releasing factors are activated by posttranslational steps involving α-amidations (52,53). Examples of hormone activation include melanotropins,

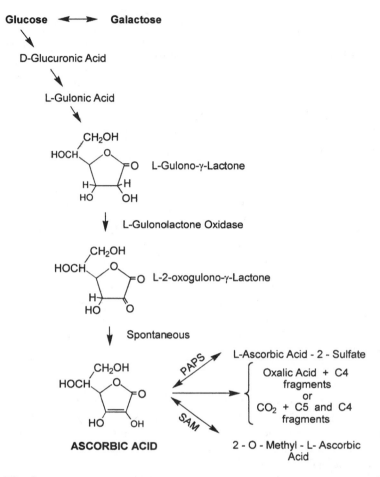

Fig. 3 Pathway for ascorbic acid synthesis from glucose in animal cells and examples of common adducts and degradation products.

calcitonin, releasing factors for growth hormone, corticotrophin and thyrotropin, pro-ACTH, vasopressin, oxytocin, cholecystokinin, and gastrin. Petidylglycine α-amidating mono-oxygenase (EC 1.14.17.3), which catalyzes α-amidations, is ascorbic acid dependent (54) and found in secretory granules of neuroendocrine cells in the brain, pituitary, thyroid, and submaxillary glands. The activation process involves C-terminal amidation and release of glyoxylate (Fig. 5) and requires copper (53).

4. Ascorbic Acid as an Antioxidant

The Food and Nutrition Board's panel on Dietary Antioxidant and Related Compounds of the NAS has defined an antioxidant as "any substance that, when present at low concentrations compared to those of an oxidizable substrate (e.g., proteins, lipids, carbohydrates, and nucleic acids), significantly delays or prevents oxidation of that substrate" (55). Ascorbic acid readily scavenges reactive oxygen and nitrogen species, such as superoxide and hydroperoxyl radicals, aqueous peroxyl radicals, singlet oxygen, ozone, peroxynitrite, nitrogen, dioxide, nitroxide radicals and hypochlorous acid. Moreover, ascorbic acid supplementation has been associated with reduced lipid, DNA, and protein oxidation in experimental systems (23,24).

Table 3 Functions of Ascorbic Acid Associated with Specific Enzymes

Function	Associated enzyme(s)	Associated mechanism and features
Extracellular matrix maturation (Collagen biosynthesis)	Prolyl-3-hydroxylase Prolyl-4-hydroxylase Lysyl hydroxylase	Dioxygenase; Fe^{+2}
Cq1 complement synthesis	Prolyl-4-hydroxylase	Dioxygenase; Fe^{+2}
Carnitine biosynthesis		Dioxygenase; Fe^{+2}
	6-N-Trimethyl-L-lysine hydroxylase	
	γ-Butyrobetaine hydroxylase	
Pyridine metabolism	Pyrimidine deoxyribonucleoside Hydroxylase (fungi)	Dioxygenase; Fe^{+2}
Cephalosporin synthesis	Deacetoxycephalosporin C Synthetase	Dioxygenase; Fe^{+2}
Tyrosine metabolism	Tyrosine-4-hydroxyphenylpyruvate hydrolase	Dioxygenase; Fe^{+2}
Norepinephrine biosynthesis	Dopamine-β-monooxygenase or hydrolase	Monooxygenase; Cu^{+1}
Peptidylglycine-α-amidation in the activation of hormones	Peptidylglycine-α-amidating Monooxygenase	Monooxygenase; Cu^{+1}

Lipids.   Although mostly inferential, numerous in vitro and in vivo studies have examined on the ability of ascorbic acid to reverse lipid peroxidation (61–69). When lipid peroxyl radicals are generated in lipoprotein fractions, vitamin C is consumed faster than other antioxidants, e.g., uric acid, bilirubins, and vitamin E (56–63). Ascorbic acid was 10^3 times more reactive than polyunsaturated fatty acids in reacting with peroxyl radicals (63), but ascorbic acid was not as effective in scavenging hydroxyl or alkoxyl radicals.

In assays for lipid peroxidation, low density lipoprotein (LDL) particles are often used as the lipid source (56–63). LDL oxidation is estimated by the lag time and propagation rate of lipid peroxidation in LDL exposed to copper ions or other catalyst. Numerous studies in smokers, nonsmokers, hypercholesterolemic subjects or healthy controls, as well as in animal models, support the observation that ascorbic acid generally retards peroxyl radical formation in LDL. The data are particularly convincing when ascorbic acid is combined with vitamin E. In studies utilizing only vitamin C, however, the effect on LDL oxidation is more varied (see Ref. 65 for excellent summary of work in this area).

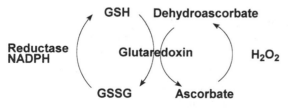

Fig. 4 Interaction between ascorbic acid and glutathione in its reduced (GSH and oxidized) forms (GSSG).

In vitro, both ascorbic acid and vitamin E may display pro-oxidant properties and promote lipid peroxidation (66). Ascorbate and tocopherol radicals at the surface of LDL particles are capable of initiating free-radical reactions (66). However, in the complexity of biological systems such as serum, vitamin C does not appear to have pro-oxidant effects (67).

DNA. In somatic cells, oxidative damage to DNA increases the risk of mutations, and in turn, is a risk factor in cancer or birth defects. Deoxyguanosine is a common target of oxidative modifications with 8-oxoguanosine and its respective nucleoside, 8-oxo-deoxyguanosine as products. These modified nucleic acid bases are found in all cells and urine (68–72). Based on measurements of 8-oxo-6-deoxyguanosine derivatives and other modified bases, tissues appear to have the potential to repair 10^4 to 10^5 oxidative lesions per cell per day.

Fraga et al. (73,74) were among the first to demonstrate a significant decrease in human sperm 8-oxo-deoxyguanosine levels following vitamin C supplementation. However, the nucleus of many cells contains relatively low concentrations of ascorbic acid, and whether ascorbic acid serves to protect DNA from oxidative damage remains unresolved (75–77).

Protein. Examples of protein oxidation include the life-long oxidation of long-lived proteins, such as the crystallins in the lens of the eye, the oxidation of α-proteinase inhibitor, and the advanced glycation end products associated with diabetes. Tyrosine, N-terminal amino acids, and cysteine are often targets of such reactions (78–81). Although data are limited, ascorbic acid supplementation may protect some proteins from oxidation. Protein carbonyl formation is reduced upon ascorbic acid repletion in guinea pigs (82).

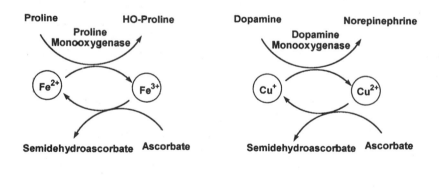

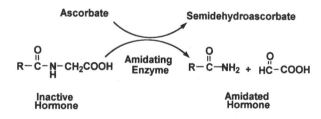

Fig. 5 Reactions catalyzed by mono- and dioxygenases for which ascorbic acid and iron or copper are catalyst. Reaction sequences for hydroxyprolyl synthesis, norepinephrine synthesis, and C-terminal amidation formation are indicated.

Carnitine Biosynthesis

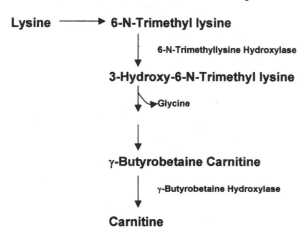

Fig. 6   Steps in carnitine biosynthesis.

Vitamin C supplementation has also been shown to reduce to formation of nitrotyrosine levels in patients with *Helicobacter pylori* gastritis (81).

5. Carnitine Biosynthesis

It has been suggested that early features of scurvy (fatigue and weakness) may be attributed to carnitine deficiency (83). Ascorbate is a cofactor of two-enzyme hydroxylation in the pathway of carnitine biosynthesis, γ-butyrobetaine hydroxylase and ε-N-trimethyllsine hydroxylase (84, Fig. 6). High doses of ascorbic acid in guinea pigs fed high-fat diets enhanced carnitine synthesis (85), and ascorbate deficiency was associated with up to a 50% decrease in heart and skeletal muscle carnitine compared to control guinea pigs (37,83,86).

6. Collagen Assembly

The role of ascorbic acid in extracellular matrix (ECM) regulation is fundamental to understanding the physiology of scurvy and is addressed in Chapter 16. In brief, ascorbic acid maintains iron in the reduced state, enhancing prolyl and lysyl hydroxylase activity and hence collagen assembly (88, Fig. 5).

IV. ASCORBIC ACID METABOLISM AND REGULATION

Many cells accumulate ascorbic acid against a concentration gradient. Intracellular concentrations of ascorbic acid are up to 40-fold higher than plasma concentrations (89, Table 4). When activated, neutrophils accumulate ascorbate with intracellular ascorbic acid levels that range from 2 mM to as much as 10 mM. Although simple diffusion accounts for some of this movement, ascorbic acid transport is primarily carrier-mediated (90–93). Ascorbic acid enters cells on a sodium- and energy-dependent transporter, and with the exception of intestine, simple diffusion is of minor importance.

Most of the intracellular ascorbate, however, is derived from rapid conversion of dehydroascorbic acid (20,94). Dehydroascorbic acid enters cells on GLUT 1, 2, or 4 transporters, and this transport is inhibited by glucose in vitro. Of clinical significance, it has

Table 4 Tissue Concentrations of Ascorbic Acid

Tissue	Rat (mg/100 g)	Human (mg/100 g)
Adrenal glands	280–400	30–40
Pituitary gland	100–130	40–50
Liver	25–40	10–16
Spleen	40–50	10–15
Lungs	20–40	7
Kidneys	15–20	5–15
Testes	25–30	3
Thyroid	22	2
Thymus	40	
Brain	35–50	13–15
Pancreas	10–16	
Eye lens	8–10	25–31
Skeletal muscle	5	3–4
Heart muscle	5–10	5–15
Bone marrow	12	
Plasma	1.6	0.4–1.0
Saliva	0.07–0.09	

been postulated that diabetics may have compromised ascorbic status due in part to the inference by glucose of dehydroascorbic acid uptake (93,94).

V. CLINICAL FEATURES INFLUENCED BY CHANGES IN ASCORBIC ACID STATUS

A. Defining Ascorbic Acid Status

Although leukocyte concentrations are generally considered to be the best indicator of vitamin C status (95,96), the measurement of leukocyte vitamin C is technically complex (95,96). The varying amounts of ascorbic acid in differing leukocyte fractions and the lack of standardized reporting procedures (95–99) also complicate interpretation of data. Hence, the measurement of plasma vitamin C concentration is currently the most widely applied test for vitamin C status. Plasma concentrations between 11–28 µmol/L represent marginal vitamin C status. At this level, there is a moderate risk for developing clinical signs of vitamin C deficiency due to inadequate tissue stores of the vitamin.

Data from the Second National Health and Nutrition Examination Survey, 1976–1980, (NHANES II) (99) indicated that the prevalence of vitamin C deficiency (plasma vitamin C concentrations less than 11 µmol/L) ranged from 0.1% in children (3–5 years of age) to 3% in females (25–44 years of age) and 7% (in males 45–64 years of age). A decade later, the more sensitive measures utilized by NHANES III indicated that the prevalency of vitamin C deficiency was 9% for adult females and 13% for adult males (100). Marginal vitamin C status (plasma vitamin C concentrations greater than 11 µmol/L and less than 28 µmol/L) was noted in 17% of adult females and 24% of adult males (100). Smokers are more likely to have marginal vitamin C status compared to nonsmoking adults. Several studies suggest that smokers required over 200 mg vitamin C daily to maintain plasma vitamin C concentrations at a level equivalent to nonsmokers consuming

60 mg vitamin C daily (cf. ref. 101 and references cited therein). The current vitamin C recommendation for smokers is about 40% greater than that for nonsmokers, 110 and 125 mg daily for females and males, respectively (102).

In its most extreme form, scurvy is characterized by subcutaneous and intramuscular hemorrhages, leg edema, neuropathy and cerebral hemorrhage, and, if untreated, the condition is ultimately fatal. Presently, even in affluent countries, scurvy should be considered when cutaneous and oral lesions are observed, particularly in alcoholics, the institutionalized elderly, or persons who live alone and consume restrictive diets containing little or no fruits and vegetables. Patients may also complain of lassitude, weakness, and vague myalgias, and seek medical attention following the appearance of a skin rash or lower extremity edema. Below are descriptions of the functional consequences of ascorbic acid deficiency, and represent the signs and symptoms of scurvy.

B. Clinical Features Influenced by Changes in Ascorbic Acid Status

1. Immune Function

Since the publication of *Vitamin C and the Common Cold* (103) by Pauling, the role of vitamin C in immune function has been the topic of lively debate. The high concentration of vitamin C in leukocytes and the rapid decline of vitamin C in plasma and leukocytes during stress and infection has been used as evidence that vitamin C plays a role in immune function (103). Descriptive data in vitro suggest that vitamin C functions as an antiviral agent (106,107), presumably by aiding in the degradation of phage/viral nucleic acids (108). Intracellular HIV replication in chronically infected T-lymphocyte cell cultures is also inhibited by ascorbic acid. Ascorbic acid suppresses reverse transcriptase activity (106), as may be the case for a number of strong reducing agents.

Dozens of clinical trials have been conducted attempting to resolve the role, if any, of vitamin C in common cold infections (see 104,105). Ultramarathon runners often suffer upper respiratory tract (URT) infections in the immediate post-race period (109). In a prospective double-blind, placebo-controlled trial, vitamin C supplementation (600 mg/ day), beginning 21 days prior to the start of the race, reduced the incidence of post-race URT infections by 50%. Sedentary control subjects were not affected by vitamin C supplementation, although the duration of symptoms was significantly less in control subjects receiving vitamin C supplements, i.e., 1-2 days. Others have also reported a decrease in the duration of URT infections in subjects consuming 1 g vitamin C daily (105). In addition, elderly patients hospitalized with acute respiratory infections (receiving 200 mg vitamin C daily) have been shown to fare better than patients receiving placebo (110).

The mechanisms by which vitamin C reduces the severity of URT infections are not well established, but must be related to any number of redox sensitive signals and sites associated with enzymes and receptors. Further, reducing agents and proton donors are needed to drive the activation of phagocytes. Vitamin C can also accelerate the destruction of histamine, a mediator of allergy and cold symptoms in vitro (111–113). Vitamin C supplementation consistently reduces blood histamine concentrations 30% to 40% in adult subjects (111–112). An acute dose of vitamin C (e.g., 2 g) can also reduce bronchial responsiveness to inhaled histamine in patients with allergy (113). Thus, the antihistamine effect of vitamin C may attenuate the severity of symptoms associated with respiratory tract infections.

Ascorbic acid can also influence neutrophil chemotaxis (114–118); however, enhancement occurs only at nonphysiological concentrations of vitamin C. Neutrophil chemotaxis is enhanced following an intravenous injection of 1 g ascorbic acid (118) and following high doses of ascorbic acid (2 or 3 g/d for one or more weeks) (119,120). Ascorbic acid supplementation (1 g/d) has also been shown to significantly improve leukocyte chemotaxis in patients with Chediak-Higashi syndrome, a disease characterized by impaired neutrophil microtubule assembly (121,122) and patients with chronic granulomatous disease, a condition characterized by depressed neutrophil activation (123). In guinea pigs, vitamin C deficiency does not affect neutrophil activity, but it does decrease neutrophil killing of internalized pathogens (124).

Data regarding the effect of vitamin C on lymphocyte proliferation in vitro are equivocal. Vitamin C depletion for a nine-week period in human subjects did not appear to alter T-cell number or T-cell proliferation in vitro (125). In a carefully controlled, metabolic depletion-repletion study, vitamin C ingested daily in amounts ranging from 5 to 250 mg for 92 days did not affect mitogen-induced lymphocyte proliferation in vitro (126). However, a delayed hypersensitivity skin response to seven recall antigens was suppressed in subjects receiving the lower levels of vitamin C. Repletion with vitamin C (60 or 250 mg/d) improved responses in about half of the subjects. In separate studies, mitogen-induced lymphocyte proliferation in vitro was significantly enhanced in subjects ingesting high levels of vitamin C (2-3 g daily) (120,127).

Vitamin C may also influence other immune system parameters. Vitamin C status in guinea pigs is directly related to serum concentrations of the complement component Clq, a protein that, in association with other complement proteins, mediates nonspecific humoral immunity (128,129). Finally, several investigators have suggested that the antihistamine effect of vitamin C may indirectly enhance immunoresponsiveness. Ascorbic acid enhances mitogen-dependent lymphocyte blastogenesis by inhibiting histamine production in spleen cell cultures (130).

2. The Progression of Selected Chronic Diseases

Atherosclerosis. Epidemiologic studies have shown that death due to cardiovascular disease is inversely related to regular use of vitamin C supplements (65,131). Males with vitamin C deficiency (plasma vitamin C < 0.2 mg/dL) were at significantly increased risk of myocardial infarction after controlling for potentially confounding variables (131). The oxidation of low-density lipoproteins has been implicated in the etiology of atherosclerosis and serum lipid peroxides were significantly reduced in patients hospitalized with acute myocardial infarction after consuming diets rich in vitamin C (132). In smokers, acute smoking (5–7 cigarettes in 90 min) increased LDL lipid peroxidation twofold; vitamin C supplementation (1.5 g daily) reversed LDL lipid smoking-induced peroxidation (133). In rabbits fed atherogenic diets for 11 weeks, the addition of vitamins C and E to diets significantly reduced the severity of atherosclerotic lesions (134). Since vitamin C can potentially regenerate vitamin E from its tocopherol radical, the protective effects of vitamin C supplementation on LDL lipid oxidation may be related to maintenance of vitamin E.

Cancer. In epidemiologic studies, cancer incidence and deaths appear inversely related to regular use of vitamin C supplements (65). Diets rich in fruits and vegetables are consistently related to reduced risk for cancers (65,135–138). Although it is likely that ascorbic acid may have some limited effect given the complexity of protective mechanism(s), the relative importance of ascorbic acid in various cancers has yet to be resolved. Ascorbic acid may simply be a marker for fruit and vegetable consumption (139).

Cataracts.   Epidemiologic studies have also shown that the risk of cataract, particularly posterior subcapsular cataract, is significantly higher in individuals with moderate to low blood concentrations of vitamin C (odds ratio, 3.3 to 11.3 after adjustment for age, gender, race, and diabetes) (140,141). After controlling for potentially confounding variables, including diabetes, smoking, sunlight exposure and regular aspirin use, taking vitamin C supplements for ~10 years was associated with reduced risk for early (odds ratio, 0.23; 95% CI, 0.99-0.60) and moderate (odds ratio, 0.17; 95% CI, 0.03-0.87) age-related lens opacities in women (141). When consumed for less than 10 years little or no association to cataract formation has been observed.

Bone Density.   Vitamin C intake is positively associated with bone mineral density (142–145). This association is independent of other nutrients correlated with dietary vitamin C, including vitamin A and β-carotene. The relationship is particularly strong at high calcium intakes in postmenopausal women (144,145). Among persons with low vitamin E intake (6 mg per day) and those with modest vitamin C intake (70 mg per day or less), the odds of sustaining a fracture were 2–4 times greater for current smokers than for women who never smoked. Among persons with low intakes of both vitamins, the odds of fracture were nearly five times greater among current smokers than among women who never smoked. Compared with those who have never smoked, the odds of fracture were not increased among smokers who also had high intakes of vitamin E and/or vitamin C, e.g., >200 mg/day (143–145).

3. Wound Healing and Connective Tissue Metabolism

As discussed, the mechanisms that link ascorbic acid intake to connective regulation and deposition are related to its role as an enzymatic cofactor (and stabilizing factor) for prolyl and lysyl hydroxylases.

VI. REQUIREMENTS, ALLOWANCES, AND UPPER LIMITS

Plasma vitamin C concentrations in people who regularly consume vitamin C supplements are 60–70% higher than those who do not take supplements (75–80 and 45–50 µmol/L, respectively (146–149). A daily intake of 500–1000 mg is necessary to maintain plasma vitamin C concentrations at 75–80 µmol/L. The recently revised RDA for vitamin C, 75 mg daily for adult females and 90 mg daily for adult males, represents a 25–50% increase over the 1989 RDA, 60 mg (102). Exclusively breastfed infants ingest approximately 10 mg vitamin C/kg body weight (149).

Serum vitamin C concentrations in newly captured vervet monkeys range from 100 to 115 µmol/L (150), and rats have serum vitamin C concentrations ranging from 60 to 100 µmol/L (150). These data indicate that high tissue levels of vitamin C are well tolerated in mammalian systems. Approximately 70% of a 500 mg dose is absorbed. However, much of the absorbed dose (>50%) is excreted unmetabolized in urine. With a dose of 1250 mg, only 50% of the dose absorbed and nearly all (>85%) of the absorbed dose is excreted (151 and references cited). These factors support the contention that ascorbic acid is relatively nontoxic.

A. Rebound Scurvy

There is some evidence that accelerated metabolism or disposal of ascorbic acid may occur after prolonged supplementation of high doses. Presumably, when vitamin C supplementation ceases abruptly, the accelerated disposal of vitamin C creates a vitamin C-deficient state, i.e., "rebound scurvy." The concerns regarding rebound scurvy come largely from work by Cochan (152). Of 42 cases of infantile scurvy at Children's Hospital in Halifax, Nova Scotia (from October 1959 to January 1961), two could not be attributed to inadequate dietary vitamin C. The possibility of rebound scurvy was considered, because the mothers of both of the infants reported taking vitamin C supplements during pregnancy, i.e., 400 mg of vitamin C daily. An intake of 400 mg per day is not remarkable and is currently consumed by 5–10% of the U.S. population. In guinea pigs fed diets containing 0.1% vitamin C by weight, intraperitoneal administration of vitamin C (1 g/kg body weight per day for four weeks) was associated with an increased rate of vitamin C turnover (153). The mean plasma vitamin C concentrations fell significantly below control values during the second and fifth week following the abrupt withdrawal of vitamin C treatment; however, plasma vitamin C concentrations measured after vitamin C treatment remained within normal ranges. Thus, the extent to which "rebound scurvy" actually occurs may be exaggerated.

B. Oxalic Acid and Uric Acid

About 75% of kidney stones contain calcium oxalate; another 5–10% are composed of uric acid. High doses of vitamin C have been shown to increase urinary excretion of both oxalic acid and uric acid; and thus, theoretically promote the formation of kidney stones (154–161). However, calculi may form in the absence of hyperoxaluria or hyperuricosuria, and conversely, patients with hyperoxaluria or hyperuricosuria often do not form stones. Unrelated factors, such as a lack of inhibitors of crystal formation in urine, changes in urinary pH, decreased urine volume, and the presence of bacteria, all influence the risk for stones.

Chronic daily ingestion of 1000 mg vitamin C increased urinary uric acid by ~30% (155). The physiologic relevance of this moderate rise in urinary uric acid is not known. More than half of patients with urate calculi do not have hyperuricemia or hyperuricosuria; rather, they have a tendency to excrete acidic urine. Although ascorbic acid may be prescribed as a urinary acidifying agent, this property is more accurately described as preventing the alkalization of urine. Vitamin C supplementation (1–6 g/day) had little effect on urinary pH in subjects with nonalkaline urine (157).

Levine et al. (161) demonstrated that mean urinary oxalate was not significantly affected by the chronic ingestion of 200 or 400 mg vitamin C in seven healthy young men. Although chronic ingestion of 1000 mg vitamin C did cause a rise in urinary oxalate, the mean oxalate concentration remained within the reference range. Early reports that attempted to connect vitamin C intake to urinary oxalate often used methods that permitted oxalate generation from the ascorbate present in urine. (Oxalate is an alkaline degradation product of ascorbic acid (159).) In preserved urine samples, no significant increase in oxalate excretion is usually noted.

Epidemiologic data do not support an association between vitamin C supplementation and kidney stones. In the Harvard Prospective Health Professional Follow-Up Study involving over 45,000 men from 40–75 years of age, 751 incident cases of kidney stones were documented over a six-year period (160). The age-adjusted relative risk for men consuming 1500 mg or more vitamin C/day compared with those consuming less than 251 mg/day was 0.78 (95% confidence interval, 0.54–1.11).

C.   Iron-Related Disorders

Red cell hemolysis, related to glucose-6-phosphate dehydrogenase (G6PD) deficiency, has been reported as a toxic effect of ascorbic acid (162). Although the mechanism is not clear, it is possible that ascorbic acid in excess can act as a pro-oxidant catalyst in the presence of available iron and the absence of an important source of reducing equivalents in red cells, i.e., the NADPH that is generated by G6DP.

Enhanced iron absorption has been associated with ascorbate intakes (163–168), and a significant relationship between serum ferritin and dietary vitamin C in elderly has been shown (166). In iron-depleted young adults, vitamin C supplementation (e.g., 500 mg at two–three times per day with meals) raised apparent iron absorption from a single test meal 30–40% (164,165). Vitamin C in meals in the range of 50–100 mg had a significant effect on improving iron absorption. However, higher intakes of vitamin C had little further effect. If hemochromatosis is recognized, a high ascorbic acid intake may not be prudent. However, omitting or reducing dietary iron sources, e.g., meat consumption, would be more effective.

D.   Vitamin B-12

Herbert et al. (169) reported that patients who received ascorbic acid in high doses had low serum vitamin B_{12}. These data have not been replicated by others, and this concern has never been confirmed. The long-term use of supplemental vitamin C is not likely to affect serum vitamin B_{12} concentration (170).

VII.   SUMMARY

Ascorbic acid is the cell's universal reducing agent. It is a dietary essential for humans and several other species that have mutations in the gene for L-gulonolactone oxidase. Ascorbic acid maintains specific enzyme activities, notably the hydroxylase enzymes involved in collagen assembly and carnitine biosynthesis. In the promotion of antioxidant defense, ascorbate metabolism is linked to the metabolism of glutathione and, probably, α-tocopherol.

Ascorbic acid deficiency results in reduced mono- and dioxygenase activities. The consequences of severe deficiency are profound, since growth, extracellular matrix, and hormonal regulation are all impaired. Low intakes of ascorbic acid may accentuate and exacerbate chronic disease, e.g., atherosclerosis and those for which defective immune responsiveness is a component. Recently, the RDA for vitamin C was set at 75 mg daily for adult females and 90 mg daily for adult males, a 25–50% increase over the previous recommendation. Gram doses (1–2 g/day) of vitamin C are well tolerated by most individuals. A number of studies suggest that optimal health benefits are achieved at intakes of 100–200 mg/day.

REFERENCES

1.   Svirbely JL, Szent-Györgyi A. Hexuronic acid as the antiscorbutic factor. Nature 129:576, 1932.
2.   Lorenz A. The conquest of scurvy. J. Amer. Diet. Ass. 30:665–70, 1954.
3.   Lorenz AJ. Scurvy in the gold rush. J. Hist. Med. Allied Sci. 7:501–510, 1957.

4. Carpenter KJ. The history of scurvy and vitamin C. Cambridge: Cambridge University Press, 1986.

5. Clemetson CAB. Vitamin C. Vol. 1. Boca Raton, FL: CRC Press, 1989.

6. Lind J. A treatise of the scurvy. Printed by Sands, Murray, and Cochran for A. Kincaid and A. Donaldson, Edinburgh, 1753.

7. Holst, Frölich T. Experimental studies relating to ship-beri beri and scurvy. II. On the etiology of scurvy. J. Hyg. 7:634–671, 1907.

8. Zilva SS. The non-specificity of the phenolindophenol reducing capacity of lemon juice and its fractions as a measure of their antiscorbutic activity. Biochem. J. 26:1624–1627, 1932.

9. Drummond JC. The nomenclature of the so-called accessory food factors (vitamins). Biochem. J. 14:660, 1920.

10. Szent-Györgyi A. Observations on the function of peroxidase systems and the chemistry of the adrenal cortex: Description of a new carbohydrate derivative. Biochem. J. 22:1387–1409, 1928.

11. Svirbely JL and Szent-Györgi A. The chemical nature of vitamin C. Biochem J 26:865–870, 1932.

12. Waugh WA and King CG. Isolation and identification of vitamin C. J. Biol. Chem. 97:325–331, 1932.

13. Svirbely JL and Szent-Györgyi A. The chemical nature of vitamin C. Biochem. J. 27:279–285, 1933.

14. Haworth WN and Hirst EL. Synthesis of ascorbic acid. J. Soc. Chem. Ind. (London) 52:645–647, 1933.

15. Tsao CS. An overview of ascorbic acid chemistry and biochemistry. In: L Packer and J Fuchs, eds. Vitamin C in Health and Disease. New York: Marcel Dekker, 1997, pp. 25–58.

16. Seib PA and Tolbert BM. Ascorbic acid: chemistry, metabolism, and uses: Advances in chemistry series. Washington, DC: American Chemical Society, 1982.

17. Crawford TC. Synthesis of L-ascorbic acid. In: Seib PA, Tolbert BM, eds. Ascorbic Acid: Chemistry, Metabolism, and Uses. Advances in Chemistry Series. Washington, D.C.: American Chemical Society, 1982, pp. 1–36.

18. Bielski BHJ. Chemistry of ascorbic acid radicals. In: Seib PA, Tolbert BM, eds. Ascorbic Acid: Chemistry, Metabolism, and Uses. Advances in Chemistry Series. Washington, DC: American Chemical Society, 1982, pp. 81–100.

19. Wang XY, Seib PA, Paukstelis JV, et al. Crystal structure of D-erythroascorbic acid. J. Carbohydr. Chem. 14:1257–1263, 1995.

20. Behrens WA and Madere R. A procedure for the separation and quantitative analysis of ascorbic acid. Dehydroascorbic acid, isoascorbic acid, and dehydroisoascorbic acid in food and animal tissue. J. Liquid Chromatogr. 17:2445–2455, 1994.

21. Smith JL, Canham JE, Kirkland WD, Wells PA. Effect of Intralipid, amino acids, container, temperature, and duration of storage on vitamin stability in total parenteral nutrition admixtures. J. Parenteral and Enteral Nut. 12:478–483, 1988.

22. Dahl GB, Jeppsson RI, Tengborn HJ. Vitamin stability in a TPN mixture stored in an EVA plastic bag. J. Clin. Hosp. Pharm. 11:271–279, 1986.

23. Halliwell B and Whiteman M Antioxidant and prooxidant properties of vitamin C. In: Packer L and Fuchs J, eds. Vitamin C in Health and Disease. New York: Marcel Dekker, 1997, pp. 25–94.

24. Bors W and Buettner GR. The vitamin C radical and its reactions. In: Packer L and Fuchs J eds. Vitamin C in Health and Disease New York: Marcel Dekker, 1997, pp. 25–94.

25. Rose RC, Bode AM. Biology of free radical scavengers: an evaluation of ascorbate. FASEB J. 7:1135–1142, 1993.

26. May JM, Qu Z, Whitesell RR. Ascorbic acid recycling enhances the antioxidant reserve of human erythrocytes. Biochem. 34:12721–12728, 1995.

27. FDA, Evaluation of the Health Aspects of Ascorbic Acid, Sodium Ascorbate, calcium ascor-

bate, erythorbic acid, sodium erythorbate, and ascorbyl palmitate as food ingredients. U.S. Department of Commerce: Springfield, VA, 1979.

28. Sauberlich HE, Tamura T, Craig CB, Freeberg Le, Liu T. Effects of erythorbic acid on vitamin C metabolism in young women. Am. J. Clin. Nutr. 64:336–346, 1996.

29. Kall MA, Andersen C. Improved method for simultaneous determination of ascorbic acid and dehydroascorbic acid, isoascorbic acid and dehydroisoascorbic acid in food and biological samples. J. Chromatogr. 730:101–111, 1999.

30. Pappa-Lovis A, Pascalidou S. Optimal conditions for the simultaneous ion-pairing HPLC determination of L-ascorbic, dehydro-L-ascorbic, D-ascorbic and uric acids with online ultraviolet absorbance and electrochemical detection. Anal. Biochem. 263:176–182, 1998.

31. Thomson CO and Trenerry VC. A rapid method for the determination of total L-ascorbic acid in fruits and vegetables by micellar electrokinetic capillary chromatography. Food Chem. 53: 43–50, 1995.

32. Kampfenkel K, Van Montagu M, Inze D. Extraction and determination of ascorbate and dehydroascorbate from plant tissue. Anal. Biochem. 225:165–167, 1995.

33. Moeslinger T, Brunner M, Spieckermann PG. Spectrophotometric determination of dehydroascorbic acid in biological samples. Anal. Biochem. 221:290–296, 1994.

34. Gensler M, Rossmann A, Schmidt H-L. Detection of added L-ascorbic acid in fruit juices by isotope ratio mass spectrometry. J. Agric. Food Chem. 43:2662–2666, 1995.

35. Moeslinger T, Brunner M, Volf I, Spieckermann PG. Spectrophotometric determination of ascorbic acid and dehydroascorbic acid. Clin. Chem. 41:1177–1181, 1995.

36. Wang S, Eide TC, Sogn EM, Berg KJ, Sund RB. Plasma ascorbic acid in patients undergoing chronic haemodialysis. Eur. J. Clin. Pharmacol. 55:527–532, 1999.

37. Schaus EE, Kutnink MA, O'Conner DK, Omaye ST. A comparison of leukocyte ascorbate levels measured by the 2,4-dinitrophenylhydrazine method with high performance liquid chromatography using electro chemical detection. Biochem. Med. Metab. Biol. 36:369–376, 1986.

38. Margolis SA and Duewer DL. Measurement of ascorbic acid in human plasma and serum: stability, intralaboratory repeatability, and interlaboratory reproducibility. Clin. Chem. 42: 1257–1262, 1996.

39. Benzie IFF and Strain JJ. The effect of ascorbic acid on the measurement of total cholesterol and triglycerides: possible artefactual lowering in individuals with high plasma concentration of ascorbic acid. Clin. Chim. Acta. 239:185–190, 1995.

40. Siest G, Appel W, Blijenberg GB, et al. Drug interference in clinical chemistry: studies on ascorbic acid. J. Clin. Chem. Clin. Biochem. 16:103–110, 1978.

41. Freemantle J, Freemantle MJ, Badrick T. Ascorbate interferences in common clinical assays performed on three analyzers. Clin. Chem. 40:950–951, 1994.

42. Nakano Y, Asada K. Hydrogen peroxide is scavenged by ascorbate-specific peroxidase in spinach chloroplasts. Plant Cell Physiol. 22:867, 1981.

43. Foyer CH. Ascorbic acid. In: Alscher RG, Hess JL, eds. Antioxidants in Higher Plants. Boca Raton, FL: CRC Press, 1993, pp 31–58.

44. Fraser PD, Miura Y, Misawa N. In vitro characterization of a staxanthin biosynthetic enzymes. J. Biol. Chem. 272:6128–6135, 1997.

45. Meister A. Glutathione-ascorbic acid antioxidant system in animals. J. Biol. Chem. 269: 9397–9400, 1994.

46. Winkler BS. Unequivocal evidence in support of the nonenzymatic redox coupling between glutathione/glutathione disulfide and ascorbic acid/dehydroascorbic acid. Biochim. Biophys. Acta. 1117:287–290, 1992.

47. Wells WW, Xu DP, Yang Y, Rocque PA. Mammalian thioltransfease (glutaredoxin) and protein disulfide isomerase have dehydroascorbate reductase activity. J. Biol. Chem. 265: 15361–15364, 1990.

48. May JM, Qu Z-C, Whitesell RR, Cobb SE. Ascorbate recycling in human erythrocytes: role of GSH in reducing dehydroascorbate. Free Radical. Biol. Med. 20:543–551, 1996.

49. Martensson J, Han J, Griffith OW, Meister A. Glutathione ester delays the onset of scurvy in ascorbate-deficient guinea pigs. Proc. Natl. Acad. Sci. USA 90:317–21, 1993.

50. Wimalasena K and Wimalasena DS. The reduction of membrane-bound dopamine beta-monooxygenase in resealed chromaffin granule ghosts. Is intragranular ascorbic acid a mediator for extragranular reducing equivalents? J. Biol. Chem. 270:27516–27524, 1995.

51. Fortin D, Coulon JF, Roberge AG. Comparative study of biochemical parameters and kinetic properties of dopamine-beta-hydroxylase activity from cat and rat adrenals. Comp. Biochem. Physiol. 104:567–575, 1993.

52. Murth ASN, Keutmann HT, Eipper BA. Further characterization of peptidylglycine α-amidating mono-oxygenase from bovine neurointermediate pituitary. Mol. Endocrinol. 1: 290–299, 1987.

53. Steiner DF. The proprotein convertases. Curr. Opin. Chem. Biol. 2:31–39, 1998.

54. Oldham CD, Girard PR, Nerem RM, May SW. Peptide amidating enzymes are present in cultured endothelial cells. Biochem. Biophys. Res. Comm. 184:323–329, 1992.

55. Food and Nutrition Board. Dietary Reference Intakes. Washington, DC: National Academy Press, 1998.

56. Niki E and Noguchi N. Protection of human low-density lipoprotein from oxidative modification by vitamin C. In: Packer L, Fuchs J, eds. Vitamin C in Health and Disease. New York: Marcel Dekker, Inc., 1997, pp 183–192.

57. Packer L. Vitamin C and redox cycling antioxidants: In: Packer L, Fuchs J, eds. Vitamin C in Health and Disease. New York: Marcel Dekker, 1997, 95–121.

58. Edge R and Truscott TG. Prooxidant and antioxidant reaction mechanisms of carotene and radical interactions with vitamins E and C. Nutrition 13:992–994, 1997.

59. Buettner GR. The pecking order of free radicals and antioxidants: Lipid peroxidation, α-tocopherol, and ascorbate. Arch. Biochem. Biophys. 300:535–543, 1993.

60. Bucala R. Lipid and lipoprotein oxidation: basic mechanisms and unresolved questions in vivo. Redox. Report 2:291–307, 1996.

61. Halliwell B. Vitamin C: antioxidant or pro-oxidant in vivo. Free. Radic. Res. 25:439–454, 1996.

62. Steinberg FM and Chait A. Antioxidant vitamin supplementation and lipid peroxidation in smokers. Am. J. Clin. Nutr. 68:319–327, 1998.

63. Niki E and Noguchi N. Protection of human low-density lipoprotein from oxidative modification by vitamin C. In: Packer L and Fuchs J, eds. Vitamin C in Health and Disease. New York: Marcel Dekker, 1997, pp 183–192.

64. Lynch SM, Gaziano JM, Frei B. Ascorbic acid and atherosclerotic cardiovascular disease. In: Harris JR, ed. Ascorbic acid: Biochemistry and Biomedical Cell Biology. New York: Plenum Press, pp 331–367, 1996.

65. Carr AC and Frei B. Toward a new recommended dietary allowance for vitamin C based on antioxidant and health effects in humans. Am. J. Clin. Nutr. 69:1086–1107, 1999.

66. Upston JM, Terentis AC, Stocker R. Tocopherol-mediated peroxidation of lipoproteins: implications for vitamin E as a potential antiatherogenic supplement. FASEB J. 13:977–994, 1999.

67. Berger TM, Polidori MC, Dabbagh A, Evans PJ, et al. Antioxidant activity of vitamin C in iron-overloaded human plasma. J. Biol. Chem. 272:15656–15660, 1997.

68. Lindahl T. Instability and decay of the primary structure of DNA. Nature 362:709–715, 1993.

69. Asami S, Manabe H, Miyake J, et al. Cigarette smoking induces an increase in oxidative DNA damage, 8-hydroxydeoxyguanosine, in a central site in the human lung. Carcinogenesis 18:1763–1766, 1997.

70. Loft S, Vistisen K, Ewertz M, Tjonneland A, Overvad K, Poulsen HE. Oxidative DNA damage estimated by 8-hydroxydeoxyguanosine excretion in humans: influence of smoking, gender and body mass index. Carcinogenesis 13:2241–2247, 1992.

71. Poulsen HE, Prieme H, Loft S. Role of oxidative DNA damage in cancer initiation and promotion. Eur. J. Cancer. Prev. 7:9–16, 1998.

72. Cross CE and Halliwell B. Nutrition and human disease: how much extra vitamin C might smokers need? Lancet 341:1091, 1993.

73. Fraga CG, Motchnik PA, Wyrobek AJ, Rempel DM, Ames BN. Smoking and low antioxidant levels increase oxidative damage to sperm DNA. Mutat. Res. 351:199–203, 1996.

74. Fraga CG, Motchnik PA, Shigenaga MK, Helbock HJ, Jacob RA, Ames BN. Ascorbic acid protects against endogenous oxidative DNA damage in human sperm. Proc. Natl. Acad. Sci. USA 88:11003–11006, 1991.

75. Green MHL, Lowe JE, Waugh APW, Aldridge KE, Cole J, Arlett CV. Effect of diet and vitamin C on DNA strand breakage in freshly isolated human white blood cells. Mutat. Res. 316:91–102, 1994.

76. Levine MA, Daruwala RC, Park JB, Rumsey SC, Wang Y. Does vitamin C have a prooxidant effect? Nature 395:231, 1998.

77. Poulsen HE, Weimann A, Salonen JT, et al. Does vitamin C have a prooxidant effect? Nature 395:231–232, 1998.

78. Berlett BS and Stadtman ER. Protein oxidation in aging, disease, and oxidative stress. J. Biol. Chem. 272:20313–20316, 1997.

79. Dean RT, Fu S, Stocker R, Davies MJ. Biochemistry and pathology of radical-mediated protein oxidation. Biochem. J. 324:1–18, 1997.

80. Ortwerth BJ and Monnier VM. Protein glycation by the oxidation products of ascorbic acid. In: Packer L, Fuchs J, eds. Vitamin C in Health and Disease. New York: Marcel Dekker, 1997, pp 123–142.

81. Mannick EE, Bravo LE, Zarama G, et al. Inducible nitric oxide synthase, nitrotyrosine, and apoptosis in *Helicobacter pylori* gastritis: effect of antibiotics and antioxidants. Cancer Res. 56:3238–3243, 1996.

82. Ghosh MK, Chattopadhyay PJ, Chatterjee IB. Vitamin C prevents oxidative damage. Free Radic. Res. 25:173–179, 1996.

83. Hughes RE, Hurley RJ, Jones E. Dietary ascorbic acid and muscle carnitine (beta-OH-gamma-(trimethylamino) butyric acid) in guinea pigs. Br. J. Nutr. 43:385–387, 1980.

84. Rebouche CJ. Ascorbic acid and carnitine biosynthesis. Am. J. Clin. Nutr. 54:1147S–1152S, 1991.

85. Otsuka M, Matsuzawa M, Ha TY, Arakawa N. Contribution of a high dose of L-ascorbic acid to carnitine synthesis in guinea pigs fed high-fat diets. J. Nutr. Sci. Vitaminol. 45:163–171, 1999.

86. Sandor A, Kispal G, Kerner J, Alkonyi I. Combined effect of ascorbic acid deficiency and underfeeding on the hepatic carnitine level in guinea pigs. Experientia 39:512–513, 1983.

87. Nelson PJ, Pruitt RE, Henderson LL, Jenness R, Henderson LM. Effect of ascorbic acid deficiency on the in vivo synthesis of carnitine. Biochim. Biophys. Acta. 672:123–127, 1981.

88. Ronchetti IP, Quaglino DJ, Bergamini G. Ascorbic acid and connective tissue. In: Harris JR, ed. Ascorbic Acid: Biochemistry and Biomedical Cell Biology, Subcellular Biochemistry 25:249–262, 1996.

89. Banhegyi G, Marcolongo P, Puskas F, Fulceri R, et al. Dehydroascorbate and ascorbate transport in rat liver microsomal vesicles. J. Biol. Chem. 273:2758–2762, 1998.

90. Goldenberg H, Schweinzer E. Transport of vitamin C in animal and human cells. J. Bioenerg. Biomembr. 26:359–367, 1994.

91. McCormick DB and Zhang Z. Cellular Assimilation of Water-Soluble Vitamins in the Mammal: Riboflavin, B$_6$, Biotin, and C. Proc. Soc. Exp. Biol. Med. 202:265–270, 1993.

92. Rose RC, Choi JL, Koch MJ. Intestinal transport and metabolism of oxidized ascorbic acid (dehydroascorbic acid). Am. J. Physiol. 254:G824–G828, 1988.

93. Rumsey SC, Kwon O, Xu G, Burant CF, Simpson I, Levine M. Glucose transporter isoforms GLUT1 and GLUT3 transport dehydroascorbic acid. J. Biol. Chem. 272:18982–18989, 1997.

94. Rumsey SC and Levine M. Absorption, transport, and disposition of ascorbic acid in humans. J. Nutr. Biochem. 9:116–130, 1998.

95. Jacob RA, Skala JH, Omaye ST. Biochemical indices of human vitamin C status. Am. J. Clin. Nutr. 46:818–826, 1987.

96. Jacob RA. Assessment of human vitamin C status. J. Nutr. 120:1480–1485, 1990.

97. Blanchard J, Conrad KA, Watson RR, Garry PJ, Crawley JD. Comparison of plasma, mononuclear, and polymorphonuclear leukocyte vitamin C levels in young and elderly women during depletion and supplementation. Eur. J. Clin. Nutr. 43:97–106, 1989.

98. Omaye ST, Schaus EE, Kutnink MA, Hawkes WC. Measurement of vitamin C in blood components by high-performance liquid chromatography. Implication in assisting vitamin C status. Ann. N.Y. Acad. Sci. 498:389–401, 1987.

99. Hematological and nutritional biochemistry reference data for persons 6 months–74 years of age: United States, 1976–80. U.S. Department of Health and Human Services; Public Health Service; National Center for Health Statistics. Hyattsville, MD: DHHS Publication No. (PHS) 83-1682; 1982:138–43.

100. Schleicher RL, Caudill SP, Yeager PR, Sowell AL. Serum vitamin C levels in the U.S. population 1988–94. Results of NHANES III. FASEB J. 12:A512, 1998.

101. Schectman G, Byrd JC, Hoffmann R. Ascorbic acid requirements for smokers: analysis of a population survey. Am. J. Clin. Nutr. 53:1466–1470, 1991.

102. Standing Committee on the Scientific Evaluation of dietary reference intakes, Food and Nutrition Board 2000 Dietary Reference intakes for vitamin C, vitamin E, selenium, and beta carotene, and other carotenoids. Washington, DC: National Academy Press.

103. Pauling L. Vitamin C and the Common Cold. San Francisco: WH Freeman & Co, 1970.

104. Hemila H. Vitamin C and the common cold. Br. J. Nutr. 67:3–16, 1991.

105. Hemila H and Herman ZS. Vitamin C and the common cold: a retrospective analysis of Chalmers' review. J. Am. Coll. Nutr. 14:116–123, 1995.

106. Harakeh S and Jariwalla RJ. Comparative study of the anti-HIV activities of ascorbate and thiol-containing reducing agents in chronically HIV infected cells. Am. J. Clin. Nutr. 54: 1231S–1235S, 1991.

107. Schwerdt PR and Schwerdt CE. Effect of ascorbic acid on rhinovirus replication in WI-38 cells (38724). Proc. Soc. Exp. Biol. Med. 148:1237–1245, 1975.

108. Kazakov SA, Astashkina TG, Mamaev SV, Vlassov VV. Site-specific cleavage of single-stranded DNA at unique sites by a copper-dependent redox reaction. Nature 335:186–188, 1988.

109. Peters EM, Goetzsche JM, Grobbelaar B, Noakes TD. Vitamin C supplementation reduces the incidence of postrace symptoms of upper respiratory tract infection in ultramarathon runners. Am. J. Clin. Nutr. 57:170–174, 1993.

110. Hunt C, Chakravorty NK, Annan G, Habibzadeh N, Schorah CJ. The clinical effects of vitamin C supplementation in elderly hospitalized patients with acute respiratory infections. Internat J. Vit. Nutr. Res. 64:212–219, 1994.

111. Johnston CS, Solomon RE, Corte C. Vitamin C depletion is associated with alterations in blood histamine and plasma free carnitine in adults. J. Am. Coll. Nutr. 15:586–591, 1996.

112. Johnston CS, Retrum KR, Srilakshmi JC. Antihistamine effects and complications of supplemental vitamin C. J. Am. Diet. Assoc. 92:988–989, 1992.

113. Bucca C, Rolla G, Oliva A, Farina JC. Effect of vitamin C on histamine bronchial responsiveness of patients with allergic rhinitis. Ann. Allergy 65:311–314, 1990.

114. Goetzl EJ. Defective responsiveness to ascorbic acid of neutrophil random and chemotactic migration in Felty's syndrome and systemic lupus erythematosus. Ann. Rheum. Dis. 35:510–515, 1976.

115. Sandler JA, Gallin JI, Vaughan M. Effects of serotonin, carbamylcholine, and ascorbic acid on leukocyte cyclic GMP and chemotaxis. J. Cell. Biol. 67:480–484, 1975.

116. Goetzl EJ, Waserman SI, Gigli I, Austen KF. Enhancement of random migration and chemotactic response of human leukocytes by ascorbic acid. J. Clin. Invest. 53:813–818, 1974.

117. Anderson R and Theron A. Effects of ascorbate on leukocytes. S. Afr. Med. J. 56:394–400, 1979.

118. Anderson R. Ascorbate-mediated stimulation of neutrophil motility and lymphocyte transformation by inhibition of the peroxidase/H_2O_2/halide system in vitro and in vivo. Am. J. Clin. Nutr. 34:1906–1911, 1981.

119. Johnston CS, Martin LJ, Cai X. Antihistamine effect of supplement ascorbic acid and neutrophil chemotaxis. J. Am. Coll. Nutr. 11:172–176, 1992.

120. Anderson R, Oosthuizen R, Maritz R, Theron A, Van Rensburg AJ. The effects of increasing weekly doses of ascorbate on certain cellular and humoral immune functions in normal volunteers. Am. J. Clin. Nutr. 33:71–76, 1980.

121. Weening RS, Schoorel EP, Roos D, van Schaik MLJ, Voetman AA, Bot AAM, Batenburg-Plenter AM, Willems C, Zeijlemaker WP, Astaldi A. Effect of ascorbate on abnormal neutrophil, platelet, and lymphocyte function in a patient with the Chediak-Hagashi syndrome. Blood 57:856–865, 1980.

122. Boxer LA, Watanabe AM, Rister M, Besch HR, Allen J, Baehner RL. Correction of leukocyte function in Chediakk-Higashi syndrome by ascorbate. N. Engl. J. Med. 295:1041–1045, 1976.

123. Anderson R. Assessment of oral ascorbate in three children with chronic granulomatous disease and defective neutrophil motility over a 2-year period. Clin. Exp. Immunol 43:180–188, 1981.

124. Goldschmidt MC. Reduced bactericidal activity in neutrophils from scorbutic animals and the effect of ascorbic acid on these target bacteria in vivo and in vitro. Am. J. Clin. Nutr. 54:1214S–1220S, 1991.

125. Kay NE, Holloway DE, Hutton SW, Bone ND, Duane WC. Human T-cell function in experimental ascorbic acid deficiency and spontaneous scurvy. Am. J. Clin. Nutr. 36:127–130, 1982.

126. Jacob RA, Kelley DS, Pianalto FS, Swendseid ME, Henning SM, Zhang JZ, Ames BN, Fraga CG, Peters JH. Immunocompetence and oxidant defense during ascorbate depletion of healthy men. Am. J. Clin. Nutr. 54:1302S–1309S, 1991.

127. Panush RS, Delafuente JC, Katz P, Johnson J. Modulation of certain immunologic responses by vitamin C. III. Potentiation of in vitro and in vivo lymphocyte responses. Int. J. Vitam. Nutr. Res. [Suppl] 23:35–47, 1982.

128. Johnston CS, Kolb WP, Haskell BE. The effect of vitamin C nutriture on complement component C1q concentrations in guinea pig plasma. J. Nutr. 117:764–768, 1987.

129. Johnston CS. Complement component C1q levels unaltered by ascorbate nutriture. J. Nutr. Biochem. 2:499–501, 1991.

130. Oh C, Nakano K. Reversal by ascorbic acid of suppression by endogenous histamine of rat lymphocyte blastogenesis. J. Nutr. 118:639–644, 1998.

131. Enstrom JE, Kanim LE, Klein MA. Vitamin C intake and mortality among a sample of the United States population. Epidemiol 3:194–202, 1992.

132. Singh RB, Niaz MA, Agarwal P, Begom R, Rastogi SS. Effect of antioxidant-rich foods on plasma ascorbic acid, cardiac enzyme, and lipid peroxide levels in patients hospitalized with acute myocardial infarction. J. Am. Diet. Assoc. 95:775–780, 1995.

133. Harats D, Ben-Naim M, Dabach Y, Hollander G, Havivi E, Stein O, Stein Y. Effect of vitamins C and E supplementation on susceptibility of plasma lipoproteins to peroxidation induced by acute smoking. Atherosclerosis 85:47–54, 1990.

134. Mahfouz MM, Kawano H, Kummerow FA. Effect of cholesterol-rich diets with and without added vitamins E and C on the severity of atherosclerosis in rabbits. Am. J. Clin. Nutr. 66:1240–1249, 1997.

135. Gandini S, Merzenich H, Robertson C, Boyle P. Meta-analysis of studies on breast cancer risk and diet: the role of fruit and vegetable consumption and the intake of associated micronutrients. Eur. J. Cancer 36:636–646, 2000.

136. Steinmaus CM, Nuänez S, Smith AH. Diet and bladder cancer: a meta-analysis of six dietary variables. Am. J. Empidemiol 151:693–702, 2000.

137. Voorrips LE, Goldbohm RA, Verhoeven DT, van Poppel GA, Sturmans F, Hermus RJ, van den Brandt PA. Vegetable and fruit consumption and lung cancer risk in the Netherlands Cohort Study on diet and cancer. Cancer Causes and Control 11:101–115, 2000.

138. Levi F, Pasche C, La Vecchia C, Lucchini F, Franceschi S. Food groups and colorectal cancer risk. Br. J. Cancer 79:1283–1287, 1999.

139. Drewnowski A, Rock CL, Henderson SA, Shore AB, Fischler C, Galan P, Preziosi P, Hercherg S. Serum β-carotene and vitamin C as biomarkers of vegetable and fruit intakes in a community-based sample of French adults. Am. J. Clin. Nutr. 55:1796–1802, 1997.

140. Jacques PF and Chylack LT. Epidemiologic evidence of a role for the antioxidant vitamins and carotenoids in cataract prevention. Am. J. Clin. Nutr. 53:325S–325S, 1991.

141. Jacques PF, Taylor A, Hankinson SE, Willet WC, Mahnken B, Lee Y, Vaid K, Lahav M. Long-term vitamin C supplement use and prevalence of early age-related lens opacities. Am. J. Clin. Nutr. 656:911–916, 1997.

142. Leveille SG, LaCroix AZ, Koepsell TD, Beresford SA, Van Belle G, Buchner DM. Dietary vitamin C and bone mineral density in postmenopausal women in Washington State, USA. J. Epidemiol. Comm. Health 51:479–485, 1997.

143. Weber P. The role of vitamins in the prevention of osteoporosis—a brief status report. Inter. J. Vit. Nutr. Res. 69:194–197, 1999.

144. Hall SL and Greendale GA. The relation of dietary vitamin C intake to bone mineral density: results from the PEPI study. Calcified Tissue International 63:183–189, 1998.

145. Melhus H, Michaelsson K, Holberg L, Wolk A, Ljunghall S. Smoking, antioxidant vitamins, and the risk of hip fracture. J. Bone. Miner. Res. 14:129–135, 1999.

146. Subar AF and Block G. Use of vitamin and mineral supplements: demographics and amounts of nutrients consumed. Am. J. Epidemiol 132:1091–1101, 1990.

147. Moss AJ. 1989. Use of vitamin and mineral supplements in the United States: current users, types of products, and nutrients. Advance Data No. 174. Washington, DC: National Center for Health Statistics, Centers for Disease Control and Prevention.

148. Dickinson VA, Block G, Russek-Cohen E. Supplement use, other dietary and demographic variables, and serum vitamin C in NHANES II. J. Am. Coll. Nutr. 13:22–32, 1994.

149. Byerley LO and Kirksey A. Effects of different levels of vitamin C concentration in human milk and the vitamin C intakes of breast-fed infants. Am. J. Clin. Nutr. 41:665–671, 1985.

150. De Klerk WA, Du Plessis JP, Van Der Watt JJ, et al. Vitamin C requirements of the vervet monkey (*Cercopithecus aethiops*) under experimental conditions. S. Afr. Med. J. 47:705–708, 1973.

151. Johnson CS. Biomarkers for establishing a tolerable upper intake level for vitamin C. Nutrition Reviews 57:71–77, 1999.

152. Cochran WA. Overnutrition in pre- and neonatal life: A problem? Cand. Med. Assoc. J. 93: 693–699, 1965.

153. Tsao CS and Leung PY. Urinary ascorbic acid levels following the withdrawal of large doses of ascorbic acid in guinea pigs. J. Nutr. 118:895–900, 1988.

154. Schrauzer GN and Rhead WJ. Ascorbic acid abuse: effects of long term ingestion of excessive amounts on blood levels and urinary excretion. Int. J. Vitam. Nutr. Res. 43:201–211, 1973.

155. Stein HB, Hasan A, Fox IH. Ascorbic acid-induced uricosuria. Ann. Intern. Med. 84:385–388, 1976.

156. Berger L and Gerson CD. The effect of ascorbic acid on uric acid excretion with a commentary on the renal handling of ascorbic acid. Am. J. Med. 62:71–76, 1977.

157. Nahata MC, Shim L, Lampman T, et al. Effect of ascorbic acid on urine pH in man. Am. J. Hosp. Pharmacy 34:1234–1237, 1977.

158. Goldfarb S. Diet and nephrolithiasis. Annu. Rev. Med. 45:235–243, 1994.

159. Wandzilak TR, D'Andre SD, Davis PA, et al. Effect of high dose vitamin C on urinary oxalate levels. J. Urol. 151:834–837, 1994.

160. Curhan GC, Willett WC, Rimm EB, et al. A prospective study of the intake of vitamins C and B_6 and the risk of kidney stones in men. J. Urol. 155:1847–1851, 1996.

161. Levine M, Conry-Cantilena C, Wang Y, Welch RW, Washko PW, Dhariwal KR, Park JB, Lazarev A, Graumlich J, King J, Cantilena LR. Vitamin C pharmacokinetics in health volunteers: evidence for a recommended dietary allowance. Proc. Natl. Acad. Sci. USA 93:3704–3709, 1996.

162. Campbell GD, Steinberg MH, Bower JD. Ascorbic acid-induced hemolysis in G6PD deficiency. Ann. Intern. Med. 82:810, 1975.

163. Hunt JR, Mullen LM, Lykken GI, et al. Ascorbic acid: effect on ongoing iron absorption and status in iron-depleted young women. Am. J. Clin. Nutr. 51:649–655, 1990.

164. Hunt Jr, Gallagher SK, Johnson LK. Effect of ascorbic acid on apparent iron absorption by women with low iron stores. Am. J. Clin. Nutr. 59:1381–1385, 1994.

165. Cook JD, Watson SS, Simpson KM, et al. The effect of high ascorbic acid supplementation on body iron stores. Blood 64:721–726, 1984.

166. Fleming DJ, Jacques PF, Dallal GE, et al. Dietary determinants of iron stores in a free-living elderly population: the Framingham Heart Study. Am. J. Clin. Nutr. 67:722–733, 1998.

167. Cohen A, Cohen IJ, Schwartz E. Scurvy and altered iron stores in thalassemia major. N. Engl. J. Med. 304:158–160, 1981.

168. Gordeuk VR, Bacon BR, Brittenham GM. Iron overload: causes and consequences. Annu. Rev. Nutr. 7:485–508, 1987.

169. Herbert V, Jacob E, Wong KTJ, et al. Low serum vitamin B_{12} levels in patients receiving ascorbic acid in megadoses: studies concerning the effect of ascorbate on radioisotope vitamin B_{12} assay. Am J Clin Nutr 31:253–258, 1978.

170. Marcus M, Prabhudesai M, Wassef S. Stability of vitamin B_{12} in the presence of ascorbic acid in food and serum: restoration by cyanide of apparent loss. Am J Clin Nutr 33:137–143, 1980.

16

Ascorbic Acid Regulation of Extracellular Matrix Expression

JEFFREY C. GEESIN

Johnson and Johnson, Skillman, New Jersey

RICHARD A. BERG

Collagen Corporation, Palo Alto, California

I. INTRODUCTION

Ascorbic acid is an essential cofactor in many biochemical reactions. Several of these reactions demand that ascorbic acid provide electrons, either directly or indirectly, to enzymes requiring prosthetic metal ions in a reduced form for activity. These reactions are accelerated by ascorbic acid but do not involve its direct participation. Included in this group of reactions are the hydroxylation of proline and lysine residues found characteristically in collagen, but also in elastin, C1q of complement, and acetylcholinesterase. The role of ascorbic acid in the synthesis of these proteins is thought to occur in these hydroxylation reactions.

Recently, the regulation of these proteins has been investigated because of results indicating that ascorbic acid regulates their expression by regulating transcription of the relevant genes. The mechanisms responsible are different from ascorbate's function as antioxidant in known hydroxylation reactions. In the following discussion, the effect of ascorbic acid on the regulation of the synthesis of extracellular matrix (ECM) and ECM-associated proteins will be described and potential mechanisms that regulate their synthesis discussed.

II. ASCORBIC ACID EFFECT ON EXTRACELLULAR MATRIX PRODUCTION IN VITRO

Collagen, which is the predominant structural protein in animals, is found in a large number of tissues, including bone, cartilage, tendon, and skin. It is also the predominant protein produced by human dermal fibroblasts, which incorporate the molecule into the ECM found in the dermis. At least 18 types of collagen have been identified to date, with more being described (1). These collagen types are often subdivided into subgroups based on common characteristics (2). Prominent in the subgroups are the fibrillar collagens, which include types I, II, III, V, and XI (1–7). The fibrillar collagens are characterized by the triple-helical ropelike structure without interruptions of the unique collagen amino acid sequence (8). Type I collagen is composed of two identical polypeptide chains, designated $\propto 1(I)$, and one $\propto 2(I)$ protein. Types II and III collagen are composed of three identical proteins, whereas types V and XI collagen are composed of three different proteins (2,8). Formation of the triple helix in these molecules requires that the helical region have the sequence of glycine-x-y where every third amino acid is glycine and many of the other residues are prolines. For the three peptides to form a stable helix, some minimum number of prolines must be hydroxylated (9–12). This hydroxylation is produced by the enzyme prolyl hydroxylase in the presence of ferrous ion, ascorbic acid, and α-ketoglutarate (8,13–15). The exact mechanism of this reaction is thought to include the production of an active oxygen species, possibly a superoxide radical, at the active site (16–19). Ascorbic acid functions in this reaction as an antioxidant that keeps the supply of iron molecules in the ferrous (+2) state, which appears to be a requirement for hydroxylation (18,20,21).

Elastin is another ECM protein that, like collagen, contains hydroxyproline; however, unlike collagen, these hydroxyproline residues represent a much smaller portion of the elastin molecule and have not been implicated in the cellular control of elastin secretion (22,23). Elastin is present in the ECM of a large number of tissues, but it is most prominent in tissues that require elasticity and are repeatedly forced to deform as part of their function. Tissues containing large amounts of elastin include lung, large blood vessels, and skin (22). Elastin is the major component of elastic fibers, which are the functional units providing elasticity to tissues (24). These fibers are complex structures composed of microfibrillar proteins, proteoglycans, and lysyl oxidase in addition to elastin.

Along with its role in proline hydroxylation, ascorbate and related analogs have been shown to stimulate the rate of synthesis of collagen by a number of different cell types, including human dermal (25–51) and lung (52) fibroblasts, bovine (53) and rat aortic smooth-muscle cells (54,55), rat perisinusoidal stellate cells (56), rabbit chondrocytes (57) and keratocytes (58), mouse (59) and porcine (60) osteoblasts, mouse BALB (61) and L1 (62) 3T3 cells, chick fibroblasts (61), neural crest cells (63), chondrocytes (64), tendon cells (65–72), and chick tibia organ culture (73,74). This process occurs with no change in the rate of degradation of the protein (37) but involves increased transcription of the collagen peptides (45,47,48,55,66). In addition, the effect of ascorbic acid on collagen synthesis appears to be independent of cell density or serum concentration, and does not involve any change in the intracellular proline pool or the rate of incorporation of radioactive proline into collagen (46).

Elastin synthesis and accumulation are also affected by ascorbic acid (54,55,75). Most of the evidence to date indicates that ascorbic acid inhibits the accumulation of insoluble, matrix-associated tropoelastin, possibly due to an increased level of hydroxyproline found in elastin after ascorbate treatment without a change in transcription or mRNA stability (54,55). Therefore, short-term treatment with ascorbate produced a shift in the partitioning of elastin from an insoluble, matrix-associated form to a soluble, media component (55). This effect has been demonstrated in both fibroblasts and smooth-muscle cells (54,55). However, recent evidence indicates that elastin production can be inhibited at both the levels of transcription and reduced mRNA stability with chronic ascorbic acid treatment (75).

A number of other molecules have been described that are regulated at the protein or mRNA level by ascorbic acid. In bovine aortic smooth-muscle cells, the levels of the minor collagen types V and VI were found to be reduced as a result of ascorbic acid treatment (53). Type IV collagen was similarly induced coincident with adipose conversion of 3T3-L1 cells (62). Cartilage maturation, along with inorganic pyrophosphate and type X collagen secretion, is induced by ascorbic acid (76–78). The production of glycosaminoglycans and gangliosides was enhanced by ascorbic acid treatment in cultured human skin fibroblasts (39,79), calf aortic smooth-muscle cells (80), and chick wing mesenchyme micromass cultures (81). The protein and mRNA levels for the acetylcholine receptor are induced in L5 muscle cells by ascorbic acid by a mechanism that is not dependent on changes in collagen synthesis (82–84).

III. POSSIBLE MECHANISMS FOR THE EFFECT OF ASCORBIC ACID ON MATRIX GENE EXPRESSION IN VITRO

There are a number of lines of evidence pointing to a role of ascorbate in stimulating collagen synthesis that is independent of its role in hydroxylation. The amount of ascorbic acid required to produce maximal activity of the prolyl hydroxylase enzyme (<0.1 μM) is at least 100-fold less than that necessary for stimulation of collagen gene expression (50 μM) (46). In addition, although D-ascorbic acid and D-isoascorbic acid can support hydroxylation reactions at equimolar concentrations to L-ascorbic acid (85), it takes at least a 10-fold greater concentration of these analogs to stimulate collagen synthesis maximally (86). Also, the effect of ascorbic acid on collagen synthesis is sensitive to the age of the donor, whereas there is no effect of age on prolyl or lysyl hydroxylase activities in culture (46). This phenomenon involves a stimulation of transcription of collagen genes (45,47,48,66), which reside on different chromosomes (1), indicating a more direct effect on gene expression than would be implied by mechanisms involving the role of ascorbate in hydroxylation. Interestingly, ascorbate treatment produces cellular remodeling of the rough endoplasmic reticulum (87,88), which may be involved in permitting increased secretion rates (61,69). In addition, calvarial bones from scorbutic guinea pigs regain normal prolyl hydroxylase activity with ascorbate treatment without a change in collagen synthesis (89). In some cells the role of ascorbate in the hydroxylation of collagen can be replaced by other cellular reductants, but collagen synthesis is still ascorbate-inducible (37,43,90).

The mechanism by which ascorbic acid stimulates collagen gene expression is unclear, although several possible explanations have been proposed, including increased transcription (32,46,66,68), increased mRNA stabilization (66,75), and release of a posttranslational block (32,46,69). One effect of high doses of ascorbic acid that has recently been considered as a means of altering collagen synthesis is the capacity to cause lipid peroxidation in cells (29,34–36,91–98). The mechanism by which ascorbic acid produces lipid peroxidation has received extensive study (97,99,100) and is outlined in Fig. 1. Ascorbic acid is potentially capable of assisting Fenton chemistry in the production of hydroxyl

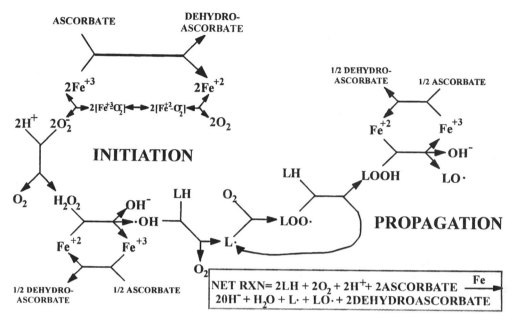

Fig. 1 Mechanism of ascorbate-induced lipid peroxidation (99).

radicals (OH·) from molecular oxygen (O_2) and ferrous ions (Fe^{2+}). Hydroxyl radical is thought to be the species that extracts a proton from available lipid molecules (LH). This step, which results in the formation of the first lipid radical (L·), is called *initiation*. In the presence of oxygen, these lipid radicals form peroxide radicals (LOO·), which can react with neighboring lipid molecules to form hydroperoxides (LOOH) and lipid radicals (L·). Since ascorbic acid has been shown to enhance the decomposition of lipid hydroperoxides to form alkoxyl radicals (LO·), ascorbic acid may also participate in this process, referred to as *propagation*.

Lipid peroxidation induces a wide range of effects. It has been associated with alterations in second-messenger pathways and cell proliferation (101–109). Also, ascorbic acid–induced cell membrane alterations have been shown to affect the ability of extracellular molecules to bind to cell surface receptors, leading to altered responses to those molecules (110–122). Aldehyde and hydroperoxide products of membrane peroxidation have been shown to modify or cross-link proteins, sugars, or DNA (123–129). Any of these changes could be involved in the regulation of collagen synthesis. With this in mind, it is noteworthy that ascorbate induction of lipid peroxidation (94) follows a concentration profile consistent with that of ascorbate-induced collagen synthesis (31,43) (Fig. 2) and that inhibitors of ascorbate-stimulated lipid peroxidation also inhibit ascorbate-induced collagen synthesis (Table 1).

Several investigators have demonstrated a link between the ability of ascorbic acid to stimulate both the production of collagen and lipid peroxidation (29,33–36,95,130–136), whereas only one report has produced results in conflict with this correlation (137). In addition, the ability of oxidized lipids (136) or aldehyde products of lipid peroxidation (29,130,131) to stimulate collagen synthesis in a number of cell types indicates that these products of lipid oxidation may be responsible for mediating the response. The mechanism by which these products of lipid peroxidation affect collagen synthesis is not clear; how-

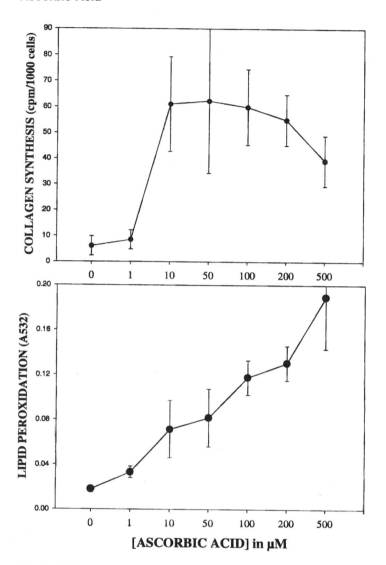

Fig. 2 Effect of ascorbic acid on collagen synthesis and lipid peroxidation in human dermal fibroblasts (34).

ever, a number of potential mechanisms have been suggested. The histone H1 has been shown to bind chromatin and regulate gene expression, particularly collagen gene expression. This binding can be prevented by modification by aldehydes of sensitive sites on the histone (138). One investigator has demonstrated the ability of malondialdehyde and oxidative stress to stimulate specifically *c-myb* expression in hepatic stellate cells. Antisense RNA to *c-myb* blocked the stimulation of collagen synthesis by ascorbic acid, indicating that this protein may be a mediator of the effect of oxidative stress on gene expression in these cells (139).

Another potential mediating mechanism involves a collagen-specific stress protein known as Hsp47 (140–148). Hsp47 has also been shown to function as a collagen-binding

Table 1 Correlation Between Various Modes of Inhibition of
Ascorbate-Induced Lipid Peroxidation and Ascorbate-Induced Collagen
Synthesis in Cultured Human Dermal Fibroblasts

	Measurement Technique	
Treatment	Collagen synthesis	Lipid peroxidation
Initiation inhibitors		
Iron chelators		
Dipyridyl	Inhibit	Inhibit
o-Phenanthroline	Inhibit	Inhibit
Iron competitors		
Cobalt chloride	Inhibit	Inhibit
Oxygen radical scavengers		
Dimethyl sulfoxide	No Effect	No Effect
Isopropanol	No Effect	No Effect
Ethanol	No Effect	No Effect
SOD and catalase	No Effect	No Effect
PEG-SOD & PEG-catalase	No Effect	No Effect
Propagation inhibitors		
Antioxidants		
Mannitol	No Effect	No Effect
Tocopherol	Inhibit	Inhibit
Propyl gallate	Inhibit	Inhibit
Naphthol	Inhibit	Inhibit
Retinoic acid	Inhibit	Inhibit
Retinol	Inhibit	Inhibit

SOD, Superoxide dismutase; PEG, polyethylene glycol
Compiled from previously published experiments (34–36).

chaperone protein located in the endoplasmic reticulum of cells. This heat shock protein has been shown to respond not only to heat shock, but is also induced by carbon tetrachloride treatment of animals (146). There have been other reports of the ability of oxygen radical–producing agents, such as carbon tetrachloride, to stimulate collagen synthesis or induce fibrosis (149–158). Recently, Hsp47 has been shown to play an unexpected role in regulating collagen gene expression because antisense RNA for Hsp47 inhibits collagen gene expression as well as collagen secretion and trafficking (159). Although no link has been established between the ability of ascorbic acid to induce oxidative stress and increased levels of Hsp47,it is an interesting possible mechanism that could regulate collagen gene expression.

A number of other potential, oxygen radical–mediated mechanisms for regulating collagen synthesis have been proposed. Poly-ADP ribosylation has recently received attention as a proposed mediator of the oxidative events involved in the stimulation of collagen synthesis by ascorbic acid. Ascorbic acid has been reported to stimulate the activity of poly-ADP-ribose synthetase, which has been reported to either inhibit (159) or stimulate (160–162) collagen synthesis and to inhibit (161,163) or stimulate (159,160,162) prolyl hydroxylase activity. ADP ribosylation is a posttranslational modification of specific proteins that has been shown to produce alterations in protein structure and function (164).

Collagen is known to be susceptible to modification or cleavage by oxygen radicals in solution (165–173) and in vivo (165,174). Consequently, collagen cleavage by ascorbic acid–induced free radicals could alter cell matrix interactions which have been shown to be capable of regulating collagen synthesis (175). Finally, the redox state within cells has been shown to be crucial for determining the activity of some transcription factors (176–182) and could be a mechanism by which ascorbic acid alters gene expression. Presumably, changes in oxidation within the cell resulting from ascorbic acid uptake or metabolism or effect on free-radical formation could alter the function of oxidation-sensitive transcription factors involved in regulating collagen synthesis.

IV. ASCORBIC ACID EFFECT ON EXTRACELLULAR MATRIX PRODUCTION IN VIVO

The role of ascorbic acid in the production of ECM synthesis in vivo is also not clearly defined and involves a number of proposed mechanisms. One factor contributing to the confusion probably involves the lack of free iron in the circulation under normal conditions. Since many of the in vitro effects of ascorbic acid are iron-dependent, it is perhaps not a surprise that these specific phenomena are not reproduced readily in vivo. Since ascorbic acid was originally discovered as the causative agent in scurvy (218), a disease that involves deficiencies in matrix production as part of its pathological process, it is clear that ascorbic acid is crucial for the production of ECM in animals. The most obvious animal model for studying the role of ascorbic acid would seem logically to be one with an ascorbic acid deficiency. However, few animals can be made ascorbic acid–deficient. Most animals synthesize their own ascorbic acid. Only humans, other primates, guinea pigs, passiformes birds, and flying mammals must obtain it from their diet. This requirement derives from their apparent inability to catalyze the conversion of L-gulonolactone to 2-keto-L-gulonolactone due to a deficiency in gulonolactone oxidase (219).

By far the largest literature in this area has been produced in guinea pigs. In scorbutic guinea pigs, ascorbic acid supplementation was shown to stimulate proline hydroxylation without increasing collagen production in healing wounds (183). The wounds contained cells that possessed an increase in the number of membrane-associated polyribosomes specific for collagen. The decrease in collagen synthesis was later shown to correlate with weight loss in the affected animals (184) and could be reproduced in fasted, ascorbic acid–supplemented guinea pigs (185). In addition to reduced collagen synthesis, proteoglycan synthesis was also reduced in cartilage and was correlated with altered levels of circulating hormones (186–188). Of particular interest are the levels of insulin-like growth factors (IGF) and their binding proteins. IGF-I and II are known to be induced during wound repair (189); however, the circulating levels of two IGF-binding proteins, IGFBP-1 and IGFBP-2,were elevated in response to fasting and vitamin C deficiency and capable of inhibiting the action of IGF-I (190–193). These differences in circulating IGFBP levels do not seem to explain the differences noted in wound repair (194). Ascorbic acid deficiency has also been associated with altered expression of iron-related proteins, which has been proposed as a potential factor contributing to altered angiogenesis during wound repair (194–195).

Another in vivo pathological condition that involves alterations in ascorbic acid utilization is diabetes mellitus. Competition for membrane transport of ascorbic acid in vitro and in vivo has been demonstrated in a number of cell types, including lymphocytes, granulocytes, and fibroblasts (196,213). Chronic reduction in ascorbic acid levels also contributed to reduced chemotaxis by affected polymorphonuclear leukocytes and mononuclear lymphocytes (196). Diabetes mellitus was further shown to be associated with reduced plasma levels of ascorbic acid without changes in dehydroascorbic acid (197–199). This altered ascorbic acid metabolism could only be partially reversed by dietary supplementation of ascorbic acid (198). In addition, although free-radical mechanisms have been implicated in the pathology of diabetes, specifically microangiopathy (197,200), supplementation with various antioxidants, such as vitamins E and C and β-carotene, produced no reduction in a marker of protein oxidation, namely, glycosylated hemoglobin (201). Other markers of diabetes, such as fasting glucose levels and cholesterol and triglyceride levels, did benefit from ascorbic acid supplementation (202), but not in all studies (203,204). Finally, ascorbic acid supplementation was also shown to inhibit erythrocyte sorbitol levels, consistent with an inhibitory effect on aldose reductase, an enzyme involved in the polyol pathway (203,204).

Experimental diabetes was induced in rats treated with streptozotocin. Animals treated in this way have shown reduced levels of hydroxyproline in nascent type I collagen isolated from skin (205,206), tendon (207), and periodontum (208). This underhydroxylated collagen correlated with an increased susceptibility to intracellular degradation (205–208) and could be reversed by ascorbic acid supplementation (205). As with clinical diabetes mellitus, streptozotocin-treated rats have deficiencies in plasma ascorbic acid levels that can be reversed by supplementation with ascorbic acid (211,212). This loss in plasma ascorbic acid can also be prevented by inhibitors of aldose reductase, consistent with the observed role of the polyol pathway in regulating ascorbic acid metabolism (211,212).

These changes in collagen production and degradation lead to reduced levels of collagen deposition in different skin wound healing models (209,210). Specifically, changes in collagen deposition have been correlated with reduced granulation tissue formation in steel-mesh cylinders implanted under the skin of diabetic rats, and the granulation tissue formed is specifically deficient in collagen content (209). This lack of collagen deposition during wound repair also correlated with altered breaking strength of incisional wounds placed in the skin of affected rats (210). The results, which indicated impaired wound repair, have not been addressed to date in clinical studies in diabetic patients; however, clinical trials in both patients undergoing tattoo removal surgery and patients with decubitus ulcers have been performed with no obvious benefit from ascorbic acid treatment (214,215).

Finally, experiments have been conducted to investigate the administration of high doses of ascorbic acid to otherwise healthy rats (216) or to mice induced to form pulmonary fibrosis by bleomycin treatment (217). In normal rats, collagen synthesis was unaffected whereas elastin accumulation was inhibited, consistent with the in vitro effect of ascorbic acid on elastin synthesis (216). In bleomycin-treated mice, ascorbic acid treatment did not prevent the effects of bleomycin on prolyl hydroxylase activity or collagen or elastin deposition (217). Interestingly, there was instead some indication that high-dose ascorbic acid treatment increased the tissue response to bleomycin (217).

V. SUMMARY

The effect of ascorbic acid on the synthesis of ECM proteins has been studied for decades and there are still many questions to be answered. It is clear that the previously accepted mechanism involving the role of ascorbic acid in the hydroxylation of collagen, which permits appropriate folding and secretion of the molecule, is certainly one of the mechanisms involved. This mechanism does not explain all of the effects of either ascorbic acid deficiency or supplementation by high doses produced both in vitro and in vivo. It is also clear that direct correlations between in vitro and in vivo observed phenomena concerning the role of ascorbic acid are not often possible due to the effect that other environmental factors play in regulating cellular and tissue responses to ascorbic acid. Consequently, additional ongoing research in the area of ascorbic acid will presumably lead to enhanced understanding of its role in ECM synthesis, degradation, and secretion. Such research is likely to provide insight into other related mechanisms for regulating matrix synthesis, such as the role of growth factors and their modulation by free radicals or other forms of cellular stress. Although much is known about ascorbic acid, there is much to be discovered.

REFERENCES

1. F. Ramirez and M. Di Liberto, *FASEB J. 4*:1616 (1990).
2. T. F. Linsenmayer, in *Cell Biology of Extracellular Matrix* (E. D. Hay, ed.), Plenum Press, New York, 1991,p. 7.
3. S. Gay, G. R. Martin, P. K. Muller, R. Timpl, and K. Kuhn, *Proc. Natl. Acad. Sci. USA 73*: 4037 (1976).
4. J. C. Geesin and R. A. Berg, in *Applications of Biomaterials in Facial Plastic Surgery* (A. I. Glasgold and F. H. Silver, eds.), CRC Press, Boca Raton, 1991, p. 7.
5. J. R. Lichtenstein, P. H. Byers, B. D. Smith, and G. R. Martin, *Biochemistry 14*:1589 (1975).
6. L. Vitellaro-Zuccarello, R. Garbelli, and V. Dal Pozzo Rossi, *Cell Tissue Res. 268*:505 (1992).
7. D. T. Woodley, V. J. Scheidt, M. J. Reese, A. S. Paller, T. O. Manning, T. Yoshiike, and R. A. Briggaman, *J. Invest. Dermatol. 88*:246 (1987).
10. G. Némethy and H. A. Scheraga, *Biochemistry 25*:3184 (1986).
11. M. Schneir, L. Golub, and N. Ramamurthy, *Ann. N. Y. Acad. Sci. 460*:500 (1985).
12. B. Steinmann, V. H. Rao, and R. Gitzelmann, *FEBS Lett. 133*:142 (1981).
13. P. L. Atreya and V. S. Ananthanarayanan, *J. Biol. Chem. 266*:2852 (1991).
14. A. De Waal and L. de Jong, *Biochemistry 27*:150 (1988).
15. A. A. Kumar, C. S. Vaidyanathan, and N. A. Rao, *J. Scient. Ind. Res. 37*:698 (1978).
16. R. S. Bhatnagar and T. Z. Liu, *FEBS Lett. 26*:32 (1972).
17. E. Holme, G. Lindstedt, S. Lindstedt, and I. Nordin, *Biochem. J. 205*:339 (1982).
18. K. I. Kivirikko and R. Myllylä, *Meth. Enzymol. 144*:96 (1987).
19. P. Sata and S. Udenfriend, *Vitam. Horm. 36*:33 (1978).
20. S. Englard and S. Seifter, *Ann. Rev. Nutr. 6*:365 (1986).
21. M. Levine, *N. Engl. J. Med. 314*:892 (1986).
22. W. C. Parks, R. A. Pierce, K. A. Lee, and R. P. Meecham, in *Advances in Molecular and Cell Biology* (H. K. Kleinman, ed.), JAI Press, Greenwich, CT, Vol. 6, 1993, p. 133.
23. J. Rosenbloom and A. Cywinski, *FEBS Lett. 65*:246 (1976).
24. R. P. Meecham and J. E. Heuser, in *Cell Biology of Extracellular Matrix* (E. D. Hay, ed.), Plenum Press, New York, 1991,p. 79.
25. S. M. L. Anderson, W. H. I. McLean, and R. J. Elliott, *Biochem. Soc. Trans. 19*:48S (1991).
26. W. D. Appling, W. R. O'Brien, D. A. Johnston, and M. Duvic, *FEBS Lett. 250*:541 (1989).
27. E. Bell, M. Rosenberg, P. Kemp, R. Gay, G. D. Green, N. Muthukumaran, and C. Nolte, *J. Biomech. Eng. 113*:113 (1991).
28. D. Chan, S. R. Lamande, W. G. Cole, and J. F. Bateman, *Biochem. J. 269*:175 (1990).
29. M. Chojkier, K. Houglum, J. Solis-Herruzo, and D. A. Brenner, *J. Biol. Chem. 264*:16957 (1989).
30. E. Fisher, S. V. McLennan, H. Tada, S. Heffernan, D. K. Yue, and J. R. Turtle, *Diabetes 40*:371 (1991).
31. H. Freiberger, D. Grove, A. Sivarajah, and S. R. Pinnell, *J. Invest. Dermatol. 75*:425 (1980).

32. J. C. Geesin, D. Darr, R. Kaufman, S. Murad, and S. R. Pinnell, *J. Invest. Dermatol. 90*: 420 (1988).
33. J. C. Geesin, J. S. Gordon, and R. A. Berg, *Skin Pharmacol. 6*:65 (1993).
34. J. C. Geesin, J. S. Gordon, and R. A. Berg, *Arch. Biochem. Biophys. 278*:350 (1990).
35. J. C. Geesin, L. J. Hendricks, P. A. Falkenstein, J. S. Gordon, and R. A. Berg, *Arch. Biochem. Biophys. 290*:127 (1991).
36. J. C. Geesin, L. J. Hendricks, J. S. Gordon, and R. A. Berg, *Arch. Biochem. Biophys. 289*: 6 (1991).
37. J. Geesin, S. Murad, and S. R. Pinnell, *Biochim. Biophys. Acta 886*:272 (1986).
38. F. Grinnell, H. Fukamizu, P. Pawelek, and S. Nakagawa, *Exp. Cell Res. 181*:483 (1989).
39. R. Hata and H. Senoo, *J. Cell. Physiol. 138*:8 (1989).
40. R. Hata, H. Sunada, K. Arai, T. Sato, Y. Ninomiya, Y. Nagai, and H. Senoo, *Eur. J. Biochem. 173*:261 (1988).
41. R. Hata, Y. Sunada, and H. Nagai, *Seikagaku 58*:823 (1986).
42. B. Lacroix, E. Didier, and J. F. Grenier, *Int. J. Vitam. Nutr. Res. 58*:407 (1988).
43. S. Murad, D. Grove, K. A. Lindberg, G. Reynolds, A. Sivarajah, and S. R. Pinnell, *Proc. Natl. Acad. Sci. USA 78*:2879 (1981).
44. S. Murad, S. Tajima, G. R. Johnson, A. Sivarajah, and S. R. Pinnell, *J. Invest. Dermatol. 81*:158 (1983).
45. C. L. Phillips, S. Tajima, and S. R. Pinnell, *Arch. Biochem. Biophys. 295*:397 (1992).
46. S. R. Pinnell, S. Murad, and D. Darr, *Arch. Dermatol. 123*:1684 (1987).
47. S. Tajima, C. Phillips, S. Murad, R. Kaufman, and S. Pinnell, *J. Invest. Dermatol. 90*:611A (1988).
48. S. Tajima and S. R. Pinnell, *Biochem. Biophys. Res. Commun. 106*:632 (1982).
49. K. Takehara, G. R. Grotendorst, M. Trojanowska, and E. C. Leroy, *Collagen Rel. Res. 6*: 455 (1986).
50. T. A. Tsvetkova, T. A. Gorokhova, and A. A. Delvig, *Vop. Med. K. H. 36*:53 (1990).
51. I. Yamamoto, N. Muto, K. Murakami, and J.-I. Akiyama, *J. Nutr. 122*:871 (1992).
52. B. Faris, R. Snider, A. Levine, R. Moscaritolo, L. Salcedo, and C. Franzblau, *In Vitro 14*: 1022 (1978).
53. J. R. A. Leushner and M. D. Haust, *Biochim. Biophys. Acta 883*:284 (1986).
54. L. M. Barone, B. Faris, S. D. Chipman, P. Toselli, B. W. Oakes, and C. Franzblau, *Biochim. Biophys. Acta 840*:245 (1985).
55. S. Tajima, H. Wachi, and A. Hayashi, *Keio J. Med. 44*:140 (1995).
56. H. Senoo and R. Hata, *Biochem. Biophys. Res. Commun. 200*:999 (1994).
57. G. C. Wright, Jr., X. Wei, C. A. McDevitt, B. P. Lane, and L. Sokoloff, *J. Orthop. Res. 6*: 397 (1988).
58. S. Saika, K. Uenoyama, K. Hiroi, and A. Ooshima, *Ophthalm. Res. 24*:68 (1992).
59. R. T. Franceschi, B. S. Iyer, and Y. Chi, *J. Bone Mineral Res. 9*:843 (1994).
60. I. Denis, A. Pointillart, and M. Lieberherr, *Bone Mineral 25*:149 (1994).
61. B. Peterkofsky, *Biochem. Biophys. Res. Commun. 49*:1343 (1972).
62. M. Ono, Y. Aratani, I. Kitagawa, and Y. Kitagawa, *Exp. Cell Res. 187*:309 (1990).
63. C. Kalcheim and V. Leviel, *Cell Different. 22*:107 (1988).
64. L. J. Sandell and J. C. Daniel, *Connective Tissue Res. 17*:11 (1988).
65. W. W. Kao, J. G. Flaks, and D. J. Prockop, *Arch. Biochem. Biophys. 173*:638 (1976).
66. B. L. Lyons and R. I. Schwarz, *Nucleic Acids Res. 12*:2569 (1984).
67. S. R. Quinones, D. S. Neblock, and R. A. Berg, *Biochem. J. 239*:179 (1986).
68. L. B. Rowe and R. I. Schwarz, *Mol. Cell. Biol. 3*:241 (1983).
69. R. I. Schwarz, *J. Biol. Chem. 260*:3045 (1985).
70. R. I. Schwarz and M. J. Bissell, *Proc. Natl. Acad. Sci. USA 74*:4453 (1977).
71. R. I. Schwarz, P. Kleinman, and N. Owens, *Ann. N. Y. Acad. Sci. 498*:172 (1987).
72. R. I. Schwarz, R. B. Mandell, and M. J. Bissell, *Mol. Cell. Biol. 1*:843 (1981).

73. J. J. Jeffrey and G. R. Martin, *Biochim. Biophys. Acta 121*:269 (1966).
74. J. J. Jeffrey and G. R. Martin, *Biochim. Biophys. Acta 121*:281 (1966).
75. J. M. Davidson, P. A. LuValle, O. Zoia, D. Quaglino, Jr., and M. G. Giro, *J. Biol. Chem. 272*:345 (1997).
76. L. M. Ryan, I. Kurup, and H. S. Cheung, *Matrix 11*:276 (1991).
77. P. S. Leboy, T. A. Sullivan, A. S. Menko, and M. Enomoto, *Bone Mineral 17*:242 (1992).
78. T. A. Sullivan, B. Uschmann, R. Hough, and P. S. Leboy, *J. Biol. Chem. 269*:22500 (1994).
79. J. Kao, G. Huey, R. Kao, and R. Stern, *Exp. Mol. Pathol. 53*:1 (1990).
80. J. A. Skrivanek, E. Schwartz, O. O. Blumenfield, and R. W. Ledeen, *In Vitro Cell. Dev. Biol. 26*:502 (1990).
81. C. W. Archer, C. P. Cottrill, and D. Lee, *In Vitro Cell. Dev. Biol. 25*:259 (1990).
82. D. Knaack, I. Shen, M. M. Salpeter, and T. R. Podleski, *J. Cell Biol. 102*:795 (1986).
83. O. Horovitz, D. Knaack, T. R. Podleski, and M. M. Salpeter, *J. Cell Biol. 108*:1823 (1989).
84. E. Liu, R. R. Minor, O. Horovitz, J. A. M. Wooten, T. R. Podleski, and M. M. Salpeter, *Exp. Cell Res. 209*:76 (1993).
85. M. A. Kutnink, B. M. Tolbert, V. L. Richmond, and E. M. Baker, *Proc. Soc. Exp. Biol. Med. 132*:440 (1969).
86. S. Murad, A. Sivarajah, and S. R. Pinnell, *Biochem. Biophys. Res. Commun. 101*:868 (1981).
87. R. Harwood, M. E. Grant, and D. S. Jackson, *Biochem. J. 142*:641 (1974).
88. M. Pacifici and R. V. Iozzo, *J. Biol. Chem. 263*:2483 (1988).
89. M. Chojkier, R. Spanheimer, and B. Peterkofsky, *J. Clin. Invest. 72*:826 (1983).
90. J. M. Mata, R. Assad, and B. Peterkofsky, *Arch. Biochem. Biophys. 206*:93 (1981).
91. G. J. Bachowski, J. P. Thomas, and A. W. Girotti, *Lipids 23*:580 (1988).
92. I. S. Chelnakova, N. D. Tron'ko, and A. S. Mikosha, *Vop. Med. K. H. 36*:32 (1990).
93. M. Dely, T. Zsoldos, A. Puppi, and A. Tigyi, *Chem.-Biol. Interact. 75*:213 (1990).
94. R. E. Heikkila and L. Manzino, *Ann. N. Y. Acad. Sci. 498*:63 (1987).
95. K. Houglum, D. A. Brenner, and M. Chojkier, *J. Clin. Invest. 87*:2230 (1991).
96. V. V. Konev and G. A. Popov, *Biokhimiya 53*:1439 (1988).
97. D. M. Miller and S. D. Aust, *Arch. Biochem. Biophys. 271*:113 (1989).
98. I. Wiswedel, L. Trümper, L. Schild, and W. Augustin, *Biochim. Biophys. Acta 934*:80 (1988).
99. B. Halliwell and J. M. C. Gutteridge, *Meth. Enzymol. 186*:1 (1990).
100. B. Halliwell, J. M. C. Gutteridge, and C. E. Cross, *J. Lab. Clin. Med. 119*:598 (1992).
101. A. Baba, E. Lee, A. Ohta, T. Tatsuno, and H. Iwata, *J. Biol. Chem. 256*:3679 (1981).
102. E. Lee, A. Baba, A. Ohta, and H. Iwata, *Biochim. Biophys. Acta 689*:370 (1982).
103. L. V. Lyzlova, V. R. Persianova, A. E. Antipenko, and S. N. Lyzlova, *Biochem. Int. 14*: 1079 (1987).
104. N. Morisaki, J. A. Lindsey, J. M. Stitts, H. Zhang, and D. G. Cornwell, *Lipids 19*:381 (1984).
105. C. A. O'Brian, N. E. Ward, I. B. Weinstein, A. W. Bull, and L. J. Marnett, *Biochem. Biophys. Res. Commun. 155*:1374 (1988).
106. L. Paradisi, C. Panagini, M. Paroloa, G. Barrera, and M. U. Dianzani, *Chem.-Biol. Interact. 53*:209 (1985).
107. C. Pieri, M. Falasca, F. Marcheselli, R. Recchioni, and F. Moroni, *Cell. Mol. Biol. 38*:437 (1992).
108. D. M. Shasby, M. Yorek, and S. S. Shasby, *Blood 72*:491 (1988).
109. A. S. Sobolev, A. A. Rosenkranz, and A. R. Kazarov, *Int. J. Radiat. Biol. 44*:31 (1983).
110. G. W. Arana, R. J. Baldessarini, and N. S. Kula, *Neuropharmacology 21*:601 (1982).
111. A. J. Bradbury, B. Costall, and R. J. Naylor, *J. Pharm. Pharmacol. 35*:738 (1983).
112. K. B. Delclos and P. M. Blumberg, *Cancer Res. 42*:1227 (1982).
113. B. J. Ebersole and P. B. Molinoff, *J. Neurochem. 58*:1300 (1992).
114. R. E. Heikkila, *Eur. J. Pharmacol. 93*:79 (1983).
115. R. E. Heikkila and F. S. Cabbat, *J. Neurochem. 41*:1384 (1983).

116. R. E. Heikkila, F. S. Cabbat, and L. Manzino, *Res. Commun. Chem. Pathol. Pharmacol. 34*: 409 (1981).
117. R. E. Heikkila, F. S. Cabbat, and L. Manzino, *J. Neurochem. 38*:1000 (1982).
118. R. E. Heikkila, F. S. Cabbat, and L. Manzino, *Life Sci. 32*:2183 (1983).
119. R. E. Heikkila, L. Manzino, F. S. Cabbat, and J. G. Hanly, *Neuropharmacology 22*:135 (1983).
120. F. P. Huger and C. P. Smith, *Fed. Proc. 41*:1327 (1982).
121. S. F. Muakkassah-Kelly, J. W. Andresen, J. C. Shih, and P. Hochstein, *J. Neurochem. 41*: 1429 (1983).
122. L. C. Tolbert, P. E. Morris, Jr., J. J. Spollen, and S. C. Ashe, *Life Sci. 51*:921 (1992).
123. B. R. Brooks and O. L. Klamerth, *Eur. J. Biochem. 5*:178 (1968).
124. A. Gòmez-Sànchez, I. Hermosin, and I. Maya, *Carbohydr. Res. 229*:307 (1992).
125. A. Haberland, A. K. Schütz, and I. Schimke, *Biochem. Pharmacol. 43*:2117 (1992).
126. A. Kautiainen, *Chem.-Biol. Interact. 83*:55 (1992).
127. J. W. Park and R. A. Floyd, *Free Rad. Biol. Med. 12*:245 (1992).
128. D. J. Tuma, T. M. Donohue, Jr., V. A. Medina, and M. F. Sorrell, *Arch. Biochem. Biophys. 234*:377 (1984).
129. J. J. M. Van den Berg, J. A. F. Op den Kamp, B. H. Lubin, B. Roelofsen, and F. A. Kuypers, *Free Rad. Biol. Med. 12*:487 (1992).
130. J. J. Maher, C. Tzagarakis, and A. Gimenez, *Alcohol Alcoholism 29*:605 (1994).
131. J. J. Maher, S. Zia, and C. Tzaragakis, *Alcoholism: Clin. Exp. Res. 18*:403 (1994).
132. P. Beossa, K. Houglum, C. Trautwein, A. Holstege, and M. Chojkier, *Hepatology 19*:1262 (1994).
133. K. S. Lee, M. Buck, K. Houglum, and M. Chojkier, *J. Clin. Invest. 96*:2461 (1995).
134. A. Holstege, P. Bedossa, T. Poynard, M. Kollinger, J. C. Chaput, K. Houglum, and M. Chojkier, *Hepatology 19*:367 (1994).
135. K. Houglum, P. Bedossa, and M. Chojkier, *Am. J. Physiol. 267*:G908 (1994).
136. S. Jimi, K. Saku, N. Uesugi, N. Sakata, and S. Takebayashi, *Atherosclerosis 116*:15 (1995).
137. D. Darr, S. Combs, and S. Pinnell, *Arch. Biochem. Biophys. 307*:331 (1993).
138. O. Niemela, R. M. Mannermaa, and J. Oikarinen, *Life Sci. 47*:2241 (1990).
139. K. Houglum, M. Buck, J. Alcorn, S. Contreas, P. Bornstein, and M. Chojkier, *J. Clin. Invest. 96*:2269 (1995).
140. K. Nagata, S. Saga, and K. M. Yamada, *J. Cell Biol. 103*:223 (1986).
141. K. Nagata, S. Saga, and K. M. Yamada, *Biochem. Biophys. Res. Commun. 152*:428 (1988).
142. K. Nagata, K. Hirayoshi, M. Obara, S. Saga, and K. M. Yamada, *J. Biol. Chem. 263*:8344 (1988).
143. A. Nakai, M. Satoh, K. Hirayoshi, and K. Nagata, *J. Cell Biol. 117*:903 (1992).
144. J. J. Sauk, T. Smith, K. Norris, and L. Ferreira, *J. Biol. Chem. 269*:3941 (1994).
145. L. R. Ferreira, K. Norris, T. Smith, C. Herbert, and J. J. Sauk, *J. Cell. Biochem. 56*:518 (1994).
146. H. Masuda, M. Fukumoto, K. Hirayoshi, and K. Nagata, *J. Clin. Invest. 94*:2481 (1994).
147. T. Smith, L. R. Ferreira, C. Hebert, K. Norris, and J. Sauk, *J. Biol. Chem. 270*:18323 (1995).
148. G. Hu, T. Gura, B. Sabsav, J. Sauk, S. N. Dixit, and A. Veis, *J. Cell. Biochem. 59*:350–367 (1995).
149. G. Chandrakasan and R. S. Bhatnagar, *Cell. Mol. Biol. 37*:751 (1991).
150. M. A. Dubick, J. A. Last, C. E. Cross, and R. B. Rucker, *Drug-Nutr. Interact. 2*:105 (1983).
151. S. N. Giri, Z.-L. Chen, W. R. Younker, and M. J. Schiedt, *Appl. Pharmacol. 71*:132 (1983).
152. S. N. Giri, H. P. Misra, D. B. Chandler, and Z. Chen, *Exp. Mol. Pathol. 39*:317 (1983).
153. K. Hiraiwa, T. Oka, and K. Yagi, *J. Biochem. 93*:1203 (1983).
154. M. Z. Hussain and R. S. Bhatnagar, *Biochem. Biophys. Res. Commun. 89*:71 (1979).
155. M. Z. Hussain, J. A. Watson, and R. S. Bhatnagar, *Hepatology 7*:502 (1987).

156. J. A. Last, J. E. Gerriets, L. G. Armstrong, T. R. Gelzleichter, and K. M. Reiser, *Chest 99*: 70S (1991).
157. G. J. Laurent and R. J. McAnulty, *Am. Rev. Respir. Dis. 128*:82 (1983).
158. S. H. Phan, R. S. Thrall, and P. A. Ward, *Am. Rev. Respir. Dis. 121*:501 (1980).
159. T. K. Hunt and Z. Hussain, in *Wound Healing; Biochemical and Clinical Aspects* (I. K. Cohen, R. F. Diegelman, and W. J. Lindblad, eds.), W. B. Saunders, Philadelphia, 1992, p. 274.
160. B. W. Duncan, J. Qian, X. Liu, and R. S. Bhatnagar, in *Fetal Wound Healing* (N. S. Adzick and M. T. Longaker, eds.), Elsevier, New York, 1991, p. 303.
161. M. Z. Hussain, S. N. Giri, and R. S. Bhatnagar, *Exp. Mol. Pathol. 43*:162 (1985).
162. J. J. Qian, B. Duncan, N. S. Adzick, and R. S. Bhatnagar, *Cell. Mol. Biol. Res. 39*:525 (1993).
163. M. Z. Hussain, Q. P. Ghani, and T. K. Hunt, *J. Biol. Chem. 264*:7850 (1989).
164. K. Ueda and O. Hayaishi, *Annu. Rev. Biochem. 54*:73 (1985).
165. J.-P. Borel and J.-C. Monboisse, *Ann. Biol. Clin. 44*:260 (1986).
166. S. F. Curran, M. A. Amoruso, B. D. Goldstein, and R. A. Berg, *FEBS Lett. 176*:155 (1984).
167. S. F. Curran, M. A. Amoruso, B. D. Goldstein, D. J. Riley, N. H. Edelman, and R. A. Berg, *Chest 85S*:43S (1984).
168. Y. Kano, Y. Sakano, and D. Fujimoto, *J. Biochem. 102*:839 (1987).
169. Y. Kato, K. Uchida, and S. Kawakishi, *J. Agric. Food Chem. 40*:373 (1992).
170. J. S. Kerr, C. U. Chae, H. Nagase, R. A. Berg, and D. J. Riley, *Am. Rev. Respir Dis. 135*: 1334 (1987).
171. J. C. Monboisse, P. Braquet, A. Randoux, and J. P. Borel, *Biochem. Pharmacol. 32*:53 (1983).
172. J. C. Monboisse, G. Poulin, P. Braquet, A. Randoux, C. Ferradini, and J. P. Borel, *Int. J. Tissue Reac. 6*:385 (1984).
173. D. J. Riley and J. S. Kerr, *Lung 163*:1 (1985).
174. D. J. Riley, M. J. Kramer, J. S. Kerr, C. U. Chae, S. Y. Yu, and R. A. Berg, *Am. Rev. Respir. Dis. 135*:441 (1987).
175. J. C. Geesin, L. J. Brown, J. S. Gordon, and R. A. Berg, *Exp. Cell Res. 206*:283 (1993).
176. C. Abate, L. Patel, F. J. Rauscher III, and T. Curran, *Science 249*:1157 (1990).
177. I. Beck, S. Ramirez, R. Weinmann, and J. Caro, *J. Biol. Chem. 266*:15563 (1991).
178. S. M. Keyse and E. A. Emslie, *Nature 359*:644 (1992).
179. C. W. Pugh, C. C. Tan, R. W. Jones, and P. J. Ratcliffe, *Proc. Natl. Acad. Sci. USA 88*: 10553 (1991).
180. R. Schreck, P. Rieber, and P. A. Baeuerle, *EMBO J. 10*:2247 (1991).
181. G. Storz, L. A. Tartaglia, and B. N. Ames, *Science 248*:189 (1990).
182. C. K. Sen and L. Packer, *FASEB J. 10*:709 (1996).
183. R. Harwood, M. E. Grant, and D. S. Jackson, *Biochem. J. 142*:641 (1974).
184. M. Chojkier, R. Spanheimer, and B. Peterkofsky, *J. Clin. Invest. 72*:826 (1983).
185. R. G. Spanheimer and B. Peterkofsky, *J. Biol. Chem. 260*:3955 (1985).
186. B. Peterkofsky, R. G. Spanheimer, and T. A. Bird, in *Development and Diseases of Cartilage and Bone Matrix* (G. Sen and T. Thornhill, eds.), Alan R. Liss, New York, 1987, p. 33.
187. T. A. Bird, N. B. Schwartz, and B. Peterkofsky, *J. Biol. Chem. 261*:11166 (1986).
188. B. Peterkofsky, *Am. Clin. Nutr. 54*:1135S (1991).
189. M. H. Gartner, J. D. Benson, and M. D. Caldwell, *J. Surg. Res. 52*:389 (1992).
190. B. Peterkofsky, J. Palka, S. Wilson, K. Takeda, and V. Shah, *Endocrinology 128*:1769 (1991).
191. A. Gosiewska, S. Wilson, D. Kwon, and B. Peterkofsky, *Endocrinology 134*:1329 (1994).
192. B. Peterkofsky, A. Gosiewska, D. E. Kipp, V. Shah, and S. Wilson, *Growth Factors 10*:229 (1994).
193. A. Gosiewska and B. Peterkofsky, *Endocrine 3*:889 (1995).
194. D. Kipp, S. Wilson, A. Gosiewska, and B. Peterkofsky, *Wound Repair Regen. 3*:192 (1995).
195. A. Gosiewska, F. Mahmoodian, and B. Peterkofsky, *Arch. Biochem. Biophys. 325*:295 (1996).

196. R. E. Pecoraro and M. S. Chen, *Ann. N. Y. Acad. Sci. 498*:248 (1987).

197. P. E. Jennings, S. Chirico, A. F. Jones, J. Lunec, and A. H. Barnett, *Diabetes Res. 6*:151 (1987).

198. A. J. Sinclair, A. J. Girling, L. Gray, C. Le Guen, J. Lunec, and A. H. Barnett, *Diabetologia 34*:171 (1991).

199. J. Lysy and J. Zimmerman, *Nutr. Res. 12*:713 (1992).

200. A. J. Sinclair, J. Lunec, A. J. Girling, and A. H. Barnett, *EXS 62*:342 (1992).

201. S. M. Shoff, J. A. Mares-Perlman, K. J. Cruickshanks, R. Klein, B. E. Klein, and L. L. Ritter, *Am. J. Clin. Nutr. 58*:412 (1993).

202. J. Eriksson and A. Kohvakka, *Ann. Nutr. Metab. 39*:217 (1995).

203. H. Wang, Z. B. Zhang, and R. R. Wen, *Chung Hua I Hsueh Tsa Chih 74*:548 (1994).

204. H. Wang, Z. B. Zhang, R. R. Wen, and J. W. Chen, *Diabetes Res. Clin. Pract. 28*:1 (1995).

205. M. Schneir, N. Ramamurthy, and L. Golub, *Coll. Relat. Res. 5*:415 (1985).

206. M. Schneir, N. Ramamurthy, and L. Goulb, *Prog. Clin. Biol. Res. 180*:667 (1985).

207. M. K. Leung, G. A. Folkes, N. S. Ramamurthy, M. Schneir, and L. M. Golub, *Biochim. Biophys. Acta 880*:147 (1986).

208. M. Schneir, M. Imberman, N. Ramamurthy, and L. Golub, *Coll. Relat. Res. 8*:221 (1988).

209. D. K. Yue, B. Swanson, S. McLennan, M. Marsh, J. Spaliviero, L. Delbridge, T. Reeve, and J. R. Turtle, *Diabet. Med. 3*:221 (1986).

210. D. K. Yue, S. McLennan, M. Marsh, Y. W. Mai, J. Spaliviero, L. Delbridge, T. Reeve, and J. R. Turtle, *Diabetes 36*:295 (1987).

211. S. McLennan, D. K. Yue, E. Fisher, C. Capogreco, S. Heffernan, G. R. Ross, and J. R. Turtle, *Diabetes 37*:359 (1988).

212. D. K. Yue, S. McLennan, E. Fisher, S. Heffernan, C. Capogreco, G. R. Ross, and J. R. Turtle, *Diabetes 38*:257 (1989).

213. E. Fisher, S. V. McLennan, H. Tada, S. Heffernan, D. K. Yue, and J. R. Turtle, *Diabetes 40*:371 (1991).

214. G. ter Riet, A. G. H. Kessels, and P. G. Knipschild, *J. Clin. Epidemiol. 48*:1453 (1995).

215. F. Vaxman, S. Olender, A. Lambert, G. Nisand, M. Aprahamian, J. F. Bruch, E. Didier, P. Volkmar, and J. F. Grenier, *Eur. Surg. Res. 27*:158 (1995).

216. J. W. Critchfield, M. Dubick, J. Last, C. E. Cross, and R. B. Rucker, *J. Nutr. 115*:70 (1985).

217. M. A. Dubick, J. A. Last. C. E. Cross, and R. B. Rucker, *Drug-Nutr. Interact. 2*:105 (1983).

218. J. Lind, in *A Treatise on the Scurvy* (C. P. Stewart and D. Cuthrie eds.), Edinburgh University Press, Edinburgh, 1953.

219. P. Sata and S. Udenfriend, *Vitam. Horm. 36*:33 (1978).

17

Nutrients and Oxidation: Actions, Transport, and Metabolism of Dietary Antioxidants

J. BRUCE GERMAN

University of California, Davis, California

MARET G. TRABER

Oregon State University, Corvallis, Oregon

I. INTRODUCTION

Free-radical–mediated oxidation reactions are ubiquitous in biological cells and fluids. Oxidation reactions that follow free-radical pathways are vital to cells in several areas: (a) Energy metabolism; the production of energy-rich intermediates that drive most of life's processes is largely accomplished by the free-radical transfer of electrons from high-energy organic substrates to oxygen, forming carbon dioxide and water. (b) Biosynthesis; many necessary macromolecules are partially synthesized by enzymes that utilize free-radical pathways to accomplish difficult reactions. (c) Detoxification; many toxic compounds are converted to inert metabolites using free-radical reactions. (d) Signaling pathways; various inter- and intracellular communication systems utilize the very rapid and energetic reactions of free-radical chemistry to produce oxidation products as signals, second messengers, and autocoids. However, oxidation reactions are also potentially catastrophic. With the ability to produce devastating cellular and extracellular damage, free-radical reactions can destroy virtually all biologically important molecules, and hence control of free-radical reactions is critical for the existence of life.

Once thought to be solely the deteriorative reactions of dead tissues and organic molecules, free-radical oxidation reactions are now recognized as important chemical pathways for a great many constitutive biological processes as well as acute response and elaboration of signaling pathways. This recognition of oxidation in the fields of biochemistry and cell biology has been followed by an appreciation that various nutrients, whose chemical structures allow them to interact with free radicals, become important in the regulation and especially the limitation of these oxidative processes. As a result of the utility and potential destructiveness of free-radical reactions, biological cells have evolved a genuinely spectacular array of defensive, preventive, responsive, and repair systems to minimize the damage associated with carrying out free-radical reactions and producing free-radical intermediates. Not surprisingly, both the extent of free-radical production and the efficacy of the defense and repair systems are influenced by an individual's nutritional status generally and the abundance of redox active nutrients specifically.

This chapter describes the basic chemistry of oxidation reactions and their prevention via readily oxidized molecules (typically referred to as antioxidants), and discusses the role of antioxidants in preventing and delimiting these reactions in human cells and tissues. This chapter reviews the chemistry of lipids as biological molecules unusually susceptible to oxidation reactions and highlights the potentially devastating effects of these reactions within certain biological environments, notably the cellular membranes. The combination of this chemical susceptibility, the solubility properties of lipids, and the structural diversity of cellular and tissue lipid components demonstrates not only the importance of the chemical properties of certain antioxidants but also the importance of biological transport systems that deliver antioxidants to the sites of oxidative reactions in cells and tissues. The chapter thus summarizes the nutritional importance of both providing and transporting these antioxidant molecules from the diet to the tissues in which they accomplish their actions.

II. CHEMISTRY OF OXIDATION

Oxidation is the chemical process by which an atom or molecule (termed the reductant) with a surfeit of outermost orbital electrons loses an electron(s) to another atom or molecule such as oxygen (termed the oxidant) that accepts the electron(s) into its orbital sphere of influence. There are always two participants in an oxidation reaction: the species that is oxidized and the species that is reduced. The oxidation–reduction reaction requires neither that the reductant and corresponding oxidant be on separate molecules nor that they dissociate into separate species after the electron transfer event. Normally, an oxidation–reduction reaction takes place when one molecule encounters another molecule and the electrons are transferred from one to another. In organic reactions, a pair of electrons are transferred. The reaction is considered a free-radical reaction when just one electron is transferred. This net transfer of electron(s) proceeds spontaneously in the direction in which the net energy of the overall system is lower. In essence, a ''poor'' oxidant requires a ''good'' reductant and vice versa. In chemical terms, this strength of molecules to undergo oxidation or reduction reactions is quantified by the oxidation–reduction potential of the molecules. Oxygen is arguably the most important biological oxidant due to the substantial energy given up when electrons are transferred from reduced carbon–based macromolecules to molecular oxygen. The paradox to modern life of humans and other animals is that the free-radical oxidative reactions involving oxygen are both life sustaining by being coupled to energy production in the mitochondria of living cells and life threatening due to the damage and deterioration of living cells that they can cause.

The evidence is compelling that throughout the course of evolution, biology has found the power of free-radical reactions to be a seductive strategy, and higher organisms use free-radical chemistry to accomplish many necessary biochemical reactions. As a result, a complex spectrum of biochemical systems that either utilize or detoxify products of free-radical chain reactions has gradually developed over the course of the evolution of organisms living in an oxygen atmosphere.

Science is only now beginning to understand the interactions between systems that generate and those that utilize oxidants. Information about these interactions will eventually lead to an understanding of where in the chemical process normal reactions begin to deviate toward pathological reactions and where antioxidants can provide protection against cell damage and disease.

A. Oxidative Stress

Oxidant production or exposure and antioxidant depletion are general phenomena that together are described as "oxidative stress." This is an important concept that is frequently poorly defined. One can consider appropriate oxidative status as the gradual oxidation of energy-rich molecules and expenditure of energy consistent with the normal functioning of cells and tissues. These various oxidative processes invariably produce a finite number of free radicals whose release and detoxification are rapidly accomplished without any apparent deleterious effects on the organism. In all cells, radicals are both made and successfully eliminated on a continual basis. Oxidative stress occurs when either the production rate increases or the elimination rate decreases, causing free-radical damage to unprotected and/or non-repairable molecules. Increases in both acute catastrophic events and natural processes compose the production side of oxidative stress (1). Oxidative stress induced by chemical toxicity, acute physical injury, or even iron overload (2–4) causes the production of excessive and frequently unusual free radicals, which results in gross tissue damage. Also, normal responses to nonradical stresses can increase oxidative stress. The most obvious example is the release of activated oxygen species by immune cells, such as macrophages, designed to kill invading pathogens. Although oxidation chemistry is an endogenously initiated process, it can result in localized damage to surrounding tissues. In this case, the cost of a defensive strategy that generates chemically toxic oxygen radicals is balanced by the benefit of protection against the threat of pathogenic infection. However, the power of free-radical oxidative chemistry to do harm is seen in the devastating consequences of chronic inflammation of several autoimmune disorders that create pathological damage due to the constant production of oxidants by the host's own immune cells. This seemingly two-edged sword underscores both the protective power of oxidative chemistry and the risks associated with these naturally evolved biological reactions.

The risks associated with acute oxidant exposure are being increasingly recognized; however, another factor that is important to oxidant balance is the overall and local concentrations of antioxidant protectors. Antioxidants are important because normal energetic metabolism, primarily through electron transport in the mitochondria, results in the continuous inadvertent release of a significant number of free-radical oxidants. Even in the absence of an acute oxidant challenge, antioxidant systems are constantly responsible for delimiting and repairing the cost of doing business, as it were, with free-radical chemistry. Given the normal production rate of free-radical oxidants, which has been estimated at 10^6 free radicals released per cell per day (5), the importance of maintaining a substantial concentration of antioxidant molecules of various types in all tissues is clear.

As will be described further below, many of the antioxidant protectors must be concentrated in areas of high oxidation and/or the production of active oxidants. This means that both the intake and delivery of antioxidants can be important. Because many of the protectors at the front line of defense are redox-active molecules that rapidly reduce damaging radicals to relatively benign radicals, their abundant presence at the sites of radical generation is important. Such redox-active molecules are typically complex phenolic molecules. Many of this class of protector molecules are not made in humans and thus are present in tissues as a function of their abundance in the diet. Vitamins C and E (ascorbate and α-tocopherol) are the most obvious examples of redox-active molecules whose abundance in the tissue sites of interest is critical to the overall balance of oxidative stress, and for which a deficiency has literally catastrophic consequences in terms of oxidative damage to tissue.

The notion of a balance between the benefits and risks of free-radical oxidative chemistry implies that this balance can be shifted by changing the predominance of pro-oxidative processes and antioxygenic protection (6). The implications of this relationship have been tested by mechanistic research, animal trials, and epidemiological studies of various populations wherein disease incidence has been related to genetic and environmental variables. The results are in general quite convincing that the balance of oxidation can be upset with deleterious consequences. The effects of acute oxidative stress or acute antioxidant deficiency are now accepted by the scientific community. In some cases, an imbalance in oxidation is accepted as a compelling rationale for recommending antioxidant therapy for overt pathologies that produce oxidative stress, such as iron overload, ischemia reperfusion injury, and certain chemical toxicities.

One additional issue emerges from the scientific perspective of oxidant–antioxidant balance. If acute oxidative stress is ameliorated by antioxidant therapy, is it possible that prolonged chronic imbalances in oxidative protection may lead to repetitive damage to sensitive biological tissues? An imbalance in oxidants–antioxidants with an excess of oxidants would, for example, lead to a buildup of potentially toxic byproducts, which over long periods leads to loss of function. Such long-term accumulation of oxidant byproducts is now believed to be an important aspect of chronic and degenerative diseases, such as coronary artery disease. This possibility has led to a consensus among the research community that antioxidant nutrients represent one of the nutritional attributes of a diet rich in plant material, i.e., fruits and vegetables.

The concept of oxidant–antioxidant balance has implications for the etiology of many chronic diseases. Atherosclerosis is an example of a disease in which oxidative damage and the molecules that result promote a metabolic dysfunction (7). LDLs become atherogenic when their basic surface structure is modified and they become substrates for uptake by macrophages located in the subendothelium of artery walls. An important mechanism by which this modification is thought to take place is oxidation. Thus, the oxidation of LDL is a part of the gradual development of atherosclerosis. Antioxidants retard oxidative modification of LDL, and it has been proposed that part of the value of diets rich in fruits and vegetables in reducing coronary disease is the provision of antioxidants as protectors against LDL oxidation.

Another example of a means by which the oxidative balance is inordinate is environmental insult to antioxidant protection. The elimination of essential protectors, such as ascorbate, through chemical insult (tobacco smoking) has been shown to increase the damage to the DNA of sperm (8). The increase in cancer incidence in the offspring of smoking male parents that was predicted by molecular results has been observed in human populations whose vitamin C intakes are low (9).

The chemistry of free-radical oxidation is highly complex and, due to its aggressive nature, tends to be nonspecific in both reaction targets and products. While all macromolecules are targets of oxidative chemistry, the chemistry of lipid oxidation is described below in detail as this is the area most understood.

B. Free-Radical Lipid Oxidation

The free-radical oxidation of lipids is a complex series of reactions with three distinct phases: initiation, propagation, and termination. Lipid oxidation should be recognized as a highly dynamic and complex result of the interplay between substrates, circumstances, and time (10).

C. Hydroperoxide Formation

During the initiation phase of oxidation, free-radical species are generated on unsaturated lipids. During the propagation phase, lipid radicals react with oxygen, forming peroxy radicals that in turn react with more lipids, generating more radicals. The propagation phase is thus a classic example of an autocatalytic chain reaction. The termination phase is the point in which two free radicals react with each other, eliminating the free radical on each.

Thermodynamic equilibrium strongly favors the net oxidation or transfer of electrons from reduced, carbon-based biomolecules to molecular oxygen. That is, biological macromolecules, proteins, polynucleotides, carbohydrates, and lipids are all reduced carbon-based molecules that contain considerably more energy than their constituent atoms in their more oxidized states, e.g., carbon dioxide. The tendency of virtually all biological molecules to oxidize is therefore driven by the strong energetic advantage to the reaction with oxygen. Life is possible because this direct oxidation reaction is so slow at the ambient temperature of the planet. This so-called kinetic stability of all biological molecules in an oxygen-rich atmosphere results from the unique spin state of the unpaired electrons in ground-state molecular (triplet) oxygen in the atmosphere. Whereas most biologically important carbon-based molecules contain an even number of electrons in which their electrons are paired in what is referred to as antiparallel arrangement. Oxygen, on the other hand, contains an even number of electrons, but in its lowest energy state the outermost two electrons occupy separate orbitals. Thus, the stable form of oxygen is a biradical in which the outermost two electrons are unpaired. This property renders atmospheric oxygen relatively inert to direct oxidation reactions with reduced, carbon-based biomolecules. Hence, even though thermodynamically favored, reactions between oxygen and protein, lipids, polynucleotides, carbohydrates, and so forth proceed at an insignificantly slow rate. The kinetic limitation can be overcome by converting oxygen to a nonradical or by converting carbon-based molecules to radicals, i.e., by removing a single electron from one of their pairs. The removal of an electron from a carbon-based molecule can be accomplished by active, single-electron oxidants.

There are various sources of free-radical oxidants, as shown in a table of free-radical redox half potentials (11). Free-radical oxidants that are important to lipid oxidation are those that will abstract a single electron from an unsaturated fatty acid and thus have a half-cell redox potential approaching 1 V. An oxidant that initiates lipid oxidation chain reactions in polyunsaturated fatty acids (PUFAs) must have sufficient redox potential to abstract the hydrogen from the most easily oxidized carbon, a methylene interrupted carbon (the carbon between two double bonds). An oxidant that can initiate the oxidation of monounsaturated fatty acids must have a slightly higher redox potential due to the greater energy required to abstract a hydrogen from a carbon that is alpha to only a single double bond.

Oxygen in the ground state is a very poor single-electron oxidant, with a half-cell redox potential of -0.33 V. Superoxide is a better oxidant but still of insufficient potential to oxidize unsaturated lipids. Hydrogen peroxide is a poor oxidant; however, in the presence of reduced metals, it oxidizes a reduced transition metal, such as cuprous or ferrous ion. The reduction of hydrogen peroxide causes it to split homolytically, yielding two free radicals. The electron from the reduced metal reduces one free radical, but the other free radical remains. This free-radical species is the hydroxyl radical. This species has a redox potential above 2 V. The hydroxyl radical will readily oxidize virtually any organic molecule, including fatty acids. This is the main reaction pathway that generates powerful oxidants in cells. The production of hydroxyl radicals then initiates free-radical chain reactions within polyunsaturated lipids.

Molecules that can remove a single electron from reduced macromolecules, such as lipids, are potentially very deleterious because as initiators of oxidation they eliminate the reactive impediments imposed by the spin restrictions of ground-state oxygen. By converting stable organic molecules, RH, to free-radical–containing molecules, R', they initiate chain reactions that continue to oxidize subsequent lipid molecules. Oxygen reacts readily with the unpaired free-radical species on fatty acids to form the peroxy radical, ROO'. The peroxy radical contains unpaired electrons and hence is active as a free radical oxidant that oxidizes other reduced molecules: ROO' + RH → ROOH + R'. Initiators of lipid oxidation are relatively ubiquitous, primarily single-electron oxidants, and they include trace metals, hydroperoxide cleavage products, and ultraviolet light. Polyunsaturated fatty acids are oxidized by the ROO' species to yield another free radical, R', and a lipid hydroperoxide, ROOH. This effectively sets up a self-propagating free-radical chain reaction, R' + O$_2$ → ROO' → ROOH + R', that can lead to the complete consumption of PUFAs (12). The ability of the peroxy radical to act as an initiating, single-electron oxidant drives the destructive and self-perpetuating reaction of PUFA oxidation.

D. Hydroperoxide Decomposition

The reduced products of the immediate reaction of a lipid radical with oxygen, the hydroperoxides, are not radicals and hence are more stable. However, their chemical formula, ROOH, is similar to that of hydrogen peroxide, HOOH, and the reactivity of the peroxides are similar. Both are readily attacked by reduced transition metals, causing a homolytic cleavage reaction. The decomposition stage of oxidation occurs when lipid hydroperoxides encounter a transition metal. From lipid hydroperoxide cleavage, the oxygen-based radical formed is termed an alkoxyl radical, RO. This free-radical species is analogous to the hydroxyl radical and has similar reactivity. The RO radical is a very strong, single-electron oxidant and will react with adjacent molecules to abstract an electron, thus oxidizing the target molecule to a free radical. In this way, hydroperoxide decomposition reactions are the most important free-radical initiators in most lipid-containing systems. The alkoxyl radical can also react internally with the lipid molecule itself, leading to a host of decomposition products (13).

Decomposition of hydroperoxides yields alcohols, aldehydes, ketones, and hydrocarbons as fragments of the original lipid hydroperoxides. While it is the reactivity of the alkoxy radical that is considered most damaging to cells and other macromolecules, the other decomposition products of lipid oxidation can also be deleterious. The release of these oxidative products provides the most discernible property of oxidized lipids, the off-flavors of rancidity. However, off-flavors are not the only consequences of lipid decomposition reactions. Aldehydes are weak electrophiles and will react with a variety of biological nucleophiles. There are many such nucleophiles in cells, including parts of proteins, vitamins, and polynucleotides. The adducts of these aldehydes with proteins in most cases render the proteins inactive.

1. Consequences of Lipid Oxidation

Lipid hydroperoxides formed during lipid oxidation decompose to various products, including short-chain aldehydes, ketones, and alcohols. These products, as well as radicals, compromise cell and tissue health in a number of ways. (a) The direct oxidation of susceptible molecules can result in loss of function. For example, oxidation of membrane lipids alters membrane integrity, promotes membrane leakage, and affects the conformation and activity of membrane-bound proteins. The direct oxidation of proteins results in a loss or alteration of enzyme catalytic activity and/or regulation. (b) Reaction of the products of lipid oxidation with protein or other macromolecules leads to adduct formation with loss of native function and, in some cases, accession to novel actions. As a well-described example, the oxidative modification of the apoB protein molecule on LDL by lipid aldehydes changes its uptake properties completely. The formation of these adducts on apoB prevents the uptake of LDL by the LDL receptor and stimulates uptake by the unregulated scavenger receptor. (c) Oxidation can cleave DNA, causing point, frame-shift, and deletion mutations, as well as base damage. This oxidative cleavage impairs or destroys normal functionality. (d) Oxidative reactions can liberate signal molecules or analogs that elicit inappropriate responses, such as activation of platelet aggregation, promotion of cell proliferation, and down-regulation of vascular relaxation.

The susceptibility and overall rate of oxidation of a lipid molecule is related to the number of double bonds on the fatty acids. First and foremost, the rate of oxidation is determined by the ease of hydrogen abstraction. Thus, oxidation of saturated fatty acids is uncommon because hydrogen abstraction is extremely rare. The addition of a single double bond renders the alpha hydrogens substantially more susceptible to oxidation. The formation of a cis interrupted pair of double bonds renders the hydrogens on this carbon much more susceptible to oxidation. Additional double bonds do not increase the susceptibility of individual hydrogens to abstraction, but they do increase the number of potential sites of abstraction and oxidation. These three considerations result in the relative oxidative rates of fatty acids. An increase in the number of double bonds increases the oxidation rate for the fatty acids 18:1, 18:2, 18:3, and 20:4 such that the relative oxidation rates for purified fatty acids are 1, 50, 100, and 200. Diets high in PUFAs require higher than normal amounts of antioxidant nutrients to prevent oxidation and rancidity. Consumption of high-PUFA diets by animals increases the antioxidant requirement. A diet enriched in highly unsaturated fatty acids increases the tendency to oxidation, and this would be predicted to increase the incidence of oxidation-associated chronic degenerative diseases. The metabolic effects of PUFA in altering lipoprotein metabolism and lowering serum cholesterol, for example, has led to their increased incorporation in human diets. This metabolic benefit does not abrogate their susceptibility to oxidation, however, and the importance of increasing dietary antioxidants to protect tissues from high intakes of polyunsaturated fat diets is implicit from their chemical susceptibility.

III. BIOLOGY OF OXIDATION: A BALANCE BETWEEN PRO- AND ANTIOXIDANTS

Organisms need to cascade and amplify chemical signals to develop appropriate responses to many stressors, including oxidants. Many of these signaling systems are oxidant-generating pathways, such as the enzymatic systems that oxidize specific PUFA moieties to form potent signaling molecules called oxylipins. These molecules, including prostaglandins, leukotrienes, etc., signal a state of stress to adjacent or responsive cells. Enzymatically produced oxidized lipids act on higher order brain functions such as pain and even sleep. Oxidation of some protein transcription factors allows binding to "oxidant response elements" within DNA and directly affects its transcription. This cascading proliferation of oxidized molecules, which accomplishes the tasks of intracellular and multicellular signaling, also places a burden on oxidant defense systems. These chemical signals clearly co-evolved with the increasingly sophisticated and necessary oxidant repair systems. Perhaps the presence of these oxidant defenses allowed the proliferation of nonlethal uses of oxidants in higher organisms.

A. Oxidant Production In Vivo

Oxidation is initiated in cells, tissues, and fluids by a host of chemical and protein, i.e., enzyme, factors. Oxidation events, such as reactions involving Fenton chemistry, mitochondrial electron transport, respiratory bursts, oxygenating enzymes, reductive cleavage of peroxides, and xenobiotic metabolism, are initiated by organisms because they are either essential for or beneficial to the success of the organism. As delineated below, the pathogenic potential for these reactions can be considerable.

1. Metal-Catalyzed Recycling of Radicals

The univalent reduction of hydroperoxides by transition metals, especially ferrous and cuprous salts, yields a radical: an unpaired electron, as either the alkoxyl or hydroxyl radical. A highly destructive free radical–initiating system has been shown by the presence of free metals and oxidized lipids in atherosclerotic plaque (7).

2. Mitochondrial Electron Transport

Mitochondria transfer billions of electrons per day to oxygen in single-electron steps that eventually converts metabolizable hydrocarbon into carbon dioxide and water. Even with transfer equal to 99.999% of total electrons transferred, leakage from this system is a significant source of reactive oxygen species. Release of free radicals has been estimated to be in excess of 10^4 free-radical species per cell per day (5). This imperfection of mitochondrial oxidative coupling is the basis of the mitochondrial damage theory of aging.

3. Respiratory Burst

The immune system has developed an oxygen-reducing system that releases superoxide, which is toxic to bacteria, to combat pathogenic organisms. Stimulation of this oxidase system in phagocytes produces a burst of oxygen consumption that stoichiometrically progresses to production of superoxide radicals. Superoxide leads to the death of bacterial cells in the immediate vicinity of the immune cells that produce it. The action of superoxide is not highly specific and can damage host cells as well as invading pathogens. This class of reactive biochemistry as an immune response is a good example of the overall cost–benefit equation that is inherent to stress responses in both animals and plants.

4. Oxygenating Enzymes

Many, perhaps all, cells respond to external stimuli by liberating PUFAs from their membranes. Arachidonic acid is a particularly active fatty acid. This fatty acid initiates a signal cascade that uses enzymatically catalyzed free-radical reactions to produce oxygenated derivatives of itself that are broadly termed *eicosanoids*. These molecules produce local signals that activate that cell and its immediate neighbors, thus acting as part of the cellular stress response system in higher organisms. Chronic activation or overactivation of this system constitutes an oxidative stress that produces inflammation and activates additional response systems.

5. Reductive Cleavage of Peroxides

A general class of enzymes called *peroxidases* eliminates peroxides as a general detoxifying mechanism. Certain peroxidases catalyze oxidant production as a result of this reaction. Peroxidases produce oxidants as a response to pathogen invasion and so are considered to be additional elements in the killing mechanisms of immune cells. Chronic activation of the oxidant-producing systems of immune cells constitutes an independent oxidative stress to many humans suffering from inflammatory diseases, autoimmune diseases, or chronic infection.

6. Xenobiotic Metabolism

The primary tissue involved in the conversion of toxic chemicals into excretable compounds is the liver. This conversion is effected by inducible enzymes that are typically oxidases. As a result, toxin metabolism can and does produce free-radical species. The poisonous properties of a host of pesticides and other chemical products are now recognized to result from the secondary byproducts produced during the metabolism of xenobiotics.

The above described use of oxidation by cells and tissues represents a clear risk–benefit relationship. Such risks are acceptable, at least in the short term, because oxidation provides a net benefit. The long-term consequences may only be relevant to aging organisms and poorly defended tissues, both of which occur in humans.

The current scientific view is that many chronic diseases exhibit secondary effects of unprotected or aberrant oxidation. Oxidant damage accumulating over a lifetime can greatly influence the health of the individual. Developments in oxidant biology indicate that antioxidant requirements need to be reevaluated in relation to cellular dysfunction. The requirement for oxidant defense will vary with oxidant stress. Antioxidant effects defined under conditions of zero stress are unlikely to be meaningful for any system either under brief, acute stress (such as viral infections, inflammation, trauma, and exposure to environmental pollutants) or during states of chronic and sustained oxidant stress (such as autoimmunity, chronic infection, chronic inflammation, elevated circulating lipoproteins, diabetes, or mild antioxidant deficiency). Thus, estimated requirements for antioxidant defense of a population would be better if based on individuals exposed to an ''average'' or ''typical'' amount of oxidant stress than based on individuals assumed to lack stress. However, the concept of what constitutes ''typical'' oxidant stress is undefined, and such definition will be difficult because, as described above, a variety of insults can elicit an oxidant stress both directly and indirectly.

Free-radical reactions are extremely powerful and absolutely necessary for a large number of biochemical and cellular processes. Cells generate energy, assemble and disassemble necessary molecules, signal within and between cells and tissues, and eliminate pathogens, all using free-radical reactions and reaction products. Biology has evolved a very complex system of catalysts whose function is to produce free-radical oxidants and reductants, and all have the potential to promote deleterious reactions if not controlled and regulated. This chemistry of free radicals was clearly a major evolutionary pathway in which successful exploitation provided selective biochemical advantage. However, this proliferation of free-radical generating systems in biology placed a substantial priority on balancing systems that could control, delimit, and repair the consequences of free-radical reactions.

B. Antioxidant Control of Oxidation

The chemistry of free-radical oxidations is multistage and complex. Oxidation is not a single catastrophic event. There is no single initiating oxidant that generates all free radicals; there are a great many sources of single-electron oxidants. Similarly, there is no single reactive product of oxidation; there are entire classes of oxidative products, many of which are either selectively or even broadly damaging. Free radicals and their products react with virtually all biological molecules, and there is no single defense against all targets of oxidative damage. Thus, organisms have evolved a spectrum of mechanisms to prevent or to respond to oxidative stresses and free radicals and their products at one or more of the many steps of oxidation. With this diversity of oxidative chemistry acting at a host of biological points, the use of the single term "antioxidant" is misleadingly simple. Nevertheless, the basic principle that an antioxidant prevents or slows the production or lowers the duration or minimizes the damage associated with free-radical oxidative reactions is still accepted. Now, however, it is possible to categorize the types of protection into several broad classes.

The potential health effects of antioxidant protectors must be considered in the context of the overall response of living organisms to oxidation. As illustrated by specific examples below, many complex biochemical pathways have evolved to prevent oxidation, to delimit the chain reactions of oxidation, and to respond to oxidation once it has occurred.

1. Scavenging of Activated Oxidants

Many free radicals are highly reactive, single-electron oxidants. Since most organic molecules contain electrons as pairs, the abstraction of an electron from an organic molecule produces another radical. This is the insidious nature of free-radical reactions, i.e., they keep going. The ability of certain organic molecules to reduce single-electron oxidants, and in so doing form a free radical that can no longer act as a strong single electron oxidant, is the major action of a classic scavenging antioxidant. Primary chain reaction–breaking antioxidants include vitamin E, coenzyme Q (reduced form, ubiquinol; oxidized form, ubiquinone), ascorbate, uric acid, polyphenolics, various flavonoids and their polymers, amino acids, and protein thiols.

There are three key ingredients to a successful scavenging antioxidant. For a free-radical reaction in which an oxidant, $O^{\cdot}$, reacts with a target, RH, to produce the reduced oxidant, OxH, and a new radical, $R^{\cdot}$, the success of an antioxidant, AH, depends on (a) its ability to successfully compete with RH for the oxidant, $Ox^{\cdot}$, (b) the effective concentration or location of AH at the site of the reaction of $Ox^{\cdot}$ with RH; and (c) the stability of the radical formed, $A^{\cdot}$, and the ease with which it reacts as an oxidant with other RH targets.

α-Tocopherol (vitamin E), because of its high biological activity and high tissue concentrations compared with other forms, is a uniquely successful antioxidant in the protection of membrane PUFAs. First, α-tocopherol reacts with the most prevalent free radical oxidant in membranes, the peroxy radical (ROO'). Even more important to its role as an antioxidant in membranes is that it reacts with the peroxy radical quite quickly and hence very competitively with the most abundant target in membranes, PUFA (RH). In fact, α-tocopherol is so competitive with PUFAs that one α-tocopherol molecule per 10,000 PUFA molecules provides effective protection. Second, α-tocopherols are insoluble molecules that partition preferentially into the membrane bilayer of cells, precisely where the abundance of PUFAs renders them particularly susceptible to free-radical chain reactions. Third, the redox potential of the α-tocopherol radical (11) is such that the significant redox couple that reduces it under biological conditions is possibly ascorbate, which in effect regenerates the α-tocopherol molecule and produces an ascorbyl radical that is even more stable than the α-tocopherol radical and that can itself be further reduced enzymatically back to ascorbate.

2. Prevention of Oxidant Formation

Compartmentation of Subcellular Reactions and Targets. Oxidative reactions are not carried out homogeneously within cells or organisms. Mitochondria and peroxisomes are specialized cellular organelles that limit the generation and transfer of electrons, and these organelles contain enzymes and free-radical scavenging activities that can dispose of toxic intermediates and the products of metabolism. This topological separation of toxic reactions from sensitive targets is an important advantage of multicellular protection; however, it carries an additional energetic requirement in that energy is required to assemble structures and concentrate reaction substrates, enzymes, and intermediates. One of the most deleterious aspects of tissue disruption and necrotic cell death is the loss of this exquisite cellular topology and the release of reactive catalysts into surrounding tissues.

Inactivation of Reactive Precursors. Several classes of molecules are capable of generating or propagating free-radical reactions. For most of these molecules that capability depends on the structures in which they are bound. Transition metals are examples of highly reactive species that, when free, readily propagate free-radical reactions. However, when they are bound within specific proteins, all of the potentially reactive metal ligands are occupied, thus inactivating the metal as an oxidation catalyst. Transferrin and ceruloplasmin are plasma proteins that not only bind and transport transition metals but actively sequester iron and copper ions, which are otherwise capable of initiating oxidation, and inactivate them. Lactoferrin similarly inactivates iron and prevents its participation in redox cycling reactions.

Reduction of Reactive Intermediates. Higher animals possess enzyme systems that scavenge active oxygen, including superoxide dismutase, catalase, and several peroxidases. Other enzymes that detoxify reactive intermediates include catalase, a scavenger of hydrogen peroxide; glutathione peroxidase, which catabolizes hydroperoxides; and superoxide dismutase, which reduces superoxide anion. Natural food constituents with antioxidant activity can also act as free-radical quenchers, antioxidants, and or protectors/regenerators of other antioxidants. Synergistic (14) and antagonistic effects among mixtures of antioxidant compounds are possible based on the nature of redox couples formed by compounds present in tissues. Phenolic antioxidants stabilize some enzymes, enhancing some activities and inhibiting others.

The complexity and interdependence of the systems described above indicate that oxidative stress could increase requirements for not only direct antioxidants but also for those nutrients essential for proper up-regulation of oxidant defense and repair mechanisms.

IV. PLANT PHENOLICS

''Antioxidant'' is a broad classification for molecules that interrupt or decelerate oxidation reactions either prior to or during a free-radical chain reaction, at initiation, propagation, termination, decomposition, or the subsequent reaction of oxidation products with sensitive targets. Antioxidants by definition are protectors against oxidation reactions. Similarly, antioxidant molecules are heterogeneous, both in chemical structures and in biochemical actions. Although various antioxidant compounds have known actions and functions apart from their ability to protect specific targets of oxidation, these functions are not the subject of this discussion. However, one should keep in mind that various molecules have dual actions, and the relative importance of these two actions in vivo depends on the immediate environment and physiological state of the tissue.

Antioxygenic compounds can participate in several of the protective strategies described above for higher animals. Differences in the chemical reactivity and the physical location between different antioxidants are not trivial, and the reactivity and location influence the efficacy of a given compound to act as a net antioxidant or protector. How different molecules act can also affect the impact of oxidation and its inhibition on biological function and damage. Plant phenolics vary in their ability to interrupt a free-radical chain reaction, with functional differences being detectable among different lipid systems, oxidation initiators, and other antioxygenic components. These chemical differences, which do not include aspects of biological structure, make it relatively clear that the most effective protection from oxidation in vivo would be afforded by a mixture of antioxidant molecules that differ in both chemical and physical properties, rather than by a single, however excellent, chemical antioxidant.

Certain plant phytochemicals also provide an interesting benefit to cells in addition to their interaction with oxidation processes in that they induce various repair pathways. Emerging from both epidemiological and mechanistic studies is the realization that induction of repair prior to an oxidative or toxic challenge can substantially decrease the damage associated with that event. ''Forewarned is forearmed,'' so to speak. Since many oxidative insults lead to DNA damage and ultimately can promote proliferative disorders, such as cancer, the ingestion of phytochemicals that induce DNA protection and repair systems is thought to explain the action of these compounds in slowing the development of some cancers.

A. Classes of Plant Phenolics

Ostensibly, phenolic molecules in animals, whether essential or not, are derived from plants. Important examples include phenolic amino acids as well as vitamins K and E. Plant secondary metabolism produces a host of phenolic molecules that are further modified and complexed during plant growth, harvesting, and processing. These molecules can be divided into three overall classes: simple phenolics, flavonoids, and high molecular weight tannins.

1. Simple Phenolics

Simple phenolics consist of various polyhydroxylated phenolics derived from further metabolism of primarily cinnamic and benzoic acids. The most abundant of these compounds in plants are chlorogenic, gallic, and caffeic acids. The ability of these and similar compounds to inhibit potentially deleterious free-radical chain reactions has been recognized by the food industry for decades. Thus, largely as extracts of the plants in which they are abundant, plant phenolics have been used to stabilize food materials for much of recent food processing history. These simple phenols are also building blocks for more complex molecules, forming polymers with many other classes of plant metabolites. The formation of more complex molecules may or may not affect their actions as free-radical scavenging antioxidants.

2. Flavonoids

The flavonoids are an extremely broad class of plant secondary metabolites that are all related to a single flavone nucleus and are responsible for much of the distinctive coloring of vascular plants. Flavonoids occur not only in multiple variants due to simple substitutions of the A, B, or C rings but in complexes with other molecules to form an astonishingly rich variety of polyphenolic structures (over 4000 different molecules) in the plant kingdom. Among the flavonoids are morin, quercetin, fisetin, myricetin, kaempferol, apigenin, and luteolin. They are an integral but highly variable part of the human diet, and their effects on human nutrition are thus of intense interest and potential importance.

3. Tannins

Tannins consist of highly complexed polymers. The monomeric units of these polymers consist either of simple phenolics, such as gallic acid, which as polymers are referred to as the hydrolyzable tannins, or of the flavonoids, which as complex polymers up to 5000 daltons are the condensed tannins. While these compounds have been pursued with intense interest for their antitumor activities, two perspectives on this class of molecules argue for some caution in applying optimistic results in animals to humans. While apparently benign when consumed, the tannins are toxic when injected. Their ability to inhibit protein synthesis is blamed for liver failures in World War II patients to whom tannins were applied as open-wound dressings. Their lack of toxicity when ingested is due to their not being absorbed or metabolized. In fact, selection appears to have responded to the need to avoid absorbing tannins by evolving modifications in salivary protein production. Both animals and humans produce a spectrum of conspicuously proline-rich proteins that actively bind tannins and prevent their absorption. Humans appear to express these proteins constitutively; however, animals require an exposure to dietary tannins to induce their production. The pre-exposure to tannins significantly lowers the toxicity of tannins to animals, implying that these induced proteins are highly protective.

The myriad structures of the polyphenolic phytochemicals from plants dictate not only their functions to the plants but their functions, biological activities, and ultimately nutritional benefits and toxicities to humans.

B. Chemical Actions of Phenolics

Polyphenolics exhibit a variety of chemical interactions with biologically important classes of molecules, particularly with potentially reactive species. Examples of these are summa-

Table 1   Actions of Polyphenolics with Biological Molecules

Action	Effect of action
Redox-active	Reduce/oxidize essential compounds, i.e., ascorbate
Free-radical scavengers	Halt propagation reactions, i.e., catechin reduces peroxy radicals
Active oxygen quenchers	Quench singlet oxygen, i.e., quercetin quenches photosensitized oxidation
Transition metal chelators	Chelation and its effects are structure-specific; redox cycling is both inhibited and stimulated by polyphenols
Form complexes with proteins, DNA, polycyclic aromatics	Binding to proteins and subsequent effects are structure-specific, both antioxidant/pro-oxidant
	Flavonoid-DNA complexes retard and promote strand cleavage depending on structure
	React with diol epoxides; complex and lower absorption of benzpyrenes
Nitrite reactants	Compete with amines for nitrosation

rized in Table 1. The phenolic compounds are primary antioxidants that act as free-radical scavengers (reductants) and chain breakers. The key to any phenolic acting as a free-radical chain-propagating inhibitor is that the phenolic radical generated as a result of scavenging a peroxy radical is kinetically stable as a free-radical oxidant. This is a logical extension of their antioxidant ability, but the actual fate of the various polyphenolic radicals that could be generated by the insertion of these compounds in a free-radical chain reaction remains to be determined. Polyphenols include salicylic, cinnamic, coumaric, and ferulic derivatives and gallic esters. In grapes alone, the following phenolics have been identified: phenolic acids (hydroxybenzoic, salicylic, cinnamic, coumaric and ferulic derivatives, and gallic esterols), flavonols (kaempferol and quercetin glycosides), flavan-3-ols (catechin, epicatechin, and derivatives), flavanonols (dihydroquercetin, dihydrokaempferol, and hamnoside), and anthocyanins (cyanidin, peronidin, petunidin, malvidin, coumarin, and caffeine glucosides). In compounds that are derivatives of benzoic and cinnamic acids as well as flavonoids, the degree and position of hydroxylation are important in determinating antioxidant efficiency. Evaluation of antioxidative activity of naturally occurring substances has been of interest; however, there is a lack of knowledge about their molecular composition, the amount of active ingredients in the source material, and relevant toxicity data.

C. Biochemical Actions of Phenolics

Largely as a consequence of the chemical interactions described above, phenolics exert a multiplicity of effects on catalytic systems and structure of biochemical systems in living cells. Summarized in Table 2 is a small subset of the multiple biochemical pathways that are influenced by the presence and/or abundance of either crude mixtures of plant phenolics or highly purified single molecules.

Based on the redox activity and potential protector or antioxidant actions of the myriad molecules discussed above, it is compelling to argue that these molecules are valuable as routine dietary constituents and that they constitute an important asset of a complex diet rich in fruits and vegetables. However, the value of polyphenolics is not

Table 2 Acions of Phenolics on Some Biochemical Pathways

Biochemical pathway	Effects of phenolics
Redox enzymes	Inhibit activity of lipoxygenases, cyclooxygenases, monooxygenases, cytochromes, etc.
	Inhibit activity of proteases
	Slow proteolysis; slow digestion
	Inhibit bleaching reactions of lipoxygenases and peroxidases
Cellular reductants	Sparing action of catechin decreases severity of tocopherol deficiency in animal models
Transcription of proteins	Various proteins exhibit selective induction by flavonoids

just nonspecific. One class of molecules is so critical to the protection of biological membranes that a protein system has evolved to ensure delivery of these compounds to sites of oxidative requirement. These are the tocopherols, or vitamin E.

V. WHY IS THE PLANT PHENOLIC, VITAMIN E, REQUIRED?

Vitamin E is the collective name for molecules that exhibit the biological activity of α-tocopherol (see Chapter 4 and Ref. 15). Vitamin E occurs naturally in eight different forms: four tocopherols and four tocotrienols, which have similar chromanol structures; trimethyl (α-), dimethyl (β- or γ-), and monomethyl (δ-). Tocotrienols differ from tocopherols in that they have an unsaturated side chain. In general, all plant phenolics inhibit lipid hydroperoxide formation catalyzed by metals, radiation, and heme compounds, and also scavenge peroxy, alkoxy, and hydroxy radicals and singlet oxygen. α-Tocopherol in oxidizing lipid systems is spared by flavonoids. If α-tocopherol is an essential antioxidant that acts where no other compound can, the sparing effect of nonessential antioxidants may be one of their most important actions.

A. Vitamin E Deficiency

The free-radical scavenging properties of vitamin E are likely the basis for its essentiality and the pathologies associated with its deficiency. That the basis for the essentiality of tocopherols lies in their ability to prevent oxidative damage raises several important nutritional questions. For example, is there value in consuming higher than simply adequate levels? Can oxidation be inhibited too much? Can other molecules with similar actions provide added benefits? Many phytochemicals have been implicated as being capable of interfering with and inhibiting free-radical chain reactions of lipids, which again raises the question, What is the function of α-tocopherol?

In humans, vitamin E deficiency results in a peripheral neuropathy. That is, the large-caliber axons in the sensory neurons die as a result of free-radical–mediated damage. This is a progressive disorder with increasing severity as the damage moves up the spinal cord and into the cerebellum (16). Initially, the disorder causes deficits in sensory perception with increasing loss of function and, ultimately, death. Patients are often diagnosed with spinocerebellar ataxia; the symptoms are nearly identical to those of Friedreich's ataxia.

Remarkably, animals fed experimental vitamin E–deficient diets are more likely to display anemia and muscle degeneration, and only at a late stage to demonstrate nerve degeneration. This suggests that the delivery mechanisms are important in determining the susceptible tissues. In experimental animals, the lack of dietary vitamin E results in continuous low plasma vitamin E concentrations. In contrast, humans with a genetic defect in the α-tocopherol transfer protein (α-TTP) have normal absorption of vitamin E but are unable to maintain plasma vitamin E concentrations (17,18). As a result, plasma concentrations fall within hours upon cessation of vitamin E consumption.

B. Delivery Systems—Intestinal Absorption

Vitamin E is present in the diet, largely in fat-containing foods. Not surprisingly, absorption of this fat-soluble vitamin is dependent on micellarization with bile acids and lipolytic products. Once taken up into the intestinal cell, vitamin E is incorporated into chylomicrons that are secreted into the lymph. This is apparently an unregulated process; all forms of vitamin E are absorbed in proportion to that in the diet. The absorption of other lipophilic dietary components is likely similar. Thus, absorption of carotenoids or other fat-soluble vitamins may be dependent on micellarization for intestinal cell uptake and chylomicron assembly for secretion into the circulation.

C. Delivery to Tissues in the Postprandial State

The mechanisms for delivery of vitamin E to tissues are largely the same mechanisms that deliver peroxidizable fats. Thus, chylomicrons that contain dietary fat, vitamin E, and other lipophilic compounds are catabolized in the circulation by lipoprotein lipase. During the initial pass through the circulation, all of the absorbed forms of vitamin E could reach peripheral tissues. This is likely an important process for delivery of vitamin E to muscles, skin, adipose tissue, and perhaps brain. Similarly, there may be distribution of other fat-soluble nutrients to these tissues as a result of the lipolytic process.

D. Role of the Liver the Antioxidant Defense: Delivering Antioxidants

Vitamin E serves as an interesting example of how the liver regulates antioxidant defenses. The intestine appears to efficiently absorb a variety of fat-soluble compounds, allowing for the potential absorption of molecules with no apparent value and that produce toxicity only. These molecules "hitch-hike" aboard the chylomicron catabolic system that serves to deliver the absorbed fatty acids to peripheral tissues. Certainly, some fat soluble molecules, including vitamin E and perhaps some carotenoids, are also deposited in peripheral tissues during the clearance of chylomicrons, but because adipose tissue contains "white fat," delivery of the fat-soluble compounds largely does not occur during lipolysis.

Chylomicron remnants deliver to the liver a variety of potentially helpful and potentially toxic compounds. In the transport of vitamin A, the retinal-binding protein serves to accomplish the distribution from the hepatocytes. This complex subject is considered in other chapters of this text.

Unlike vitamin A, vitamin E is transported in the plasma in all lipoproteins. The ready solubility of vitamin E in membranes and the apparent ease of its exchange between membranes and lipoproteins made it appear likely that vitamin E distribution was a nonspecific process. Thus, the description of patients who were vitamin E–deficient without having fat malabsorption were critical to the recognition of a specific vitamin E transport process (17). Now it is clear that tocopherols are not easily exchanged between membranes, and specific protein-based mechanisms are necessary for efficient quantitative transport of tocopherol between cellular and tissue membranes.

The hepatic α-TTP (30–35 kDa) was purified from rat (19,20) and human liver (21,22), and the amino acid sequence of both has been reported (19,22). All of the following features of vitamin E are required for recognition and transfer by α-TTP (19,23,24): (a) the fully methylated, intact chromane ring with a free 6-OH group; (b) the presence of the phytyl side chain; and (c) the stereochemical configuration of the phytyl tail in the 2R position.

In vitro, purified α-TTP transfers α-tocopherol between liposomes and microsomes (24,25). This ability to recognize α-tocopherol and to transfer it suggests that α-TTP is likely responsible for the export of α-tocopherol from the liver, and nascent lipoproteins isolated from perfused monkey livers were preferentially enriched in *RRR*-α-tocopherol (26). However, hepatocytes overexpressing α-TTP were able to export α-tocopherol even when lipoprotein synthesis was inhibited by Brefeldin A. Thus, the mechanism by which α-TTP exports α-tocopherol from the liver remains unknown.

Once chylomicron remnants deliver vitamin E to the liver, the α-tocopherol transfer protein is necessary for its export to the plasma (17,18). Recent discoveries of humans with vitamin E deficiency–like diseases demonstrate that defects in α-TTP result in vitamin E deficiency (27,28). Thus, delivery of tocopherols to the liver is nonspecific, as it is for many of the fat-soluble compounds ingested. With respect to vitamin E form, delivery to the liver is nonspecific for the different isomers, but export from the liver is selective. An exquisite advantage of such a system acting within the liver is that, because of the continuous recycling of vitamin E with lipoprotein metabolism, the entire plasma pool of vitamin E is replaced daily, thereby assuring sustained vitamin E concentrations.

E. Liver-Detoxifying Systems

Because of the relative nonselectivity of intestinal absorption of fat-soluble molecules, the liver becomes the site of clearance of unneeded, unwanted, and potentially toxic compounds. That excess vitamin E is metabolized to a chain-shortened metabolite, α-tocopherol, 2,5,7,8-tetramethyl-2(2′carboxyethyl)-6-hydroxychroman, is also known (29). This metabolite is sulfated and glucuronidated to form a water-soluble product that is excreted in urine. Excretion of this metabolite increases linearly with the dose of supplemental vitamin E. Where or how the metabolite is synthesized is unknown, but the liver seems a likely site.

F. Non-antioxidant Functions of Vitamin E

In addition to its antioxidant function, reported structure-specific effects of α-tocopherol on specific enzyme activities or on membrane properties have been reviewed (30). The most convincing work in this respect is on the suppression of arachidonic acid metabolism via phospholipase A_2 inhibition (31) and on the regulation of vascular smooth-muscle cell proliferation and protein kinase C (PKC) activity (32–35). Reportedly, α-tocopherol, but not β-tocopherol, prevents the phosphorylation of PKC-α, and this is the mechanism by which α-tocopherol decreases PKC-α activity (32).

Interestingly, PKCα inhibition by α-tocopherol has been demonstrated not only in smooth-muscle cells but also in monocytes (36) and platelets (37)—three cell types involved in atherosclerosis. Thus, these observations begin to provide a mechanism for the beneficial effects reported for vitamin E in heart disease and perhaps a mechanism for the specific vitamin requirement for α-tocopherol.

VI. CONCLUSIONS

Free radical–mediated redox reactions are a fact of life, especially in an oxygen-rich atmosphere. Oxidation reactions are essential not only to the continuous transduction of energy from carbon-rich molecules to cellular processes, but also to the successful responses of living organisms to potentially catastrophic environmental, toxic, and pathogenic challenges. However, oxidation reactions are also fundamentally damaging, with the potential to destroy the functions of ostensibly all biologically important molecules.

The control of free-radical reactions by biological cells and tissues is effected by overlapping and redundant biochemical processes that prevent excessive oxidant production, protect susceptible molecular targets, and repair or replace damaged molecules. However, the key protective molecules as a portion of this overall control are largely phenolic molecules that are not made by cells but that must be consumed in the diet. Thus, in the biological battle against oxidation in humans, the quality and abundance of these phenolic molecules in the diet is an important variable to ongoing and especially long-term success.

Although not recognized until relatively recently, oxidation is not only an ongoing chemical insult but is also a highly localized stress. Thus, biological processes that deliver protective molecules to the sites of oxidation reactions are as important as the presence of these molecules in the diet. The identification of populations of humans with perfectly adequate intakes of vitamin E but who suffer from devastating vitamin E deficiency as they age, solely because they have a defect in the protein that is responsible for successfully delivering vitamin E to peripheral tissues, emphasizes the importance of this new dimension to lipid metabolism to the nutritional value of antioxidants and the overall diet.

REFERENCES

1. Halliwell B., Gutteridge JMC. Free Radicals in Biology and Medicine. Oxford: Clarendon Press, 1989.
2. Stohs S. J., Bagchi D. Free Radical *Biol Med* 1995; *18*:321–336.
3. Gutteridge J. M. *Chem Biol Interact* 1994; *91*:133–140.
4. Halliwell B. *Am J Med* 1991; *91(3C)*:14S–22S.
5. Ames B. N., Shigenaga MK, Hagen TM. Mitochondrial decay in aging. *Biochim Biophys Acta* 1995; *1271*:165–170.
6. Halliwell B., Gutteridge J. M. The definition and measurement of antioxidants in biological systems [letter; comment]. *Free Rad Biol Med* 1995; *18*:125–126.
7. Steinberg D. Antioxidants in the prevention of human atherosclerosis. Summary of the proceedings of a National Heart, Lung, and Blood Institute Workshop, September 5–6, 1991, Bethesda, Maryland. Circulation 1992; *85*:2337–2344.
8. Fraga C. G., Motchnik P. A., Shigenaga M. K.; Helbock H. J., Jacob R. A., Ames B. N. Ascorbic acid protects against endogenous oxidative DNA damage in human sperm. *Proc Natl Acad Sci* USA 1991; *88*:11003–11006.
9. McCredie, M. Maisonneuve, P. Boyle, P. Antinatal risk factors for malignant brain tumors in New South Wales children. *Int J Cancer* 1994; *56*:6–10.
10. Frankel E. N. Chemistry of free radical and singlet oxidation of lipids. *Prog Lipid Res* 1985; *23*:197–221.

11. Buettner G. R. The pecking order of free radicals and antioxidants: lipid peroxidation, alpha-tocopherol, and ascorbate. *Arch Biochem Biophys* 1993; *300*:535–543.

12. Kanner J., German J. B., Kinsella J. E. *Crit Rev Food Sci Nutr* 1987; *25*:317–364.

13. Frankel E. N., Volatile lipid oxidation products. *Prog Lipid Res* 1982; *22*:1–33.

14. Tappel, A. L. Will antioxidants slow the aging process? *Geriatrics* 1968; *23*:97–105.

15. Sheppard A. J., Pennington J. A. T., Weihrauch J. L. Analysis and distribution of vitamin E in vegetable oils and foods. In: Packer L, Fuchs J, eds. *Vitamin E in Health and Disease.* New York: Marcel Dekker, 1993:9–31.

16. Sokol R. J. 1993. Vitamin E deficiency and neurological disorders. In: Packer L, Fuchs J, eds. *Vitamin E in Health and Disease.* New York: Marcel Dekker, 1993: 815–849.

17. Traber M. G., Sokol R. J., Burton G. W., Ingold K. U., Papas A. M., Huffaker J. E., Kayden H. J. Impaired ability of patients with familial isolated vitamin E deficiency to incorporate α-tocopherol into lipoproteins secreted by the liver. *J Clin Invest* 1990; *85*:397–407.

18. Traber M. G., Sokol R. J., Kohlschütter A., Yokota T., Muller D. P. R., Dufour R, Kayden H. J. Impaired discrimination between stereoisomers of α-tocopherol in patients with familial isolated vitamin E deficiency. *J Lipid Res* 1993; *34*:201–210.

19. Sato Y., Arai H., Miyata A., Tokita S., Yamamoto K., Tanabe T., Inoue K. Primary structure of alpha-tocopherol transfer protein from rat liver. Homology with cellular retinaldehyde-binding protein. *J Biol Chem* 1993; *268*:17705–17710.

20. Yoshida H., Yusin M., Ren I., Kuhlenkamp J., Hirano T., Stolz A., Kaplowitz N. Identification, purification and immunochemical characterization of a tocopherol-binding protein in rat liver cytosol. *J Lipid Res* 1992; *33*:343–350.

21. Kuhlenkamp J., Ronk M., Yusin M., Stolz A., Kaplowitz N. Identification and purification of a human liver cytosolic tocopherol binding protein. *Prot Exp Purific* 1993; *4*:382–389.

22. Arita M., Sato Y., Miyata A., Tanabe T., Takahashi E., Kayden H. J., Arai H., Inoue K. Human alpha-tocopherol transfer protein: cDNA cloning, expression and chromosomal localization. *Biochem J* 1995; *306*:437–443.

23. M. G. Traber M. G. Determinants of plasma vitamin E concentrations. *Free Rad Biol Med* 1994; *16*:229–239.

24. Hosomi A., Arita M., Sato Y., Kiyose C., Ueda T., Igarashi O., Arai H., Inoue K. Affinity for alpha-tocopherol transfer protein as a determinant of the biological activities of vitamin E analogs. *FEBS Lett* 1997; *409*:105–108.

25. Sato Y., Hagiwara K., Arai H., Inoue K. Purification and characterization of the α-tocopherol transfer protein from rat liver. *FEBS Lett* 1991; *288*:41–45.

26. Traber M. G., Rudel L. L., Burton G. W., Hughes L., Ingold K. U., Kayden H. J. Nascent VLDL from liver perfusions of cynomolgus monkeys are preferentially enriched in RRR compared with SRR-a tocopherol: studies using deuterated tocopherols. *J Lipid Res* 1990; *31*:687–694.

27. Ouahchi K, Arita M., Kayden H., Hentati F., Ben Hamida M., Sokol R., Arai H., Inoue K., Mandel J.-L., Koenig M. Ataxia with isolated vitamin E deficiency is caused by mutations in the α-tocopherol transfer protein. *Nature Gene* 1995; *9*:141–145.

28. Cavalier L., Ouahchi K., Kayden H. J., DiDonato S., Ataxia with isolated vitamin E deficiency: Heterogeneity of mutations and phenotypic variability in a large number of families. *Am J Hum Genet* 1998; *62*:301–310.

29. Schultz M., Leist M., Petrzika M., Gassmann B., Brigelius-Flohé R. A novel urinary metabolite of α-tocopherol, 2,5,7,8-tetramethyl-2(2′ carboxyethyl)-6-hydroxychroman (α-CEHC) as an indicator of an adequate vitamin E supply? *Am J Clin Nutr* 1995; *62*(Suppl):1527S–1534S.

30. Traber M. G., Packer L. Vitamin E: beyond antioxidant function. *Am J Clin Nutr* 1995; *62*(Suppl):1501S–1509S.

31. Pentland A. P., Morrison A. R., Jacobs S. C., Hruza L. L., Hebert J. S., Packer L. Tocopherol analogs suppress arachidonic acid metabolism via phospholipase inhibition. *J Biol Chem* 1992; *267*:15578–15584.

32. Azzi A., Boscoboinik D., Clement S., Marilley D., Ozer N., Ricciarelli R., Tasinato A. Alpha-tocopherol as a modulator of smooth muscle cell proliferation. *Prostaglandins Leukotrienes Essential Fatty Acids* 1997; *57*:507–514.

33. Boscoboinik D., Szewczyk A., Hensey C., Azzi A. Inhibition of cell proliferation by alpha-tocopherol. Role of protein kinase C. *J Biol Chem* 1991; *266*:6188–6194.

34. Boscoboinik D. O., Chatelain E., Bartoli G. M., Stauble B., Azzi A. Inhibition of protein kinase C activity and vascular smooth muscle cell growth by δ-alpha-tocopherol. *Biochim Biophys Acta* 1994; *1224*:418–426.

35. Tasinato A., Boscoboinik D, Bartoli G. M., Maroni P., Azzi A. d-Alpha-tocopherol inhibition of vascular smooth muscle cell proliferation occurs at physiological concentrations, correlates with protein kinase C inhibition, and is independent of its antioxidant properties. *Proc Natl Acad Sci USA* 1995; *92*:12190–12194.

36. Devaraj S, Li D., Jialal I. The effects of alpha tocopherol supplementation on monocyte function. Decreased lipid oxidation, interleukin 1 beta secretion, and monocyte adhesion to endothelium. *J Clin Invest* 1996; *98*:756–763.

37. Freedman J. E., Farhat J. H., Loscalzo J., Keaney J. F. J. Jr. Alpha-tocopherol inhibits aggregation of human platelets by a protein kinase C–dependent mechanism. *Circulation* 1996; *94*: 2434–2440.

Index